# TechOne: Automatic Transmissions

# TechOne: Automatic Transmissions

Jack Erjavec



THOMSON
DELMAR LEARNING
Australia Canada Mexico Singapore Spain United Kingdom United States

**TechOne: Automatic Transmissions**

Jack Erjavec

**Vice President, Technology and Trades SBU:**
Alar Elken

**Editorial Director:**
Sandy Clark

**Sr. Acquisitions Editor:**
David Boelio

**Developmental Editor:**
Matthew Thouin

**Marketing Director:**
Dave Garza

**Channel Manager:**
William Lawrensen

**Marketing Coordinator:**
Mark Pierro

**Production Director:**
Mary Ellen Black

**Production Editor:**
Barbara L. Diaz

**Art/Design Specialist:**
Cheri Plasse

**Technology Project Specialist:**
Kevin Smith

**Editorial Assistant:**
Andrea Domkowski



Printed in the United States of America
1 2 3 4 5 XX 06 05 04

For more information contact
Thomson Delmar Learning
Executive Woods
5 Maxwell Drive, PO Box 8007,
Clifton Park, NY 12065-8007
Or find us on the World Wide
Web at www.delmarlearning.com



Library of Congress Cataloging-in-Publication Data:

Erjavec, Jack.
TechOne. Automotive transmissions / Jack Erjavec.
p. cm.
Includes index.
ISBN 0-7668-1169-7
1. Automobiles—Transmission devices, Automatic—Maintenance and repair. I. Title.
TL263.E76 2005
629.2'446'0288—dc22 2004057994

## NOTICE TO THE READER

Publisher does not warrant or guarantee any of the products described herein or perform any independent analysis in connection with any of the product information contained herein. Publisher does not assume, and expressly disclaims, any obligation to obtain and include information other than that provided to it by the manufacturer.

The reader is expressly warned to consider and adopt all safety precautions that might be indicated by the activities herein and to avoid all potential hazards. By following the instructions contained herein, the reader willingly assumes all risks in connection with such instructions.

The publisher makes no representation or warranties of any kind, including but not limited to, the warranties of fitness for particular purpose or merchantability, nor are any such representations implied with respect to the material set forth herein, and the publisher takes no responsibility with respect to such material. The publisher shall not be liable for any special, consequential, or exemplary damages resulting, in whole or part, from the readers' use of, or reliance upon, this material.

# Contents

# Preface

## THE SERIES

Welcome to Thomson Delmar Learning's *TechOne*, a state-of-the-art series designed to respond to today's automotive instructor and student needs. *TechOne* offers current, concise information on the ASE certification and other specific subject areas, combining classroom theory, diagnosis, and repair into one easy to use volume.

You'll notice several differences from a traditional textbook. First, a large number of short chapters divide complex material into chunks. Instructors can give tight, detailed reading assignments that students will find easier to digest. These shorter chapters can be taught in almost any order, allowing instructors to pick and choose the material that best reflects the depth, direction, and pace of their individual classes.

*TechOne* also features an art-intensive approach to suit today's visual learners—images drive the chapters. From drawings to photos, you will find more art to better understand the systems, parts, and procedures under discussion. Look also for helpful graphics that draw attention to key points in features such as "You Should Know" and "Interesting Fact."

Just as important, each *TechOne* starts off with a section on Safety and Communication, which stresses safe work practices, tool competence, and familiarity with workplace "soft skills," such as customer communication and the roles necessary to succeed as an automotive technician. From there, learners are ready to tackle the technical material in successive sections, ultimately leading them to the real test—an ASE practice exam in the Appendix.

## THE SUPPLEMENTS

*TechOne* comes with an **Instructor's Manual** that includes answers to all chapter-end review questions and a complete correlation of the text to NATEF standards. A **CD-ROM**, included with each Instructor's Manual, contains **PowerPoint Slides** for classroom presentations, a **Computerized Testbank** with hundreds of questions to aid in creating tests and quizzes, and an electronic version of the Instructor's Manual. Chapter-end review questions from the text also have been redesigned into adaptable **Electronic Worksheets**, so instructors can modify questions if desired to create in-class assignments or homework.

Flexibility is the key to *TechOne*. For those who would like to purchase job sheets, Thomson Delmar Learning's NATEF Standards Job Sheets are a good match. Topics cover the eight ASE subject areas and include:

- Engine Repair
- Automatic Transmissions and Transaxles
- Manual Drive Trains and Axles
- Suspension and Steering
- Brakes
- Electrical and Electronic Systems
- Heating and Air Conditioning
- Engine Performance

Visit **http://www.autoed.com** for a complete catalog.

## OTHER TITLES IN THIS SERIES

*TechOne* is Thomson Delmar Learning's latest automotive series. We are excited to announce these future titles:

- Engine Repair
- Suspension and Steering
- Heating and Air Conditioning
- Advanced Automotive Electronic Systems
- Advanced Engine Performance
- Automotive Fuels & Emissions

Check with your sales representative for availability.

## A NOTE TO THE STUDENT

There are now more computers on a car than aboard the first spacecraft, and even gifted backyard mechanics long ago turned their cars over to automotive professionals for diagnosis and repair. That's a statement about the nation's need for the knowledge and skills you'll develop as you continue your studies. Whether you eventually choose a career as a certified or licensed technician, service writer or manager, automotive engineer—or even decide to open your own shop—hard work will give you the opportunity to become one of the 840,000 automotive professionals providing and maintaining safe and efficient automobiles on our roads. As a member of a technically proficient, cutting-edge workforce, you'll fill a need, and, even better, you'll have a career to feel proud of.

Best of luck in your studies,
The Editors of Thomson Delmar Learning

# About the Author

Jack Erjavec, a master-certified ASE technician and professor emeritus at Columbus State Community College in Columbus, Ohio, has become a fixture in the automotive textbook world, with more than thirty works to his credit. In addition to his bestselling comprehensive text, *Automotive Technology*, Jack also is series editor of Thomson Delmar Learning's popular *Today's Technician* series and the new *TechOne* automotive series.

Prior to assuming these editorships, Jack was the Product Development Manager at Thomson Delmar Learning for business, industry, government and retail automotive materials, as well as a professional consultant for several automotive manufacturers. He is a long-time affiliate of the North American Council of Automotive Teachers, is currently the Vice President of Marketing, and has served on the board of directors and as executive vice president. Jack also is associated with a number of professional organizations, including the Society of Automotive Engineers, and remains active in the industry. In his free time, he most cherishes spending time with his family.

# Acknowledgments

I'd like to thank the following reviewers, whose technical expertise was invaluable in creating this text:

C. Neel Flannagan
Aiken Technical College
Aiken, SC

James Haun
Walla Walla Community College
Walla Walla, WA

Donald Lumsdon
Ivy Tech State College
Terre Haute, IN

Pat Paul
Kenner, LA

Scott Sanford
Monroe Community College
Rochester, NY

# Features of the Text

*TechOne* includes a variety of learning aids designed to encourage student comprehension of complex automotive concepts, diagnostics, and repair. Look for these helpful features:

**Section Openers** provide students with a **Section Table of Contents** and **Objectives** to focus the learner on the section's goals.

**Interesting Facts** spark student attention with industry trivia or history. Interesting facts appear on the section openers and are then scattered throughout the chapters to maintain reader interest.

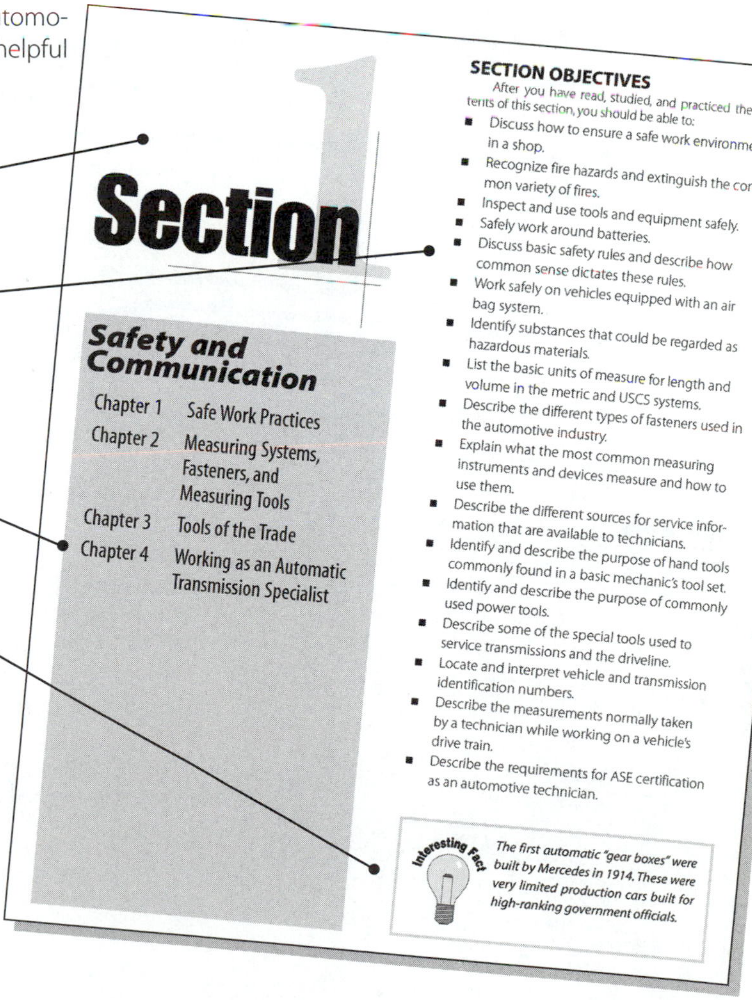

An **Introduction** orients readers at the beginning of each new chapter. **Technical Terms** are bolded in the text upon first reference and are defined.

# Torque Converter Operation

## Introduction

A torque converter uses fluid to smoothly transfer engine torque to the transmission. The torque converter is a doughnut-shaped unit, located between the engine and the transmission and filled with ATF **(Figure 1)**. Internally, the torque converter has three main parts: the impeller, turbine, and stator **(Figure 2)**. Each of these has blades, which are curved to increase torque converter efficiency.

The impeller is driven by the engine and directs fluid flow against the turbine blades, causing them to rotate and drive the turbine shaft, which is the transmission's input shaft. The stator is located between the impeller and the turbine and returns fluid from the turbine to the impeller, so that the cycle can be repeated.

During certain operating conditions, the torque converter multiplies torque. It provides extra reduction to meet the driveline needs while under a heavy load. When the vehicle is operating at cruising speeds, the torque converter operates as a **fluid coupling** and transfers engine torque to the transmission. It also absorbs the shock from gear changing in the transmission. Not all of the engine's power is transferred through the fluid to the transmission; some is lost. To reduce the amount of power lost through the converter, especially at cruising speeds, manufacturers equip most of their current transmissions with a torque converter clutch.

The engagement of the converter clutch is based on both engine and vehicle speeds and the clutch are controlled by transmission hydraulics and on-board computer electronic controls. When the clutch engages, a mechanical

**Figure 1.** A typical torque converter.

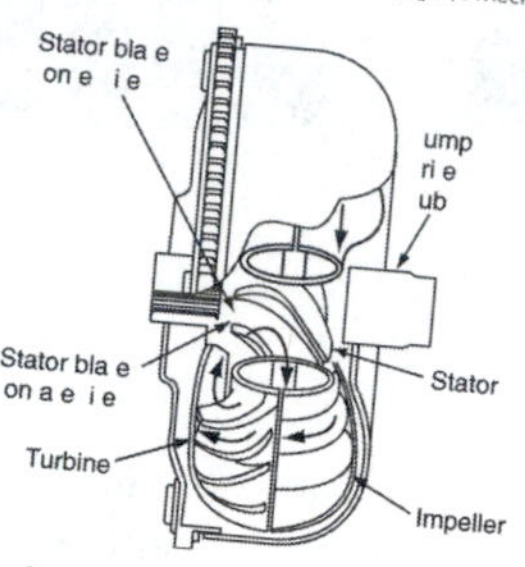


**Figure 2.** A torque converter's major internal parts are its impeller, turbine, and stator.

103

**Chapter 28 Transmission/Transaxle Removal • 343**

can completely remove the starter from the vehicle to get it totally out of the way.

Now pull the transaxle away from the engine. It may be necessary to use a pry bar between the transaxle and engine block to separate the two units. Make sure the converter comes out with the transmission. This prevents bending the input shaft, damaging the oil pump, or distorting the drive hub. After separating the transaxle from the engine, retain the torque converter in the bellhousing. This can be simply done by bolting a small combination wrench to a bellhousing bolthole across the outer edge of the converter.

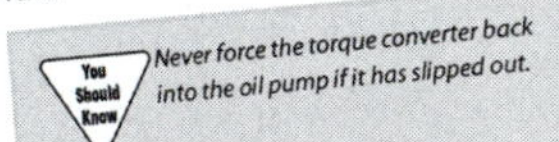


## Summary

- Normally, transmission removal begins with placing the vehicle on a hoist so that you can easily work under the vehicle and under the hood.
- Always disconnect the negative battery cable and place it away from the battery before beginning to remove the transmission or transaxle.
- Disconnect and remove anything that may get in the way when removing the transmission or transaxle.
- Unbolt the torque converter from the flexplate.
- Place a transmission jack under the transmission and secure the transmission to it before loosening and removing mounts and transmission-to-engine bolts.
- When lowering the transmission, make sure the converter hub and any associated shafts have a clear path.
- On some vehicles, the recommended removal procedure may include removing the engine with the transaxle.
- The drive or axle shafts must be removed before a transaxle can be removed.
- Engine support bars must be installed before attempting to remove the transaxle in most FWD vehicles.

## Review Questions

1. What is the best way to drain the fluid from a transmission?
2. What type of special tool may be needed to unbolt the torque converter from the flexplate?
3. What is the typical procedure for removing the axle shafts from a FWD vehicle?
4. *True or False:* On some vehicles, the engine must be removed before the transaxle can be removed.
5. Technician A says that a transmission jack is a good workstand for overhauling a transmission after it has been removed from the vehicle. Technician B says that the engine in most FWD vehicles needs to be supported when removing the transaxle. Who is correct?
   A. Technician A only
   B. Technician B only
   C. Both Technician A and Technician B
   D. Neither Technician A nor Technician B

**You Should Know** informs the reader whenever special safety cautions, warnings or other important points deserve emphasis.

A **Summary** concludes each chapter in short, bulleted sentences. **Review Questions** are structured in a variety of formats, including ASE style, challenging students to prove they've mastered the material.

An **ASE Practice Exam** is found in the **Appendix** of every *TechOne* book, followed by a **Bilingual Glossary**, which offers Spanish translations of technical terms alongside their English counterparts.

# Appendix A

## ASE PRACTICE EXAM FOR AUTOMATIC TRANSMISSIONS AND TRANSAXLES

1. When diagnosing a no engine braking condition in manual low operation, Technician A suspects a defective oil pump. Technician B suspects a bad drive link. Who is correct?
   A. Technician A only
   B. Technician B only
   C. Both Technician A and Technician B
   D. Neither Technician A nor Technician B
2. Which of the following is the MOST likely cause of slipping in all forward gear ranges?
   A. Faulty governor
   B. Clogged oil filter
   C. Faulty band or clutch
   D. Sticking valve in the valve body
3. When diagnosing the cause of no forced downshifts during full throttle operation, Technician A suspects a misadjusted manual linkage. Technician B suspects a dirty valve body and valves. Who is correct?
   A. Technician A only
   B. Technician B only
   C. Both Technician A and Technician B
   D. Neither Technician A nor Technician B
4. When diagnosing the cause of gear slipping in second gear only, Technician A suspects a worn or damaged clutch assembly. Technician B suspects a damaged drive link. Who is correct?
   A. Technician A only
   B. Technician B only
   C. Both Technician A and Technician B
   D. Neither Technician A nor Technician B
5. When conducting a pressure test, Technician A says that the cause of low pressure in all operating ranges could be a clogged filter. Technician B says that the cause of high pressure in all operating ranges could be a defective throttle valve. Who is correct?
   A. Technician A only
   B. Technician B only
   C. Both Technician A and Technician B
   D. Neither Technician A nor Technician B
6. When diagnosing the cause of no torque converter clutch engagement on a transmission that seems to shift fine, Technician A suspects a damaged clutch pressure plate. Technician B suspects a severely worn input clutch. Who is correct?
   A. Technician A only
   B. Technician B only
   C. Both Technician A and Technician B
   D. Neither Technician A nor Technician B
7. When diagnosing the cause of transmission overheating, Technician A checks for contaminated fluid. Technician B suspects a damaged flexplate. Who is correct?
   A. Technician A only
   B. Technician B only
   C. Both Technician A and Technician B
   D. Neither Technician A nor Technician B
8. When diagnosing the cause of sluggish acceleration, Technician A suspects a faulty governor. Technician B checks the condition of the fluid. Who is correct?
   A. Technician A only
   B. Technician B only
   C. Both Technician A and Technician B
   D. Neither Technician A nor Technician B

431

# Bilingual Glossary

**Abrasion** Wearing or rubbing away of a part.
***Erosión*** *cuando una pieza se desgasta.*

**Accumulator** A device used in automatic transmissions to cushion the shock of shifting between gears, providing a smoother feel inside the vehicle.
***Acumulador*** *Dispositivo que se usa en las transmisiones automáticas para amortiguar los cambios y provoca que se sienta el movimiento más suave dentro del vehículo.*

**Actuator** A control device that delivers mechanical action in response to an electrical signal.
***Biela de accionamiento*** *Dispositivo de control que envía una acción mecánica en respuesta a una señal eléctrica.*

**Adaptive Learning** The ability of a computer to monitor the drivers' habits and the operating conditions of its system and make adjustments to its program to correct for them.
***Aprendizaje adaptable*** *Habilidad de una computadora para inspeccionar los hábitos del conductor y las condiciones operacionales de su sistema y hacer ajustes a su programa para corregirlos.*

**Adhesives** Chemicals used to hold gaskets in place during the assembly of an engine. They also aid the gasket in maintaining a tight seal by filling in the small irregularities on the surfaces and by preventing the gasket from shifting because of engine vibration.
***Adhesivos*** *Químicos que se usan para detener el empaque en su lugar durante el montaje del motor. También ayudan al empaque a mantenerse firmemente sellado al llenar las pequeñas irregularidades en la superficie y al prevenir que el empaque se mueva con la vibración del motor.*

**Aeration** The process of mixing air into a liquid.
***Ventilación*** *Proceso de mezclar aire con un líquido.*

**Alignment** An adjustment to a line or made to bring into a line.
***Alineación*** *Ajuste a una línea o hecho para regular la posición de una línea.*

**Alternating Current** Electrical current that changes direction between positive and negative.
***Corriente alterna*** *Corriente eléctrica que cambia de dirección entre positivo y negativo.*

**Ammeter** The instrument used to measure electrical current flow in a circuit.
***Amperímetro*** *Instrumento que se usa para medir el flujo de la corriente eléctrica en un circuito.*

**Ampere** The unit for measuring electrical current; usually called an amp.
***Amperio*** *Unidad para medir la corriente eléctrica.*

**Amplitude** The height of a waveform.
***Amplitud*** *Altura de una onda.*

**Analog Signal** A voltage signal that varies within a given range (from high to low, including all points inbetween).
***Señal análoga*** *Señal de voltaje que varía en un cierto campo (de alto a bajo, incluyendo todos los puntos intermedios.)*

**Annulus gear** Another name for the ring gear of a planetary gear set.
***Palanca circular*** *Otro nombre para el eje de anillo de un conjunto de ejes planetarios.*

**Apply devices** Devices that hold or drive members of a planetary gear set. They may be hydraulically or mechanically applied.
***Dispositivo de ajuste*** *Dispositivos que detienen o conducen a los miembros de un conjunto de ejes planetarios. Se pueden ajustar hidráulica o mecánicamente.*

**ATF** Automatic Transmission Fluid.
***LTA*** *Líquido para la transmisión automática.*

**AWG** American Wire Gauge System. The system used to designate wire size.

440

A comprehensive **Index** helps instructors and students pinpoint information in the text.

# Index



463

# Section 1

## Safety and Communication



## SECTION OBJECTIVES

After you have read, studied, and practiced the contents of this section, you should be able to:

- Discuss how to ensure a safe work environment in a shop.
- Recognize fire hazards and extinguish the common variety of fires.
- Inspect and use tools and equipment safely.
- Safely work around batteries.
- Discuss basic safety rules and describe how common sense dictates these rules.
- Work safely on vehicles equipped with an air bag system.
- Identify substances that could be regarded as hazardous materials.
- List the basic units of measure for length and volume in the metric and USCS systems.
- Describe the different types of fasteners used in the automotive industry.
- Explain what the most common measuring instruments and devices measure and how to use them.
- Describe the different sources for service information that are available to technicians.
- Identify and describe the purpose of hand tools commonly found in a basic mechanic's tool set.
- Identify and describe the purpose of commonly used power tools.
- Describe some of the special tools used to service transmissions and the driveline.
- Locate and interpret vehicle and transmission identification numbers.
- Describe the measurements normally taken by a technician while working on a vehicle's drive train.
- Describe the requirements for ASE certification as an automotive technician.

*The first automatic "gear boxes" were built by Mercedes in 1914. These were very limited production cars built for high-ranking government officials.*

# Safe Work Practices

## Introduction

Safety is an important topic—important to you and those around you. Although the safety issues presented in this chapter are categorized, true safe work practices are based on nothing else but common sense and a knowledge of safety equipment.

### SAFE WORK AREAS

Your work area should be kept clean and safe. The floor and bench tops should be kept clean, dry, and orderly. Any oil, coolant, or grease on the floor can make it slippery. Slips can result in serious injuries. To clean up oil, use commercial oil absorbent. Keep all water off the floor. Water is slippery on smooth floors, and electricity flows well through water. Aisles and walkways should be kept clean and wide enough to easily move through. Make sure the work areas around machines are large enough to safely operate the machines.

Make sure all drain covers are snugly in place. Open drains or covers that are not flush to the floor can cause toe, ankle, and leg injuries.

Shop safety is the responsibility of everyone in the shop. Everyone must work together to protect the health and welfare of all who work in the shop.

### PERSONAL SAFETY

Personal safety simply involves those precautions you take to protect yourself from injury. Your eyes can become infected or permanently damaged by many things in a shop. Eye protection should be worn whenever you are working in the shop. There are many types of eye protection available. To provide adequate eye protection, safety glasses have lenses made of safety glass. They also offer some sort of side protection. Driveline and transmission work is either performed on the vehicle or at a bench. For nearly all services performed on the vehicle, eye protection **(Figure 1)** should be worn. This is especially true while working under the vehicle.

Some procedures may require that you wear other eye protection in addition to safety glasses. For example, when you are cleaning parts with a pressurized spray you should wear a face shield. A face shield not only gives added protection to your eyes but also protects the rest of your face.

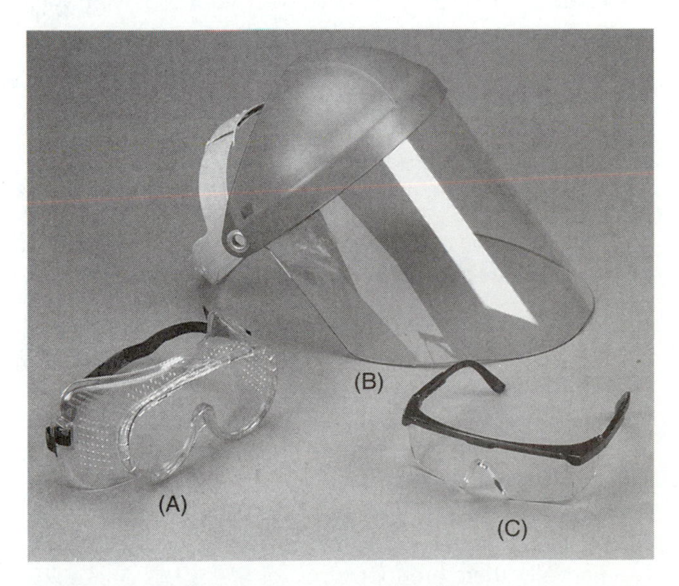


**Figure 1.** Different types of eye protection worn by automotive technicians: (A) goggles, (B) a face shield, and (C) safety glasses.

If chemicals such as battery acid, fuel, or solvents get into your eyes, flush them continuously with clean water. Have someone call a doctor and get medical help immediately. Many shops have eyewash stations or safety showers that should be used whenever you or someone else has been sprayed or splashed with a chemical.

Your clothing should be well fitted and comfortable but made with strong material. If you have long hair, tie it back or tuck it under a cap. Never wear rings, watches, bracelets, and neck chains. These can easily get caught in moving parts and cause serious injury.

Automotive work involves the handling of many heavy objects, which can be accidentally dropped on your feet or toes. Always wear shoes or boots of leather or similar material with no slip soles. Steel-tipped safety shoes can give added protection to your feet.

Good hand protection is often overlooked. A scrape, cut, or burn can limit your effectiveness at work for many days. A well-fitted pair of heavy work gloves should be worn during operations such as grinding and welding or when handling high-temperature components. Always wear approved rubber gloves when handling strong and dangerous caustic chemicals. They can easily burn your skin. Many technicians wear thin, surgical-type latex gloves whenever they are working on vehicles. These offer little protection against cuts but do offer protection against disease and grease buildup under and around your fingernails. These gloves are comfortable and are quite inexpensive.

## LIFTING AND CARRYING

When lifting a heavy object like a transmission, use a hoist or have someone else help you. If you must work alone, always lift heavy objects with your legs, not your back. Bend down with your legs, not your back, securely grasp the object you are lifting, and then stand up, keeping the object close to you **(Figure 2)**. Trying to "muscle" something with your arms or back can result in severe damage to your back and may end your career and limit what you do for the rest of your life.

**Figure 2.** When lifting a heavy object, always use your legs and keep the object close to your body.

## FIRE HAZARDS AND PREVENTION

Many items around a typical shop are a potential fire hazard. These include: gasoline, diesel fuel, cleaning solvents, and dirty rags. Each of these should be treated as a potential firebomb and handled and stored properly.

In case of a fire you should know the location of the fire extinguishers and fire alarms in the shop and also how to use them. You also should be aware of the different types of fires and the fire extinguishers used to put out these types of fires.

Basically, there are four types of fires. Class A fires are those in which wood, paper, and other ordinary materials are burning, Class B fires are those involving flammable liquids, such as gasoline, diesel fuel, paint, grease, oil, and other similar liquids. Class C fires are electrical fires. Class D fires are a unique type of fire, for the material burning is a metal. An example of this is a burning "mag" wheel; the magnesium used in the construction of the wheel is a flammable metal and will burn brightly when subjected to high heat.

## USING A FIRE EXTINGUISHER

Remember: during a fire never open doors or windows unless it is absolutely necessary; the extra draft will only make the fire worse. Make sure the fire department is contacted before or during your attempt to extinguish a fire. To extinguish a fire, stand six to ten feet from the fire **(Figure 3)**. Hold the extinguisher firmly in an upright position. Aim the nozzle at the base and use a side-to-side motion, sweeping the entire width of the fire. Stay low to avoid inhaling the

**Figure 3.** Stand away from the fire before activating the extinguisher.

smoke. If it gets too hot or too smoky, get out. Remember: never go back into a burning building for anything. When using an extinguisher, remember the word "PASS":

**P**ull the pin from the handle of the extinguisher.
**A**im the extinguisher's nozzle at the base of the fire.
**S**queeze the handle.
**S**weep the entire width of the fire with the contents of the extinguisher.

If there is not a fire extinguisher handy, a blanket or fender cover may be used to smother the flames. You must be careful when doing this because the heat of the fire may burn you and the blanket. If the fire is too great to smother, move everyone away from the fire and call the local fire department. A simple under-the-hood fire can cause the total destruction of the car and the building and can take some lives. You must be able to respond quickly and precisely to avoid a disaster.

## SAFE TOOLS AND EQUIPMENT

Whenever you are using any equipment, make sure you use it properly and that it is set up according to the manufacturer's instructions. All equipment should be properly maintained and periodically inspected for unsafe conditions. Frayed electrical cords or loose mountings can cause serious injuries. All electrical outlets should be equipped to allow for the use of three-pronged electrical cords. All equipment with rotating parts should be equipped with safety guards **(Figure 4)** that reduce the possibility of the parts coming loose and injuring someone.

Do not depend on someone else to inspect and maintain equipment. Check it out before you use it! If you find the equipment unsafe, put a sign on it to warn others and notify the person in charge.

Never use tools and equipment for purposes other than those for which they are designed. Using the proper

**Figure 4.** Equipment such as fixtures for wire brushes and grinding wheels should be equipped with safety shields and tool guards.

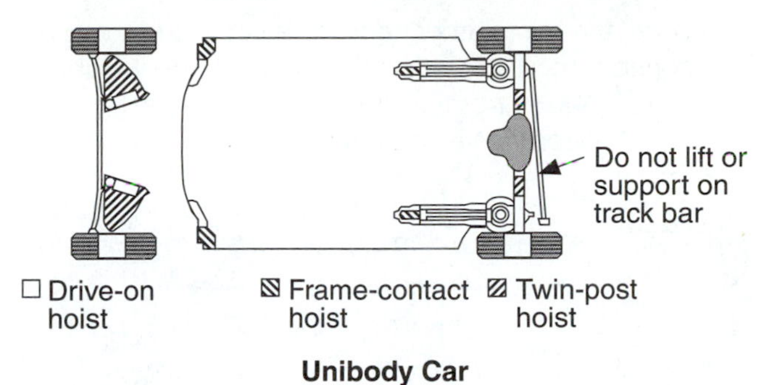


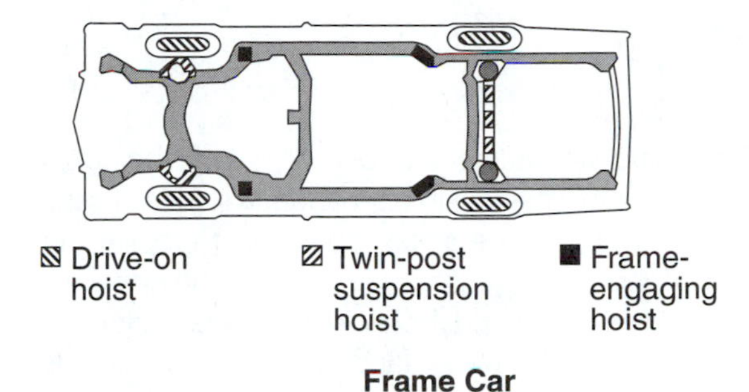


**Figure 5.** Typical lift points for (A) unibody and (B) frame/body vehicles.

tool in the correct way will not only be safer but also will allow you to do a better job.

## LIFT SAFETY

Always be careful when raising a vehicle on a lift or a hoist. Adapters and hoist plates must be positioned correctly on twin post and rail-type lifts to prevent damage to the underbody of the vehicle. There are specific lift points. These points allow the weight of the vehicle to be evenly supported by the adapters or hoist plates. The correct lift points can be found in the vehicle's service manual. **Figure 5** shows typical lift locations for unibody and frame cars. These diagrams are for illustration only. Always follow the manufacturer's instructions. Before operating any lift or hoist, carefully read the operating manual and follow the operating instructions.

Once you feel the lift supports are properly positioned under the vehicle, raise the lift until the supports contact the vehicle. Then, check the supports to make sure they are in full contact with the vehicle. Shake the vehicle to make sure it is securely balanced on the lift, and then raise the lift to the desired working height.

## JACK AND JACK STAND SAFETY

A vehicle also can be raised off the ground with a hydraulic jack. The lifting pad of the jack must be positioned under an area of the vehicle's frame or at one of the manufacturer's recommended lift points. Never place the pad under the floor pan or under steering and suspension components; these are easily damaged by the weight of the vehicle. Always position the jack so the wheels of the vehicle can roll as the vehicle is being raised.

> **You Should Know** *Never use a lift or jack to move something heavier than it is designed for. Always check the rating before using a lift or jack. If a jack is rated for two tons, do not attempt to use it for a job requiring five tons. It is dangerous for you and the vehicle.*

Safety (jack) stands are supports of different heights that sit on the floor. They are placed under a sturdy chassis member, such as the frame or axle housing, to support the vehicle. Once the safety stands are in position, the hydraulic pressure in the jack should be slowly released until the weight of the vehicle is on the stands. Like jacks, jack stands also have a capacity rating. Always use a jack stand of the correct rating.

Never move under a vehicle when it is only supported by a hydraulic jack; rest the vehicle on the safety stands first. The jack should be removed after the jack stands are set in place. This eliminates a hazard, such as a jack handle sticking out into a walkway. A jack handle that is bumped or kicked can cause a tripping accident or can cause the vehicle to fall.

## BATTERIES

When possible, you should disconnect the **battery** of a car before you disconnect any electrical wire or component. This prevents the possibility of a fire or electrical shock. It also eliminates the possibility of an accidental short, which can ruin the car's electrical system. This is especially true of newer cars that are equipped with many electronic and computerized controls. Any electrical arcing can cause damage to the components. To properly disconnect the battery, disconnect the negative or ground cable first, then disconnect the positive cable. Because electrical circuits require a ground to be complete, by removing the ground cable you eliminate the possibility of accidentally completing a circuit. When reconnecting the battery, connect the positive cable first, then the negative.

> **You Should Know** *The active chemical in a battery, the* ***electrolyte****, is basically sulfuric acid. Sulfuric acid can cause severe skin burns and permanent eye damage, including blindness, if it gets in your eye. If some battery acid gets on your skin, wash it off immediately and flush your skin with water for at least 5 minutes. If the electrolyte gets into your eyes, immediately flush them out with water, then immediately see a doctor. Never rub your eyes; just flush them well and go to a doctor. It shows common sense to wear safety glasses or goggles while working with and around batteries.*

## ELECTRICAL SYSTEM REPAIRS

Some electronic replacement parts are very sensitive to **static electricity**. These parts will be labeled as such. Whenever you are handling a part that is sensitive to static, you should follow these guidelines to reduce any possible electrostatic charge buildup on your body and the electronic part:

1. Do not open the package until it is time to install the component.
2. Before removing the part from the package, ground the package to a known good ground on the car.
3. Always touch a known good ground before handling the part. This should be repeated while handling the part and more frequently after sliding across the seat, sitting down from a standing position, or walking a distance.
4. Never touch the electrical terminals of the component.

## ACCIDENTS

Make sure you are aware of the location and contents of the shop's first aid kit **(Figure 6)**. There should be an eyewash station in the shop so that you can rinse your eyes thoroughly should you get acid or some other irritant into them. If there are specific first aid rules in your

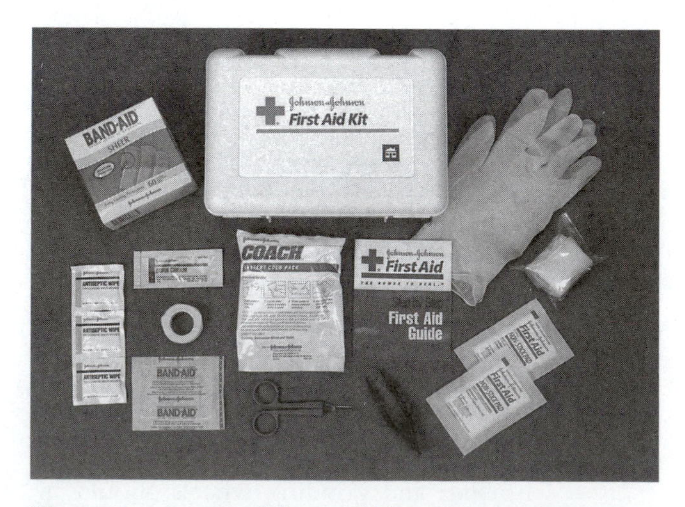


**Figure 6.** A typical first aid kit and its contents.

school or shop, make sure you are aware of them and follow them. Some first aid rules apply to all circumstances and are normally included by everyone. If someone is overcome by carbon monoxide, get him or her fresh air immediately. Burns should be cooled immediately by rinsing them with water. Whenever there is severe bleeding from a wound, try to stop the bleeding by applying pressure with clean gauze on or around the wound, and get medical help. Never move someone who may have broken bones unless the person's life is otherwise endangered. Moving that person may cause additional injury. Call for medical assistance.

Your supervisor should be informed immediately of all accidents that occur in the shop. It is a good idea to keep a list of up-to-date emergency telephone numbers posted next to the telephone. The numbers should include those of a doctor, a hospital, and of the fire and police departments.

## AIR BAG SAFETY AND SERVICE WARNINGS

Transmission controls and other transmission related components are located in the proximity of the dash. The dash and steering wheel contain the circuits that control the **air bag system**. Whenever working on or around air bag systems, it is important to follow some safety warnings. There are safety concerns with both **deployed** and live (undeployed) air bag modules.

1. Wear safety glasses when servicing the air bag system.
2. Wear safety glasses when handling an air bag module.
3. Wait at least 30 minutes after disconnecting the battery before beginning any service on or around the air bag system. The reserve energy module is capable of storing enough power to deploy the air bag for up to ten minutes after battery voltage is lost.
4. Handle all air bag sensors with care. Do not strike or jar a sensor in such a manner that deployment may occur.
5. When carrying a live air bag module, face the trim and bag away from your body.
6. Do not carry the module by its wires or connector.
7. When placing a live module on a bench, face the trim and air bag up.
8. Deployed air bags may have a powdery residue on them. Sodium hydroxide is produced by the deployment reaction and is converted to sodium carbonate when it comes into contact with atmospheric moisture. It is unlikely that sodium hydroxide will still be present. However, wear safety glasses and gloves when handling a deployed air bag. Wash your hands immediately after handling the bag.
9. A live air bag module must be deployed before disposal. Because an air bag is deployed through an explosive process, improper disposal may result in injury and in fines. A deployed air bag should be disposed of in a manner consistent with EPA and manufacturer procedures.
10. Do not use a battery- or AC-powered voltmeter, ohmmeter, or any other type of test equipment not specified in the service manual. Never use a test light to probe for voltage.

## HAZARDOUS MATERIALS

Many solvents and other chemicals used in an auto shop have warning and caution labels that should be read and understood by everyone that uses them. These solvents and chemicals are typically considered to be hazardous materials. Also, many service procedures generate what are known as **hazardous wastes**. Dirty solvents and liquid cleaners are good examples of these.

Every employee in a shop is protected by **"Right-to-Know" Laws** concerning hazardous materials and wastes. The general intent of these laws is for employers to provide a safe working place as it relates to hazardous materials. All employees must be trained about their rights under the legislation, the nature of the hazardous chemicals in their workplace, the labeling of chemicals, and the information about each chemical listed and described on **Material Safety Data Sheets (MSDS)**. These sheets are available from the manufacturers and suppliers of the chemicals. They detail the chemical composition and precautionary information for all products that can present health or safety hazards.

Employees must be familiar with the intended purposes of the substance, the recommended protective equipment, the accident and spill procedures, and any other information regarding the safe handling of hazardous materials. This training must be given annually to employees and be provided to new employees as part of their job orientation. The Canadian equivalents to the MSDS are called Workplace Hazardous Materials Information Systems (WHMIS).

You Should Know *When handling any hazardous material, always wear the appropriate safety protection. Always follow the correct procedures while using the material and be familiar with the information given on the MSDS for that material.*

All hazardous materials should be properly labeled, indicating what health, fire, or reactivity hazards they pose and what protective equipment is necessary when handling each chemical. The manufacturers of the hazardous materials must provide all warnings and precautionary information, which must be read and understood by all users before they use the materials. You should pay great attention to the label information. By doing so, you will use the substance in the proper and safe way, thereby preventing hazardous conditions.

A list of all hazardous materials used in the shop should be posted for the employees to see. Shops must maintain documentation on the hazardous chemicals in the workplace, proof of training programs, records of accidents or spill incidents, satisfaction of employee requests for specific chemical information via the MSDS, and a general right-to-know compliance procedure manual utilized within the shop.

There are many government agencies charged with ensuring safe work environments for all workers. These include the **Occupational Safety and Health Administration (OSHA)**, the Mine Safety and Health Administration (MSHA), and the National Institute for Occupational Safety and Health (NIOSH). These, in addition to state and local governments, have instituted regulations that must be understood and followed. Everyone in a shop has the responsibility to adhere to these regulations.

## OSHA

In 1970, OSHA was formed by the federal government to "assure safe and healthful working conditions for working men and women; by authorizing enforcement of the standards developed under the Act; by assisting and encouraging the States in their efforts to assure safe and healthful working conditions by providing research, information, education, and training in the field of occupational safety and health."

Safety standards have been established that will be consistent across the country. It is the employers' responsibility to provide a place of employment that is free from all recognized hazards and that will be inspected by government agents knowledgeable in the law of working conditions. OSHA controls all safety and health issues of the automotive industry.

OSHA and the EPA have other strict rules and regulations that help to promote safety in the auto shop. These are described throughout this text whenever they are applicable. Maintaining a vehicle involves handling and managing a wide variety of materials and wastes. Some of these wastes can be toxic to fish, wildlife, and humans when improperly managed. No matter the amount of waste produced, it is to the shop's legal and financial advantage to manage the wastes properly and, even more importantly, to prevent pollution.

## HANDLING SHOP WASTES

- Recycle engine oil. Set up equipment, such as a drip table or a screen table with a used oil collection bucket, to collect oils dripping off parts. Place drip pans underneath vehicles that are leaking fluids onto the storage area. Never allow oil to enter into the shop's sewage or drain system.
- Used oil filters should be drained for at least 24 hours, then crushed and recycled.
- Old batteries should be recycled by sending them to a reclaimer or back to the distributor.
- Collect metal filings when machining metal parts. If possible, keep them separate from other dirt and recycle the metal. Prevent metal filings from falling into a storm sewer drain.
- Recover and/or recycle refrigerants during the service and disposal of motor vehicle air conditioners and refrigeration equipment. It is illegal to knowingly vent refrigerants to the atmosphere. Recovery and/or recycling during service must be performed by an EPA-certified technician using certified equipment and following specified procedures.
- Replace hazardous chemicals with less toxic alternatives that have equal performance. For example, substitute water-based cleaning solvents for petroleum-based solvent degreasers. Hire a hazardous waste management service to clean and recycle solvents. Store solvents in closed containers to prevent evaporation. Properly label spent solvents.
- Store materials such as scrap metal, old machine parts, and worn tires under a roof or tarpaulin to protect them from the elements and to prevent the potential for contaminated runoff. Consider recycling tires by retreading them.
- Collect and recycle coolants from radiators. Store transmission fluids, brake fluids, and solvents containing chlorinated hydrocarbons separately, and recycle or dispose of them properly.
- Keep waste towels in a closed container marked "Contaminated Shop Towels Only." To reduce costs and liabilities associated with disposal of used towels, which can be classified as hazardous wastes, investigate using a laundry service that is able to treat the wastewater generated from cleaning the towels.

# *Summary*

- True safe work practices are based on nothing else but common sense and knowledge of safety equipment.
- A safe work area needs to be kept clean and orderly.
- Shop safety is the responsibility of everyone in the shop, and everyone must work together to protect the health and welfare of everyone in the shop.
- Eye protection should be worn whenever you are working in the shop.
- If chemicals or solvents get into your eyes, flush them continuously with clean water. Have someone call a doctor and get medical help immediately.

- Dress to protect yourself and to avoid accidents. Wear well-fitted and comfortable clothing and foot and hand protection. Never wear anything that may get caught in moving parts or that can conduct electricity.
- When lifting a heavy object, use a hoist or have someone help you, and always lift with your legs, not your back.
- Gasoline, diesel fuel, cleaning solvents, and dirty rags should be treated as potential firebombs and handled and stored properly.
- Know where the fire extinguishers and fire alarms are located in the shop and know how to use them (think of the word PASS.).
- Inspect all tools and equipment before using them. Check them for unsafe conditions.
- Never use tools and equipment for purposes other than those for which they are designed.
- Make sure the contacts of a lift are properly positioned on the vehicle before lifting it.
- Use jack stands to secure a vehicle after it has been raised by a jack.
- When possible, disconnect the battery of a car before you begin to work around it or before you disconnect any electrical wire or component.
- Disconnect the negative cable of the battery first and reconnect it last.
- Follow the proper procedures for eliminating static when handling parts that are static sensitive.
- Know the location and contents of the shop's first aid kits and eye wash stations and know where the list of emergency telephone numbers is.
- Whenever working on or around air bag systems, it is important to follow certain procedures.
- Many solvents and other chemicals used in an auto shop have warning and caution labels that should be read and understood by everyone that uses them.
- Handle and dispose of all hazardous wastes according to local laws and your own common sense.
- Know where the MSDS can be found in the shop and know how to quickly find the appropriate information in them.

## Review Questions

1. When should you wear eye protection?
2. What is the correct procedure for putting out a fire with an extinguisher?
3. What do all employees of a shop have the right to know?
4. The general intent of __________ ___ __________ laws is for employers to provide a safe working place as it relates to an awareness of hazardous materials.
5. Technician A says that accidents can be prevented by not having anything dangle near rotating equipment and parts. Technician B says that most accidents can be prevented by using common sense. Who is correct?
   A. Technician A only
   B. Technician B only
   C. Both Technician A and Technician B
   D. Neither Technician A nor Technician B
6. Technician A says that unsafe equipment should have its power disconnected and marked with a sign to warn others not to use it. Technician B says that all equipment should be inspected for safety hazards before being used. Who is correct?
   A. Technician A only
   B. Technician B only
   C. Both Technician A and Technician B
   D. Neither Technician A nor Technician B
7. While discussing ways to create an accident-free work environment Technician A says that everyone in the shop should take full responsibility for ensuring safe work areas. Technician B says that the appearance and work habits of technicians can help prevent accidents. Who is correct?
   A. Technician A only
   B. Technician B only
   C. Both Technician A and Technician B
   D. Neither Technician A nor Technician B
8. While discussing the car's electrical system Technician A says that you should always disconnect the negative or ground battery cable first, then disconnect the positive cable. Technician B says that you should always connect the positive battery cable first, then the negative. Who is correct?
   A. Technician A only
   B. Technician B only
   C. Both Technician A and Technician B
   D. Neither Technician A nor Technician B
9. If a chemical gets into someone's eye, what should you do?
10. What should you think about before lifting a heavy object?

# Measuring Systems, Fasteners, and Measuring Tools

## *Introduction*

Servicing automatic transmissions requires the use of various tools. Many of these tools are used to remove and install fasteners. Fasteners of different sizes and shapes are used on today's cars. Basically, the correct tool for the job is the tool that fits the fastener. The size of both tools and fasteners is expressed according to increments defined by a measuring system.

In this chapter, the most common measuring systems will be explained. These systems will then be used to describe various fasteners. Also covered in this chapter are the most commonly used measuring instruments. The proper use of these instruments ensures that parts are installed properly and according to specifications.

### MEASURING SYSTEMS

Two different systems of weights and measures are used: the United States Customary System (USCS) and the International System (SI), commonly referred to as the metric system. The USCS units of measurement were brought to the United States by the original English settlers and are sometimes referred to as the English or Imperial System. The United States is slowly changing over to the metric system; during the changeover, cars produced in the United States are being made with both English and metric fasteners and specifications. Most of the world outside the United States uses the metric system.

Linear USCS units are inches, feet, and yards. The inch can be broken down into fractions, such as 1/64, 1/32, 1/16, 1/8, 1/4, and 1/2 inch. The inch is also commonly expressed in decimals. When an inch is divided into tenths, each part is a tenth of an inch (0.1 in.). Tenths of an inch can be further divided by ten into hundredths of an inch (0.01 in.). The division after hundredths is thousandths (0.001 in.); these are followed by ten-thousandths (0.0001 in.).

In the metric system, the basic measurement unit of length is the meter. For exact measurements, the meter is divided into units of ten. The first division is called a decimeter (dm), the second division is the centimeter (cm), and the third—and most commonly used—division is the millimeter (mm). One dm equals 0.1 meter, 1 cm equals 0.01 m, and 1 mm equals 0.001 m.

Because some vehicles have metric fasteners, some have USCS, and others have both, automotive technicians must have both English and metric tools. Vehicle specifications are normally listed in both meters and inches; therefore, measuring tools are available in both measuring systems. Appendix C contains some of the common equivalents of the two systems.

### FASTENERS

Fasteners are those things that are used to secure or hold parts of something together. Many types and sizes of fasteners are commonly used. Each fastener is designed for a specific purpose. One of the most commonly used types of fastener is the threaded fastener. Threaded fasteners include bolts, nuts, screws, and similar items that allow a technician to install or remove parts easily.

Threaded fasteners are available in many sizes, designs, and threads. The threads can be either cut or rolled into the fastener. Rolled threads are 30 percent stronger than cut threads. They also offer better fatigue resistance because there are no sharp notches to create stress points. Fasteners are made to Imperial or metric measurements. There are

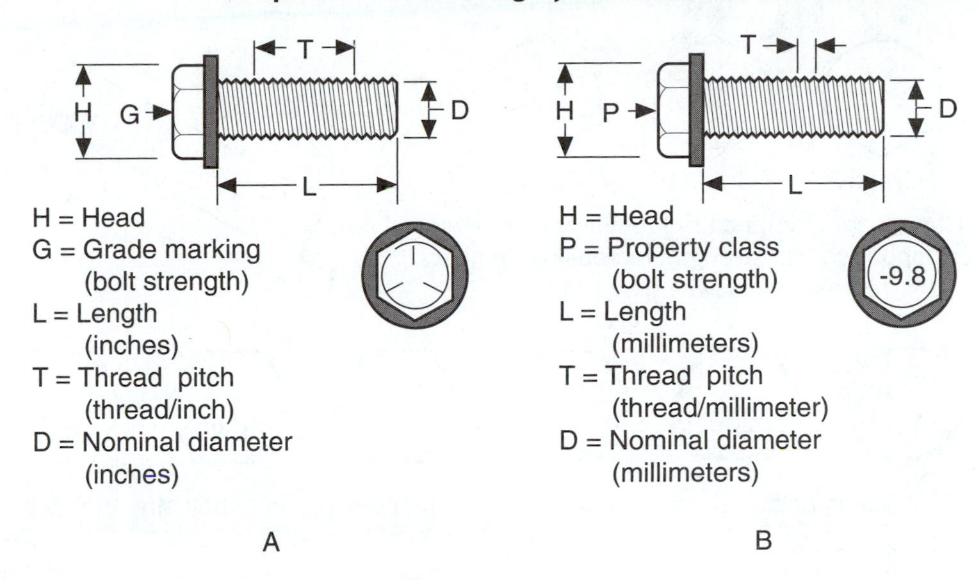


**Figure 1.** (A) English and (B) metric bolt terminology.

four classifications for the threads of Imperial fasteners: Unified National Coarse (UNC), Unified National Fine (UNF), Unified National Extra Fine (UNEF), and Unified National Pipe Thread (UNPT or NPT). Metric fasteners are also available in fine and coarse threads.

Coarse threads are used for general-purpose work, especially where rapid assembly and disassembly is required. Fine threaded fasteners are used when greater holding force is necessary. They also are used when greater resistance to vibration is desired.

Bolts have a head on one end and threads on the other. Bolts are identified by their head size, shank diameter, thread pitch, length, and grade **(Figure 1)**. Bolts have a shoulder below the head and the threads do not travel all the way from the head to the end of the bolt.

Studs are rods with threads on both ends. Most often, the threads on one end are coarse, while the other end is fine thread. One end of the stud is screwed into a threaded bore. A hole in the part to be secured is fitted over the stud and held in place with a nut screwed over the stud. Studs are used when the clamping pressures of a fine thread are needed and a bolt will not work. If the material the stud is being screwed into is soft (such as aluminum) or granular (such as cast iron), fine threads will not withstand a great amount of pulling force on the stud. Therefore, a coarse thread is used to secure the stud in the work piece and a fine threaded nut is used to secure the other part to it. Doing this combines the clamping force of fine threads and the holding power of coarse threads.

Nuts are used with other threaded fasteners when the fastener is not threaded into a piece of work. Many different designs of nuts are found on today's cars. The most common one is the hex nut, which is used with studs and bolts and is tightened with a wrench.

Setscrews are used to prevent rotary motion between two parts, such as a pulley and shaft. Headless setscrews require an Allen wrench or screwdriver to loosen and tighten them. Setscrews can have a square head.

## BOLT IDENTIFICATION

The **bolt head** is used to loosen and tighten the bolt; a socket or wrench fits over the head and is to screw the bolt in or out. The size of the bolt head varies with the diameter of the bolt and is available in USCS and metric wrench sizes. Many confuse the size of the head with the size of the bolt. The size of a bolt is determined by the diameter of its shank. The size of the bolt head determines what size wrench is required to screw it.

Bolt diameter is the measurement across the major diameter of the threaded area or across the bolt shank. The length of a bolt is measured from the bottom surface of the head to the end of the threads.

The **thread pitch** of a bolt in the Imperial system is determined by the number of threads that are in one inch of the threaded bolt length and is expressed in number of threads per inch. A UNF bolt with a 3/8-inch diameter would be a 3/8 x 24 bolt. It would have 24 threads per inch. Likewise, a 3/8-inch UNC bolt would be called a 3/8 x 16.

The distance in millimeters between two adjacent threads determines the thread pitch in the metric system. This distance will vary between 1.0 and 2.0 and depends on the diameter of the bolt. The lower the number, the closer the threads are placed and the finer the threads are.

The bolt's tensile strength, or grade, is the amount of stress or stretch it is able to withstand before it breaks. The type of material the bolt is made of and the diameter of the bolt determine its grade. In the Imperial system, the tensile

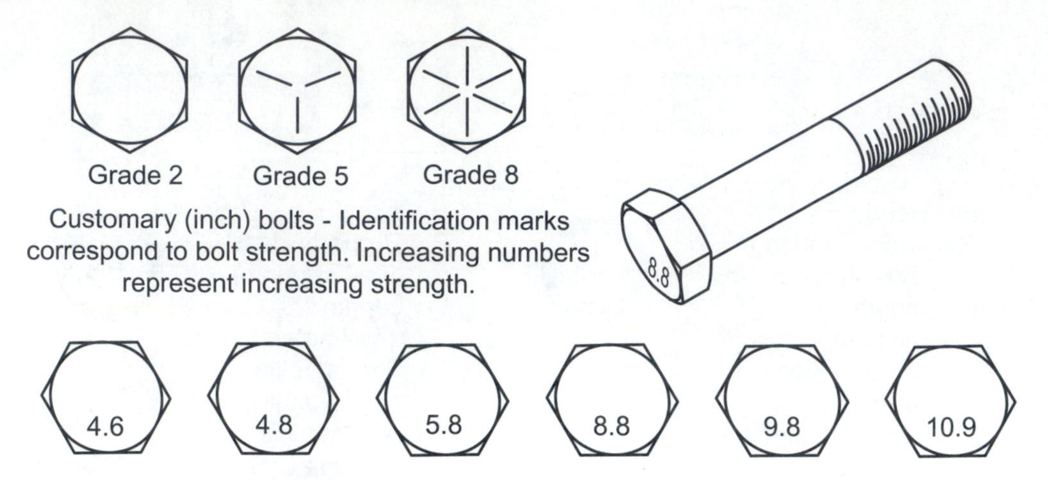


**Figure 2.** Bolt head markings indicating the grade of a bolt.

strength of a bolt is identified by the number of radial lines (**grade marks**) on the bolt's head. More lines mean higher tensile strength. Count the number of lines and add two to determine the grade of a bolt **(Figure 2)**.

A property class number on the bolt head identifies the grade of metric bolts. This numerical identification is comprised of two numbers. The first number represents the tensile strength of the bolt. The higher the number, the greater the tensile strength. The second number represents the yield strength of the bolt. This number represents how much stress the bolt can take before it is not able to return to its original shape without damage. The second number represents a percentage rating. For example, a 10.9 bolt has a tensile strength of 1000 MPa (145,000 psi) and a yield strength of 900 MPa (90 percent of 1000). A 10.9 metric bolt is similar in strength to an SAE grade 8 bolt.

Nuts are graded to match their respective bolts. For example, a grade 8 nut must be used with a grade 8 bolt. If a grade 5 nut were used, a grade 5 connection would result. Grade 8 and critical applications require the use of fully hardened flat washers. These will not dish out when torqued like soft washers will.

Bolt heads can pop off because of **fillet** damage. The fillet is the smooth curve, where the shank flows into the bolt head **(Figure 3)**. Scratches in this area introduce stress to the bolt head, causing failure. Removing any burrs around the edges of holes can protect the bolt head. Also place flat washers with their rounded, punched side against the bolt head and their sharp side to the work surface.

Fatigue breaks are the most common type of bolt failure. A bolt becomes fatigued from working back and forth when it is too loose. Undertightening the bolt causes this problem. Bolts also can be broken or damaged by overtightening, being forced into a nonmatching thread, or bottoming out, which happens when the bolt is too long.

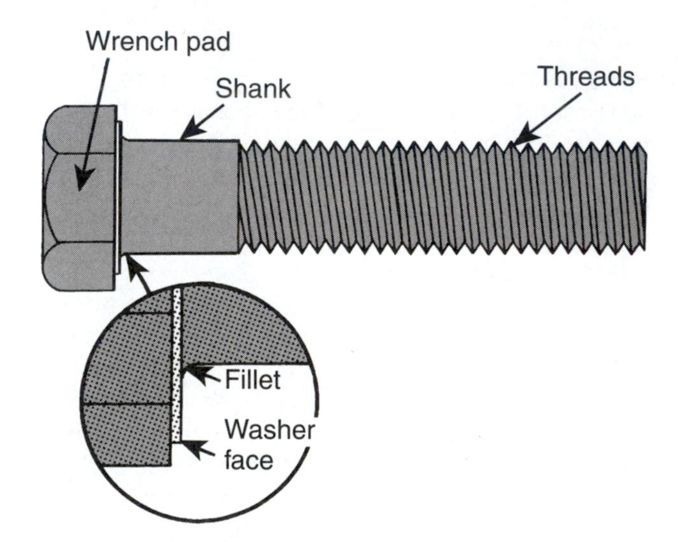


**Figure 3.** Bolt fillet detail.

## TIGHTENING BOLTS

Any fastener is near worthless if it is not as tight as it should be. When a bolt is properly tightened, it will be "spring loaded" against the part it is holding. This spring effect is caused by the stretch of the bolt when it is tightened. Normally a properly tightened bolt is stretched to 70 percent of its elastic limit. The elastic limit of a bolt is that point of stretch at which the bolt will not return to its original shape when it is loosened. Not only will an overtightened or stretched bolt not have sufficient clamping force, but also it will have distorted threads. The stretched threads will make it more difficult to screw and unscrew the bolt or a nut on the bolt. Always check the service manual to see if there is a torque specification for a bolt before tightening it. If there is, use a torque wrench and tighten the bolt properly.

## MEASURING TOOLS

Many of the procedures discussed in this manual require exact measurements of parts and clearances. Accurate measurements require the use of precision measuring devices that are designed to measure things in very small increments. Measuring tools are delicate instruments and should be handled with great care. Never strike, pry, drop, or force these tools. Also make sure you clean them before and after every use.

### Machinist's Rule

The **machinist's rule (Figure 4)** looks very much like an ordinary ruler. Each edge of this basic measuring tool is divided into increments based on a different scale. A typical machinist's rule based on the Imperial system of measurement may have scales based on 1/8-, 1/16-, 1/32-, and 1/64-inch intervals. Of course, metric machinist's rules are also available. Metric rules are usually divided into 0.5-mm and 1-mm increments.

Some machinist rules may be based on decimal intervals. These are typically divided into 1/10-, 1/50-, and 1/1,000-inch (0.1-, 0.03-, and 0.01-mm) increments. Decimal machinist's rules are very helpful when measuring dimensions that are specified in decimals; they make such measurements much easier.

### Feeler Gauge

A **feeler gauge** is a thin strip of metal or plastic of known and closely controlled thickness. Several of these metal strips are often assembled together as a feeler gauge set that looks like a pocketknife **(Figure 5)**. The desired thickness gauge can be pivoted away from others for convenient use. A steel feeler gauge pack usually contains strips or leaves of 0.002- to 0.010-inch thickness (in steps of

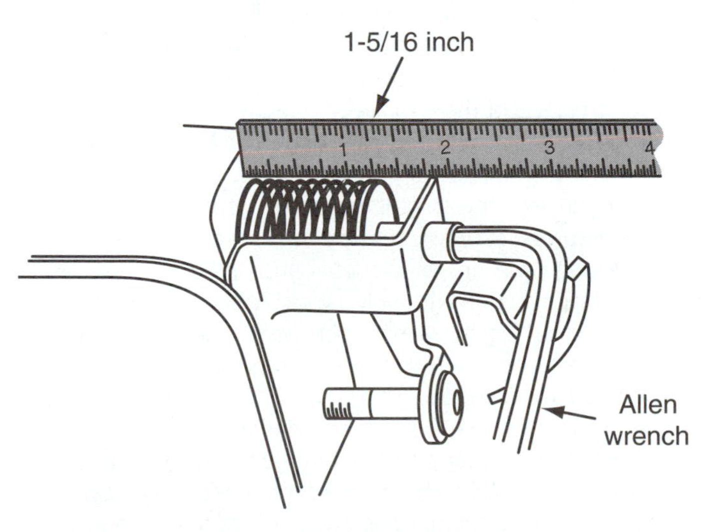


**Figure 4.** Using a machinist's rule to make an adjustment to line pressure.

**Figure 5.** Using a feeler gauge and straightedge to check the flatness of a transmission pump.

0.001 inch) and leaves of 0.012- to 0.024-inch thickness (in steps of 0.002 inch).

A feeler gauge can be used by itself to measure clearances and gaps or it can be used with a precision straightedge to measure alignment and surface warpage.

### Screw Pitch Gauge

The use of a **screw pitch gauge** provides a quick and accurate method of checking the thread pitch of a fastener. The leaves of this measuring tool are marked with the various pitches. To check the pitch of threads, simply match the teeth of the gauge with the threads of the fastener. Then, read the pitch from the leaf. Screw pitch gauges are available for the various types of fastener threads used by the automotive industry.

### Vernier Caliper

A **vernier caliper** is a measuring tool **(Figure 6)**. It can make inside, outside, or depth measurements. It is marked in both USCS and metric divisions called a vernier scale. A vernier scale consists of a stationary scale and a movable scale—in this case, the vernier bar to the vernier plate. The length is read from the vernier scale.

A vernier caliper has a movable scale that is parallel to a fixed scale. These precision measuring instruments are capable of measuring outside and inside diameters and most will even measure depth. Vernier calipers are available in both Imperial and metric scales. The main scale of the caliper is divided into inches; most measure up to six inches. Each inch is divided into ten parts, each equal to 0.100 inch. The area between the 0.100 marks is divided into four. Each of these divisions is equal to 0.025 inch.

The vernier scale has 25 divisions, each one representing 0.001 inch. Measurement readings are taken by combining the main and vernier scales. At all times, only one division line on the main scale will line up with a line on the vernier scale. This is the basis for accurate measurements.

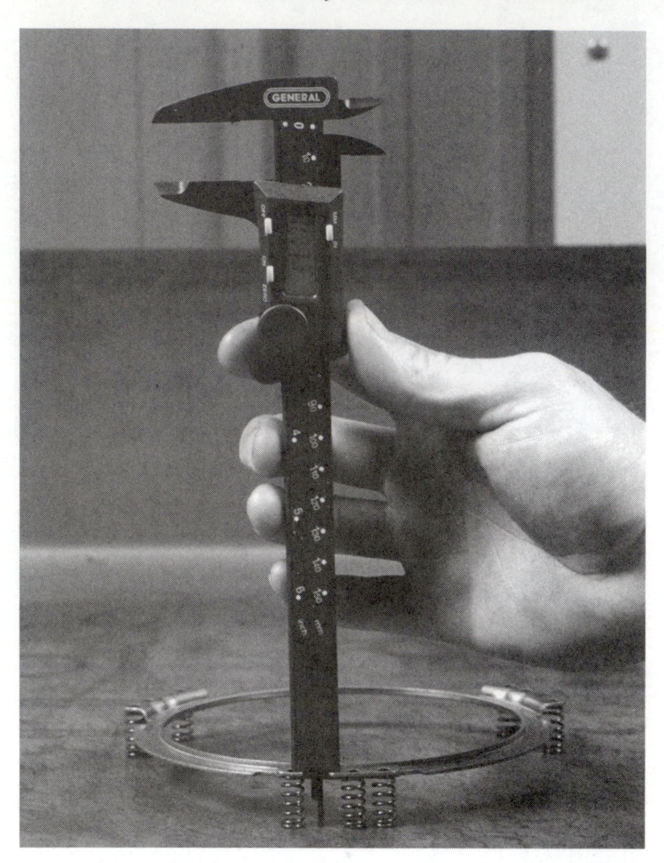

Figure 6. Vernier calipers are commonly used to measure depth. This dimension is read in the same way as thickness or width.

To read the caliper, locate the line on the main scale that lines up with the zero (0) on the vernier scale. If the zero lined up with the 1 on the main scale, the reading would be 0.100 inch. If the zero on the vernier scale does not line up exactly with a line on the main scale, then look for a line on the vernier scale that does line up with a line on the main scale.

## Dial Caliper

The **dial caliper (Figure 7)** is an easier to use version of the vernier caliper. USCS calipers commonly measure dimensions from 0 to 6 inches. Metric dial calipers typically measure from 0 to 150 mm in increments of 0.02 mm. The dial caliper features a depth scale, bar scale, dial indicator, inside measurement jaws, and outside measurement jaws.

The main scale of an Imperial dial caliper is divided into one-tenth (0.1-) -inch graduations. The dial indicator is divided into one-thousandth (0.001-) -inch graduations. Therefore, one revolution of the dial indicator needle equals one-tenth inch on the bar scale.

A metric dial caliper is similar in appearance but the bar scale is divided into 2-mm increments. Additionally, on a metric dial caliper, one revolution of the dial indicator needle equals 2 mm.

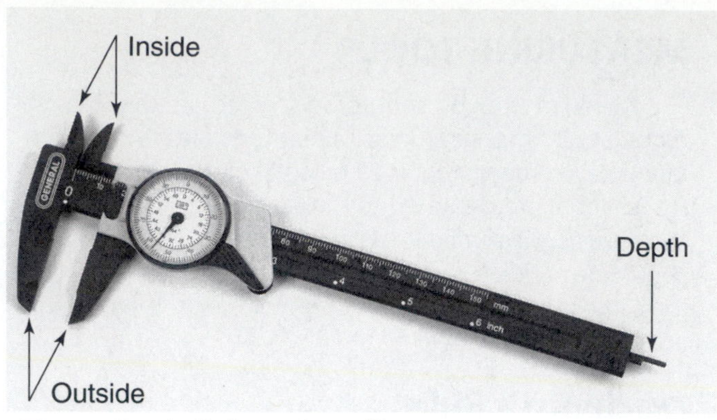


Figure 7. A dial vernier caliper.

Both English and metric dial calipers use a thumb-operated roll knob for fine adjustment. When you use a dial caliper, always move the measuring jaws backward and forward to center the jaws on the object being measured. Make sure the caliper jaws lay flat on or around the object. If the jaws are tilted in any way, you will not obtain an accurate measurement.

Although dial calipers are precision measuring instruments, they are only accurate to plus or minus two thousandths (0.002) of an inch. Micrometers are preferred when extremely precise measurements are desired.

## MICROMETERS

The **micrometer** is used to measure linear outside and inside dimensions. Both outside and inside micrometers are calibrated and read in the same manner. Measurements on both are taken with the measuring points in contact with the surfaces being measured.

The major components and markings of a micrometer include the frame, anvil, spindle, locknut, sleeve, sleeve numbers, sleeve long line, thimble marks, thimble, and ratchet **(Figure 8)**. Micrometers are calibrated in either inch or metric graduations and are available in a range of sizes.

To measure small objects with an outside micrometer, open the jaws of the tool and slip the object between the spindle and the anvil **(Figure 9)**. While holding the object against the anvil, turn the thimble using your thumb and forefinger until the spindle contacts the object. Use only enough pressure on the thimble to allow the object to just fit between the tips of the anvil and spindle. The object should slip through with only a very slight resistance. When a satisfactory feel is reached, lock the micrometer. Because each graduation on the sleeve represents 0.025 inch **(Figure 10)**, begin reading the measurement by counting the visible lines on the sleeve and multiply that number by 0.025. The graduations on the thimble assembly define the area between the lines on the sleeve; therefore, the number indicated on the thimble should be added to the measurement shown on the sleeve. The sum is the outside diameter of the object.

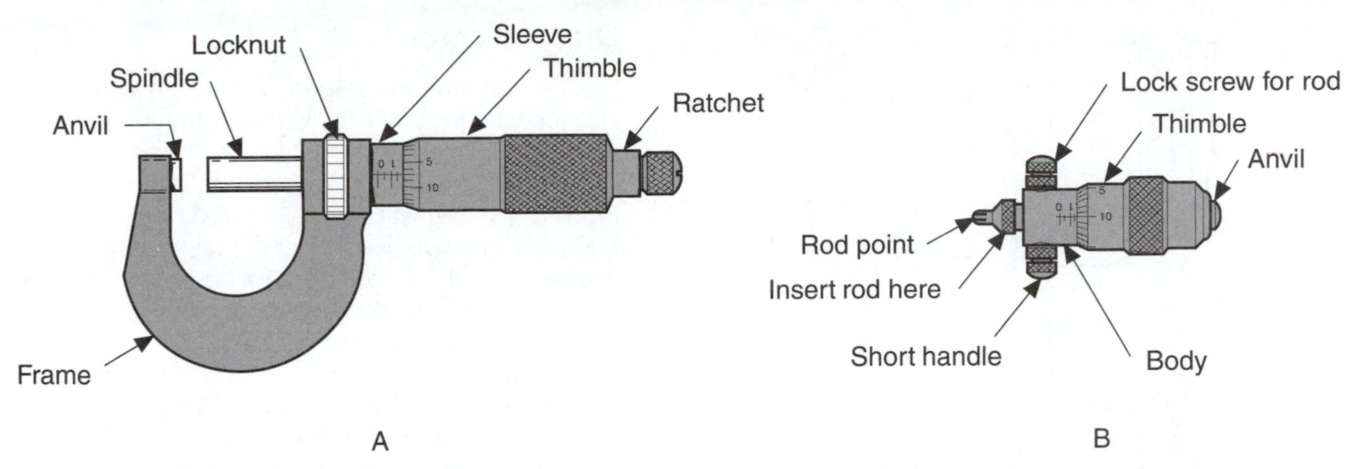


**Figure 8.** Major components of (A) an outside and (B) an inside micrometer.

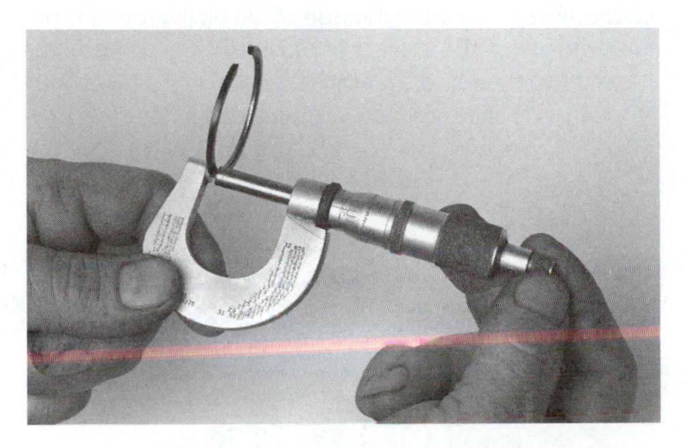

**Figure 9.** To measure an object, place it between the anvil and spindle of the micrometer and move the spindle until the object has good contact with both the anvil and the spindle.

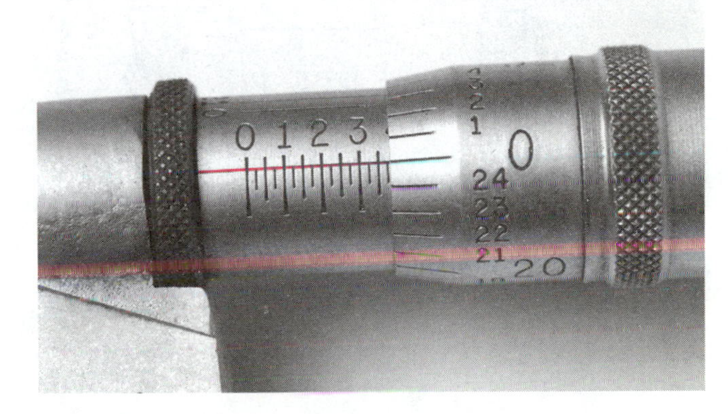

**Figure 10.** This micrometer has a reading of 0.375 inch.

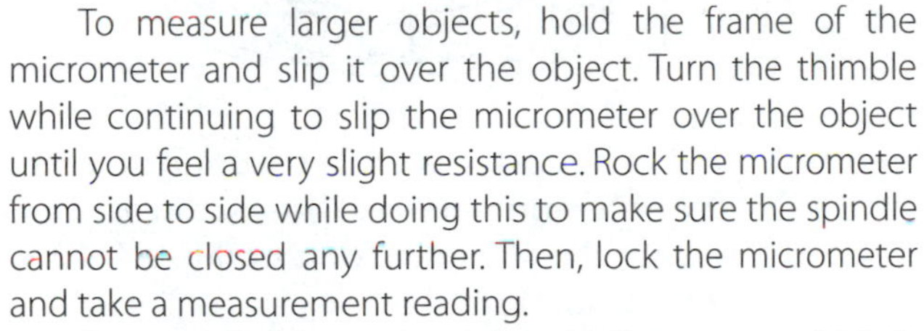

To measure larger objects, hold the frame of the micrometer and slip it over the object. Turn the thimble while continuing to slip the micrometer over the object until you feel a very slight resistance. Rock the micrometer from side to side while doing this to make sure the spindle cannot be closed any further. Then, lock the micrometer and take a measurement reading.

Some technicians use a digital micrometer, which is easier to read. These tools do not have the various scales; rather, the measurement is displayed and read directly off the micrometer.

Micrometers are available in a number of different sizes. The size is dictated by the smallest measurement it can make to the largest. Examples of these sizes are the 0- to 1-inch, 1- to 2-inch, 2- to 3-inch, and 3- to 4-inch micrometers.

## Reading a Metric Outside Micrometer

A metric micrometer is read in the same manner as the inch-graduated micrometer, except the graduations are expressed in the metric system of measurement. Readings are obtained as follows.

1. Each number on the sleeve of the micrometer represents 5 millimeters (mm) or 0.005 meter (m) **(Figure 11A)**.
2. Each of the ten equal spaces between each number, with index lines alternating above and below the horizontal line, represents 0.5 mm or five tenths of a millimeter. One revolution of the thimble changes the reading one space on the sleeve scale or 0.5 mm **(Figure 11B)**.
3. The beveled edge of the thimble is divided into 50 equal divisions with every fifth line numbered: 0, 5, 10, and so on up to 45. Since one complete revolution of the thimble advances the spindle 0.5 mm, each graduation on the thimble is equal to one-hundredth of a millimeter **(Figure 11C)**.
4. As with the inch-graduated micrometer, the three separate readings are added together to obtain the total reading.

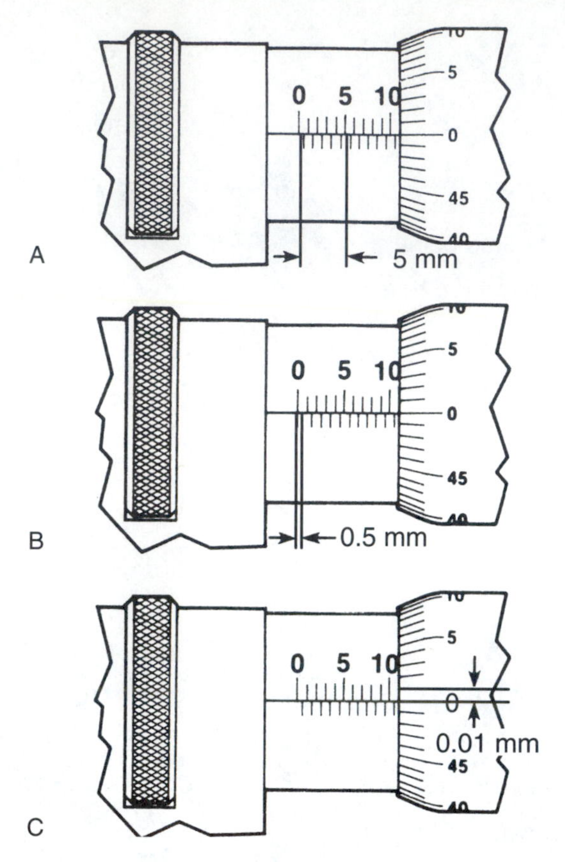


**Figure 11.** Reading a metric micrometer: (A) 5 mm plus (B) 0.5 mm plus (C) 0.01 mm equals 5.51 mm.

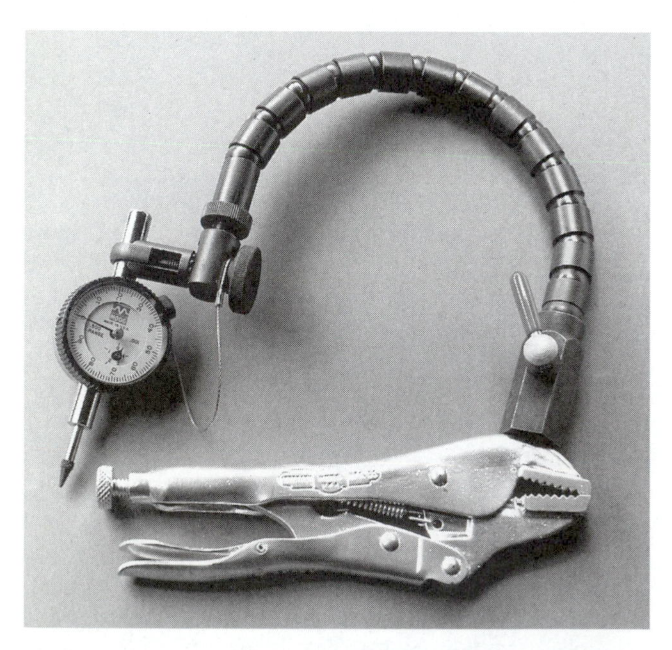

**Figure 12.** A dial indicator with a highly adaptive holding fixture.

## Dial Indicator

The **dial indicator (Figure 12)** is calibrated in 0.001-inch (one-thousandth-inch) increments. Metric dial indicators are also available. Both types are used to measure movement. Common uses of the dial indicator include measuring movement, journal concentricity, flywheel or brake rotor runout, gear backlash, clutch pack clearance **(Figure 13)**, and shaft end play. Dial indicators are available with various face markings and measurement ranges to accommodate many measuring tasks.

To use a dial indicator, position the indicator rod against the object to be measured. Then, push the indicator toward the work until the indicator needle travels far enough around the gauge face to permit movement to be read in either direction. Zero the indicator needle on the gauge. Always be sure the range of the dial indicator is sufficient to allow the amount of movement required by the measuring procedure. For example, never use a 1-inch indicator on a component that will move two inches.

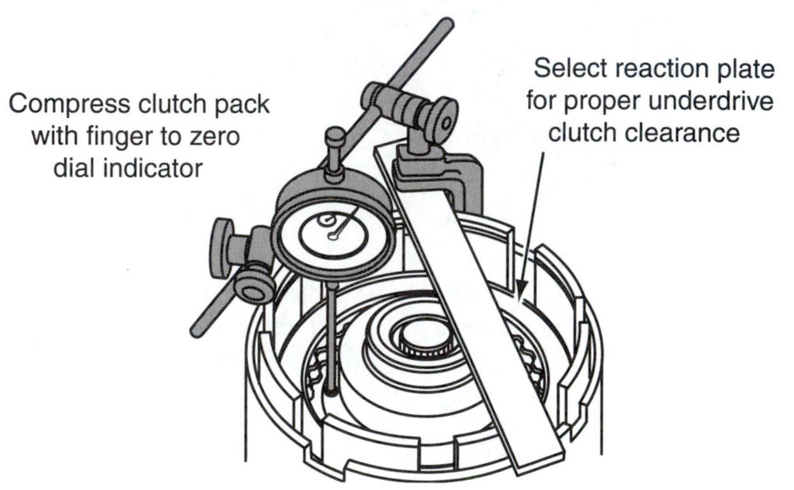


**Figure 13.** A dial indicator is often used in automatic transmission service. This setup is measuring the clearance of a multiple friction disc (clutch) pack.

**You Should Know** *Like all tools, measuring tools should only be used for the purpose for which they were designed. Some instruments are not accurate enough for very precise measurements; others are too accurate to be practical for less critical measurements.*

## Summary


- In the United States Customary System (USCS), the inch is also commonly broken down into fractions or decimals.
- In the metric system, the basic measurement unit of length is the meter, which for exact measurements is divided into units of ten. The most commonly used division is the millimeter (mm).
- One of the most commonly used types of fastener is the threaded fastener. This type includes bolts, nuts, screws, and similar items that allow a technician to install or remove parts easily.
- Bolts are identified by their head size, shank diameter, thread pitch, length, and grade.
- The size of a bolt is determined by the diameter of its shank. The size of the bolt head determines what size wrench is required.
- The length of a bolt is measured from the bottom surface of the head to the end of the threads.
- The bolt's tensile strength, or grade, is the amount of stress or stretch it is able to withstand before it breaks.
- Measuring instruments commonly used to service automatic transmissions include the machinist's rule, feeler gauge, vernier caliper, dial indicator, and micrometer.


## Review Questions

1. Technician A says that the USCS measuring system is based on the inch and that fine measurements are expressed in fractions of an inch or decimals. Technician B says that the metric system is based on the meter and that fine measurements are expressed in tenths, hundredths, and thousandths. Who is correct?
   A. Technician A only
   B. Technician B only
   C. Both Technician A and Technician B
   D. Neither Technician A nor Technician B
2. Technician A says that one millimeter is equal to 0.001 meters. Technician B says that one meter is equal to 0.03937 inches. Who is correct?
   A. Technician A only
   B. Technician B only
   C. Both Technician A and Technician B
   D. Neither Technician A nor Technician B
3. When discussing automotive fasteners Technician A says that bolt sizes are listed by their appropriate wrench size. Technician B says that whenever bolts are replaced, they should be replaced with exactly the same size as the manufacturer had installed. Who is correct?
   A. Technician A only
   B. Technician B only
   C. Both Technician A and Technician B
   D. Neither Technician A nor Technician B
4. When discussing the purpose of micrometers Technician A says that micrometers are used to measure the diameter of an object. Technician B says outside micrometers are used to measure the outside diameter of an object, whereas inside micrometers are used to measure the inside diameter. Who is correct?
   A. Technician A only
   B. Technician B only
   C. Both Technician A and Technician B
   D. Neither Technician A nor Technician B
5. Technician A says that dial indicators are commonly used to measure the backlash or movement of a set of gears. Technician B says that dial indicators are commonly used to measure the endplay of a set of gears. Who is correct?
   A. Technician A only
   B. Technician B only
   C. Both Technician A and Technician B
   D. Neither Technician A nor Technician B
6. Technician A says that a vernier caliper can be used to measure the outside diameter of something. Technician B says that a vernier caliper can be used to measure the inside diameter of a bore. Who is correct?
   A. Technician A only
   B. Technician B only
   C. Both Technician A and Technician B
   D. Neither Technician A nor Technician B
7. Technician A uses a dial caliper to take inside and outside measurements. Technician B uses a dial caliper to take depth measurements. Who is correct?
   A. Technician A only
   B. Technician B only
   C. Both Technician A and Technician B
   D. Neither Technician A nor Technician B
8. List five features of a bolt that are commonly used to identify it.
9. Explain the two primary purposes of a feeler gauge.
10. One millimeter is equal to how many inches?

# Tools of the Trade

## Introduction

Although every technician's toolbox contains many different tools, there are certain hand tools that are a must. These are discussed briefly in this chapter, as are the power and special tools commonly used to work on automatic transmissions and transaxles.

### HAND TOOLS

A basic tool set should include a set of box-end and open-end wrenches. Box-end wrenches are available in either 6 or 12 points. Twelve-point box-end wrenches allow you to work in tighter areas than do 6-point wrenches. An open-end wrench is normally the best tool for turning a nut down or holding a bolt head.

Older domestic cars were built with bolts and nuts that require USCS wrenches. Most imported and newer domestic cars require metric-sized wrenches. To be able to work on today's transmissions, you should have complete sets of both USCS and metric wrenches.

> **You Should Know** *Metric and USCS wrenches are not interchangeable. For example, a 9/16-inch wrench is 0.02 inch larger than a 14-mm nut. If the 9/16-inch wrench is used to turn or hold a 14-mm nut, the wrench will probably slip. This may cause the points of the bolt head or nut to round off and can possibly cause skinned knuckles.*

To loosen or tighten line or tubing fittings, flare nut wrenches (also called line wrenches) should be used rather than open-end wrenches. Using an open-end wrench will tend to round the corners of the nut, which typically are made of soft metal and can distort easily. Line wrenches surround the nut and provide a better grip on the fitting **(Figure 1)**.

Your tool set should include both blade and Phillips screwdrivers in a variety of lengths. Some vehicles may require special screwdrivers, such as those with a Torx head design.

You also should have a good variety of pliers and at least three hammers: one 8-ounce and one 12- to 16-ounce ball-peen hammer and a small sledgehammer. You also should have a plastic and lead or brass-faced mallet. Hammers are used with punches and chisels, and mallets are used for tapping parts apart or aligning parts. A soft-faced

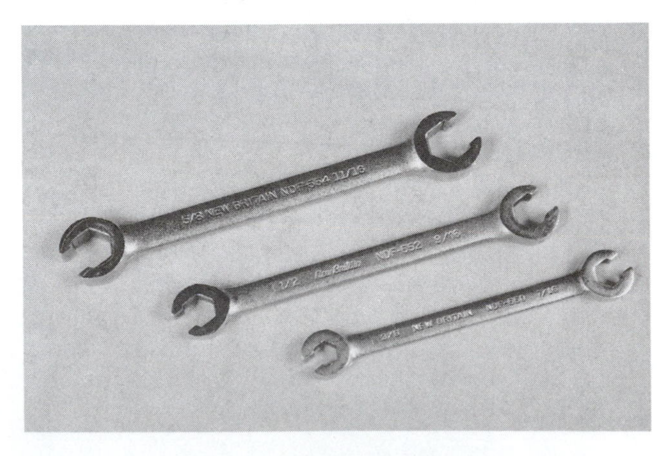

**Figure 1.** A flare nut (line) wrench set.

mallet will not harm the part it is hitting against, whereas a hammer will.

A variety of punches and chisels are used by transmission technicians. Your tool set should include several types of drift and starter punches and flat, cape, round-nose cape, and diamond point chisels.

## Ratchets and Sockets

A set of USCS and metric sockets combined with a ratchet handle and a few extensions also should be included in your tool set. These sockets should be 3/8-inch drive, although 1/4- and 1/2-inch drive sets are also handy. You also may want a long breaker bar to fit your sockets. Breaker bars offer increased leverage when loosening very tight bolts. Extensions allow you to put the handle in the best position while working. They range from 1 inch to 3 feet long. Universal joints that allow a technician to work a bolt or nut at a slight angle also are available.

A 6-point socket has stronger walls and improved grip on a bolt when compared to a normal 12-point socket. Eight-point sockets are available to use on square nuts or square-headed bolts. Some axle and transmission assemblies use square-headed plugs in the fluid reservoir. Deep-well sockets that are used to reach a nut when it is on a bolt or stud with long threads also are available.

## Torque Wrenches

Many nuts and bolts should be tightened a certain amount and have a torque specification that typically is expressed in pounds-feet (USCS) or Newton-meters (metric). Torque wrenches are available with drives that correspond with sockets: 1/4, 3/8, and 1/2 inch. Automatic transmission work often requires the use of a pound-inch torque wrench. A well-equipped transmission technician will have one pound-inch torque wrench as well as at least one pound-foot wrench.

You also should have an OSHA approved air blowgun **(Figure 2)**. These are used not only for blowing off parts during cleaning but also for testing transmission assemblies. Never point a blowgun at yourself or someone else.

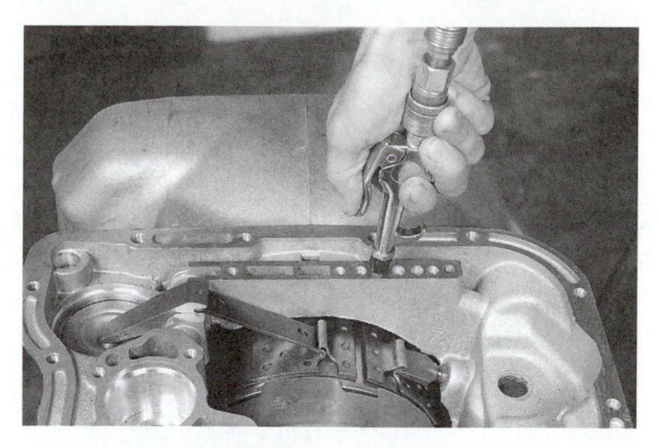

**Figure 2.** An OSHA-approved air blowgun being used to test the hydraulic passages in a transmission.

# SERVICE INFORMATION AND MANUALS

Service manuals are a necessary part of automatic transmission service. There is no way a technician can remember all of the procedures and specifications needed to repair an automobile correctly. Thus, a good technician relies on service manuals and other information sources **(Figure 3)**. Good information and knowledge allow a technician to fix a problem with the least bit of frustration and at the lowest expense to the customer. Service manuals also provide drawings and photographs that show where and how to perform certain procedures on the particular car on which you are working. Special tools or instruments, if required, are listed and shown. Precautions also are given to prevent injury or damage to parts. Perhaps the most important tools you will use are service manuals.

**Figure 3.** Breakdown of the information about a vehicle that may be contained in a service manual.

Most automobile manufacturers publish a service manual or set of manuals for each model and year of their cars. These manuals provide the best and most complete information for those cars. The most commonly used specifications and procedures are compiled in comprehensive service manuals. Various editions are available, covering different ranges of model years for both domestic and imported cars. Specialized shop manuals also are produced for special areas, such as transmissions. Although these manuals are similar to manufacturers' manuals in many ways, they do not provide as much information or detail.

Although the manuals from different publishers vary in presentation and arrangement of topics, all service manuals are easy to use after you become familiar with their organization. To use a service manual:

1. Select the appropriate manual for the vehicle being serviced.
2. Use the table of contents to locate the section that applies to the work being done.
3. Use the index at the front of that section to locate the required information.
4. Carefully read the information and study the applicable illustrations and diagrams.
5. Follow all of the required steps and procedures given for that service operation.
6. Adhere to all of the given specifications and perform all measurement and adjustment procedures with accuracy and precision.

Throughout this book, you will be told to refer to the appropriate shop manual to find the correct procedures and specifications. Although the various systems of all automobiles function in much the same way, there are many variations in design. Each design has its own set of repair and diagnostic procedures. Therefore, it is important that you always follow the recommendations of the manufacturer to identify and repair problems.

Because many technical changes occur on specific vehicles each year, manufacturers' service manuals need to be updated constantly. Updates are published as service bulletins (often referred to as Technical Service Bulletins or TSBs) that show the changes in specifications and repair procedures during the model year. These changes do not appear in the service manual until the next year. The car manufacturer provides these bulletins to dealers and repair facilities on a regular basis.

Automotive manufacturers also publish a series of technician reference books. The publications provide general instructions on the service and repair of their vehicles with their recommended techniques.

A vehicle's service history also can be a valuable diagnostic aid. This information may be found in a shop's records or through on-line services with the manufacturer or another provider of information. The history records can show the vehicle's maintenance record, recall actions, and previous related services.

**Figure 4.** The use of CD-ROMs, DVDs, and computers makes accessing information quick and easy.

## Aftermarket Suppliers' Guides and Catalogs

Many of the larger parts manufacturers have excellent guides on the various parts they manufacture or supply. They also provide updated service bulletins on their products. Other sources for up-to-date technical information are trade magazines and trade associations.

## Computer-based Information

The same information that is available in service manuals and bulletins is also available electronically: on CD-ROMs **(Figure 4)**, DVDs, and the Internet. A single compact disk can hold a quarter million pages of text. This eliminates the need for a huge library to contain all of the printed manuals. Using electronics to find information is also easier and quicker. The disks are normally updated monthly and not only contain the most recent service bulletins but also engineering and field service fixes. Online data can be updated instantly and requires no space for physical storage.

# POWER TOOLS

Power tools make a technician's job easier. They operate faster and with more torque than hand tools. However, power tools require greater safety measures. Power tools do not stop unless they are turned off.

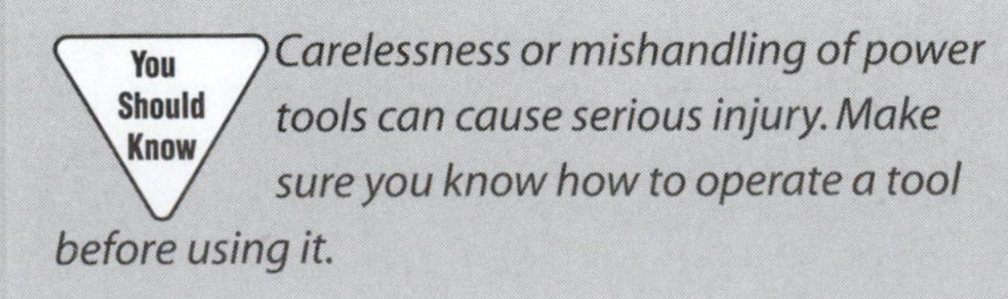


*Carelessness or mishandling of power tools can cause serious injury. Make sure you know how to operate a tool before using it.*

## Air Wrenches

An impact wrench uses compressed air or electricity to hammer or drive a nut or bolt loose or tight. **Impact sockets** should be used with impact wrenches. These sockets are designed to withstand the force of the impact. Ordinary sockets must not be used with impact wrenches—they will crack or shatter because of the force and can cause injury.

You Should Know *Impact wrenches should not be used to tighten critical parts or parts that may be damaged by the hammering force of the wrench.*

Air ratchets are often used during disassembly or reassembly work to save time. Because the ratchet turns the socket without an impact force, these wrenches can be used on most parts and with ordinary sockets. Air ratchets usually have a 3/8-inch drive. Air ratchets are not torque sensitive; therefore, a torque wrench should be used on all fasteners after snugging them up with an air ratchet.

# SHOP TOOLS

Some tools required to work on transmissions are not part of a technician's tool set; rather, these are general shop tools. A brief discussion of the common shop tools used during an automatic transmission service follows.

## Bench Grinder

A bench grinder is generally bolted to a workbench, but it may be freestanding. In either case, the grinder should have safety shields and guards. Three types of wheels are commonly used on a bench grinder:

- Wire wheel brush. Used for general cleaning and buffing, removing rust and scale, paint removal, deburring, and so forth.
- Grinding wheel. For a wide variety of grinding jobs from sharpening cutting tools to deburring.
- Buffing wheel. For general purpose buffing, polishing, and light cutting.

## Presses

Many jobs require the use of force to assemble or disassemble parts that are **press fit** together. Presses can be hydraulic, electric, air, or hand driven. Capacities range up to 150 tons of pressing force, depending on the size and design of the press. Smaller arbor and C-frame presses can be bench or pedestal mounted, whereas high capacity units are freestanding or floor mounted **(Figure 5)**.

**Figure 5.** A floor-mounted hydraulic press.

## Jacks and Lifts

The most common jacks are hydraulic floor jacks, which are classified by the weights they can lift: 1½, 2, and 2½ tons, and so on. These jacks are controlled by moving a handle up and down. The other design of portable floor jack uses compressed air. Pneumatic jacks are operated by controlling air pressure at the jack.

When a vehicle is raised by a jack, it should be supported by safety stands. Never work under a car with only a jack supporting it—you always should use safety stands. Seals in the jack can leak and allow the vehicle to drop.

A floor lift is the safest lifting tool, as it can raise the vehicle high enough to allow you to walk and work under it. Various safety features prevent a hydraulic lift from dropping if a seal does leak or if air pressure is lost. Before lifting a vehicle, make sure the lift is correctly positioned.

## Transmission Jack

A transmission jack **(Figure 6)** is designed to help you while removing a transmission from under the vehicle. The weight of the transmission makes it difficult and unsafe to

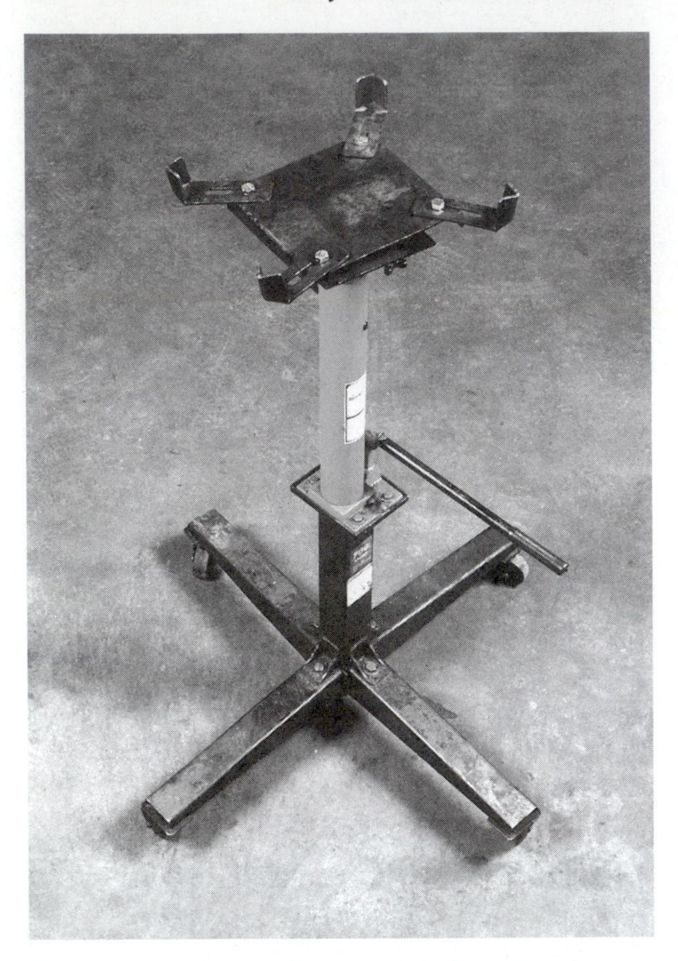

**Figure 6.** Typical hydraulic transmission jack.

remove it without assistance and/or a transmission jack. These jacks fit under the transmission and typically are equipped with hold down chains. These chains are used to secure the transmission to the jack. The transmission's weight rests on the jack's saddle.

Transmission jacks are available in two basic styles. One is used when the vehicle is raised by a hydraulic jack and setting on jack stands. The other style is used when the vehicle is raised on a lift.

## Engine Hoist

An engine hoist allows you to lift an engine, with or without the transmission attached, from a car. Hydraulic pressure converts power to a mechanical advantage and lifts the engine from the vehicle. After the engine has been removed, the engine should be separated from the transmission and mounted to an engine stand. The transmission also should be mounted to a holding fixture.

## Transaxle Removal and Installation Equipment

Removing and installing a transaxle from a vehicle with a transversely mounted engine may require other tools. The

**Figure 7.** An engine holding fixture for a FWD vehicle.

engine and transaxle in some front wheel drive (FWD) vehicles are removed by lifting them from the top. In other vehicles, they must be removed from the bottom, and this procedure requires different equipment. The required equipment varies with manufacturer and vehicle model; however, most types accomplish the same thing.

To remove the engine and transmission from under the vehicle, the vehicle must be raised. A crane and/or support fixture is used to hold the engine and transaxle assembly in place while the assembly is being readied for removal. When everything is set for removal of the assembly, the crane is used to lower the assembly onto a cradle. The cradle is similar to a hydraulic floor jack and is used to lower the assembly further so it can be rolled out from under the vehicle. The transaxle can be separated from the engine once it has been removed from the vehicle.

When a transaxle is removed unattached from the engine, the engine must be supported while it is in the vehicle before, during, and after transaxle removal. Special fixtures mount to the vehicle's upper frame or suspension parts **(Figure 7)**. These supports have a bracket that is attached to the engine. With the bracket in place, the engine's weight is now on the support fixture and the transmission can be removed.

## Transmission/Transaxle Holding Fixtures

Holding fixtures should be used to support the transmission or transaxle after it has been removed from the vehicle. These holding fixtures may be stand-alone units **(Figure 8)** or may be bench mounted and allow the transmission to be easily repositioned during repair work.

# TRANSMISSION TOOLS

There is a large variety of special tools designed for automatic transmission work. Often these tools are for a

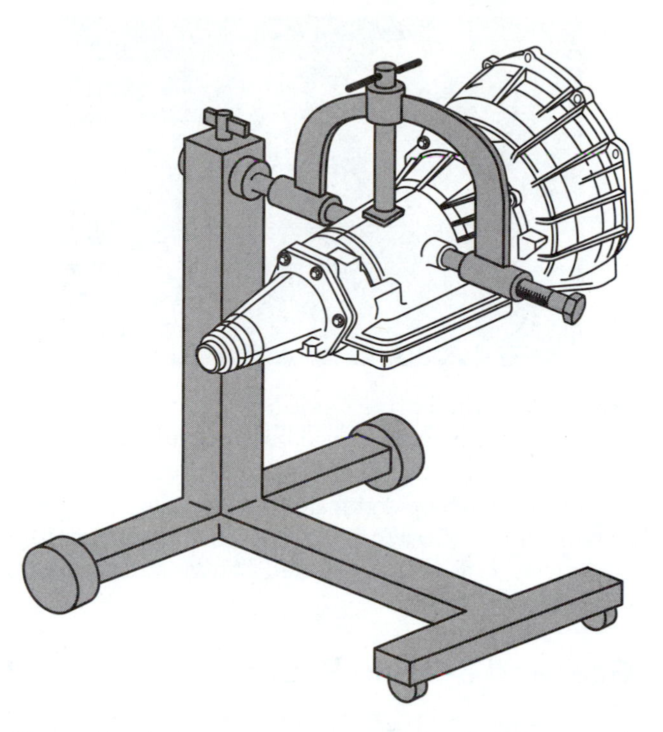

**Figure 8.** A stand-alone transmission holding stand.

specific model of transmission. Most of these tools are listed in the appropriate service manuals and are part of each manufacturer's essential tool kit.

Other less model-specific tools are used primarily for servicing automatic transmissions. These are discussed here.

## Scan Tools

The introduction of computer-controlled systems brought with it the need for tools capable of troubleshooting electronic control systems. There is a variety of computer scan tools available today that do just that. A **scan tool (Figure 9)** is a microprocessor designed to communicate with the vehicle's computer. A scan tool is connected to the computer through diagnostic connectors and can access trouble codes, run tests to check system operations, and monitor the activity of the system. Trouble codes and test results are displayed on an LED screen or printed out on the scanner printer.

Scan tools will retrieve fault codes from a computer's memory and digitally display these codes on the tool. A scan tool also may perform other diagnostic functions, depending on the year and make of the vehicle.

Scan tools are capable of testing many on-board computer systems—such as transmission controls, engine computers, antilock brake computers, air bag computers, and suspension computers—depending on the year and make of the vehicle and the type of scan tester. In many cases, the technician must select the computer system to be tested with the scan tool after it has been connected to the vehicle.

**Figure 9.** A scan tool connected to a vehicle's DLC.

The scan tool is connected to specific diagnostic connectors on various vehicles. Most manufacturers have one diagnostic connector. This connects the data wire from each on-board computer to a specific terminal in this connector. Other vehicle manufacturers have several different diagnostic connectors on each vehicle, and each of these connectors may be connected to one or more on-board computers. A set of connectors is supplied with the scan tool to allow tester connection to various diagnostic connectors on different vehicles.

The scan tool must be programmed for the model year, make of vehicle, and type of engine. With some scan tools, this selection is made by pressing the appropriate buttons on the tester, as directed by the digital tester display. On other scan tools, the appropriate memory card must be installed in the tester for the vehicle being tested.

As automotive computer systems become more complex, the diagnostic capabilities of scan testers continue to expand. Many scan testers now have the capability to store, or "freeze" data into the tester during a road test **(Figure 10)**, and then play back this data when the vehicle is returned to the shop.

Some scan testers now display diagnostic information based on the fault code in the computer memory. Service bulletins published by the scan tester manufacturer may be indexed by the tester after the vehicle information is entered in the tester. Other scan testers will display sensor specifications for the vehicle being tested.

Trouble codes are set by the vehicle's computer only when an input signal is entirely out of its normal range. The codes help technicians identify the cause of the problem when this is the case. If a signal is within its normal range but is still not correct, the vehicle's computer may not display a trouble code. However, a problem will still exist. As an aid to identify this type of problem, most manufacturers recommend that the signals to and from the computer be looked at carefully. This is done by checking voltage and resistance between specific points within the computer's wiring harness.

Figure 10. Using a scan tool during a road test. (Courtesy of OTC Tool and Equipment, Division of SPX Corporation)

With on-board diagnostics (OBD) II, the diagnostic connectors look the same on all vehicles. Also, any scan tool designed for OBD II will work on all OBD II systems; therefore, the need to have designated scan tools or cartridges is eliminated. The OBD II scan tool has the ability to run diagnostic tests on all emission-related systems and has "freeze frame" capabilities.

Figure 11. A final drive's side bearing being removed with an universal bearing puller.

## Gear and Bearing Pullers

Some gears and bearings in a transmission often have a slight interference fit (press fit) when they are installed on a shaft or in housing. For example, the inside diameter of a bore is 0.001 inch smaller than the outside diameter of a shaft. When the shaft is fitted into the bore, it must be

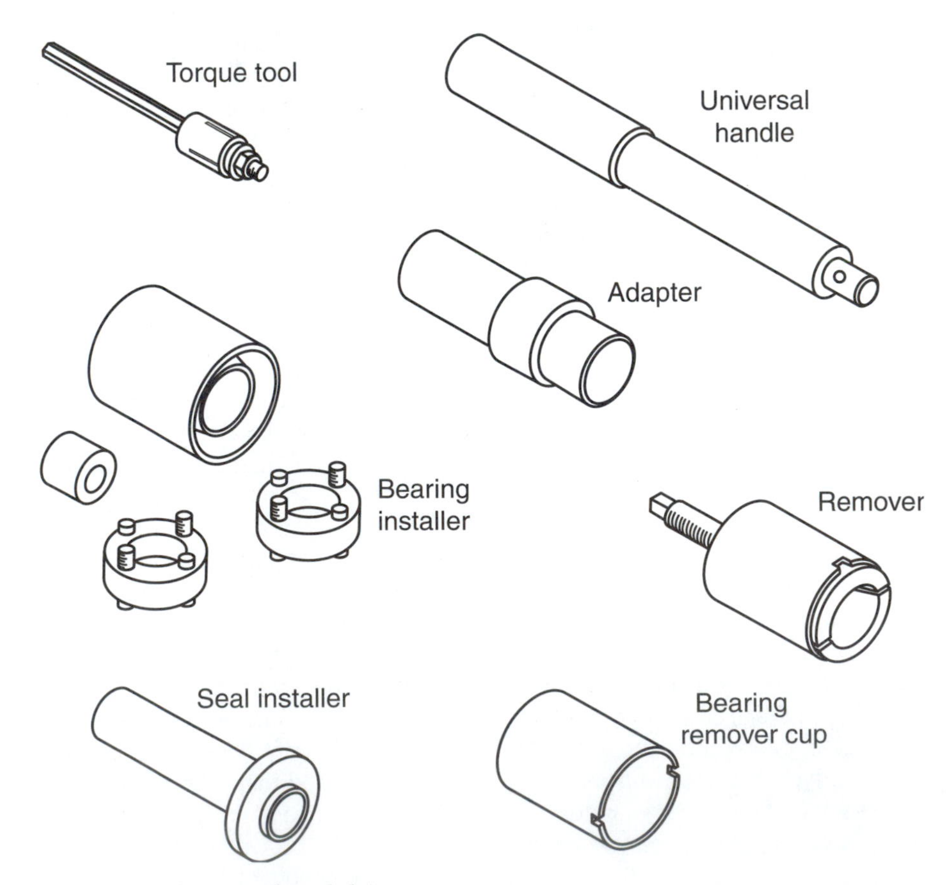


Figure 12. An assortment of bushing and seal drivers.

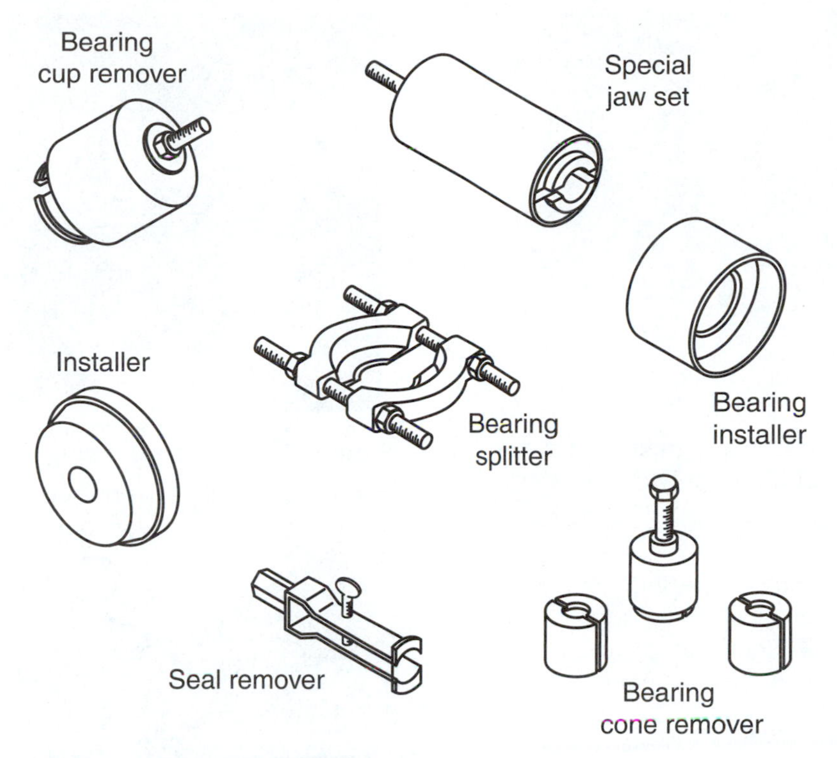


**Figure 13.** An assortment of pullers and removal tools.

pressed in to overcome the 0.001 inch interference fit. This press fit prevents the parts from moving on each other. The removal of these gears and bearings must be done carefully to prevent damage to the gears, bearings, or shafts. Prying or hammering can break or bind the parts. A gear and bearing puller **(Figure 11)** with the proper jaws and adapters should be used to remove gears and bearings. Using the proper puller, the force required to remove a gear or bearing can be applied with a slight and steady motion.

## Bushing and Seal Pullers and Drivers

Another commonly used group of special tools is the various designs of **bushing** and seal drivers **(Figure 12)** and pullers. Pullers either are a threaded or slide hammer-type tool **(Figure 13)**. Always make sure you use the correct tool for the job—bushings and seals are damaged easily if the wrong tool or procedure is used.

## Special Pliers

Snap or lock ring pliers are made with a linkage that allows the movable jaw to stay parallel throughout its range of opening. The jaw surface is usually notched or toothed to prevent slippage. Often, a transmission technician will run into many different styles and sizes of retaining rings that hold subassemblies together or keep them in a fixed location **(Figure 14)**. Using the correct tool to remove and install these rings is the only safe way to work with them. All transmission technicians should have an assortment of retaining ring pliers **(Figure 15)**.

**Figure 14.** This retaining ring is easily removed with the correct type of pliers.

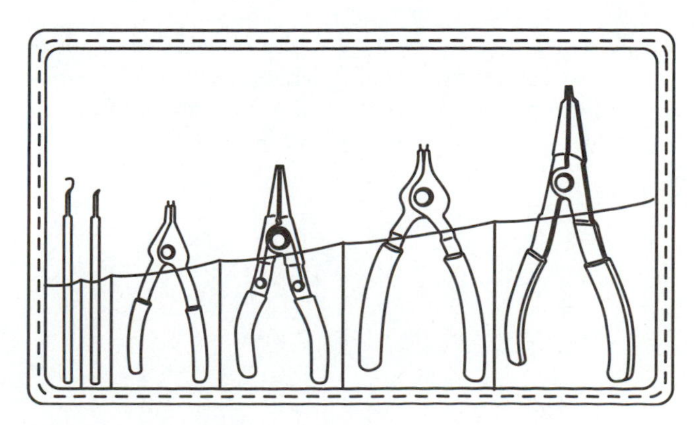

**Figure 15.** Retaining ring pliers set for transmission service.

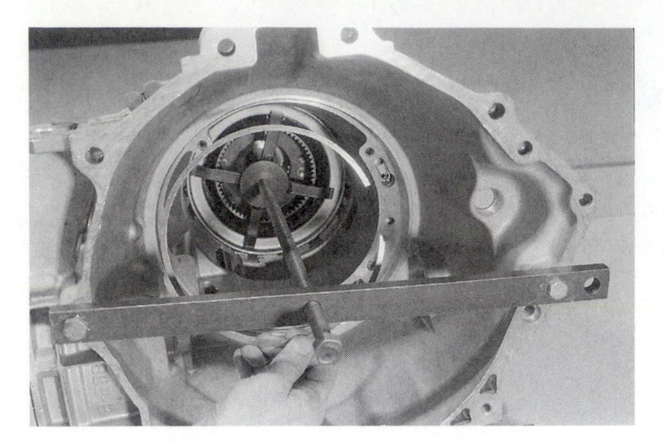

**Figure 16.** A spring compressor designed for a particular multiple-disc assembly.

**Figure 17.** A universal spring compressor.

## Spring Compressors

Small hand-operated spring compressors are used to disassemble and reassemble multiple-disc clutch and brake assemblies **(Figure 16)**. Tool manufacturers have compressors of specific sizes, as well as universal spring compressor tools **(Figure 17)**.

## Hydraulic Pressure Gauge Set

A common diagnostic tool for automatic transmissions is a hydraulic pressure gauge. A pressure gauge measures pressure in pounds per square inch (psi) and/or kilopascals (kPa). The gauge is normally part of a kit that contains various fittings and adapters so that it can be connected to the test ports on a transmission **(Figure 18)**.

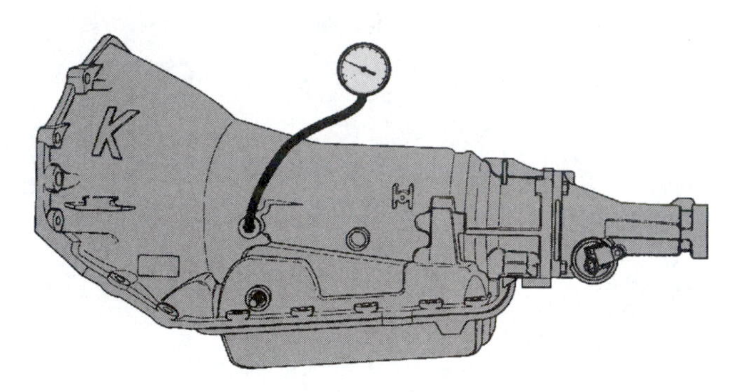

**Figure 18.** A hydraulic pressure gauge connected to a transmission.

# *Summary*

- An automatic transmission technician should have a good assortment of hand tools.
- Service manuals provide the procedures and specifications needed to repair an automobile correctly.
- Blowguns are used for blowing off parts during cleaning and to test transmission assemblies.
- A bench grinder may have a wire wheel brush, a grinding wheel, and/or a buffing wheel.
- Many jobs require the use of a press to assemble or disassemble parts that are press fit together.
- A transmission jack is used to support a transmission while it is being removed or installed.
- Gear and bearing pullers are used to separate gears or bearings from a shaft or housing. They are designed to do this without damaging any components.
- A spring compressor is a tool that is used to compress a spring or set of springs while assembling a unit.

## Review Questions

1. What is the primary purpose of the wire wheel on a bench grinder unit?
2. Why should you not use an air impact wrench to tighten critical components?
3. What is the purpose of a blowgun and what type should always be used?
4. When discussing the fit of gears and bearings, Technician A says that an interference fit allows the parts to be easily disassembled. Technician B says that parts that are press fit have an interference fit. Who is correct?
   A. Technician A only
   B. Technician B only
   C. Both Technician A and Technician B
   D. Neither Technician A nor Technician B
5. When discussing the removal of gears from a shaft, Technician A says that prying or hammering on the gears may cause them to bind on the shaft and damage the shaft or the gears. Technician B says that by using the proper driver and press, gears can be installed on a shaft without damage to either part. Who is correct?
   A. Technician A only
   B. Technician B only
   C. Both Technician A and Technician B
   D. Neither Technician A nor Technician B
6. Why should you not use regular sockets with an air impact wrench?
7. What type of socket should be used on square nuts or square-headed bolts?
8. Describe how most service manuals are divided and how information is organized.
9. When discussing the purpose of torque wrenches, Technician A says that they are used to tighten fasteners to a specified torque. Technician B says that they are used for added leverage while loosening or tightening a bolt. Who is correct?
   A. Technician A only
   B. Technician B only
   C. Both Technician A and Technician B
   D. Neither Technician A nor Technician B
10. Technician A says that one of the most important tools for a technician is a service manual. Technician B says that service manuals are not needed if a good selection of TSBs is available. Who is correct?
    A. Technician A only
    B. Technician B only
    C. Both Technician A and Technician B
    D. Neither Technician A nor Technician B

# Chapter 4

# Working as an Automatic Transmission Specialist

## *Introduction*

To be a successful automotive technician, you need to have good training, a desire to succeed, and to be committed to becoming a good technician and a good employee. A good employee works well with others and strives to make the business successful. The required training is not just in the field of automotive. Good technicians **(Figure 1)** need to have good reading, writing, and math skills. These skills will allow you to better understand and use the material found in service manuals and textbooks, as well as provide you with the basics for good communications with customers and others.

**Figure 1.** A good technician has good training, a desire to succeed, is a good employee, works well with others, strives to make the business successful, and has good reading, writing, and math skills.

### COMPENSATION

Technicians typically are paid according to their abilities. Most often, new or apprentice technicians are paid by the hour. Apprentice technicians learn the trade and the business while they are paid. Their time is usually spent working with a master technician or doing low-skilled jobs. As an apprentice learns more, he or she can earn more and take on more complex jobs. Once technicians have demonstrated a satisfactory level of skills, they are responsible for their own work. Then they can receive higher hourly wages or they can go on **flat rate**.

Flat rate is a pay system in which a technician is paid for the amount of work done. Each job has a flat rate time. Pay is based on that time, regardless of how long it took to complete the job. To explain how this system works, let us look at a technician who is paid $15.00 per flat rate. If a job has a flat rate time of 3 hours, the technician will be paid $45.00 for the job, regardless of how long it took to complete it. Experienced technicians beat the flat rate time nearly all of the time. Their weekly pay is based on the time "turned in," not on the time spent. If the technician turns in 60 hours of work in a 40-hour workweek, he or she actually earned $22.50 each hour worked. However, if he or she turned in only 30 hours in the 40-hour week, the hourly pay is $11.25.

The flat rate system favors good technicians who work in a shop with a large volume of work. The use of flat rate times allows for more accurate repair estimates to the customers. It also rewards skilled and productive technicians.

## EMPLOYER-EMPLOYEE RELATIONSHIPS

When you begin a job, you enter into a business agreement with your employer. When you become an employee, you sell your time, skills, and efforts. In return, your employer pays you for these resources. As part of the employment agreement, your employer also has certain responsibilities:

- Instruction and Supervision—You should be told what is expected of you. A supervisor should observe your work, tell you if it is satisfactory, and offer ways to improve your performance.
- Clean, Safe Place to Work—An employer should provide a clean and safe work area as well as a place for personal cleanup.
- Wages—You should know how much you are to be paid, what your pay will be based on, and when you will be paid, before accepting a job.
- Fringe Benefits—When you are hired, you should be told what benefits to expect, such as paid vacations and employer contributions to health insurance and retirement plans.
- Opportunity—You should be given a chance to succeed and possibly advance within the company.
- Fair Treatment—All employees should be treated equally, without prejudice or favoritism.

On the other side of this business transaction, employees have responsibilities to their employers. Your obligations as an employee to the employer include the following:

- Regular Attendance—A good employee is reliable.
- Following Directions—As an employee, you are part of a team; doing things your way may not serve the best interests of the company.
- Responsibility—You must be willing to answer for your behavior and work. You also need to realize that you are legally responsible for the work you do.
- Productivity—Remember, you are paid for your time as well as your skills and effort.
- Loyalty—Loyalty is expected. By being loyal, you will act in the best interests of your employer, both on and off the job.

## CUSTOMER RELATIONS

Another responsibility you have is good customer relations. Learn to listen and communicate clearly. Be polite and organized, particularly when dealing with customers. Always be as honest as you possibly can.

Look and present yourself as a professional, which is what automotive technicians are. Professionals are proud of what they do and they show it. Always dress and act appropriately and watch your language, even when you think no one is near.

Respect the vehicles on which you work. They are important to the lives of your customers. Always return the vehicle to the owner in a clean, undamaged condition. Remember, a car is the second largest expense a customer has. Treat it that way. It doesn't matter if you like the car. It belongs to the customer; treat it respectfully.

Explain the repair process to the customer in understandable terms. Whenever you are explaining something to a customer, make sure you do this in a simple way without making the customer feel unintelligent. Always show the customers respect and be courteous to them. Not only is this the right thing to do, but also it leads to loyal customers.

## ASE CERTIFICATION

An obvious sign of your knowledge and abilities—in addition to your dedication to the trade—is ASE certification. The National Institute for Automotive Service Excellence (ASE) has established a voluntary certification program for automotive, heavy-duty truck, auto body repair, and engine machine shop technicians. In addition to these programs, ASE also offers individual testing in some specialty areas. This certification system combines voluntary testing with on-the-job experience to confirm that technicians have the skills needed to work on today's vehicles **(Figure 2)**. ASE recognizes two distinct levels of service capability—the automotive technician and the master automotive technician.

To become ASE certified, you must pass one or more tests that stress diagnostic and repair problems. Automatic Transmissions and Transaxles is one of the eight basic automotive certification areas.

After passing at least one exam and providing proof of 2 years of hands-on work experience, you become ASE certified. Retesting is necessary every 5 years to remain certified. A technician who passes one examination receives an

**Figure 2.** The ASE certification shoulder patch worn by automotive technicians certified in one or more areas is shown on the left. When technicians are certified in all eight areas, they become Master Automotive Technicians and can wear the patch shown on the right.

automotive technician shoulder patch. The master automotive technician patch is awarded to technicians who pass all eight of the basic automotive certification exams.

You may receive credit for one of the 2 years by substituting relevant formal training in one, or a combination, of the following:

- High school training. Three years of training may be substituted for 1 year of experience.
- Post-high school training. Two years of post-high school training in a trade school, technical institute, or community college may be counted as 1 year of work experience.
- Short courses. For shorter periods of post-high school training, you may substitute 2 months of training for 1 month of work experience.
- Apprenticeship Programs. You may receive full credit for the experience requirement by satisfactorily completing a 3- or 4-year apprenticeship program.

Each certification test consists of 40 to 80 multiple-choice questions. The questions are written by a panel of technical service experts, including domestic and import vehicle manufacturers, repair and test equipment and parts manufacturers, working automotive technicians, and automotive instructors. All questions are pre-tested and quality-checked on a sample of national technicians before they are included in the actual test. Many test questions force the student to choose between two distinct repair or diagnostic methods. The test questions focus on basic technical knowledge, repair knowledge, and skills, and testing and diagnostic knowledge and skills.

## DUTIES OF A TRANSMISSION SPECIALIST

As a transmission specialist, you will be responsible for servicing, troubleshooting, and repairing automatic transmissions and all of the components affecting the operation of a transmission and transaxle. An automatic transmission or transaxle requires little maintenance other than periodic fluid and filter changes. The condition and level of the fluid should be checked on a regular basis. Always follow the recommended services and time intervals of the manufacturer.

## VEHICLE IDENTIFICATION

Before any service is done to a vehicle, it is important for you to know exactly what type of vehicle you are working on. The best way to do this is to refer to the **vehicle's identification number (VIN)**. The VIN is always on a plate behind the lower corner of the driver's side of the windshield as well as in other locations in the vehicle. It may be stamped into various body parts or other main components of the vehicle. Check the service manual for those locations. The VIN is made up of seventeen characters and contains all pertinent information about the vehicle. The use of a standard VIN code **(Figure 3)** became mandatory beginning with vehicles manufactured in 1981 and is used by all manufacturers of vehicles, both domestic and foreign.

The first character identifies the country in which the vehicle was manufactured:

- 1 or 4—U.S.A.
- 2—Canada
- 3—Mexico
- J—Japan
- K—South Korea
- S—England
- W—Germany

The second character identifies the manufacturer, for example:

- A—Audi
- B—BMW
- C—Chrysler
- D—Mercedes Benz
- F—Ford
- G—General Motors
- H—Honda
- N—Nissan
- T—Toyota

The third character identifies the vehicle type or manufacturing division (passenger car, truck, bus, etc.). The fourth through eighth characters identify the features of the vehicle, such as the body style, vehicle model, engine type, and so on.

The ninth character is used to identify the accuracy of the VIN and is a check digit. The 10th character identifies the model year, for example:

- S—1995
- T—1996
- V—1997
- W—1998
- X—1999
- Y—2000
- 1—2001
- 2—2002
- 3—2003
- 4—2004
- 5—2005

The 11th character identifies the plant in which the vehicle was assembled, and the 12th to 17th characters identify the production sequence of the vehicle as it rolled off the manufacturer's assembly line.

The specifics needed for decoding the characters of the VIN can be found in the service manual for the vehicle.

In addition to the VIN, all vehicles produced since 1972 have an under-hood emission control label and/or calibration decal. This label gives useful information, including engine specifications, a description of the emission control devices, and vacuum hose routing.

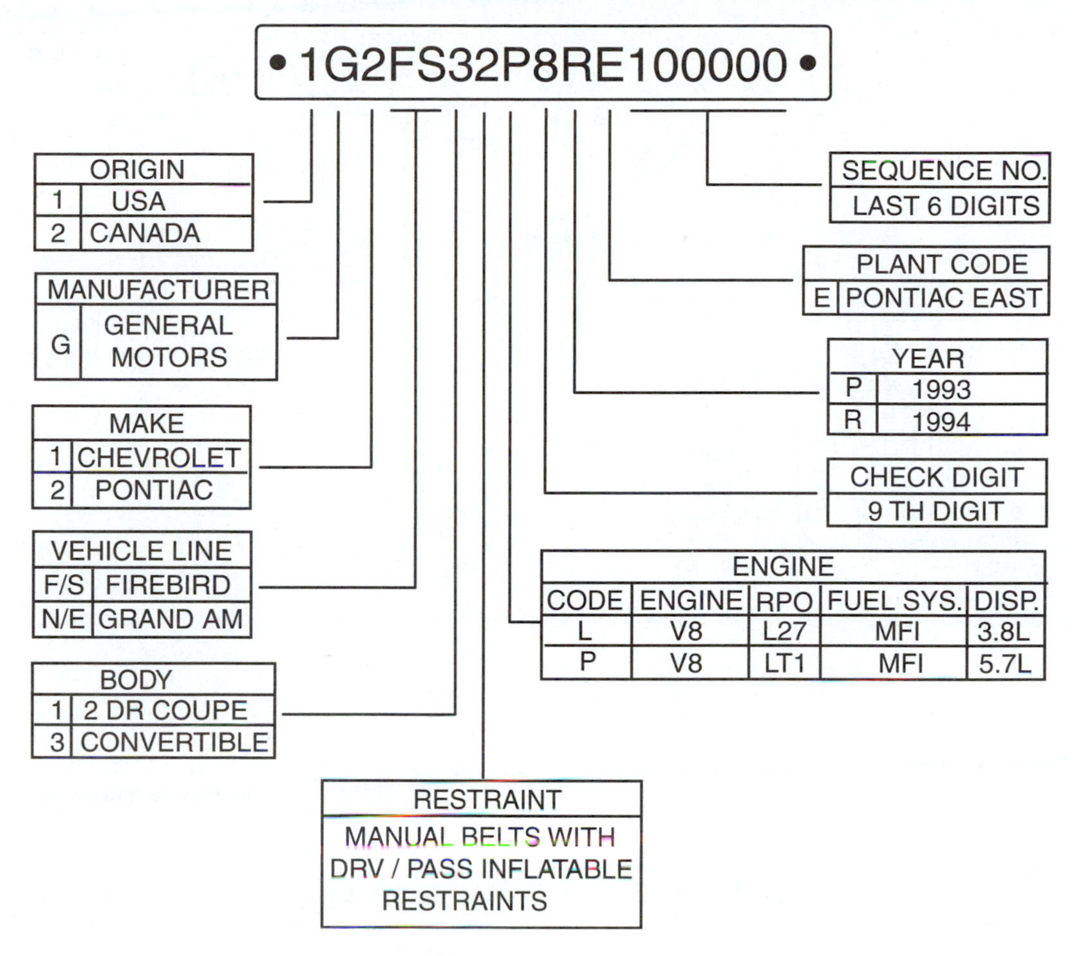


**Figure 3.** Basic interpretation of a VIN.

## Transmission Identification

You also must make sure you know exactly which transmission you are working on. This will ensure that you are following the correct procedures and specifications and are installing the correct parts. Proper identification can be difficult, because transmissions cannot be identified accurately just by the way they look. The only exception to this is the shape of the oil pan, which can be used to help identify some transmissions.

The only positive way to identify the exact design of the transmission is by its identification numbers. Transmission identification numbers are found either as stamped numbers in the case, a label adhered to the case **(Figure 4)**, or on a metal tag held by a bolt head **(Figure 5)**. Use a service

50
M6TM-DK
NIE-WB
00102068
MH-9K33
6248 00102068 4X4

**Figure 4.** An example of a transmission identification sticker. These typically are stuck to the transmission is an easily viewed spot.

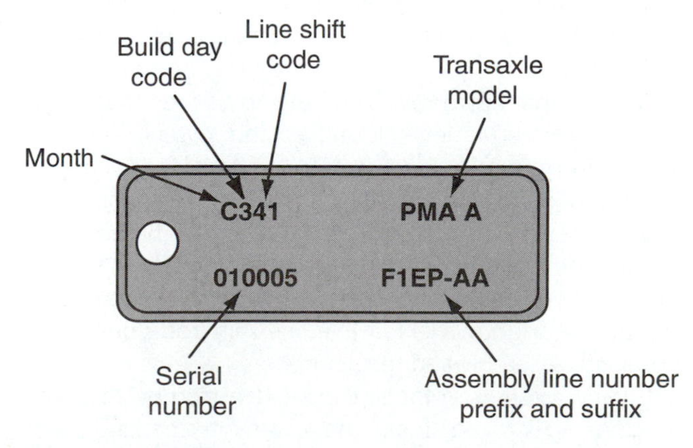


**Figure 5.** A typical identification tag for a Ford Motor Company automatic transmission.

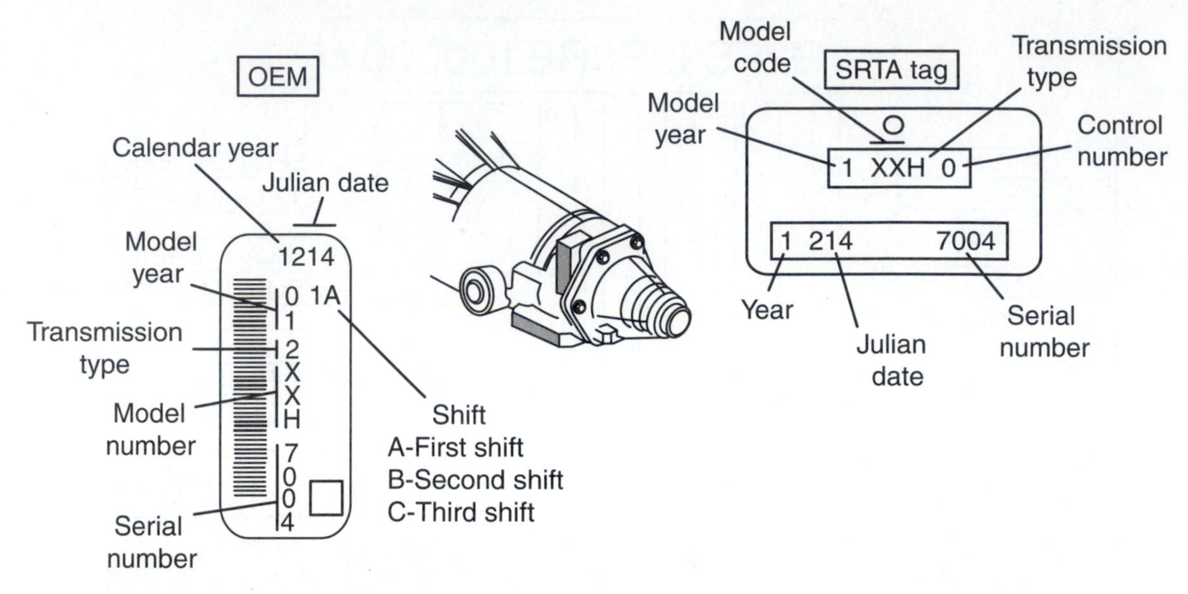


**Figure 6.** Location of and information contained on a General Motors' transaxle identification plate.

manual to decipher the identification number. Most identification numbers include the model, manufacturer, and assembly date **(Figure 6)**.

The following gives some examples of the location of and the information contained on the identification tags of common transmissions.

On Chrysler front-wheel drive (FWD) vehicles, the transaxle can be identified by a Transaxle Identification Number (TIN) stamped on a boss located on the transaxle housing just above the oil pan flange. On rear-wheel drive (RWD) vehicles, the TIN is stamped on a pad on the left side of the transmission case's oil pan flange. In addition to the TIN, each transmission carries an assembly part number that must be referenced when ordering transaxle replacement parts. Transmission operation requirements are different for each vehicle and engine combination. Some internal parts will differ among models. Always refer to the seven-digit part number for positive transmission identification when replacing parts.

Late-model Ford RWD vehicles can be identified by an identification code letter found on the lower line of the Vehicle Certificate label under "TR." This label is attached to the left (driver's) side door lock post. An RWD transmission can be identified by a metal tag attached to the transmission by the lower extension housing retaining bolt. The tag on a transaxle is attached to the valve body cover and gives the transmission model number, line shift code, build date code, and assembly and serial numbers.

Early Ford RWD models can be identified by an identification tag located under the lower intermediate servo cover bolt or attached to the lower extension housing retaining bolt. A number appearing after the suffix indicates internal parts in the transmission have been changed after initial production startup. The top line of the tag shows the transmission model number and build date code.

The transaxle in General Motors FWD vehicles can be identified by an identification number stamped on a plate attached to the rear face of the transaxle. The transaxle model is printed on the Service Parts Identification label located inside the vehicle.

The transmission in early RWD General Motors products can be identified by the production number located on the ID plate attached to the right side of the case near the modulator. Production number consists of a year code, a two-character model code, and a build date code.

The transmission in late model RWD GM products can be identified by a letter code contained in the identification number. The ID number is stamped on the transmission case above the oil pan rail on the right rear side. The identification number contains information, which must be used when ordering replacement parts.

The transaxle in Hondas and Acuras can be identified by a number stamped on a metal pad on top of the transmission. The first two characters indicate the transmission model.

Aisin-Warner transmissions—used by Jeep, Isuzu, and Volvo—can be identified by a plate attached to a side of the transmission case. The plate shows the transmission model number and serial number.

The transmission used in some models from Audi, Porsche, and Volkswagen can be identified by numbers cast into the top rear of the case. The transaxle model code is identified by figures stamped into the torque converter housing. These figures consist of model code and build date code. Some models also will have code letters and the date of manufacture stamped into the machined flat on the bellhousing rim. The valve body also may be stamped on the machined boss on the valve body. The valve body identification tag is secured with valve body mounting screws. A torque converter code letter is stamped on the side of the attaching lug.

Models that use Borg-Warner transmissions have an identification number stamped on a plate attached to the torque converter housing near the throttle cable and the distributor.

Jatco transmissions, manufactured by the Japan Automatic Transmission Company, are used by Chrysler, Mitsubishi, Mazda, and Nissan. The model may be identified by a stamped metal plate attached to the right side of the transmission case. The plate lists the model code on the second line and the serial number on the bottom line.

Some Mazda, Mercedes Benz, Toyota, and Subaru (Gunma) transmissions are best identified by the 11th character in the VIN, which is located at the top left of the instrument panel and on the transaxle flange on the exhaust side of the engine or the driver's doorpost. However, most Mercedes Benz transmissions have the identification number stamped into the pan rail on the right side of the case.

BMW, Peugeot, and some models of Volvo use ZF transmissions, which have an identification plate fixed to the left side of the transmission case. The lower left series of numbers on the plate indicate the number of gears, type of controls, type of gears, and torque capacity.

## DIAGNOSTICS

The true measure of a good technician is an ability to find and correct the cause of problems. Service manuals and other information sources will guide you through the diagnosis and repair of problems. But those guidelines will not always lead you to the exact cause of the problem. To do this, you must use your knowledge and take a logical approach when troubleshooting. Diagnosis is not guessing; it is more than following a series of interrelated steps in order to find the solution to a specific problem. Diagnosis is a way of looking at systems that are not functioning the way they should to find out why. You should know how the system should work and decide if it is working correctly. Through an understanding of the purpose and operation of the system, you can accurately diagnose problems.

Most good technicians use the same basic diagnostic approach. Simply because this is a logical approach, it can quickly lead to the cause of a problem. Logical diagnosis follows these steps:

1. Gather information about the problem.
2. Verify that the problem exists.
3. Thoroughly define what the problem is and when it occurs.
4. Research all available information and knowledge to determine the possible causes of the problem.
5. Isolate the problem by testing.
6. Continue testing to pinpoint the cause of the problem.
7. Locate and repair the problem, then verify the repair.

# Summary

- Technicians are typically paid according to their abilities. New or apprentice technicians are paid by the hour. Once technicians have demonstrated a satisfactory level of skills, they can earn a higher hourly wage or go on flat rate.
- When you begin a job, you enter into a business agreement with your employer. When you become an employee, you sell your time, skills, and efforts. In return, your employer pays you for these resources.
- As part of the employment agreement, your employer also has certain responsibilities: instruction and supervision; a clean, safe place to work; wages, fringe benefits; opportunity; and fair treatment.
- Your obligations as an employee to the employer include regular attendance, following directions, responsibility, productivity, and loyalty.
- When dealing with customers, be polite, respectful, organized, and honest.
- An obvious sign of your knowledge and abilities, in addition to your dedication to the trade, is Automotive Service Excellence (ASE) certification.
- The true measure of a good technician is an ability to find and correct the cause of problems.
- Diagnosis is not guessing; it also is more than following a series of interrelated steps in order to find the solution to a specific problem.

# Review Questions

1. What must a technician do in order to become certified as an ASE Master Automobile Technician?
2. Explain the flat rate pay system.
3. What are your responsibilities as an employee to your employer?

4. List the steps that should be followed when logically diagnosing a problem.
5. Technician A says that after an individual passes a particular ASE certification exam, he or she is certified in that test area. Technician B says that all of the questions on an ASE certification exam are written as Technician A and Technician B questions. Who is correct?
   A. Technician A only
   B. Technician B only
   C. Both Technician A and Technician B
   D. Neither Technician A nor Technician B

## Basic Theories and Services



## SECTION OBJECTIVES

After you have read, studied, and practiced the contents of this section, you should be able to:

- Identify the major components of a vehicle's drive train.
- State the purpose of a transmission.
- Describe the major differences between a transmission and a transaxle.
- Explain how a set of gears can increase torque.
- Define and determine the ratio between two meshed gears.
- Describe the basic operation of a planetary gear set.
- State the purpose of a torque converter assembly.
- Describe the differences between a typical FWD and RWD car.
- Identify and describe the various gears used in modern drive trains.
- Identify and describe the various bearings used in modern drive trains.
- Explain the basic principles of electricity.
- Define the terms voltage, current, and resistance.
- Name the various electrical components and their uses in electrical circuits.
- Diagnose electrical problems by logic and symptom description.
- Perform troubleshooting procedures using meters, test lights, and jumper wires.
- Repair electrical wiring.
- Replace electrical connectors.
- Locate, test, adjust, and replace electrical switches and sensors on a transmission.

*Until recently the only electrical part connected to a transmission was the backup switch. This normally was positioned so the external shift linkage opened and closed the switch.*

# Complete Drive Train Theory

## Introduction

An automobile can be divided into four major systems or basic components: (1) the engine, which serves as a source of power; (2) the drive train, which transmits the engine's power to the car's wheels; (3) the chassis, which supports the engine and body and includes the brake, steering, and suspension systems; and (4) the car's body, interior, and accessories, which include the seats, heater and air conditioner, lights, windshield wipers, and other comfort and safety features.

Basically, a drive train has four main purposes: to connect and disconnect the engine's power to the wheels, to select different speed ratios, to provide a way to move the car in reverse, and to control the power to the drive wheels for safe turning of the vehicle. The main components of the drive train are the transmission, differential, and drive axles **(Figure 1)**.

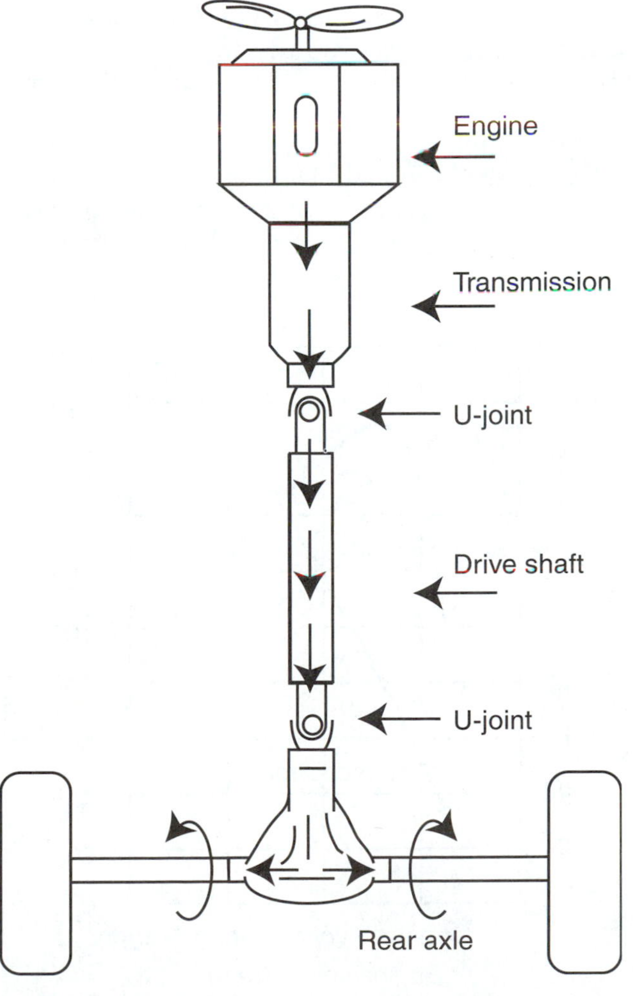


**Figure 1.** Typical drive train components for a RWD car.

### ENGINE

Although the engine is a major system by itself, its output can be considered a component of the drive train. An engine develops a rotary motion or **torque** that, when multiplied by the transmission gears, will move the car under a variety of conditions. The engine produces power by burning a mixture of fuel and air in its combustion chambers. Combustion causes a high pressure in the cylinders, which forces the pistons downward. Connecting rods transfer the downward movement of the pistons to the crankshaft, which rotates by the force on the pistons.

Most automotive engines are four-stroke cycle engines. The opening and closing of the intake and exhaust valves are timed to the movement of the pistons. As a result, the engine passes through four different events or strokes during one combustion cycle. The four strokes are called the intake, compression, power, and exhaust strokes. As long as the engine is running, this cycle of events repeats itself, resulting in the production of engine torque.

## Engine Torque

The rotating or turning effort of the engine's crankshaft is called **engine torque**. Engine torque is measured in foot-pounds (ft.-lb.) or in the metric system Newton-meters (N • m). Most engines produce a maximum amount of torque while operating within a range of engine speeds and loads. When an engine reaches the maximum speed of that range, torque is no longer increased. This range of engine speeds is normally referred to as the engine's **torque curve (Figure 2)**. For maximum efficiency, the engine should operate within its torque curve at all times.

As a manual transmission equipped car is climbing up a steep hill, its driving wheels slow down, which causes engine speed to decrease and reduces the engine's output. The driver must downshift the transmission, which increases engine speed and allows the engine to produce more torque. When the car reaches the top of the hill and begins to go down, its speed and the speed of the engine rapidly increase. The driver can now upshift, which allows the engine's speed to decrease and places it back in the torque curve.

As a car equipped with an automatic transmission is climbing up a steep hill, the speed of its driving wheels and the engine also decreases. However, the driver does not need to downshift the transmission. The transmission senses the increased load and automatically downshifts. This downshifting increases engine speed and allows the engine to produce more torque. When the car reaches the top of the hill and begins to go down, the transmission will automatically upshift to decrease the engine's speed.

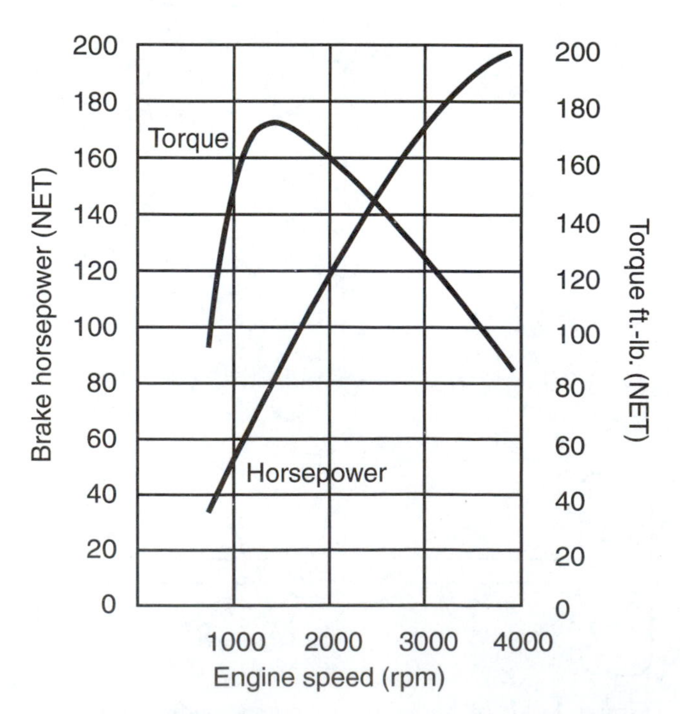


**Figure 2.** The amount of torque produced by an engine varies with the speed of the engine.

## Torque Multiplication

Measurements of **horsepower** indicate the amount of work being performed and the rate at which it is being done. The term **power** actually means a force at work, that is, doing work over a period of time. The driveline can transmit power and multiply torque, but it cannot multiply power. When power flows through one gear to another, the torque is multiplied in proportion to the different gear sizes. Torque is multiplied, but the power remains the same, as the torque is multiplied at the expense of rotational speed.

# BASIC GEAR THEORY

The main components of all drive train are gears. Gears apply torque to other rotating members of the drive train and are used to multiply torque. As gears with different numbers of teeth mesh, each rotates at a different speed and torque.

Torque is calculated by multiplying the force by the distance from the center of the shaft to the point at which the force is exerted **(Figure 3)**. For example, if you tighten a bolt with a wrench that is 1 foot long and apply a force of 10 pounds to the wrench, you are applying 10 pounds per foot of torque to the bolt. Likewise, if you apply a force of 20 pounds to the wrench, you are applying 20 ft.-lbs. of torque. You could also apply 20 ft.-lbs. of torque by applying only 10 pounds of force if the wrench were 2 feet long.

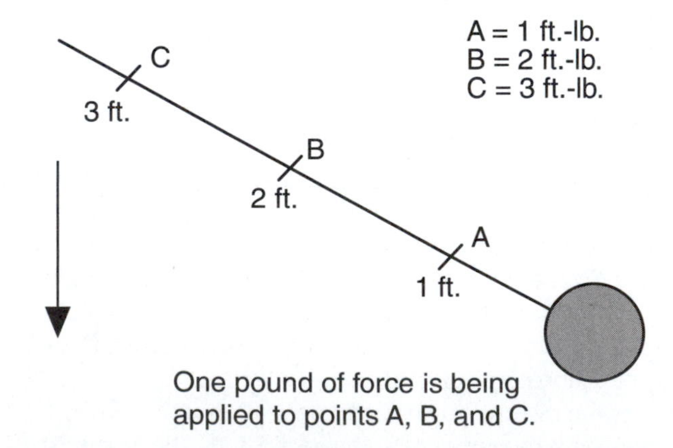


**Figure 3.** Torque is calculated by multiplying the force (1 pound) by the distance from the center of the shaft to the point (Points A, B, and C) where force is exerted.

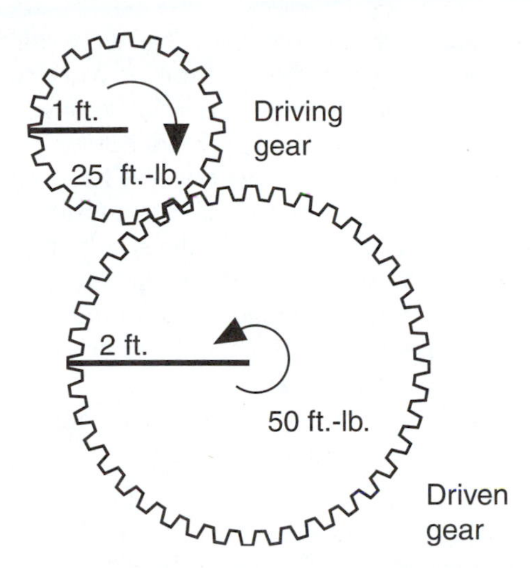


**Figure 4.** The driven gear will turn at half the speed but twice the torque, because it is two times larger than the driving gear.

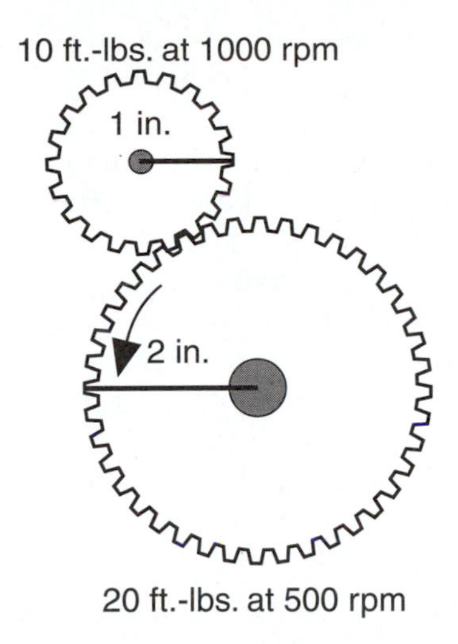


**Figure 5.** The one-inch gear will turn the two-inch gear at half its speed but twice the torque.

The distance from the center of a circle to its outside edge is called the radius. On a gear, the radius is the distance from the center of the gear to the point on its teeth at which force is applied.

If a tooth on the driving gear is pushing against a tooth on the driven gear with a force of 25 pounds and the force is applied at a distance of 1 foot (the radius of the driving gear) a torque of 25 ft.-lb. is applied to the driven gear. The 25 pounds of force from the teeth of the smaller (driving) gear is applied to the teeth of the larger (driven) gear. If that same force were applied at a distance of 2 feet from the center, the torque on the shaft at the center of the driven gear would be 50 ft.-lbs. The same force is acting at twice the distance from the shaft center **(Figure 4)**.

The amount of torque that can be applied from a power source is proportional to the distance from the center at which it is applied. If a fulcrum or pivot point is placed closer to object being moved, more torque is available to move the object, but the lever must move farther than if the fulcrum was farther away from the object. The same principle is used for gears in mesh: A small gear will drive a large gear more slowly but with greater torque. It is important to realize that the driven gear rotates in the opposite direction as the drive gear.

A gear set consisting of a driving gear with 24 teeth and a radius of 1 inch and a driven gear with 48 teeth and a radius of 2 inches will have a torque multiplication factor of 2 and a speed reduction of $\frac{1}{2}$. It doubles the amount of torque applied to it at half the speed **(Figure 5)**. The radii between the teeth of a gear act as a lever; therefore, a gear that is twice the size of another has twice the lever arm length of the other.

**Gear ratios** express the mathematical relationship of one gear to another. Gear ratios can be varied by changing the diameter and number of teeth of the gears in mesh. A gear ratio also expresses the amount of torque multiplication between two gears. The ratio is obtained by dividing the diameter or number of teeth of the driven gear by the diameter or teeth of the drive gear. If the smaller driving gear had 11 teeth and the larger gear had 44 teeth, the ratio would be 4:1 **(Figure 6)**. The gear ratio tells you how many times the driving gear must turn to rotate the driven gear once. With a 4:1 ratio, the smaller gear must turn four times to rotate the larger gear once.

The larger gear turns at one fourth the speed of the smaller gear but has four times the torque of the smaller

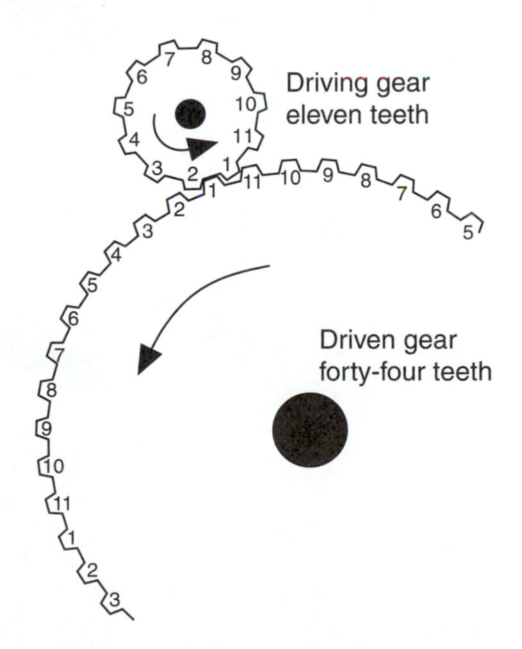


**Figure 6.** The driving gear must rotate four times to rotate the driven gear once. The ratio of the gear set is 4:1.

gear. In gear systems, speed reduction means torque increase. For example, when a typical five-speed transmission is in first gear, there is a speed reduction of 12:1 from the engine to the drive wheels, which means the crankshaft turns 12 times to turn the wheels once. The resulting torque is 12 times the engine's output; therefore, if the engine produces 100 ft.-lbs. of torque, a torque of 1200 ft.-lb. is applied to the drive wheels.

Pulleys also can be used to change speed and torque. Because a drive belt typically connects them, the direction of the driven pulley is the same as the direction of the drive pulley. However, the difference in size has the same effect as the size of gears. When the drive pulley is the same diameter as the driven pulley, the two will rotate at the same speed and with the same torque. When the drive pulley is smaller than the driven pulley, the driven pulley will turn at a lower rotational speed but with greater torque. Likewise, when the drive pulley is larger than the driven pulley, the driven pulley will rotate faster but with less torque. Pulleys are used with drive belts to operate some engine components such as generators, power steering pumps, and air conditioning compressors. Pulleys also are the basis for the operation of constant variable ratio transmissions **(Figure 7)**.

Constantly variable ratio transmissions are found on some cars. These transmissions automatically change

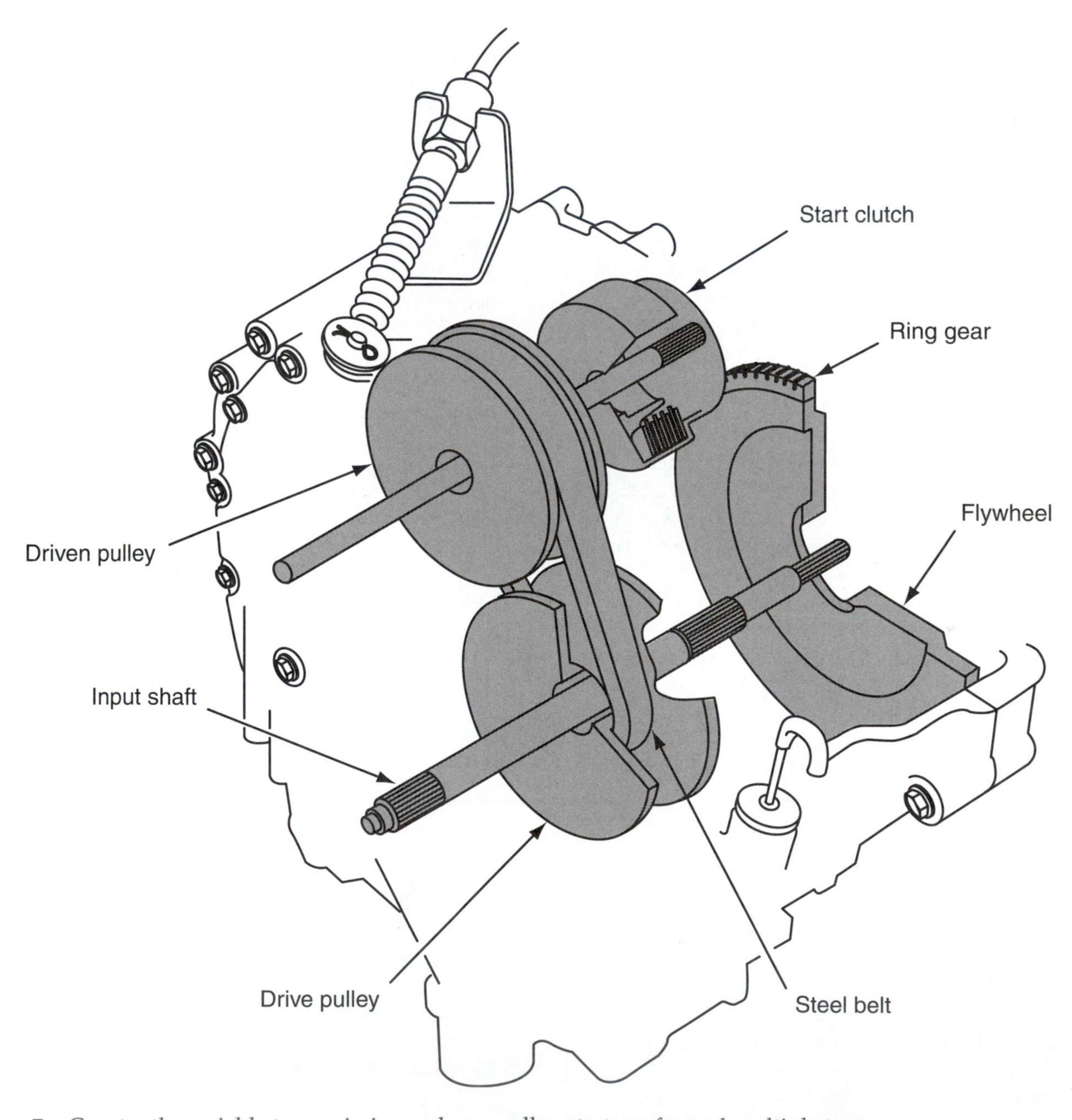


**Figure 7.** Constantly variable transmissions rely on pulleys to transfer and multiple torque.

torque and speed ranges without causing a change in engine speed. The premise behind this transmission design is to keep the engine operating within a fixed speed range. This allows for improved fuel economy and decreased emission levels.

## TRANSMISSIONS

The transmission is mounted to the rear of the engine and is designed to allow the car to move forward and in reverse. It also has a neutral position. In this position, the engine can run without applying power to the drive wheels. Therefore, although there is input to the transmission when the vehicle is in neutral, there is no output from the transmission because the driving gears are not engaged to the output shaft.

There are two basic types of transmissions: automatic and manual transmissions. Automatic transmissions use a combination of a torque converter and planetary gear sets to change gear ratios automatically. A manual transmission is an assembly of gears and shafts that transmits power from the engine to the drive axle and changes in gear ratios is controlled by the driver. By moving the shift lever on a manual transmission and depressing the clutch pedal, various gear and speed ratios can be selected. The gears in a transmission are selected by the driver.

An automatic transmission or transaxle selects gear ratios according to engine speed, power train load, vehicle speed, and other operating factors. Little effort is needed on the part of the driver, because both upshifts and downshifts occur automatically. There is no need for a manually operated clutch to assist in the change of gears, as is the case for a manual transmission. Also, an automatic transmission can remain engaged in a gear without stalling the engine while the vehicle is stopped. The driver also can manually select a lower forward gear, reverse, neutral, or park. Depending on the forward range selected, the transmission also can provide engine braking during deceleration.

Today, most automatic transmissions have four or five forward speeds. These speeds or gears are identified as first, second, third, fourth, and fifth. Different gear ratios are necessary because an engine develops relatively little power at low engine speeds. Without the aid of gears, the engine must turn at a fairly high speed before it can deliver enough power to get the car moving. Through selection of the proper gear ratio, torque applied to the drive wheels can be multiplied. This eliminates the need to run the engine at high speeds to initiate and maintain movement.

## TRANSMISSION GEARS

Transmissions contain several combinations of large and small gears. In low or first gear, a small gear drives a large gear on another shaft. This reduces the speed of the larger gear but increases its torque and offers the proper gear ratio for starting movement or for pulling heavy loads. First gear has the lowest gear ratio of any gear in a transmission.

The ratio of second gear does not offer the same amount of torque multiplication as does first gear; however, it does offer a substantial amount. Because the car is already in motion, less torque is needed to move the car.

Third gear allows for a further decrease in torque multiplication, while increasing vehicle speed and encouraging fuel economy. This gear may provide a direct drive (1:1) ratio so that the amount of torque that enters the transmission is also the amount of torque that passes through and out of the transmission output shaft. This gear is used at cruising speeds and promotes fuel economy. While the car is in third gear, it lacks the performance characteristics of the lower gears.

Many of today's transmissions have a fourth and/or fifth gear, called overdrive gears. Overdrive gears have ratios of less than 1:1. These ratios are achieved by using a small driving gear meshed with a smaller driven gear. Output speed is increased and torque is reduced. The purpose of overdrive is to promote fuel economy and reduce operating noise while maintaining highway cruising speed.

Through the use of an additional gear in mesh with two other speed gears, vehicles can move in reverse. Because reverse gear ratios are typically based on the drive and driven gears used for first gear, only low speeds can be obtained in reverse.

The transmission's gear ratios are further increased by the gear ratio of the final drive gears in the drive axle assembly. Typical axle ratios are between 2.5 and 4.5:1. The total final drive gear ratio is calculated by multiplying the transmission gear ratio by the final drive ratio. If a transmission is in first gear with a ratio of 3.63:1 and has a final drive ratio of 3.52:1, the overall gear ratio is 12.87:1.

Although a manual transmission must be disconnected from the engine briefly each time the gears are shifted, by disengaging the clutch; an automatic transmission does its gear shifting while it is engaged to the engine. This is accomplished through the use of constantly meshing planetary gears.

### Planetary Gears

A planetary gear set consists of a ring gear, a sun gear, and several planet gears all mounted in the same plane **(Figure 8)**. The ring gear has its teeth on its inner surface and the sun gear, concentric with the ring gear, has its teeth on its outer surface. The planet gears are spaced evenly around the sun gear and mesh with both the ring and sun gears. The ring, sun, and planet gears each have their own shaft or carrier on which to rotate.

By applying the engine's torque to one of the gears in a planetary gear set and preventing another member of the set from moving, torque change is available on the

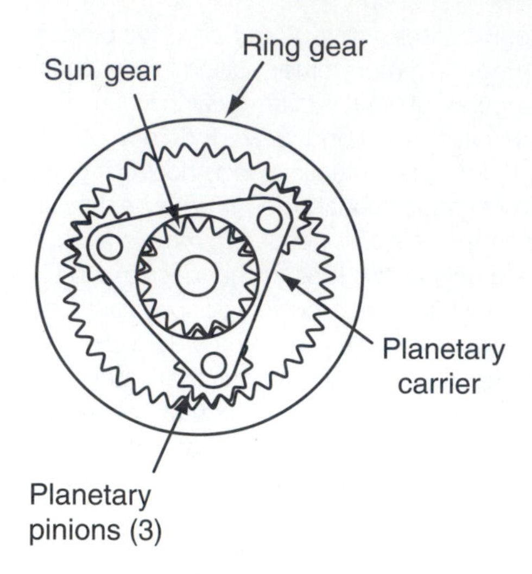


**Figure 8.** A planetary gearset.

third set of gears. Brake bands **(Figure 9)** or clutch packs **(Figure 10)** attached to the individual gear carriers and shafts are hydraulically activated to direct power flow from the engine to any of the gears and to hold any of the gears from rotating. This allows gear ratio changes and the reversing of power flow while the engine is running.

## Compound Planetary Gear Sets

A limited number of gear ratios are available from a single planetary gear set. To increase the number of available gear ratios, gear sets can be added. The typical automatic transmission with four forward speeds has at least two planetary gear sets.

There are two common designs of compound gear sets: the Simpson gear set, in which two planetary gear sets share a common sun gear, and the Ravingeaux gear set, which has two sun gears, two sets of planet gears, and a common ring gear. Some transmissions are fitted with an additional single planetary gear set, which is used to provide additional gear ratios.

## Planetary Gear Ratios

Calculating the gear ratios of a planetary gear set requires the use of different formulas than used to calculate the ratio of two gears in mesh. In a planetary gear set, three sets of gears are in mesh and some of the gears have internal teeth. The resulting ratio is based on the number of teeth on the drive and driven gears. In most cases, the carrier walks around a held gear. Because the carrier links the drive and driven gears, the total number of teeth on the drive gear is added to the number of teeth on the driven gear. This sum is then divided by the number of teeth on the drive gear. Therefore, the formula for calculating the gear ratio of a planetary gear set is:

$$\text{Gear ratio} = \frac{\text{Drive gear} + \text{Driven gear}}{\text{Drive gear}}$$

For example, let us look at a gear set with a sun gear that has 25 teeth and a ring gear with 75 teeth. In low or first gear, the sun gear is the drive gear. The ratio of this combination is calculated as follows:

$$\text{Gear ratio} = \frac{25 + 75}{25}$$

$$\text{Gear ratio} = \frac{100}{25} = 4{:}1$$

When the ring gear is the drive gear, such as during second gear, the ratio changes:

$$\text{Gear ratio} = \frac{25 + 75}{75}$$

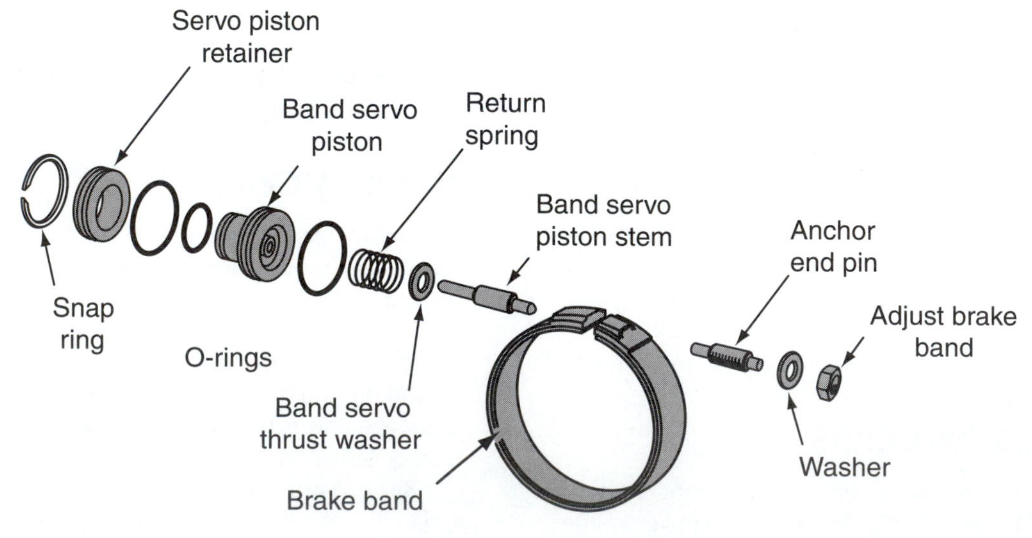


**Figure 9.** A typical band assembly.

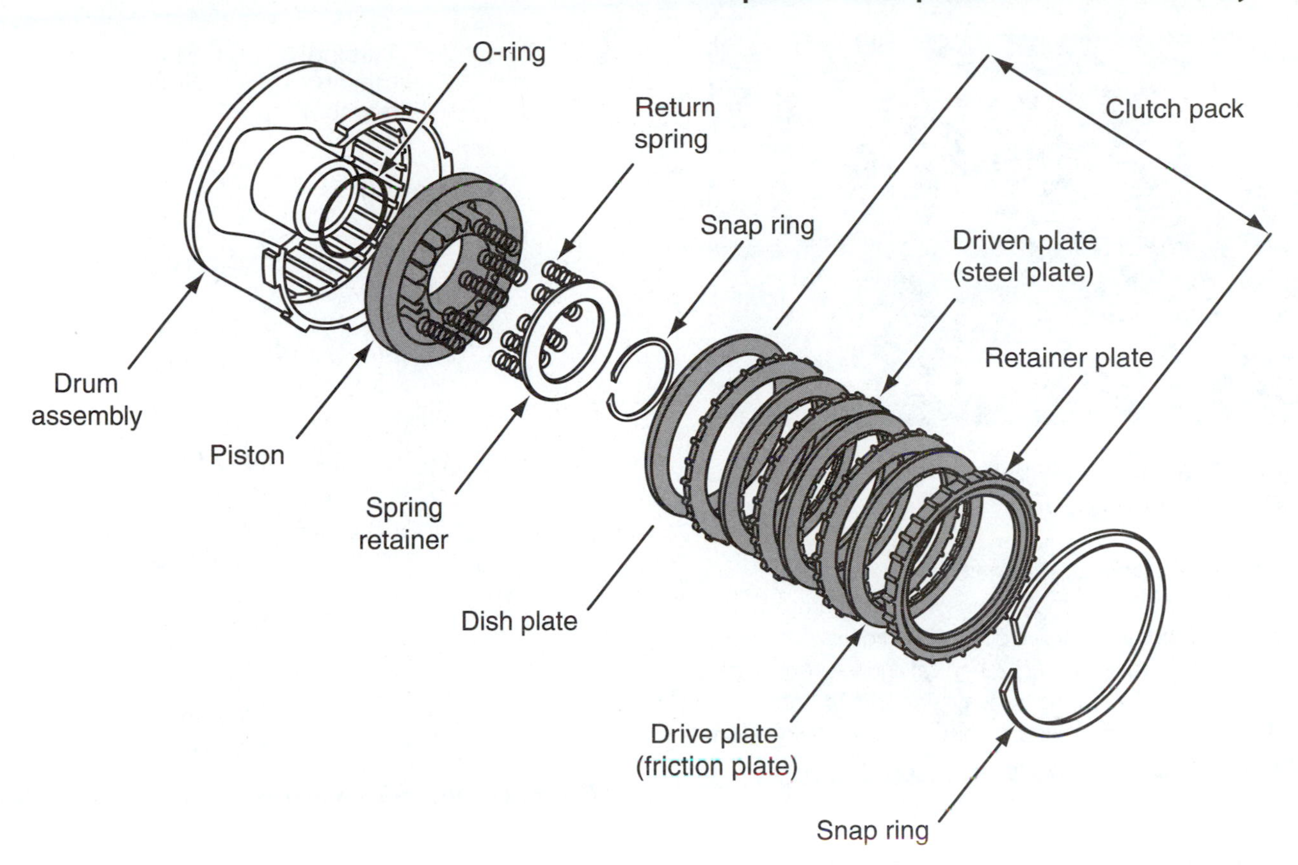


**Figure 10.** A typical multiple friction disc pack.

$$\text{Gear ratio} = \frac{100}{75} = 1.33{:}1$$

When either the ring gear or sun gear is held (this happens in overdrive), the ratio is determined by adding the number of teeth on the gears. Use this formula when the driven gear is the ring gear:

$$\text{Gear ratio} = \frac{\text{Driven gear}}{\text{Drive gear} + \text{Driven gear}}$$

$$\text{Gear ratio} = \frac{75}{100} = 0.75{:}1$$

When the gear set is in direct drive, the members of the gear set are locked together and rotate as a single member. Because of this action, no gear reduction occurs and the ratio is simply 1:1.

When the transmission is in reverse, a simple driven to drive relationship exists. Using the sun and ring gears from the previous examples, the ratio of reverse gear would be 3:1 when the sun gear is the drive gear:

$$\text{Gear ratio} = \frac{\text{Driven gear}}{\text{Drive gear}}$$

$$\text{Gear ratio} = \frac{75}{25} = 3{:}1$$

## Shift Control

The engine's power is transmitted to the transmission through a fluid coupler, called the torque converter **(Figure 11)**. The torque converter drives an oil pump, which transmits fluid to a control-valve assembly. This valve assembly provides the hydraulic fluid needed to activate the various brake bands and multiple-friction disc packs. The **valve body (Figure 12)** controls the flow of the fluid throughout the transmission in response to the inputs it receives about engine and vehicle speeds and loads.

**Figure 11.** A typical torque converter.

Figure 12. A valve body being removed from a transmission case.

Most new automatic transmissions rely on data received from electronic sensors and use an electronic control unit to operate solenoids in the valve body to shift gears. Older automatic transmissions rely on mechanical and vacuum signals, which determine when the transmission should shift.

Electronically controlled automatics have many advantages over the older designs, including more precise shifting of gears, greater fuel economy, and increased reliability. When shifting gears, older designs rely on the action of a cable-operated or vacuum-controlled modulator to adjust fluid pressure in the transmission's lines to activate spring-loaded valves. This action took a small amount of time and delays in shifting and slippage resulted. The shifting of electronic transmissions is precisely controlled by a computer, which gathers information from many sensors, including those for throttle position, temperature, engine load, and vehicle speed. The computer processes this information every few milliseconds and sends electrical signals to the shift solenoids, which control the shift valves of the valve body. Electrically controlled solenoids also match transmission line-pressure to engine torque for better shift feel than mechanically controlled transmissions.

## Torque Converter

An automatic transmission is connected to the engine by a fluid filled torque converter. The rotary motion of the engine's crankshaft is transferred from the **flywheel**, through the torque converter, to the transmission. This rotary motion is then delivered by the transmission to the **differential** and transferred by axle shafts to the tires, which push against the ground to move the car.

The torque converter consists of an impeller, which is attached to the engine's crankshaft, a mating turbine, which is attached to the transmission's input shaft, and a torque-multiplying stator that is mounted between the turbine and the impeller **(Figure 13)**.

The torque converter operates by hydraulic force generated by automatic transmission fluid. The torque converter

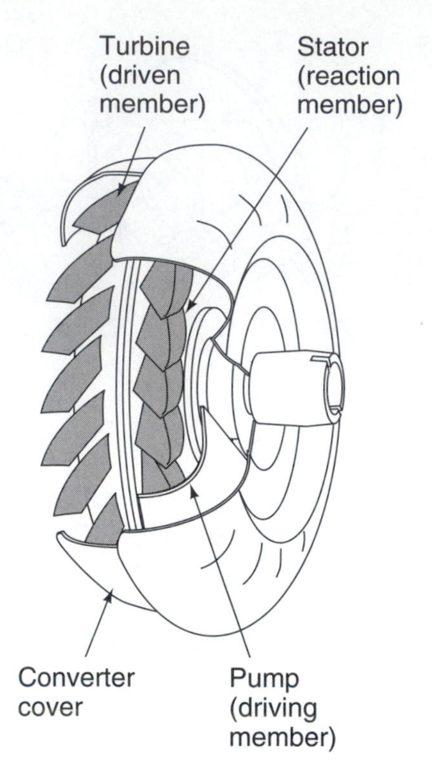


Figure 13. A torque converter.

changes or multiplies torque transmitted by the engine's crankshaft and directs it through the transmission. The torque converter also automatically engages and disengages engine power to the transmission in response to engine speed.

As the engine rotates, the impeller throws the transmission fluid at the blades of the turbine. The turbine spins in response to the force exerted by the moving fluid. Because the turbine is connected to the transmission's input shaft, engine output is transferred to the transmission.

While operating at normal idle speeds, the engine does not rotate fast enough to allow the impeller to throw fluid against the turbine with enough force to cause it to spin. This lack of hydraulic force enables the vehicle to stand still, without stalling the engine, when the gears are engaged. The hydraulic engaging and disengaging action of the impeller to the turbine performs a similar duty as the operation of the clutch in a manual transmission equipped vehicle.

The clutch assembly of a vehicle equipped with a manual transmission is mounted to a flywheel, which is a large and heavy disc, attached to the rear end of the crankshaft. In addition to providing a friction surface and mounting for the clutch, the flywheel also dampens crankshaft vibrations, adds inertia to the rotation of the crankshaft, and serves as a large gear for the starter motor. Automatic transmissions do not require the use of a heavy flywheel; rather the weight of the fluid-filled torque converter mounted to a lightweight **flexplate** is used for the same purposes.

## DRIVE LINE CONFIGURATIONS

Today's cars are designed to transfer the engine's power to either the front or rear wheels. In a FWD car, the transmission and final drive unit are located in a single cast aluminum housing called a transaxle assembly. All of the driving components are located compactly at the front of the vehicle. One of the major advantages of FWD is that the weight of the power train components is placed over the driving wheels, which provides for improved traction on slippery road surfaces.

RWD cars have the drive train, with the exception of the engine, located beneath the body. The engine is mounted at the front of the chassis and the related components extend to the rear driving wheels. The transmission's internal parts are located within an aluminum or cast iron housing called the transmission case. The driving axle is located at the rear of the vehicle, in a separate housing called the rear axle assembly. A drive shaft connects the output of the transmission to the rear axle.

## TYPES OF GEARS

Gears are used to transmit torque from one shaft to another. These shafts may operate in line, parallel to each other, or at an angle to each other. These different applications require a variety of gear designs that vary primarily in the size and shape of the teeth. In order for gears to mesh, they must have teeth of the same size and design. Meshed gears have at least one pair of teeth engaged at all times.

Automobiles use a variety of gear types to meet the demands of speed and torque. The most basic type of gear is the spur gear, which has its teeth parallel to and in alignment with the center of the gear. Early manual transmissions used straight-cut spur gears, which were easier to machine but were noisy and difficult to shift. Today these gears are commonly used in simple devices such as hand or powered winches.

**Helical gears** are like spur gears, except that their teeth have been twisted at an angle from the gear centerline. Engagement of these gears begins at the tooth tip of one gear and rolls down the trailing edge of the teeth. Helical spur gears are quieter in operation and have greater strength and durability than straight spur gears, simply because the contacting teeth are longer. Helical spur gears are widely used in transmissions because they are quieter at high speeds and are durable.

Bevel gears are shaped like a cone with its top cut off. The teeth point inward toward the peak of the cone. These gears permit the power flow to "turn a corner." Spiral bevel gears have their teeth cut obliquely on the angular faces of the gears. The most commonly used spiral beveled gear set is the ring and pinion gears used in heavy-duty truck differentials.

The **hypoid gear set** have a pinion drive gear located below the center of the ring gear. Its teeth and general construction are the same as the spiral bevel gear. The most common use for hypoid gears is in modern differentials.

In a worm gear set, the mating gear has teeth, which are curved at the tips to permit a greater contact area. Power is supplied to the worm gear, which drives the mating gear. Worm gears usually provide right-angle power flows.

**Internal gears** have their teeth pointing inward and are commonly found in the planetary gear sets used in automatic transmissions and transfer cases. These gear sets have an outer ring gear with internal teeth and mate with smaller planetary gears. These gears, in turn, mesh with a center or sun gear. In a planetary gear set, one gear is normally the input, another is prevented from moving or held, and the third gear is the output gear. In planetary gear sets, the load is spread over several gears, reducing stress and wear.

## BEARINGS

The ease with which gears rotate on the shaft or the shaft rotates with the gears partially determines the amount of power needed to rotate them. If they rotate with great difficulty because of high friction, much power is lost. High friction also will cause excessive wear to the gears and shaft. To reduce the friction, bearings are fitted to the shaft and/or gears **(Figure 14)**.

The simplest type of **bearing** is a cylindrical hole formed in a piece of material, into which the shaft fits freely. The hole is usually lined with a brass or bronze lining, or **bushing**, which not only reduces the friction but also allows for easy replacement when wear occurs. Bushings usually have a tight fit in the hole in which they fit.

Ball or roller bearings are used wherever friction must be minimized. With these types of bearings, rolling friction replaces the sliding friction that occurs in plain bearings. Typically, two bearings are used to support a shaft instead of a single long bushing. Bearings have three purposes: they support a load, maintain alignment of a shaft, and reduce rotating friction.

Most bearings are capable of withstanding only loads that are perpendicular to the axis of the shaft. Such loads are called **radial loads**, and bearings that carry them are called radial or journal bearings.

To prevent the shaft from moving in the **axial** direction, shoulders or collars may be formed on it or secured to it. If both collars are made integral with the shaft, the bearings must be split or made into halves, and the top half or cap bolted in place after the shaft has been put in place. The collars or shoulders withstand any end thrusts, and bearings designed this way are termed **thrust bearings**.

Some thrust bearings look similar to a thick washer fitted with needle bearings on its flat surface. These typically are called thrust needle bearings or Torrington bearings and commonly are used in automatic transmissions. Automatic

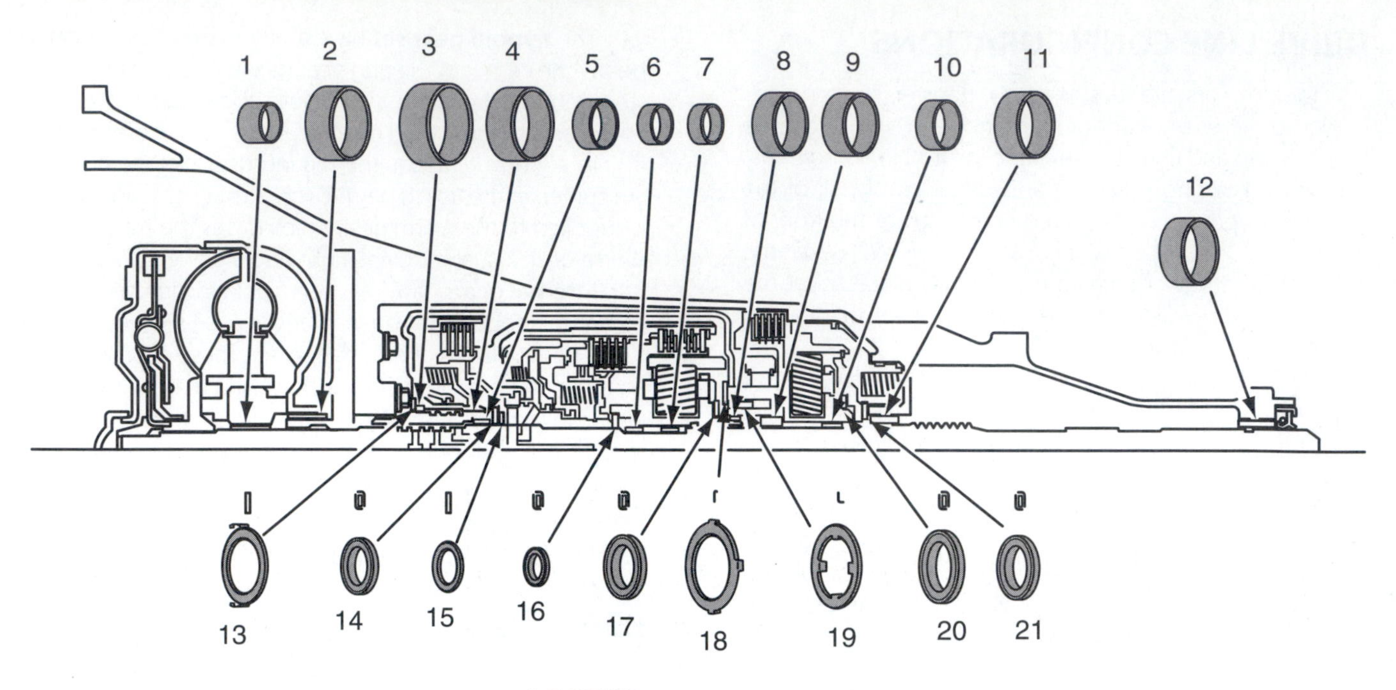


**LEGEND**

1. Bushing, Stator shaft (front)
2. Bushing, Oil pump body
3. Bushing, Reverse input clutch (front)
4. Bushing, Reverse input clutch (rear)
5. Bushing, Stator shaft (rear)
6. Bushing, Input sun gear (front)
7. Bushing, Input sun gear (rear)
8. Bushing, Reaction carrier shaft (front)
9. Bushing, Reaction gear
10. Bushing, Reaction carrier shaft (rear)
11. Bushing, Case
12. Bushing, Case extension
13. Washer, Thrust (pump to drum)
14. Bearing assembly, Stator shaft/selective washer
15. Washer, Thrust (selective)
16. Bearing assembly, Input sun gear
17. Bearing assembly, Thrust (input carrier to reaction shaft)
18. Washer, Thrust (reaction shaft/shell)
19. Washer, Thrust (race/reaction shell)
20. Bearing assembly, Thrust (reaction carrier/support)
21. Bearing, Reaction gear support to case

**Figure 14.** Typical location of the various bushings and bearings used in an automatic transmission. (Courtesy of General Motors Corporation, Service Operations)

transmissions also use items that look like thrust needle bearings without the needle bearings. These are thrust washers and are used to control end clearance.

A single row journal or radial **ball bearing** has an inner race made of a ring of case-hardened steel with a groove or track formed on its outer circumference for a number of hardened steel balls to run upon. The outer race is another ring, which has a track on its inner circumference. The balls fit between the two tracks and roll around in the tracks as either race turns. The balls are kept from rubbing against each other by some form of cage. These bearings can withstand radial loads and can also withstand a considerable amount of axial thrust. Therefore, they are often used as combined journal and thrust bearings.

A bearing designed to take only radial loads has only one of its races machined with a track for the balls. Other bearings are designed to take thrust loads in only one direction. If this type of bearing is installed wrong, the slightest amount of thrust will cause the bearing to come apart.

Another type of ball bearing uses two rows of balls. These are designed to withstand considerable amounts of radial and axial loads. Constructed as two single-row ball bearings joined together, these bearings are often used in rear axle assemblies.

**Roller bearings** are used wherever it is desirable to have a large bearing surface and low amounts of friction. Large bearing surfaces are needed in areas of extremely heavy loads. The rollers are usually fitted between a journal of a shaft and an outer race. As the shaft rotates, the rollers turn and rotate in the race. Tapered roller bearings are commonly used in drive axle and transfer shaft assemblies.

## Summary


- The drive train has four primary purposes: to connect the engine's power to the drive wheels, to select different speed ratios, to provide a way to move the vehicle in reverse, and to control the power to the drive wheels for safe turning of the vehicle.
- The main components of the drive train are the transmission, differential, and drive axles.
- The rotating or turning effort of the engine's crankshaft is called engine torque. Gears are used to apply torque to other rotating parts of the drive train and to multiply torque.
- Torque is calculated by multiplying the applied force by the distance from the center of the shaft to the point where the force is exerted. Torque is measured in either inch- or foot-pounds and Newton-meters.
- Gear ratios express the mathematical relationship, in size and number of teeth, of one gear to another.
- Gear ratios are determined by dividing the number of teeth on the driven gear by the number of teeth on the driving gear.
- Transmissions offer various gear ratios through the meshing of various sized gears.
- Like manual transmissions, automatic transmissions provide various gear ratios that match engine speed to the vehicle's speed. However, an automatic transmission is able to shift between gear ratios by itself and there is no need for a manually operated clutch to assist in the change of gears.
- A planetary gear set consists of a ring gear, a sun gear, and several planet gears all mounted in the same plane.
- The ring gear has its teeth on its inner surface and the sun gear has its teeth on its outer surface. The planet gears are spaced evenly around the sun gear and mesh with both the ring and sun gears.
- By applying the engine's torque to one of the gears in a planetary gear set and preventing another member of the set from moving, torque multiplication, speed increase, or change of rotational direction is available on the third set of gears.
- Brake bands or multiple-friction disc packs attached to the individual gear carriers and shafts are hydraulically activated to direct engine power to any of the gears and to hold any of the gears from rotating. This allows gear ratio changes and the reversing of power flow while the engine is running.
- An oil pump in the transmission provides the hydraulic fluid needed to activate the various brake bands and clutch packs.
- The valve body controls the flow of the fluid throughout the transmission and acts on the vacuum and mechanical signals it receives about engine and vehicle speeds and loads.
- Most new automatic transmissions rely on data received from electronic sensors and use an electronic control unit to operate solenoids in the valve body to shift gears.
- In FWD cars, the transmission and drive axle is located in a single assembly called a transaxle. In RWD cars, the drive axle is connected to the transmission through a drive shaft.
- The primary purpose of the differential is to allow a difference in driving wheel speed when the vehicle is rounding a corner or curve. The ring and pinion in the drive axle also multiples the torque it receives from the transmission.
- An understanding of gears and bearings is the key to effective troubleshooting and repair of driveline components.


## Review Questions

1. What are primary purposes of a vehicle's drive train?
2. Why does torque increase when a smaller gear drives a larger gear?
3. The rotating or turning effort of the engine's crankshaft is called ______________________________.
4. Gear ratios are determined by dividing the number of teeth on the ______________ gear by the number of teeth on the ______________ gear.
5. The torque converter assembly is comprised of a ________________, an ________________________, and a ________________.
6. While discussing the purposes of a drive train, Technician A says that it connects the engine's power to the drive wheels. Technician B says that it controls the power to the drive wheels for safe turning of the vehicle. Who is correct?
   A. Technician A only
   B. Technician B only
   C. Both Technician A and Technician B
   D. Neither Technician A nor Technician B
7. Technician A says that gears are used to apply torque to other rotating parts of the drive train. Technician B says that gears are used to multiply torque. Who is correct?
   A. Technician A only
   B. Technician B only
   C. Both Technician A and Technician B
   D. Neither Technician A nor Technician B

8. While discussing gear ratios, Technician A says that they express the mathematical relationship, according to the number of teeth, of one gear to another. Technician B says that they express the size difference of two gears by stating the ratio of the smaller gear to the larger gear. Who is correct?
   A. Technician A only
   B. Technician B only
   C. Both Technician A and Technician B
   D. Neither Technician A nor Technician B
9. While discussing torque converters, Technician A says that they are used to transfer engine torque to the transmission. Technician B says that they rotate at engine speed at all times. Who is correct?
   A. Technician A only
   B. Technician B only
   C. Both Technician A and Technician B
   D. Neither Technician A nor Technician B
10. Technician A says that a valve body may rely on vacuum signals to determine the best time to cause a change in gear ratios. Technician B says that the valve body's action may be controlled by electrical solenoids. Who is correct?
   A. Technician A only
   B. Technician B only
   C. Both Technician A and Technician B
   D. Neither Technician A nor Technician B

# Chapter 6 Hydraulics

## *Introduction*

An automatic transmission shifts automatically through the gears during its forward range. The forward range is selected by the driver through the gearshift lever. The selection of park, neutral, and reverse are controlled by the driver as well. In order for the transmission to operate in the desired range, a series of other controls are needed. These controls are mechanical, electrical (electronic), or hydraulic.

The mechanical controls are comprised of the gear selector linkage and other linkages that work with the electronic and hydraulic control devices.

### Electronic Controls

Electrical and electronic circuits are used to perform work or control the operation of a device. An electrical switch is a simple electrical control that turns a circuit on or off, and consequently turns the electrical device connected to the circuit on or off. Switches can be opened or closed by the driver, mechanical linkages, or a predetermined condition. The latter typically is a low or high hydraulic pressure.

Electrical sensors are also a type of control; however, rather than merely opening and closing a circuit, sensors vary the flow of electricity through the circuit. Sensors typically are potentiometers or variable resistors that respond to changes in conditions. The sensor's resistance reflects the condition it is monitoring.

Solenoids are commonly used in transmissions. These devices convert electrical energy into mechanical energy. A solenoid provides for a linear movement. Solenoids in automatic transmissions are used to control the direction and flow of hydraulic fluid pressure.

Until recently, all automatic transmissions were controlled by hydraulic circuits. However, nearly all transmissions now control the operation of the torque converter and transmission through a computer and hydraulics. Based on information received from various electronic sensors and switches, a computer can control the operation of the torque converter and transmission's shift points. Computer-controlled electrical solenoids typically are used to control shifting **(Figure 1)**.

### Hydraulic Controls

Automatic transmission fluid (ATF) is special oil designed to allow for proper transmission operation. Transmissions are equipped with a fluid cooler, which prevents the overheating of the fluid that can result in damage to the transmission. The transmission's pump is the source of all fluid flow in the hydraulic system. It provides a constant supply of fluid under pressure to operate, lubricate, and cool the transmission. **Pressure-regulating valves** change the fluid's pressure to control the shift quality of a transmission and the shift points of the transmission equipped with a governor. **Flow-directing valves** direct the pressurized fluid to the appropriate apply device to cause a change in gear ratios. The hydraulic system also keeps the torque converter filled with fluid.

### LAWS OF HYDRAULICS

An automatic transmission is a complex hydraulic circuit. To better understand how an automatic transmission works, a good understanding of how basic hydraulic circuits work is needed. A simple hydraulic system has liquid, a pump, lines to carry the liquid, control valves, and an output device. The liquid must be available from a

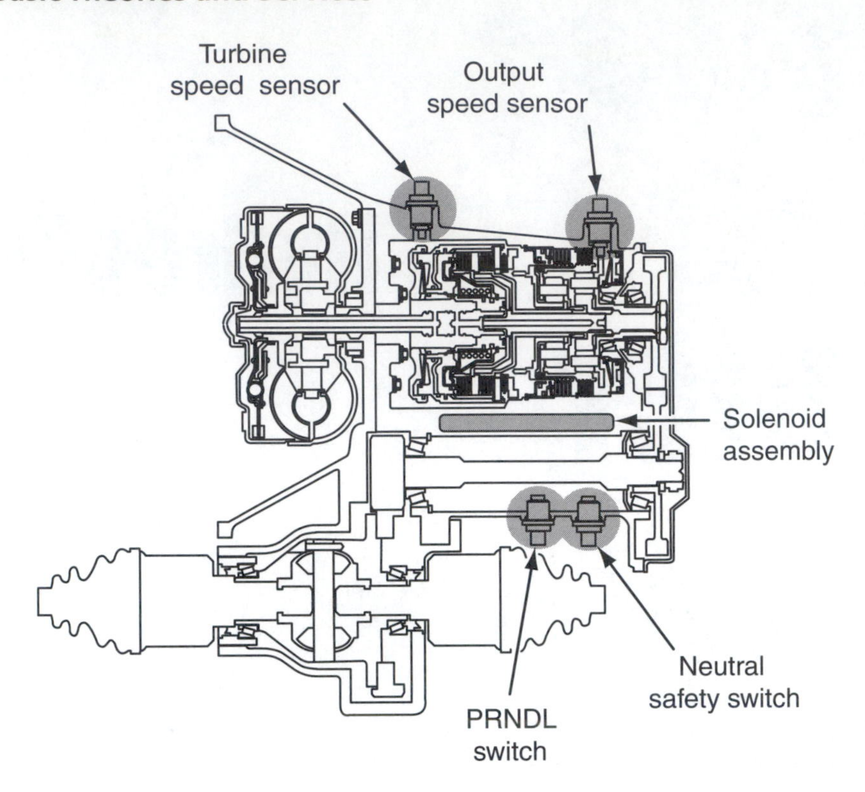


**Figure 1.** An electronically controlled automatic transaxle.

continuous source, such as an oil pan or sump. An oil pump is used to move the liquid through the system. The lines to carry the liquid may be pipes **(Figure 2)**, hoses, or a network of internal bores or passages in a single housing **(Figure 3)**. Control valves are used to regulate hydraulic pressure and direct the flow of the liquid. The output device is the unit that uses the pressurized liquid to do work.

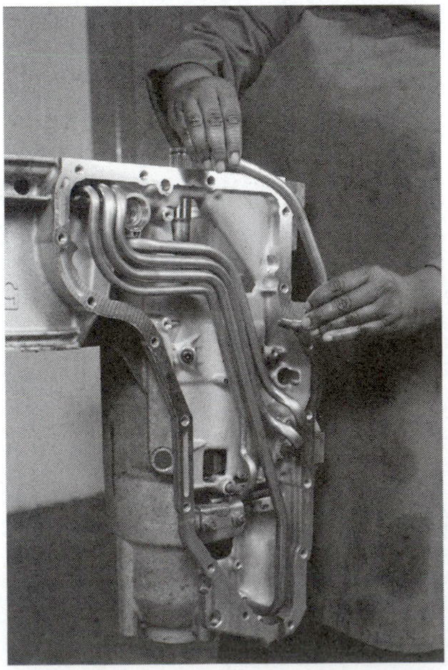

**Figure 2.** An example of oil lines within a transmission.

As can be seen, hydraulics involves the use of a liquid or fluid. Hydraulics is the study of liquids in motion. All matter, everything in the universe, exists in three basic forms: solids, liquids, and gases. A fluid is something that does not have a definite shape; therefore, liquids and gases are fluids. A characteristic of all fluids is that they will conform to the shape of their container. A major difference between a gas and a liquid is that a gas will always fill a sealed container, whereas a liquid may not. A gas also will readily expand or compress according to the pressure exerted on it. A liquid

**Figure 3.** Oil passages cast into a transmission housing.

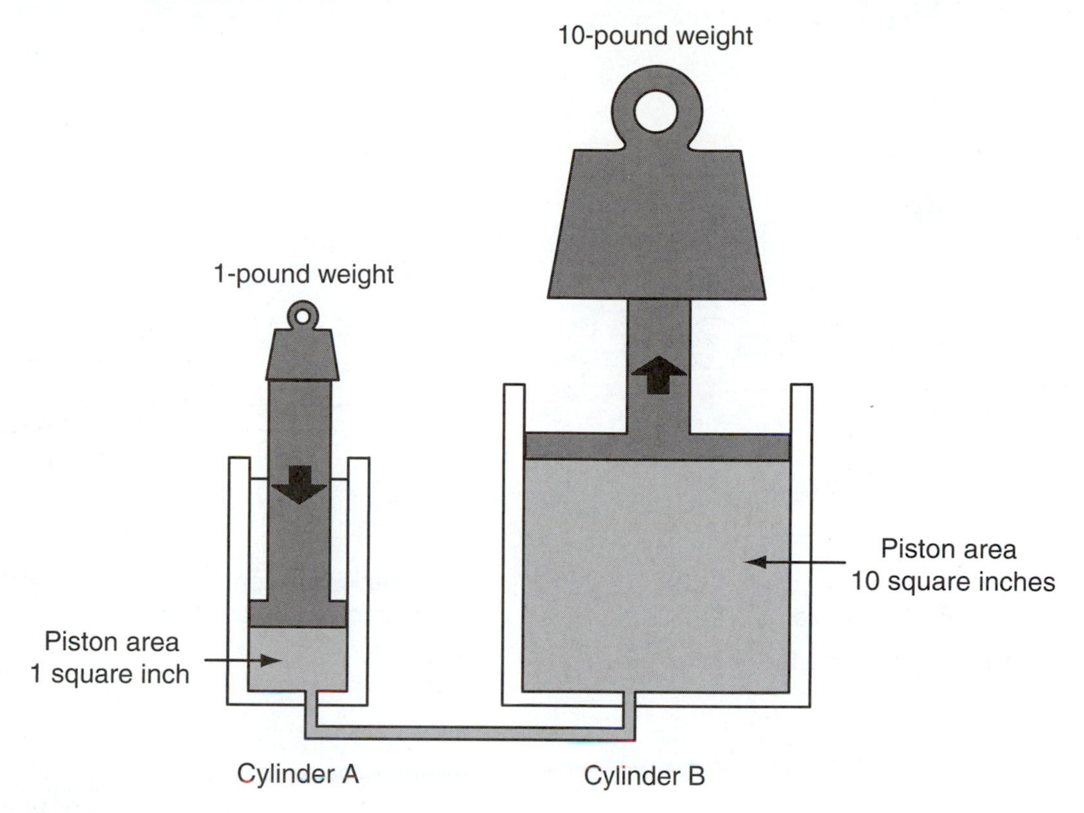


**Figure 4.** Hydraulics can be used to increase the amount of work available from a force.

will typically not compress, regardless of the pressure on it. Therefore, liquids are considered noncompressible fluids.

Liquids will, however, predictably respond to pressures exerted on them. Their reaction to pressure is the basis of all hydraulic applications. This fact allows hydraulics to do work.

## Pascal's Law

Over 300 years ago, a French scientist, Blaise Pascal, determined that if you had a liquid-filled container with only one opening and applied force to the liquid through that opening, the force would be evenly distributed throughout the liquid. This explains how pressurized liquid is used to operate and control an automatic transmission. A pump pressurizes the transmission fluid and the fluid is delivered to the various apply devices. When the pressure increases enough to activate the apply device, it holds a gear set member until the pressure decreases. The valve body is responsible for the distribution of the pressurized fluid to the appropriate reaction member according to the operating conditions of the vehicle.

Pascal constructed the first known hydraulic device, which consisted of two sealed containers connected by a tube. The pistons inside the cylinders seal against the walls of each cylinder, prevent the liquid from leaking out of the cylinder, and prevent air from entering into the cylinder. When the piston in the first cylinder has a force applied to it, the pressure moves everywhere within the system. The force is transmitted through the connecting tube to the second cylinder. The pressurized fluid, in the second cylinder, exerts force on the bottom of the second piston, moving it upward and lifting the load on the top of it. By using this device, Pascal found he could increase the force available to do work **(Figure 4)**, just as could be done with levers or gears.

Pascal determined that force applied to liquid creates pressure **(Figure 5)**, or the transmission of force through the liquid. These experiments revealed two important

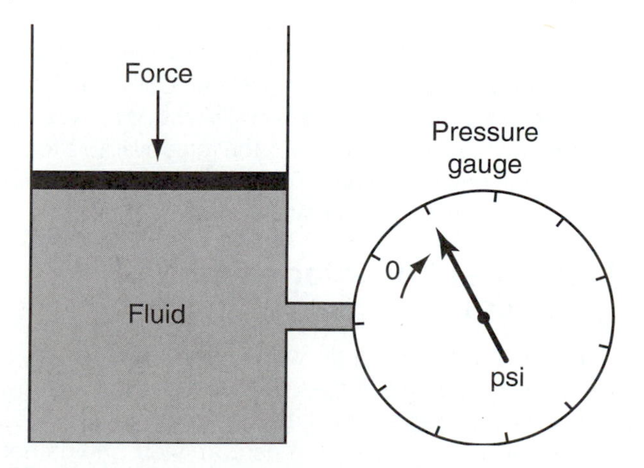


**Figure 5.** When a force is applied to liquid, it pressurizes the liquid.

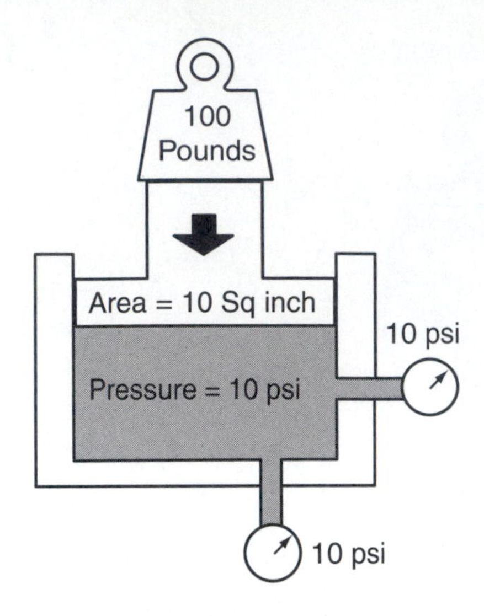


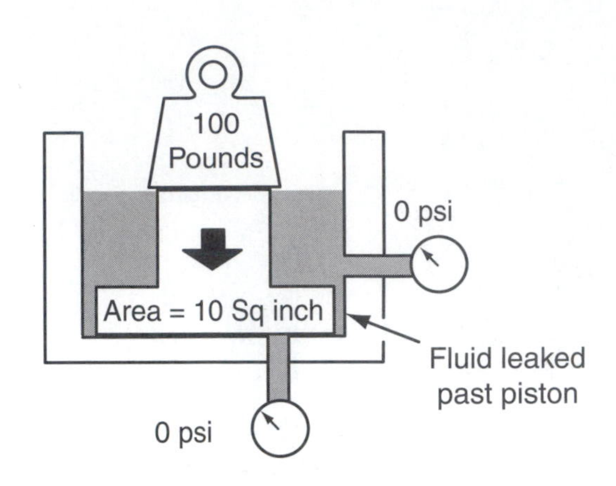


**Figure 6.** The effect of a leak on a fluid under pressure.

aspects of a liquid when it is confined and put under pressure **(Figure 6)**. The pressure applied to it is transmitted equally in all directions and this pressure acts with equal force at every point in the container.

## Fluid Characteristics

If a liquid is confined and a force applied, pressure is produced. In order to pressurize a liquid, the liquid must be in a sealed container. Any leak in the container will decrease the pressure.

The basic principles of hydraulics are based on certain characteristics of liquids. Liquids have no shape of their own; they acquire the shape of the container they are put in. They also always seek a common level. Therefore, oil in a hydraulic system will flow in any direction and through any passage, regardless of size or shape. Liquids are basically incompressible, which gives them the ability to transmit force. The pressure applied to a liquid in a sealed container is transmitted equally in all directions and to all areas of the system and acts with equal force on all areas. As a result, liquids can provide great increases in the force available to do work. A liquid under pressure also may change from a liquid to a gas in response to temperature changes.

## Mechanical Advantage with Hydraulics

Hydraulics is used to do work in the same way as a lever or gear does. All of these systems transmit energy. Because energy cannot be created or destroyed, these systems only redirect energy to perform work and do not create more energy. **Work** is actually the amount of force applied and the distance over which it is applied.

When a pressure is applied to a confined liquid, the pressure of the liquid is the same everywhere within the hydraulic system. If the hydraulic pump provides 100 psi, there will be 100 pounds of pressure on every square inch of the system **(Figure 7)**. If the system has a piston with

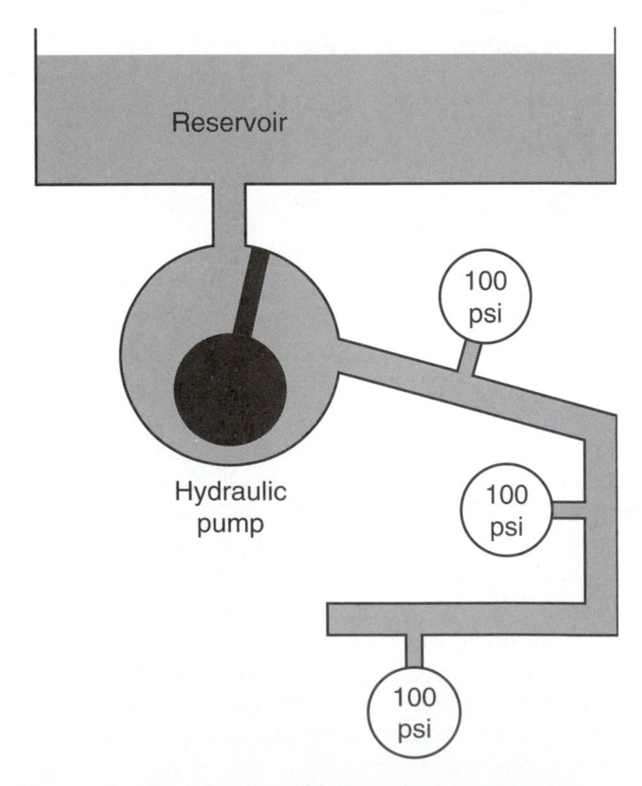


**Figure 7.** If a hydraulic pump provides 100 psi, there will be 100 pounds of pressure on every square inch of the system.

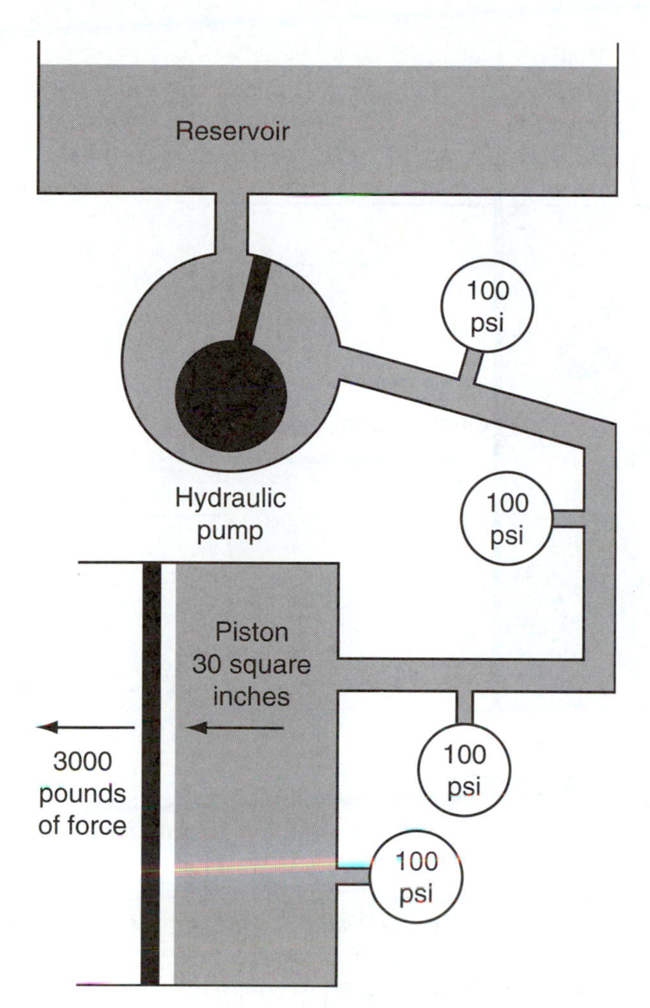


**Figure 8.** If the system included a piston with an area of 30 square inches, each square inch receives 100 pounds of pressure; therefore, there will be 3000 pounds of force applied by that piston.

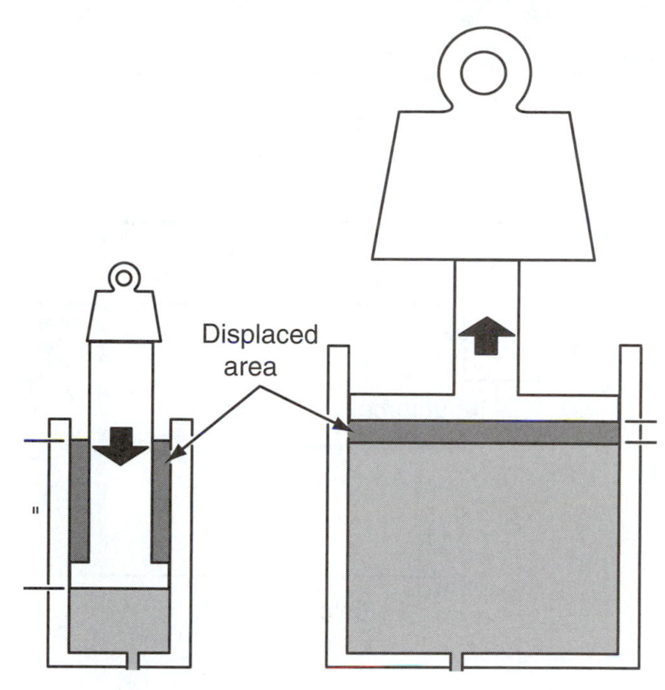


**Figure 9.** The difference in piston travel between a smaller cylinder and a larger one. (Courtesy of General Motors Corporation, Service Operations)

an area of 30 square inches, each square inch receives 100 pounds of pressure. This means there will be 3000 pounds of force applied to that piston **(Figure 8)**. According to Pascal's law, force is equal to the pressure multiplied by the area of piston (F = P × A). The use of the larger piston gives the system a mechanical advantage or increase in the force available to do work. The multiplication of force through a hydraulic system is directly proportional to the difference in the piston sizes throughout the system.

By changing the size of the pistons in a hydraulic system, force is multiplied, and, as a result, low amounts of force are needed to move heavy objects. The mechanical advantage of a hydraulic system can be further increased by the use of levers to increase the force applied to a piston.

Although the force available to do work is increased by using a larger piston in one cylinder, the total movement of the larger piston is less than that of the smaller one **(Figure 9)**. A hydraulic system with two cylinders, one with a 1-inch piston and the other with a 2-inch piston, will double the force at the second piston. However, the total movement of the larger piston will be half the distance of the smaller one **(Figure 10)**, simply because the amount of fluid in the system has not changed.

The use of hydraulics to gain a mechanical advantage is similar to the use of levers or gears. All of these systems can increase force but with an increase of force comes a decrease in the distance moved. Although hydraulic systems, gears, and levers can accomplish the same results, hydraulics is preferred when size and shape of the system is of concern. In hydraulics, the force applied to one piston will transmit through the fluid and the opposite piston will have the same force on it. The distance between the two pistons in a hydraulic system does not affect the force in a static system. Therefore, the force applied to one piston can be transmitted without change to another piston located somewhere else.

A hydraulic system responds to the pressure or force applied to it. The mere presence of different sized pistons does not always result in fluid power. Either the pressure applied to the pistons must be different or the size of the pistons must be different in order to cause fluid power. If an equal amount of pressure is exerted onto both pistons in a system and both pistons are the same size, neither piston will move, the system is balanced or is at equilibrium. The pressure inside the hydraulic system is called **static pressure** because there is no fluid motion.

When an unequal amount of pressure is exerted on the pistons, the piston with the least amount of pressure on

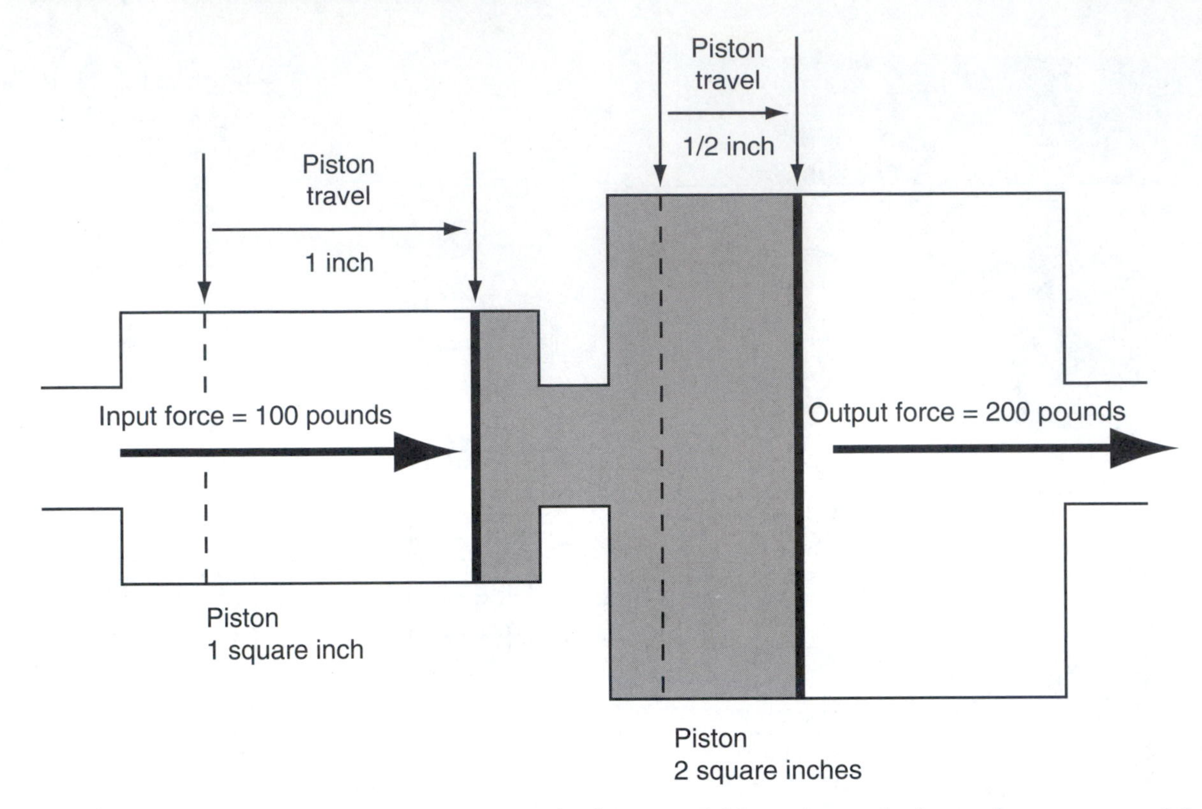


**Figure 10.** Although using a larger piston increases the force available to do work, the total movement of the larger piston will be half the distance of the smaller one.

it will move in response to the difference between the two pressures. Likewise, if the size of the two pistons is different and an equal amount of pressure is exerted on the pistons, the fluid will move. The pressure of the fluid while it is in motion is called **dynamic pressure**.

## APPLICATION OF HYDRAULICS IN TRANSMISSIONS

A common hydraulic system within an automatic transmission is the servo assembly. The servo assembly is used to control the application of a band. The band must hold the drum it surrounds tightly when it is applied. The holding capacity of the band is determined by the construction of the band and the pressure applied to it. This pressure or holding force is the result of the action of a servo. The servo multiplies the force through hydraulic action.

If a servo has an area of 10 square inches and has a pressure of 70 psi applied to it, the apply force of the servo is 700 pounds. The force exerted by the servo is increased further by its lever-type linkage and the self-energizing action of the band. The total force applied by the band is sufficient to stop and hold the rotating drum connected a planetary gear set member.

A multiple-disc assembly is also used to stop and hold gear set members. This assembly also uses hydraulics to increase its holding force. If the fluid pressure applied to the clutch assembly is 70 psi and the diameter of the clutch piston is 5 inches, the force applying the clutch pack is 1374 pounds. If the clutch assembly uses a **Belleville spring** or piston spring **(Figure 11)**, which adds a mechanical advantage of 1.25, the total force available to engage the clutch will be 1374 pounds multiplied by 1.25, or 1717 pounds.

## AUTOMATIC TRANSMISSION FLUID

The ATF circulating through the transmission and torque converter and over the parts of the transmission cools the transmission. The heated fluid moves to the transmission fluid cooler, where the heat is removed. As the fluid lubricates and cools the transmission, it also cleans the parts. The dirt is carried by the fluid to a filter, where the dirt is removed.

Another critical job of ATF is its role in shifting gears. ATF moves under pressure throughout the transmission and causes various valves to move. The pressure of the ATF changes with changes in engine speed and load.

ATF is also used to operate the various apply devices (clutches and brakes) in the transmission. At the appropriate time, a switching valve opens and sends pressurized fluid to the apply device, which engages or disengages a gear.

### Description of ATF

The fluid used in an automatic transmission's hydraulic system is called ATF. ATF is an hydraulic oil designed

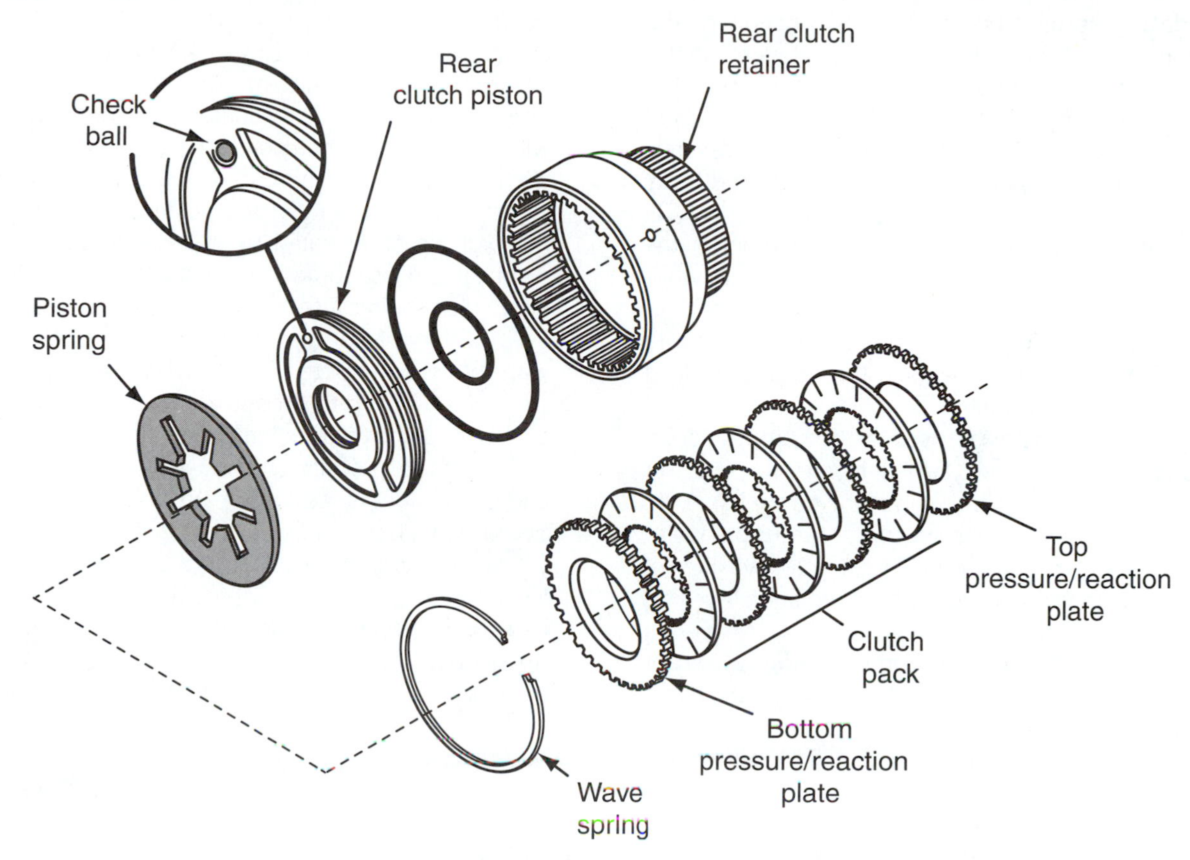


**Figure 11.** A clutch assembly with a Belleville (piston) spring.

specifically for automatic transmissions. Its primary purpose is to transmit pressure to activate the transmission's brakes and clutches. It also serves as a fluid connector between the engine and the transmission, removes heat from the transmission, and lubricates the transmission's moving parts.

ATF can be a petroleum-based, partially synthetic- or totally synthetic-based oil. Nearly all domestic automobile manufacturers require a petroleum-based fluid in their automatic transmissions; however, some imported vehicles require the use of a partially synthetic fluid.

ATF is a compound liquid, which also includes special additives, and it allows the lubricant to better meet the flow and friction requirements of an automatic transmission. ATF is normally dyed red, primarily to help distinguish it from engine oil when determining the source of fluid leaks.

Petroleum-based ATF typically has a clear red color and will darken when it is burned or will become milky when contaminated by water. Synthetic ATF is normally a darker red than petroleum-based fluid. Synthetic fluids tend to look and smell burned after normal use; therefore, the appearance and smell of these fluids is not a good indicator of the fluid's condition.

The various chemicals added to ATF ensure the durability and overall performance of the fluid. Zinc, phosphorous, and sulfur commonly are added to reduce friction. Detergent additives are added to ATF to help keep the transmission parts clean. Also added are dispersants, which keep contaminants suspended in the fluid so they can be trapped by the filter.

All certified types and brands of ATF have been tested to ensure they meet the criteria set by the manufacturers. Some of these test standards apply to all types of ATF, such as oxidation resistance, corrosion and rust inhibition, flash and flame points, and resistance to foaming. Other standards are specific for a particular fluid rating or type.

The following is a summary of the common additives blended into the various types of ATF:

**Antifoam agents:** These minimize foaming caused by the movement of the planetary gears and the movement inside the torque converter. These movements tend to cause fluid foaming.

**Antiwear agents:** Zinc is blended into the fluid to minimize gear, bushing, and thrust washer wear.

**Corrosion inhibitors:** These are added to prevent corrosion of the transmission's bushings and thrust washers, in addition to preventing fluid cooler line corrosion.

**Dispersants:** These keep dirt suspended in the fluid, which helps prevent the buildup of sludge inside the transmission.

**Friction modifiers:** Additives are blended into the base fluid to provide for an intentional amount of clutch and band slippage. This improves or softens the feel of the shift.

**Oxidation stabilizers:** To control oxidation of the fluid, additives are used to allow the ATF to absorb and dissipate heat. If the fluid is not designed to handle the high heat that is normally present in a transmission, the fluid will burn or oxidize. Oxidized fluid will cause severely damaged friction materials, clogged fluid filters, and sticky valves.

**Seal swell controllers:** These additives control the swelling and hardening of the transmission's seals, while maintaining their normal pliability and tensile strength.

**Viscosity index improvers:** These are blended into the fluid as an attempt to maintain the viscosity of fluid regardless of its temperature.

Because some chemicals used in the composition of transmission fluid may react adversely with the fibers or synthetic materials used in the seals of the transmission, the compatibility of the fluids with specific transmissions is also tested. Incompatibility can result in external and internal transmission fluid leaks because of deterioration, swelling, and/or shrinking of the seals.

All brands of ATF also are tested for their miscibility or compatibility with other brands of ATF. Although the different brands of transmission fluids must meet the same set of standards, they may differ in their actual chemical composition and be incompatible with others. There should be no fluid separation, color change, or chemical breakdown when two different brands of ATF are mixed together. This level of compatibility is important to the service life of a transmission, as it allows for the maintenance of fluid levels without the worry about switching to or mixing different brands of fluid.

## Recommended Applications

There are several ratings or types of ATF available; each type is designed for a specific application. The different classifications of transmission fluid have resulted from the inclusion of new or different additives, which enhance the operation of the different transmission designs. Each automobile manufacturer specifies the proper type of ATF that should be used in their transmissions. Both the design of the transmission and the desired shift characteristics are considered when a specific ATF is chosen.

To reduce wear and friction inside a transmission, most commonly used transmission fluids are mixed with friction modifiers. Transmission fluids with these additives allow for the use of lower friction disc and brake application pressures, which, in turn, provide for a very smooth feeling shift. Transmission fluids without a friction modifier tend to have a firmer shift because higher friction disc and brake application pressures are required to avoid excessive slippage during gear changes.

If an ATF without friction modifiers is used in a transmission designed for friction-modified fluid, the service life of the transmission is not normally affected. However, firmer shifting will result and the driver may not welcome this change in shifting quality. Transmission durability is affected by using friction-modified fluid in a transmission designed for nonmodified fluids. This incorrect use of fluid will cause slippage, primarily when the vehicle is working under a load. Any amount of slippage can cause the clutches and brakes to wear prematurely. Also, because of the high heat generated by the slippage, the fluid may overheat and lose some of its lubrication and cooling qualities, which could cause the entire transmission to fail.

The formulation of an ATF also must be concerned with the viscosity of the fluid. Although the fluids are not selected according to viscosity numbers, proper flow characteristics of the fluid are important in operation of a transmission. If the viscosity is too low, the chances of internal and external leaks increases, parts can wear prematurely because of a lack of adequate lubrication, system pressure will be reduced, and overall control of the hydraulics will be less effective. If the viscosity is too high, internal friction will increase resulting in an increase in the chance of building up sludge, hydraulic operation will be sluggish, and the transmission will require more engine power for operation.

The viscosity of a fluid is directly affected by temperature. Viscosity increases at low temperatures and decreases with higher temperatures. A transmission operates at many different temperatures. Because the fluid is used for lubricating and for shifting it must be able to flow well at any temperature. ATF has a low viscosity but it is viscous enough to prevent deterioration at higher temperatures. High temperature performance is improved by many additives, such as friction modifiers. Low temperature fluid flow of ATF is enhanced by mixing pour point depressants into the base fluid. These additives are normally referred to as "Viscosity Index Improvers."

The use of the correct ATF is critical to the operation and durability of automatic transmissions. We have already discussed the differences between friction modified and nonfriction modified fluids. Certainly this is one aspect of ATF that must be considered when putting fluid into a transmission. There are other considerations. Each type of ATF is specifically blended for a particular application. Each one has a unique mixture of additives that make it suitable for certain types of transmissions. Always fill a transmission with the fluid recommended by the transmission's manufacturer. Through the years, automatic transmissions have changed and so have their fluid requirements. In many cases, the development of new fluids has allowed automobile manufacturers to improve their transmission designs. In other cases, changes in the transmission have mandated the development of new fluid types.

Each manufacturer recommends the use of a particular type of ATF. Each of these is specially blended for their transmissions. Some manufacturers, such as Volkswagen, have their own brand of fluid that should be used in their transmissions.

Figure 12. A typical fluid filter for an automatic transmission.

## Filtering

To trap dirt and metal particles from the circulating ATF, automatic transmissions have an oil filter **(Figure 12)** normally located inside the transmission case between the oil pump pickup and the bottom of the oil pan. If dirt, metal, and friction materials are allowed to circulate, they can cause valves to stick and/or cause premature transmission wear.

You Should Know *Some automatic transmissions are equipped with an extra deep oil sump, which allows for improved cooling and increased capacity. If the transmission is equipped with a deep pan, a special filter must be used that will reach into the bottom of the pan.*

Current automatic transmissions are fitted with one of three types of filter: a screen filter, paper filter, and felt filter. Screen filters use a fine wire mesh to trap the contaminants in the ATF. This type filter is considered a surface filter because it traps the contaminants on its surface. As a screen filter traps dirt, metal, and other materials, fluid flow through the filter is reduced. The openings in the screen are relatively large so that only larger particles are trapped and small particles remain in the fluid. Although this does not remove all of the contaminants from the fluid, it does prevent quick clogging of the screen and helps to maintain normal fluid flow.

Another surface filter that is commonly used is the paper filter. This type of filter is more efficient than the screen type because it can trap smaller sized particles. Paper filters typically are made from a cellulose or Dacron fabric. Although this type filter is quite efficient, it can quickly clog and cause a reduction in fluid flow through the transmission. Therefore, some older transmissions equipped with a paper filter have a bypass circuit, which allows contaminated fluid to circulate through the transmission if the filter becomes clogged and greatly restricts fluid flow.

The most commonly used filter in current model transmissions is the felt-type. These are not surface filters; rather, they are considered depth filters because they trap contaminants within the filter and not just on its surface. Normally made from randomly spaced polyester materials, felt filters are able to trap both large and small particles and are less likely susceptible to clogging.

To protect vital transmission circuits and components, most transmissions are equipped with a secondary filter located in a hydraulic passage, which helps keep dirt out of the pump, valves, and solenoids. Secondary filters are simply small screens fit into a passage or bore.

## Reservoir

All hydraulic systems require a reservoir to store fluid and to provide a constant source of fluid for the system. In an automatic transmission, the reservoir is the pan, typically located at the bottom of the transmission case **(Figure 13)**. Transmission fluid is forced out of the pan by atmospheric pressure into the pump and returned to it after it has circulated through the selected circuits. A transmission dipstick and filler tube is used to check the level of the fluid and to add ATF to the transmission. The tube and dipstick is normally located at the front of the transaxle housing and the right front of a transmission housing. Other transmissions have a side plug on the pan or the transmission to check and replenish fluid level.

## Venting

In order to allow the fluid to be pumped through the transmission by the pump, all reservoirs must have an air vent that allows atmospheric pressure to force the fluid into the pump when the pump creates a low pressure at its inlet port. The pans of many automatic transmissions vent through the handle of the dipstick. Transmissions also must

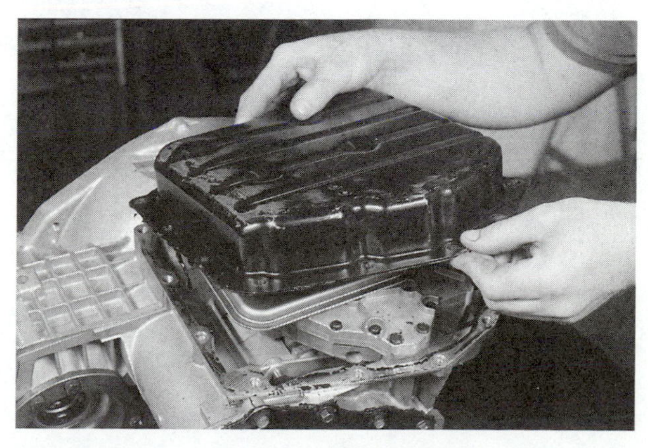

Figure 13. The oil pan being removed from a transaxle as it sits bottom up on a workbench.

be vented to allow for the exhaust of builtup air pressure that results from the moving components inside the transmission. The movement of these parts can force air into the ATF, which would not allow it to cool or lubricate the transmission properly.

## Transmission Coolers

The removal of heat from ATF is extremely important to the durability of the transmission. Excessive heat causes the fluid to break down. Once broken down, ATF no longer

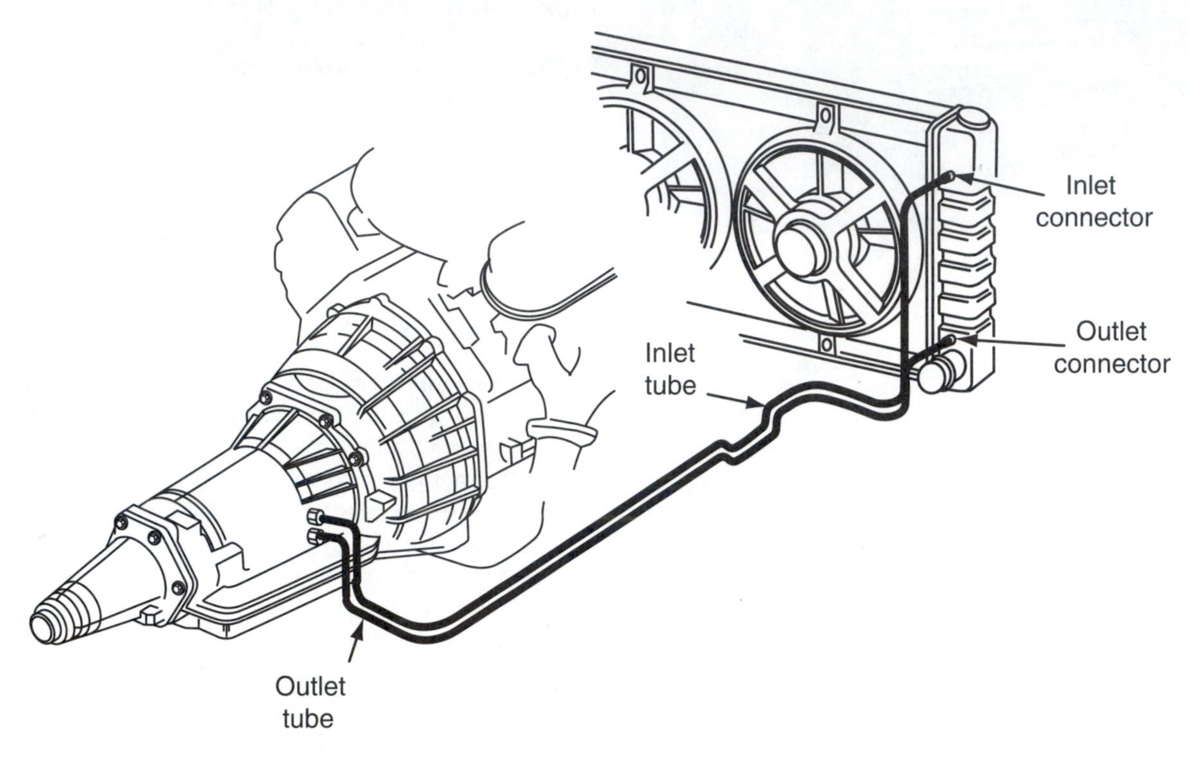


**Figure 14.** Routing of transmission cooler lines.

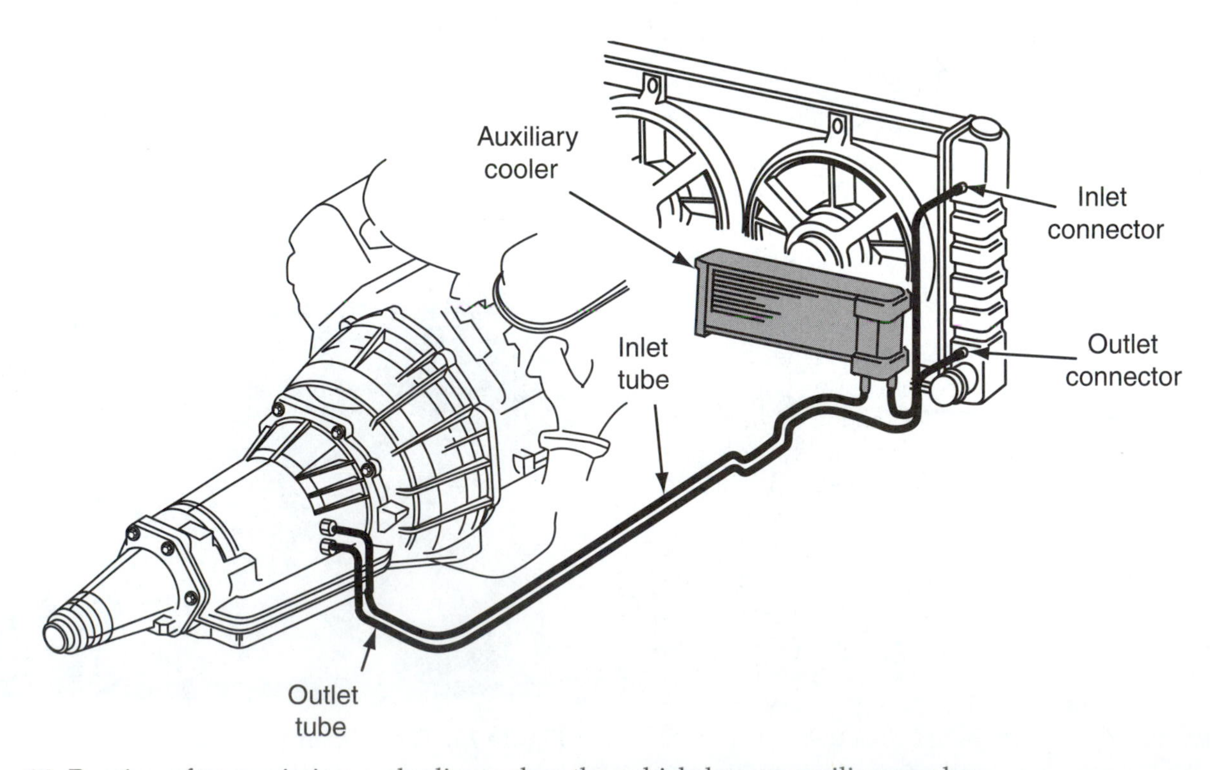


**Figure 15.** Routing of transmission cooler lines when the vehicle has an auxiliary cooler.

lubricates well and has poor resistance to oxidation. Oxidized ATF may damage transmission seals. When a transmission is operated for some time with overheated ATF, varnish is formed inside the transmission. Varnish buildup on valves can cause them to stick or move slowly. The result is poor shifting and glazed or burnt friction surfaces. Continued operation can lead to the need for a complete rebuilding of the transmission.

It is important to note that ATF is designed to operate at 175°F (80°C). At this temperature, the fluid should remain effective for 100,000 miles. However, when the operating temperature increases, the useful life of the fluid quickly decreases. A 20° increase in operating temperature will decrease the life of ATF by one half!

Transmission housings are fitted with ATF cooler lines **(Figure 14)**. These lines direct the hot fluid from the torque converter to the transmission cooler, located in the vehicle's radiator. The heat of the fluid is reduced by the cooler and the cool ATF returns to the transmission. In some transmissions, the cooled fluid flows directly to the transmission's bushings, bearings, and gears. Then, the fluid is circulated through the rest of the transmission. The cooled fluid in other transmissions is returned to the oil pan where it is drawn into the pump and circulated throughout the transmission.

Some vehicles are equipped with an auxiliary fluid cooler **(Figure 15)**, in addition to the one in the radiator. This cooler serves to remove additional amounts of heat from the fluid before it is sent back to the transmission. Auxiliary coolers are most often found on heavy duty and performance vehicles.

## Summary

- Pressure-regulating valves change the pressure of the oil to control the shift points of the transmission. Flow-directing valves direct the pressurized oil to the appropriate reaction members, which cause a change in gear ratios.
- A simple hydraulic system has liquid, a pump, lines to carry the liquid, control valves, and an output device.
- Pascal's Law states that "pressure exerted on a confined liquid or fluid is transmitted undiminished and equally in all directions and acts with equal force on all areas."
- All fluids conform to the shape of their container.
- Liquids have no shape of their own; they acquire the shape of the container into which they are put.
- Liquids are basically incompressible.
- The pressure applied to a liquid in a sealed container is transmitted equally in all directions and to all areas of the system and acts with equal force on all areas.
- When a pressure is applied to a confined liquid, the pressure of the liquid is the same everywhere within the hydraulic system.
- ATF is hydraulic oil designed specifically for automatic transmissions. Its primary purpose is to transmit pressure to activate the transmission's brakes and clutches. It also serves as a fluid connector between the engine and the transmission, removes heat from the transmission, and lubricates the transmission's moving parts.
- To remove the heat from the fluid, the ATF is directed through a cooler where it dissipates heat.

## Review Questions

1. What is the purpose of a transmission fluid cooler?
2. Which of the following is not a true statement?
   A. Force applied to liquid creates pressure.
   B. Force applied to a liquid causes it to compress.
   C. The pressure applied to a liquid is transmitted equally in all directions.
   D. The pressure on a liquid acts with equal force at every point in its container.
3. True or False: Petroleum-based ATF typically has a clear red color and will darken when it is burned or will become milky when contaminated by water.
4. Why do transmissions need an air vent?
5. True or False: By decreasing the size of the pistons in a hydraulic system, force is multiplied, and, as a result, low amounts of force are needed to move heavy objects.

# Chapter 7 Basic Electrical Theory

## *Introduction*

Much of the operation of today's automatic transmissions is electrically operated or controlled. In addition, the driveline is also fitted with sensors that give vital information to a computer that controls other systems of the automobile. To understand the operation of these, you must have a good understanding of electricity and electronics. Although the subject is normally covered in a separate course, a quick overview of electricity and its principles is presented here.

### BASIC ELECTRICITY

Electricity is caused by the flow of electrons from one atom to another **(Figure 1)**. The release of energy as one electron leaves the orbit of one atom and jumps into the orbit of another is **electricity**. The key behind creating electricity is to give a reason for the electrons to move.

There is a natural attraction of electrons to protons. Electrons have a negative charge and are attracted to something with a positive charge. An electron moves from one atom to another because the atom next to it appears to be more positive than the one it is orbiting around. An electrical power source provides for a more positive charge and in order to allow for a continuous flow of electricity, it supplies free electrons. In order to have a continuous flow of electricity, three things must be present: an excess of electrons in one place, a lack of electrons in another place, and a path between the two places.

Two power or energy sources are used in an automobile's electrical system—the vehicle's battery **(Figure 2)** and AC Generator **(Figure 3)**. To provide power, a chemical reaction in the battery provides for an excess of electrons and a lack of electrons in another place. Batteries have two terminals—a positive and a negative. Basically, the negative terminal is the outlet for the electrons and the positive terminal in the inlet for the electrons to get to the protons. The chemical reaction in a battery causes a lack of electrons at the positive (+) terminal and an excess at the negative (−) terminal. This creates an electrical imbalance, causing the electrons to flow through the path provided by a wire.

The chemical process in the battery continues to provide electrons until the chemicals become weak. At that

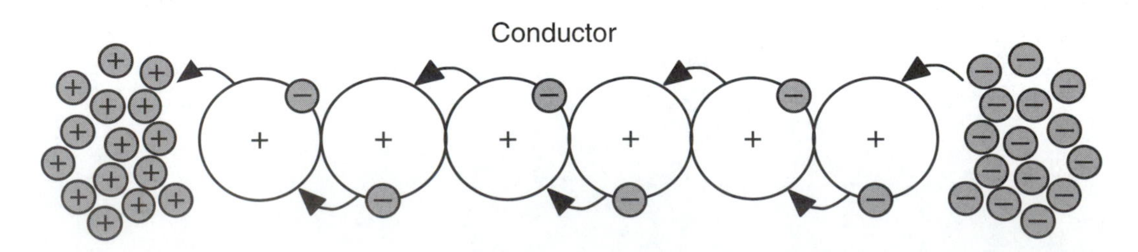


**Figure 1.** Electricity is the flow of electrons from one atom to another.

**Figure 2.** A typical automotive battery.

**Figure 3.** An AC generator (alternator).

time, either the battery has run out of electrons or all of the protons are matched with an electron. When this happens, there is no longer a reason for the electrons to want to move to the positive side of the battery. It no longer looks more positive. Fortunately, the vehicle's charging system restores the battery's supply of electrons. This allows the chemical reaction in the battery to continue indefinitely.

Moving a wire through an already existing magnetic field (such as a permanent magnet) can produce electricity. This process of producing electricity through magnetism is called **induction**. In a generator, a magnetic field is moved through a coil of wire and electricity is induced. The amount of induced electricity depends on a number of factors: the strength of the magnetic field, the number of wires that pass through the field, and the speed at which the wire moves through the magnetic field.

## ELECTRICAL TERMS

Electrical **current** is a term used to describe the movement or flow of electricity. The greater the number of electrons flowing past a given point in a given amount of time, the more current there is in the circuit. **Voltage** is the force developed by the attraction of the electrons to the protons. The more positive one side of the circuit is, the more voltage is available to the circuit. Voltage does not flow; rather, it is the pressure that causes current flow. When any substance flows, it meets resistance. The resistance to electrical flow can be measured.

### Electrical Current

The unit for measuring electrical current is the **ampere**, usually called an amp. The instrument used to measure electrical current flow in a circuit is called an **ammeter**.

When electricity flows, millions of electrons are moving past any given point at the speed of light. The electrical charge from any one moving electron is extremely small. It takes millions of electrons to make a charge that can be measured.

There are two types of current: **direct current** (DC) and **alternating current** (AC). In direct current, the electrons flow in one direction only. In alternating current, the electrons change direction at a fixed rate. Most automobile circuits operate on DC current, whereas the current in homes and buildings is AC.

### Resistance

In every atom, the electrons resist being moved out of their shell. The amount of resistance depends on the type of atom. As explained earlier, in some atoms (such as those in copper) there is very little resistance to electron flow, because the outer electron is loosely held. In other substances there is more resistance to flow, because the outer electrons are tightly held.

The resistance to current flow produces heat. This heat can be measured to determine the amount of resistance. A unit of measured resistance is called an **ohm**. Resistance can be measured by an instrument called an **ohmmeter**.

### Voltage

In electrical flow, some force is needed to move the electrons from one atom to another. This force is the pressure that exists between a positive and negative point within an electrical circuit. This force, also called electromotive force **(EMF)**, is measured in units called **volts**. One volt is the amount of pressure (force) required to move 1 ampere of current through a resistance of 1 ohm. Voltage is measured by an instrument called a **voltmeter**.

## ELECTRICAL CIRCUITS

When electrons are able to flow along a path (wire) between two points, an electrical circuit is formed. An electrical circuit is considered a **complete circuit** when there is a path that connects the positive and negative terminals of the electrical power source. In a complete circuit, resistance must be low enough to allow the available voltage to push electrons between the two points. Most automotive circuits contain four basic parts.

- A power source.
- Conductors, such as copper wires that provide a path for the electrons.
- **Loads** or devices that use electricity to perform work, such as lights and motors.
- **Controllers**, such as switches or relays that direct the flow of electrons.

A functioning electrical circuit must have a complete path from the power source to the load and back to the source. With the many circuits in an automobile, this would require hundreds of wires connected to both sides of the battery. To avoid this, vehicles are equipped with power distribution centers or fuse blocks that distribute battery voltage to the various circuits.

As a common return circuit, the vehicle's metal frame **(Figure 4)** is used as part of the return circuit. The load is often grounded directly to the metal frame. Current passes from the battery, through the load, and into the frame. The frame is connected to the negative terminal of the battery through the battery's ground cable.

An electrical component, such as an alternator is often mounted directly to the engine block, transmission case, or frame. This direct mounting effectively grounds the component without the use of a separate ground wire. In other cases, however, a separate ground wire must be run from the component to the frame or another metal part to ensure a sound return path. The increased use of plastics and other nonmetallic materials in body panels and engine parts has made electrical grounding more difficult. To assure good grounding back to the battery, some manufacturers now use a network of common grounding terminals and wires.

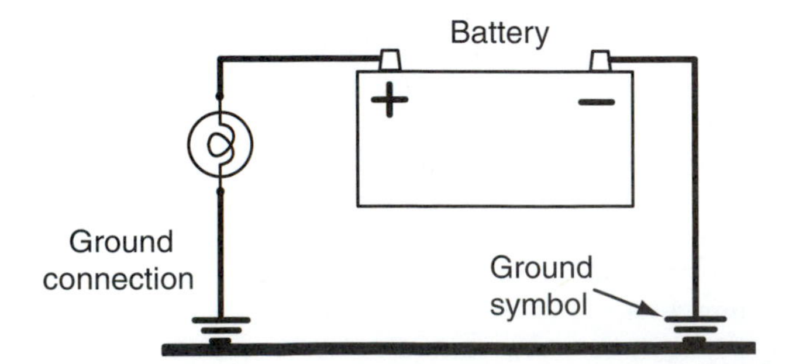


**Figure 4.** To eliminate the need to run separate return wires, many accessories and automotive components are grounded to the vehicle's chassis.

### Conductors and Insulators

Controlling and routing electricity requires the use of materials known as conductors and insulators. **Conductors** are materials with a low resistance to the flow of current. Many metals, such as copper, silver, and aluminum, are good conductors.

Copper wire is by far the most used conductor. Where flexibility is required, the copper wire will be made of a large number of very small strands of wire woven together.

**Insulators** resist the flow of current. Wire wound inside of electrical units, such as ignition coils and generators, usually has a very thin baked-on insulating coating. External wiring often is covered with a plastic-type insulating material that is highly resistant to environmental factors like heat, vibration, and moisture.

## OHM'S LAW

To understand the relationship among current, voltage, and resistance in a circuit, you should become familiar with Ohm's Law. This law states that it takes 1 volt of electrical pressure to push 1 ampere of electrical current through 1 ohm of resistance. As such, the law provides a mathematical formula for determining the amount of current, voltage, or resistance in a circuit when two of these are known. The basic formula is: Voltage (E) equals the Current (I) multiplied by the Resistance (R).

Although this formula is used to calculate unknown values in an electrical circuit **(Figure 5)**, it also helps to define the behaviors of electrical circuits.

If voltage does not change but there is a change in the resistance of the circuit, the current will change. If resistance increases, current decreases. If resistance decreases, current will increase. If voltage changes, so must the current or resistance. If the resistance stays the same and current decrease, so will voltage. Likewise, if current increases, so will the voltage.

In a complete circuit, the flow of electricity is controlled and applied to do useful work, such as light a headlamp or turn over a starter motor. Components that use electrical power put a load on the circuit and consume electrical energy. The energy used by a load is measured in volts and is called **voltage drop**.

Voltage (E) = Current (I) times Resistance (R), therefore

$$E = I \times R.$$

Current (I) = Voltage (E) divided by Resistance (R), therefore

$$I = \frac{E}{R}.$$

Resistance (R) = Voltage (E) divided by Current (I), therefore

$$R = \frac{E}{I}.$$

**Figure 5.** Ohm's Law.

# CIRCUIT COMPONENTS

Automotive electrical circuits contain a number of different types of electrical devices.

## Resistors

Resistors are used to limit current flow (and thereby voltage) in circuits in which full current flow and voltage are not needed. Resistors are devices specially constructed to introduce a measured amount of electrical resistance into a circuit. Fixed value resistors are designed to have only one rating, which should not change. Some electrical loads use resistance to produce heat. An electric window defroster is a specialized type of resistor that produces heat. Electric lights are resistors that get so hot they produce light.

Tapped or stepped resistors are designed to have two or more fixed values, available by connecting wires to the several taps on the resistor. Heater motor resistor packs, which provide for different heater fan speeds, are an example of this type of resistor.

Variable resistors are designed to have a range of resistances available through two or more taps and a control. Two examples of this type of resistor are rheostats and potentiometers. **Rheostats** have two connections **(Figure 6)**, one to the fixed end of a resistor and one to a sliding contact with the resistor. Turning the control moves the sliding contact away from or toward the fixed end tap, increasing or decreasing the resistance. **Potentiometers** have three connections **(Figure 7)**, one at each end of the resistance and one connected to a sliding contact with the resistor. Turning the control moves the sliding contact away from one end of the resistance but toward the other end.

Another type of variable resistor is the **thermistor**. Thermistors are designed to change in value as its temperature changes. Thermistors are used to provide compensating voltage in components or to determine temperature. As a temperature sender, the thermistor is connected to a voltmeter calibrated in degrees. As the temperature rises or falls, the resistance also changes. This changes the reading on the meter.

**Figure 6.** A rheostat.

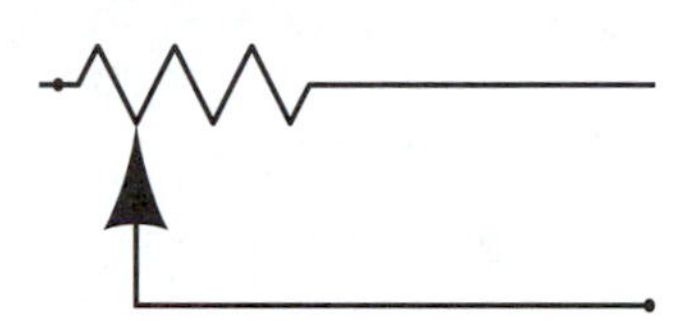

**Figure 7.** A potentiometer.

## Circuit Protective Devices

When overloads or shorts in a circuit cause excessive current flow, the wiring in the circuit heats up, the insulation melts, and a fire can result, unless the circuit has some kind of protection device. Fuses, fuse links, maxi-fuses, and circuit breakers are designed to prevent circuit damage caused by high current. Protection devices open the circuit when high current is present. As a result, the circuit no longer works but the wiring and the components are saved from damage.

## Switches

Electrical circuits are usually controlled by some type of switch. Switches turn the circuit on or off or direct the flow of current in a circuit. Switches can be controlled by the driver or can be self-operating through a condition of the circuit, the vehicle, or the environment.

A simple switch either makes or breaks (completes or opens) a single conductor or circuit. This is a single-pole, single-throw **(SPST)** switch. The throw refers to the number of output circuits, and the pole refers to the number of input circuits made by the switch.

Switches can be designed with a great number of poles and throws. Single-pole, double-throw switches have one wire in and two wires out. This type of switch allows the driver to select between two circuits, such as high-beam or low-beam headlights. The transmission neutral start switch may have two poles and six throws, and is referred to as a multiple-pole, multiple-throw **(MPMT)** switch. It contains two movable wipers that move in unison across two sets of terminals. The wipers are mechanically linked, or ganged. The switch closes a circuit to the starter in either P (park) or N (neutral) and to the backup lights in R (reverse) **(Figure 8)**.

A temperature-sensitive switch usually contains a bimetallic element heated either electrically or by some component where the switch is used as a sensor. When engine coolant is below or at normal operating temperature,

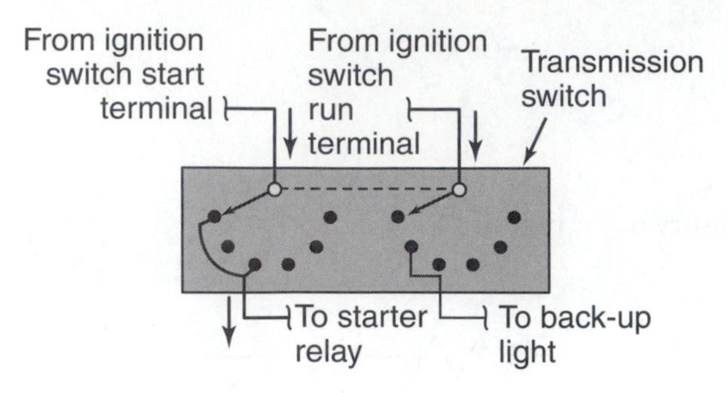


Figure 8. A multiple-pole, multiple-throw neutral start safety switch.

the engine coolant temperature sensor is in its normally open condition. If the coolant exceeds the temperature limit, the bimetallic element bends the two contacts together and the switch is closed to the indicator or the instrument panel. As a result of this action, the indicator lamp is illuminated. Other applications for heat-sensitive switches are time delay switches and flashers.

## Relays

A **relay (Figure 9)** is an electric switch that allows a small amount of current to control a circuit with high current. The low current circuit is called the control circuit. When the control circuit switch is open, no current flows to the coil in the relay, so the coil windings are deenergized. When the switch is closed, the coil is energized, turning the soft iron core into an electromagnet and drawing the armature down. This closes the high current circuit contacts, and connects power to the load circuit. When the control switch is opened, the current stops slowing in the coil, the electromagnet field disappears, and the armature is released, which opens the power circuit contacts.

## Solenoids

**Solenoids (Figure 10)** are electromagnetic devices with movable cores and are used to translate electrical current flow into mechanical movement. The movement of the core causes something else to move, such as lever. They also can close electrical contacts, acting as a relay at the same time.

## ELECTROMAGNETISM BASICS

A substance is said to be a magnet if it has the property of magnetism—the ability to attract substances such as iron, steel, magnetite, or cobalt. A magnet has two points of maximum attraction, one at each end of the magnet. These points are called poles, with one being designated the North Pole and the other the South Pole. When two magnets are brought together, opposite poles attract, while similar poles repel each other.

A magnetic field, called a **field of flux**, exists around every magnet **(Figure 11)**. The field consists of imaginary lines along which a magnetic force acts. These lines emerge from the North Pole and enter the South Pole, returning to the North Pole through the magnet itself. All lines of force leave the magnet at right angles to the magnet. None of the lines cross each other. All lines are complete.

Magnets can occur naturally in the form of a mineral called magnetite. Artificial magnets also can be made by inserting a bar of magnetic material inside a coil of insulated wire and passing a heavy direct current through the coil. This principle is very important in understanding certain automotive electrical components. Another way of creating a magnet is by stroking the magnetic material with a bar magnet. Both methods force the randomly arranged molecules of the magnetic material to align themselves along North and South poles.

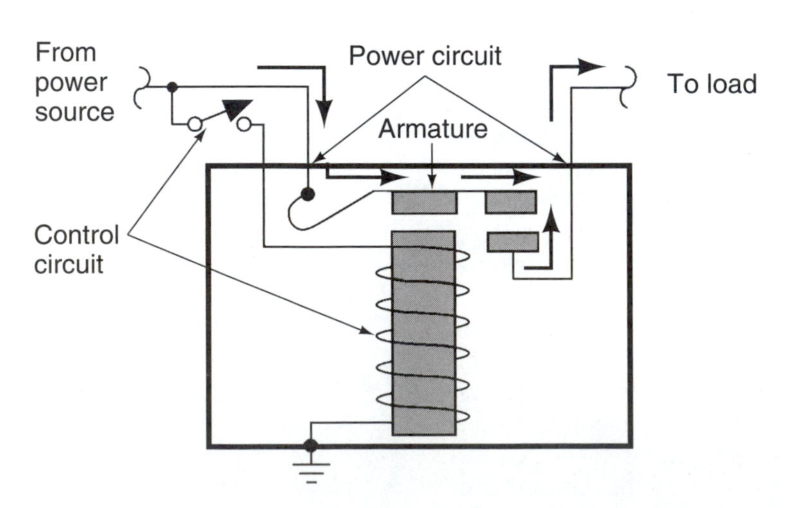


Figure 9. A typical relay.

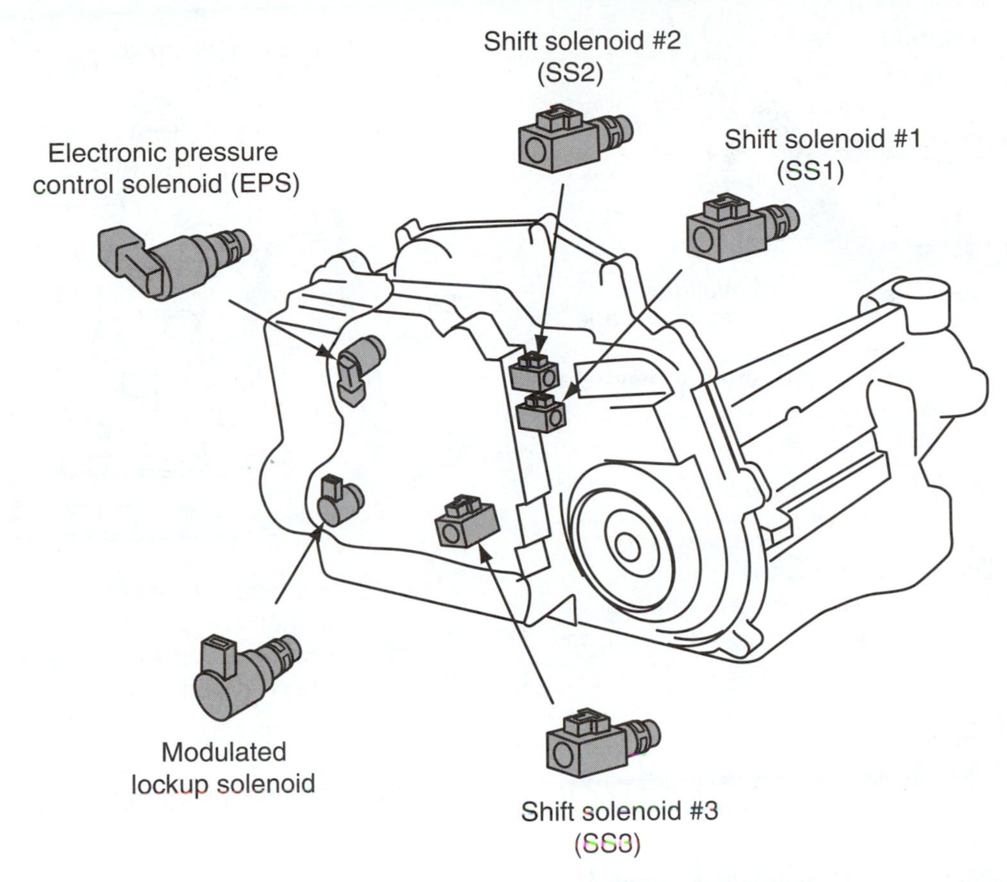


**Figure 10.** Today's automatic transmissions are fitted with many different solenoids.

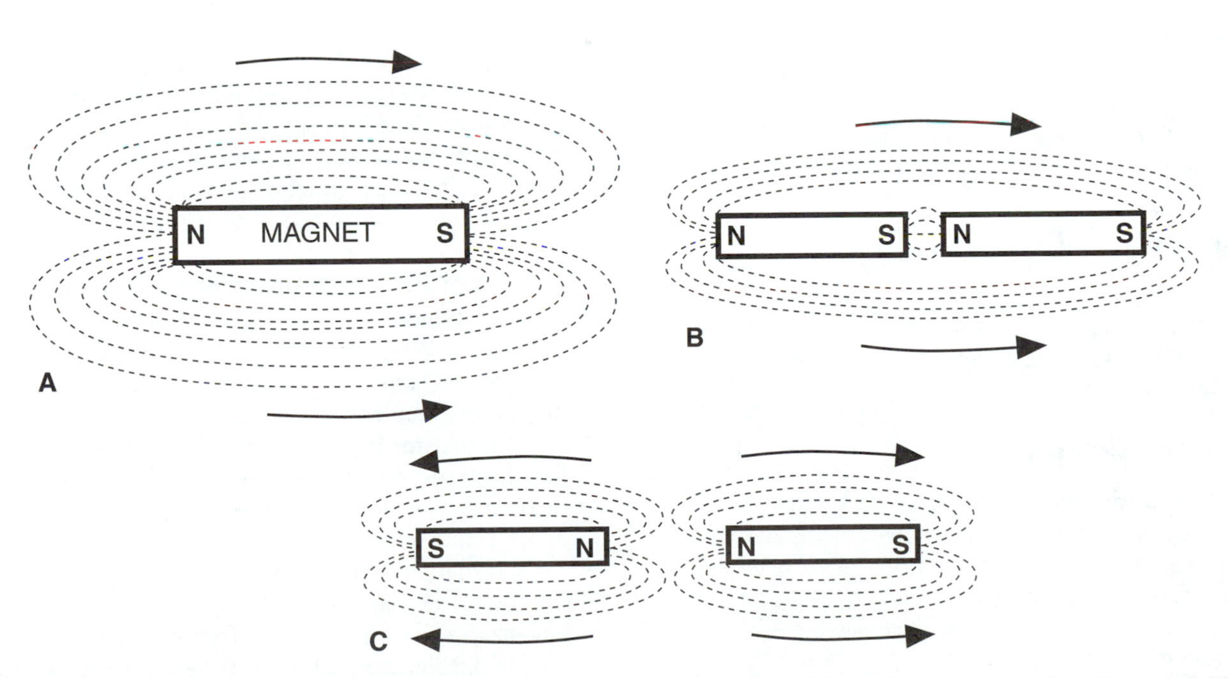


**Figure 11.** Magnetic principles: (A) field of flux around a magnet, (B) unlike poles attract each other, and (C) like poles repel.

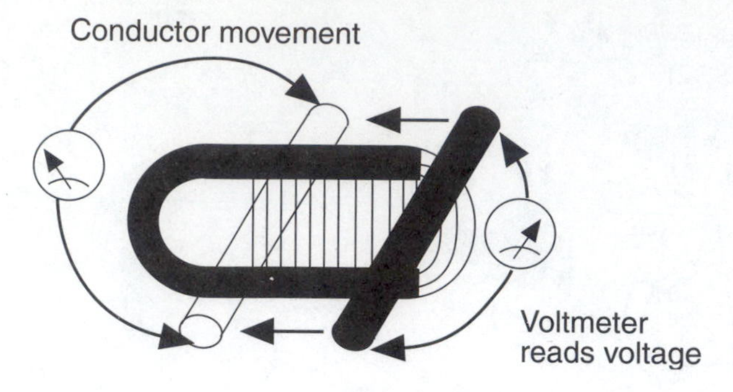


**Figure 12.** Moving a conductor through a magnetic field induces a voltage.

Artificial magnets can be either temporary or permanent. Temporary magnets are usually made of soft iron. They are easy to magnetize but quickly lose their magnetism. Permanent magnets are difficult to magnetize but, once magnetized, they retain this property for very long periods.

## Induced Voltage

Now that we have explained how current can be used to generate a magnetic field, it is time to examine the opposite effect of how magnetic fields can produce electricity. Consider a straight piece of conducting wire with the terminals of a voltmeter attached to both ends. If the wire is moved across a magnetic field, the voltmeter registers a small voltage reading **(Figure 12)**. A voltage has been induced in the wire.

It is important to remember that the conducting wire must cut across the flux lines to induce a voltage. Moving the wire parallel to the lines of flux does not induce voltage. Holding the wire still and moving the magnetic field at right angles to it also induces voltage in the wire.

# BASICS OF ELECTRONICS

Computerized engine controls and other features of today's cars would not be possible if it were not for electronics. Electronics is best defined as the technology of controlling electricity.

## Semiconductors

A **semiconductor** is a material or device that can function either as a conductor or an insulator, depending on its structure. Semiconductor materials have less resistance than an insulator but more resistance than a conductor. Some common semiconductor materials include silicon (Si) and germanium (Ge).

In semiconductor applications, materials have a crystal structure. This means that their atoms do not lose and gain electrons as the atoms in conductors do. Instead, the atoms

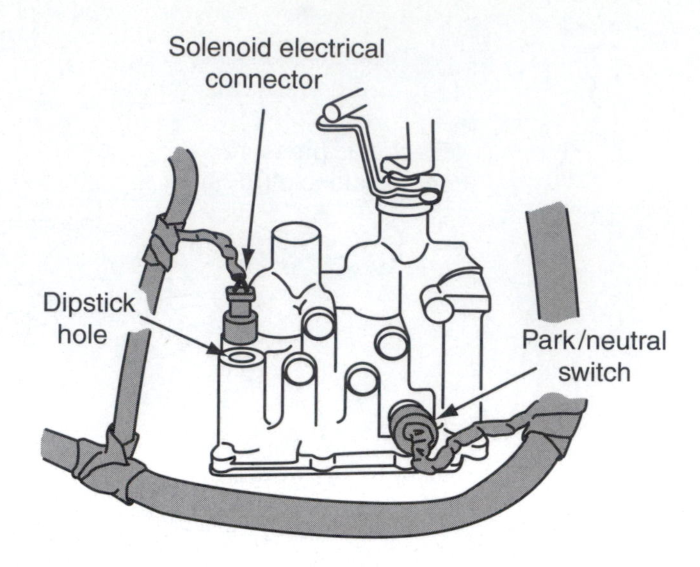


**Figure 13.** A diode connected in parallel with a coil to prevent voltage spikes that normally occur when the circuit's switch is opened. This is called a clamping diode.

in these semiconductor materials share outer electrons with each other. In this type of atomic structure, the electrons are tightly held and the element is stable.

Because the electrons are not free, crystals cannot conduct current. These materials are called electrically inert materials. In order to function as semiconductors, a small amount of trace element must be added. The addition of these traces, called impurities, allows the material to function as a semiconductor. The type of impurity added determines what type of semiconductor will be produced.

The **diode** is the simplest semiconductor device. A diode **(Figure 13)** allows current to flow in only one direction. Therefore, it can function as a switch, acting as either a conductor or an insulator, depending on the direction of current flow.

A variation of the diode is the zener diode, which functions like a standard diode until a certain voltage is applied to the diode. When the voltage reaches this point, the zener diode allows current to flow in the reverse direction. Zener diodes are often used in electronic voltage regulators.

A **transistor** is an electronic device produced by joining three sections of semiconductor materials. Like the diode, it is very useful as a switching device, functioning as either a conductor or an insulator.

One transistor or diode is limited in its ability to do complex tasks. However, when many semiconductors are combined into a circuit, they can perform complex functions.

An **integrated circuit** is simply a large number of diodes, transistors, and other electronic components such as resistors and capacitors, all mounted on a single piece of semiconductor material. This type of circuit is extremely small. Circuitry that used to take up entire rooms can now fit into a pocket. The principles of semiconductor operation remain the same in integrated circuits; only the size has changed.

The increasingly small size of integrated circuits is very important to automobiles. Many transistors, diodes, and other solid-state components are installed in a car to make logic decisions and issue commands to other areas of the engine. This is the foundation of computerized control systems.

The computer has taken over many of the tasks in cars and trucks that were formerly performed by vacuum, electromechanical, or mechanical devices. When properly programmed, they can carry out explicit instructions with blinding speed and almost flawless consistency. A typical electronic control system **(Figure 14)** is made up of sensors, actuators, and related wiring connected to a computer.

## Sensors

All sensors perform the same basic function. They detect a mechanical condition (movement or position), chemical state, or temperature condition and change it into an electrical signal that can be used by the computer to make decisions. The computer makes decisions based on information it receives from sensors. Each sensor used in a particular system has a specific job to do. Although there are a variety of sensor designs, they all fall under one of two

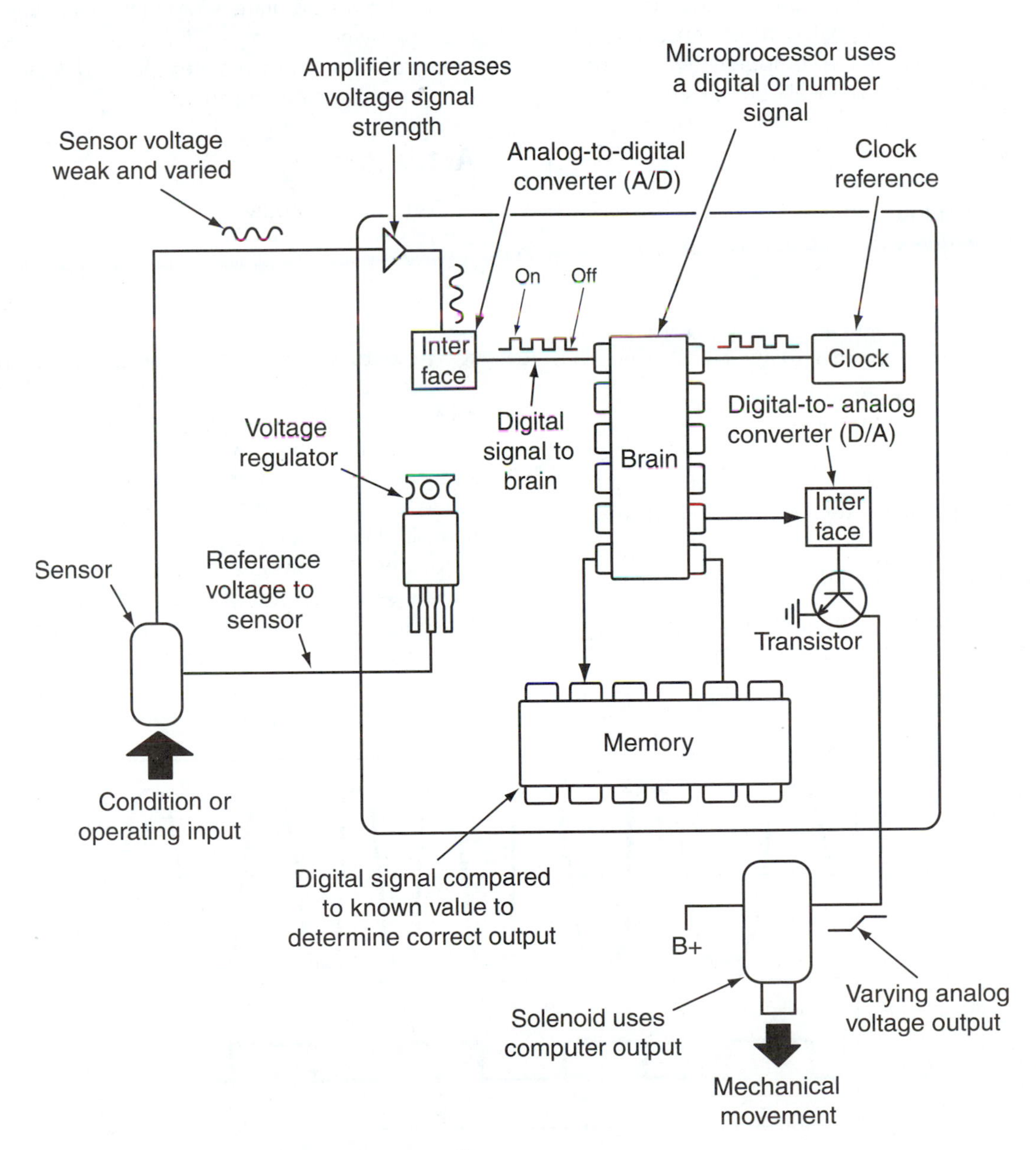


**Figure 14.** A simplified look at an electronic control system.

operating categories: reference voltage sensors or voltage generating sensors.

**Reference voltage (Vref) sensors** provide input to the computer by modifying or controlling a constant, predetermined voltage signal. This signal, which can have a reference value from 5 to 9 volts, is generated and sent out to each sensor by a reference voltage regulator located inside the **processor**. The term "processor" is used to describe the actual metal box that houses the computer and its related components. Because the computer knows that a certain voltage value has been sent out, it can indirectly interpret things like motion, temperature, and component position, based on what comes back.

Most sensors are variable resistors or potentiometers. They modify a voltage to or from the computer, indicating a constantly changing status that can be calculated, compensated for, and modified. That is, most sensors simply control a voltage signal from the computer. The monitored signal line may be the output signal from the computer to the sensor (one- and two-wire sensors), or the computer may use a separate return line from the sensor to monitor voltage changes (three-wire sensors).

**Voltage-generating sensors** are commonly Hall-effect switches, oxygen sensors, and knock sensors, which are capable of producing their own input voltage signal. This varying voltage signal, when received by the computer, enables the computer to monitor and adjust for changes in the computerized control system.

In addition to variable resistors, two other commonly used reference voltage sensors are switches and thermistors. Regardless of the type of sensors used in electronic control systems, the computer is incapable of functioning properly without input signal voltages from the sensors.

## Communication Signals

Most input sensors are designed to produce a voltage signal that varies within a given range (from high to low, including all points in between). A signal of this type is called an **analog signal**. Unfortunately, the computer does not understand analog signals. It can only read a **digital signal**, a signal that has only two values—on or off **(Figure 15)**.

To overcome this communication problem, all analog voltage signals are converted to a digital format by a device known as an analog-to-digital converter (A/D converter), located in a section of the processor. Some sensors, such as the Hall-effect switch, produce a digital or square wave signal that can go directly to the computer as input.

A computer's memory holds the programs and other data, such as vehicle calibrations, which the microprocessor refers to in performing calculations. The program is a set of instructions or procedures that the computer must follow. Included in the program is information that tells the microprocessor when to retrieve input (based on temperature, time, etc.), how to process the input, and what to do with it once it has been processed.

## Actuators

After the computer has processed the information, it sends output signals to control devices called actuators or outputs. These actuators—solenoids, switches, relays, or motors—physically act or carry out a decision made by the computer **(Figure 16)**.

**Actuators** are electromechanical devices that convert an electrical current into mechanical action. This mechanical action can then be used to open and close valves, control vacuum to other components, or open and close switches. When the computer receives an input signal indicating a change in one or more of the operating conditions, it determines the best strategy for handling the conditions. The computer then controls a set of actuators to achieve the desired effect or strategy goal. In order for the computer to control an actuator, it must rely on a component called an **output driver**.

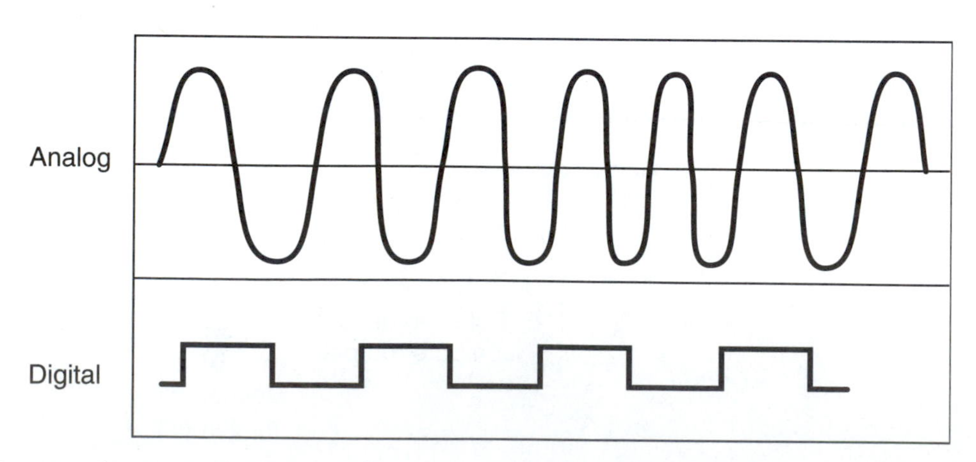


**Figure 15.** Analog signals are constantly variable, whereas digital signals are either on or off, or high or low.

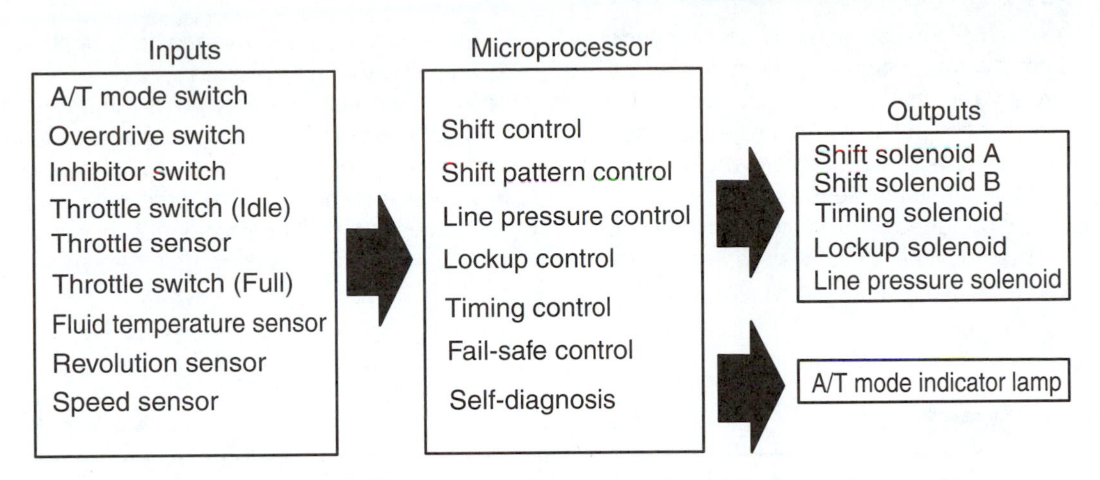


**Figure 16.** The pattern for the information that controls the system's outputs.

Output drivers are located in the processor and operate by the digital commands issued by the computer. Basically, the output driver is nothing more than an electronic on/off switch used to control the ground circuit of a specific actuator.

## Summary

- In order to have a continuous flow of electricity, there must be an excess of electrons in one place, a lack of electrons in another place, and a path between the two places.
- Electrical current is a term used to describe the movement or flow of electricity. The unit for measuring electrical current is the ampere, usually called an amp. The instrument used to measure electrical current flow in a circuit is called an ammeter.
- Voltage is electrical pressure. Voltage is the force developed by the attraction of the electrons to the protons. The more positive one side of the circuit is, the more voltage is available to the circuit. An instrument called a voltmeter measures voltage.
- When any substance flows, it meets resistance. The resistance to electrical flow can be measured. A unit of measured resistance is called an ohm. An instrument called an ohmmeter can measure resistance.
- An electrical circuit is considered complete when there is a path that connects the positive and negative terminals of the electrical power source.
- Conductors are materials with a low resistance to the flow of current.
- Insulators resist the flow of current.
- The basic formula for Ohm's Law is: Voltage = Current multiplied by Resistance.
- Resistors are used to limit current flow (and thereby voltage) in circuits in which full current flow and voltage are not needed. Some electrical loads use resistance to produce heat.
- Rheostats have two connections, one to the fixed end of a resistor and one to a sliding contact with the resistor. Turning the control moves the sliding contact away from or toward the fixed end tap, increasing or decreasing the resistance.
- Potentiometers have three connections, one at each end of the resistance and one connected to a sliding contact with the resistor.
- Thermistors change their resistance value as their temperature changes.
- Fuses, fuse links, maxi-fuses, and circuit breakers are designed to prevent circuit damage caused by high current.
- Switches turn the circuit on or off or direct the flow of current in a circuit.
- A relay is an electric switch that allows a small amount of current to control a circuit with high current.
- Solenoids are electromagnetic devices with movable cores and are used to translate electrical current flow into mechanical movement.
- When two magnets are brought together, opposite poles attract, while similar poles repel each other.
- A diode is the simplest semiconductor device. It allows current to flow in only one direction and can function as a switch, depending on the direction of current flow.
- A transistor is an electronic device produced by joining three sections of semiconductor materials. It is very useful as a switching device, functioning as either a conductor or an insulator.

- Sensors detect a mechanical condition (movement or position), chemical state, or temperature condition and change it into an electrical signal that can be used by the computer to make decisions. All sensors fall under one of two operating categories: reference voltage sensors or voltage generating sensors.
- Most sensors are designed to produce an analog voltage signal that varies within a given range. Because the computer can only read a digital signal, all analog voltage signals are converted to a digital format by an analog-to-digital converter.
- A computer's memory holds the programs and other data, which the microprocessor refers to in performing calculations.
- After the computer has processed the information, it sends output signals to control devices called actuators.

## *Review Questions*

1. Name the two energy sources used in automobile electrical systems.
2. What is the difference between voltage and current?
3. What is the difference between a fixed resistor and a variable resistor?
4. What types of sensors are typically used in an automotive computer system?
5. Current is measured in ________________, electrical voltage is measured in ________________, and electrical resistance is measured in ________________.
6. ________________, ________________, ________________, and ________________ ________________ are used to protect circuits against current overloads.
7. A computerized circuit depends on two types of signals: ________________ and ________________.
8. While discussing the behavior of electricity, Technician A says that if voltage does not change but there is a change in the resistance of the circuit, the current will change. Technician B says that if resistance increases, current decreases. Who is correct?
   A. Technician A only
   B. Technician B only
   C. Both Technician A and Technician B
   D. Neither Technician A nor Technician B
9. Technician A says that rheostats have three connections, one at each end of the resistance and one connected to a sliding contact with the resistor. Technician B says that potentiometers have two connections, one at the power end of the resistance and one connected to a sliding contact with the resistor. Who is correct?
   A. Technician A only
   B. Technician B only
   C. Both Technician A and Technician B
   D. Neither Technician A nor Technician B
10. Technician A says that electrical resistance is the pressure that causes current to flow in a circuit. Technician B says that if there is zero resistance in a circuit, a maximum amount of current will flow in the circuit. Who is correct?
    A. Technician A only
    B. Technician B only
    C. Both Technician A and Technician B
    D. Neither Technician A nor Technician B

# Basic Electrical Service

## *Introduction*

Certain diagnostic basics apply to all electrical systems. Many who have a difficult time diagnosing electrical problems do not understand what electricity is, what electrical meters can and do tell you, and how the different types of electrical problems affect a circuit. The purpose of this chapter is to help you understand these things. Also covered in this chapter is the repair of basic electrical systems.

## ELECTRICAL CIRCUITS

In a **series circuit (Figure 1)** current follows only one path and the amount of current flow through the circuit depends on the total resistance of the circuit. To calculate the total resistance in a series circuit, all resistance values are added together. At each resistor, voltage is dropped and the total amount of voltage dropped in a series circuit is equal to the voltage of the source (battery). Regardless of the possible differences and resistance values of the loads, current in a series circuit is always constant and is determined by the total resistance in the circuit.

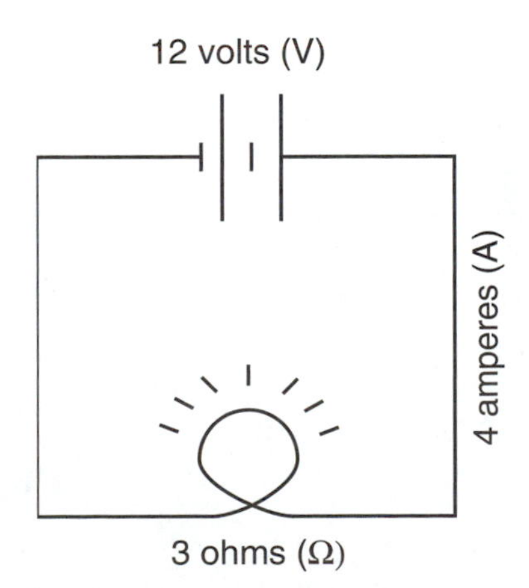


**Figure 1.** In a series circuit, the same amount of current flows through the entire circuit.

**Parallel circuits (Figure 2)** are designed to allow current to flow in more than one path. This allows one power source to power more than one circuit or load. A car's accessories and other electrical devices can be individually controlled through the use of parallel circuits. Within a parallel circuit, there is a common path to and from the power source. Each branch or leg of a parallel circuit behaves as if it were an individual circuit. Current flows only through the individual circuits when each is closed or completed. Not all legs of the circuit need to be complete in order for current to flow through one of them.

In parallel circuits, the total amperage of the circuit is equal to the sum of the amperages in all of the legs of the circuit. No voltage is dropped when the circuit splits into its branches; therefore, equal amounts of voltage are applied to each branch of the circuit. The total resistance of a parallel circuit is always less than the resistance of the leg with the smallest amount of resistance. The total resistance of

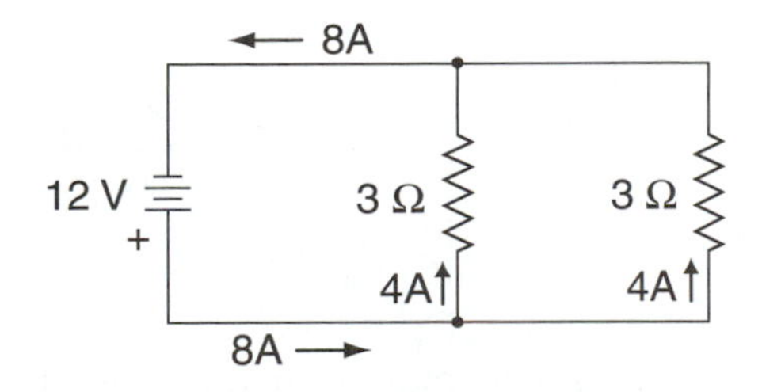


**Figure 2.** A simple parallel circuit.

two resistors in parallel can be calculated by dividing the product of the two by their sum.

The legs of a parallel circuit may contain a series circuit. To determine the resistance of that leg, the resistance values are added together. The resistance values of each leg are used to calculate the total resistance of a parallel circuit. Total circuit current flows only through the common power and ground paths; therefore, a change in a branch's resistance not only will affect the current in the branch but also will affect total circuit current.

## Effects of Resistance

The amount of resistance in a circuit determines how much current will flow in the circuit. Any device that uses electricity has some resistance. This resistance converts the electrical energy into another form of energy. For example, a lamp illuminates because the filament inside the bulb resists the flow of electricity. The bulb shines from the heat given off, as the circuit's voltage pushes the current through the resistance of the filament. The electrical energy converted to heat can be measured as voltage drop.

Voltage pushes current through the circuit's resistors or loads. The amount of voltage within a load decreases as the current passes through the load. By the time the current passes through the last bit of resistance in a circuit, all of the voltage is lost.

In a 12-volt light series circuit, when the resistance of the bulb is 3 ohms, 4 amps of current will flow through the circuit **(Figure 3)**. However, if there is a bad connection at the bulb that creates an additional 1 ohm of resistance, the current will decrease to 3 amps. As a result of the added resistance and decreased current, the bulb will lose about half of its brightness.

This lack of brightness is also caused by the amount of voltage dropped by the bulb. In the normal circuit, the bulb drops 12 volts. However, in the circuit with the added resistance, 3 volts will be dropped by the bad connection and the bulb will drop the remaining 9 volts.

The way resistance affects a circuit depends on the placement of the resistance in the circuit. If the resistances (wanted or unwanted) are placed directly into the circuit, the resistances are said to be in series. If a resistor is placed so that it allows an alternative path for current, it is a parallel resistor.

# ELECTRICAL PROBLEMS

All electrical problems can be classified into one of three categories: open, short, or high-resistance. Identifying

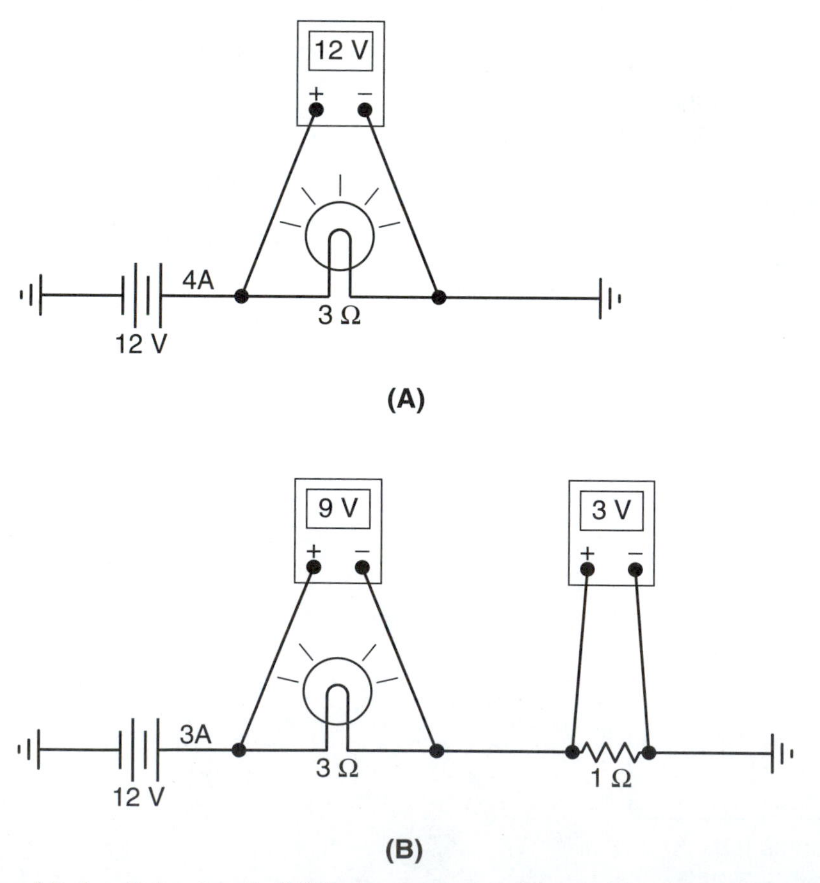


**Figure 3.** "A" is a normal 3 ohm light circuit. "B" is the same circuit but with a high-resistance ground. Notice that the voltage drop across the lamp decreased, as did the circuit current. The bulb will burn dimmer.

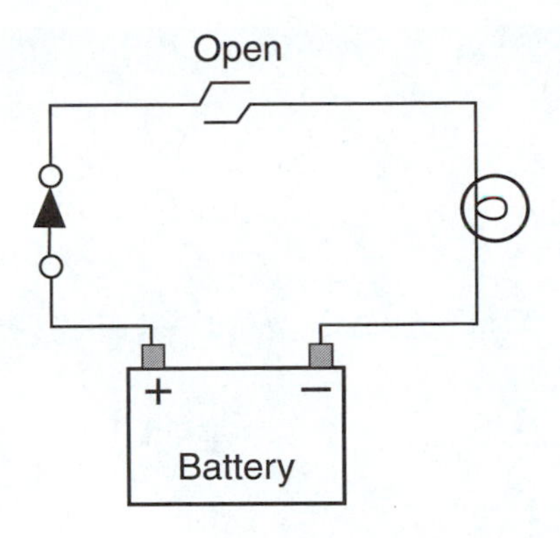


Figure 4. An open prevents current flow.

the type of problem helps identify the correct tests to conduct when diagnosing an electrical problem.

An **open** occurs when a circuit has a break in the wire **(Figure 4)**. Without a completed path, current cannot flow and the load or component cannot work. A disconnected or broken wire or a switch in the OFF position can cause an open circuit. Although voltage will be present up to the open point, there is no current flow. Without current flow, there are no voltage drops across the various loads. If there is an open in one leg of a parallel circuit, the remaining part of that parallel circuit will operate normally.

A **short** results from an unwanted path for current. Shorts cause an increase in current flow, which can burn wires or components. Sometimes two circuits become shorted together resulting in one circuit powering another. This can result in strange happenings, such as the horn sounding every time the brake pedal is depressed. In this case, the brake light circuit is shorted to the horn circuit **(Figure 5)**.

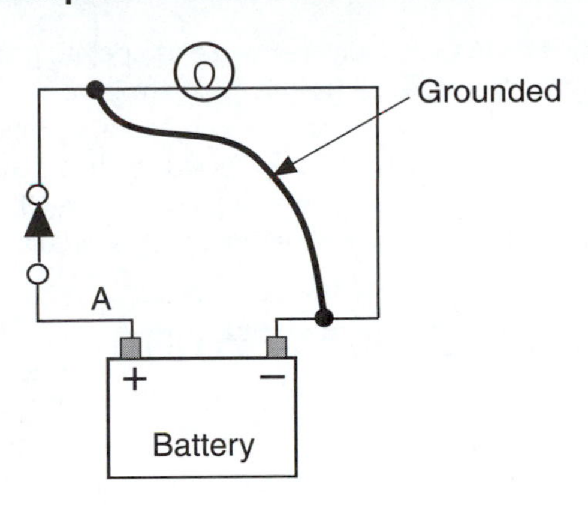


Figure 6. A short to ground causes a large increase in circuit current.

A short to ground can be present before the load in the circuit or internally within the load. This problem provides a low resistance path to ground **(Figure 6)**. Improper wiring and damaged insulation are the two major causes of short circuits.

High-resistance problems occur when there is unwanted resistance in the circuit. The higher than normal resistance causes current to be lower than normal and the

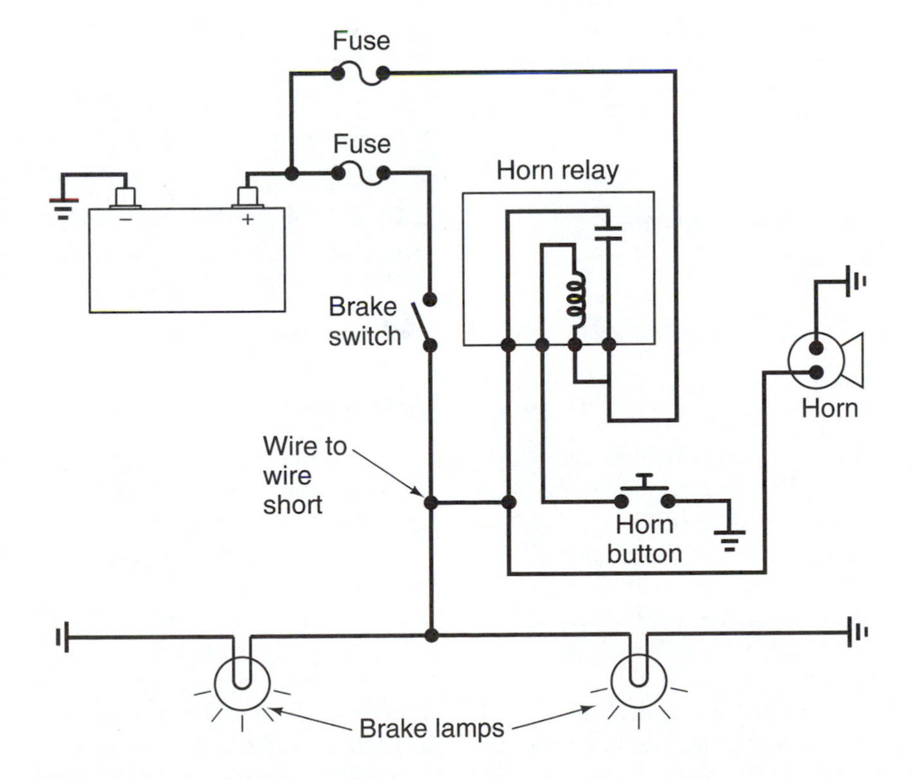


Figure 5. This horn circuit is shorted to the brake light circuit.

components in the circuit are unable to operate properly, if at all. A common cause for this problem is corrosion at a connector. The corrosion becomes an additional resistance in the circuit. This load not only decreases the circuit's current but also uses some of the circuit's voltage, which prevents full voltage to the normal loads in the circuit.

## BASIC ELECTRICAL TESTING

To troubleshoot a problem, always begin by verifying the customer's complaint. Then operate the system and others, to get a complete understanding of the problem. Often there are other problems that are not as evident or bothersome to the customer that will provide helpful information for diagnostics. Refer to the correct wiring diagram and study the circuit that is affected. From the diagram, you should be able to identify testing points and probable problem areas. Then test and use logic to identify the cause of the problem. Several meters are used to test and diagnose electrical systems. These are the ammeter, voltmeter, and ohmmeter. These can be used along with test lights and jumper wires.

### Ammeters

An ammeter must be placed into or in series with the circuit being tested. Normally, this requires disconnecting a wire or connector from a component and connecting the ammeter between the wire or connector and the component. The red lead of the ammeter always should be connected to the side of the connector closest to the positive side of the battery and the black lead should be connected to the other side.

You Should Know *Never place the leads of an ammeter across the battery or a load. This puts the meter in parallel with the circuit and will blow the fuse in the ammeter or possibly destroy the meter.*

It is much easier to test with an ammeter if it has an inductive pickup, commonly referred to a current probe. The pickup clamps around the wire **(Figure 7)** or cable being tested and measures amperage by the strength of the magnetic field created by the current flowing through the wire. This type of pickup eliminates the need to separate the circuit to insert the meter. When the circuit is activated, current flow will be read on the meter.

Assume that a 6-amp fuse protects a circuit with four identical lights wired in parallel. If the circuit constantly blows the fuse, a short exists somewhere in the circuit. To find the short, disconnect all lights by removing them from their sockets. Then, close the switch and read the ammeter. With

**Figure 7.** A current probe clamped around a wire will measure the current flow in that wire.

the load disconnected, the meter should read 0 amperes. If there is any reading, the wire between the fuse block and the socket is shorted to ground.

If zero amps were measured, reconnect each light in sequence; the reading should increase with each bulb. If, when making any connection, the reading is higher than expected, the problem is in that part of the light circuit.

You Should Know *When testing for a short, always use a fuse. Never by-pass the fuse with a wire. The fuse should be rated at no more than 50 percent higher capacity than specifications. This offers circuit protection and provides enough amperage for testing. After the problem is found and corrected, be sure to install the specified rating of fuse for circuit protection.*

### Voltmeters

Electrical circuits and components cannot operate properly if the proper amount of voltage is not available. Not only must the battery be able to deliver the proper amount of voltage, the proper amount of voltage also must be available to operate the intended component. Because of this, voltage is measured at the source, at the electrical loads, and across the loads and circuits.

A voltmeter has two leads: a positive and a negative lead. Connecting the positive lead to any point within a circuit and connecting the negative lead to a ground will measure the voltage at the point where the positive lead was connected. If battery voltage is being measured, the positive lead is connected to the positive terminal of the

**Figure 8.** Probing a circuit with the red voltmeter lead will show the voltage present at that point in the circuit.

battery and the negative lead to the negative post or to a clean spot on the common ground.

To measure the amount of available voltage, the negative lead can be connected to any point in the common ground circuit and the positive lead connected any point within the circuit. This will measure the voltage present at that point **(Figure 8)**.

To measure the voltage drop of a light bulb in a simple circuit, connect the positive lead to the battery or power lead at the bulb and the other lead to the groundside of the load **(Figure 9)**. The meter will read the amount of voltage drop across the bulb. If no other resistances are present in the circuit, the amount of voltage drop will equal the amount of source voltage. However, if there is a resistance present in the ground connection, the bulb will have a voltage drop that is less than battery voltage. If the voltage drop across the ground is measured, it will equal the source voltage minus the amount of voltage dropped by the light. In order to measure voltage drop and available voltage at various points within a circuit, the circuit must be activated.

The maximum allowable voltage drop across wires, connectors, and other conductors in an automotive circuit is 10 percent of the system voltage. Therefore, in a 12-volt automotive electrical system, this maximum loss is 1.2 volts.

A voltmeter also can be used to check for proper circuit grounding. Connect the positive meter lead to the groundside of the bulb and the negative lead to the ground. Any voltage reading indicates there a voltage drop across that connection and the connection is faulty.

## Ohmmeters

Ohmmeters use their own power to measure the resistance between two points. They never should be connected into an activated circuit. External power sources can damage the meter.

The amount of resistance, measured by the meter, is based on the voltage dropped across the component being tested. The scale of an ohmmeter reads from 0 to infinity ( ∞ ). A 0 reading means there is no resistance in the circuit and may indicate a short in a component that should show some resistance. An infinite reading indicates a number higher than the meter can measure. Most digital multimeters display an infinite reading as "O.L" **(Figure 10)**. This usually is an indication of an open circuit.

Most ohmmeters have a variety of ranges it measures within. The expected resistance should be used to select the appropriate range. After the range has been selected, the meter must be zeroed in that scale. To do this, the leads of the meter are held together and the needle or number display is adjusted until it reads zero.

To measure the resistance of a circuit or component, connect one of the meter's leads to the power side of the component and the other lead to the groundside. If the resistance is greater than the selected range, the meter will show a reading of "infinity." When this happens, the next

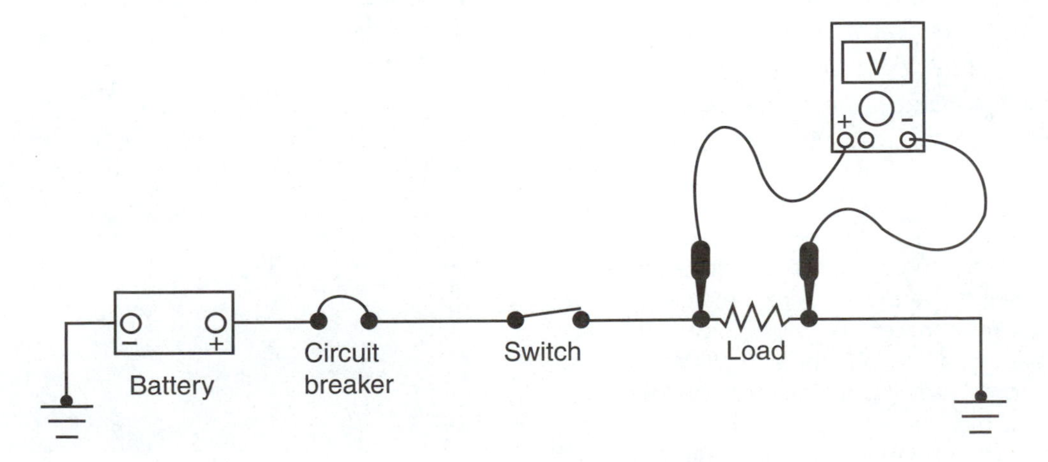


**Figure 9.** To measure voltage drop, the voltmeter is connected across the item being tested.

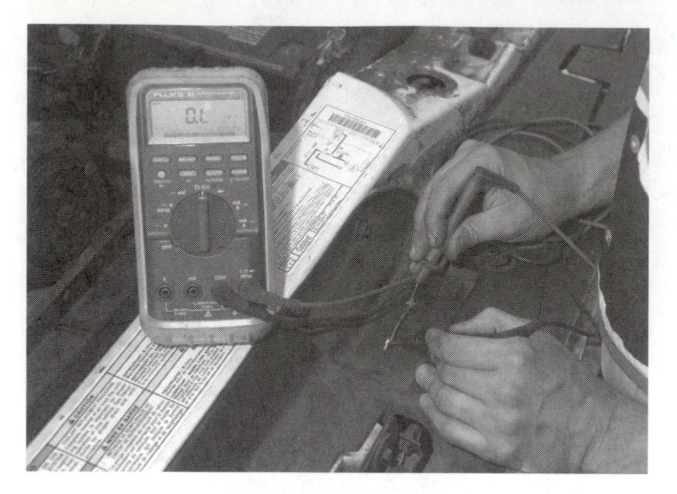

**Figure 10.** Most DVOMs display an infinite resistance reading as "O.L."

highest range should be selected and the component retested. Whenever the range has been changed, the meter must be zeroed for that range prior to measuring the resistance. If subsequent range changes result in continued readings of infinity, it can be assumed that there is no continuity between the two measured points.

An ohmmeter is often used to test circuit completeness or continuity. If a circuit is complete, the meter will display a low resistance, whereas if the circuit is open, the meter will give an infinite reading.

Ohmmeters also can be used to compare the resistance of a component to the value it should have. Many electrical components have a specified resistance value that is listed in the service manuals. This resistance value is important as it controls the amount of voltage dropped and the amount of current that will flow in the circuit. If a component does not have the proper amount of resistance, the circuit will not operate properly.

**You Should Know** *Prior to testing a component or circuit with an ohmmeter, the service manual should be checked for precautions regarding the impedance of the meter.*

## Multimeters

It is not necessary to own separate voltmeters, ohmmeters, and ammeters. These meters are combined in a single tool called a **multimeter**. The most commonly used multimeter is the digital volt/ohmmeter **(DVOM)**, which is often referred to as a digital multimeter **(DMM)**. These meters display the measurements digitally on the meter.

Most digital meters have high input **impedance**, usually at least 10 megaohms (10 million ohms). The high impedance reduces the risk of damaging sensitive components and delicate computer circuits. Analog meters have low input impedance. The low input impedance causes a large amount of current to flow through circuits and should not be used on delicate electronic devices.

Most multimeters test DC and AC volts, ohms, and amperes. Usually there are several test ranges provided for each of these functions. In addition to the basic electrical tests, some multimeters also measure engine revolutions per minute (rpm), ignition dwell, diode condition, distributor conditions, frequency, and even temperature.

DMMs have either an "auto range" feature, in which the meter automatically selects the appropriate scale, or they must be set to a particular range. In either case, you should be familiar with the ranges and the different settings available on the meter you are using. To designate particular ranges and readings, meters display a prefix before the reading or range. For example, if the meter has a setting for mAmps, this means the readings will be given in milliamps or $\frac{1}{1000}$th of an amp.

## Test Lights

There are two types of test lights commonly used in diagnosing electrical problems: nonpowered and self-powered. A **test light (Figure 11)** looks like a stubby ice pick. Its handle is transparent and contains a light bulb. A probe extends from one end of the handle and a ground clip and wire from the other end. With the wire lead connected to ground and the tester's probe at a point of voltage, the light turns on with the presence of voltage. The brightness of the bulb is an indication of the amount of voltage present at the test point.

A self-powered test light is often used to test for continuity instead of an ohmmeter. A self-powered test light does not rely on the power of the circuit to light its bulb. Rather, it contains a battery for a power supply and when connected across a completed circuit, the bulb will light.

**Figure 11.** A nonpowered test light.

This type test light is connected to the circuit in the same way as an ohmmeter.

## Jumper Wires

**Jumper wires** are used to bypass individual wires, connectors, components, or switches. Bypassing a component or wire helps to determine if that part is faulty. If the problem is no longer evident after the jumper wire is installed, the part bypassed is probably faulty. Technicians typically have jumper wires of various lengths and contain a fuse or circuit breaker in them to protect the circuits being tested.

# ELECTRICAL WIRING

Two types of wire are used in automobiles: solid and stranded. Solid wires are single-strand conductors. Stranded wires are the most commonly used wire type and are made up of a number of small solid wires twisted together to form a single conductor. Computers and other electronic units use specially shielded, twisted cable for protection from unwanted induced voltages that can interfere with computer functions. In addition, some use printed circuits.

The current-carrying capacity of a wire is determined by its length and gauge (size). Based on the American Wire Gauge **(AWG)** system, wire size is identified by a numbering system ranging from 0 to 20 gauge, with 0-gauge wire having the largest cross-sectional area and 20-gauge the smallest. Most automotive wiring is 10- to 18-gauge, and battery cables are at least 4-gauge.

## Electrical Wiring Diagrams

Wiring diagrams, sometimes called **schematics**, show how circuits are wired and how the components are connected. A wiring diagram does not show the actual position of the parts on the vehicle or their appearance, nor does it indicate the length of the wire that runs between components. It usually indicates the color of the wire's insulation, and sometimes the wire gauge size. The first letter of the color-coding is a combination of letters usually indicating the base color. The second letter usually refers to the strip color (if any). Tracing a circuit through a vehicle is basically a matter of following the colored wires.

Many different symbols also are used to represent components, such as motors, batteries, switches, transistors, and diodes. Common symbols are shown in **Figure 12**.

Wiring diagrams can become quite complex. To avoid this, the vehicle's electrical system may be divided into many diagrams each illustrating only one system, such as the backup light circuit, oil pressure indicator light circuit, or wiper motor circuit.

# TROUBLESHOOTING GUIDELINES

Troubleshooting electrical problems involves the use of meters, test lights, and jumper wires. Two other basic

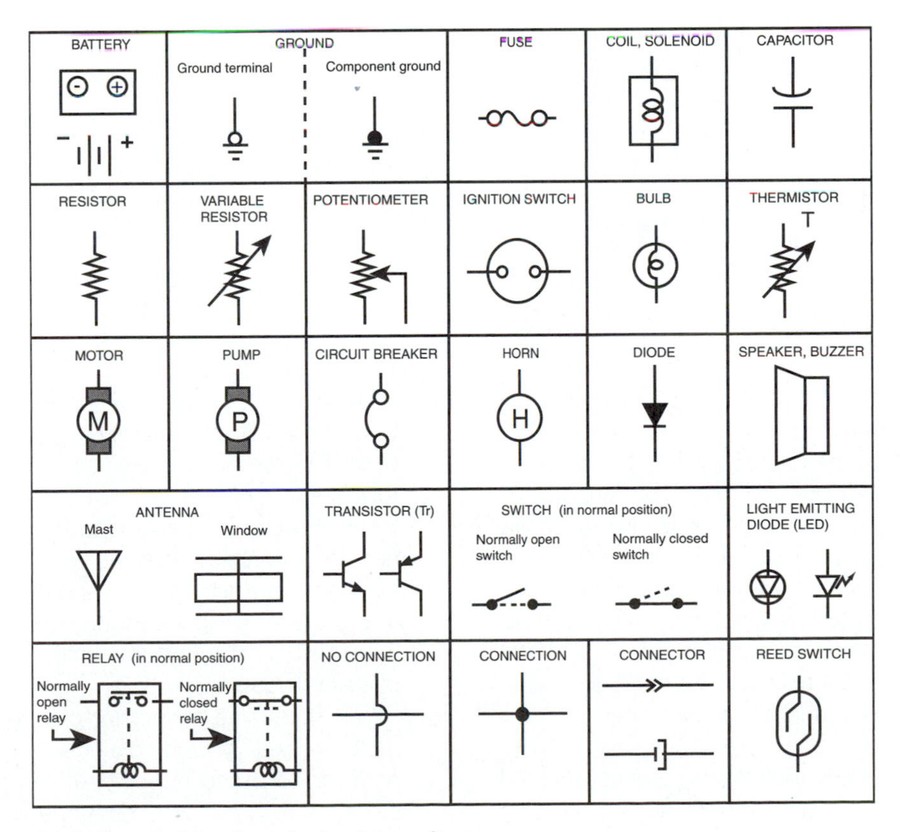


**Figure 12.** Common symbols used in electrical wiring diagrams.

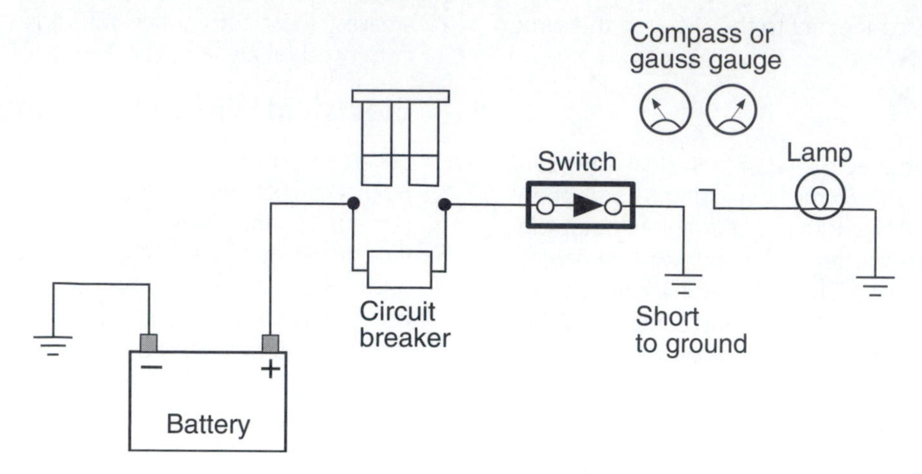


**Figure 13.** The needle of a compass or gauss gauge will fluctuate over the portion of the circuit that has current flowing through it. Once the gauge has passed the spot where the wire is broken and shorted to ground, the needle will stop fluctuating.

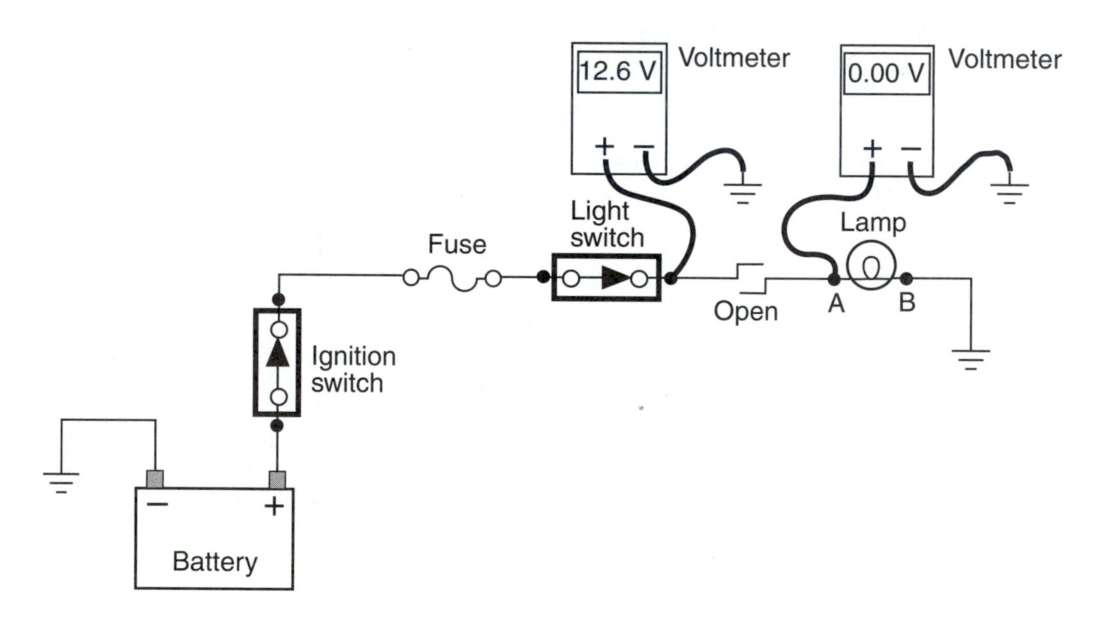


**Figure 14.** Locating an open by testing for voltage.

tools are also quite useful and need to be discussed: circuit breakers and short detection devices.

A circuit breaker is a protection device that resets itself after it has been tripped by high current. Circuit breakers open due to the heat of high current. While open, the breaker cools and closes again to complete the circuit. When a fuse is blown or burned out by high current, it must be replaced to reactivate the circuit. To diagnose a circuit with a short, a completed circuit makes locating the fault easier. By inserting a circuit breaker in place of the fuse, the circuit can be activated and still protected.

A number of techniques and test instruments are used to locate a short, the most common of which is called a short detector. This device is also known as a gauss gauge and looks like a compass **(Figure 13)**. It uses the magnetic field formed by current flow to identify the location of a short. The cycling of the circuit breaker causes the detector's needle to sweep or move back and forth as the detector is moved along the length of the circuit. When the needle no longer sweeps in response to the breaker, the location of break in the wire and the short can be assumed to be before that point in the circuit. Carefully moving the detector back through the circuit should locate the exact place of the short.

To determine the location of an open, a test-light or voltmeter is used. By probing along the circuit, the completed path can be traced. The point at which voltage is no longer present is the point where the open is **(Figure 14)**. If the circuit is open, there will be no current flow. This means that if the open is after the load, source voltage will be measured after the load because there is no voltage drop in the circuit. Often a jumper wire is used to verify or isolate the problem area.

Semiconductors allow current flow only in certain directions and that function can be tested with an ohmmeter. By placing the leads of the meter to the semiconductor connections, continuity should be observed. Reversing the meter leads to the same connections should result in different readings of continuity. For example, a diode should show good continuity when the leads are connected to it and poor continuity when the leads are reversed.

You Should Know

*Similar electrical components can be tested in similar way, regardless of the circuit or system they are part of. Although the following discusses how to test some of the more common ways to test components, always refer to the appropriate service manual before you begin your tests.*

## Testing Circuit Protection Devices

**Fuses.** There are three basic types of **fuses** found in automobiles: cartridge, blade, and ceramic. Late-model vehicles use blade or spade fuses. To check a blade fuse, pull it from the fuse panel and look at the fuse element through the transparent plastic housing. Look for internal breaks and discoloration. The current rating for blade fuses is indicated by the color of the plastic case. It may also be marked on the top.

**Maxi-Fuses.** Many late-model vehicles use **maxi-fuses** instead of fusible links. Maxi-fuses **(Figure 15)** look and operate like two-prong, blade, or spade fuses, except they are much larger and can handle more current. Maxi-fuses are located in their own under hood fuse block.

To check a maxi-fuse, look at the fuse element through the transparent plastic housing. If there is a break in the element, the maxi-fuse has blown. To replace it, pull it from its fuse box or panel. Always replace a blown maxi-fuse with a new one with the same ampere rating.

**Circuit Breakers.** Some circuits are protected by **circuit breakers**. Each circuit breaker conducts current through an arm made of two types of metal bonded together (bimetal arm). If the arm starts to carry too much current, it heats up. As one metal expands faster than the other, the arm bends, opening the contacts. Current flow is broken. A circuit breaker will either automatically reset or it must be manually reset by depressing a button.

Figure 15. Maxi-fuses have replaced fusible links on many vehicles.

## Testing Switches

Switches can be tested with a voltmeter, test light, or ohmmeter. To check the operation of a switch with a voltmeter or a test light, connect the meter's positive lead to the battery side of the switch. With the negative lead attached to a good ground, voltage should be measured at this point. Without closing the switch, move the positive lead to the other side of the switch. If the switch is open, no voltage will be present at that point. The amount of voltage present at this side of the switch should equal the amount on the other side when the switch is closed. If the voltage decreases, the switch is causing a voltage drop because of excessive resistance. If no voltage is present on the ground-side of the switch with it closed, the switch is not functioning properly and should be replaced.

If a switch has been removed from the circuit, it can be tested with an ohmmeter or a self-powered test light. By connecting the leads across the switch connections, the action of the switch should open and close the circuit **(Figure 16)**.

## Testing Solenoids

Solenoids are commonly found on automatic transmissions. A problem with a solenoid or solenoid circuit can cause shift problems. For example, if there is an open in the power wire to a shift solenoid, the transmission will not shift in particular gears. To diagnose a problem like this, check for voltage at the solenoid. Then work away from the solenoid until voltage is present. The open will be between the last point in the circuit where voltage was not present and the point at which voltage was measured. To diagnose this problem, you should refer to the wiring diagram for the vehicle first.

If less than battery voltage is available at the solenoid, check the circuit for excessive voltage drops. If a solenoid is bad, it will either be open, have a short, or have high internal

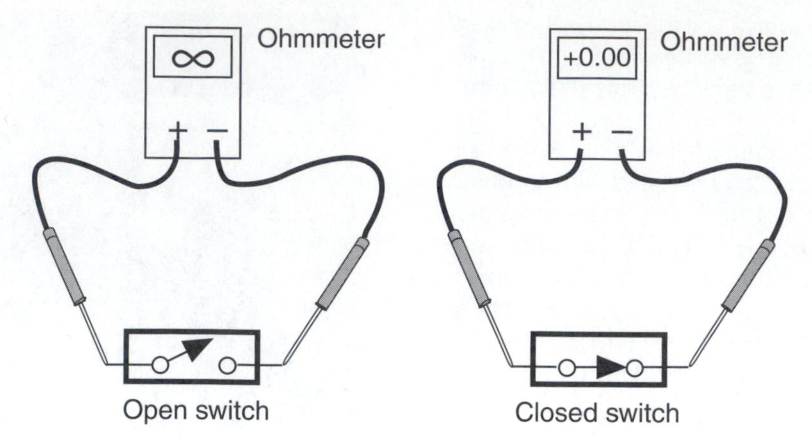


**Figure 16.** After a switch has been removed or disconnected from the circuit, it can be checked with an ohmmeter.

resistance. The exact problem can be identified by checking the solenoid with an ohmmeter. Resistance checks should be made across the terminals of the solenoid and between the terminals and the case. A good solenoid will have low resistance only across the terminals. A resistance reading between the terminals and the case indicates that the windings are shorted to the case. If there is high resistance across the terminals, there is excessive resistance in the solenoid. Also, if there is no continuity across the terminals, the windings are open.

## Testing Electronic Circuits

The introduction of powertrain computer controls brought with it the need for tools capable of troubleshooting electronic control systems. There is a variety of computer scan tools available today that do just that. A scan tool is a microprocessor designed to communicate with the vehicle's computer. Connected to the computer through diagnostic connectors, a scan tool can access diagnostic trouble codes (DTCs), run tests to check system operations, and monitor the activity of the system.

A scan tool receives its testing information from one of several sources. Some scan tools have a programmable read-only memory **(PROM)** chip that contains all the information needed to diagnose specific model lines. This chip may be contained in a cartridge that is plugged into the tool. The type of vehicle being tested determines the appropriate cartridge that should be inserted. These cartridges contain the test information for that particular car. A cartridge typically is needed for each make and model vehicle. As new systems are introduced on new car models, a new cartridge is made available.

LED displays are generally only large enough to display four short lines of information. This feature limits the technician's ability to compare test data. However, most scan tools overcome this inadequacy by storing the test data in a random access memory **(RAM)**. The scan tool then can be interfaced with a printer, personal computer, or larger engine analyzer that can retrieve the information stored in the memory.

The vehicle's computer only sets trouble codes when a voltage signal is entirely out of its normal range. The codes help technicians identify the cause of the problem when this is the case. If a signal is within its normal range but it still not correct, the vehicle's computer will not display a trouble code. However, a problem will still exist. As an aid to identify this type of problem, most manufacturers recommend that the signals to and from the computer be carefully looked at.

Most diagnostic work on computer control systems should be based on a description of symptoms. With this description, you can locate any technical service bulletins that refer to the problem. You also can use the description to locate the appropriate troubleshooting sequence in the manufacturer's service manuals.

# BASIC ELECTRICAL REPAIRS

Faulty wiring can cause many electrical problems. Loose or corroded terminals, frayed, broken, or oil-soaked wires, and faulty insulation are the most common causes. Wires, fuses, and connections should be checked carefully during troubleshooting.

Wire end terminals are connecting devices. They are generally made of tin-plated copper and come in many shapes and sizes. They may be either soldered or crimped in place.

When working with wiring and connectors, never pull on the wires to separate the connectors. This can loosen the connector and cause an intermittent problem that may be very difficult to find later. Always follow the correct procedures and use the tools designed for separating connectors.

Nearly all connectors have pushdown release type locks **(Figure 17)**. Make sure these are not damaged when disconnecting the connectors. Many connectors have covers over them to protect them from dirt and moisture. Make sure these are properly installed to provide for that protection.

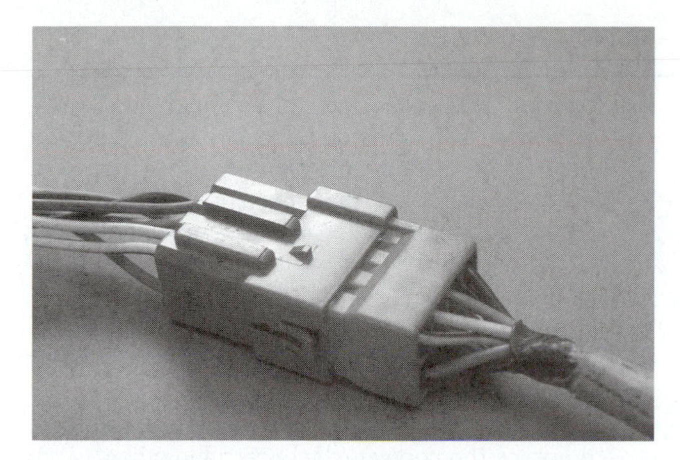

**Figure 17.** A connector with a pushdown release lever.

Never reroute wires when making repairs. Rerouting wires can result in induced voltages from nearby components. These stray voltages can interfere with the function of electronic circuits.

Dielectric grease should be used to moisture proof and to protect connections from corrosion. If the manufacturer specifies that a connector be filled with grease, make sure it is. If the old grease is contaminated, replace it. Some car manufacturers suggest using petroleum jelly to protect connection points.

When installing a terminal, select the appropriate size and type of terminal. Be sure it fits the unit's connecting post and has enough current carrying capacity for the circuit. Also, make sure it is heavy enough to endure normal wire flexing and vibration.

To crimp a connector to a wire, make sure you use the correct size-stripping opening on the crimping tool. Remove enough insulation to allow the wire(s) to completely penetrate the connector. Place the wire(s) into the connector and crimp the connector **(Figure 18)**. To get a proper crimp, place the open area of the connector facing toward the anvil. Make sure the wire is compressed under the crimp. Use electrical tape or heat shrink tubing to tightly seal the connection. This will provide good protection for the wire and connector.

**Figure 18.** Use the correct tool and opening on the tool when crimping a connector.

The preferred way to connect wires or to install a connector is by soldering **(Figure 19)**. Some car manufacturers have used aluminum in their wiring. Aluminum cannot be soldered. Follow the manufacturer's guidelines and use the proper repair kits when repairing aluminum wiring.

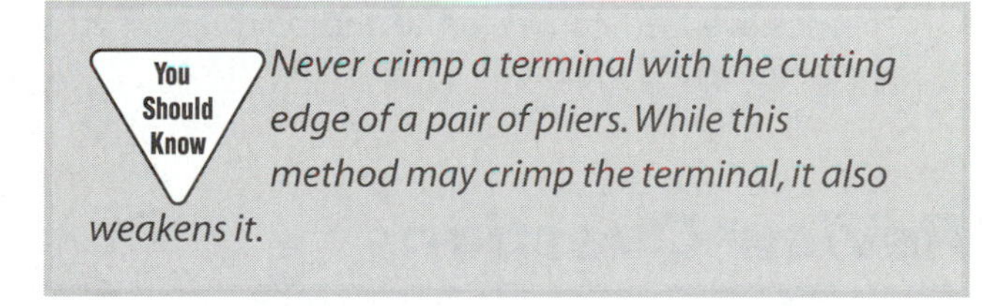


*Never crimp a terminal with the cutting edge of a pair of pliers. While this method may crimp the terminal, it also weakens it.*

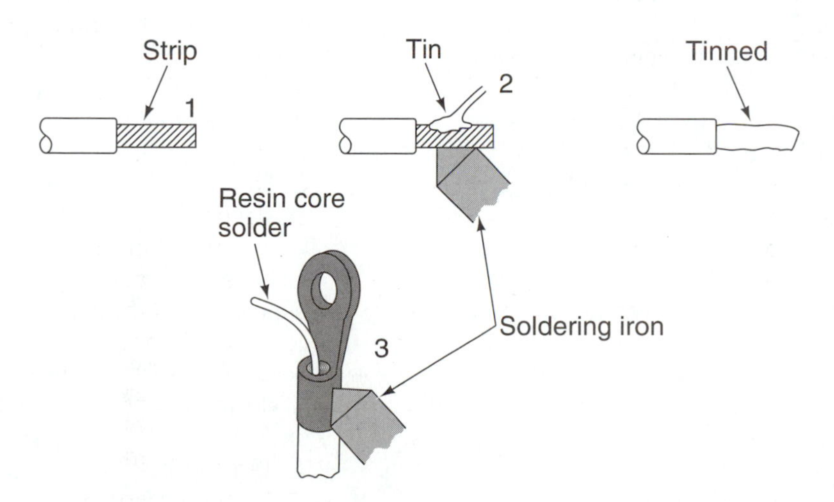


**Figure 19.** Proper procedure for soldering a terminal to a wire.

## Summary

- The amount of voltage within a load decreases as the current passes through the load. By the time the current passes through the last bit of resistance in a circuit, all of the voltage is lost.
- In a series circuit, current follows only one path and the amount of current flow through the circuit depends on the total resistance of the circuit. The total resistance in a series circuit is equal to the sum of the resistances in the circuit.
- The total amount of voltage dropped in a series circuit is equal to the voltage of the source.
- Current in a series circuit is always constant.
- In parallel circuits, the total amperage of the circuit is equal to the sum of the amperages in all of the legs of the circuit. Equal amounts of voltage are applied to each branch of the circuit.
- The total resistance of a parallel circuit is always less than the resistance of the leg with the smallest amount of resistance.
- An open occurs when a circuit is not complete and therefore there is no current flow through the circuit.
- A short results from an unwanted path for current and cause an increase in current flow.
- High-resistance problems occur when there is unwanted resistance in the circuit, which causes lower than normal current flow and unwanted voltage drops.
- To measure available voltage, connect the negative lead to any point in the common ground circuit and the positive lead connected to the point at which a measurement is sought.
- To measure the voltage drop of a component, connect the positive lead to the power lead of the component and the other lead to the groundside.
- On an ohmmeter, a 0 reading means there is no resistance in the circuit, and an infinite reading indicates a number higher than the meter can measure, usually this indicates an open circuit.
- A DMM is the preferred electrical tester because it can measure volts, amps, and ohms, as well as conduct many other tests.
- Begin diagnostic work by verifying the customer's complaint. Then operate the system and others, to get a complete understanding of the problem. Refer to the wiring diagram to identify testing points and probable problem areas. Then test and use logic to identify the cause of the problem.
- To check the operation of a switch with a voltmeter or a test light, check the voltage in and out of the switch when it is in the open and closed positions.
- Connected to the computer through diagnostic connectors, a scan tool can access diagnostic trouble codes, run tests to check system operations, and monitor the activity of the system.
- The vehicle's computer only sets trouble codes when a voltage signal is entirely out of its normal range.
- Faulty wiring can cause many electrical problems. Loose or corroded terminals, frayed, broken, or oil-soaked wires, and faulty insulation are the most common causes.

## Review Questions

1. Name the three types of electrical problems and their effect on current flow.
2. Describe what happens to the circuit when a bulb in a parallel circuit burns out.
3. Explain why impedance is an important consideration when choosing a multimeter.
4. Describe one way the location of an open can be found.
5. While discussing measuring resistance, Technician A says that an ohmmeter can be used to measure resistance of a component before disconnecting it from the circuit. Technician B says that a voltmeter can be used to measure voltage drop. Something that has very little resistance will drop a zero or very little voltage. Who is correct?
   A. Technician A only
   B. Technician B only
   C. Both Technician A and Technician B
   D. Neither Technician A nor Technician B
6. When using a voltmeter, Technician A connects it across the circuit being tested. Technician B connects the meter's red lead to the most positive side of the circuit. Who is correct?
   A. Technician A only
   B. Technician B only
   C. Both Technician A and Technician B
   D. Neither Technician A nor Technician B
7. If you were probing along with a voltmeter on a circuit that has a bad ground, how many volts would you read before and after the load? Why?
8. Technician A says that current will increase with a decrease in resistance. Technician B says that current will decrease with an increase in resistance. Who is correct?
   A. Technician A only
   B. Technician B only
   C. Both Technician A and Technician B
   D. Neither Technician A nor Technician B

9. While discussing how to test a switch, Technician A says that the action of the switch can be monitored by a voltmeter. Technician B says that continuity across the switch can be checked by measuring the resistance across the switch's terminals when the switch is in its different positions. Who is correct?
   A. Technician A only
   B. Technician B only
   C. Both Technician A and Technician B
   D. Neither Technician A nor Technician B

10. While discussing electricity, Technician A says that an open causes unwanted voltage drops. Technician B says that high-resistance problems cause increased current flow. Who is correct?
    A. Technician A only
    B. Technician B only
    C. Both Technician A and Technician B
    D. Neither Technician A nor Technician B

# Basic Operation

*In 1939, Chrysler introduced the Fluid Drive, a three-speed synchromesh transmission with a fluid coupling between the engine and a conventional friction clutch. In 1941, this was replaced by the Vacamatic, a manually shifted two-speed in series with a vacuum-actuated, automatic overdrive. The result was four forward speeds, only half of which were engaged automatically. Ford launched its similar Liquamatic Drive on 1942 Lincolns and Mercurys. Few were sold, and many owners had standard transmissions retrofitted during the war, when replacement parts were hard to find.*

## SECTION OBJECTIVES

After you have read, studied, and practiced the contents of this section, you should be able to:

- List the factors that determine when an automatic transmission will automatically shift.
- Describe the four basic systems of all automatic transmissions.
- Describe the operation and purpose of a torque converter.
- Identify the major components in a transmission's hydraulic circuit and describe how they provide fluid flow and pressurization.
- List and describe the various load-sensing devices used in an automatic transmission's hydraulic system.
- Describe the basic construction of automatic transmission housings, including the purpose of the various mechanical and electrical connections.
- Identify the major components in a torque converter and explain their purpose.
- Explain the operation of typical electronic controls for a converter clutch.
- Explain the advantages of using electronic controls for transmission shifting.
- Identify the input and output devices in a typical automatic transmission electronic control system and briefly describe the function of each.
- Describe the construction and operation of typical Simpson and Ravigneaux geartrain-based transmissions.
- Describe the construction and operation of transaxles that use planetary gearsets in tandem.
- Describe the construction and operation of automatic transmissions that use helical gears in constant mesh.

# Chapter 9 Automatic Transmission Operation

## *Introduction*

An automatic transmission/transaxle shifts automatically through the forward gears without the driver moving a gearshift lever or depressing a clutch pedal. Transmissions are primarily used in RWD and 4WD vehicles. Transaxles are used in most 4WD and mid-engined rear wheel drive cars. The major components of a transaxle are the same as those in a transmission, except the transaxle assembly includes the final drive and differential gears, in addition to the transmission.

Most of today's automatic transmissions and transaxles have electronic controls in addition to hydraulic controls. This chapter focuses on how an automatic transmission works and on the basic transmission systems. Electronic controls are mentioned, but detailed discussions of these are in other chapters.

### BASIC OPERATION

An automatic transmission receives engine power through a torque converter, which is driven by the engine's crankshaft. Hydraulic pressure in the converter allows power to flow from the torque converter to the transmission's input shaft. The only time a torque converter is mechanically connected to the transmission is during the application of the torque converter clutch, if the converter is fitted with one.

The input shaft drives a planetary gear set, which provides the various forward gears, a neutral position, and one reverse gear. Power flow through the gears is controlled by multiple-friction disc packs, one-way brakes and clutches, and/or brake friction bands. These connect one member of the gear set to another member or hold a member of the gear set when they are activated. By holding different members of the planetary gear set **(Figure 1)**, different gear ratios are possible.

Hydraulic pressure is routed to the correct driving or holding element by the transmission's valve body, which contains many hydraulic valves and controls the pressure and direction of the hydraulic fluid.

An automatic transmission or transaxle selects gear ratios according to engine speed, engine load, vehicle speed, and other operating conditions. Both upshifts and downshifts occur automatically. The transmission also can be manually selected into a lower forward gear, reverse, neutral, or park. Depending on the forward gear range selected, the transmission may provide engine braking during deceleration.

The most commonly used transmissions today have four forward speeds and neutral, reverse, and park positions. Five- and six-speed automatics also are becoming quite common. These typically have two overdrive gears (fourth and fifth or fifth and sixth). An overdrive condition occurs when the transmission's input shaft rotates less than one full revolution for every one full revolution of the output shaft. Overdrive improves fuel economy and reduces engine noise. Overdrive also allows the engine to run at slower speeds at high vehicle speeds, which increases engine life. Most automatic transmissions also feature a torque converter clutch, which reduces the power lost through the operation of a conventional torque converter.

A typical transmission has a seven-position shift selector **(Figure 2)**. The selector lever is linked to the manual lever on the transmission to select the desired gear range. These ranges are as follows:

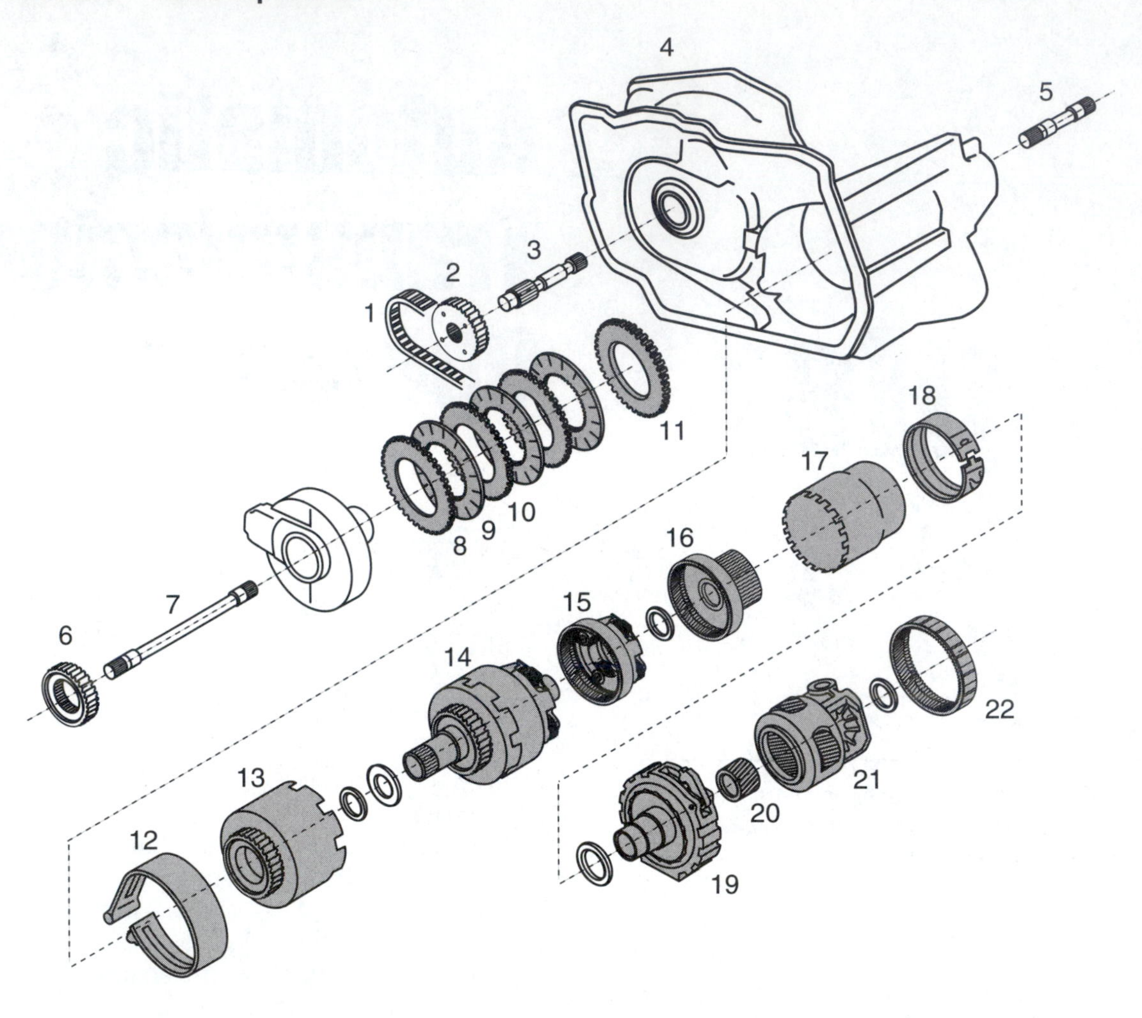


**LEGEND**

1. Drive link assembly
2. Drive sprocket
3. Turbine shaft
4. Transaxle case
5. Output stub shaft
6. Driven sprocket
7. Output shaft
8. 2nd clutch waved plate
9. 2nd clutch steel plate
10. 2nd clutch fiber plate
11. 2nd clutch backing plate
12. Intermediate 4th band
13. Reverse input clutch assembly
14. Direct and coast clutch assembly
15. Input carrier assembly
16. Input flange and forward clutch hub assembly
17. Forward clutch assembly
18. Lo/Reverse band
19. Forward clutch support assembly
20. Final drive sun gear
21. Differential and final drive assembly
22. Final drive internal gear

**Figure 1.** An example of planetary gears with their associated brakes and clutches.

**Figure 2.** Gear selection pattern for a seven-position shift selector.

"P" (Park)—The transmission is in neutral with its output shaft locked to the case by engaging a parking pawl into the parking gear. The engine can be started in this selector position, and this is the only position in which the ignition key can be removed.

"R" (Reverse)—The transmission is in reverse at a reduced or lower gear ratio.

"N" (Neutral)—Like Park, the transmission is in neutral and the engine can be started; however, the output shaft is not locked to the case and the parking brake should be applied.

"OD" (Overdrive)—This is the normal driving gear range. Selection of this position provides for automatic shifting into all forward gears and allows for application and release of the converter clutch.

"D" (Drive)—Selection of this gear range provides automatic shifting into all forward gears, except overdrive gear, and allows for the application and release of the converter clutch. This position is selected when overdrive is not desired such as when traveling on hilly or mountainous roads or when towing a trailer. Operating in overdrive during these conditions places an extraordinary amount of load on the engine and can result in severe damage.

"2" (Manual Second)—The selection of this gear position provides only second gear operation regardless of vehicle speed. When Manual Second is selected, some transmission will start off in first gear then upshift to second. Others will start and remain in second gear. The selection of this gear range is wise for acceleration on slippery surfaces or for engine braking while going down steep hills.

"1" (Manual Low)—In most transmissions, the shift selector allows only first gear operation. Some transmissions are designed to shift into second gear when engine speed is high and Manual Low is selected. Selection of this gear at high speeds results in a downshift into second gear. An automatic downshift to first will occur only when the vehicle's speed decreases to a predetermined level, normally at about 30 mph. Once the transmission is in first gear, it will stay in low until the selector lever is moved to another position.

An automatic transmission automatically selects the gear ratio and torque output best suited for the existing operating conditions, vehicle load, and engine speed. A typical automatic transmission consists of four basic systems: the torque converter, planetary gear sets, the hydraulic system, and **reaction members**.

## Manual Shifting

Some late-model cars feature a manual shift option that allows the driver to control the shifting of the transmission. Chrysler calls this option "Autostick." The benefit of this option is simply driver control of gear changes without the use of a clutch pedal. When the shifter is moved into the autostick position, the transaxle remains in whatever gear it was using before Autostick was activated. Moving the shifter toward the driver causes the transmission to upshift. Moving the shifter to the passenger side causes the transaxle to downshift. The selected gear is illuminated on the instrument panel.

The car can be launched in first, second, or third gear while in the Autostick mode. Shifting into overdrive cancels the autostick option and the transmission resumes its normal overdrive operation. Although the driver has near full control of transmission shifting, there are limitations as to when shifts can be made. A shift into third or fourth gear cannot be made if the vehicle is going below 15 mph, nor can the transmission be downshifted into first if the car is traveling at speeds greater than 41 mph. These safeguards prevent transmission and engine damage possible if the gears were selected.

## Torque Converters

The torque converter automatically engages and disengages power from the engine to the transmission in relation to engine speed. When the engine is running at the correct idle speed, there is an insufficient amount of fluid flow in the converter to allow for power transfer through the torque converter. As engine speed increases, fluid flow increases and creates a sufficient amount of force to transmit engine power through the torque converter to the input shaft of the transmission. This action couples the engine to the transmission's planetary gear set and provides some torque multiplication during some driving conditions.

*In 1941, Chrysler introduced a four-speed, semi-automatic transmission with a hydraulic coupling.*

The torque converter also absorbs the shock of the transmission while it is changing gears. The main components of a torque converter are the cover, turbine, impeller, and stator **(Figure 3)**. The **converter cover** is connected to the flywheel and transmits power from the engine into the torque converter. The impeller is driven by the converter cover. Its rear hub drives the transmission's oil pump. The turbine is splined to the **input shaft**, and is driven by the force of the fluid from the impeller. The stator aids in redirecting the flow of fluid from the turbine to the impeller and is equipped with a one-way clutch that allows it to remain stationary during maximum torque development.

Most torque converters are fitted with a clutch mechanism that provides a mechanical connection between the transmission and the engine. The clutch is only engaged when prescribed conditions are met. The torque converter clutch is typically controlled by a computer.

# PLANETARY GEARING

Planetary gearsets provide for the different gear ratios in most automatic transmissions. A single planetary gearset

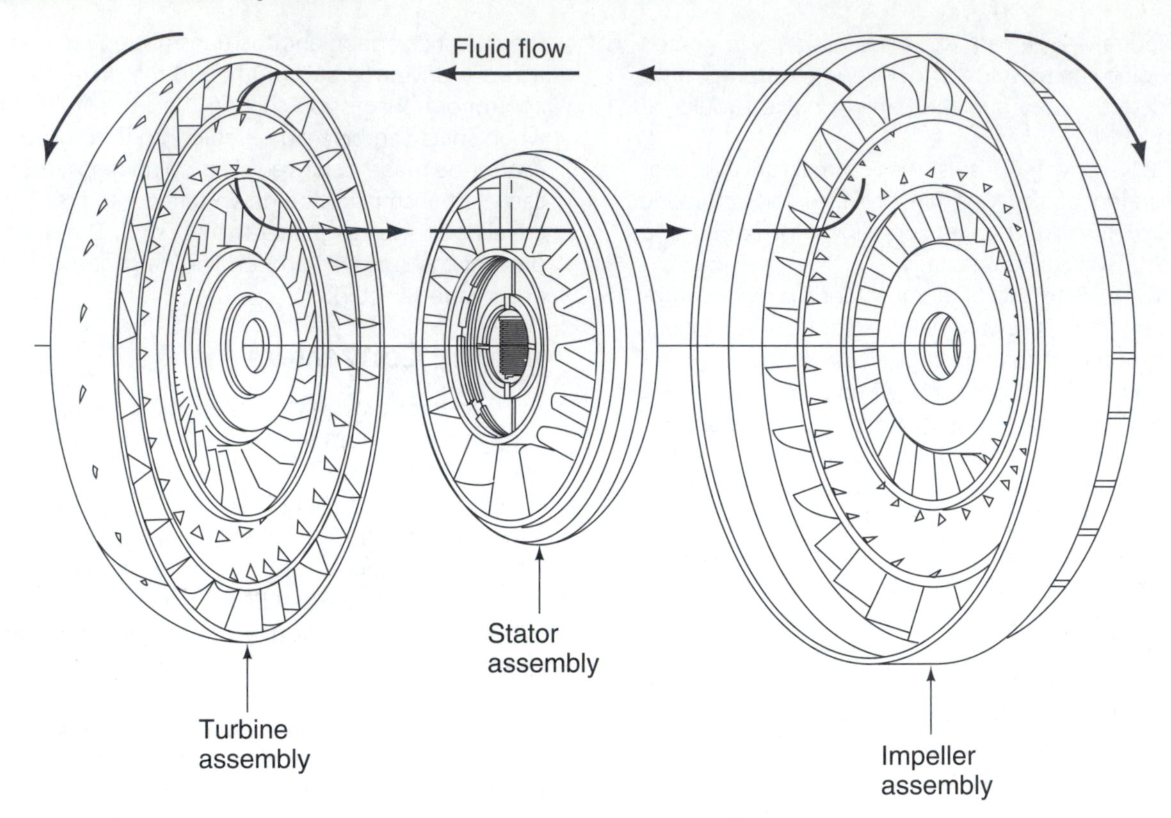


**Figure 3.** The main components of a torque converter. The converter cover and impeller are shown as a solid unit.

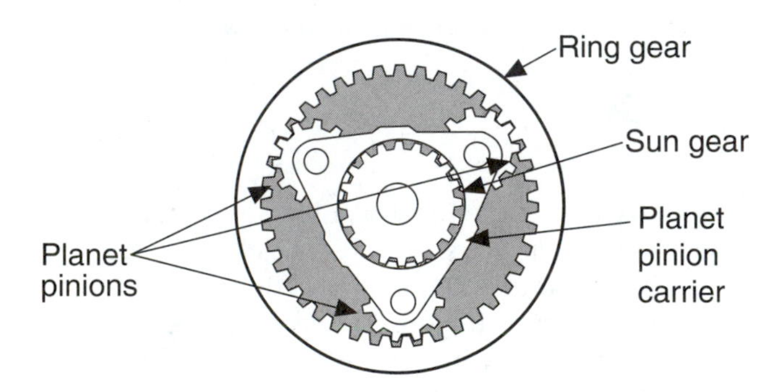


**Figure 4.** A planetary gearset.

consists of a sun gear, a planet carrier with two or more planet pinion gears, and an internal ring gear **(Figure 4)**. These gears typically are helically cut gears, which offer quiet operation. From a single planetary gearset, two reverses and five forward speeds are possible. The five forward speeds are two reduction ratios, a direct drive, and two overdrive ratios. These different speeds are obtained by holding one or more gear set members and driving another.

At the center of the planetary gearset is the sun gear. The placement of this gear in the gearset is the reason for its name; the earth's sun is at the center of our solar system. Planet gears surround the sun gear, just like the earth and other planets in our solar system. These gears are mounted and supported by the planet carrier and each gear spins on its own separate shaft. The planet gears are in constant mesh with the sun and ring gears. The ring gear is the outer gear of the gear set. It has internal teeth and surrounds the rest of the gear set. Its gear teeth are in constant mesh with the planet gears. The number of planet gears used in a planetary gearset varies according to the loads the transmission is designed to face. For heavy loads, the number of planet gears is increased to spread the workload over more gear teeth.

The planetary gearset can provide a gear reduction or overdrive, direct drive or reverse, or a neutral position. Because the gears are in constant mesh, gear changes are made without engaging or disengaging gears, as is required in a manual transmission. Rather, clutches and bands are used to either hold or release different members of the gearset to get the proper direction of rotation and/or gear ratio.

## Compound Planetary Gear Sets

A limited number of gear ratios are available from a single planetary gearset. To increase the number of available gear ratios, gear sets can be added. The typical automatic transmission with four forward speeds has at least two planetary gearsets.

In automatic transmissions, two or more planetary gear sets **(Figure 5)** are connected together to provide the various gear ratios needed to efficiently move a vehicle. There are two common designs of compound gear sets: the Simpson gearset, in which two planetary gearsets share a common sun gear, and the Ravingeaux gear set, which

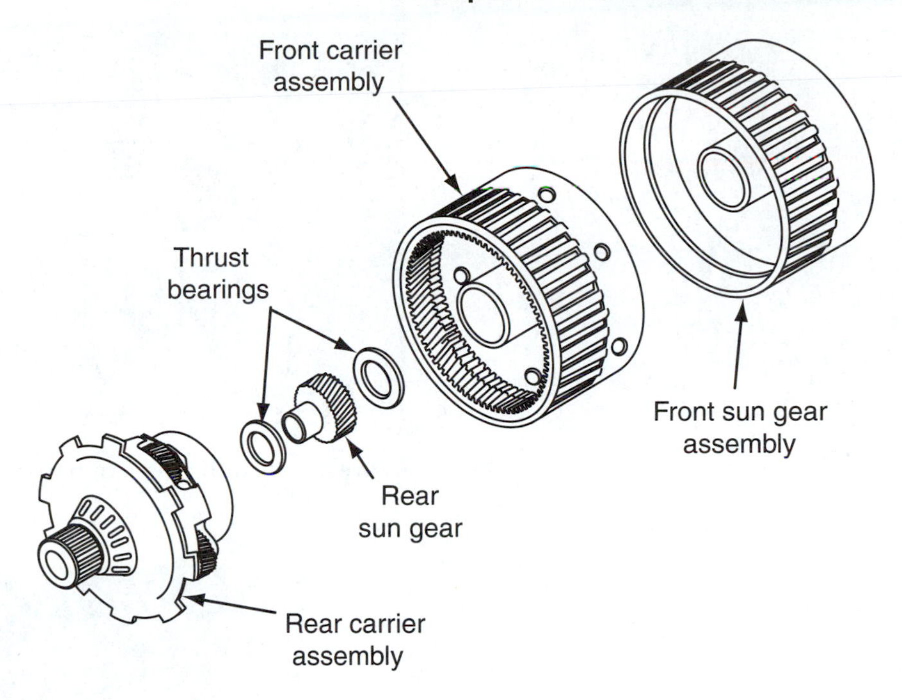


**Figure 5.** A compound gear set.

has two sun gears, two sets of planet gears, and a common ring gear. Some transmissions are fitted with an additional single planetary gearset, which is used to provide additional overdrive gear ratios.

## INTERNAL COMPONENTS

Contained in the transmission/transaxle housing, or case, are the basic parts that allow for the transfer of power and the change of gears. One end of the housing encases the torque converter and bolts to the engine. The torque converter is bolted to the engine and not the transmission. However, the transmission is designed to support the unbolted end. This end rides on a bearing or bushing in the transmission.

The torque converter not only transfers engine power to the transmission, it also drives the transmission pump. The pump is a critical component as it supplies fluid flow for the entire transmission, including the torque converter.

The valve body contains the valves and orifices that control the operation of a transmission. The valve body is typically bolted to the bottom of the transmission housing and is covered by the oil pan. In response to the movement of the valves, fluid is routed to the different hydraulic apply devices. This action causes a change in gears.

Although the pump is the main source of fluid flow, there are many devices involved in the flow and pressurization of the fluid. All of these work together to enable the transmission to shift at the correct time and with the correct feel.

### Planetary Gear Controls

Certain parts of the planetary geartrain must be held, while others must be driven to provide the needed torque multiplication and direction for vehicle operation. "Planetary gear controls" is the general term used to describe transmission bands, servos, and clutches. Reaction members are those parts of an automatic transmission that hold members of the planetary gear set to the case in order to change gears. One-way overrunning clutches, bands, and multiple disc packs are examples of reaction members. One-way overrunning clutches are purely mechanical devices, whereas clutches and brakes are hydraulically controlled mechanical devices. Most automatic transmissions use more than one type of these members; some use all three.

**Transmission bands** A **band (Figure 6)** is a braking assembly positioned around a stationary or rotating drum.

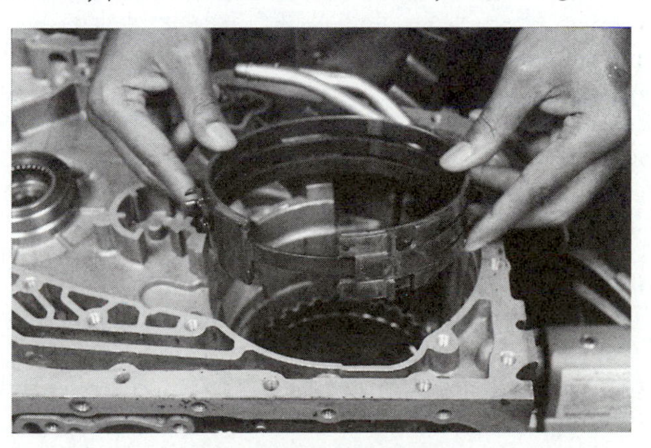

**Figure 6.** A typical band assembly.

Figure 7. A typical one-way clutch assembly.

Figure 8. A multiple-friction disc assembly.

The band brings a drum to a stop by wrapping itself around the drum and holding it. The band is applied hydraulically by a servo assembly. Connected to the drum is a member of the planetary gear train. The purpose of a band is to hold a member of the planetary gear set by holding the drum and connecting planetary gear member stationary. Bands provide excellent holding characteristics and require a minimum amount of space within the transmission housing.

The **servo** assembly converts hydraulic pressure into a mechanical force that applies a band to hold a drum stationary. Simple and compound servos are used to engage bands in modern transmissions.

In contrast to a band, which can only hold a planetary gear member, transmission clutches are capable of both holding and driving members.

**Overrunning clutches** In an automatic transmission operation, both sprag and roller overrunning clutches **(Figure 7)** are used to hold or drive members of the planetary gear set. These clutches operate mechanically. An overrunning clutch allows rotation in only one direction and operates at all times.

**Multiple-disc clutches** A multiple-disc clutch uses a series of friction discs to transmit torque or apply braking force. The discs have internal teeth that are sized and shaped to mesh with splines on the clutch assembly hub. In turn, this hub is connected to a member of the planetary gear set that will receive the desired braking or transfer force when the clutch is applied or released.

Multiple-disc clutches are enclosed in a large drum-shaped housing that can be either a separate casting **(Figure 8)** or part of the existing transmission housing. This drum housing also holds the other clutch components: cylinder, hub, piston, piston return springs, seals, pressure plate, friction plates, and snap rings.

Figure 9. A typical input shaft protruding from a transaxle.

## Shafts

Transmissions have at least two shafts: an input shaft and an output shaft. The input shaft **(Figure 9)** connects the output of the torque converter to the driving members inside the transmission. Each end of the input shaft is externally splined to fit into the internal splines of torque converter's turbine and the driving member in the transmission. Normally, the front clutch pack's hub is the driving member. In many transmissions there is a tube, called the stator shaft, that surrounds the input shaft. The stator shaft is splined to the torque converter's stator and is a stationary shaft.

The output shaft **(Figure 10)** connects the driven members of the gear sets to the final drive gear set. The rotational torque and speed of this shaft varies with input speed and the operating gear. On RWD vehicles, the output shaft is connected to the rear drive axle by the drive shaft. The output shaft may be splined to any one member of the planetary gear set.

Some transaxles have additional shafts. These shafts are actually a continuation of the input and output shafts.

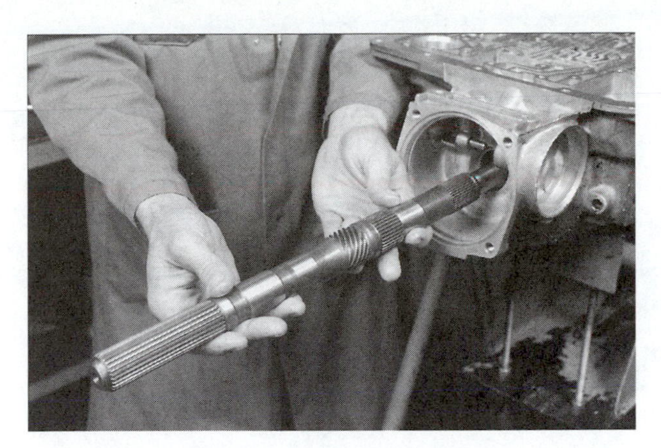

Figure 10. A typical output shaft removed from the rear of a transmission case.

Figure 12. A typical valve body.

They are placed in parallel where the rotating torque can be easily transferred from one shaft to another. The shafts are divided to keep the transaxle unit compact.

## Valve Body

For efficient operation of the transmission, the bands and multiple-disc packs must be released and applied at the proper time. It is the responsibility of the hydraulic control system to control the hydraulic pressure being sent to the different hydraulic members. Central to the hydraulic control system is the valve body assembly. This assembly is made of two or three main parts: a valve body, **separator plate**, and **transfer plate (Figure 11)**. These parts are bolted as a single unit to the transmission housing. The valve body **(Figure 12)** is machined from aluminum or iron and has many precisely machined bores and fluid passages. Various valves are fit into the bores and the passages direct fluid to various valves and other parts of the transmission. The separator and transfer plates are designed to seal off some of these passages and to allow fluid to flow through specific passages.

The purpose of a valve body is to sense and respond to engine and vehicle load, as well as to meet the needs of the driver. Valve bodies are normally fitted with three different types of valves: spool valves, check ball valves, and poppet valves. The purpose of these valves is start, stop, or use movable parts to regulate and direct the flow of fluid throughout the transmission.

**Check ball valve** The check ball valve is a ball that operates on a seat located on the valve body. The check ball operates by having a fluid pressure or manually operated linkage force it against the ball seat to block fluid flow **(Figure 13)**. Pressure on the opposite side unseats the check ball. Check balls and poppet valves can be normally open, which allows free flow of fluid pressure, or normally closed, which blocks fluid pressure flow.

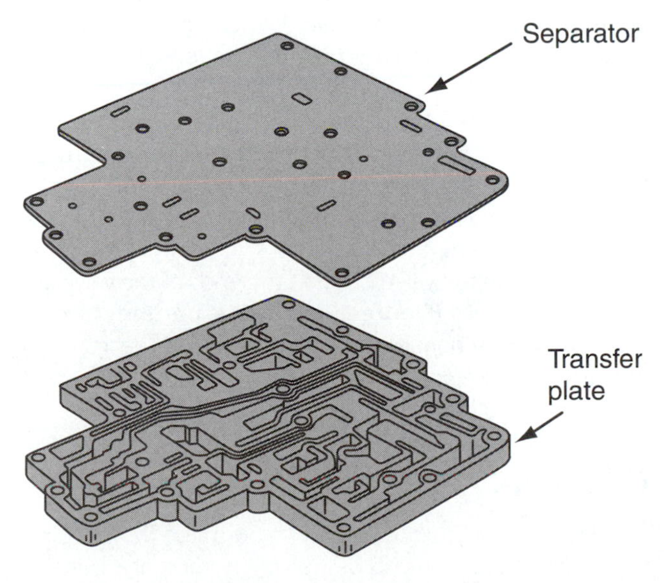


Figure 11. The separator and transfer plates of a valve body.

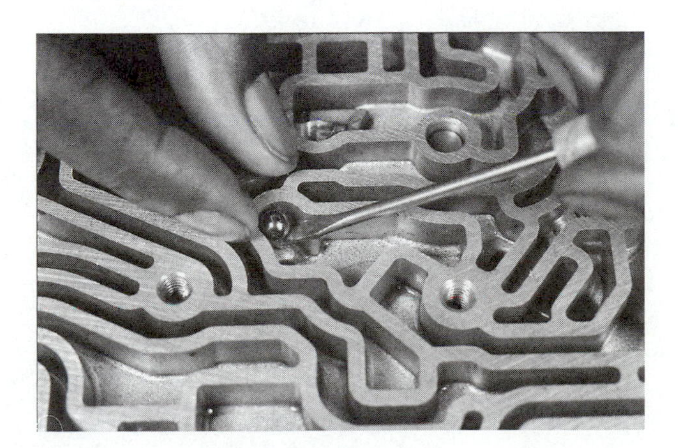

Figure 13. A check ball.

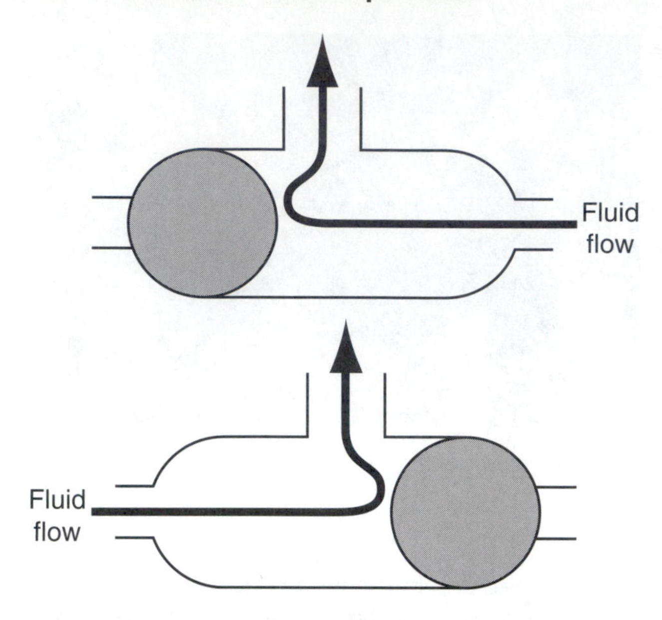


**Figure 14.** Fluid can be directed by the movement of a check ball.

At times, the check ball will have two seats to check and direct fluid flow from two directions: that is, seated and unseated by pressures from either source shown in **Figure 14**.

**Poppet valve** A poppet valve **(Figure 15)** can be a ball or a flat disc. In either case, the poppet valve acts to block fluid flow. Often the poppet valve has a stem to guide the valve's operation. The stem normally fits into a hole acting as a guide to the valve's opening and closing. Poppet valves tend to pop open and close, hence their name.

**Spool valve** The most commonly used valve in a valve body is the spool valve. A spool valve **(Figure 16)** looks similar to a sewing thread spool. The large circular parts of the valve are called the lands. There are a minimum of two lands per valve. Each land of the assembly is connected by a stem. The space between the lands and stem is called the

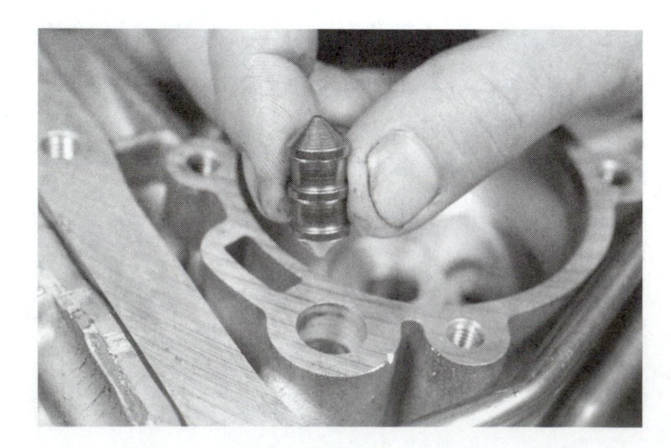

**Figure 15.** A poppet valve.

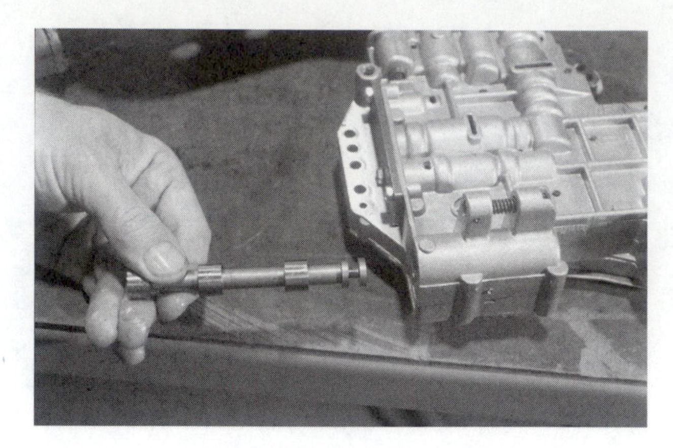

**Figure 16.** A manual valve, which is a spool valve controlled by the shift linkage.

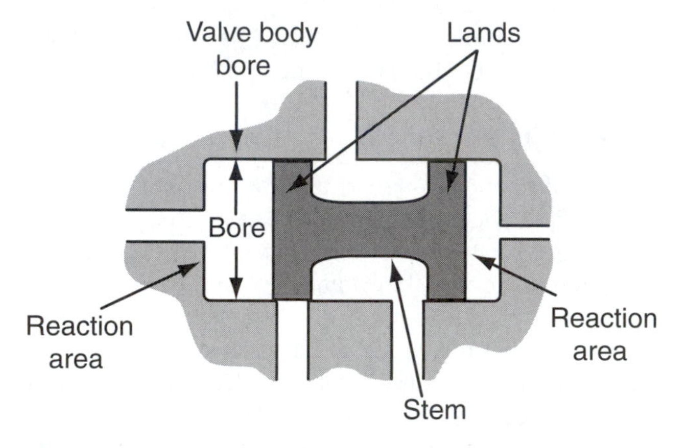


**Figure 17.** A spool valve.

valley. Valleys form a fluid pressure chamber between the lands and valve body bore. Fluid flow can be directed into other passages depending on the location of the spool valve and the design of the valve body.

The lands are precisely machined and ride on a very thin film of fluid in the valve body bore. The land must be treated very carefully because any damage, even a small score or scratch, can impair smooth valve operation. As the spool valve moves, the land covers (closes) or uncovers (opens) ports in the valve body. At the ends of the valve are the reaction areas **(Figure 17)**, which cause the valve to move within its bore. Forces acting against the reaction area to move the spool valve include spring tension, fluid pressure, or mechanical linkage.

## Oil Pump

The only source of fluid flow through the transmission is the oil pump. Three types of oil pumps are used in automatic transmission: the **gear-type**, **rotor-type (Figure 18)**, and **vane-type (Figure 19)**. Oil pumps are driven by

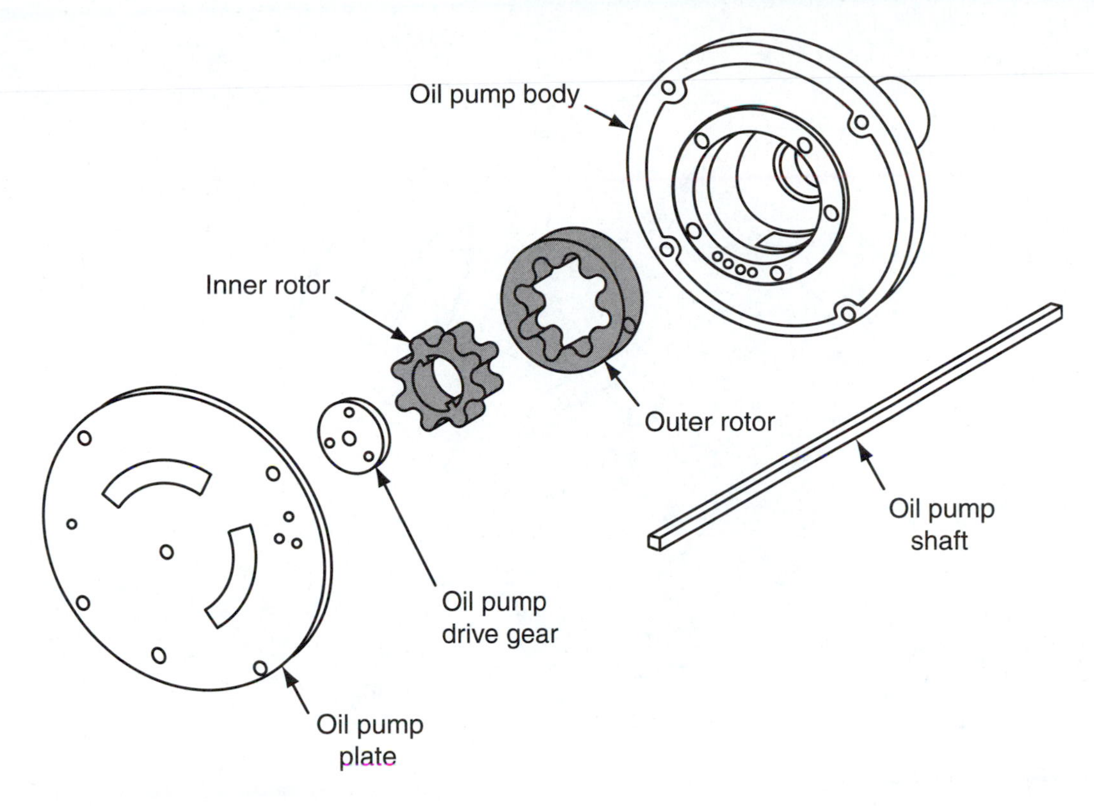


**Figure 18.** A typical fixed displacement, rotor-type oil pump.

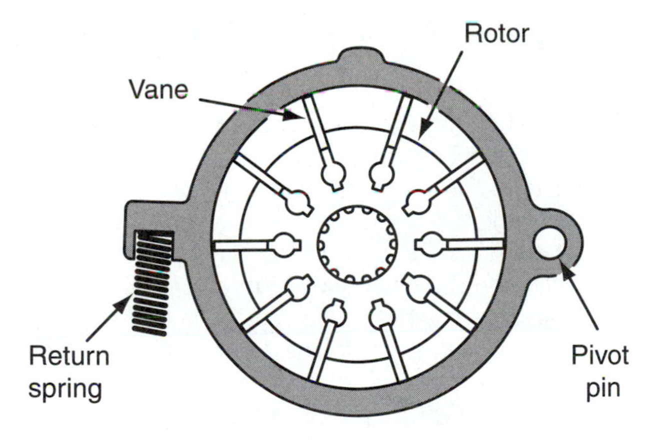


**Figure 19.** A typical variable displacement vane-type oil pump.

the pump drive hub of the torque converter or oil pump shaft and/or converter cover on transaxles. Therefore, whenever the torque converter cover is rotating, the oil pump is driven. The oil pump creates fluid flow throughout the transmission. The valve body regulates and directs the fluid flow to meet the needs of the transmission.

Transmissions are capable of creating excessive amounts of pressure, which may cause damage. Therefore, the transmission is equipped with a pressure regulator valve.

### Pressure Regulator Valve

The **pressure regulator valve** is located in the valve body or the pump. It maintains basic fluid pressure. Pressure regulating valves are typically spool valves that toggle back and forth in their bore to open and close an exhaust passage. By opening the exhaust passage, the valve decreases the pressure of the fluid. As soon as the pressure decreases to a predetermined amount, the spool valve moves to close off the exhaust port and pressure again begins to build. The action of the spool valve regulates the fluid pressure.

Many late-model transmissions use an electronic pressure control (EPC or PC) solenoid to regulate system pressure.

## TRANSMISSION/TRANSAXLE HOUSINGS

The basic shape and size of a transmission/transaxle reflects its design to some degree. Automatic transmission housings are typically aluminum castings. The torque converter and transmission housings are normally cast as a single unit; however, in some designs, the torque converter housing is a separate casting **(Figure 20)**. By bolting the torque converter housing to the transmission housing, engineers can use the same transmission with different sized torque converters to meet the specific needs of particular types of vehicles. Also, at the rear of some

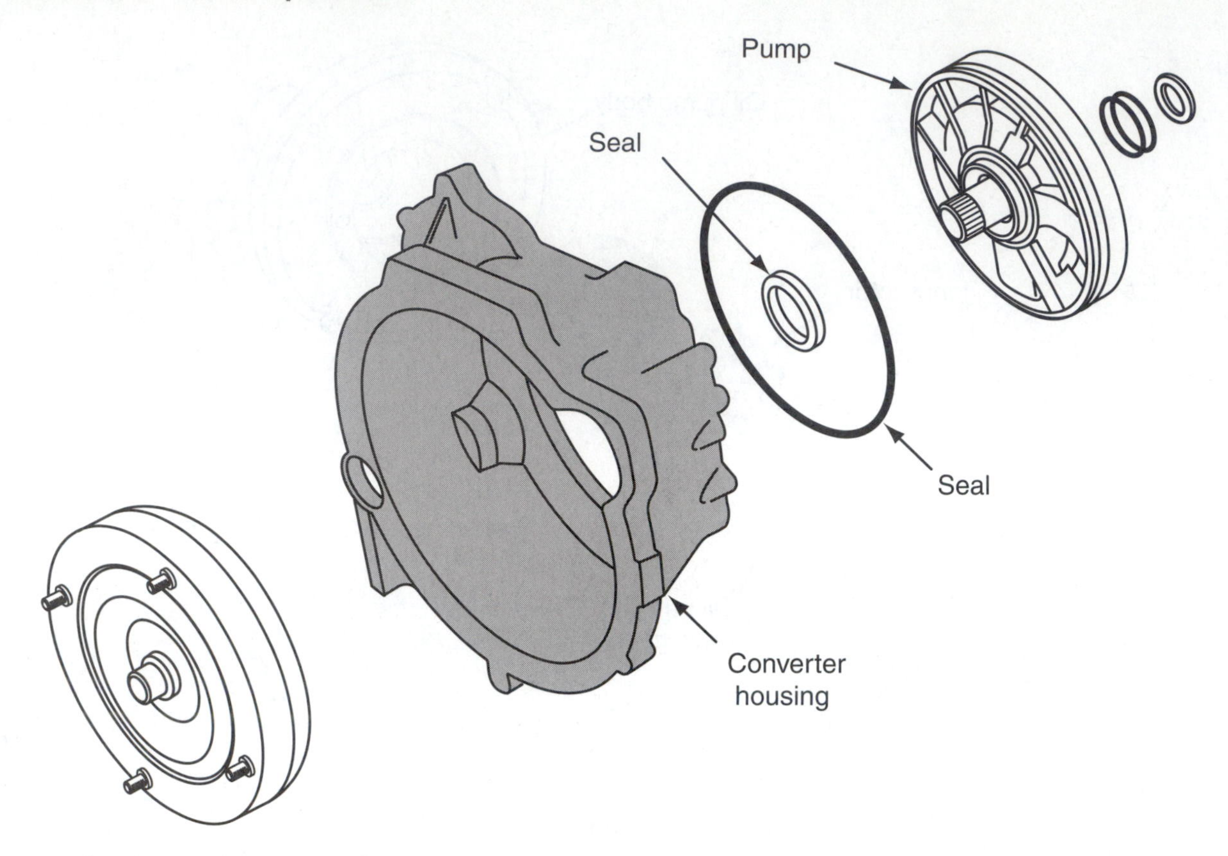


Figure 20. A converter housing with associated components separated from the transaxle.

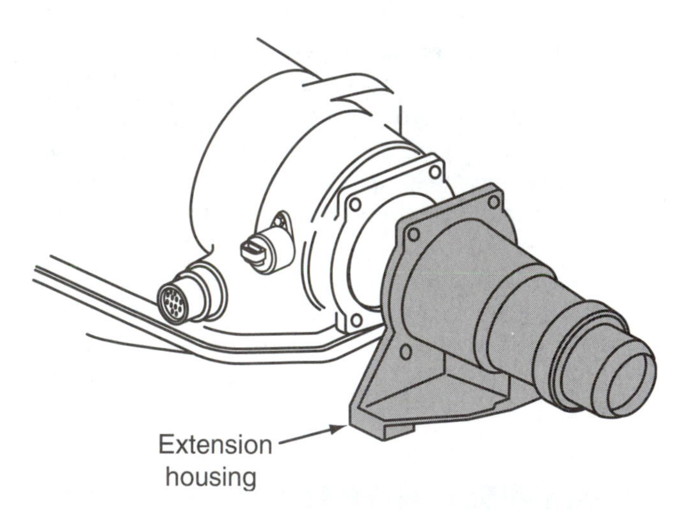


Figure 21. A transmission housing with a bolted-on extension housing.

transmission housings is a bolted-on extension housing **(Figure 21)**. This allows for the use of housings of different lengths; therefore, the transmission can be used on vehicles with different wheelbases.

Transaxle housings are typically comprised of two or three separate castings bolted together; one of these castings is the torque converter housing. Although they serve many of the same purposes as a transmission, transaxle housings are considerably different in appearance from transmission housings. Because FWD vehicles do not use a drive shaft to convey power to a drive axle, many transaxles are not fitted with an extension housing. Although they serve many of the same purposes as a transmission, transaxle housings are considerably different in appearance from transmission housings **(Figure 22)**.

The surfaces at which the sections of a transmission/transaxle mate are critical to the operation of the unit. Proper mounting surfaces are necessary to keep the various shafts aligned and to provide a good seal. It is important to remember that a poor seal not only will cause fluid leaks but also will allow dirt to enter into the unit. Dirt is a transmission's most feared enemy.

Transmission housings are cast to secure and accommodate the following components:

- Multiple friction disc assemblies—inside the housing are linear keyways designed to hold a multiple-disc clutch assembly in position in the housing.
- Fluid passages **(Figure 23)**.
- Threaded bores and/or studs to fasten mounts.
- Bores for the fluid level dipstick and filler tube **(Figure 24)**.
- Various threaded bores for a variety of sensors and switches **(Figure 25)**.
- Bores for gear selector attachment.

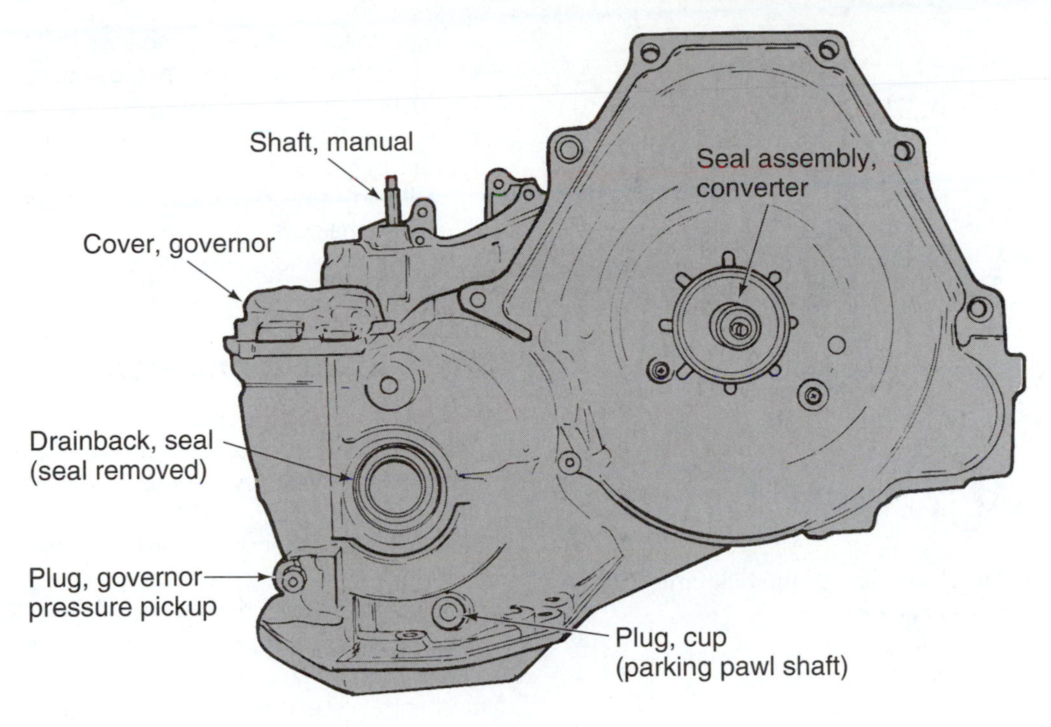


**Figure 22.** A side view of a transaxle housing. Note that the converter housing is an integral part of the housing.

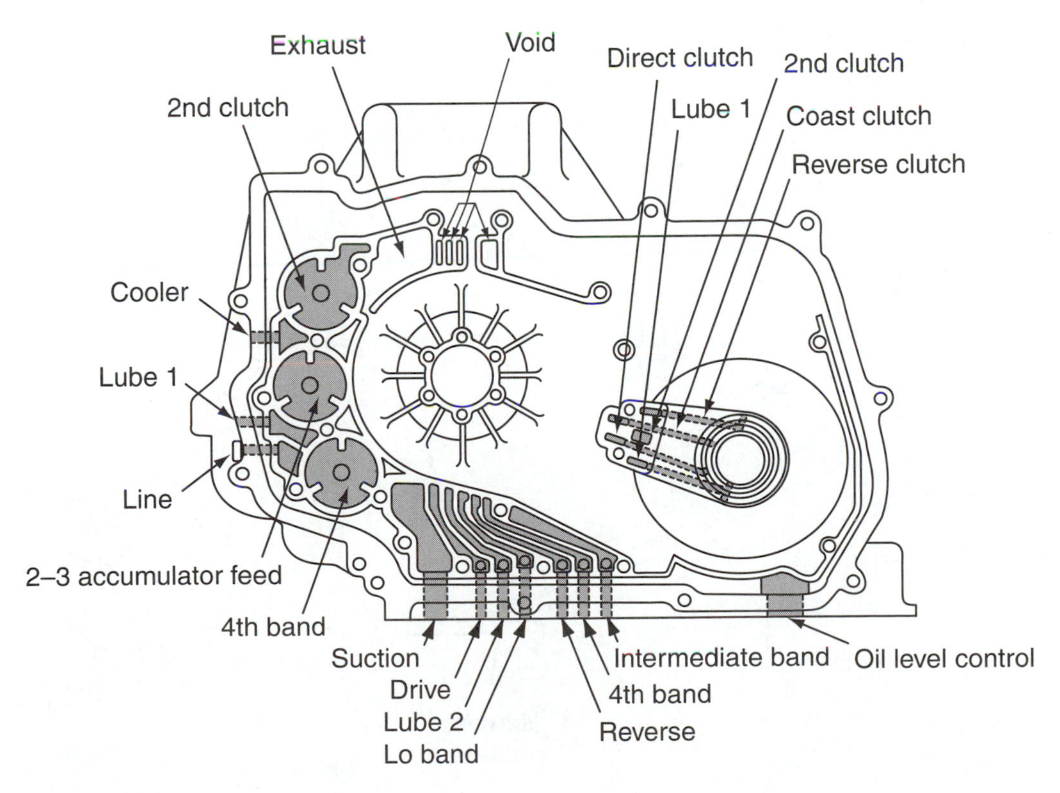


**Figure 23.** An example of the fluid passages cast into a transmission housing.

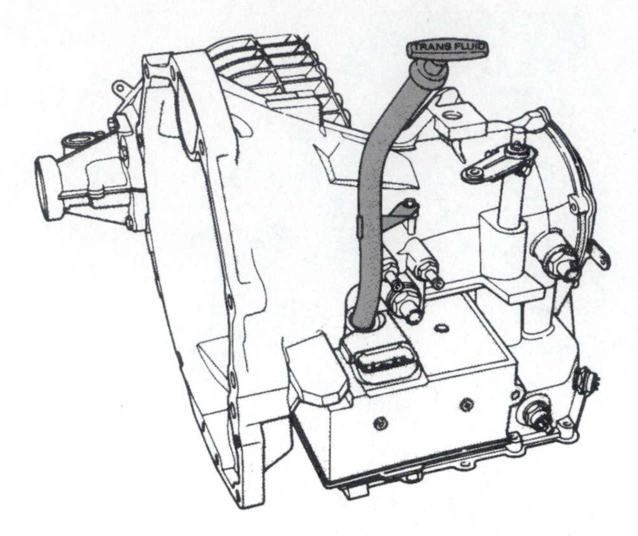


**Figure 24.** Location of oil level dipstick and filler tube on a transaxle.

- Fittings for the fluid cooling lines.
- Mounting points for the pump and valve body.
- Mounting for the oil pan.
- Round structures projecting from the side to serve as the cylinders for the servo and accumulator assemblies. In some designs, there also are projections to house the governor assembly.
- Some housings have bores for the band adjusting screws.

## Mechanical Connections

To monitor vehicle speed, vehicles are equipped with a speedometer. A speedometer can operate mechanically or electrically. On most new vehicles, it is an electrical device. A Vehicle Speed Sensor (VSS) is most commonly found mounted in the housing near the transmission's **output shaft**. The shaft is fitted with a trigger tooth. As the output shaft rotates, the trigger rotates past the VSS. The VSS then sends an electrical signal to processing unit. This unit then translates the signal into vehicle speed. Vehicles with a mechanical speedometer have a speedometer drive assembly geared to the output shaft. A flexible drive cable connected the drive assembly to the speedometer in the instrument panel.

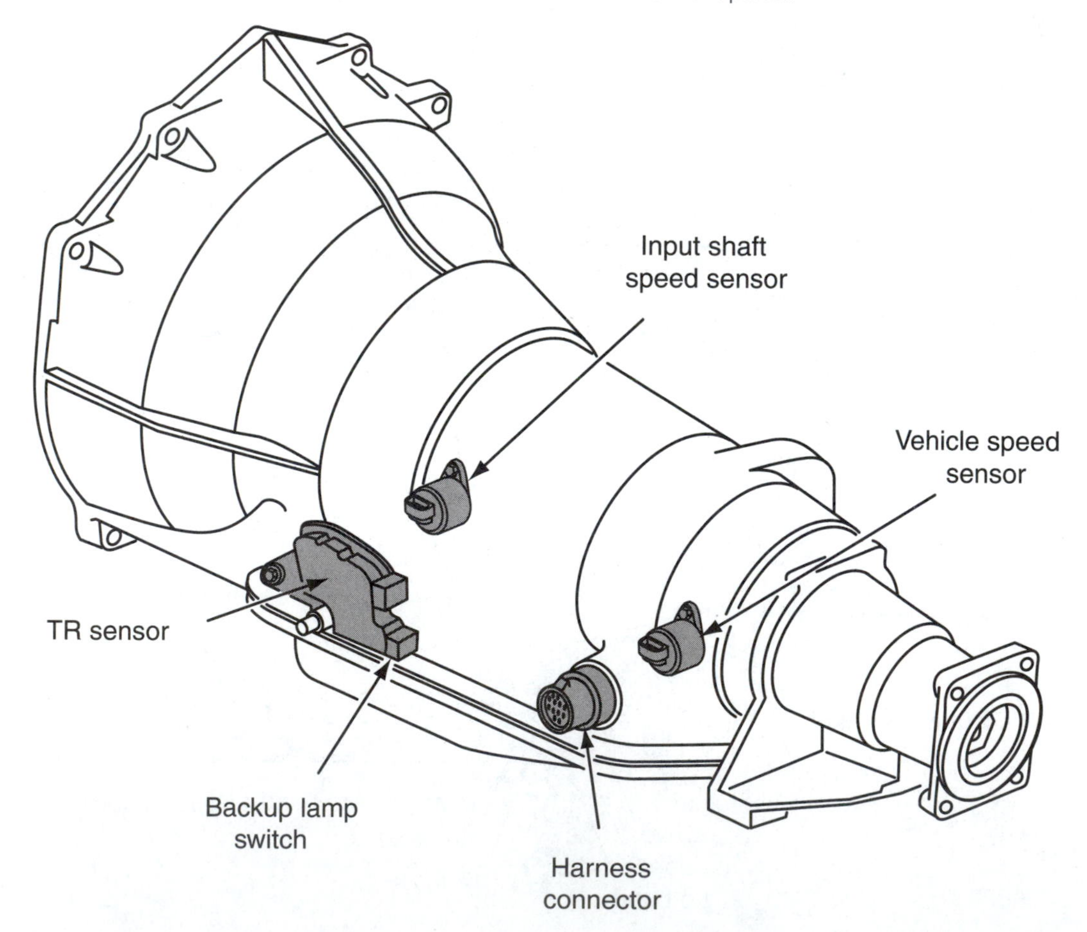


**Figure 25.** Various sensors and switches attached to a transmission housing.

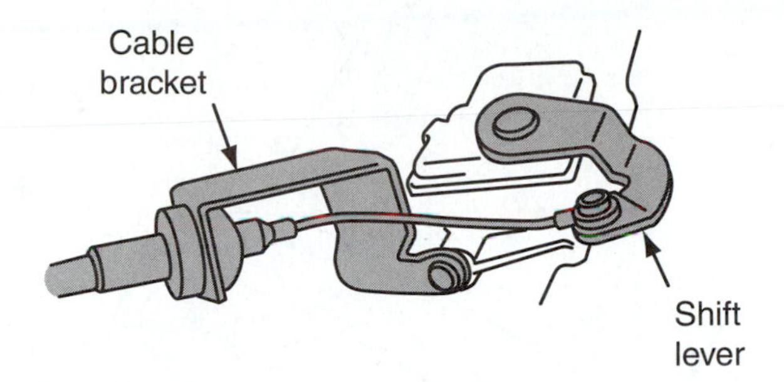


Figure 26. As the gearshift selector is moved into its various positions, the cable moves the manual shift lever at the transmission.

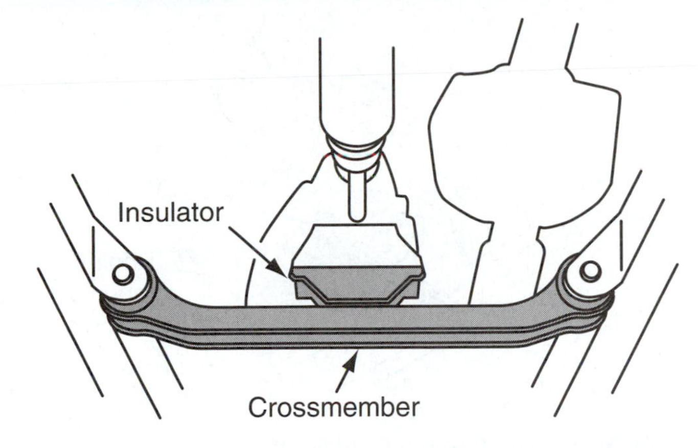


Figure 27. Typical rear transmission to frame mount.

## Linkages

In addition to the throttle valve (TV) linkage, there may be another cable or rod linkage connected to the housing. This could be the gear selector linkage. As the gearshift selector is moved into its various positions, the linkage moves the manual shift lever, which in turn moves the **manual shift valve** in the transmission's valve body **(Figure 26)**. Each gear position of the lever and/or shift valve is held in place by the internal linkage seating itself into the various seats or detents for the gear range selected.

## Electrical Connections

Many different electrical switches, sensors, and connectors may be connected to the housing. As more electronics are used to control and monitor the functioning of the transmission, more connectors and electrical devices will be found at the housing. One of the more basic switches is the neutral safety switch, which allows the engine to start only when the transmission is in neutral or park. The backup light switch is sometimes incorporated with the neutral safety switch, as are other shift position switches. These switches may be individual switch assemblies. Connectors for shift solenoids also may be present on the outside of the case.

# MOUNTS

The weight of the transmission or transaxle is supported by the engine and its mounts and by a transmission mount **(Figure 27)**. These mounts not only are critical for proper operation of the transmission but also isolate transmission noise and vibrations from the passenger compartment. The mountings for the engine and transmission keep the powertrain in proper alignment with the rest of the drive train and help to maintain proper adjustment of the various linkages attached to the housing.

An engine has vibrates or oscillates as it runs. Because the transmission is directly connected to the engine, those vibrations carry through to the transmission. Faulty mounts not only will allow these vibrations and the resulting noise to transfer into the vehicle but also can cause internal transmission problems. When an engine is mounted transversely, the inherent vibrations of the engine are easily transmitted to the vehicle's suspension and wheels. Also, most FWD vehicles use compact in-line four-cylinder or V6 engines that do not run as smoothly as larger engines. For these reasons, the manufacturers of FWD vehicles have developed many different mounting systems for their engines and transmissions.

The typical mount bolts to the transaxle and is connected to a plate or to the surface on the transaxle housing. A bolt passes through a rubber insulator and connects the mount to the transaxle. This type of mount is used on the side and toward the top of the engine/transaxle assembly. With this mount is a lower mount located between the subframe and the transaxle. These two mounts keep the assembly in place but do little to control noise and vibration; therefore, additional mounts are used.

Another common way to suppress vibrations and noise is the use of an engine mount strut. This strut limits the rocking motion of the engine and connects the top of the engine to the frame of the vehicle **(Figure 28)**. Again, the connecting points between the two parts of the mount are made through an insulator. Some vehicles use more than one of these struts or have an additional strut mounted to the side of the engine.

Some models have an upper engine mount that connects the lower front of the engine **(Figure 29)** to the frame of the vehicle and an upper engine mount of the rear of the assembly.

A few models use an adjustable mount that responds to vibrations **(Figure 30)**. This type system relies on the action of a solenoid on a hydraulic mount. The solenoid responds to the commands from the PCM and decreases and increases the fluid pressure at the mount.

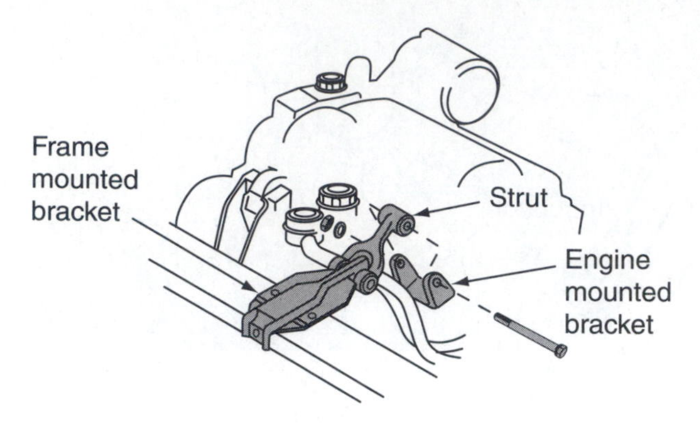


Figure 28. An engine mount strut.

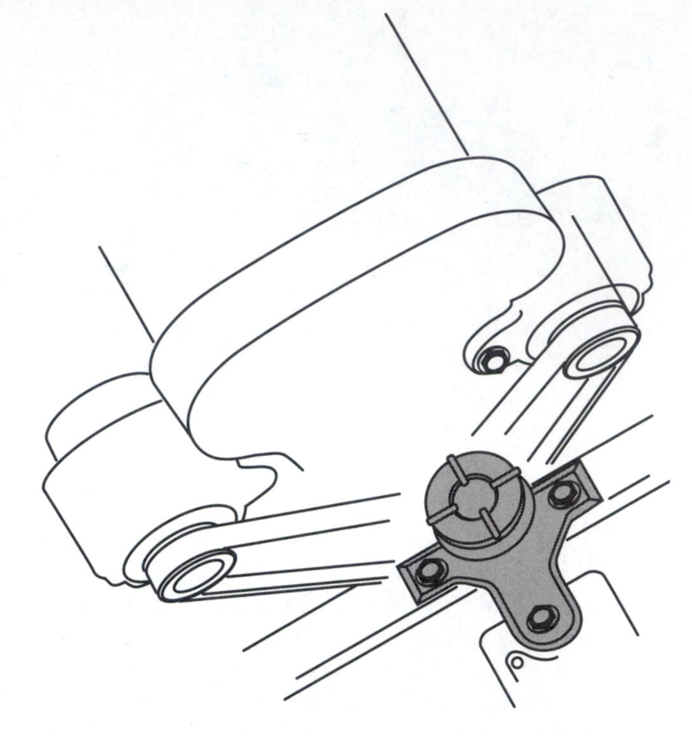

Figure 29. A lower engine mount.

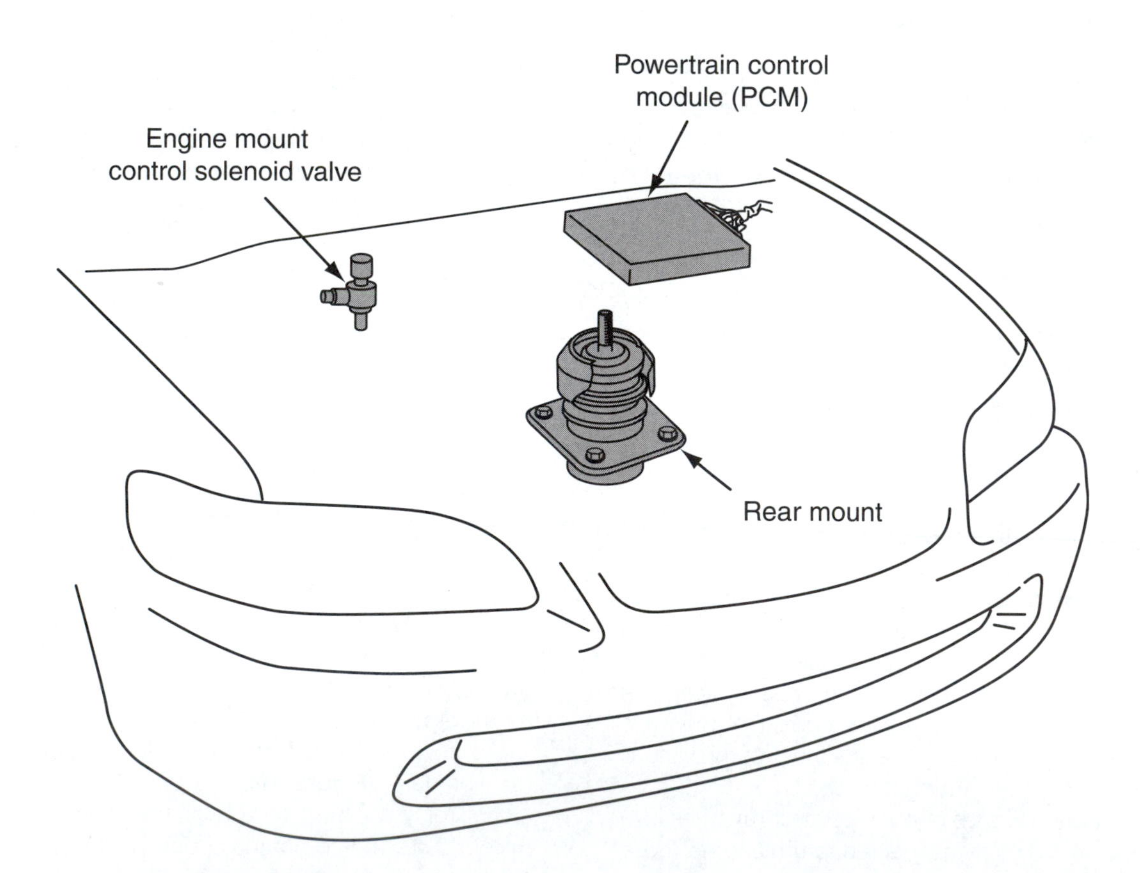


Figure 30. An electronically controlled engine mount.

## Summary

- An automatic transmission or transaxle selects gear ratios according to engine speed, engine load, vehicle speed, and other operating conditions.
- The torque converter automatically engages and disengages power from the engine to the transmission in relation to engine rpm.
- The main components of a torque converter are the cover, turbine, impeller, and stator. The cover is connected to the flywheel and transmits power from the engine to the impeller, which allows for torque multiplication and drives the transmission's oil pump. The turbine is splined to the input shaft and is driven by the force of the fluid from the impeller. The stator aids in directing the flow of fluid from the impeller to the turbine.
- Power flow through the gears is controlled by multiple friction disc packs, one-way clutches, and/or friction bands. These hold a member of the gear set when force is applied to them, thereby making different gear ratios possible.
- Many transmissions rely on information sent from various electronic sensors and switches to a computer that controls the engagement of the torque converter clutch and the transmission's shift points.
- A band is a braking assembly positioned around a stationary or rotating drum that is connected to a member of the planetary gearset.
- A servo assembly hydraulically applies the band. It converts hydraulic pressure into a mechanical force that applies a band to hold a drum stationary.
- In an automatic transmission operation, both sprag and roller overrunning clutches are used to hold or drive members of the planetary gear set. These clutches operate mechanically.
- A multiple-disc pack uses a series of friction discs to transmit torque or apply braking force.
- Multiple-disc clutches are enclosed in a large drum-shaped housing that can be either a separate casting or part of the existing transmission housing.
- Central to the hydraulic control system is the valve body assembly, which is typically made of two or three main parts: a valve body, separator plate, and transfer plate.
- Valve bodies are normally fitted with three different types of valves: spool valves, check ball valves, and poppet valves. The purpose of these valves is start, stop, or use movable parts to regulate and direct the flow of fluid throughout the transmission.
- The transmission's oil pump provides a constant supply of oil under pressure to operate, lubricate, and cool the transmission.
- Three types of oil pumps are used in automatic transmission: the gear-type, rotor-type, and vane-type.
- The pressure regulator valve is located in the valve body or front pump and maintains basic fluid pressure.
- Transmission housings are cast to secure and accommodate many important components and assemblies.
- The weight of the transmission or transaxle is supported by the engine and its mounts and by a transmission mount.
- Transmission mounts also isolate transmission noise and vibrations from the passenger compartment.
- Proper mounting surfaces of a transmission/transaxle are critical in order to keep the various shafts aligned and to provide a good seal.

## Review Questions

1. Transmission housings are cast to secure and accommodate many important components. List at least five of them.
2. What are the major components of all automatic transmissions?
3. In an automatic transmission operation, both ____________ and ______________ overrunning clutches are used to hold or drive members of the planetary gearset.
4. Briefly explain how a torque converter works.
5. What are the primary purposes of the valves in the valve body?
6. What are purposes of the reaction members of a transmission?
7. Power flow through the gears is controlled by ________-______ ___________, ___-___ ________, and/or ______________, which hold a member of the gear set when hydraulic pressure is applied to them.
8. Technician A says that the torque converter's impeller is driven by the converter cover and its rear hub drives the transmission's oil pump. Technician B says that the impeller is splined to the input shaft, and is driven by the force of the fluid from the turbine. Who is correct?
   A. Technician A only
   B. Technician B only
   C. Both Technician A and Technician B
   D. Neither Technician A nor Technician B
9. Technician A says that oil pumps are driven by the pump drive hub of the torque converter. Technician B says that the oil pump is indirectly driven by the torque converter's stator. Who is correct?

A. Technician A only
B. Technician B only
C. Both Technician A and Technician B
D. Neither Technician A nor Technician B

10. Technician A says that engine/transaxle mounts isolate the inherent vibrations of the assembly from the passenger compartment. Technician B says that improper mounting of an engine and/or transmission can cause problems with the transmission. Who is correct?
A. Technician A only
B. Technician B only
C. Both Technician A and Technician B
D. Neither Technician A nor Technician B

# Chapter 10 Torque Converter Operation

## *Introduction*

A torque converter uses fluid to smoothly transfer engine torque to the transmission. The torque converter is a doughnut-shaped unit, located between the engine and the transmission and filled with ATF **(Figure 1)**. Internally, the torque converter has three main parts: the impeller, turbine, and stator **(Figure 2)**. Each of these has blades, which are curved to increase torque converter efficiency.

The impeller is driven by the engine and directs fluid flow against the turbine blades, causing them to rotate and drive the turbine shaft, which is the transmission's input shaft. The stator is located between the impeller and the turbine and returns fluid from the turbine to the impeller, so that the cycle can be repeated.

During certain operating conditions, the torque converter multiplies torque. It provides extra reduction to meet the driveline needs while under a heavy load. When the vehicle is operating at cruising speeds, the torque converter operates as a **fluid coupling** and transfers engine torque to the transmission. It also absorbs the shock from gear changing in the transmission. Not all of the engine's power is transferred through the fluid to the transmission, some is lost. To reduce the amount of power lost through the converter, especially at cruising speeds, manufacturers equip most of their current transmissions with a torque converter clutch.

The engagement of the converter clutch is based on both engine and vehicle speeds and the clutch are controlled by transmission hydraulics and on-board computer electronic controls. When the clutch engages, a mechanical

**Figure 1.** A typical torque converter.

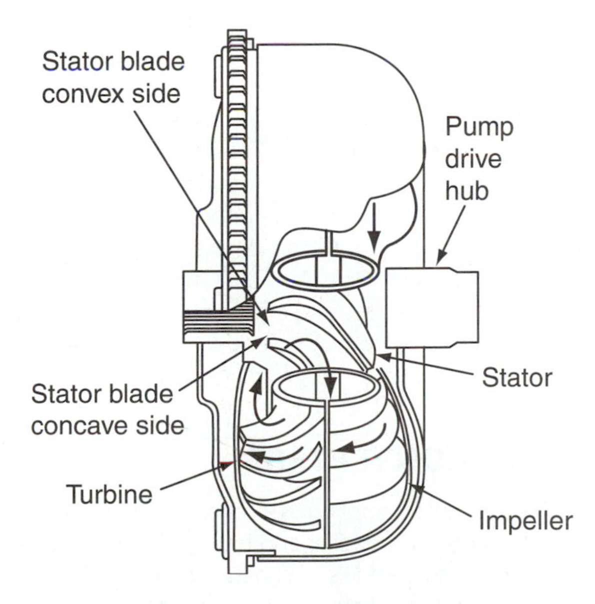


**Figure 2.** A torque converter's major internal parts are its impeller, turbine, and stator.

connection exists between the engine and the drive wheels. This improves overall efficiency and fuel economy.

## CVT Units

All automatic transmissions, except some designs of the Continously Variable Transmission (CVT), use a torque converter to transfer power from the engine to the transmission. The CVT is a compact transmission that uses belts and pulleys to match the transmission's torque multiplication with the needs of the vehicle. These transmissions do not have fixed gear ratios; rather, the size of the drive and driven pulleys changes in accordance to engine speed and load. This provides for constantly changing torque multiplication factors.

The same basic design of the CVT transmission is used by different manufacturers. The primary design difference among the manufacturers relates to the method used to transmit engine torque to the transmission. Some manufacturers use a torque converter; however, some do not.

Late model Hondas are available with a CVT; these vehicles are not equipped with a torque converter. The transaxles have an internal start clutch that allows the vehicle maintain an idle speed while it is in gear.

To transfer torque from the engine to the transmission, an electromagnetic clutch is used. Its operation is controlled by a clutch control unit. The clutch control unit switches current to energize and deenergize the electromagnetic clutch in response to inputs from various sensors. These sensors include a brake switch, accelerator pedal switches, and an inhibitor switch. The brake switch deenergizes the clutch when the vehicle is slowing or coming to a stop. The control unit uses inputs from the accelerator pedal switches to vary the amount of current to the clutch and to signal ratio changes to the adjustable pulleys inside the transmission. The inhibitor switch prevents clutch engagement when the gear selector is in the P or N position.

When the electromagnetic clutch is energized, engine torque is transferred to the transmission's input shaft and the drive pulley. At the end of the transmission's input shaft is the transmission's oil pump; therefore, the clutch control unit has indirect control of the oil pressure in the transmission. Pressure from the pump is used to control the **sheaves** of both the drive and driven pulleys. Line pressure is controlled by a three-way solenoid in the transmission valve body.

## TORQUE CONVERTERS

A simple fluid coupling is comprised of three basic members: a housing, impeller, and turbine. The impeller and the turbine are shaped like two halves of a doughnut. The impeller and turbine have internal vanes radiating from their centers. Both the **impeller** and **turbine** are enclosed in the housing. ATF is forced into the housing by the transmission's oil pump. The housing is sealed and filled with fluid. The impeller is driven by the engine. The impeller acts like a pump and moves the fluid in the direction of its rotation. The moving fluid hits against the turbine vanes. This causes the turbine to rotate, which brings an input into the transmission via the input shaft.

When the transmission is in gear, the fluid coupling allows the vehicle to stop without stalling the engine because the impeller does not rotate fast enough to drive the turbine. As engine speed increases, the force from the fluid flow off the impeller increases and forces the turbine to rotate. Based on the fluid flow, a fluid coupling can transmit very little to very much force. Maximum efficiency is approximately 90 percent, but that figure depends upon a number of factors: fluid type, impeller and turbine vane design, vehicle load, and engine torque.

*The use of fluid couplings actually began in the 1900s with steamship propulsion and later was used to help dampen the vibrations of large diesel engines. Just before World War II, Chrysler was the first American automobile manufacturer to use a fluid coupling. The fluid coupling was added to a manual transmission driveline, which allowed the engine to idle in gear, but a foot-operated clutch was still needed to shift gears.*

**Fluid Flow** The operation of a fluid coupling is totally based on the flow of fluid inside the coupling. Two types of fluid flow take place inside the fluid coupling: rotary and vortex flow **(Figure 3)**. These different flows complement each other, depending on the difference in speed between the impeller and the turbine. **Rotary flow** is the movement of the fluid in the direction of impeller rotation and results from the paddle action of the impeller vanes against the fluid. As this rotating fluid hits against the blades of the slower turning or stationary turbine, it exerts a turning force onto the turbine.

**Vortex flow** is the fluid flow circulating between the impeller and turbine as the fluid moves from the impeller to the turbine and back to the impeller. This type fluid flow is only present when there is a difference in rotational speeds between the impeller and the turbine.

Both types of flow can occur within a fluid coupling at the same time. When the impeller is rotating fast but turbine speed is restricted because of a load, most of the fluid within the housing moves with a vortex flow. However, as the load is overcome and turbine speed increases, more rotary flow is occurring.

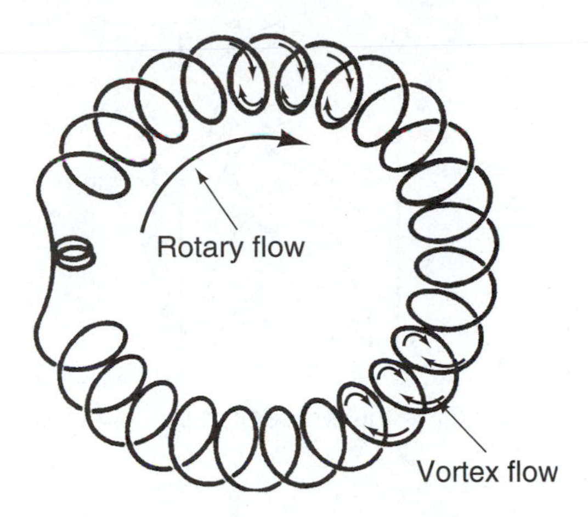


Figure 3. Difference between rotary and vortex flow. Note that vortex flow spirals its way around the converter.

*Variations of the basic three-element torque converter have been used. The Buick Dynaflow had two impellers, two stators, and one turbine in 1948. In 1953, the Twin-Turbine Dynaflow was released with two turbines, one impeller, and one stator. In 1956, Buick, again, introduced a multiple-turbine torque converter, which had a variable pitch stator. By the late 1960s, the industry, including Buick, had returned to the basic three-element converter.*

## Basic Torque Converters

*Interesting Fact* *The most important improvement to automatic transmissions was the result of Buick's development of a torque converter for tanks during the war. The converter was incorporated into the Dynaflow transmission in 1948, and eventually into other automatics.*

A torque converter is a type of fluid coupling or connector that connects the engine's crankshaft to the transmission's input shaft. The torque converter, working as a fluid connector, smoothly transmits engine torque to the transmission. The torque converter allows some slippage between the engine and the transmission so that the engine will remain running when the vehicle is stopped and is in gear. A torque converter also multiplies torque when the vehicle is under load to improve performance. Torque converters are more efficient than fluid couplings because they allow for increased fluid flow, which multiplies the torque from the engine when it is operating at low speeds.

## Torque Converter Construction

A typical torque converter consists of three elements sealed in a single housing: the impeller, the turbine, and the stator. The impeller is the drive member of the unit and its fins are attached directly to the converter cover. Therefore, the impeller is the input device for the converter and it always rotates at engine speed.

The turbine is the converter's output member and is coupled to the transmission's input shaft **(Figure 4)**. The turbine is driven by the fluid flow from the impeller and always turns at its own speed. The fins of the turbine face

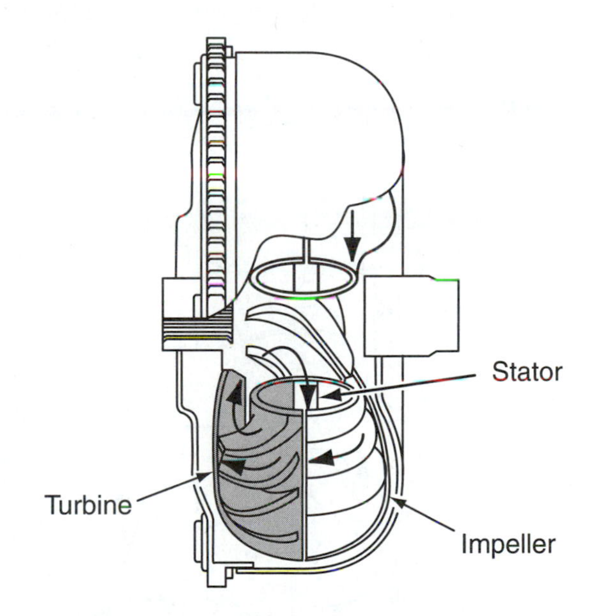


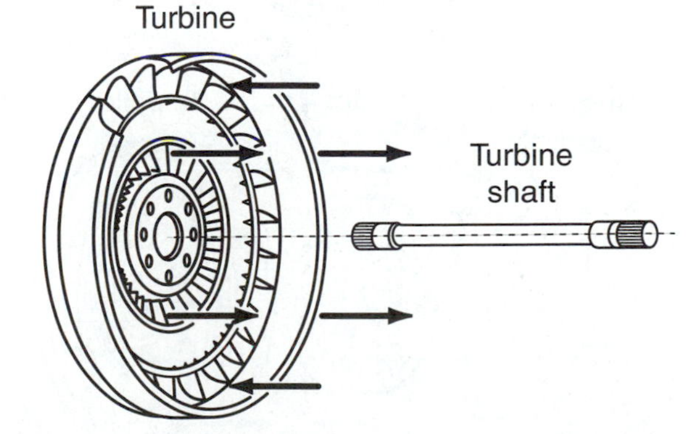


Figure 4. Location of turbine. Note that the turbine shaft extends from the turbine to the transmission and serves as the input shaft of the transmission.

toward the fins of the impeller. The impeller and the turbine have internal fins, but the fins point in opposite directions.

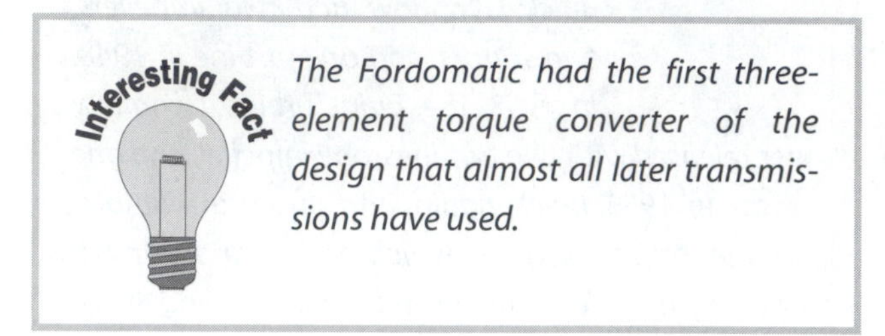


*The Fordomatic had the first three-element torque converter of the design that almost all later transmissions have used.*

The **stator** is the reaction member of the converter **(Figure 5)**. This assembly is about one-half the diameter of the impeller or turbine and is positioned between the impeller and turbine. The stator is not mechanically connected to either the impeller or turbine; rather it fits between the turbine outlet and the inlet of the impeller. All of the fluid returning from the turbine to the impeller must pass through the stator. The stator redirects the fluid, leaving

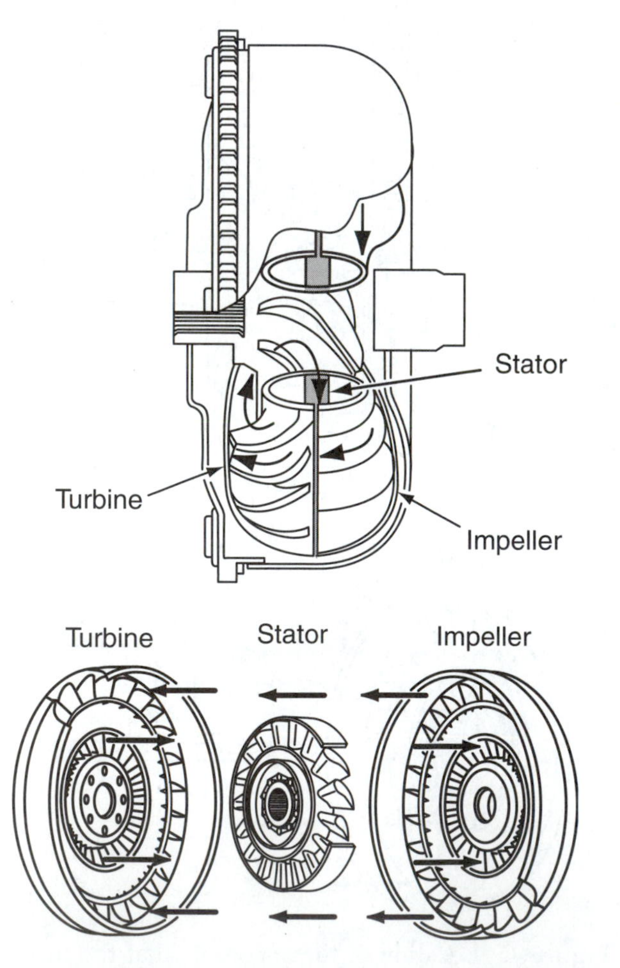


**Figure 5.** Location of stator assembly. Note that the stator is not attached to the turbine or the impeller.

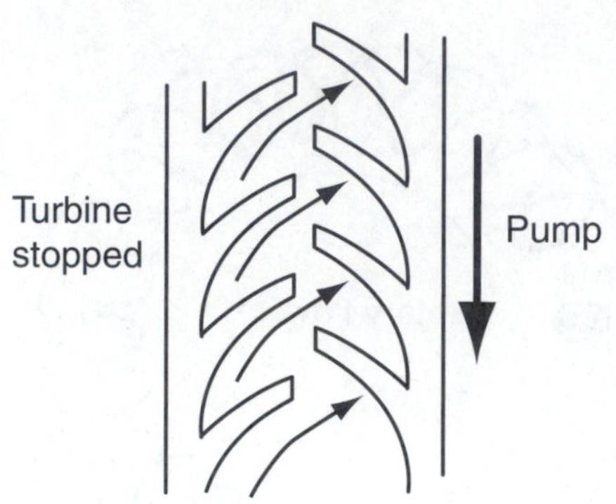


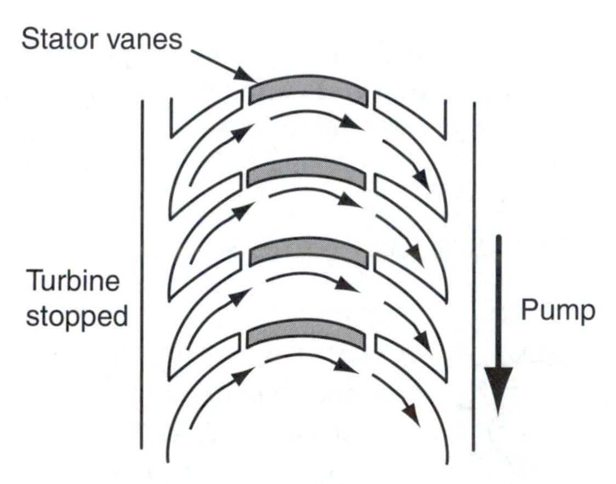


**Figure 6.** (A) Without a stator, fluid leaving the turbine works against the direction in which the impeller is rotating. (B) With a stator in its lock mode, fluid is directed to help push the impeller in its rotating direction.

the turbine, back to the impeller **(Figure 6)**. By redirecting the fluid so that it is flowing in the same direction as engine rotation, it allows the impeller to rotate more efficiently.

## Torque Converter Mountings

On all RWD and most FWD vehicles, the torque converter is mounted inline with the transmission's input shaft. Some transaxles use a drive chain to connect the converter's output shaft with the transmission's input shaft **(Figure 7)**. Offsetting the input shaft from the centerline of the transmission gives the manufacturers more flexibility in the positioning of the transaxle.

The torque converter fits between the transmission and the engine and is supported by the transmission housing's oil pump bushing and the engine's crankshaft. The rear hub of the converter's cover fits into the transmission's pump. The hub or a connecting shaft drives the pump whenever the engine is running.

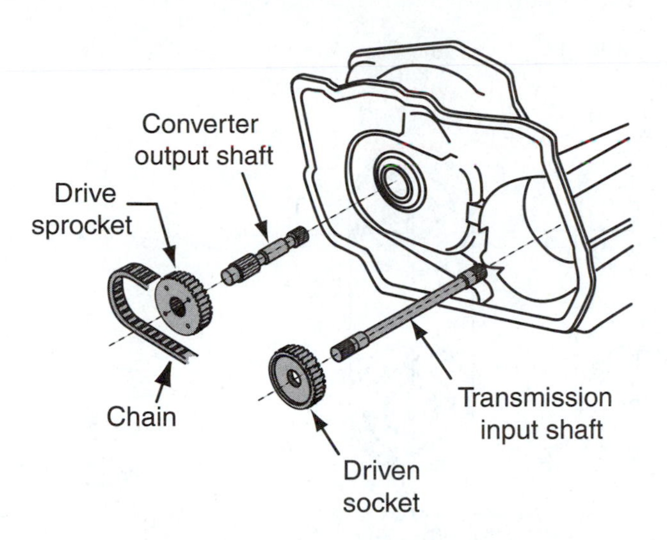


**Figure 7.** Typical drive chain setup for offsetting the torque converter and the input shaft.

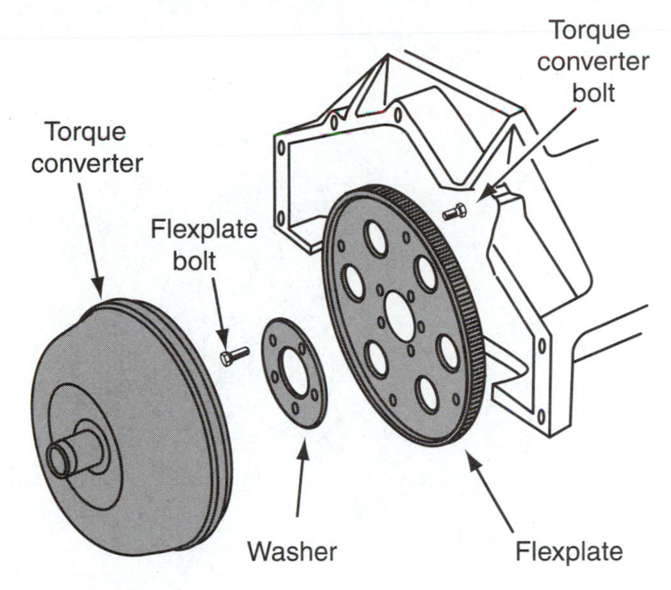


**Figure 8.** A torque converter is bolted to the flexplate, which is bolted to the engine's crankshaft.

The front of a torque converter is mounted to a flexplate, which is bolted to the end of the engine's crankshaft **(Figure 8)**. Normally, the converter cover is fitted with studs that are used to tighten the cover to the flexplate. The flexplate is designed to be flexible enough to allow the front of the converter to move forward or backward if it expands or contracts because of heat or pressure. The centerline of the converter and crankshaft are matched and this alignment is maintained by a pilot on the converter cover that fits into a recess in the crankshaft.

An externally toothed ring gear is normally pressed or welded to the outside diameter of the flexplate. The starter motor's drive gear meshes with the ring gear to crank the engine for starting.

The transmission's input shaft is supported by bushings in the stator support inside the torque converter **(Figure 9)**. There is no mechanical link between the output of the engine and the transmission. The fluid connects the power from the engine to the transmission. The combined weight of the fluid, torque converter, and flexplate serve as the flywheel for the engine.

## TORQUE CONVERTER OPERATION

The impeller rotates at engine speed whenever the engine is running. The impeller is comprised of many curved vanes radiating out of an inner ring, which form fluid passages for the fluid. As soon as the converter begins to rotate, the vanes of the impeller begin to circulate fluid.

The fluid in the torque converter is supplied by the transmission's oil pump and enters through the converter's hub, then flows into the passages between the vanes. As the impeller rotates, the fluid is moved outward and upward through the vanes by centrifugal force because of the curved shape of the impeller. The faster the impeller rotates, the greater the centrifugal force becomes.

The fluid moves from the outer edge of the vanes into the turbine. As the fluid strikes the curved vanes of the turbine, it attempts to push the turbine into a rotation. Because the impeller is turning in a clockwise direction, the fluid also rotates in a clockwise direction as it leaves the vanes of the impeller. However, the turbine vanes are

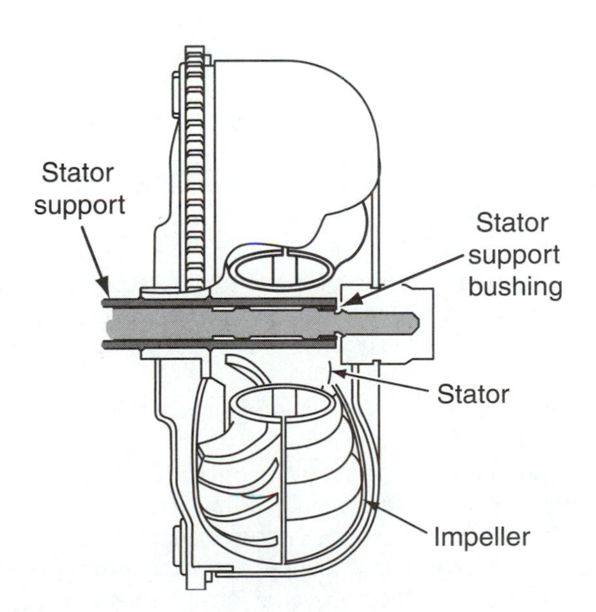


**Figure 9.** The transmission's input shaft is supported by the stator support inside the torque converter.

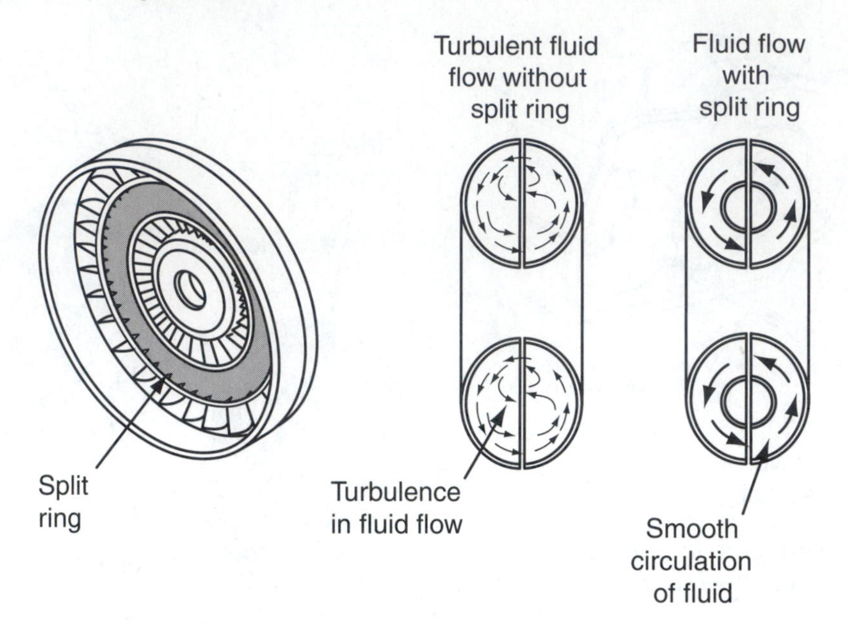


**Figure 10.** Location and action of the split guide rings.

curved in an opposite direction as the impeller and the fluid turns the turbine in the same direction as the impeller.

The higher the engine speed, the faster the impeller turns, and more force is transferred from the impeller to the turbine by the fluid. This explains why a torque converter allows the engine to idle in gear. When engine speed is low, the fluid does not have enough force to turn the turbine against the load on the drive train. The movement of the fluid in the converter is very weak; it just circulates from the impeller to the turbine and back to the impeller. Therefore, little, if any, power is transmitted through the torque converter to the transmission.

As engine speed increases, the fluid is thrown at the turbine with a greater force causing the turbine to begin to rotate. Once the turbine begins to turn, engine power is transmitted to the transmission. However, the force from the fluid must be great enough to overcome the load of the vehicle before the turbine can rotate. Some of the energy in the moving fluid returns to the impeller as the torque converter responds to the torque requirements of the vehicle and to vortex flow. At low speeds, most of the energy in the fluid is lost as the fluid bounces back away from the turbine vanes.

Vortex flow is a continuous circulation of the fluid, outward in the impeller and inward in the turbine, around the **split guide rings** attached to the turbine and the impeller **(Figure 10)**. The guide rings direct the vortex flow to provide for a smooth and turbulence-free fluid flow.

As the vortex flow continues, the fluid leaving the turbine to return to the impeller is moving in the opposite direction as crankshaft rotation. If the fluid were allowed to continue in this direction, it would enter the impeller as an opposing force and some of the engine's power would be used to redirect the flow of fluid. To prevent this loss of power, torque converters are fitted with a stator.

## Torque Multiplication

The stator receives the fluid thrown off by the turbine and redirects the fluid so that it reenters the impeller in the same direction as crankshaft rotation **(Figure 11)**. The redirection of the fluid by the stator not only prevents a torque loss but also provides for a multiplication of torque.

The stator is attached to a circular hub, which is mounted on a one-way clutch **(Figure 12)**. This clutch

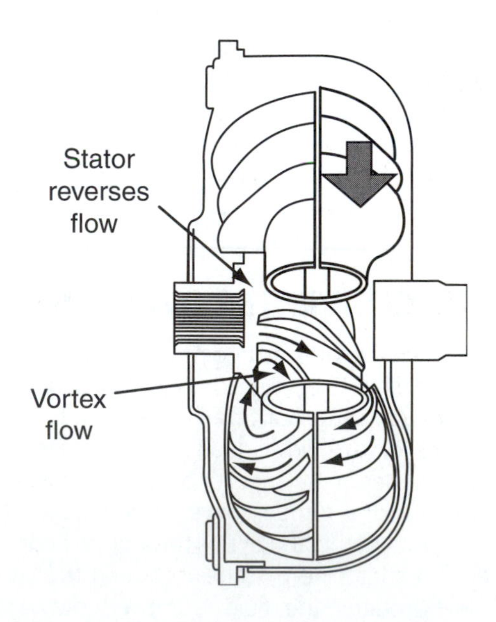


**Figure 11.** Action of the stator on fluid flow.

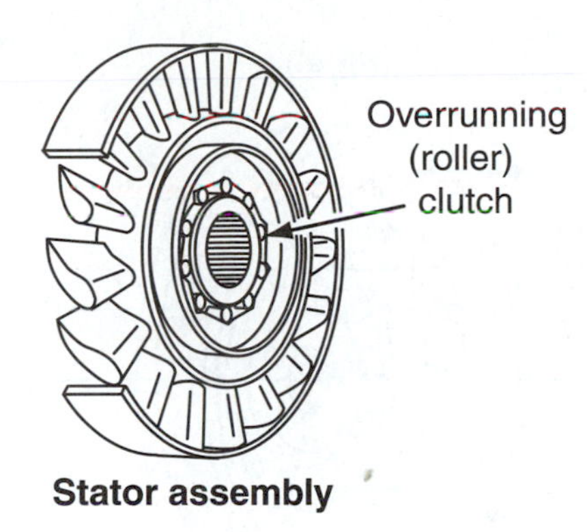


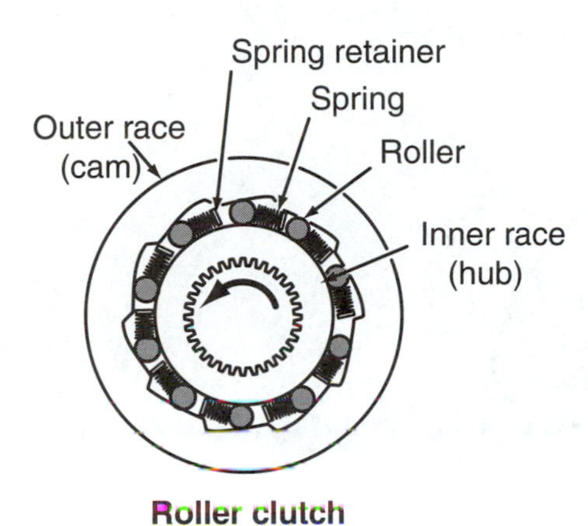


**Figure 12.** The one-way overrunning clutch in a stator assembly. (Courtesy of General Motors Corporation, Service Operations)

assembly has an inner and outer race separated by spring-loaded roller bearings or sprags. The inner race is splined to the stator support; therefore, it cannot turn. The outer race is fitted into the stator. The rollers or sprags are fitted between the two races and will allow the outer race to rotate in one direction only. The stator locks in the opposite direction of turbine rotation. When the stator is turned in the same direction as turbine rotation, the rollers are free between the races and the stator is able to turn **(Figure 13)**.

The fluid leaving the turbine has to pass through the stator blades before reaching the impeller. In passing through the stator, the direction of fluid flow is reversed by the curvature of the stator blade. The fluid now moves in a direction that aids in the rotation of the impeller. The impeller accelerates the movement of the fluid and it now leaves with nearly twice the energy and exerts a greater force on the turbine. This action results in a torque multiplication.

It is vortex flow that allows for torque multiplication. Torque multiplication occurs when there is high impeller speed and low turbine speed. Low turbine speed and the stator cause the returning fluid to have a high velocity vortex flow. This allows the impeller to rotate more efficiently and increase the force of the fluid pushing the turbine in rotation. When the vehicle's torque requirements become greater than the output of the engine, the turbine slows down and causes an increase in vortex flow velocity. This causes an increase in torque multiplication.

As the vortex flow slows down, torque multiplication is reduced. Torque multiplication is obtained anytime the turbine turns at less than 90 percent of the impeller speed. Most automotive torque converters are capable of maximum torque multiplication factors ranging from 1.7:1 to 2.8:1.

## Coupling Phase

Torque multiplication occurs because of the redirection of the fluid flow by the stator. This takes place only when the impeller is rotating faster than the turbine. As

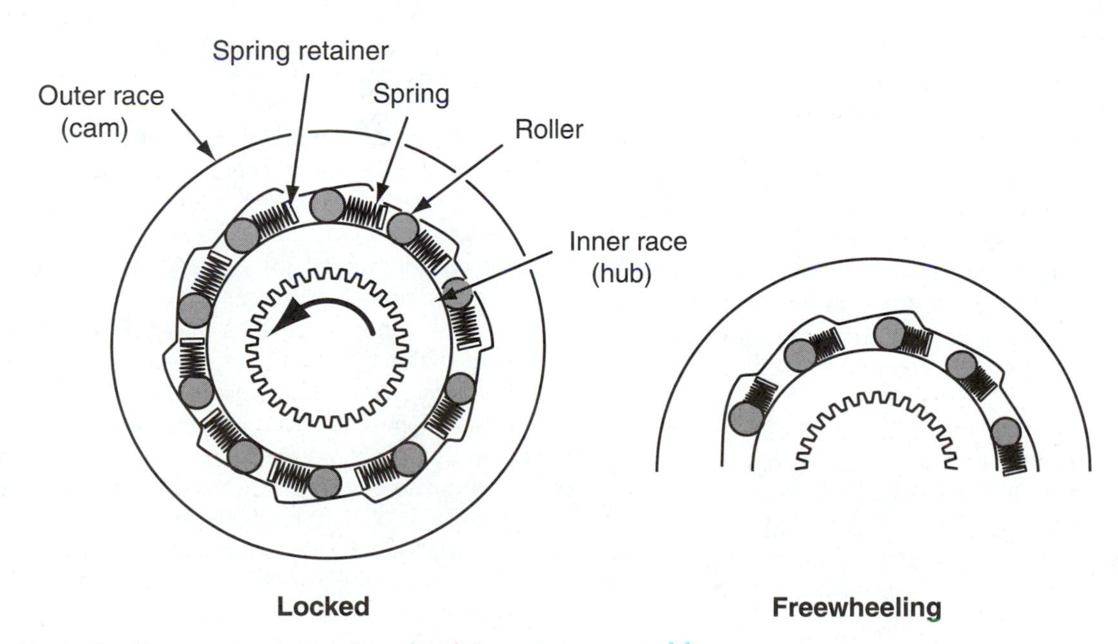


**Figure 13.** Typical roller-type overrunning clutch in a stator assembly.

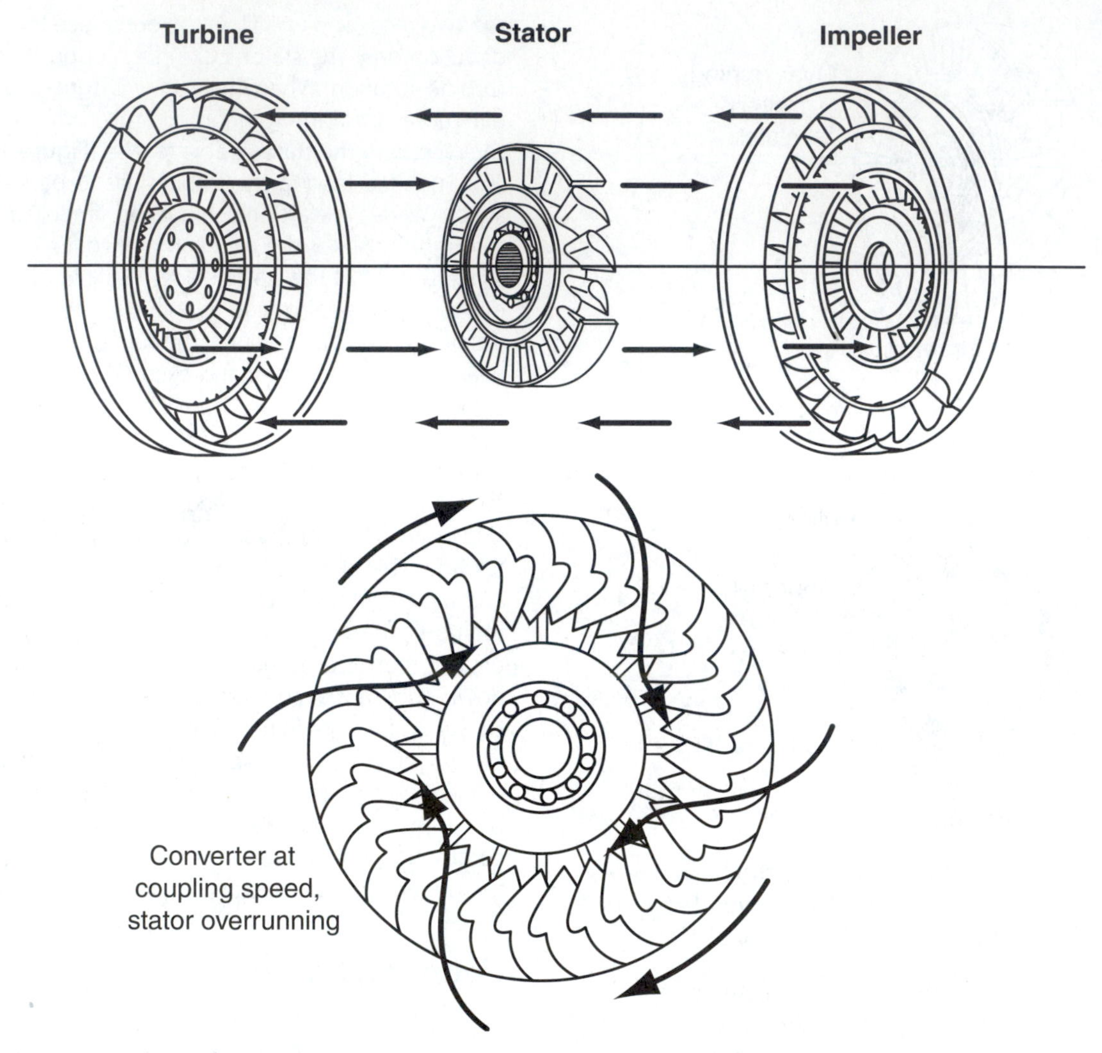


**Figure 14.** Fluid flow through the converter when the coupling phase has been achieved and the stator is overrunning.

the speed of the turbine increases, the direction of the flow changes and there is less multiplication of torque. When the speed of the turbine nearly equals the speed of the impeller, fluid flows against the stator vanes in the same direction as the fluid from the impeller **(Figure 14)**. This releases the one-way clutch and allows the stator to rotate freely. At this point, there is little vortex flow and the engine's torque is carried through the converter by the rotary flow of the fluid. This is the **coupling phase** of the torque converter and no torque multiplication takes place.

When the speed of the turbine reaches approximately 90 percent of the impeller's speed, coupling occurs. During coupling, the converter is acting as a fluid coupling transmitting engine torque to the transmission. The coupling phase does not occur at a specific speed or condition; rather, it occurs whenever the speed of the turbine nearly equals the speed of the impeller.

## Stall Speed

A torque converter is not a very efficient device. This is especially true at stall speeds when all of the engine torque that enters the torque converter is lost as heat and no power is inputted into the transmission. From stall speed to the coupling speed, the efficiency of a converter increases to approximately 90 percent.

The condition called stall occurs when the turbine is held stationary and the impeller is spinning. **Stall speed** is the fastest speed an engine can reach while the turbine is at stall. Some stall occurs every time vehicle starts moving forward or backward, as well as each time the vehicle is brought to a stop. Today, most torque converters have a stall speed of 1200 to 2800 rpm. Torque converters with a high stall speed are normally used with the less powerful engines and those with a low speed are used with more powerful engines.

## Converter Capacity

Although the basic construction of most torque converters is similar, the actual design of a converter is dictated by its application. The desired converter capacity and required stall speed are the two primary design considerations.

The capacity of a torque converter is an expression of the converter's ability to absorb and transmit engine torque with a limited amounted of slippage. A low-capacity converter has a high stall speed but allows for a relatively large amount of slippage during the coupling phase. However, it also provides for more torque multiplication and better acceleration.

A high-capacity converter allows for less slippage but has a low stall speed. This type of converter is very efficient at highway speeds but offers low torque multiplication. If a vehicle is equipped with a torque converter with the correct capacity, there will be a sort of balance between the amount of slippage, torque multiplication, and the stall speed.

The stall speed of a torque converter changes with a change in the diameter of the converter and the angles of the impeller, turbine, and/or stator vanes. The diameter of a torque converter directly affects the stall speed. Small diameter torque converters have high stall speeds and multiply torque at high engine speeds, but they do not couple until high engine speeds. Large diameter converters offer low stall speeds and torque multiplication at lower speeds. Typically, low output engines are fitted with small diameter converters to allow those engines to run at higher speeds to allow the engine to operate when it is making most of its power.

Stall speed is actually determined by the circumference of the converter because this is the distance the fluid travels inside the converter. As the circumference increases, the speed of the fluid increases, at a given engine speed. The speed of the fluid in the impeller almost matches the speed of the converter's circumference.

The angle of the impeller and turbine vanes also determines the stall speed of a converter. The vanes can be angled forward or backward. A forward angle produces higher fluid speeds; therefore, the stall speed tends to be lower. When the vanes have a backward angle, the stall speed increases. The angle of the stator vanes also changes capacity and stall of the converter. As the angle of the vanes increases, more fluid will return to the impeller and stall speed will increase. Changing the vane angle of an impeller, turbine, or stator is a common way to change torque converter capacity. This, of course, is not done by the technician; rather, it is done during manufacturing to match the converter to a vehicle.

## Cooling the Torque Converter

Slippage in a torque converter results from a loss of energy and most of these losses are in the generation of frictional heat. Heat is produced as the fluid hits and pushes its internal members. To maintain the efficiency that it has, the fluid must be cooled. Excessive fluid heat would result in even more inefficiencies.

The transmission's oil pump continually delivers ATF into the torque converter through a hollow shaft in the center of the torque converter assembly. A seal is used to prevent fluid from leaking at the point where the shaft enters the converter **(Figure 15)**. The fluid is circulated through the converter and exits past the turbine through the turbine shaft, which is located within the hollow fluid feed shaft. From there, the fluid is directed to an external oil cooler and back to the transmission's oil pan. This cooling circuit ensures that the fluid flowing into the converter is cooled, thereby maintaining the converter's normal efficiency.

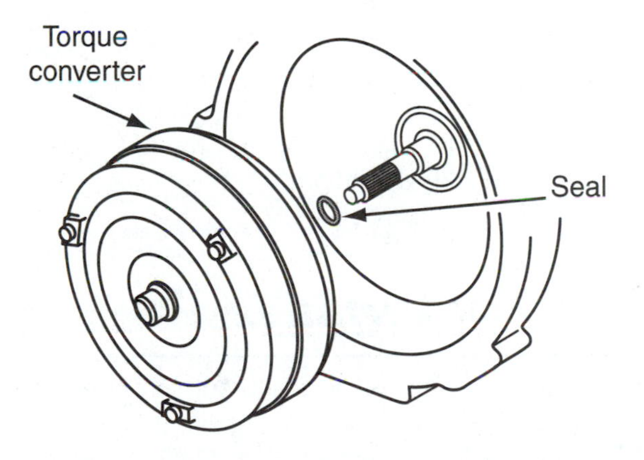


**Figure 15.** The turbine shaft-to-converter seal.

## Summary


- All automatic transmissions use a torque converter to transfer power from the engine to the transmission, except for some designs of the Constant Variable Transmission.
- A simple fluid coupling is comprised of three basic members: a housing, impeller, and turbine.
- The impeller acts like a pump and moves the fluid in the direction of its rotation. The moving fluid hits against the turbine vanes. This causes the turbine to rotate which brings an input into the transmission.
- Two types of fluid flow take place inside the fluid coupling: rotary and vortex flow.
- On all RWD and most FWD vehicles, the torque converter is mounted inline with the transmission's input shaft. Some transaxles use a drive chain to connect the converter's output shaft with the transmission's input shaft.
- The torque converter fits between the transmission and the engine and is supported by the transmission's oil pump housing and the engine's crankshaft. The rear hub of the converter's cover fits into the transmission's pump. The hub drives the oil pump whenever the engine is running.
- The front of a torque converter is mounted to a flexplate. The flexplate is designed to be flexible enough to allow the front of the converter to move forward or backward if it expands or contracts because of heat or pressure.
- A typical torque converter consists of three elements sealed in a single housing: the impeller, the turbine, and the stator.
- The impeller is the drive member of the unit and its fins are attached directly to the converter cover.
- The turbine is the converter's output member and is coupled to the transmission's input shaft. The turbine is driven by the fluid flow from the impeller and always turns at its own speed.
- The stator is the reaction member of the converter. This assembly is about one half the diameter of the impeller or turbine and is positioned between the impeller and turbine.
- The stator redirects the fluid leaving the turbine, back to the impeller. By redirecting the fluid so that it is flowing in the same direction as engine rotation, it allows the impeller to rotate more efficiently.
- A stator is supported by a one-way clutch, which is splined to the stator shaft and allows the stator to rotate only in the same direction as the impeller.
- The outer edge of the stator fins normally forms the inner edge of a three-piece fluid guide ring. This split guide ring directs fluid into a smooth, turbulence-free flow.
- It is vortex flow that allows for torque multiplication. Torque multiplication occurs when there is high impeller speed and low turbine speed.
- The condition called stall occurs when the turbine is held stationary and the impeller is spinning. Stall speed is the fastest speed an engine can reach while the turbine is at stall.
- The fluid is circulated through the converter and exits past the turbine through the turbine shaft, which is located within the hollow fluid feed shaft. From there the fluid is directed to an external oil cooler and back to the transmission's oil pan.


## Review Questions

1. What does a CVT use instead of a torque converter and briefly describe how it works?
2. What is the difference between rotary and vortex oil flow?
3. Name the three major components of a torque converter and describe the purpose of each.
4. What determines the vortex flow in a torque converter?
5. What is meant by stall speed?
6. A torque converter has reached its coupling phase when turbine speed is about _______ percent of the speed of the impeller. At this point, there is no _________________ _________________.
7. It is __________ flow that allows for torque multiplication. Torque multiplication occurs when there is ___________ impeller speed and ___________ turbine speed.
8. Technician A says that when the speed of the turbine nearly equals the speed of the impeller, fluid flows against the stator vanes in the same direction as the fluid from the impeller. Technician B says that when there is little vortex flow and the engine's torque is carried through the converter by the rotary flow of the fluid, the coupling phase of the torque converter occurs and no torque multiplication takes place. Who is correct?
   A. Technician A only
   B. Technician B only
   C. Both Technician A and Technician B
   D. Neither Technician A nor Technician B

9. Technician A says that the rear hub of the torque converter is bolted to the flexplate. Technician B says that the flexplate is designed to be flexible enough to allow the front of the converter to move forward or backward if it expands or contracts because of heat or pressure. Who is correct?
   A. Technician A only
   B. Technician B only
   C. Both Technician A and Technician B
   D. Neither Technician A nor Technician B

10. When discussing the operation of the stator, Technician A says that the stator redirects the fluid leaving the turbine back to the impeller. Technician B says that only a portion of the fluid returning from the turbine to the impeller passes through the stator. Who is correct?
    A. Technician A only
    B. Technician B only
    C. Both Technician A and Technician B
    D. Neither Technician A nor Technician B

# Chapter 11 Basic Hydraulic Controls

## Introduction

In order for the transmission to operate in the desired range, a series of other controls are needed. These controls are mechanical, electronic, or hydraulic. This chapter looks at the components that control the activity of the hydraulic fluid in a transmission. Many of these hydraulic controls have been replaced with or are used in conjunction with electronic devices in late-model transmissions. Although some of these components can be considered obsolete, an understanding of their purpose and how they function are important to gaining an understanding of electronic controls.

Hydraulic controls stop, send, and change the hydraulic pressure within the transmission. The total control of a transmission depends on how much fluid pressure is applied and where it is applied. This is true for both hydraulically controlled and electronically controlled transmissions.

## PRESSURES

All automatic transmissions use three basic pressures to control their operation: mainline pressure, throttle pressure, and governor pressure. The valves that regulate these pressures are the pressure regulator valve, **throttle valve**, and governor valve. Mainline pressure is regulated pump pressure that is the source of all other circuits in the transmission. Mainline pressure may fill the torque converter and typically lubricates the transmission, supplies fluid to the valve body, and is used to apply the brakes and clutches. Governor pressure is a regulated pressure that varies with vehicle speed. Throttle pressure is regulated pressure that varies with engine load or throttle position. Throttle pressure interacts with governor pressure to control shifting.

### Mainline Pressure

The transmission's pump provides a flow of fluid to the mainline circuit. Engine speed and the pressure regulator valve limit this flow. The pressure regulator valve develops mainline pressure when it blocks or resists the flow of fluid.

**Mainline pressure**, also called line pressure, is the source of all other pressures used by the transmission **(Figure 1)**. It is the pressure used to engage or apply the clutches and bands within the transmission.

The pressure regulator valve controls line pressure to meet the needs of the transmission regardless of operating speeds. As engine speed increases, the pump works faster and delivers a greater flow than when the engine is running at low speeds. The pressure regulator keeps line pressure from building to a point at which it could damage the transmission. Line pressure also increases as the pump fills all of the circuits in the hydraulic system. Again, the pressure regulator valve prevents excessive pressure buildup by exhausting some of the fluid flow from the pump when the pressure reaches a predetermined limit.

### Throttle Pressure

**Throttle pressure** is a signal pressure. The amount of throttle pressure for many transmissions depends directly on the engine's load. Engine load is sensed by mechanical linkages relaying the amount of throttle opening, the effect of engine vacuum on a modulator, a combination of these two, or various sensors tied to a PCM. Many late-model transmissions use a pulse-modulated solenoid, controlled

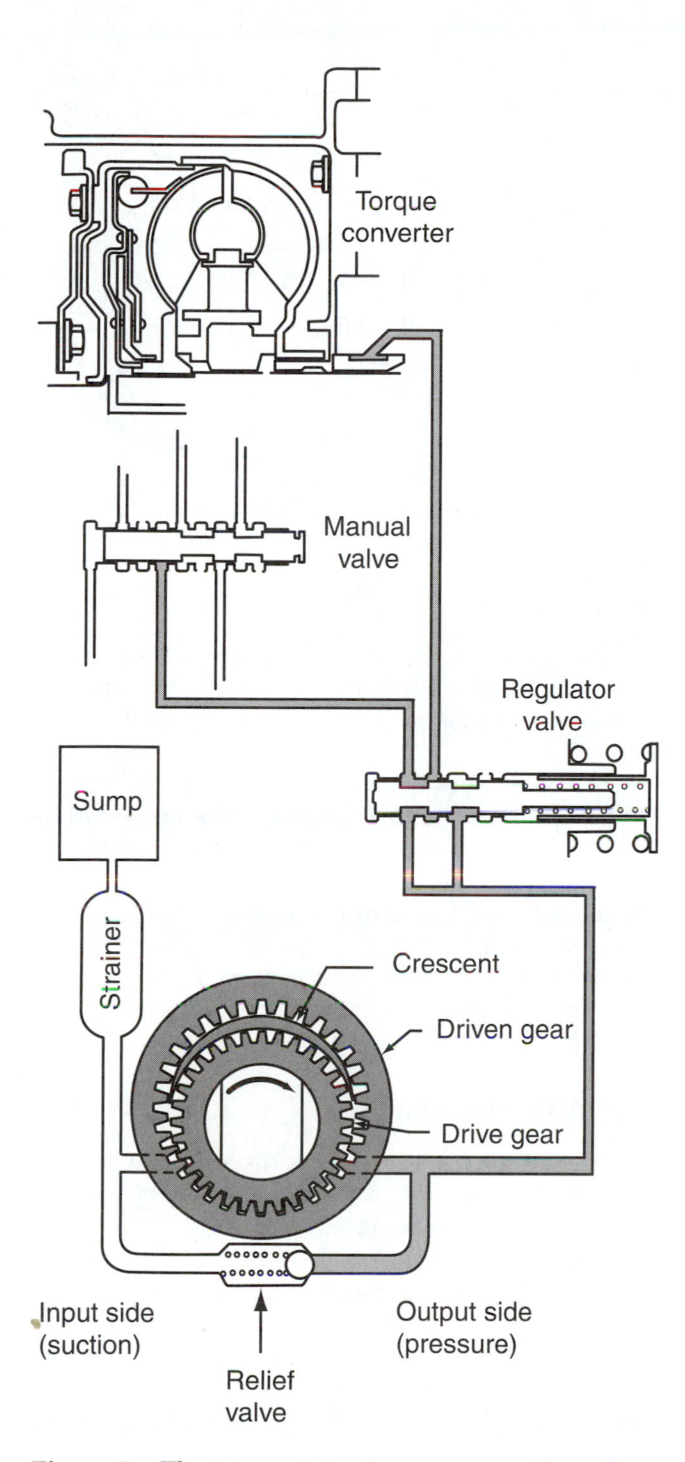


**Figure 1.** The transmission's pump provides a flow of fluid to the mainline circuit.

by the computer, to control throttle pressure in response to inputs to the computer.

## Governor Pressure

**Governor pressure** increases with an increase of vehicle speed. When governor pressure is higher than throttle pressure, an upshift occurs. Early transmission designs used a mechanical governor made of springs, weights, and a spool valve to create this signal. Late-model transmissions use a computer-controlled solenoid to control this pressure.

A transmission's change of gears is actually caused by the movement of a **shift valve**. The shift valve is simply a control valve that directs fluid flow. When the shift valve moves, line pressure is directed to the appropriate apply device. Because throttle and governor pressures just control the movement of the shift valve, they are only signal pressures. Signal pressures control the direction of the line pressure by their action on shift valves.

Most electronically controlled transmissions do not use governors; rather, a speed sensor is used to help the PCM or TCM decide when to activate a shift solenoid. These transmissions also receive the signals from a load sensor, such as a MAP sensor, to determine when to shift.

The **governor** assembly is driven by the transmission's output shaft and senses road speed and sends a fluid pressure signal to the valve body to either upshift or downshift. When vehicle speed is increased, the pressure developed by the governor is directed to the shift valve. As the speed (and therefore the pressure) increases, the spring tension on the shift valve is overcome and the valve moves **(Figure 2)**. This action causes an upshift. Likewise, a decrease in speed will result in a decrease in pressure and a downshift.

Although the governor sends a signal that will force an upshift, engine load may cause a delay in the shift. This allows for operation in a lower gear when there is a heavy load and the vehicle needs the gear reduction. During heavy load operation, the governor pressure must be strong enough to overcome the high throttle pressure plus the spring tension on the shift valve before it can force an upshift. Because of this, the transmission will remain in a particular gear range until a higher than normal engine speed is reached.

## Shift Timing

Shift timing is determined by throttle pressure and governor pressure acting on opposite ends of the shift valve. When a vehicle is accelerating from a stop, throttle pressure is high and governor pressure is low. As vehicle speed increases, the throttle pressure decreases and the governor pressure increases. When governor pressure overcomes throttle pressure and the spring tension at the shift valve, the shift valve moves to direct pressure to the appropriate apply device and the transmission upshifts.

# PRESSURE BOOSTS

When the engine operates under heavy load conditions, fluid pressure must be increased to increase the holding capacity of a hydraulic member. Increasing the fluid pressure holds the brake and clutch control units tightly to

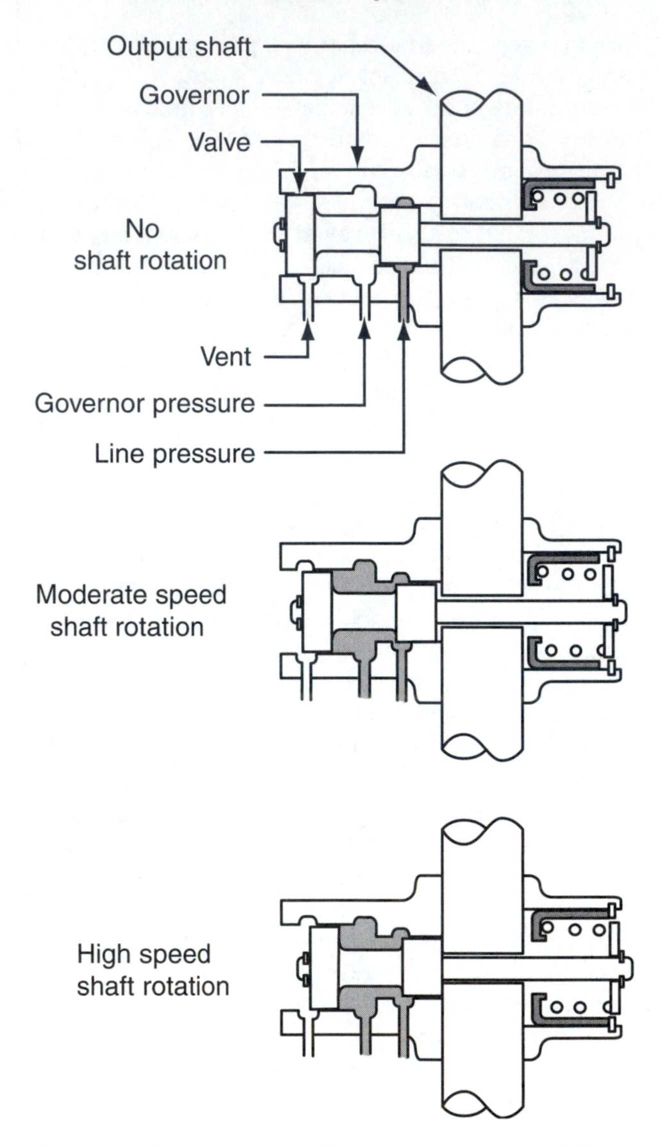


Figure 2. The action of a governor at different speeds.

reduce the chance of slipping while under heavy load. This is accomplished by sending pressurized fluid to one side of the pressure regulator's spool valve. This pressure works against the spool valve's normal movement to open the exhaust port and allows pressure to build to a higher point than normal.

When the vehicle is placed under heavy load, throttle pressure is applied to a booster valve at the pressure regulator. This pressure acting on the booster valve assists the pressure regulator valve's spring in pushing the regulator valve up against the line pressure. Line pressure is able to continue to increase until the pressure on the regulator valve overcomes the spring in the pressure regulator and throttle pressure. The pressure regulator valve now opens its exhaust port with a boosted line pressure, which is used to hold the reaction members tightly to resist slippage.

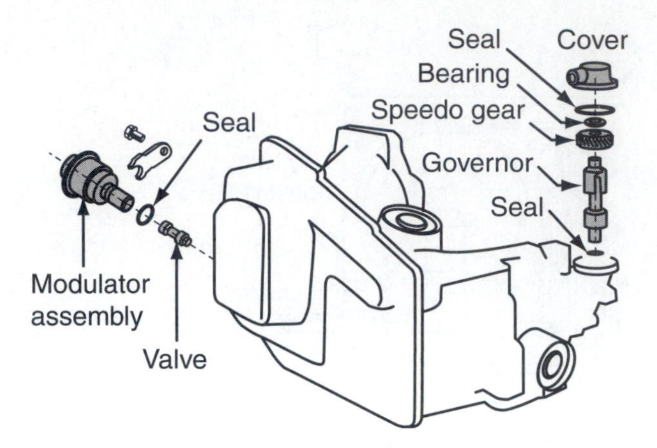


Figure 3. A vacuum modulator and governor assembly.

Some transmissions are equipped with two boost valves. The second boost valve is used for improved shifting into reverse gear and for greater holding power when operating in reverse.

Engine load can be monitored electronically through the use of various electronic sensors that send information to an electronic control unit, which in turn controls the pressure at the valve body. Load also can be monitored by throttle pressure. Many early transmissions were equipped with a **vacuum modulator**, which uses **engine vacuum** to change transmission pressure **(Figure 3)**. The vacuum modulator allowed for an increase in pressure when vacuum was low and decreased it when vacuum was high.

## Vacuum Modulators

A vacuum modulator is a **load-sensing device** that regulates fluid pressure in response to a vacuum signal, which varies with throttle opening and vehicle load **(Figure 4)**. Engine vacuum is low when there is a heavy load and high when the load is low. When high vacuum is present at the modulator, the pressure regulator valve works normally and maintains normal pressure. However, when the vacuum is low, the modulator allows fluid flow to enter onto the spool valve in the pressure regulator, which allows for an increase in pressure.

Usually, a vacuum modulator is a small round metal canister that is threaded or push fit into the transmission's housing. A vacuum line connects the vacuum modulator to the engine's intake manifold. This supplies vacuum to operate the modulator.

The vacuum modulator is basically a canister divided into two chambers by a diaphragm. The diaphragm forms a seal between the two chambers. The chamber closest to the transmission housing is open to atmospheric pressure and the other chamber is closed to atmospheric pressure. A pushrod is connected to the open side of the diaphragm.

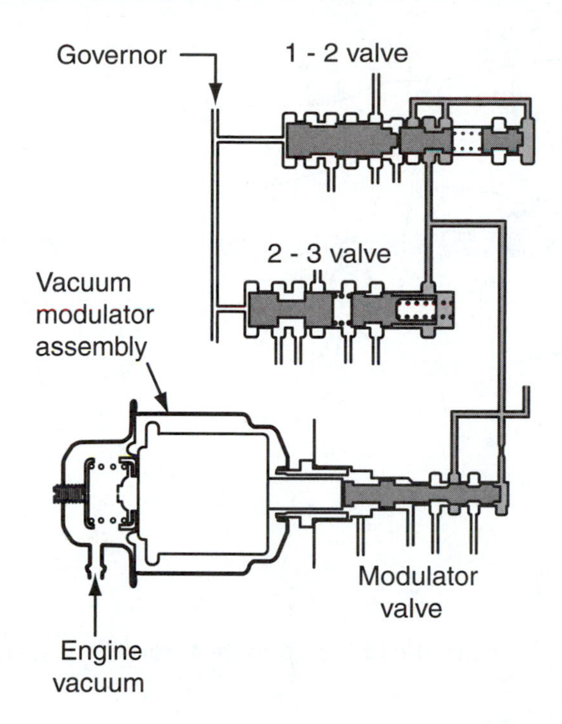


**Figure 4.** A vacuum modulator is a load-sensing device that regulates fluid pressure in response to a vacuum signal, which varies with throttle opening and vehicle load.

This rod connects the diaphragm to the **modulator valve**. The closed chamber of the vacuum modulator contains a coil spring positioned between the diaphragm and the end of the canister. Engine vacuum is sensed on the spring side of the diaphragm.

The modulator valve directs hydraulic pressure to other control devices at low vacuum and withholds it at high vacuum. When engine load is low and the throttle closed, high manifold vacuum retracts the diaphragm and pushrod against spring tension to reduce throttle pressure. As engine load increases and the throttle is opened wider, manifold vacuum drops and the diaphragm spring forces the push rod into firmer contact with the modulator valve to increase fluid flow in the modulator circuit.

A vacuum diaphragm responds to differences in atmospheric pressure on one side and vacuum on the other side. Normal vacuum modulators are less effective at higher altitudes because atmospheric pressure and engine vacuum are less than they are at sea level. Some vacuum modulators use an altitude compensating vacuum valve with a spring-type bellows added to the atmospheric side of the valve. When the bellows units are made, a vacuum is drawn in the bellows, which causes atmospheric pressure to compress the bellows. The bellows unit is then sealed to hold the vacuum. As altitude increases and atmospheric pressure decreases, the bellows unit expands because of the decrease in pressure differential. This allows the bellows unit to exert pressure against the atmospheric side of the diaphragm to compensate for the lower atmospheric pressure at higher altitudes.

Some vacuum modulators use two diaphragms; one diaphragm is exposed to intake vacuum while the other receives a ported vacuum signal. Ported vacuum increases with throttle plate opening; therefore, ported vacuum is available to assist the lower engine vacuum during times of heavy load. Transmissions not equipped with a vacuum modulator use a throttle cable or electronics to sense engine load and change fluid pressures.

## Throttle Linkages

Some transmissions use mechanical linkages to control throttle valve pressure and increase line pressure. The throttle cable is normally connected to a lever located at the left front of the transmission housing. The **throttle valve** is moved by the throttle lever, which is moved by the throttle linkage at the throttle body. The throttle lever moves according to the position of the throttle pedal. Some older automatic transmissions use a rod-type linkage assembly, not a cable, to move the throttle lever.

Normally, the throttle valve converts mainline pressure into a variable throttle pressure based on the position of the throttle plate **(Figure 5)**. The linkage moves the throttle valve against spring pressure, the more the throttle is opened, the higher the throttle pressure, until throttle pressure equals mainline pressure. The throttle cable is connected between

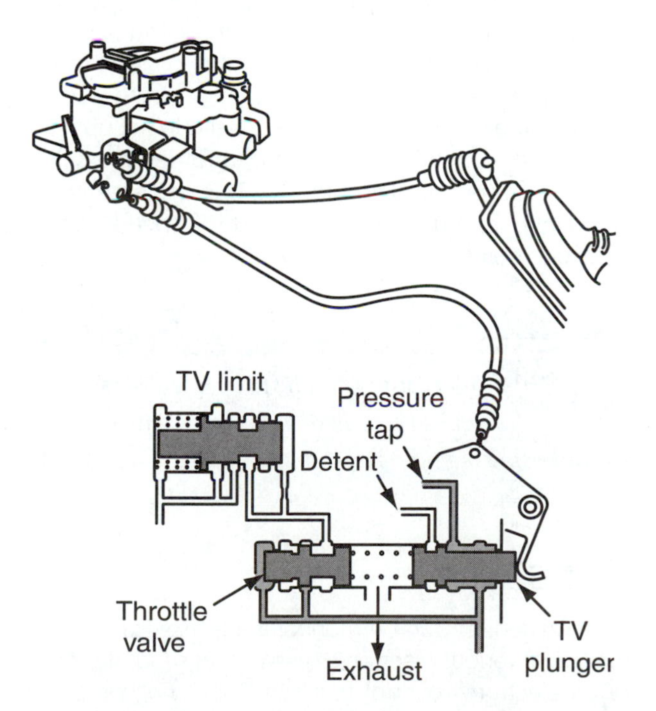


**Figure 5.** The throttle linkage moves the throttle valve against spring pressure; the more the throttle is opened, the higher the throttle pressure, until throttle pressure equals line pressure.

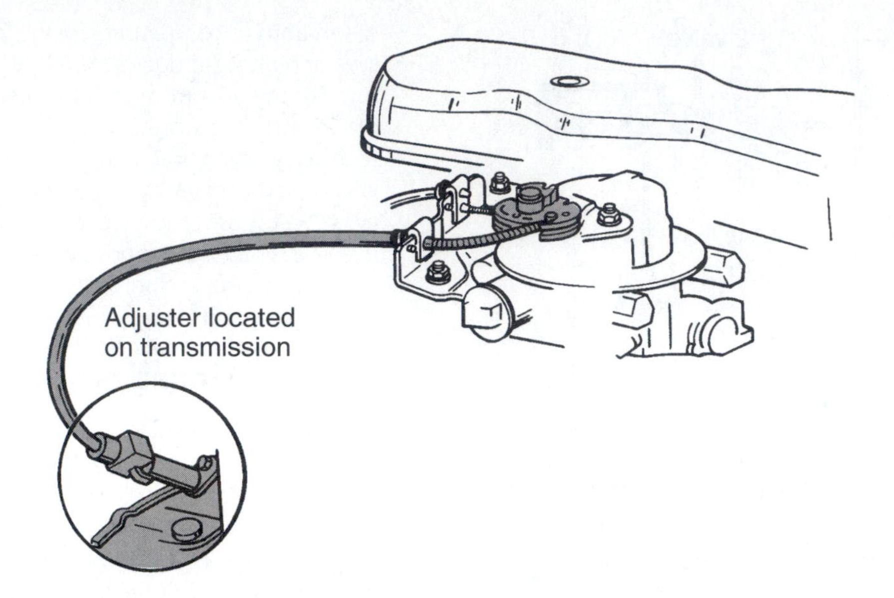


**Figure 6.** The throttle cable is connected between the throttle body or throttle linkage and the throttle lever, which moves with the movement of the throttle pedal.

the throttle body or throttle linkage and a lever located at the transmission housing **(Figure 6)**. The throttle lever moves with the movement of the throttle pedal.

When the throttle pedal is depressed, the throttle valve opens to produce throttle pressure, which is directed to the pressure regulator valve. This pressure helps the pressure regulator valve spring hold the pressure regulator valve in position to close the exhaust port, which causes the pressure to increase. This increased pressure is used to hold the apply devices and causes a delay in the upshift. As the throttle pedal is released, throttle pressure is decreased, as is mainline pressure. This decrease in pressure allows for a downshift when vehicle speed reaches a particular point.

You Should Know *Improper throttle valve cable adjustment can cause damage to a transmission very quickly. Never drive a vehicle with a disconnected or improperly adjusted cable.*

## MAP Sensor

Engine load can be monitored electronically through the use of various electronic sensors that send information to an electronic control unit, which in turn controls the pressure at the valve body. The most commonly used sensor is the **Manifold Absolute Pressure (MAP) sensor**. The MAP sensor **(Figure 7)** senses air pressure in the intake manifold. The control unit uses this information as an indication of engine load. A pressure-sensitive ceramic or

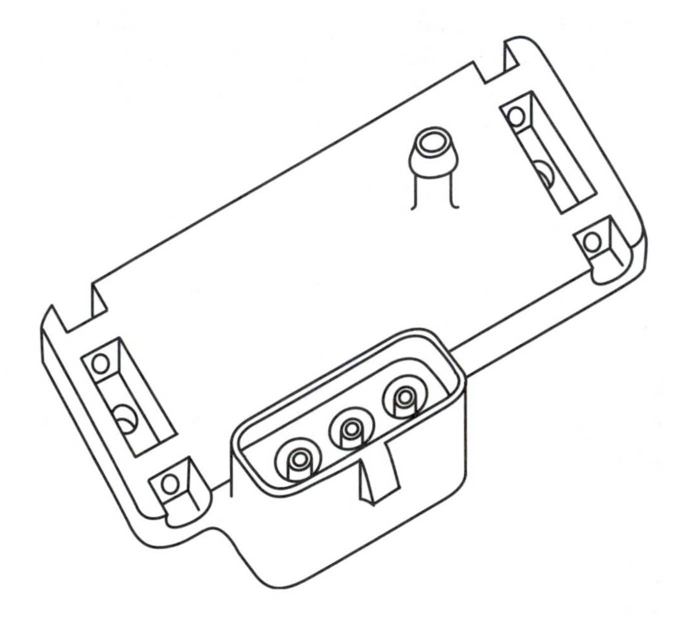

**Figure 7.** A MAP sensor.

silicon element and electronic circuit in the sensor generates a voltage signal that changes in direct proportion to pressure. A MAP sensor measures manifold air pressure against a precalibrated absolute pressure; therefore, the readings from these sensors are not adversely affected by changes in operating altitudes or barometric pressures.

## Kickdown Valve

The valve body is also fitted with a **kickdown** circuit, which provides a downshift when the driver requires additional power. When the throttle pedal is opened

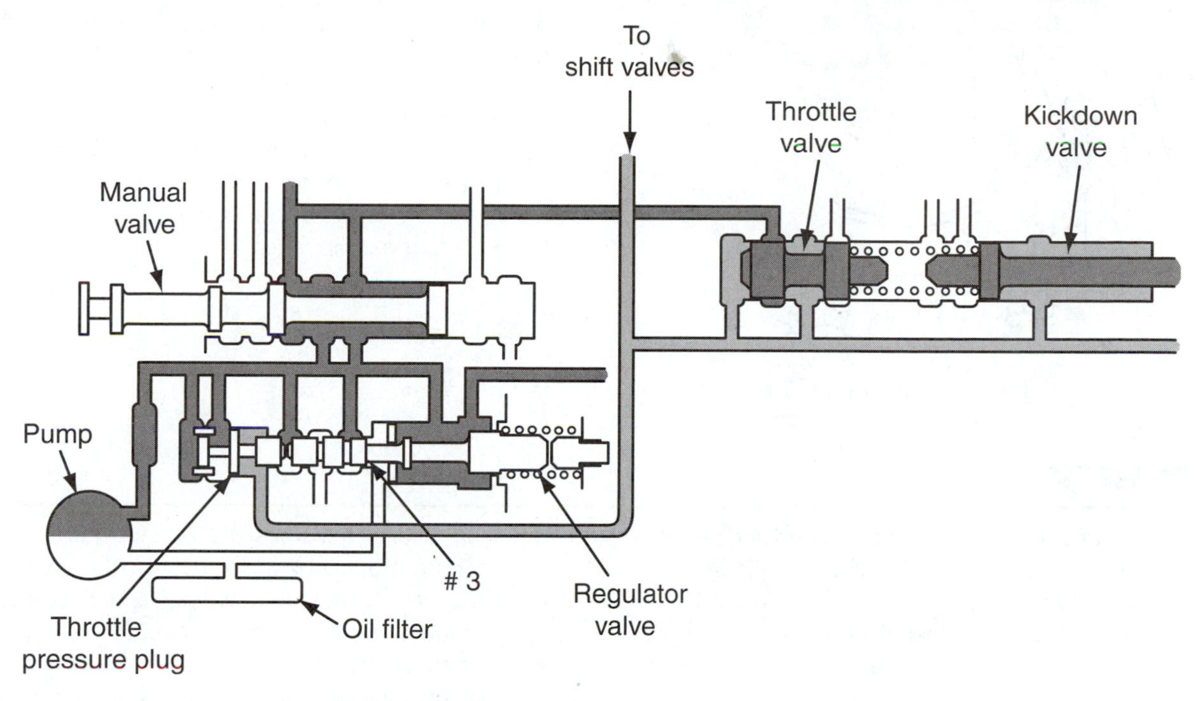


**Figure 8.** Fluid flow for the throttle and kickdown valves.

quickly, throttle pressure rapidly increases and directs a large amount of pressure onto the kickdown valve. This moves the kickdown valve, which opens a port and allows mainline pressure to flow against the shift valve. The spring tension on the shift valve, the kickdown pressure, and throttle pressure will push on the end of the shift valve causing it to move to the downshift position **(Figure 8)**. The transmission automatically downshifts to the next lower gear. The position of the shift valve blocks fluid at its inlet port, which prevents an upshift during this time.

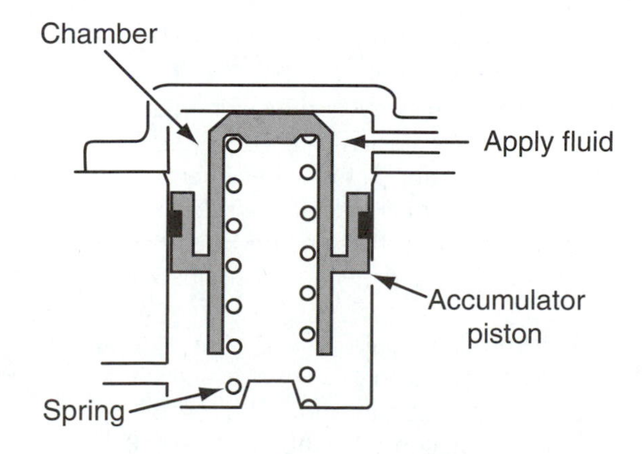


**Figure 9.** An accumulator assembly.

## SHIFT QUALITY

All transmissions are designed to change gears at the correct time, according to engine speed and load and driver intent. However, transmissions also are designed to provide for positive change of gear ratios without jarring the driver or passengers. If a band or clutch is applied too quickly, a harsh shift will occur. **Shift feel** is controlled by the pressure at which each hydraulic member is applied or released, the rate at which each is pressurized or exhausted, and the relative timing of the apply and release of the members.

To improve shift feel during gear changes, a band is often released while a multiple friction disc pack is being applied. The timing of these two actions must be just right or both components will be released or applied at the same time, which would cause engine flareup or driveline binding. Several other methods are used to smoothen gear changes and improve shift feel.

Multiple friction disc packs sometimes contain a wavy spring-steel separator plate that helps smoothen the application of the clutch. Shift feel can also be smoothened out by using a restricting **orifice** or an **accumulator** piston **(Figure 9)** in the band or clutch apply circuit. A restricting orifice or check ball **(Figure 10)** in the passage to the apply piston restricts fluid flow and slows the pressure increase at the piston by limiting the quantity of fluid that can pass in a given time. An accumulator piston slows pressure buildup at the apply piston by

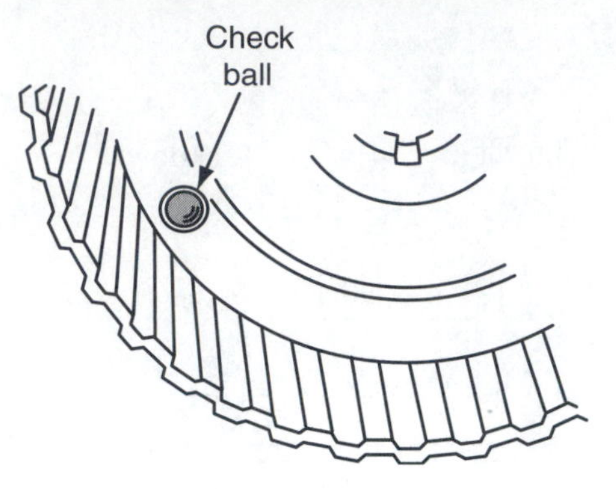


**Figure 10.** A check ball in the cylinder of a clutch pack.

diverting a portion of the pressure to a second piston in the same hydraulic circuit. This delays and smoothens the application of a clutch or band.

Manufacturers also have applied electronics to get the desired shift feel. One of the most common techniques is the pulsing (turning on and off) of the shift solenoids. Doing this prevents the immediate engagement of a gear, by allowing some slippage.

## Summary

- The governor assembly senses road speed and sends a fluid pressure signal to the valve body to shift gears.
- Throttle pressure increases the fluid pressure applied to the apply devices of the planetary units to hold them tightly to reduce the chance of slipping while the vehicle is operating under heavy load.
- Shift timing is determined by throttle pressure and governor pressure acting on opposite ends of the shift valve.
- The vacuum modulator is a load-sensing device that increases or decreases fluid pressure in response to the vacuum signal.
- Transmissions not equipped with a vacuum modulator use a throttle cable or electronics to sense engine load and change fluid pressures.
- The throttle cable is connected to the throttle lever at the valve body and is moved according to the position of the throttle pedal.
- Many early transmissions were equipped with a vacuum modulator, which allowed for an increase in pressure when vacuum was low and a decrease when vacuum was high.
- The throttle valve converts mainline pressure into a variable throttle pressure based on the position of the throttle plate.
- Engine load can be monitored electronically through the use of a Manifold Absolute Pressure (MAP) sensor. The MAP sensor senses air pressure in the intake manifold.
- Shift feel is controlled by the pressure at which each hydraulic member is applied or released, the rate at which each is pressurized or exhausted, and the relative timing of the apply and release of the members.
- An accumulator is a hydraulic piston assembly that helps a servo apply a band or clutch smoothly. They do this by absorbing sudden pressure surges in the hydraulic circuit to the servo.

## Review Questions

1. How can shift feel be controlled?
2. What is the purpose of a vacuum modulator?
3. What determines the timing of the shifts in an automatic transmission?
4. How does engine load affect engine vacuum?
5. Why must hydraulic line pressures increase when there is an increase of load on the engine?
6. The governor assembly senses ____ speed and sends a fluid pressure signal to the ____ ____ to either upshift or downshift.
7. Engine load can be monitored ______________, by ________ ________ or by a ______ _________.
8. An automatic transmission or transaxle selects gear ratios according to engine _____, engine ____, vehicle _____, and other operating conditions.
9. When discussing load sensing devices, Technician A says that transmissions not equipped with a vacuum modulator use a throttle cable or electronics to sense engine load. Technician B says that increasing the fluid pressure holds the planetary control units tightly to reduce the chance of slipping while under heavy load. Who is correct?
   A. Technician A only
   B. Technician B only
   C. Both Technician A and Technician B
   D. Neither Technician A nor Technician B
10. Technician A says that changes in engine load cause changes in hydraulic pressure inside the transmission. Technician B says that engine load is monitored by throttle pedal movement or by engine vacuum. Who is correct?
   A. Technician A only
   B. Technician B only
   C. Both Technician A and Technician B
   D. Neither Technician A nor Technician B

# Chapter 12 Electronic Controls

## Introduction

Electronic transmission control has become increasingly more common on today's cars. Although these transmissions function in the same way as earlier hydraulically based transmissions, their shift points are determined by a computer **(Figure 1)**. The computer uses inputs from several different sensors and matches this information to a predetermined schedule.

Hydraulically controlled transmissions rely on signals from a governor and throttle pressure device to force a shift in gears. They also rely on pressure differentials at the sides of a shift valve to hold or change a gear. Electronically controlled transmissions do not have governors or throttle pressure devices. The required pressure differential to shift or hold a gear is provided by the action of computer-controlled shift solenoids that allow for changes in pressure on the side of a shift valve.

*In the mid-1980s, Toyota introduced the A140E transaxle. This was the first automatic transmission with electronic shift controls. This transmission began the trend in which, by 1990, all domestic manufacturers offered at least one electronic automatic transmission (EAT). Today, EATs are found in nearly all new vehicles.*

Older hydraulically controlled automatic transmissions waste a good amount of the torque produced by the engine through the heat generated by the moving fluid. Also, because gear changes were dependent on the movement of fluid, upshifts and downshifts were somewhat lazy. Manufacturers could not get the transmissions to respond immediately to the needs of the vehicle without jarring the driver and the vehicle. By using electronic controls for transmission operation, the amount of wasted power was reduced and the overall operation of the transmission is smoother and more responsive.

### BASICS OF ELECTRONIC CONTROLS

The basic part of all electronic control systems is a computer. A **computer** is an electronic device that receives information, stores information, processes information, and communicates information. All of the information it works with is really nothing more than electricity. To a computer, certain voltage and current values mean something and based on these values, the computer becomes informed.

Computers receive information from a variety of input devices, which send voltage signals to the computer. These signals tell the computer the current condition of a particular part or the conditions that a particular part is operating in. After the computer receives these signals, it stores them and interprets the signals by comparing the values to data it has in its memory. By processing this data, the computer knows what conditions the input data represents. It also can search its memory to identify any actions it should take in response to those conditions. If an action is required, the computer will send out a voltage signal to the device that should take the action causing it to respond and correct the situation.

This entire process describes the operation of an electronic control system. Information is received by a microprocessor from some input sensors; the computer processes the information, and then sends commands out

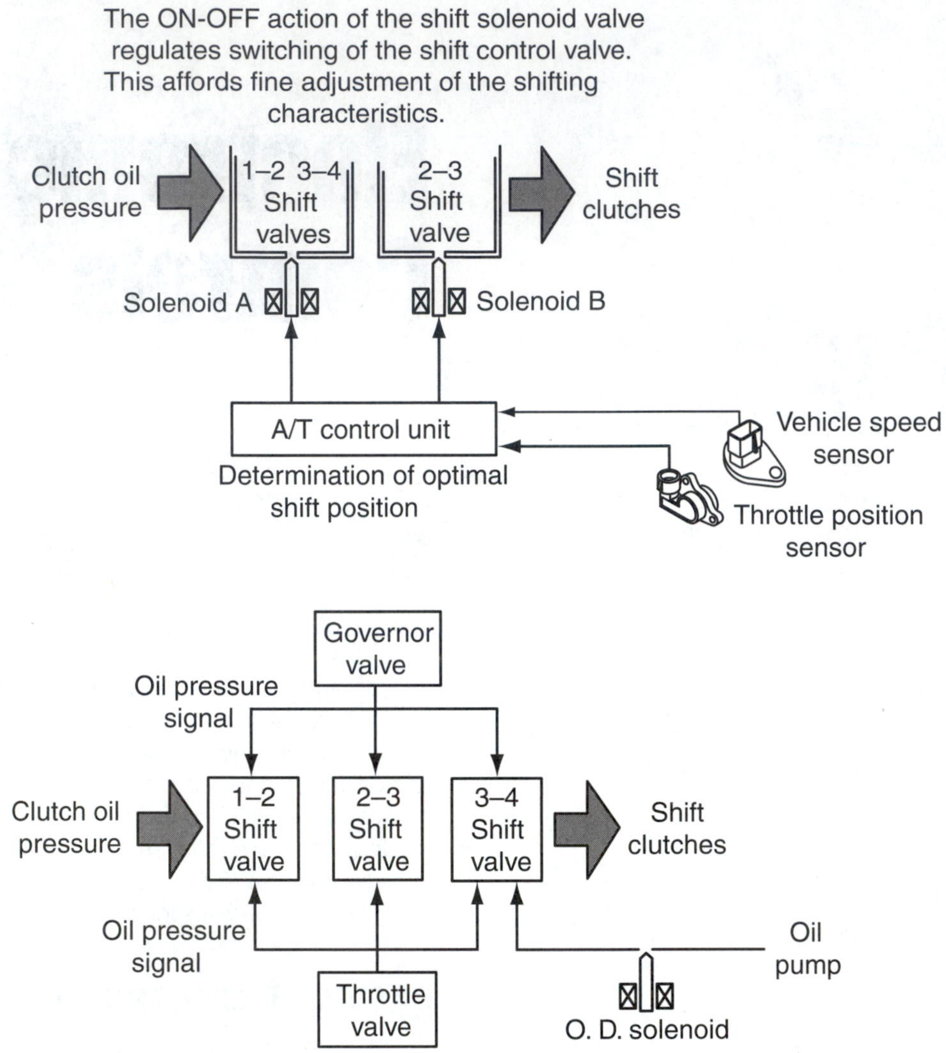

**Figure 1.** A comparison of electronic (top) and hydraulic (bottom) transmission controls.

to actuators **(Figure 2)**. Most control systems also are designed to monitor their own work; they will check to see if their commands resulted in the expected results. If the result is not what was desired, the computer will alter its command until the desired outcome is achieved.

## Inputs

The input devices used in electronic control systems vary with each system; however, they can be grouped into distinct categories: **reference voltage sensors** and **voltage generators**. Voltage generation devices typically are used to monitor rotational speeds, the most common of which is the PM generator, used as a Vehicle Speed Sensor. A speed sensor is a magnetic pickup that senses and transmits low-voltage pulses **(Figure 3)**. These pulses are generated each time a small magnet, normally located on the circumference of a shaft or rotating component, passes by the wire coil of the pickup unit. The pulse of voltage is the signal sent to the computer. The computer then compares the signal to its clock or counter, which converts the signal into speed.

Common voltage reference sensors include on/off switches, potentiometers, thermistors, and pressure sensors. These sensors normally complete a circuit to and from the computer. The computer sends a reference voltage to them and reads the voltage sent back to it by them. The change in voltage represents a condition that the computer can again identify by looking at its program and comparing values. Potentiometers **(Figure 4)**, thermistors, and pressure sensors are designed to change their electrical resistance in response to something else changing. Normally, a potentiometer is linked to devices such as the throttle linkage. As the throttle is moved, the resistance of the potentiometer in the throttle position (TP) sensor changes. This causes a varied return signal to the computer as the throttle opening changes.

Thermistors **(Figure 5)** also change their resistance values in response to conditions. However, they respond to changes in heat. Pressure sensors respond to pressure

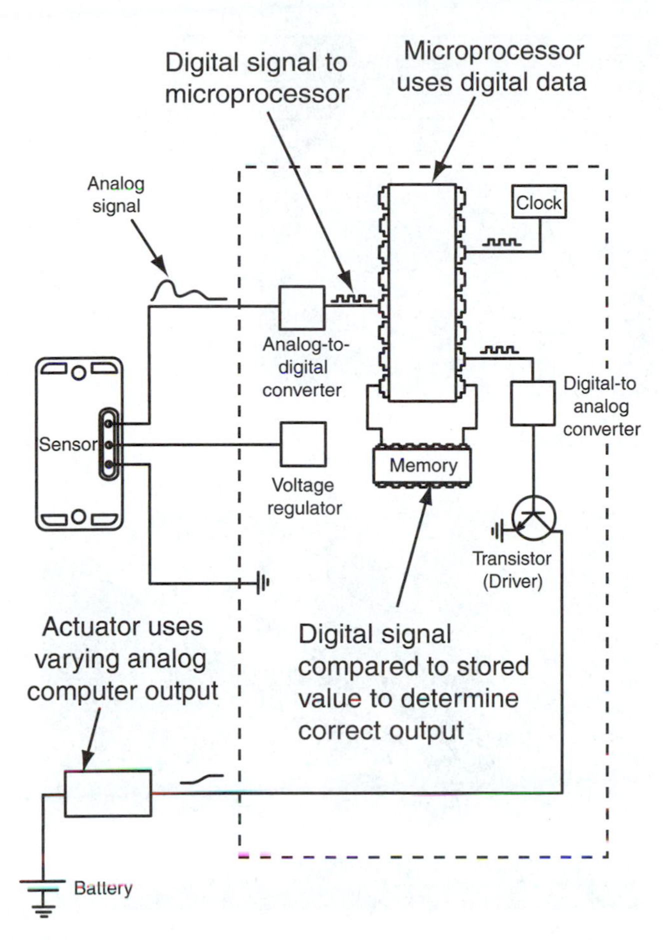


**Figure 2.** A simple schematic of an electronic control system and the voltage signals present throughout the circuit.

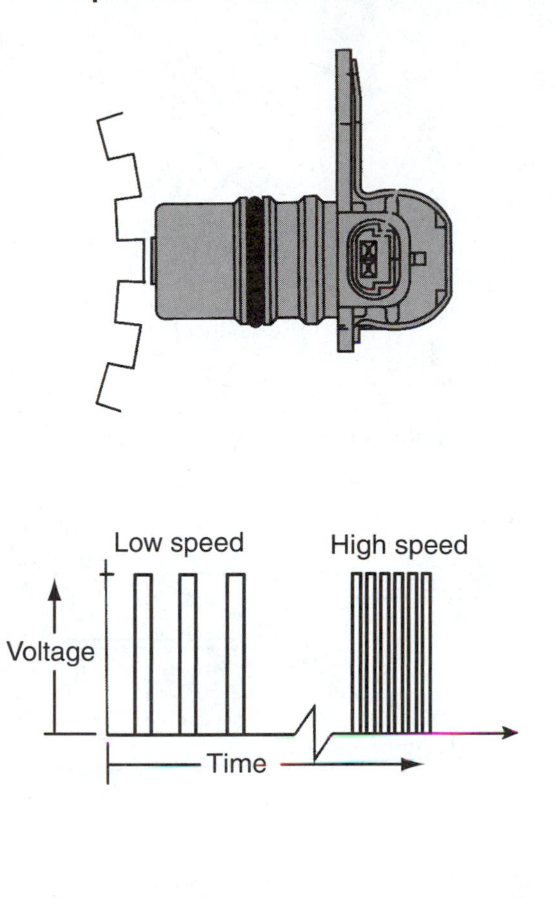


**Figure 3.** (Top) A typical speed sensor. (Bottom) The signal a speed sensor sends to the computer.

applied to a movable diaphragm in the sensor. A pressure increases, so does the movement of the diaphragm and the amount of resistance in the sensor **(Figure 6)**.

On/off switches are simple in operation. When a particular condition exists, the switch is either forced closed or open. When the switch is closed, the computer receives a return signal from the switch. When it is open, there is no return signal.

The return signal from a switch represents circuit on and off times. If a switch is cycled on/off very rapidly, the return signal is a rapid on/off one. This describes a digital

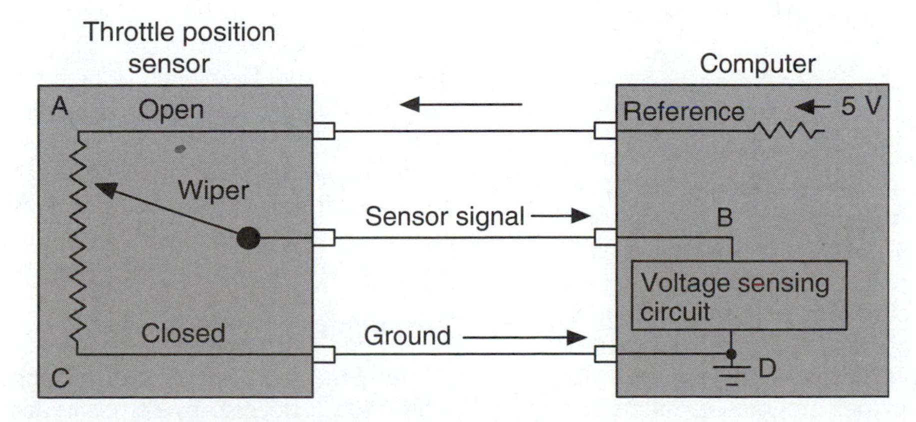


**Figure 4.** A potentiometer sensor circuit measures the amount of voltage drop to determine the position of the device attached to the sensor's wiper.

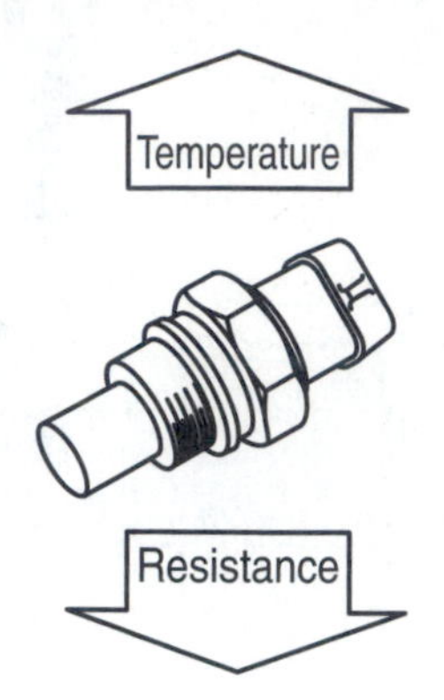


| Thermistor temperature to resistance valves | | |
|---|---|---|
| °F | °C | Ohms |
| 210 | 100 | 185 |
| 160 | 70 | 450 |
| 100 | 38 | 1,800 |
| 40 | 4 | 7,500 |
| 0 | –18 | 25,000 |
| –40 | –40 | 100,700 |

**Figure 5.** A typical thermistor's resistance decreases as the temperature rises.

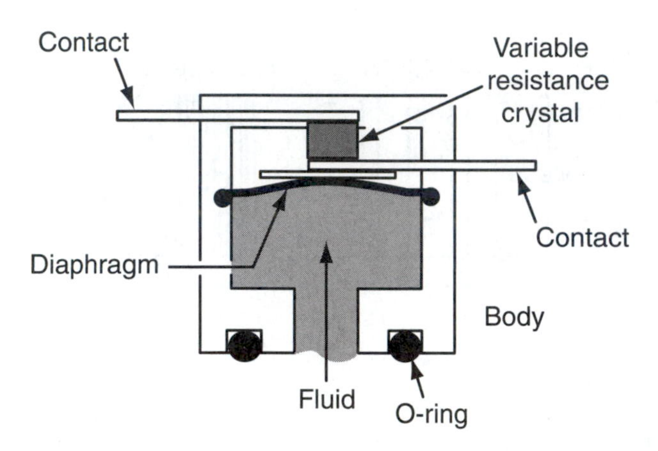


**Figure 6.** The working parts of a pressure switch.

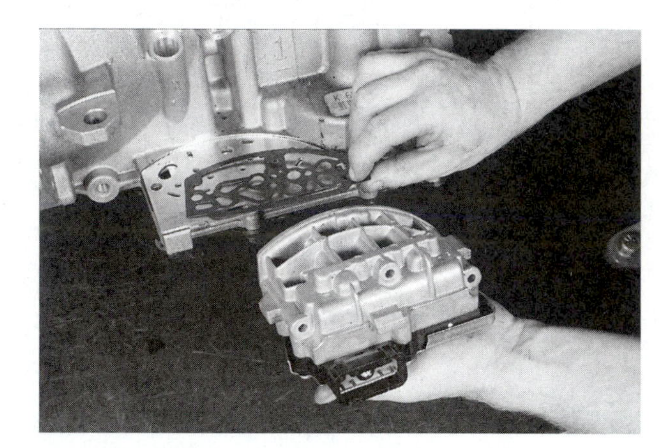

**Figure 7.** A typical solenoid assembly.

signal, a series of on/off pulses. A computer is a digital device that processes information through a series of on/off signals. In order for a computer to process information, it must receive its data digitally. The inputs from voltage reference sensors arrive to the computer as a change in voltage not as a digitized signal. Therefore, in order for the computer to analyze the data, first it must be changed to a digital signal. This is the first processing task of a computer, to translate the analog signals into a digital signal.

## Outputs

After the information has been processed, the computer sends out commands to its output devices or actuators. Typical output devices are solenoids **(Figure 7)** and motors, which cause something mechanical or hydraulic to change. The movement of these outputs is controlled by the commands of the computer. Sometimes the command is merely the application of voltage to operate or energize the device. Other times it is a variable signal that causes the device to cycle in response to the signal. Most often, the solenoids are controlled on their ground side. Battery voltage is applied to the solenoids whenever the ignition is on. The computer controls the ground of the solenoid.

The typical electronically controlled transmission uses hydraulically operated clutches and brakes. The computer receives information from various inputs and controls a solenoid assembly. The solenoid assembly consists of two to five solenoids that control hydraulic pressure and flow to the various apply devices and to the clutch of the torque converter.

Most solenoid-operated valves are ball-type valves that open and close a hydraulic passage. They are designed to block fluid flow when voltage is applied to the solenoid and to allow fluid flow when voltage is not applied. Most four-speed transmissions have two shift solenoid valves that control the shifting through all of the forward gears. By controlling which solenoid is energized, the computer controls the shift timing. Many transmissions energize one of the solenoids for first gear, the other solenoid for third gear, both solenoids for second gear, and neither for fourth gear.

## Processing

The primary purpose of using a computer is to process information. It does so by comparing all data it receives against data it has stored. All stored data remains in the computer's memory and is used, as needed, to analyze and correct operating conditions. The computers used to control transmission functions rely on programming stored in their memory to provide gear shifting at the optimum time. The decision to shift or not to shift is based on shift schedules and logic programmed into the memory of the computer.

In order for a computer in any electronic control system to determine when to initiate a gearshift change, it must be able to refer to shift schedules that it has stored in its memory. A **shift schedule** contains the actual shift points to be used by the computer according to the input data it receives from the sensors. Shift schedule logic chooses the proper shift schedule for the current conditions of the transmission. It uses the shift schedule to select the appropriate gear, and then determines the correct shift schedule or pattern that should be followed.

The first input a computer looks at to determine the correct shift logic, is the position of the gearshift lever. All shift schedules are based on the gear selected by the driver. The choices of shift schedules are limited by the type and size of engine that is coupled to the automatic transmission. Each engine/transmission combination has a different set of shift schedules. These schedules are coded by selector lever position and current gear range, and use throttle angle and vehicle speed as primary determining factors. The computer also looks at different temperature, load, and engine operation inputs for more information.

The basic shift logic of the computer allows the releasing apply device to slip slightly during the engagement of the engaging apply device. Once the apply device has engaged and the next gear is driven, the releasing apply device is pulled totally away from its engaging member and the transmission is fully into its next gear. This allows for smooth shifting into all gears.

The shift schedules set the conditions that need to be met for a change in gears. Since the computer frequently reviews the input information, it can make quick adjustments to the schedule if needed and as needed. The result of the computer's processing of this information and commanding outcomes according to a logically program is optimum shifting of the automatic transmission. This results in improved fuel economy and overall performance.

The electronic control systems used by the manufacturers differ with the various transmission models and the engines they are attached to. The components in each system and the overall operation of the system also vary with the different transmissions. However, all operate in a similar fashion and use basically the same parts.

## TORQUE CONVERTER CLUTCHES

Up to 10 percent of the engine's energy is wasted by torque converter slippage and the speed difference between the impeller and turbine during the coupling phase. To eliminate this slippage, late-model vehicles have a torque converter clutch (TCC) that mechanically links the engine to the input of the transmission during certain operating conditions. This results in improved fuel economy and reduced transmission fluid temperatures.

When applied, the TCC connects the turbine to the cover of the converter. Most converters equipped with a clutch consist of the three basic elements: impeller, turbine, and stator, plus a piston and clutch plate assembly, special thrust washers, and roller bearings.

The piston and clutch plate assembly has friction material on the outer portion of the plate with a spring cushioned damper assembly in the center. The clutch disc (plate) is splined to the turbine shaft and when the piston is applied, the plate locks the turbine to the converter **(Figure 8)**. The thrust washers and roller bearings control the movements of and provide bearing surfaces for the components of the converter.

Converter clutch apply or engagement only occurs when the PCM (or other control) determines that conditions are right for direct drive. The converter clutch does not connect the engine to the transmission when torque multiplication from the torque converter is needed, such as during acceleration. The clutch is disengaged during braking; this prevents the engine from stalling. Likewise, the clutch is disengaged when the vehicle is at a standstill while the transmission is in gear. Normally, during deceleration, the clutch is not applied. If the clutch is applied during deceleration, fuel will be wasted and there may be high exhaust emissions. The clutch is applied only when the engine is warm enough and it is running at a great enough speed to prevent a shudder or stumble when it engages. The control systems use inputs from many different sensors to determine when the conditions are suitable for clutch engagement.

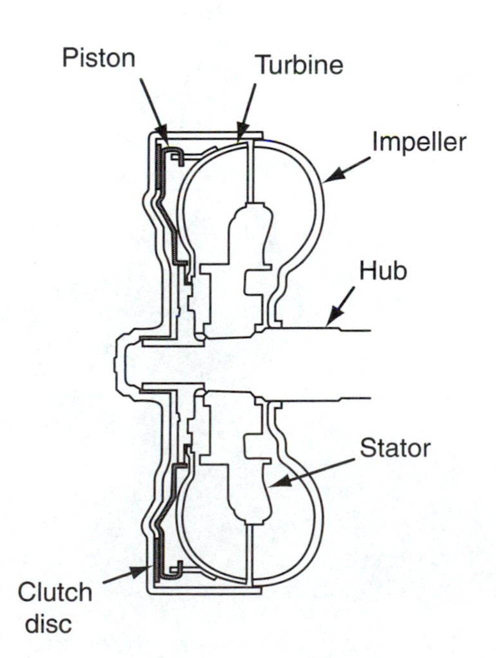


**Figure 8.** Piston-type converter clutch assembly.

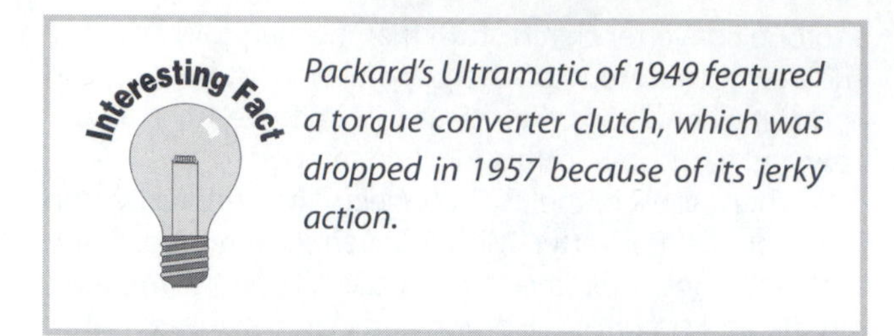


*Packard's Ultramatic of 1949 featured a torque converter clutch, which was dropped in 1957 because of its jerky action.*

## Hydraulic Converter Clutch

Early converter clutch systems relied on a hydraulically operated clutch. This design of converter clutch was controlled and operated totally by hydraulics and provided a mechanical link only when the transmission was in high gear. Today's transmissions apply the clutch in more than one forward gear and rely on computer-controlled electrical solenoids to control the hydraulic pressure to the clutch assembly **(Figure 9)**.

The operation of the hydraulic clutch is rather simple. However, the systems that control it can be fairly complex. The converter clutch is applied when the fluid flow through the torque converter is reversed by a valve. When the torque converter clutch control valve moves, the fluid begins to flow in a reversed direction **(Figure 10)**. This forces the clutch disc or pressure plate against the front of the torque converter's cover. The position of the clutch now blocks the fluid flow through the converter and a mechanical link exists between the impeller and the turbine.

Normally, when the converter clutch is not engaged, fluid flows down the turbine shaft, past the clutch assembly, through the converter, and out past the outside of the turbine shaft. Normal torque converter pressures keep the clutch firmly against the turbine. There is no mechanical link and only a fluid coupling exists.

To engage the clutch, mainline pressure is directed between the plate and the turbine. This forces the plate into contact with the front inner surface of the cover and

Figure 9. A typical torque converter clutch (TCC) solenoid.

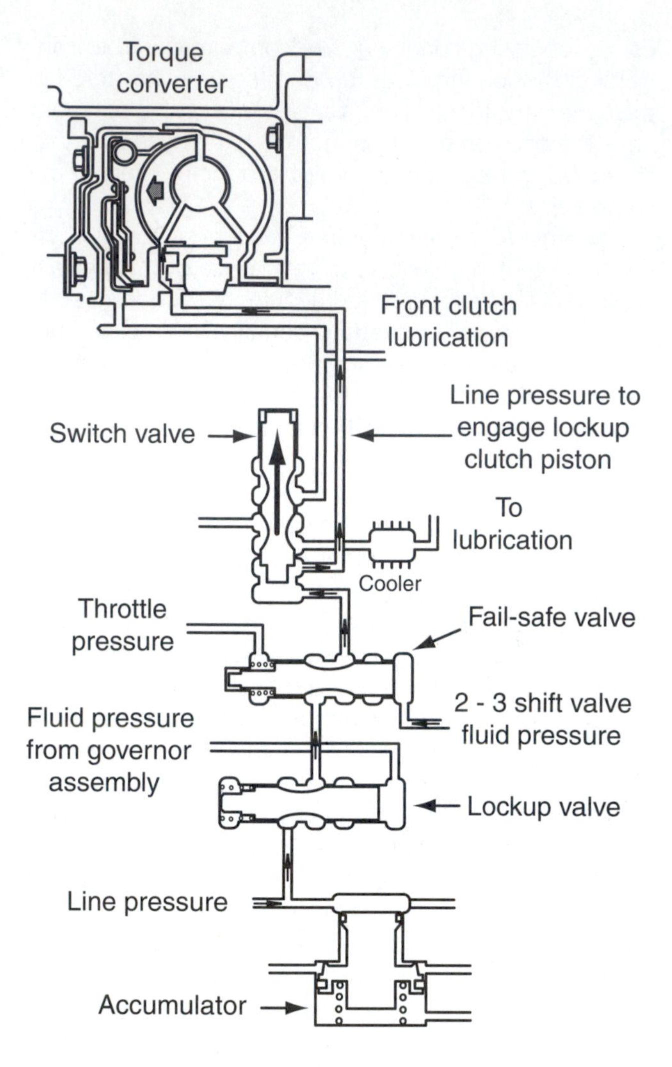


Figure 10. Lockup control valve operation to produce total clutch engagement in a piston-type converter clutch assembly.

locks the turbine to the impeller. With the engagement of the clutch, the fluid in front of the clutch is squeezed out before the clutch is totally engaged. The presence of this fluid softens the engagement of the clutch and acts much like an accumulator.

## Viscous Clutch

A **viscous converter clutch (VCC)** is used on some vehicles to provide torque converter clutch engagement. This design allows the clutch to engage in a very smooth manner with no engagement shock. It operates in the same way as an electronically controlled, hydraulically applied clutch, except that it uses a viscous clutch assembly to connect the impeller to the turbine. The viscous clutch assembly consists of a rotor, body, clutch cover, and silicone fluid **(Figure 11)**. The silicone fluid is sealed between the cover and the body of the clutch assembly. It

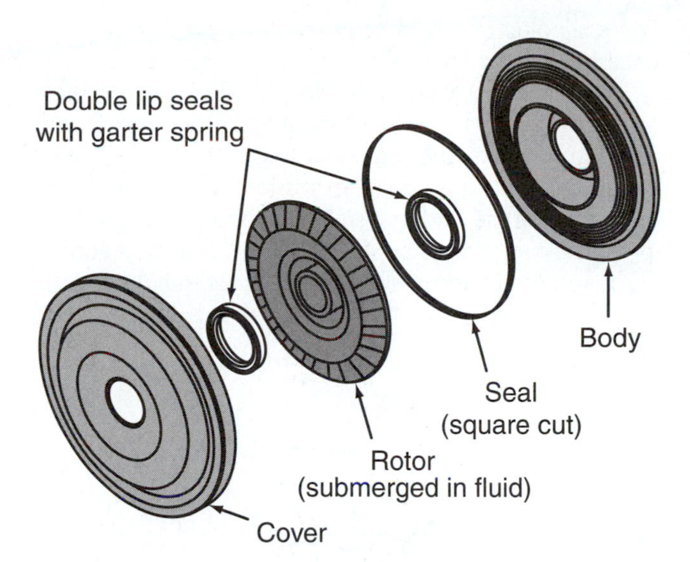


**Figure 11.** A viscous converter clutch assembly.

is the viscous silicone fluid that cushions the feel of clutch application.

The viscous converter clutch is a self-contained fluid coupling with a built-in friction faced pressure plate. When the clutch is engaged, the pressure plate is forced against the converter cover. Engine power is transmitted from the pressure plate through the fluid coupling to the transaxle's input shaft. The clutch's fluid coupling uses the viscous properties of thick silicone fluid, between the closely spaced pressure plate and cover plate, to transmit the power.

When the clutch is applied, there is a constant but minor amount of slippage in the viscous unit, about 40 rpm at 60 mph. However, this slippage is nothing compared to a conventional torque converter without lockup. Engagement of the viscous clutch allows for improved fuel economy and reduced fluid operating temperatures. When the clutch is disengaged, the assembly operates in the same way as a conventional torque converter.

## Clutch Assembly

A typical hydraulic converter clutch has a plate that acts as a clutch piston and is splined to the front of the turbine. Friction material is usually bonded to either the forward face of the clutch plate or the inner front surface of the converter cover.

A converter clutch disc is splined to the turbine so that it can drive the turbine when the friction material is forced against the torque converter's cover. A damper assembly directs the power flow through a group of coil springs, which are designed to absorb the normal torsional vibrations of an engine. The damper springs are placed evenly from the center of the disc and are sandwiched between two steel plates. One of the plates is attached directly to the clutch assembly's hub and the other to the clutch disc. The two plates will move as a single unit after the plates have moved against the tension of the springs. During clutch engagement, the sudden application of torque to one plate is absorbed by the springs as they compress and start the other plate turning. The damper assembly acts as a shock absorber to the pulses of the engine's vibrations and softens the engagement of the clutch. When the converter clutch is not applied, the fluid inside the torque converter absorbs the torsional vibrations. Therefore, a damper assembly is not needed until the clutch mechanically connects the engine to the input of the transmission.

## Converter Control Circuits

Many different systems have been used to control the application of the converter clutch. The systems vary from simple hydraulic controls at the valve body to computer-controlled solenoids. Simple systems limit the engagement of the torque converter clutch in high gear only and when the vehicle is traveling above a particular speed. Systems that are more complex allow the clutch to engage at any time efficiency would be improved by doing so.

Most of the computer-controlled systems base clutch engagement on inputs from various sensors, which give information about the engine's fuel system, ignition system, vacuum, operating temperature and vehicle speed. This information allows the computer to engage the clutch at exactly the right time according to operating conditions.

The clutch is typically engaged by a clutch piston controlled by an electric solenoid and one or more spool valves. The solenoid controls pressurized fluid that moves the spool valve to move the clutch piston. The movement of the piston engages or disengages the converter clutch. When oil pressure moves the piston against the clutch disc, engagement occurs. As the piston moves away from the disc, the clutch disengages.

# ELECTRONICALLY CONTROLLED TRANSMISSIONS

Electronic transmission control has become increasingly more common on today's cars. These controls provide automatic gear changes when certain operating conditions are met. Through the use of electronics, transmissions have better shift timing and quality. As a result, the transmissions contribute to improved fuel economy, lower exhaust emission levels, and improved driver comfort. Although these transmissions function in the same way as earlier hydraulically based transmissions, a computer determines their shift points. The computer uses inputs from several different sensors and matches this information to a predetermined schedule.

Hydraulically controlled transmissions relied on pressure differentials at the sides of a shift valve to hold or change a gear. Electronic transmissions still do. However, the pressure differential is caused by the action of shift solenoids that allow for changes in pressure on the side of a shift valve **(Figure 12)**. The computer controls these solenoids. The solenoids do not directly control the transmission's clutches and bands. These are engaged or disengaged in the same way as hydraulically controlled units. The solenoids simply control the fluid pressures in the transmission and do not perform a mechanical function.

Most electronically controlled systems are complete computer systems. There is a central processing unit, inputs, and outputs. Often the central processing unit is a separate computer designated for transmission control. This computer may be the TCM (transmission control module), PCM (powertrain control module), or the BCM (body control module). When transmission control is not handled by the

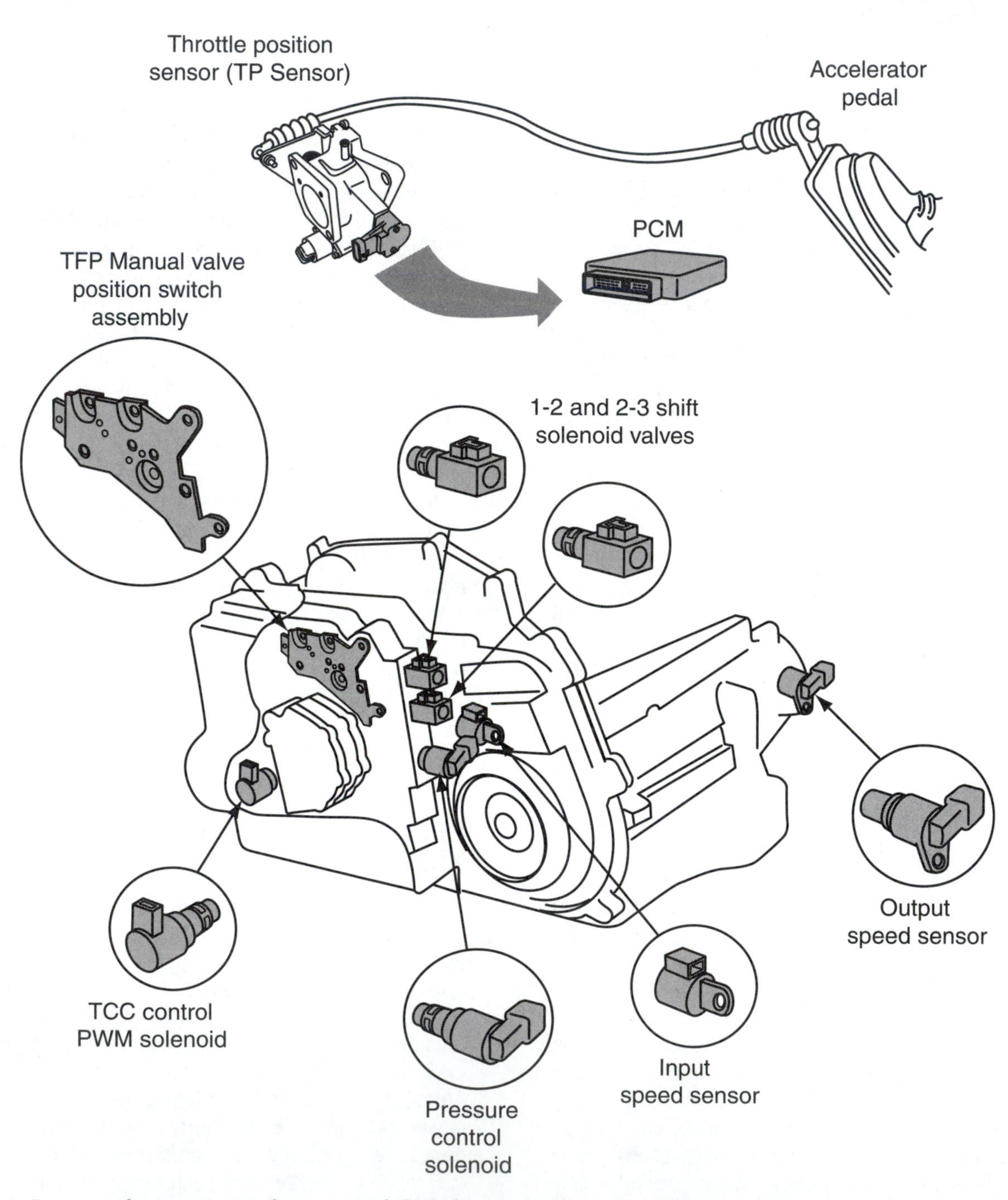


**Figure 12.** Layout of components for a typical GM electronically controlled transmission.

PCM, the controlling unit communicates with the PCM. In this chapter, the controlling computer for the transmission will be referred to as the TCM whether it is a separate computer or if it is integrated into another computer.

## Inputs

The inputs include transmission operation monitors plus the some of the sensors used by the PCM. Input sensors, such as the TP sensor, supply information for many different control systems. Various control modules may share this information.

The computer may receive information from two different sources: directly from a sensor, or through a twisted-pair bus circuit, which connects all of the vehicle computer systems **(Figure 13)**. This modulated bidirectional bus system allows the various computers in the vehicle to share information; this called **multiplexing**. Bus wires are twisted to reduce the chance of the signals being disrupted by radio frequency interference. This interference can cause the sensitive voltage signals to be altered and send false information to the PCM. The computers that share the bus have unique frequencies that serve as identification. The frequency may also be altered by radio frequency interferences. Radio frequency interference is a form of electromagnetic interference or electrical noise caused by the secondary ignition, high current flow, and the operation of devices, such as motors and solenoids.

Engine speed, throttle position, engine temperature, engine load, and other typical engine-related inputs are used by the computer to determine the best shift points. Many of these inputs are available through and are inputted from the common bus. Other information, such as engine and body identification, the TCM's target idle speed, and speed control operation are not the result of monitoring by sensors; rather, these have been calculated or determined by the TCM and made available on the bus **(Figure 14)**.

Typical bus inputs used by the TCM are from the Mass Airflow (MAF), Intake Air Temperature (IAT), Manifold Absolute Pressure (MAP), Barometric Pressure (BARO), Engine Coolant Temperature (ECT), and Crankshaft Position (CKP) sensors. These provide the TCM with information about the operating condition of the engine. Through these, the TCM is able to control shifting and TCC operation according to the temperature, speed, and load of the engine.

The inputs from the ECT are critical to the operation of the transmission. If the engine's coolant temperature is cold, the computer may delay upshifts to improve driveability. The computer also may engage the converter clutch in second or third gear if the coolant temperature rises.

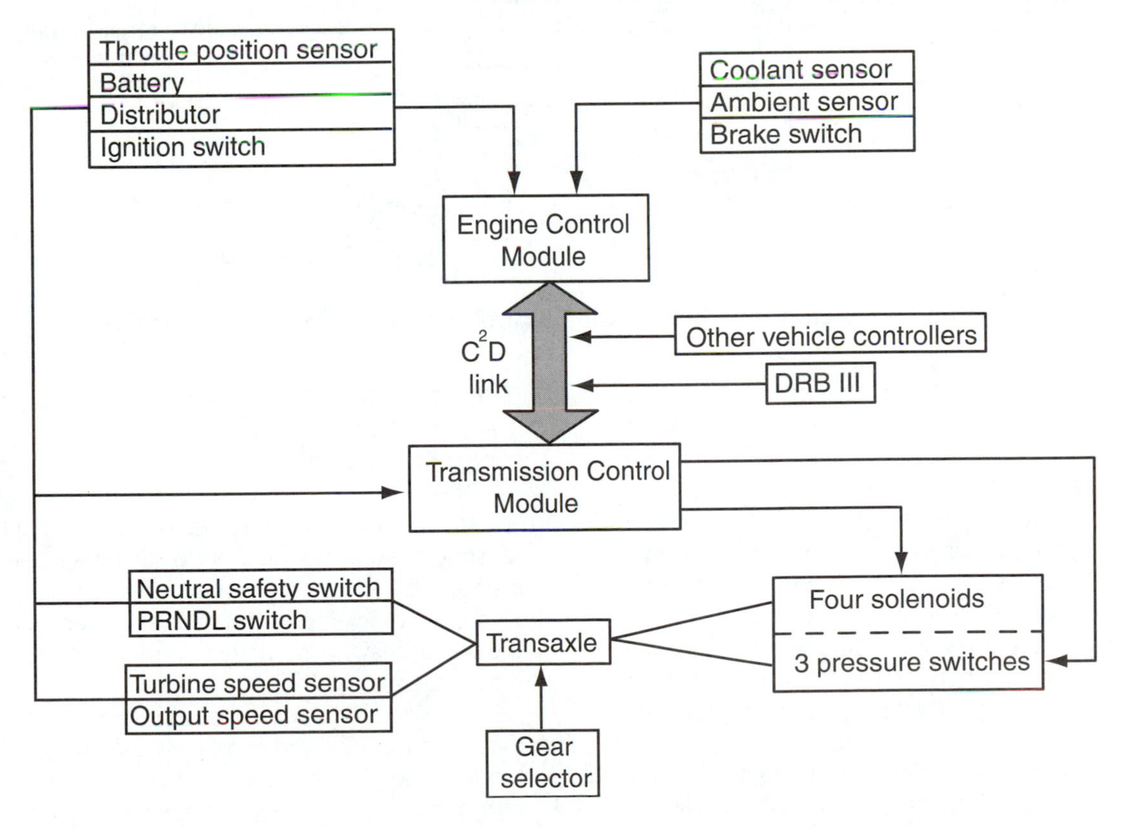


**Figure 13.** Schematic of the electronic control system using multiplexing. Note the CCD ($C^2D$) bus (link).

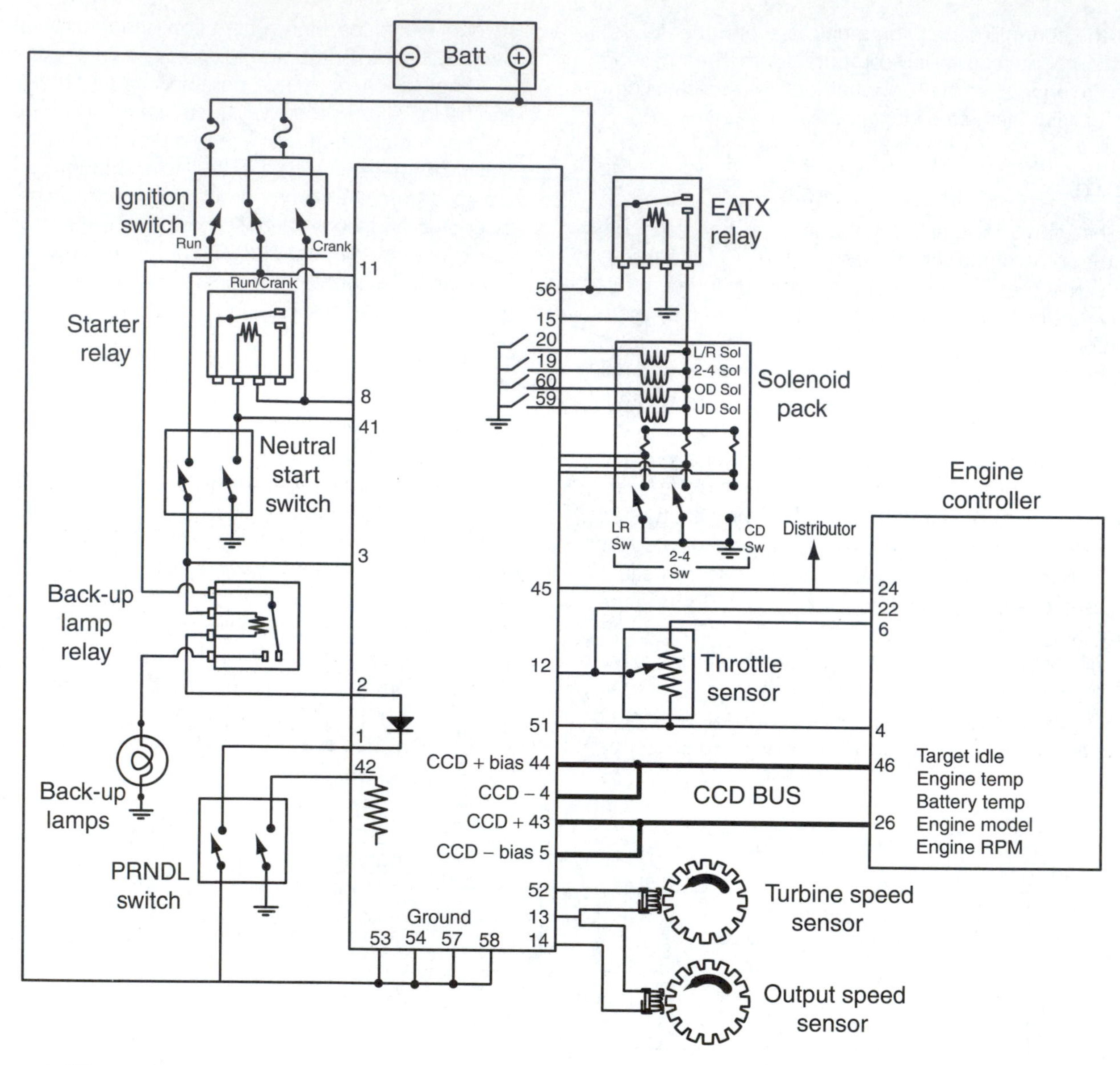


**Figure 14.** Wiring diagram for the electronic transmission controls on a Chrysler 41TE transaxle.

The MAP sensor keeps the computer informed of changes in engine load. This signal, combined with the signals from the TP and MAF sensors, allow the computer to have a good idea of the load and the driver's intent. The TP sensor sends a voltage signal to the TCM in response to throttle position. Not only is this signal used to inform the TCM of the driver's intent; it is also used in place of the hydraulic throttle pressure linkage.

The TCM may use the signal from the IAT to calculate the temperature of the battery. The TCM then uses this temperature calculation to estimate transmission fluid temperature.

The CKP sensor provides engine speed to the TCM. Although engine speed information is available at the bus, the computer receives this signal directly from the distributor pick-up coil or CKP sensor. With the direct feed, any time delay at the bus circuit is avoided and the computer is aware of current engine speeds. This input is used to determine shift timing and TCC apply and release.

The signals from the BARO are used by the TCM to adjust line pressures according to changes in altitude. This sensor input may not be used; its use depends on the type of intake air monitoring system the vehicle is equipped with. On those vehicles using the BARO sensor as an input, the sensor may be integrated on the PCM circuit board or mounted externally.

The direct inputs are those sensors that provide information to the TCM and do not use the bus circuit. Many of these sensors produce an analog signal that must be changed to a digital signal before the TCM can respond. This conversion is handled by an analog-to-digital (A/D) converter, the PCM, or a digital radio adapter controller

(DRAC). These typically convert an analog AC signal to a digital 5-volt square wave.

**On/Off Switches** The brake switch is used to disengage the torque converter clutch when the brakes are applied. Its input has little to do with the up and down shifting of gears, except in some systems it signals a need for engine braking. An A/C Request switch informs the TCM that the A/C has been turned on. The TCM then changes line pressure and shift timing to accommodate the extra engine load created by A/C compressor operation.

The Transmission Range (TR) sensor informs the TCM of the gear selected by the driver. This sensor normally also contains the neutral safety switch and the reverse light switch. The TR sensor is typically a multiple pole type on/off switch **(Figure 15)**.

**Transmission Fluid Temperature (TFT) Sensor** The TFT sensor monitors the temperature of the transmission's fluid. When the signal from this sensor is normal, the transmission will operate within its normal range. However, when the signal indicates that the fluid is overly hot, the TCM will allow the transmission to only operate in such a way that will allow the transmission to cool down. This prevents damage to the transmission. When the TFT signal indicates that the fluid is cooler than normal, the TCM will alter the shift schedule.

The TFT sensor is a thermistor located in the valve body. Its electrical resistance varies with ATF temperature. The TCM integrates this input with others to control TCC operation. Typically, the TCM prevents TCC engagement until fluid temperatures reach about 68°F (20°C). If fluid temperature reaches about 250°F (122°C), the TCM applies the TCC in second, third, or fourth gears. If mechanically connecting the engine to the input shaft of the transmission does not reduce fluid temperature, and it reaches 300°F (150°C), the TCM will release the TCC to prevent damage to the converter clutch from excessive temperatures. If the fluid reaches about 310°F (154°C), the TCM sets a fluid temperature trouble code and uses a fixed value as the fluid temperature input signal.

**Figure 15.** A TR range sensor.

Some vehicles are equipped with a low transmission fluid level indicator. This type of warning light is normally a simple circuit consisting of a sensor or sending unit, wires and the lamp. The sending unit provides a path for ground when the fluid is low, thereby lighting the warning lamp. When the fluid level is high enough to keep the switch open, the lamp remains off.

**Transmission Pressure Switches** Various transmission pressure switches can be used to keep the TCM informed as to which hydraulic circuits are pressurized and which clutches and brakes are applied. These input signals can serve as verification to other inputs and as self-monitoring or feedback signals.

**Voltage-Generating Sensors** The Vehicle Speed Sensor (VSS) and **Output Shaft Speed (OSS)** sensors are used to monitor output of the transmission and/or vehicle speed. In some electronic control systems, only one of these sensors is used. When a vehicle has both sensors, the OSS signal is used as a verification signal for the VSS by the engine control system. The VSS **(Figure 16)** is used as a verification signal for the OSS by the TCM. Some transmissions use these speed-related inputs in place of a governor. These signals are used to regulate hydraulic pressure and shift points and to control TCC operation. 4WD vehicles may use a third speed sensor installed in the transfer case. The TCM determines vehicle speed from this sensor, rather than the OSS.

Some transmissions have an input speed sensor. This sensor and its operation are identical to the OSS and its signal is used by the TCM to calculate converter turbine speed. Input and output speeds provided by the two sensors are used by the TCM to help determine line pressure, shift patterns, and TCC apply pressure and timing.

## Adaptive Controls

Many late-model transmissions have systems that allow the computer to change transmission behavior in response to the operating conditions and to the habits of the driver. The system monitors the condition of the engine and compensates for any changes in the engine's performance. It also monitors and memorizes the typical driving style of the driver and the operating conditions of the vehicle. With this information, the computer adjusts the timing of shifts and converter clutch engagement to provide good shifting at the appropriate time.

These systems are constantly learning about the vehicle and driver. The computer adapts its normal operating procedures to best meet the needs of the vehicle and the driver. When systems are capable of doing this, they are said to have **adaptive learning** capabilities. To store this information, the computer includes a long-term adaptive memory.

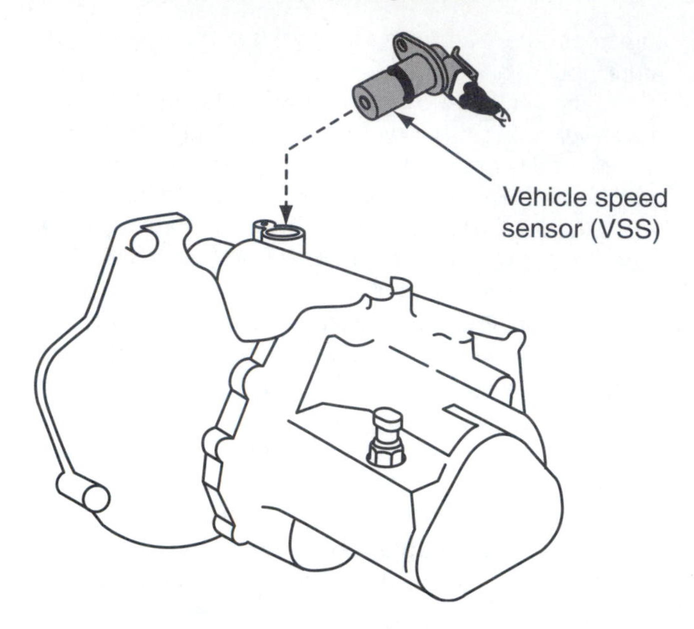


Figure 16. Typical location of a VSS mounted in a transaxle.

## Outputs

A transmission control system uses two to five solenoids **(Figure 17)**, controlled by the TCM, to regulate shift timing, feel, and TCC application. The number and purpose of each depends on the model of transmission. Normally, two solenoids are used as shift solenoids. They control the delivery of fluid to the manual shift valve. An **EPC (Electronic Pressure Control) solenoid** is used to control hydraulic pressures throughout the transmission. This solenoid operates on a duty cycle controlled by the TCM **(Figure 18)**. Its purpose is to regulate line pressure according to engine running conditions and engine torque. An additional solenoid is used to provide for engine braking during coasting. This solenoid operates when the vehicle is slowing down and the throttle is closed **(Figure 19)**. The other solenoid controls the operation of the converter clutch.

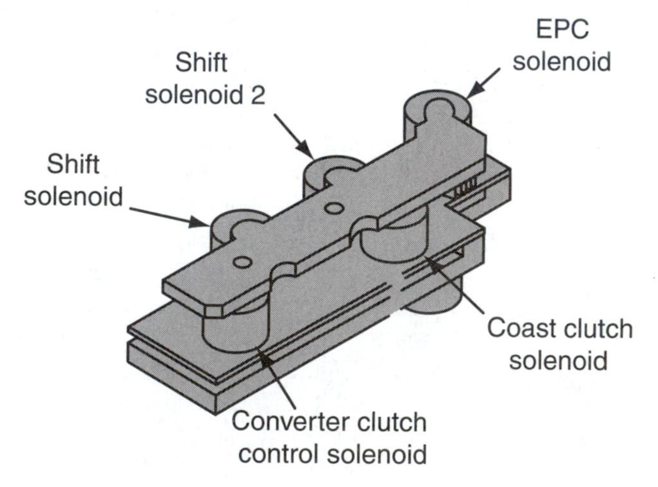


Figure 17. Typical transmission solenoid assembly.

The computer also learns the characteristics of the transmission and changes its programming accordingly. It learns the release and application rates of various transmission components during various operating conditions. Adaptive learning allows the computer to compensate for wear and other events that might occur and cause the normal shift programming to be inefficient. Doing this, the adaptive learning capability of the transmission computer allows for this smooth shifting throughout the life of the transmission. As component wear and shift overlap times increase, the TCM adjusts line pressure to maintain proper shift timing calibrations.

The adaptive learning takes place as the TCM reads input and output speeds more than 140 times per second. The computer responds to each new reading. This learning process allows the TCM to make adjustments to its program so that quality shifting always occurs.

Direct battery voltage is supplied to the TCM. If the computer loses source voltage, the transmission on some vehicles will enter into a default or "limp-in" mode. The transmission will also enter into the default mode if the TCM senses a transmission failure. At this point, a fault code will be stored in the memory of the computer and the transmission will remain in default until the transmission is repaired. While in the default mode, the transmission will operate only in Park, Neutral, Reverse, and Second gears. The transmission will not upshift or downshift. This mode allows the vehicle to be operated, although its efficiency and performance is hurt.

Figure 18. A solenoid.

The shift solenoids offer four possible on/off combinations to control fluid to the various shift valves. These solenoids are on/off solenoids that are normally off and in the open position. Being open, the solenoid valves allow line pressure to the bore of the shift valve and keep the shift valve closed. When the shift solenoids are energized, they block the line pressure and exhaust line pressure from the valve. This allows the shift valve to open.

The **EPC solenoid** replaces the conventional TV cable setup to provide changes in pressure in response to engine load. This solenoid is a **variable force solenoid (VFS)** and contains a spool valve and spring. To control fluid pressure, the PCM sends a varying signal to the solenoid. This varies the amount the solenoid will cause the spool valve to move. When the solenoid is off, the spring tension keeps the valve in place to maintain maximum pressure. As more current is applied to the solenoid, the solenoid moves the spool valve more, which moves to uncover the exhaust port more, thereby causing a decrease in pressure **(Figure 20)**. The EPC solenoid controls line pressure, at all times, based on the programming of the system's computer, and it is able to match shift timing and feel with the current needs of the vehicle.

Typically, the operation of the converter clutch is also totally controlled by the computer. The only exception to this is during first gear and reverse gear operation when the clutch is disabled hydraulically to prevent engagement regardless of the commands by the computer. Normally, the converter clutch is hydraulically applied and electrically controlled through a pulse width modulated (PWM) solenoid **(Figure 21)**, which is controlled by the TCM. Modulat-

| SHIFT SOLENOID OPERATION CHART | | | | | |
|---|---|---|---|---|---|
| Transaxle range selector lever position | Powertrain Control Module Gear commanded | Eng braking | AX4N solenoids | | |
| | | | SS 1 | SS 2 | SS 3 |
| P / N [a] | P / N | NO | OFF [b] | ON [b] | OFF |
| R (Reverse) | R | YES | OFF | ON | OFF |
| Overdrive | 1 | NO | OFF | ON | OFF |
| | 2 | NO | OFF | OFF | OFF |
| | 3 | NO | ON | OFF | ON |
| | 4 | YES | ON | ON | ON |
| D (Drive) | 1 | NO | OFF | ON | OFF |
| | 2 | NO | OFF | OFF | OFF |
| | 3 | YES | ON | OFF | OFF |
| Manual 1 | 2 [c] | YES | OFF | ON | OFF |
| | 3 [c] | YES | OFF | OFF | OFF |
| | | YES | ON | OFF | OFF |

a When transmission fluid temperature is below 50° then SS1=OFF, SS2=ON, SS3=ON to prevent cold creep.

b Not contributing to powerflow

c When a manual pull-in occurs above calibrated speed the transaxle will downshift from the higher gear until the vehicle speed drops below this calibrated speed.

**Figure 19.** Shift solenoid operation for an AX4N transaxle.

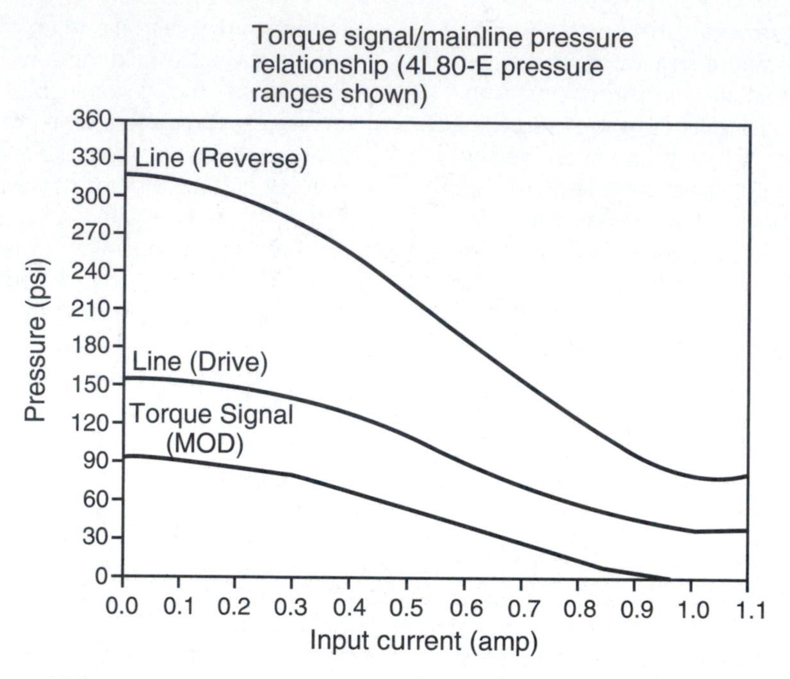


**Figure 20.** The EPC solenoid controls hydraulic pressures throughout the transmission by controlling the current going to the solenoid.

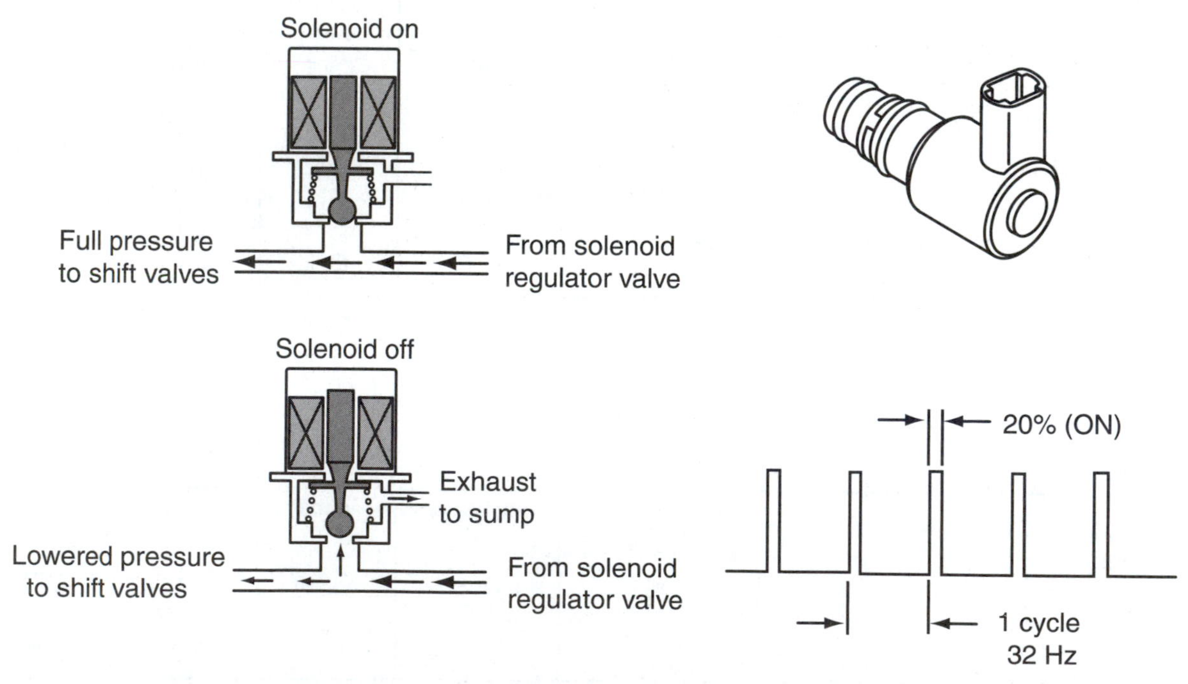


**Figure 21.** (Left) A typical PWM solenoid. (Right) The signal representing the control or ordered duty cycle from the computer.

ing the pressure to the converter clutch allows for smooth engagement and disengagement and also allows for partial engagement of the clutch.

The PWM solenoid is installed in the valve body. It controls the position of the TCC apply valve. When the solenoid is off, TCC signal fluid exhausts and the converter clutch remains released. Once the solenoid is energized, the plunger moves the metering ball to allow TCC signal fluid to pass to the TCC regulator valve. The TCM cycles the PWM solenoid on and off 32 times per second, but varies the length of time it is energized in each cycle.

## Operational Modes

With electronic controls, automatic transmissions can be programmed to operate in different modes of operation. The desired mode is selected by the driver **(Figure 22)**. The mode selection switch can be located on the center console or the instrument panel. Most transmissions with this feature have two selective modes, "Normal" and "Power." During the normal mode, the transmission operates according to the shift schedule and logic set for normal or regular operation. In the power mode, the TCM uses different logic and shift schedules to provide for better acceleration and performance with heavy loads. Normally this means delaying upshifts.

If three modes are available, the third mode is called the "Auto" mode. The auto mode is a mixture between the normal and power modes. While in this mode, the TCM will control the shifts in a normal way. However, if the throttle is quickly opened, the shift pattern will switch to the power mode.

## Manual Shifting

Some electronically controlled transmissions are available with manual shift controls. These systems allow the driver to manually upshift and downshift the transmission at will, much like a manual transmission. Unlike a manual transmission, the driver does not need to depress a clutch pedal. All the driver does is move a gear selector or hit a button and the transmission changes gears. If the driver does not change gears and engine speed is high, the transmission shifts on its own. If the driver elects to let the transmission shift automatically, a switch disconnects the manual control and the transmission operates automatically.

**Figure 22.** The transmission mode selector switch for Toyota's A-140E electronic transmission.

**BMW's Steptronic** This is a five-speed automatic transmission with a manual control option. Manual shifting is performed by moving a control on the console. Moving the selector forward provides for an upshift. A movement down allows for a downshift. The selected gear is displayed on the instrument panel to keep the driver informed of the selected gear. There also is a sport position for the gearshift. In this position, automatic gear changes are delayed to maximize acceleration.

**Chrysler AutoStick** This is the most familiar system. Manual shifting is performed by moving a control on the console. Moving the selector to the right provides for an upshift. A movement to left allows for a downshift.

The transaxle is not modified for this option; rather, it is fitted with a special gear selector and switch assembly. The driver either can manually shift the gears or allow the transaxle to shift automatically. The selected gear is displayed on the instrument panel to keep the driver informed of the selected gear.

The Autostick feature also will be deactivated if the TCM senses problems and/or sets a trouble code that relates to the TR sensor or Autostick switch or when there is a high engine and transmission temperature code.

**Honda's Sequential Sport Shift** Manual shifting is performed by moving a control on the console. Moving the selector forward provides for an upshift. A movement down allows for a downshift. This transmission is unique in that it will not automatically upshift if the driver brings the engine's speed too high. All other transmissions of this design will upshift automatically at a predetermined engine speed to prevent damage to the engine.

**Tiptronic** This is a five-speed transmission available in Porsches. Audi also has a Tiptronic transmission that is different from Porsche's. Manual shifting on the Audi is performed by moving a control on the console. Moving the selector forward provides for an upshift. A movement down allows for a downshift. To shift Porsche's transmission, the driver moves the gear selector into the manual gate, next to the automatic ranges, and depresses buttons on the steering wheel.

**Toyota** Taking electronics one step further, Toyota has a series of Lexus High Performance cars. These cars have many features designed to mix traditional Lexus luxury with performance. One of these is the five-speed transmission that can operate in either of two modes providing fully automatic shifting or electronic manual control. The TCM is programmed to allow for rapid shifts in response to the driver's commands. It also will prevent shifting during conditions that may cause engine or transmission failure. The transmission also may be shifted by its gated console-mounted shift lever **(Figure 23)**. The shift lever allows the driver to select individual gear ranges as well as the full-automatic mode.

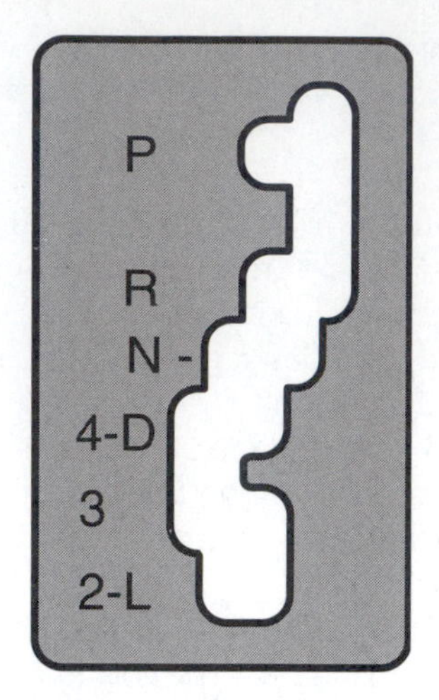


**Figure 23.** The gated shifter for Lexus' automatic–manual transmission.

Manual shifting may also be controlled by fingertip shifting buttons located on both horizontal spokes of the steering wheel. Touching a button on the front of the steering wheel triggers downshifts. Contacting the buttons on the backside of the steering wheel controls upshifts. The buttons are located so that either thumb can be used to downshift and either index finger can be used to upshift.

## Honda's CVT Controls

The electronic control system for Honda's CVT consists of a TCM, various sensors, three linear solenoids, and an inhibitor solenoid. Pulley ratios are always controlled by the control system. Input from the various sensors determines which linear solenoid the TCM will activate **(Figure 24)**. Activating the shift control solenoid changes the shift control valve pressure, causing the shift valve to move. This changes the pressure applied to the driven and drive pulleys, which changes the effective pulley ratio. Activating the start clutch control solenoid moves the start clutch valve. This valve allows or disallows pressure to the start clutch assembly. When pressure is applied to the clutch, power is transmitted from the pulleys to the final drive gear set.

The start clutch allows for smooth starting. Since this transaxle does not have a torque converter, the start clutch is designed to slip just enough to get the car moving without stalling or straining the engine. The slippage is controlled by the hydraulic pressure applied to the start

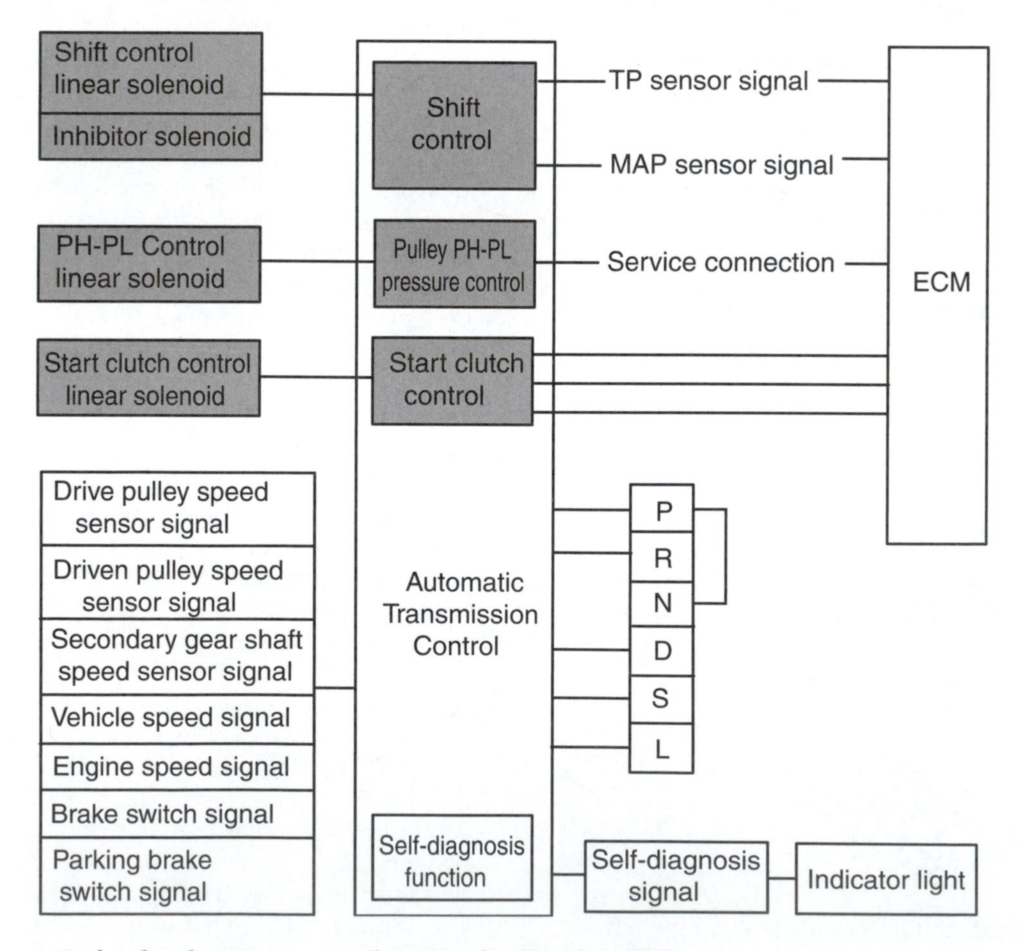


**Figure 24.** Schematic for the electronic control system for Honda's CVT.

clutch. To compensate for engine loads, the TCM monitors the engine's vacuum and compares it to the measured vacuum of the engine while the transaxle was in Park or Neutral.

The TCM controls the pulley ratios to reduce engine speed and maintain ideal engine temperatures during acceleration. If the car is continuously driven at full throttle acceleration, the TCM causes an increase in pulley ratio. This reduces engine speed and maintains normal engine temperature while not adversely affecting acceleration. After the car has been driven at a lower speed or not accelerated for a while, the TCM lowers the pulley ratio. When the gear selector is placed into reverse, the TCM sends a signal to the PCM. The PCM then turns off the car's air conditioning and causes a slight increase in engine speed.

## Summary

- By using electronic controls for transmission operation, the amount of wasted power can be reduced and the overall operation of the transmission can be more responsive with better reliability.
- A computer is an electronic device that receives information, stores information, processes information, and communicates information.
- When applied, the torque converter clutch connects the turbine to the cover of the converter.
- Most torque converters with a clutch consist of the three basic elements—impeller, turbine, and stator—plus a piston and clutch plate assembly, special thrust washers, and roller bearings.
- The piston and clutch plate assembly has friction material on the outer portion of the plate with a spring cushioned damper assembly in the center. The clutch plate is splined to the turbine shaft and when the piston is applied, the plate connects the turbine to the converter.
- To engage the clutch, mainline pressure is directed between the plate and the turbine. This forces the plate into contact with the front inner surface of the cover and locks the turbine to the impeller.
- In an electronically controlled transmission, once the information has been processed, the computer sends out commands to its output devices or actuators. Typical output devices are solenoids and motors, which cause something mechanical or hydraulic to change.
- Common voltage reference sensors include on/off switches, potentiometers, thermistors, and pressure sensors.
- Pressure sensors respond to pressure applied to a movable diaphragm in the switch. As pressure increases, so does the movement of the diaphragm and the sensor's resistance.
- If a switch is cycled on/off very rapidly, the return signal is a rapid on/off one. This describes a digital signal, a series of on/off pulses.
- A shift schedule contains the actual shift points to be used by the computer according to the input data it receives from the sensors. Shift schedule logic chooses the proper shift schedule for the current conditions of the transmission.
- Adaptive learning allows the computer to compensate for wear and other events that might occur and cause the normal shift programming to be inefficient.
- The typical output devices are solenoids and motors, which cause something mechanical or hydraulic to change.
- Engine speed, throttle position, temperature, engine load, and other typical engine-related inputs are also used by the computer to determine the best shift points. Many of these inputs are available through multiplexing and are inputted from the common bus.
- The adaptive learning takes place as the computer reads input and output speeds over 140 times per second.
- The EPC solenoid replaces the conventional TV cable setup to provide changes in pressure in response to engine load.
- The TCM is programmed to adjust its operating parameters in response to changes within the system, such as component wear. As component wear and shift overlap times increase, the TCM adjusts line pressure to maintain proper shift timing calibrations.
- Some electronically controlled transmissions are available with manual shift controls that allow the driver to manually upshift and downshift the transmission at will.
- On some models, an operational mode selector switch is located on the center console or instrument panel. This switch allows the driver to select different modes to change transmission upshift characteristics.

# Review Questions

1. A computer relies on many different reference voltage sensors. These can be divided into types. Name the types and give a brief description of their operation.
2. Although computers receive different information from a variety of sensors, the decisions for shifting are actually based on more than the inputs. What are they based on?
3. Some transmissions receive information through multiplexing. How does this work?
4. How does a solenoid in an electronically controlled transmission work?
5. What is adaptive learning?
6. Technician A says that voltage generation devices are typically used to monitor rotational speeds. Technician B says that the most common voltage generation device used in electronic transmission control systems is the Vehicle Speed Sensor. Who is correct?
   A. Technician A only
   B. Technician B only
   C. Both Technician A and Technician B
   D. Neither Technician A nor Technician B
7. Technician A says that throttle position is an important input in most electronic shift control systems. Technician B says that vehicle speed is an important input for most electronic shift control systems. Who is correct?
   A. Technician A only
   B. Technician B only
   C. Both Technician A and Technician B
   D. Neither Technician A nor Technician B
8. Technician A says that shift solenoids direct fluid flow to and away from the various apply devices in the transmission. Technician B says that shift solenoids are used to mechanically apply a friction brake or multiple-disc clutch assembly. Who is correct?
   A. Technician A only
   B. Technician B only
   C. Both Technician A and Technician B
   D. Neither Technician A nor Technician B
9. When discussing shift logic, Technician A says that the logic must be based on the current engine and transmission combination. Technician B says that the logic must be based on the driver's driving habits. Who is correct?
   A. Technician A only
   B. Technician B only
   C. Both Technician A and Technician B
   D. Neither Technician A nor Technician B
10. Technician A says that multiplexing allows information to be shared with many computers. Technician B says that multiplexing reduces the number of wires and components needed in current vehicles. Who is correct?
    A. Technician A only
    B. Technician B only
    C. Both Technician A and Technician B
    D. Neither Technician A nor Technician B

# Chapter 13 Common Transmissions

## Introduction

Throughout the years, the automatic transmission has evolved into a complex machine with electronic, hydraulic, and mechanical components. It is safe to say this evolution will continue for years to come. This chapter looks at the different designs of automatic transmissions. Transmission designs vary from manufacturer to manufacturer and within each manufacturer. These variations may be simply different electronic controls for shifting or for the torque converter or the number of forward speeds a transmission has. Variations also result from the way a particular model of transmission is constructed and the vehicle it will be installed in.

> **Interesting Fact**
> *In 1939, the big automotive news was the Hydra-Matic Drive introduced by Oldsmobile. The Hydra-Matic was a combination of a liquid flywheel and fully automatic transmission. The headlines read "The '40 Olds shifts without a clutch? What will they think of next?"*

### TRANSMISSION VERSUS TRANSAXLE

As you know, automobiles are propelled in one of three ways: by the rear wheels, by the front wheels, or by all four wheels. The type of driveline helps determines whether a transmission or a transaxle will be used.

Vehicles propelled by the rear wheels normally use a transmission. Transmission gearing is located within an aluminum or iron housing. A few front engined vehicles have the transmission located with the final drive unit at the rear axle **(Figure 1)**. Typically, the transmission case is attached to the rear of the engine, at the front of the vehicle. The transmission case may be made of two **(Figure 2)** or three separate castings. A drive shaft links the output shaft of the transmission with the differential and drive axles located in a separate housing at the rear of the vehicle.

The front wheels propel FWD vehicles. For this reason, they must use a drive design different from that of a RWD vehicle. FWD vehicles are typically equipped with a transaxle and have no need for a separate differential and drive axle housing. A transaxle is a compact unit that combines the transmission gearing, differential, and drive axle connections into an aluminum housing located in front of the vehicle **(Figure 3)**. This design has two primary advantages: good traction on slippery roads due to the weight of

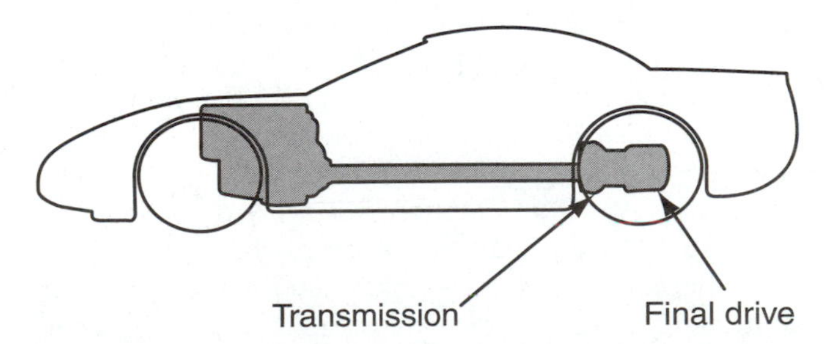


**Figure 1.** A front-engined car with the transmission mounted directly to the final drive unit at the rear axle.

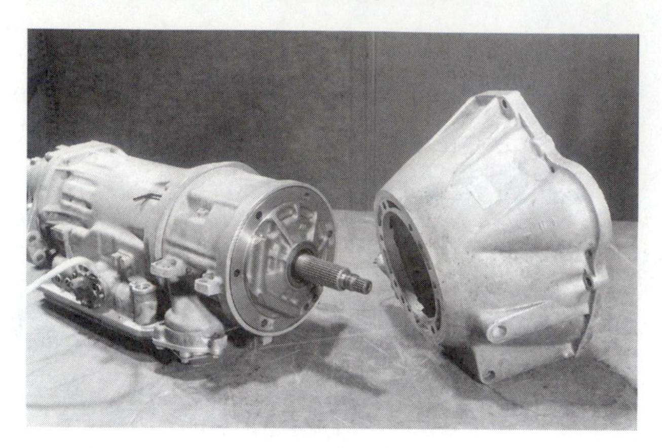

**Figure 2.** A transmission housing with a separate torque converter housing.

**Figure 3.** A typical transaxle.

the power train components being directly over the driving axles of the vehicle and transverse engine and transaxle configurations also allow for lower hood lines, thereby improving the vehicle's aerodynamics.

Not all FWD vehicles have a transversely mounted engine. Some have the engine mounted longitudinally and use a transaxle that looks like a conventional transmission design modified to drive the front wheels directly.

4WD vehicles typically use a transmission and transfer case. The transfer case mounts on the side or back of the transmission. A chain or gear drive inside the transfer case receives power from the transmission and transfers it to two separate drive shafts. One drive shaft connects to a differential on the front drive axle. The other drive shaft connects to a differential on the rear drive axle.

There are a few mid-engined and rear-engined cars out there. These can use a transmission or a transaxle. To simplify things, you can look at the driveline and tell whether or not the vehicle has a transmission or a transaxle. If there is a separate drive axle unit with a differential, a transmission is used. If the drive axles extend from the transmission unit, it is a transaxle.

## BASIC DESIGNS

All automotive transmissions/transaxles are equipped with a varied number of forward speed gears, a neutral gear, and one reverse speed. Transmissions can be divided into groupings based on the number of forward speed gears they have. Most current designs have four or five forward speeds. A few have six or seven forward speeds.

Transmissions are designed and built for particular applications. Things that must be considered in the design of a transmission is the power output of the engine, the weight of the vehicle the transmission will be installed in, and the typical workload of the vehicle. All of these play a part in the size and strength of the parts and materials used in a transmission. Obviously, high-powered and hard working vehicles need larger and stronger materials.

Overall transmission/transaxle size is also an important design consideration. This often dictates the placement of the various shafts inside the transmission and the size of the components. Transmission designs also vary according to the type of compound planetary gear sets used.

Transaxle design is most affected by the intended application. Because these units attach directly to the engine and that combination must fit between the two drive wheels, size is important. A variety of shaft arrangements can be found in today's transaxles. Long shafts may be divided into two parts and the ends of each connected by gears or chains. Final drive units vary in design as well. Common final drive setups for FWD vehicles use helical gears, planetary gears, hypoid gears, and/or a chain drive.

It is important to know that no two transmission models are exactly alike, regardless of their outward appearance. Keep in mind that these variations are based on external and internal components.

Although there are many different designs, all transmissions and transaxles rely on gears, shafts, bearings, and apply devices to function. All automatic transmissions, except CVTs, have a torque converter to connect and disconnect the engine's power to the transmission. All have an input shaft to transfer power from the torque converter to the internal gear sets and drive members. They also have gear sets to provide the different gear ratios, a reverse gear, and a neutral position. All have an output shaft to transfer power from the transmission to the final drive unit. They also have a hydraulic system that includes the pump, valve body, pistons, servos, brakes, and clutches.

### Model Numbers

In past, manufacturers identified their transmission models with internal codes that represented the design of the transmission. Although technicians who worked on transmissions learned to know the differences between the models, there was little logic used to denote the features of a particular design.

Manufacturers now use more definitive codes for model identification. Most use an alphanumeric code and the features of each transmission model are explained in their service manuals. Let us take a look at the transmission model codes for some manufacturers.

Interpreting a General Motors' transmission code will reveal the work capacity of the transmission, the number of forward gears, its directional placement in the vehicle, and if it has electronic controls. A 4T40-E transmission is a four-speed, transversely mounted, light-to-medium duty, electronically controlled unit. The "4" in the model number designates the number of forward gears, the "T" shows that it is a transverse mounted unit, the "40" is the product series, and the "E" means it has electronic controls.

Other commonly used GM transmissions are the 4L60-E and the 4L80-E. Both of these are found in pickup trucks and SUVs. Both of these models are longitudinally mounted, four-speed units with electronic controls. The internal construction of these units is different, as noted by the different product series number. The 4L80-E unit is designed for heavier duty than the 4L60-E.

Chrysler uses a similar system **(Figure 4)**. The first character denotes the number of forward gears. The second character is the duty rating. In contrast to GM's use of a two-digit code for duty rating, Chrysler uses a single digit code.

Ford Motor Company uses similar logic for model designations; however, their duty rating is more specific. The 4F27E transmission is used in the Focus. The model code is broken into four character groups. The "4" indicates the transmission is a four-speed unit. The "F" denotes that this is a transaxle for FWD vehicles. The "27" is the duty rating and represents the maximum input torque, after the torque converter, this transaxle was designed for. In this case, the 27 represents 270 lbs.-ft. (365 N•m). The "E" means the transaxle is fully electronically controlled.

The importance of using the service manual to decipher the codes is apparent when considering the model code for another commonly used Ford transmission. The 4R70W is a four-speed transmission for RWD pickups and SUVs. This transmission is electronically controlled, but the model does not contain an "E"; rather, there is a "W." The W indicates this heavy-duty (maximum input torque of 700 lbs.-ft.) unit has wide ratio gears.

## SIMPSON GEAR TRAIN

The Simpson gear train is an arrangement of two separate planetary gear sets with a common sun gear, two ring gears, and two planetary pinion carriers. One half of the compound set or one planetary unit is referred to as the front planetary and the other planetary unit is the rear planetary. The two planetary units do not need to be the same size or have the same number of teeth on their gears. The size and number of gear teeth determine the actual gear ratios obtained by the compound planetary gear assembly.

The typical power flow through a Simpson gear train when it is in Neutral has engine torque being delivered to the transmission's input shaft by the torque converter's turbine. No planetary gear set member is locked to the shaft; therefore, engine torque enters the transmission but goes nowhere else.

When the transmission is shifted into first gear, the input shaft is locked to the front planetary ring gear. The front ring gear drives the front planet gears, which drive the sun gear. The rear planet carrier is locked; therefore, the sun gear spins the rear planet gears. The planet gears drive rear ring gear, which is locked to the output shaft, in a clockwise direction. The result is a forward gear reduction.

When the transmission is operating in second gear, the input shaft is locked to the front planetary ring gear. The front ring gear drives the front planet gears. The front planet gears walk around the sun gear because it is held. The walking of the planets forces the planet carrier to turn. Because

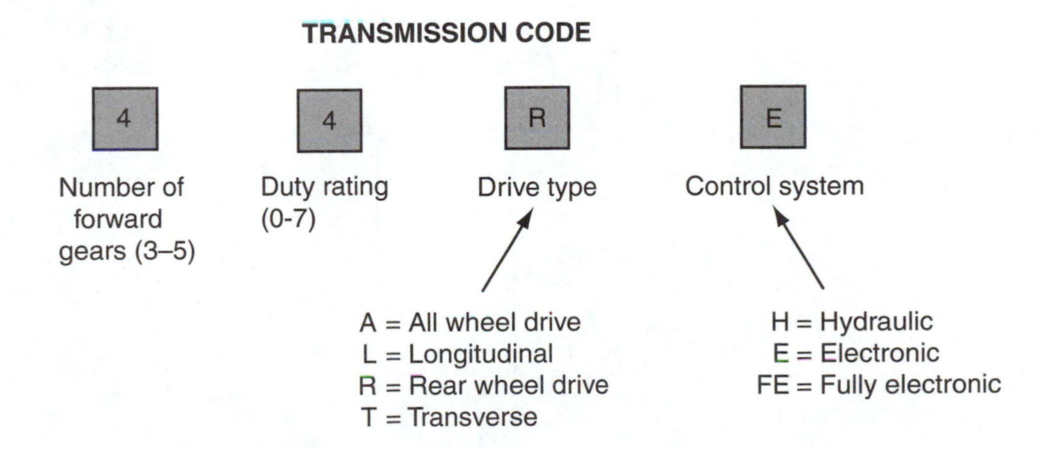


**Figure 4.** Transmission model designation code interpretation for Chrysler vehicles.

the carrier is locked to the output shaft, it causes the shaft to rotate in a forward direction with some gear reduction.

When operating in third gear, the front ring gear and the sun gear receive the input. Because the sun and ring gear are rotating at the same speed and in the same direction, the front planet carrier is locked between the two and is forced to move with them. Because the front carrier is locked to the output shaft, direct drive results.

To obtain reverse gear, input is received by the sun gear. The sun gear is driven in a clockwise direction and drives the rear planet gears in a clockwise direction. The rear planet carrier is held and the planet gears drive the rear ring gear in a counterclockwise direction. The result is a reverse gear with gear reduction.

## CHRYSLER TRANSMISSIONS

Chrysler Corporation introduced the Torqueflite transmission in 1956. This transmission was the first modern three-speed automatic transmission with a torque converter and the first to use the Simpson two-planetary compound gear train. All Torqueflite-based transmissions and transaxles use a Simpson gear train **(Figure 5)**.

Chrysler has used many different models and designs of transmissions. Because they were based on a Simpson gear set, they operate similarly. Three-speed models relied on one compound gear set, and many four-speed models use that same planetary unit with an additional simple planetary unit for fourth gear.

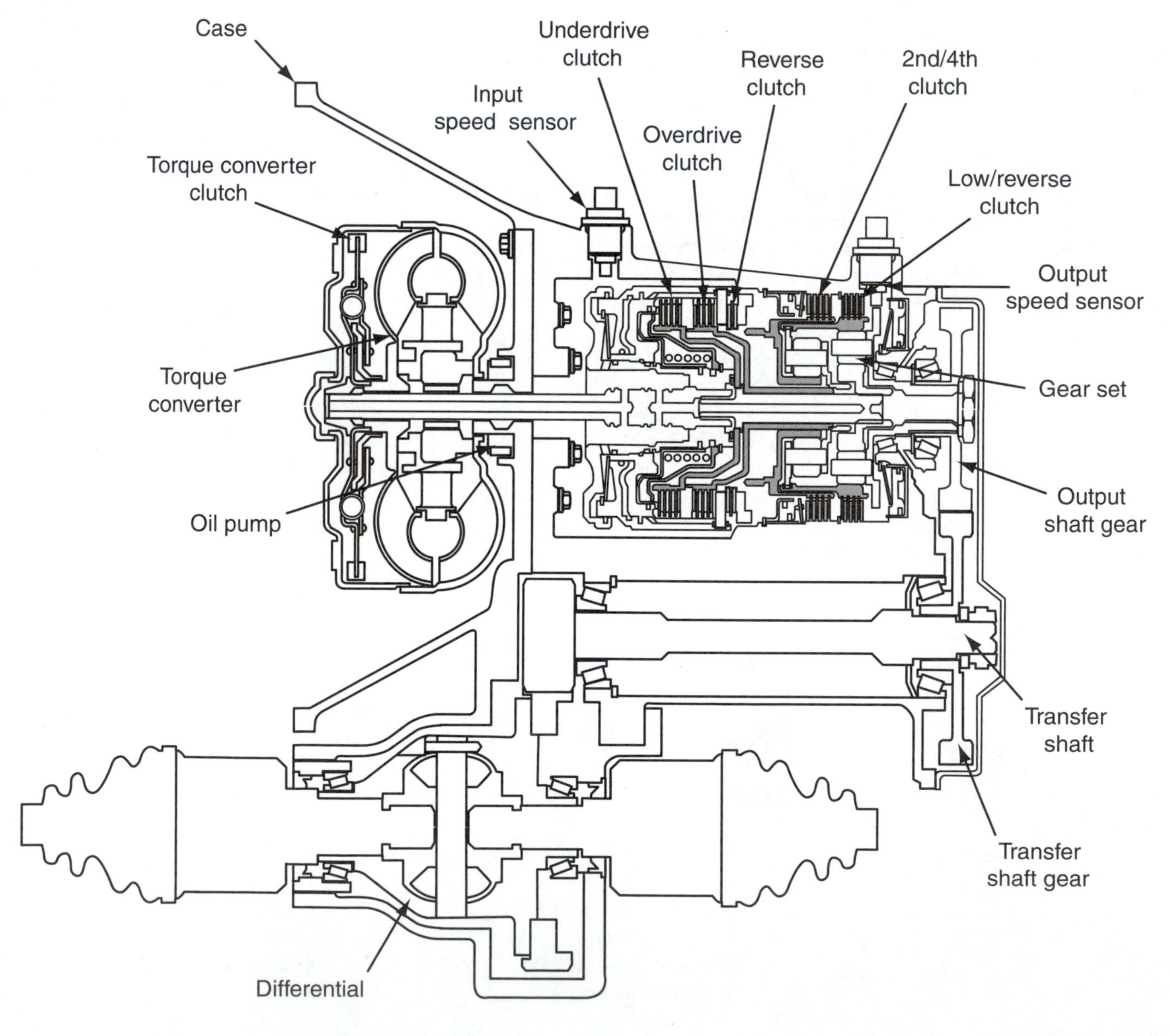


**Figure 5.** A Chrysler Simpson-based transaxle.

The different models vary by application. Transmissions designed for large automobiles have evolved into the transmissions for today's Chrysler pickups and SUVs. Nearly all Chrysler cars are FWD, so the basic design of the early transmissions was modified to become a transaxle. Chrysler also modified this basic design for transverse and longitudinal mounting in FWD cars. Through the years, these transmissions were lightened or strengthened depending on their application, fitted with converter clutches, received hydraulic control enhancements, and, of course, electronic controls were added.

## Input Devices

The front and rear multiple-disc clutches serve as the input devices for these transmissions. The turbine of the torque converter rotates the transmission's input shaft clockwise. Because the front clutch hub and rear clutch drum are splined together at one end of the input shaft, they also rotate clockwise **(Figure 6)**. The rear ring gear carries the output of the transmission to the final drive unit.

The inner edges of the front clutch friction discs are splined to the outer edges of the front clutch hub **(Figure 7)** and therefore turn with the hub. The outer edges of the front clutch steel plates are splined to the inner edges of the front clutch drum. The drum rotates with the input shaft whenever the front clutch is applied.

An input shell is splined to the front clutch drum and the common sun gear, which rotates on the output shaft but is not splined to it. Because the front and rear pinion gears mesh with the sun gear, the drum, the input shell, and the sun gear rotate with the input shaft when the clutch is applied.

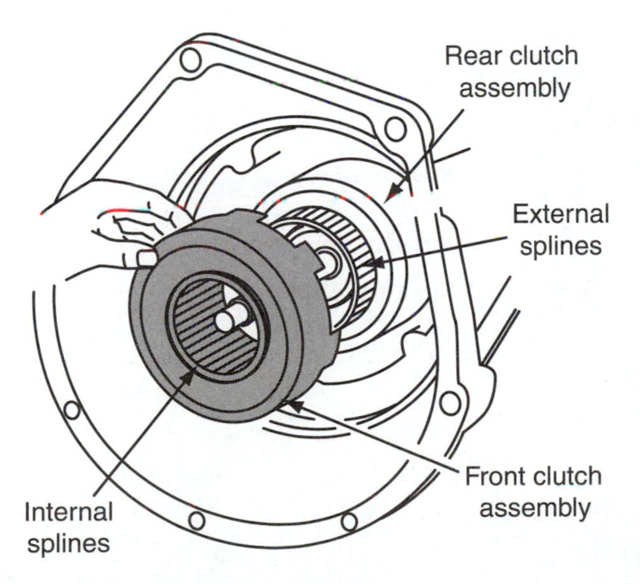


**Figure 6.** The front clutch hub and rear clutch drum are splined together.

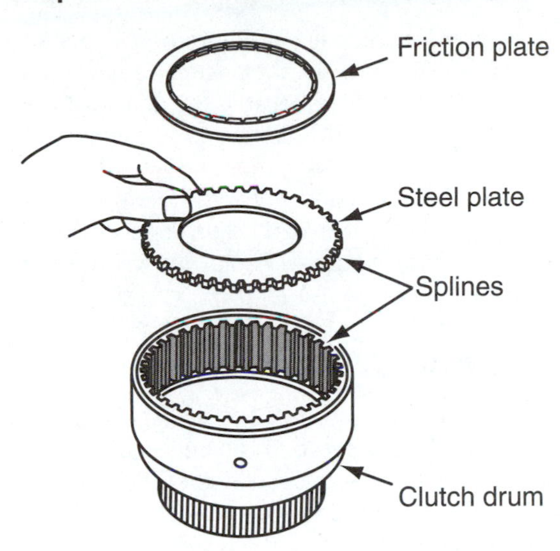


**Figure 7.** The friction plates are splined to the front clutch hub, and the steel plates are splined to the front clutch drum.

The front clutch is applied in third and reverse gears to drive the sun gear. The front clutch is released by either one large coil spring or several small coil springs when hydraulic pressure at the clutch is released.

The rear clutch is applied in all forward gears. It uses a Belleville spring to multiple the applying force of the piston and to help the piston retract into its bore.

## Holding Devices

The front and rear bands and the one-way overrunning clutch are the holding devices for the transmission. The front or **kickdown band** is used only in second gear and holds the input shell and the sun gear stationary **(Figure 8)**.

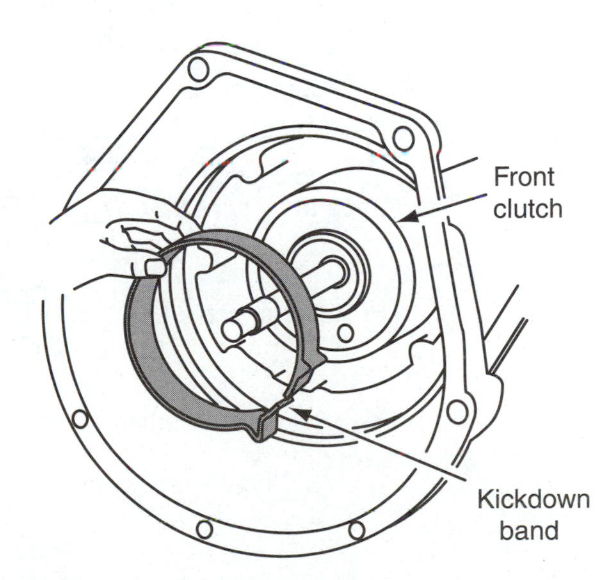


**Figure 8.** The kickdown band holds the front clutch drum; this keeps the sun gear stationary.

The rear or low/reverse band is applied in reverse and manual low and holds the rear planetary carrier. In manual low, the band is ineffective, because the one-way overrunning clutch is holding. The rear band is effective during coasting or deceleration when the one-way clutch begins to freewheel; the rear band holds the drum to allow for engine braking. If the rear band slips or cannot hold the drum, the one-way clutch will hold but will not provide engine braking.

## Apply Devices

Most Torqueflite transmissions are equipped with a **controlled load front servo**. This type servo has two pistons and allows for the quick release of the band during shifts from second to third gear and second to first shifts. The servo also allows for smooth engagement of the band during first to second and third to second gearshifts.

Some Chrysler transmissions are not equipped with a controlled-load front servo. Rather, these use a conventional servo, which firmly applies and releases the band. Most Chrysler transmissions have an external adjustment screw for the front band.

Power flow through these transmissions occurs by the engagement and disengagement of the clutches and bands.

## Add-On Overdrive

Early Chrysler four-speed automatic transmissions had an additional planetary gearset, controlled by an overdrive clutch, direct clutch, and overrunning clutch to provide for the fourth gear. The power flow for the first three forward gear ratios is the same as the single compound planetary equipped units. An intermediate shaft is locked to the output shaft whenever the overrunning clutch is locked. This locking results in bypassing the overdrive planetary and provides for direct drive. The intermediate shaft also drives the planetary carrier of the overdrive gear set. When the transmission shifts from third to fourth gear, the overdrive clutch piston moves the clutch's hub to relieve the spring tension on the direct clutch assembly. It also apples pressure to the overdrive clutch, which locks the sun gear to the transmission case. With the sun gear held, the planet carrier forces the ring gear and output shaft to rotate in an overdrive condition.

The direct clutch locks the sun and ring gears together to prevent freewheeling of the overrunning clutch during coasting and deceleration. This provides for engine braking.

## Electronic Controls

Late-model Chrysler transmissions and transaxles used in most Chrysler/Mitsubishi vehicles rely on electronics to control the shifting into forward gears, including the overdrive gear. Electronic controls are also involved in the control of the converter clutch.

Most transmissions use hydraulically operated clutches, which are controlled by solenoids that are activated by the transaxle's computer. The computer receives information from various inputs and controls a solenoid assembly through the Electronic Automatic Transaxle **(EATX)** relay **(Figure 9)**. The solenoid assembly consists of

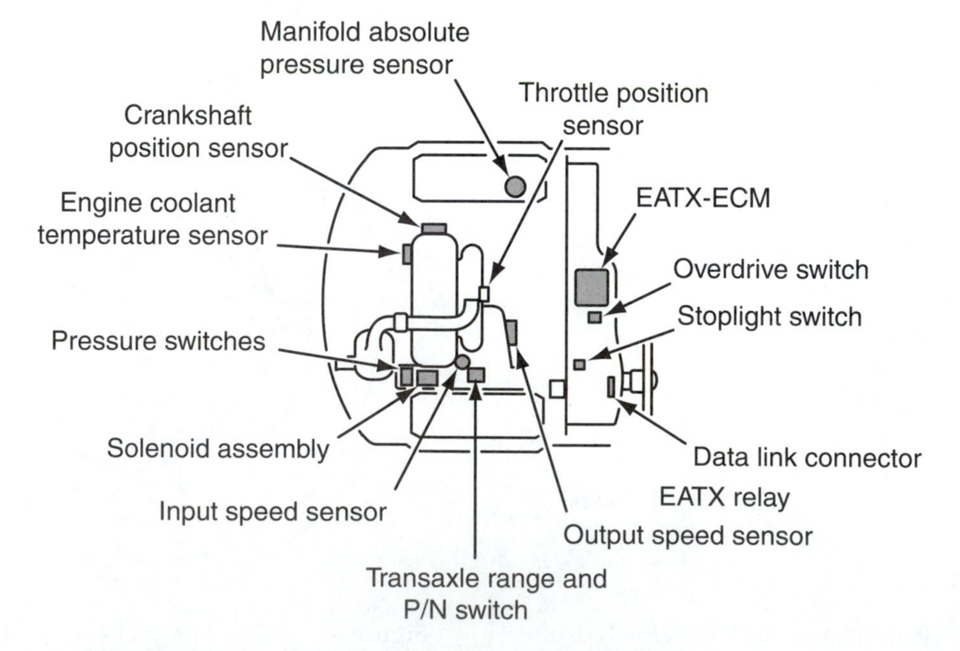


**Figure 9.** The layout of the electrical system for Chrysler's 42LE transaxle.

| GEAR SELECTOR POSITION | OPERATING GEAR | UNDERDRIVE CLUTCH | OVERDRIVE CLUTCH | 2/4 CLUTCH | LOW/REVERSE CLUTCH | REVERSE CLUTCH |
|---|---|---|---|---|---|---|
| OD | 1st gear | x | | | x | |
| | 2nd gear | x | | x | | |
| | 3rd gear | x | x | | | |
| | 4th gear | | x | x | | |
| D | 1st gear | x | | | x | |
| | 2nd gear | x | | x | | |
| | 3rd gear | x | x | | | |
| L | 1st gear | x | | | x | |
| | 2nd gear | x | | x | | |
| R | Reverse | | | | x | x |

**Figure 10.** Clutch and band application chart for late-model Chrysler transaxles.

four solenoids that control hydraulic pressure to four of the five clutches in the transaxle and to the clutch of the torque converter **(Figure 10)**. The solenoids act directly on steel poppet and ball valves. Two of the solenoid valves are normally venting and the other two are normally applying.

The computer has an adaptive learning that allows it to compensate for wear and other events that might occur and cause the normal shift programming to be inefficient. The adaptive learning takes place as the computer reads input and output speeds over 140 times per second. The computer responds to each new reading. This learning process allows the transaxle controller to make adjustments to its program so that quality shifting always occurs.

The computer may receive information from two different sources: directly from a sensor; or through a twisted-pair bus circuit, which connects all of the vehicle computer systems **(Figure 11)**. This modulated bidirectional bus system is called Chrysler Collision Detection **(CCD)**

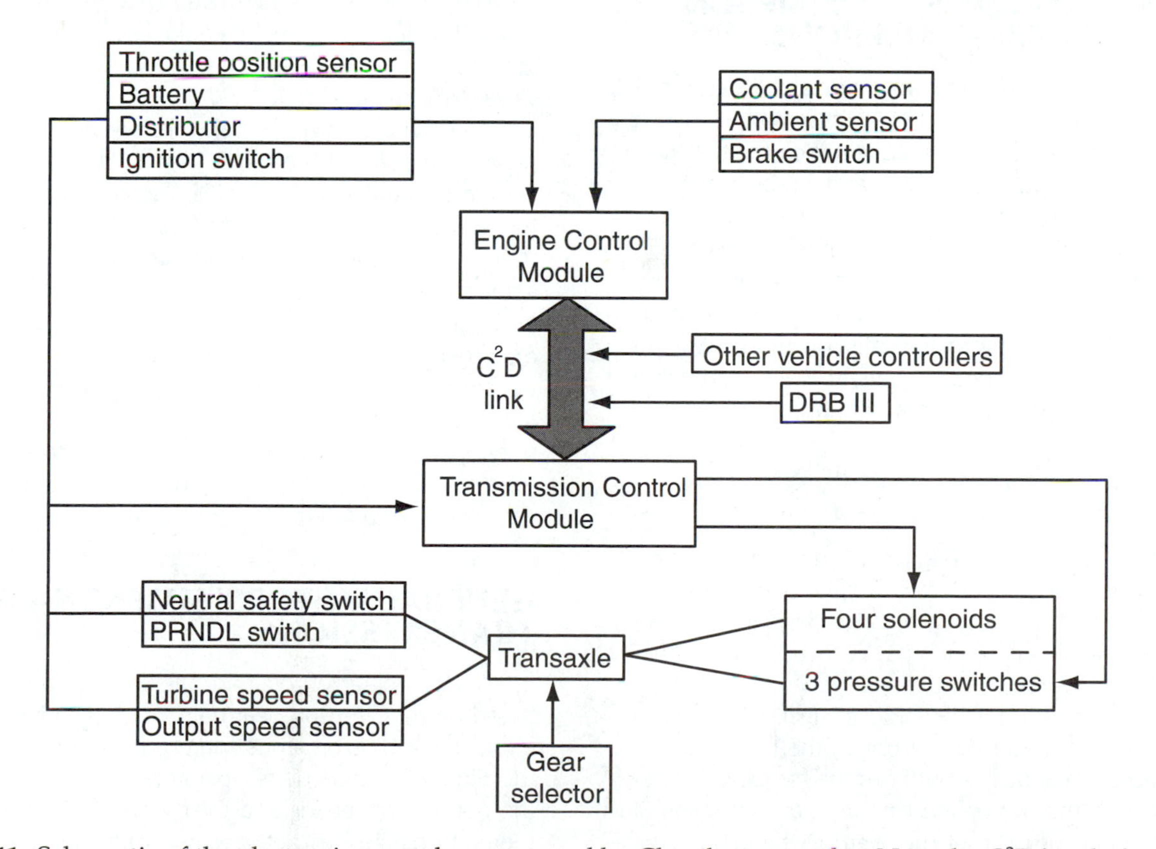


**Figure 11.** Schematic of the electronic control system used by Chrysler transaxles. Note the $C^2D$ bus link.

bus and allows the vehicle's various computers to share information.

Direct battery voltage is supplied to the computer. If the computer loses source voltage, the transaxle will enter into limp-in mode. The transaxle will also enter into limp-in if the system senses a transmission failure. At this point, a fault code will be stored in the computer's memory and the transaxle will remain in limp-in until the transaxle is repaired. While in limp-in, the transaxle will operate only in Park, Neutral, Reverse, and Second gears. The transaxle will not up or downshift. This allows the vehicle to be operated, although its efficiency and performance is hurt.

The computer completes the ground circuit of a solenoid when it should be activated. The computer also monitors the operation of the solenoids through inputs from a turbine speed sensor, output shaft speed sensor, manual shift valve position sensor, low/reverse switch, 2–4 pressure sensor, and/or an overdrive switch. These switches and sensors are located in the solenoid assembly.

The basic shift logic of the controller allows the releasing apply device to slip slightly during the engagement of the engaging apply device. Once the apply device has engaged and the next gear is driven, the releasing apply device is pulled totally away from its engaging member and the transmission is fully into its next gear. This allows for smooth shifting into all gears.

## FORD MOTOR COMPANY SIMPSON GEAR TRAIN TRANSMISSIONS

Ford has used many different models and designs of transmissions. Many recent models are based on a Simpson gear set and operate similarly. In appearance, the transmissions are similar and have a separate and removable extension housing. The major external difference between the models is the size and the reinforcements of the case. To control planetary gear action, the transmissions typically use two multiple-disc clutches, two bands, and one overrunning clutch. The different transmission models vary in their control of the planetary units, therefore on some models a multiple-disc pack may be used instead of a band, or vice versa.

### Overdrive

The A4LD transmission **(Figure 12)** is a fully automatic transmission with four speeds forward, and one reverse. The transmission consists of a torque converter, planetary gear train, three multiple disc clutch packs, three bands, a one-way clutch, and hydraulic control system.

To provide for an overdrive fourth gear, an additional planetary gear set is placed in front of the normal three-speed Simpson gear train. Input from the torque converter passes through the overdrive planetary gear set before it reaches the Simpson gear set. The overdrive band holds the overdrive sun gear to provide for fourth gear overdrive. The overdrive clutch locks the overdrive sun gear to the overdrive planetary carrier to provide for engine braking while the transmission is operating in Drive. The overdrive one-way roller clutch locks the input from the torque converter's turbine shaft directly to the input shaft of the Simpson gear train and provides direct drive through the planetary gear set by locking the carrier to the ring gear.

Direct drive through the overdrive planetary unit is present whenever the transmission is operating in first, second, third, and reverse gears. During these conditions, all input power flows through the overdrive one-way clutch, also the overdrive clutch is applied to provide for engine braking.

When overdrive is selected, the overdrive band is applied and holds the sun gear. The overdrive clutch is released and the planet carrier becomes the input member driving the ring gear at an overdrive ratio of approximately 0.75:1.

### Electronic Controls

The A4LD is a four-speed automatic transmission with a torque converter clutch. The engine control system controls the operation of the converter clutch and the operation of a shift solenoid for third to fourth gear shifting. This solenoid is normally off and permits line pressure flow through the solenoid valve, which inhibits the 3–4 shift. When the PCM receives information that the time is right to allow the 3–4 shift, the solenoid is energized and the inhibitor oil is exhausted. This allows the hydraulic shift valve to make the shift.

The 4R44E and 4R55E transmissions are based on the previous models of the A4LD. The 4R55E is a heavy duty, version of the 4R44E. The 5R55E is a five-speed version of the 4R55E. This five-speed version is similar to the other designs, except for the controls necessary to provide the additional gear. A planetary gear set was not added; rather, new combinations of apply and holding devices were programmed into the design. The electronic controls for these transmissions are based on the A4LD and other Ford systems. These transmissions are equipped with five or six solenoids mounted to the valve body. These solenoids control line pressure, shift quality and timing, and torque converter clutch operation.

## GENERAL MOTORS' SIMPSON-BASED TRANSMISSIONS

Most General Motors transmissions and transaxles are based on the Simpson gear train. The transmission model found in a vehicle depends entirely on its application. All of the Simpson-based units operate similarly, regardless of the number of speeds and their work capacity. Although there are differences between the various models of GM

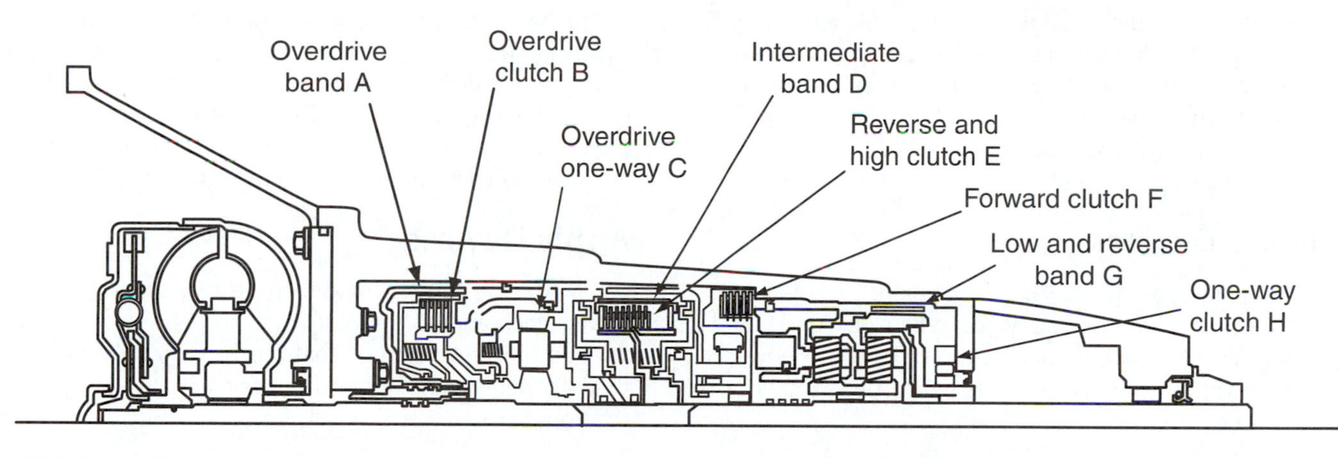


| Gear | Overdrive band A | Overdrive clutch B | Overdrive one-way clutch C | Intermediate band D | Reverse and high clutch E | Forward clutch F | Low and reverse band G | One-way clutch H | Gear ratio |
|---|---|---|---|---|---|---|---|---|---|
| 1- Manual first gear (low) | | Applied | Holding | | | Applied | Applied | Holding | 2.47:1 |
| 2- Manual second gear | | Applied | Holding | Applied | | Applied | | | 1.47:1 |
| D- Drive auto-1st gear | | Applied | Holding | | | Applied | | Holding | 2.47:1 |
| D- O/D auto-1st gear | | | Holding | | | Applied | | Holding | 2.47:1 |
| D- Drive auto-2nd gear | | Applied | Holding | Applied | | Applied | | | 1.47:1 |
| D- O/D auto-2nd gear | | | Holding | Applied | | Applied | | | 1.47:1 |
| D- Drive auto-3rd gear | | Applied | Holding | | Applied | Applied | | | 1.0:1 |
| D- O/D auto-3rd gear | | | Holding | | Applied | Applied | | | 1.0:1 |
| D- Overdrive automatic fourth gear | Applied | | | | Applied | Applied | | | 0.75:1 |
| R- Reverse | | Applied | Holding | | Applied | | Applied | | 2.1:1 |

**Figure 12.** A Ford A4LD transmission.

transmissions with a Simpson gear train, most use three or four multiple-disc clutches, one band, and a single one-way roller clutch to provide the various gear ratios.

## Input Devices

None of the members of the Simpson gear train are directly attached to the input shaft. However, two members are splined to the output shaft. The rear carrier or the rear ring gear will always be the output member of the gear set. The outside of the rear ring gear serves as the parking gear, which is locked to the transmission/transaxle case by the parking pawl when the gear selector is placed into the Park position.

The input shaft is splined to the direct clutch hub and the forward clutch drum; therefore, the hub, drum, and input shaft rotate as a unit **(Figure 13)**. The inside edges of the direct clutch friction discs are splined to the outside of the direct clutch hub. The outside edges of the steel discs are splined to the inside of the direct clutch drum. When the direct clutch is applied, the input shaft rotates the drum clockwise.

The drum is splined to the input shell, which is splined to the common sun gear. The sun gear, therefore, rotates with the input shell. The sun gear is the input member for the gear set when the direct clutch is applied.

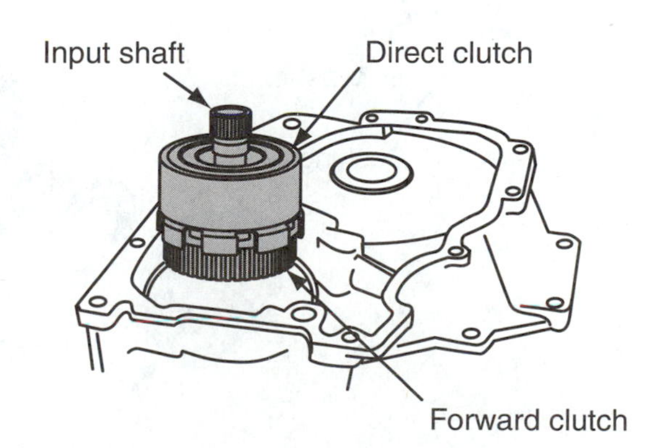


**Figure 13.** The input shaft is splined to the direct clutch hub and the forward clutch drum.

The inside edges of the forward clutch's friction discs are splined to the outside edge of the front planetary ring gear. The outside edges of the steel discs are splined to the inside of the forward clutch drum, which is splined to the input shaft. When the forward clutch is applied, the front ring gear rotates with the input shaft.

## Holding Devices

The different models of GM transmissions are based on which members of the planetary gear set are held and how they are held. The model specifications cause the power flows through the various transmissions to be slightly different from each other.

Some models use front and rear bands plus a low roller clutch, intermediate one-way clutch, and intermediate clutch as the holding devices. Most other models use a low/reverse clutch, low roller clutch, and an intermediate band as the holding devices **(Figure 14)**. Some other models may have an intermediate clutch brake and intermediate roller clutch in addition to the other holding devices.

## Apply Devices

Many transmission models have two servos: the rear servo, which applies the rear band and also contains the 1–2 accumulator piston that cushions application of the intermediate clutch; and the front servo, which applies the front band.

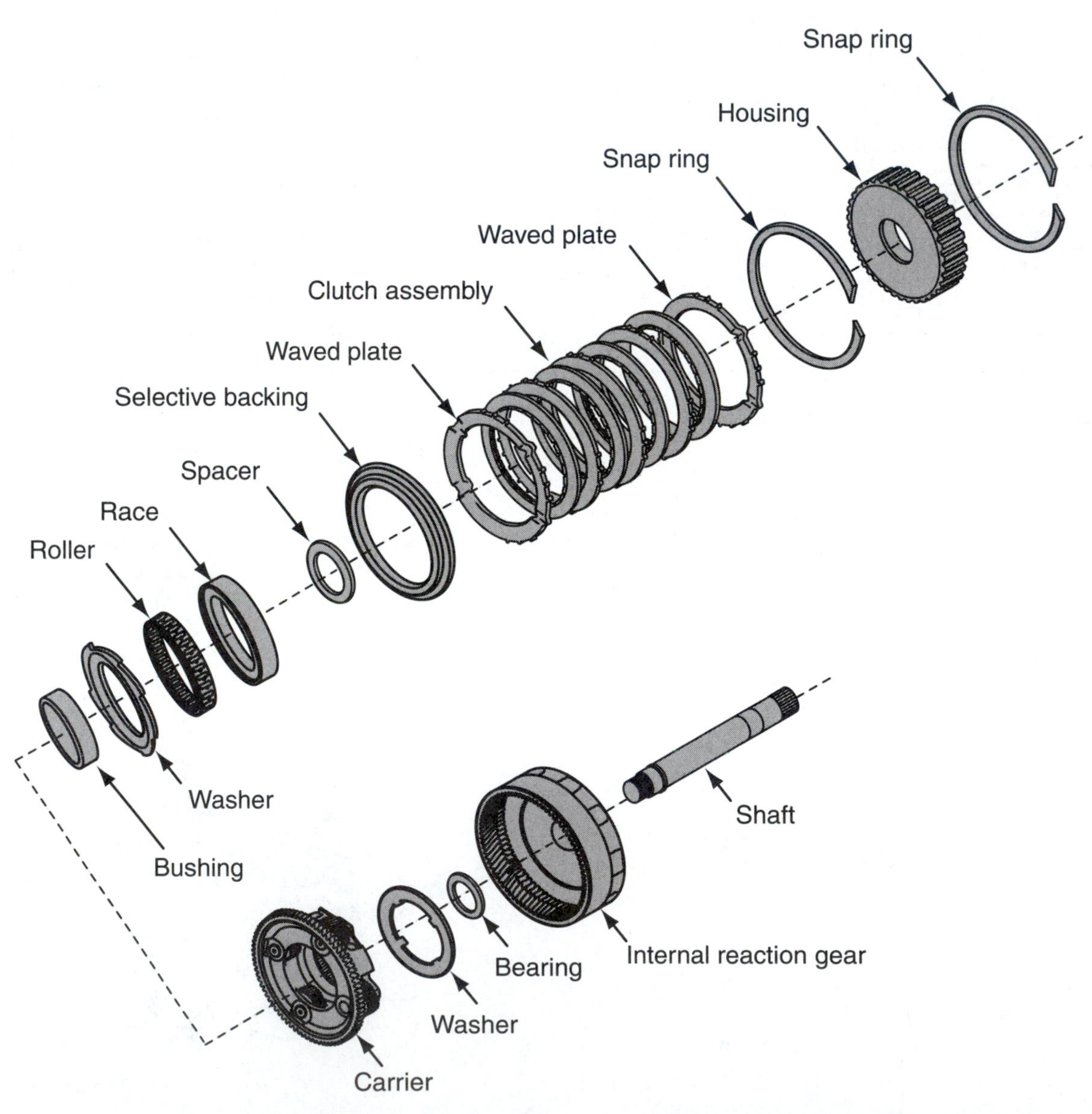


**Figure 14.** The low/reverse clutch and low one-way clutch, in addition to the intermediate band, are used as the holding devices in most THM transmissions.

| GEAR SELECTOR POSITION | OPERATING GEAR | 2–4 BAND | REVERSE INPUT CLUTCH | OVERRUN CLUTCH | FORWARD CLUTCH | FORWARD SPRAG CLUTCH | 3–4 CLUTCH | LOW ROLLER CLUTCH | LOW/REVERSE CLUTCH |
|---|---|---|---|---|---|---|---|---|---|
| OD | 1st gear | | | | x | x | | x | |
| | 2nd gear | x | | | x | x | | | |
| | 3rd gear | | | | x | x | x | | |
| | Overdrive | x | | | x | | x | | |
| D | 1st gear | | | x | x | x | | x | |
| | 2nd gear | x | | x | x | x | | | |
| | 3rd gear | | | x | x | x | x | | |
| 2 | 1st gear | | | x | x | x | | x | |
| | 2nd gear | x | | x | x | x | | | |
| R | Reverse | | x | | | | | | x |

**Figure 15.** Clutch and band application chart for a THM 4L60 transmission.

Other transmission models use one servo assembly to apply the front band or intermediate overrun band. The servo bore is in the transmission case and consists of a piston, an apply pin, cushion spring, and a cover.

## Overdrive

When GM first introduced a four-speed automatic transmission, they simply added a planetary unit to the existing transmissions and transaxles. By adding a planetary unit in front of the Simpson gear set, a 3T40 became a 4T40 and a 3L80 became the 4L80, and so on.

The 4L60/4L80 is a fully automatic transmission consisting of two planetary gear sets, five multiple disc clutches, one sprag clutch, one roller clutch, and a band. (Refer to the clutch and band application chart, **Figure 15**, for more details.) These provide four forward gears, including an overdrive. The five multiple-disc clutches are applied hydraulically and released by several small coil springs.

## Electronic Controls

General Motors is using several different transmission models, but the electronic control systems for each are very similar. All automatic upshifts and downshifts are electronically controlled by the PCM for gasoline engines **(Figure 16)** or the TCM for diesel engines.

The selected and operating gear ranges are hydraulically controlled according to the position of the manual shift valve, as are selector-initiated and forced downshifts. A **variable force motor (VFM)** controlled by the PCM is used to change line pressure in response to component wear, engine speed, and vehicle load. By responding to current operating conditions, the VFM is able to match shift timing and feel with the current needs of the vehicle. The torque converter clutch is hydraulically applied and electrically controlled by the PCM through a **pulse width modulated (PWM) solenoid (Figure 17)**.

The two shift solenoids are attached to the valve body and are normally open. There are four possible on/off combinations of the solenoids, which determine fluid flow to the shift valves in the valve body.

When a solenoid is grounded by the PCM, a check ball held in place by the solenoid plunger blocks the fluid pressure feed. This closes the exhaust passage and causes signal fluid pressure to increase. When the solenoid is deenergized, fluid pressure moves the check ball and plunger off the check ball seat. This allows fluid to flow past the check ball and exhaust through the solenoid, decreasing signal pressure.

# OTHER SIMPSON GEAR–BASED TRANSMISSIONS

There are many different automatic transmissions that are based on the basic Simpson gear train. Most have similar power flows and major components. The primary differences between most of the transmissions are in the nomenclature and holding devices used by the different manufacturers. The following transmissions are presented to illustrate the similarities of all Simpson-based transmissions and to expose you to some of the nomenclature used by the manufacturers.

## Aisin-Warner Transmissions

Aisin-Warner is a manufacturer, largely owned by Toyota, that supplies transmissions to many different vehicle manufacturers including Chrysler, Isuzu, Mitsubishi, GM, and Toyota. The most common, the Aisin-Warner (AW) 4, is a four-speed overdrive; this is an electronically controlled automatic transmission that has a converter clutch, three

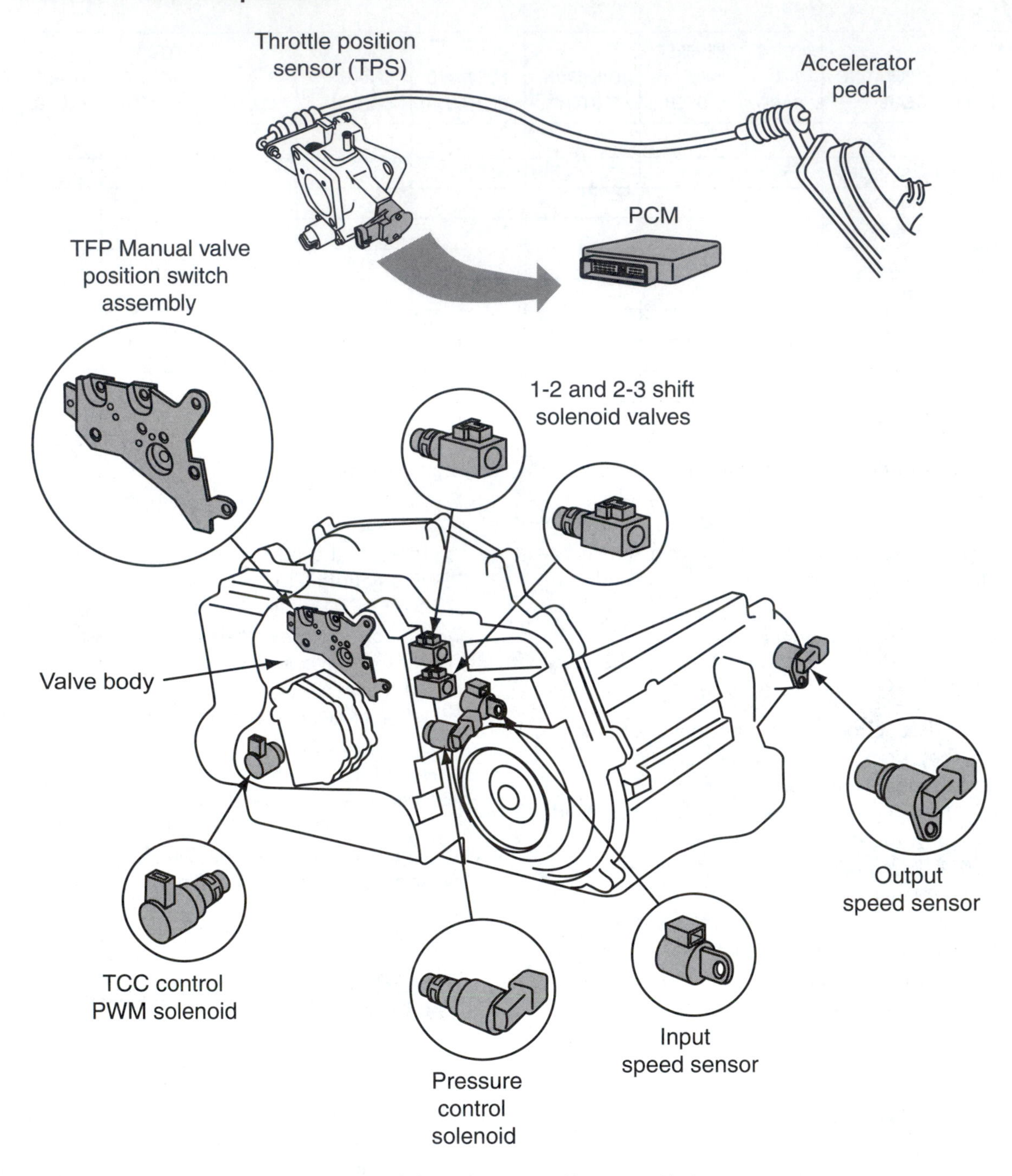


**Figure 16.** Layout of components for a typical GM electronically controlled transmission.

planetary gear sets, apply devices, and a valve body fitted with three solenoids.

The solenoids are controlled by a TCM and are used to control the operation of the transmission. Solenoids #1 and #2 are used to control shifting and solenoid #3 is used for torque converter clutch operation.

On some models, a POWER/COMFORT switch is mounted on the instrument panel. This switch is an automatic transmission mode selection switch, which allows the driver to select different modes to change transmission upshift characteristics.

## Nissan Motor Company Transmissions

Most Nissan RWD vehicles use a transmission that provide four forward gears through the use of a Simpson gear set and an additional planetary unit mounted in front of the Simpson. This series provides electronic control of

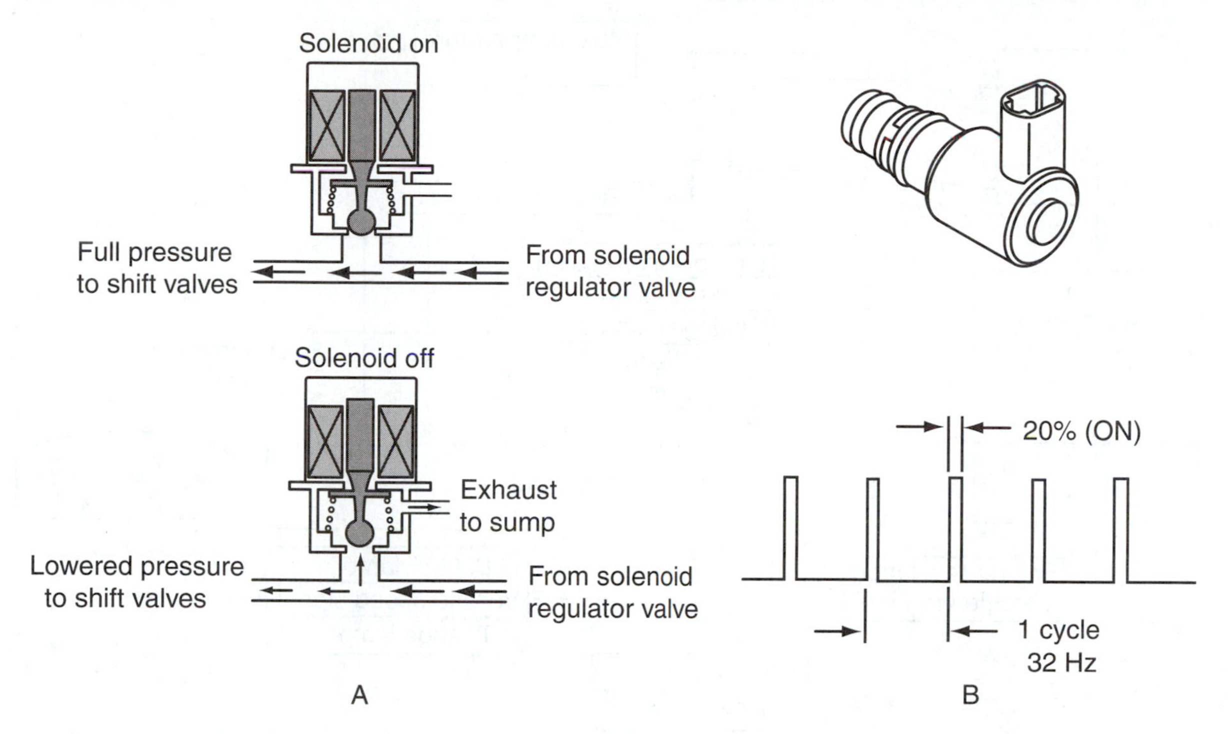


**Figure 17.** (A) A typical PWM solenoid. (B) The signal representing the control or ordered duty cycle from the computer.

shifting and torque converter clutch engagement. These transmissions use four multiple-disc clutches, three servos and bands, and two one-way clutches to provide for the different ranges of gears **(Figure 18)**.

Nissan uses different transaxles in their vehicles. The typical transaxle has a converter clutch and electronic shift controls **(Figure 19)**. The PCM/TCM operates, primarily, in response to throttle position and vehicle speed. Based on inputs, the PCM provides the appropriate shift schedule for the current operating conditions. The control unit also has a fail-safe feature, which allows the vehicle to be driven even if an important input fails.

This system relies on two shift solenoids controlled by the PCM. In addition to these solenoids, two other solenoids are incorporated into the system. One of these solenoids, called the timing solenoid, provides for smooth downshifting. The other, called the line pressure solenoid, provide for smooth upshifting. Both of these solenoids control the engagement and disengagement of the transmission's apply units. The system also has a fifth solenoid, which is used to control converter clutch activity. The PCM relies on inputs from various sensors to control the solenoids.

## Toyota Motors Transmissions

Many different transmissions and transaxles have been used by this manufacturer. A common early design is the A-140E/A-140L. This transaxle combines a three-speed

| GEAR SELECTOR POSITION | OPERATING GEAR | REAR CLUTCH | ONE-WAY CLUTCH | BAND | FRONT CLUTCH | LOW/REVERSE BAND | DIRECT CLUTCH | OVERDRIVE BAND | OVERDRIVE ONE-WAY CLUTCH |
|---|---|---|---|---|---|---|---|---|---|
| D | 1st gear | x | x | | | | x | | x |
| | 2nd gear | x | | x | | | x | | x |
| | 3rd gear | x | | | x | | x | | x |
| | 4th gear | x | | | x | | | x | |
| 2 | 2nd gear | x | | x | | | x | | x |
| 1 | 1st gear | x | x | | | x | x | | x |
| R | Reverse | | | | x | x | x | | x |

**Figure 18.** Clutch and band application chart for a Nissan L4N71B transmission.

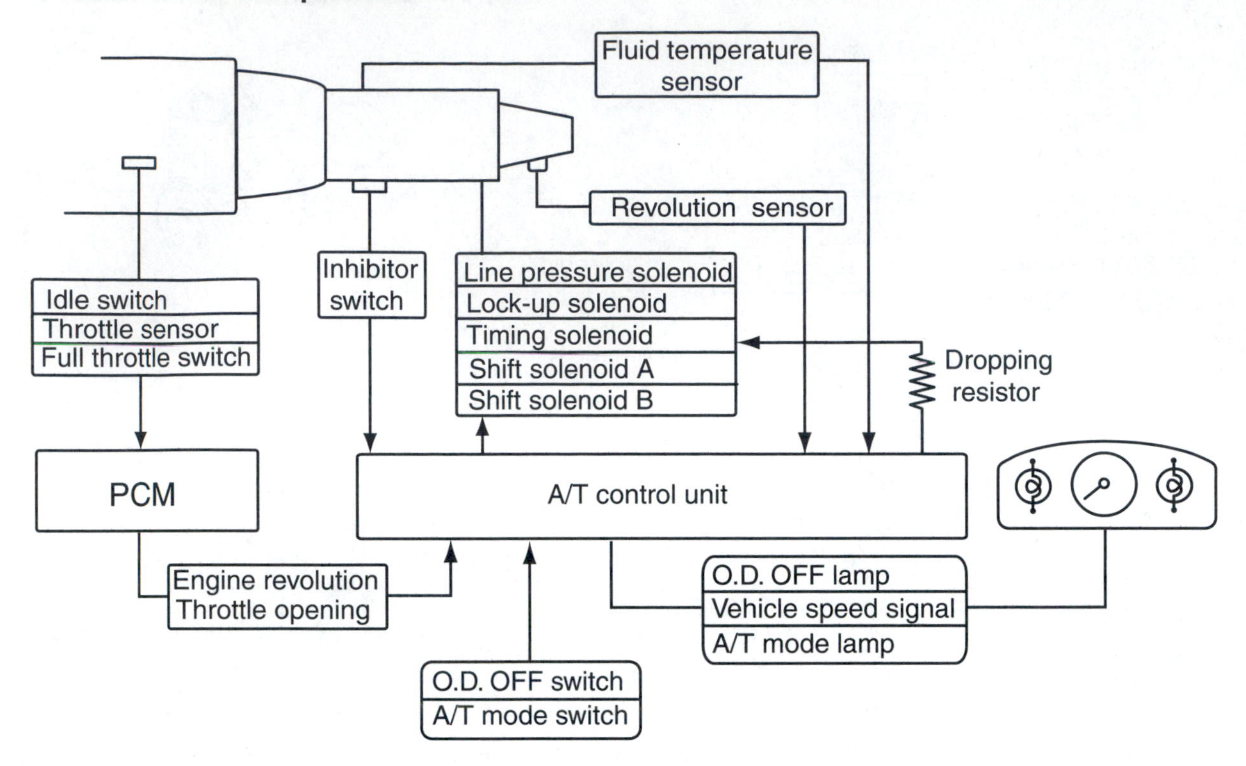


**Figure 19.** Schematic of a Nissan RE4FO2A transaxle's electronic control system.

transmission with an overdrive assembly. The primary differences between the L and E type transaxles are the main valve body, operating mechanism, and electronic control. The E refers to an ECT. The ECT is different from the oil pressure control transaxle (A140L) in that it is controlled by a microprocessor located behind the glove box. The A140L uses an electronic overdrive solenoid system.

Although these transaxles are electronically controlled, the basis for operation is a Simpson gear train in line with a single overdrive planetary gear set. The transaxles uses four multiple-disc clutches, two band and servo assemblies, and three one-way clutches to provide the various gear ranges **(Figure 20)**.

The A541E transaxle is a revised copy of the A140E. The biggest change was in the electronic controls, which now had adaptive learning. This transaxle has been used in many different models of Toyotas, especially the Camry and Avalon. This transaxle uses six multiple-disc clutches, three one-way clutches, and one brake band. The operation of the transaxle and the lockup converter is totally controlled

| GEAR SELECTOR POSITION | OPERATING GEAR | OD CLUTCH | FORWARD CLUTCH | DIRECT CLUTCH | 2ND COAST BAND | 2ND COAST DRUM | 1ST & REVERSE BRAKE | OD ONE-WAY CLUTCH | #1 ONE-WAY CLUTCH | #2 ONE-WAY CLUTCH |
|---|---|---|---|---|---|---|---|---|---|---|
| D-Drive | 1st gear | x | x | | | | | x | | x |
| | 2nd gear | x | x | | | x | | x | x | |
| | 3rd gear | x | x | x | | x | | x | | |
| | Overdrive | | x | x | | x | | | | |
| 2-Second | 1st gear | x | x | | | | | x | | x |
| | 2nd gear | x | x | | x | x | | x | x | |
| | 3rd gear | x | x | x | | x | | x | | |
| L-Low | 1st gear | x | x | | | | x | x | | x |
| | 2nd gear | x | x | | x | x | | x | x | |
| R | Reverse | x | | x | | | x | | | |
| N/P | Neutral | x | | | | | | | | |

**Figure 20.** Clutch and band application chart for a Toyota A140 transmission.

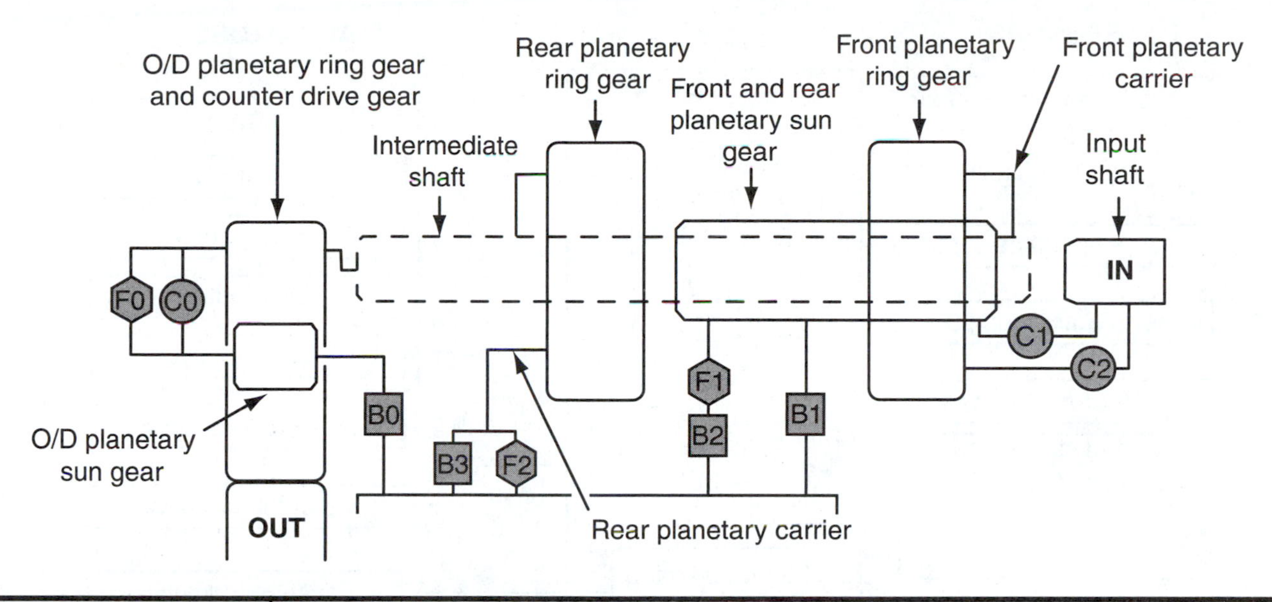


| COMPONENT | | FUNCTION |
|---|---|---|
| Forward clutch | C1 | Connects input shaft and front planetary ring gear. |
| Direct clutch | C2 | Connects input shaft and front and rear planetary sun gear. |
| 2nd Coast brake | B1 | Prevents front and rear planetary sun gear from turning either clockwise or counterclockwise. |
| 2nd Brake | B2 | Prevents outer race of F1 from turning either clockwise or counterclockwise, thus preventing front and rear planetary sun gear from turning counterclockwise. |
| 1st & Reverse brake | B3 | Prevents rear planetary carrier from turning either clockwise or counterclockwise. |
| #1 One-way clutch | F1 | When B2 is operating, prevents front and rear planetary sun gear from turning counterclockwise. |
| #2 One-way clutch | F2 | Prevents rear planetary carrier from turning counterclockwise. |
| O/D Direct clutch | C0 | Connects overdrive sun gear and overdrive planetary carrier. |
| O/D Brake | B0 | Prevents overdrive sun gear from turning either clockwise or counterclockwise. |
| O/D One-way clutch | F0 | When transaxle is being driven by engine, connects overdrive sun gear and overdrive carrier. |
| Planetary gears | | These gears change the route through which driving force is transmitted in accordance with the operation of each clutch and brake in order to increase or reduce the input and output speed. |

**Figure 21.** Purpose of the clutches and bands in Toyota's A541E transaxle.

by the PCM. However, a throttle pressure cable is used to mechanically modulate line pressure. The band and clutch location, as well as the application of each **(Figure 21)**, is the same as the previous models.

## Electronic Controls

Toyota led the way into electronically controlled automatic transmissions. The A-140E was the first fully electronic transmission. This transaxle is actually a three-speed unit with an add-on fourth or overdrive gear. The control module receives input signals from the water temperature-sensing switch, throttle position switch and the shift pattern selector switch **(Figure 22)**. The module also receives signals from two speed sensors; one is located at the speedometer and the other at the transaxle. The backup lamp/neutral safety switch signals the module for starting and backup lamp circuits. The module controls and sends output signals to the stop and backup lamps. The module also controls two shift solenoids, located within the transaxle.

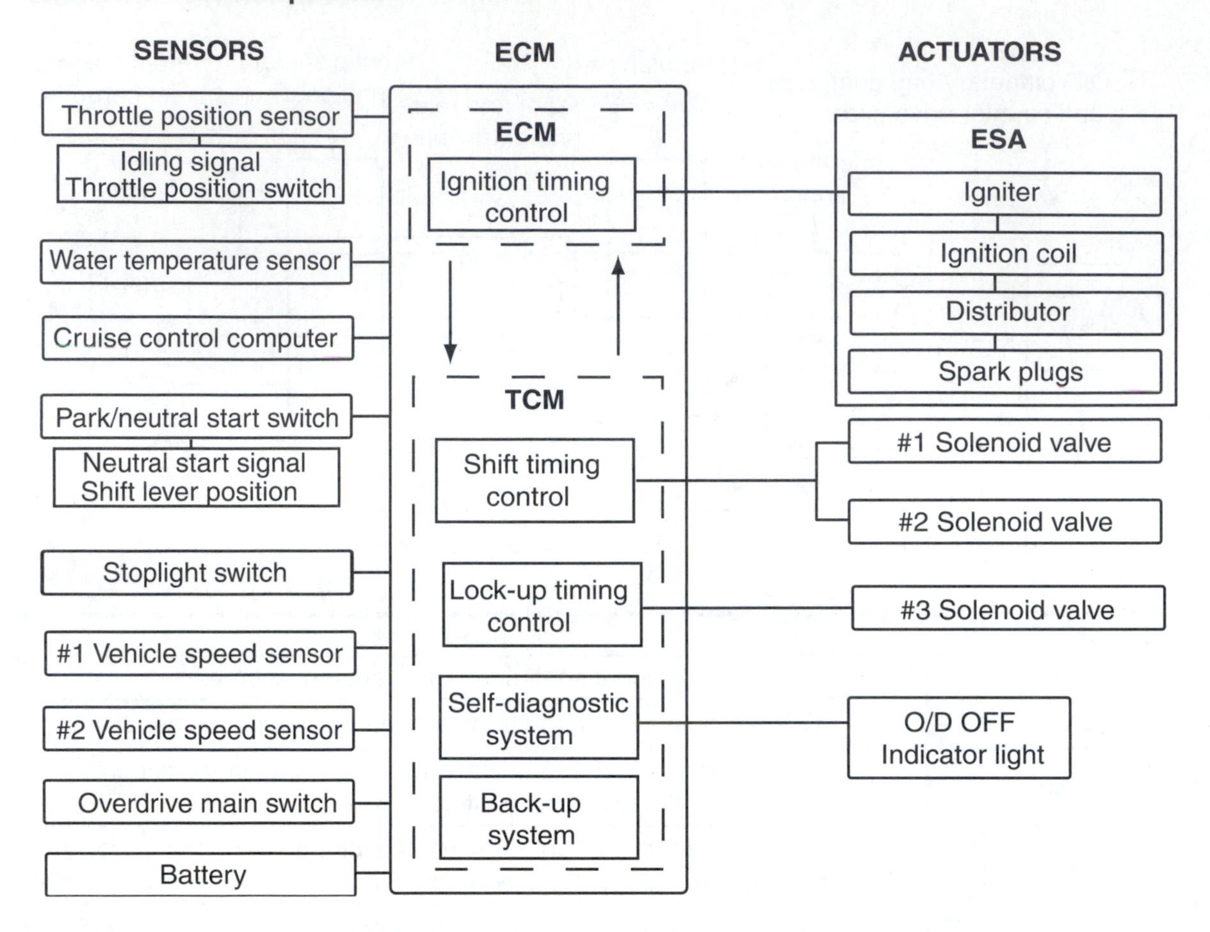


**Figure 22.** Schematic of the electronic controls for a typical electronically-controlled transmission.

## RAVIGNEAUX GEAR TRAIN

The Ravigneaux gear train uses two sun gears, one small and one large, and two sets of planetary pinion gears, three long pinions and three short pinions **(Figure 23)**. The planetary pinion gears rotate on their own shafts, which are fastened to a common planetary carrier. The small sun gear is meshed with the short planetary pinion gears. These short pinions act as idler gears to drive the long planetary pinion gears. The long planetary pinion gears mesh with the large sun gear and the ring gear. A single ring gear surrounds the complete assembly.

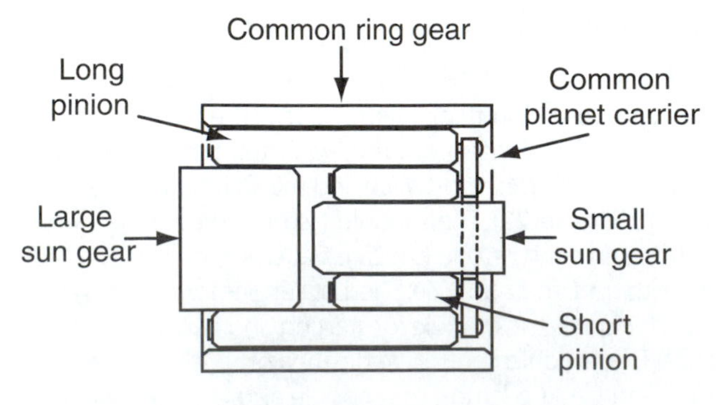


**Figure 23.** A typical Ravigneaux gear set.

With a Ravigneaux gear set in neutral, the engine drives the forward clutch drum. Because the forward clutch is not applied, the power is not transmitted through the gear train and there is no power output.

When the transmission is operating in first gear, engine torque drives the small sun gear clockwise. The planetary carrier is prevented from rotating counterclockwise by a one-way clutch or another brake; therefore, the small sun gear drives the short planetary pinion gears counterclockwise. The direction of rotation is reversed as the short pinion gears drive the long pinion gears, which drive the ring gear and output shaft in a clockwise direction with greater torque but at a lower speed than the input.

In second gear operation, the large sun gear is held. The small sun gear receives the input and rotates in a clockwise direction. The small sun gear drives the short pinion gears counterclockwise. The direction of rotation is reversed as the short pinion gears drive the long pinion gears, which walk around the stationary large sun gear. The output on the ring gear and output shaft is in a clockwise direction and at a gear reduction.

In third gear, there are two inputs into the planetary gear train, the small sun gear and the planetary gear carrier. Because two members of the gear train are being driven at the same time, the planetary gear carrier and the small sun gear rotate as a unit. The long pinion gears transfer the torque, in a clockwise direction, through the gear set to the ring gear and output shaft. This results in direct drive.

In overdrive or fourth gear, input drives the planetary carrier. The long pinion gears walk around the stationary large sun gear and drive the ring gear and output shaft. This results in an overdrive condition.

During reverse gear, input is received at the large sun gear and the planetary gear carrier is held. The clockwise rotation of the large sun gear drives the long pinion gears in a counterclockwise direction. The long pinions then drive the ring gear and output shaft in a counterclockwise direction with a gear reduction.

Several transmissions/transaxle use a Ravigneaux gear set. Primarily because of the increased tooth contact area, it can withstand heavy torque loads and has a compact size. Through the years, Ford has probably used the Ravigneaux gear set more than other manufacturers. The 4R70W is a commonly used transmission for RWD and 4WD vehicles and will be the focus of the discussion of Ravigneaux transmissions.

## Ford's 4R70W Transmission

The 4R70W is a four-speed electronically controlled transmission that uses the common ring gear as the output member. It uses four multiple-disc clutches, two one-way clutches, and two bands to obtain the various gear ranges. See the clutch and band application chart **(Figure 24)**.

## Input Devices

The gear train receives power through the application of the forward, direct, or reverse clutches, which serve as the input devices for the gear train.

The forward clutch drives the forward sun gear and is applied in first, second, and third gears. The clutch is applied hydraulically and is released by the tension of a single large coil spring when hydraulic pressure at the clutch is exhausted.

The direct clutch is applied in third and fourth gears. This clutch connects the input from the converter cover to the planetary carrier. The direct clutch is released by several small coil springs when hydraulic pressure to the clutch is relieved.

The reverse clutch is applied only in reverse and drives the reverse sun gear. The reverse clutch is released by a Belleville spring, which also increases the clamping force of the clutch.

## Holding Devices

The gear train relies on the intermediate clutch, both one-way clutches, the low/reverse band, and the overdrive band for holding gear set members. The intermediate clutch is applied in second, third, and fourth gears. This clutch prevents the reverse sun gear from turning counterclockwise but is only effective in second gear because the intermediate one-way clutch freewheels in third and fourth gear. The intermediate clutch is released by several small coil springs.

The intermediate one-way clutch locks and prevents the reverse sun gear from turning counterclockwise during acceleration in second gear. During deceleration, the intermediate one-way clutch freewheels and prevents engine braking. The intermediate one-way clutch is also locked during first gear operation.

| GEAR SELECTOR POSITION | OPERATING GEAR | INTERMEDIATE CLUTCH | INTERMEDIATE ONE-WAY CLUTCH | OVERDRIVE BAND | REVERSE CLUTCH | FORWARD CLUTCH | PLANETARY ONE-WAY | LOW/REVERSE BAND | DIRECT CLUTCH |
|---|---|---|---|---|---|---|---|---|---|
| OD | 1st gear | | | | | x | x | | |
| | 2nd gear | x | x | | | x | | | |
| | 3rd gear | x | | | | x | | | x |
| | 4th gear | x | | x | | | | | x |
| 3 | 1st gear | | | | | x | x | | |
| | 2nd gear | x | x | | | x | | | |
| | 3rd gear | x | | | | x | | | x |
| 1 | 1st gear | | | | | x | x | x | |
| | 2nd gear | x | x | x | | x | | | |
| R | Reverse | | | | x | | | x | |

**Figure 24.** Clutch and band application chart for a Ford 4R70W transmission.

The low one-way clutch locks and prevents the planetary carrier from turning counterclockwise while operating in the drive range and accelerating in first gear. The low one-way clutch is also locked during manual low operation but does not provide engine braking.

The low/reverse band is applied during manual low and reverse gear operation. The low/reverse band holds the planetary carrier stationary. The overdrive band holds the reverse sun gear to provide for an overdrive condition in fourth gear.

## Apply Devices

The low/reverse and the overdrive servos are located in bores in the bottom of the transmission case. Neither band is adjustable. Each is only adjusted during transmission assembly through the selection of different length apply stems.

## Electronic Controls

The operation of a 4R70W is controlled by the PCM. Input sensors provide information to the PCM, which, in turn, controls actuators, which control transmission operation. The PCM uses logic to control shift scheduling, shift feel, and converter lockup. The PCM relies on information from the engine control system, as well as information from the transmission to determine the optimum shift timing.

The transmission system includes four solenoids: two shift solenoids, an EPC solenoid, and a modulated converter clutch (MCCC) solenoid. The EPC solenoid controls line pressure, at all times, based on the programming of the system's computer. Likewise, the operation of the converter clutch is also totally controlled by the computer through the MCCC. The only exception to this is during first gear and reverse gear operation when the clutch is disabled hydraulically to prevent lockup regardless of the commands by the computer.

The EPC solenoid is a VFS and contains a spool valve and spring. To control fluid pressure, the PCM sends a varying signal to the solenoid. This varies the amount the solenoid will cause the spool valve to move. When the solenoid is off, the spring tension keeps the valve in place to maintain maximum pressure. As more current is applied to the solenoid, the solenoid moves the spool valve more, which moves to uncover the exhaust port more, thereby causing a decrease in pressure **(Figure 25)**. The shift solenoids offer four possible on/off combinations to control fluid to the various shift valves.

# PLANETARY GEARSETS IN TANDEM

Rather than relying on the use of a compound gear set, some automatic transmissions use two or three simple planetary units in series. In this type of arrangement, gear set members are not shared, instead, the holding devices are used to lock different members of the planetary units together. GM's 4T65-E is a common example of this design and will be discussed here.

## 4T65-E Transaxle

The 4T60 was the first domestically produced four-speed automatic transaxle built for FWD vehicles. Overdrive

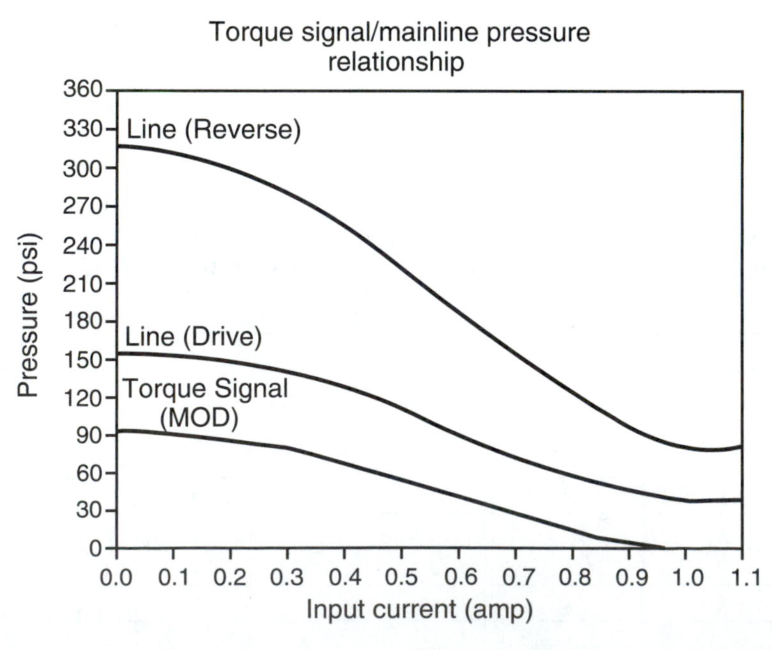


**Figure 25.** The EPC solenoid controls hydraulic pressures throughout the transmission by controlling the current going to the solenoid.

| GEAR SELECTOR POSITION | OPERATING GEAR | 4TH CLUTCH | REVERSE BAND | 2ND CLUTCH | 3RD CLUTCH | 3RD ROLLER CLUTCH | INPUT CLUTCH | INPUT SPRAG CLUTCH | 1–2 BAND |
|---|---|---|---|---|---|---|---|---|---|
| D | 1st gear | | | | | | x | x | x |
| | 2nd gear | | | x | | | x | | x |
| | 3rd gear | | | x | x | x | | | |
| | Overdrive | x | | x | x | | | | |
| 3 | 3rd gear | | | x | x | x | x | x | |
| 2 | 2nd gear | | | x | | | x | | x |
| Low | 1st gear | | | | x | x | x | x | x |
| R | Reverse | | x | | | | x | x | |

**Figure 26.** Clutch and band application chart for a GM 4T65 transaxle.

fourth gear is obtained by using the same planetary units as is used for the first three speeds, instead of adding a planetary unit just for overdrive. The 4T65-E is a modified version of the 4T60-E transaxle. The vacuum modulator used on the 4T60 was replaced with a **pressure control solenoid (PCS)** and a pressure switch was added as an input for the PCM.

Although the gear train is based on two simple planetary gear sets operating in tandem, the combination of the two planetary units does function much like a compound unit. The front planetary carrier is locked to the rear ring gear and the front ring gear is locked to the rear planetary carrier. The transaxle houses a third planetary unit, which is used only as the final drive unit and not for overdrive.

The 4T65 uses four multiple-disc clutches, two bands, and two one-way clutches to provide the various gear ranges **(Figure 26)**.

## Input Devices

Engine torque is transferred to the transaxle through the torque converter and a sprocket and drive link assembly. The torque converter does not directly drive the gear train of the transaxle; rather, the converter drives a sprocket that is linked, by a drive chain, to an input sprocket at the gear train. This arrangement allows the gear train to be positioned away from the centerline of the engine's crankshaft. Input through the gear train is controlled by the input clutch, input one-way clutch, and second clutch **(Figure 27)**.

## Holding Devices

The third clutch, third roller clutch, fourth clutch, 1–2 band, and reverse band are the holding devices for the gear train. The third clutch is applied in third, fourth, and manual low gears. The third roller clutch is locked in third gear and in the manual low position. The third roller clutch overruns in fourth gear. Both the third clutch and third roller clutch prevent the drive wheels from rotating the front sun gear at a speed faster than input shaft speed.

The fourth clutch holds the front sun gear and is applied in fourth gear. The 1–2 band is applied in first and second gears, whereas the reverse band is only applied in reverse.

## Apply Devices

Like many other GM transmissions, the 4T65 uses servos fitted with two pistons. These servos not only apply the bands but also act as an accumulator for the application of the clutch involved in upshifting. Band adjustment is controlled by the length of the servo rod or apply pin, which is selected during transaxle assembly. External adjustment is possible after assembly but requires the use of a special tool that helps select the proper length rod or apply pin.

## Electronic Controls

The shifting of this four-speed transaxle is controlled by two shift solenoids that are controlled by the PCM. The transaxle relies on two solenoids to cause a change of gears. A summary of the solenoid activity and power flow is shown in **(Figure 28)**.

The transaxle also has a PCS to control line pressure in response to normal clutch, seal, and spring wear. The PCS is controlled by a varying the duty cycle of the solenoid, which varies the current flow through its windings. As current flow increases, the magnetic field around the windings also increases. This increased strength moves the solenoid's plunger farther away from the fluid exhaust port. Allowing less fluid to exhaust will cause the pressure of the fluid to increase.

The torque converter clutch is controlled by two electronic solenoids, one for apply and release and the other to control the feel of the apply and release of the TCC. Some transaxles use a PWM solenoid to control converter clutch engagement.

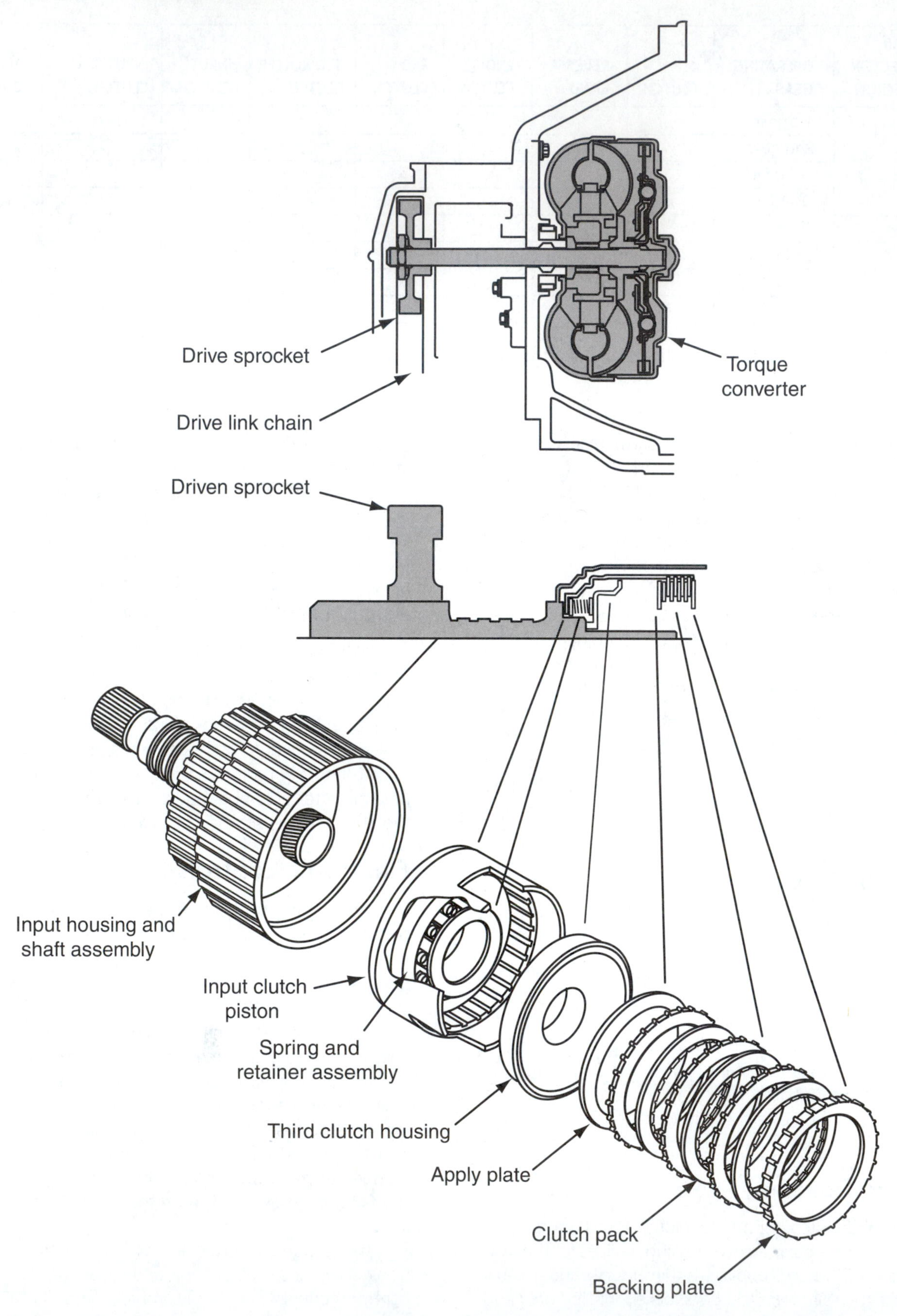


**Figure 27.** Input through the gear train is controlled by the direct clutch, direct one-way clutch, and intermediate/high clutch.

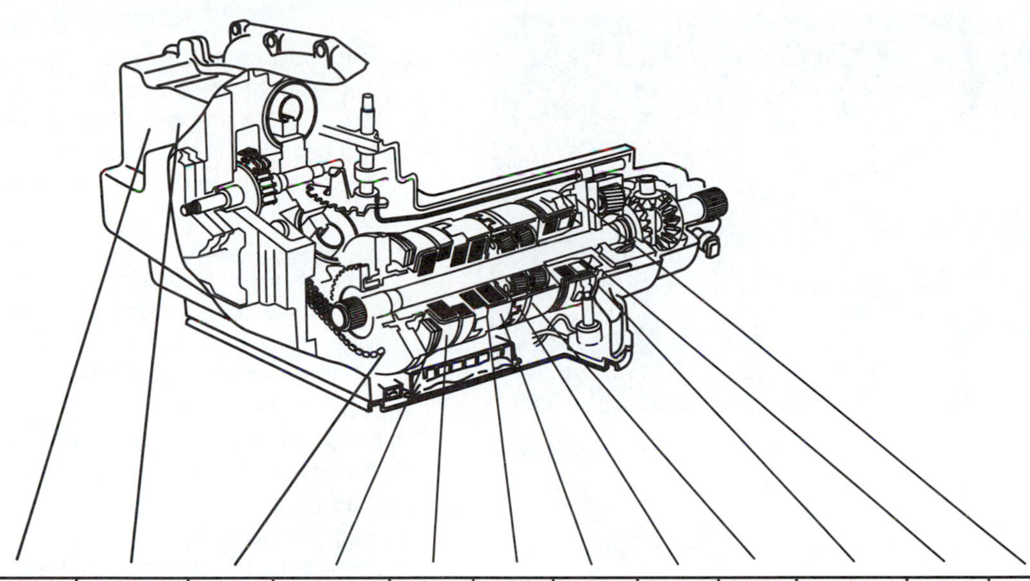

| Range | Gear | A solenoid | B solenoid | 4th clutch | Reverse band | 2nd clutch | 3rd clutch | 3rd roller clutch | Input clutch | Input sprag | Forward band | 1-2 roller clutch | 2-1 band |
|---|---|---|---|---|---|---|---|---|---|---|---|---|---|
| P - N | | On | On | | | | | | * | * | | | |
| D | 1st | On | On | | | | | | Apply | Hold | Apply | Hold | |
| | 2nd | Off | On | | | Apply | | | * | Orun | Apply | Hold | |
| | 3rd | Off | Off | | | Apply | Apply | Hold | | | Apply | | |
| | 4th | On | Off | Apply | | Apply | * | Orun | | | Apply | | |
| D | 3rd | @Off | @Off | | | Apply | Apply | Hold | Apply | Hold | Apply | | |
| | 2nd | @Off | @On | | | Apply | | | * | Orun | Apply | Hold | |
| | 1st | @On | @On | | | | | | Apply | Hold | Apply | Hold | |
| 2 | 2nd | @Off | @On | | | Apply | | | * | Orun | Apply | Hold | Apply |
| | 1st | @On | @On | | | | | | Apply | Hold | Apply | Hold | Apply |
| 1 | 1st | @On | @On | | | | Apply | Hold | Apply | Hold | Apply | Hold | Apply |
| R | Rev | On | On | | Apply | | | | Apply | Hold | | | |

* Applied but not effective
On - Solenoid energized
Off - Solenoid de-energized

@ The solenoid's state follows a shift pattern that depends upon vehicle speed and throttle opening. It does not depend upon the selected gear.

**Figure 28.** Range reference chart for a 4T60-E transaxle.

## NONPLANETARY BASED TRANSMISSIONS

Most transaxles used in Honda and Acura cars (as well as Saturn vehicles) are unique in that they do not use a planetary gear set to provide for the different gear ranges. Constant-mesh helical and square-cut gears are used in a manner similar to that of a manual transmission **(Figure 29)**.

These transaxles have a mainshaft and countershaft on which the gears ride. To provide the four forward and one reverse gear, different pairs of gears are locked to the shafts by hydraulically controlled clutches **(Figure 30)**. Four multiple-disc clutches, the sliding reverse gear, and a one-way clutch are used to control the gears. These are designated by the gear they activate: first gear clutch, first gear one-way clutch, second gear clutch, third gear clutch, fourth gear clutch, and reverse gear. Refer to the application chart **(Figure 31)** for details on the power flow through this transaxle.

Reverse gear is obtained through the use of a shift fork, which slides the reverse gear into position. The power flow through these transaxles is also similar to that of a manual transaxle. The action of the clutches is much the same as the action of the synchronizer assemblies in a manual

Figure 29. An example of the helical and spur gears used in a Honda automatic transaxle.

transmission. When Honda's transaxle is in Park or Neutral, none of the clutches are applied and no power is transmitted to the countershaft.

When "drive" is selected and the transaxle is in first gear, hydraulic pressure is applied to the first clutch, which rotates with the mainshaft. This causes first gear to rotate. Power is transmitted from first gear on the mainshaft to first gear on the countershaft, then it energizes the first one-way clutch. From there, power moves through the final drive unit to the drive wheels.

When the correct conditions are present, the transaxle shifts into second gear. To do this, hydraulic pressure is applied to the second clutch on the mainshaft. This allows for power on the second gear, which is meshed with second gear on the countershaft. Power is transmitted across the gears, driving the countershaft, which is connected to the final drive unit.

The first clutch is still hydraulically applied; however, the first one-way clutch is now freewheeling. This allows the first clutch to remain on all the way through the gears without transmitting any torque.

Third gear engagement is made after the right operating conditions are met. To engage third gear, hydraulic pressure is applied to the third clutch on the countershaft. Power from the mainshaft is transmitted to the countershaft third gear. Hydraulic pressure is exhausted from the second clutch. From the third gear on the countershaft power is then transmitted to the final drive gear.

When engine load and vehicle speeds permit, the transmission upshifts into fourth gear. Hydraulic pressure is applied to the fourth clutch on the mainshaft, which rotates together with the mainshaft causing the mainshaft fourth gear to rotate. Hydraulic pressure is exhausted from the third clutch. Power is transmitted to the fourth countershaft gear, which drives the countershaft, which is connected to the final drive unit.

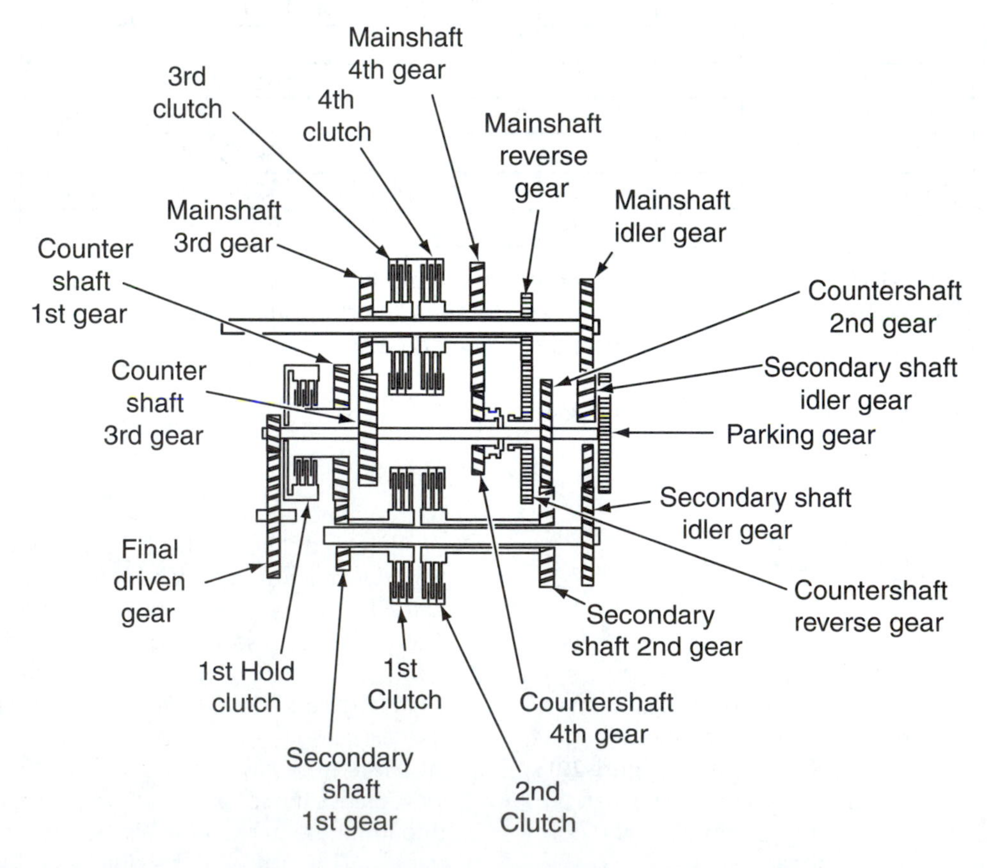


Figure 30. Arrangement of gears and reaction devices in a typical Honda transaxle.

| GEAR SELECTOR POSITION | OPERATING GEAR | 1ST CLUTCH | 1ST ONE-WAY CLUTCH | 2ND CLUTCH | 3RD CLUTCH | 4TH CLUTCH | REVERSE GEAR |
|---|---|---|---|---|---|---|---|
| D4 | 1st gear | X | X | | | | |
| | 2nd gear | X | | X | | | |
| | 3rd gear | X | | | X | | |
| | 4th gear | X | | | | X | |
| D3 | 1st gear | X | X | | | | |
| | 2nd gear | X | | X | | | |
| | 3rd gear | X | | | X | | |
| 2 | 2nd gear | | | X | | | |
| R | Reverse | | | | | X | X |

**Figure 31.** Clutch and band application chart for a typical Honda non-planetary transaxle.

When the driver selects reverse gear, the manual valve switches hydraulic pressure to the servo valve, which moves the reverse shift fork to the reverse position. The reverse shift fork engages with the reverse selector, reverse selector hub, and the countershaft reverse gear. Hydraulic pressure is also applied to the fourth clutch. Power is transmitted from the mainshaft reverse gear via the reverse idler gear to the countershaft reverse gear. The rotational direction of the countershaft reverse gear is changed by the reverse idler gear.

## Electronic Controls

The electronic control system consists of the TCM, sensors, and four solenoid valves **(Figure 32)**. Shifting and converter clutch engagement are electronically controlled. Activating a shift solenoid valve changes modulator pressure, causing a shift valve to move. This allows pressurized fluid to flow through a line to engage the appropriate clutch and its corresponding gear set. Converter clutch engagement is controlled by the action of the other two solenoid valves.

## Honda's Continuously Variable Transmission

Another unconventional automatic transmission found in some late-model Hondas and an increasing number of other vehicles is the CVT. Basically a CVT is a transmission with no fixed forward speeds. The gear ratio varies with engine speed and temperature. These transmissions are equipped with a one-speed reverse gear. These transaxles do not have a torque converter; instead a manual transmission-type flywheel is used with a start clutch. Rather than rely on planetary or helical gear sets to provide drive ratios, most CVTs are equipped with two pulleys and a steel belt **(Figure 33)**. CVT transaxles may have a planetary gearset that sets (forward or reverse) directional gear ranges.

One pulley is the driven member and the other is the drive. Each pulley has a moveable face and a fixed face. When the moveable face moves, the effective diameter of the pulley changes. The change in effective diameter changes the effective pulley (gear) ratio. The driven and drive pulleys are linked by a steel belt.

To achieve a low pulley ratio, high hydraulic pressure works on the moveable face of the driven pulley to make it larger. Because the belt connects the two and proper tension of the belt is critical, a reduction of pressure at the drive pulley allows it to decrease in diameter. If the change in the drive pulley is the equivalent to the change of the driven pulley, the belt will have the correct stretch or tension. This means the increase of hydraulic pressure at the driven pulley is proportional to the decrease of pressure at the drive pulley. The opposite is true for high pulley ratios. Low hydraulic pressure causes the driven pulley to decrease in size, while high pressure increases the size of the drive pulley.

The input shaft is in line with the engine's crankshaft. One end of it is the sun gear for a planetary gear set. The drive pulley shaft holds the drive pulley and the forward clutch. The driven pulley shaft includes the start clutch and the secondary drive gear. The secondary gear shaft is located between the driven pulley shaft and the final driven gear **(Figure 34)**. The secondary gear shaft includes the secondary driven gear. The drive and driven pulleys rotate in the same direction, so the secondary shaft is necessary to correct the direction of rotation for reverse.

The sun gear of the planetary gear set is part of the input shaft. The forward clutch on the drive pulley shaft is mounted to the carrier on the forward clutch drum. The carrier assembly includes pinion gears that mesh with the sun gear and the ring gear. The ring gear has a hub-mounted reverse brake multiple-disc assembly and is only used for switching the rotational direction of the pulley shafts. The parking gear is an integral part of the secondary drive gear.

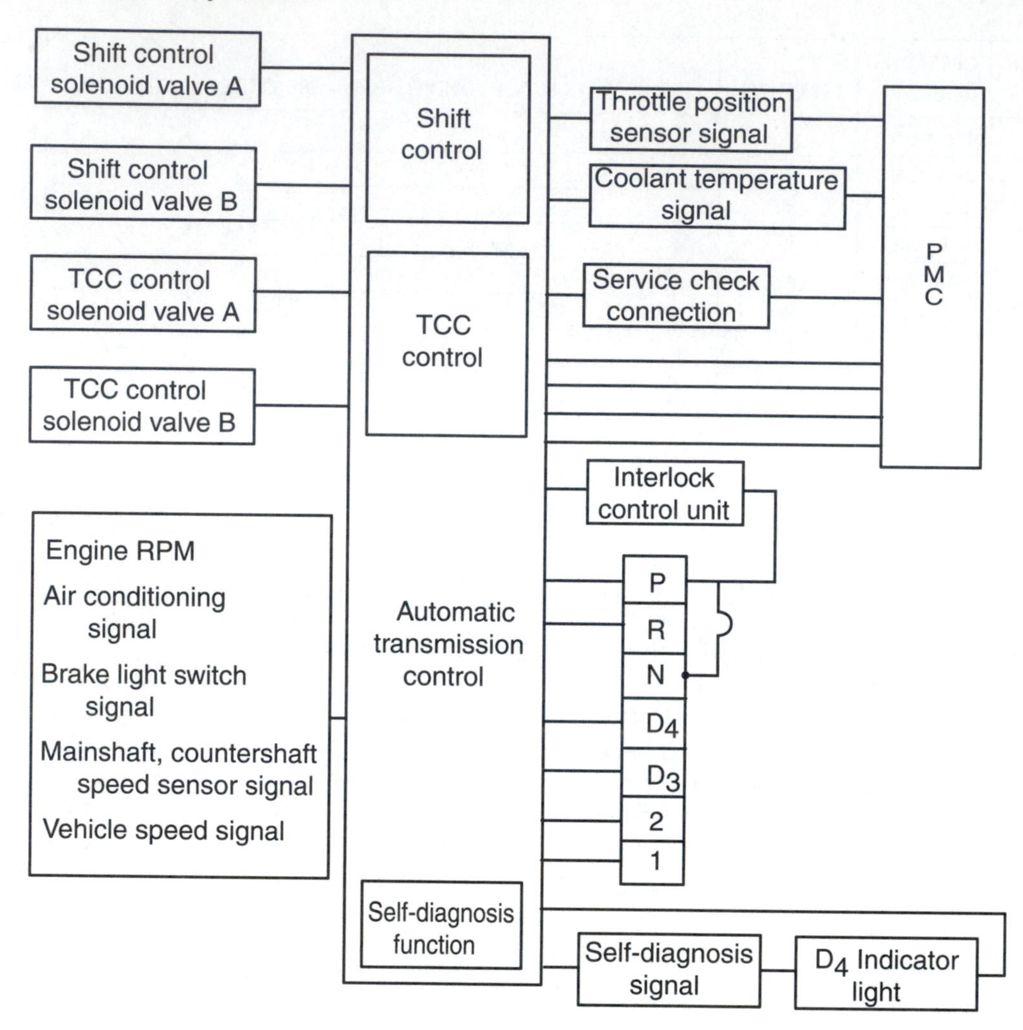


**Figure 32.** Schematic of Honda's electronically controlled transaxle.

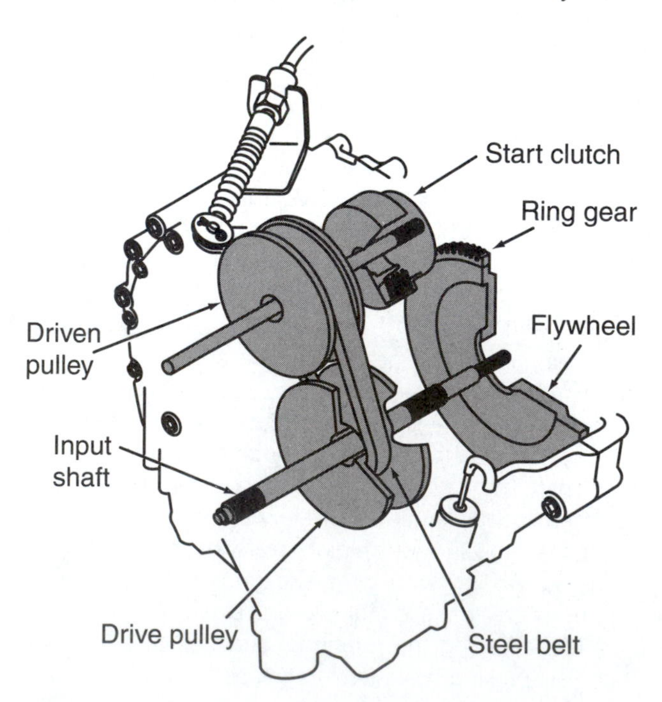


**Figure 33.** The main components of a typical CVT.

The gear selector has six positions: Park, Reverse, Neutral, Drive, Second, and Low. Although these positions seem to indicate fixed gear ratios, this is not true of the forward ranges on this type of transmission. When Low is selected, the transaxle shifts into the lowest range of available pulley ratios. This gear is intended for engine braking and power for climbing steep grades or for very heavy loads. Second gear is designed for quicker acceleration while the car is moving at highway speeds and for moderate engine braking. When this range is selected, the transaxle shifts into a lower range of pulley ratios. When Drive is selected, the transaxle automatically adjusts the pulley ratios to keep the engine at the best speed for the current driving conditions. During these forward ranges, the start and forward clutches are engaged.

When the selector is in Park, the parking pawl is engaged to the parking gear on the driven pulley shaft. This locks the front wheels. While the transaxle is in Park and Neutral, the start clutch and the forward clutch are released. When the selector is placed into reverse. The reverse brake clutch is engaged, as is the start clutch.

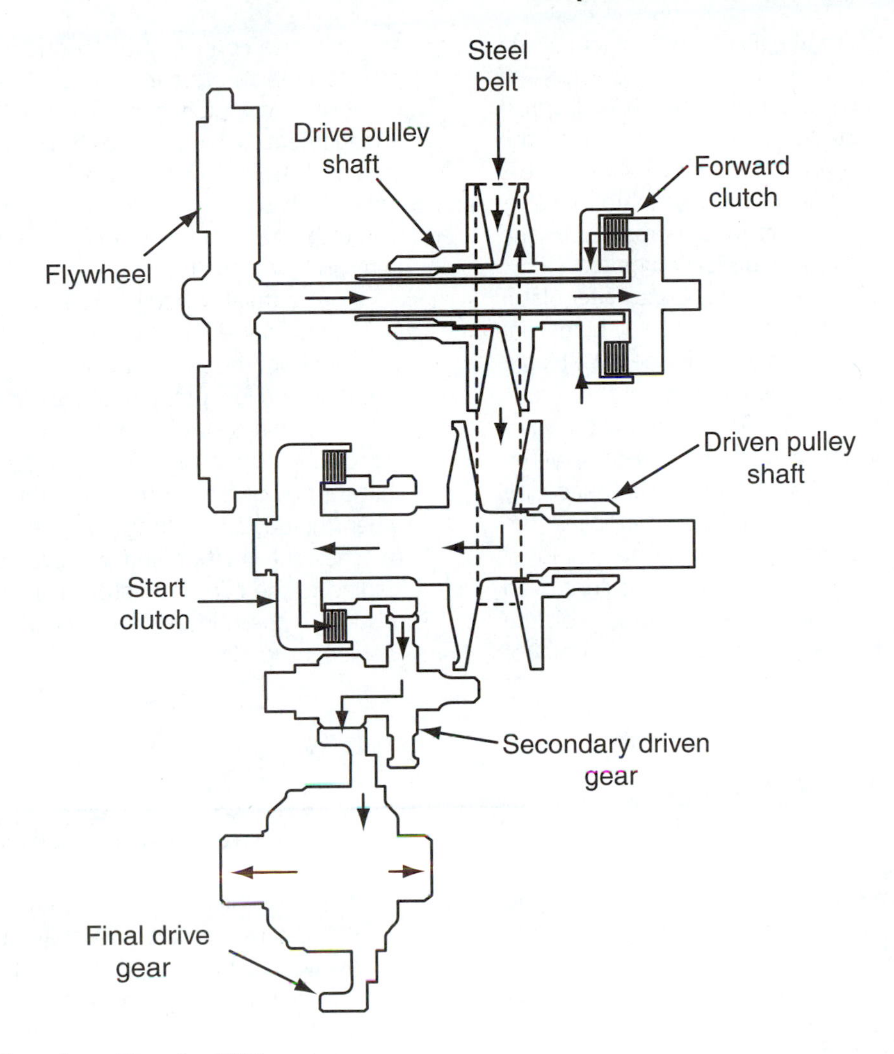


**Figure 34.** Power flow for a Honda CVT.

In the forward range, the pinion gears do not rotate on their shafts. They revolve with the sun gear. This causes the carrier and forward clutch drum to rotate. The input shaft drives the sun gear. The carrier outputs the power to the drive pulley shaft. In reverse, the reverse clutch locks the ring gear to the housing. The pinion gears revolve around the sun gear causing the carrier to rotate in the opposite direction from the rotation of the sun gear. The start clutch engages and disengages the secondary drive gear to transfer power to the final drive unit.

**Hydraulic Circuits** Shifting (or pulley ratio changes) in Honda's CVT is totally controlled by electronics. Three solenoids control the flow of hydraulic pressure, which in turn control the action of the transaxle. The transaxle's oil pump is connected to the input shaft by sprockets and a chain. There are two valve body assemblies within the transaxle. The primary valve body, called the lower valve body, is mounted to the lower part of the transaxle case. The manual shift valve body is bolted to the intermediate housing of the transaxle.

The lower valve body includes the main valve body, secondary valve body, the low pressure (PL) regulator valve body, start clutch control valve body, and shift valve body. The main valve body consists of the high pressure (PH) control valve, the lubrication valve, and the pilot regulator valve. The PH control valve supplies high control pressure to the PH regulator valve, which also regulates high pressure. The lubrication valve controls and maintains the pressure of the ATF going to the shafts. The pilot regulator valve controls the start clutch pressure in relationship to engine speed when a fault is detected in the electronic system.

The secondary valve body contains the PH regulator valve, clutch reducing valve, start clutch valve accumulator, and shift inhibitor valve. The PH regulator valve maintains the pressure of the fluid from the oil pump and supplies fluid to the shafts and the rest of the hydraulic control circuit. The clutch reducing valve receives high pressure from the PH regulator valve and controls the clutch reducing pressure. The clutch reducing valve supplies fluid to the manual valve and the start clutch control valve. It also supplies signal pressure to the PH–PL pressure control valve, shift control valve, and inhibitor solenoid valve. The start clutch valve accumulator stabilizes the hydraulic pressure applied to the start clutch. The switching of the hydraulic passages to the start clutch control when there is an electronic failure is the duty of the shift inhibitor valve. This valve also supplies clutch reducing pressure to the pilot regulator valve and the pilot lubrication pipe.

The low pressure valve body contains the PL regulator valve and the PH–PL control valve. Two solenoids are mounted to this portion of the valve body: the PH–PL control solenoid and the inhibitor solenoid, both of these solenoids are controlled by the TCM. The PL regulator valve supplies low pressure to the pulleys when necessary. The PH–PL control valve controls the PL regulator valve according to engine torque. The PH–PL control valve supplies control pressure to the PH control valve to regulate PH higher than PL. The action of the PH–PL control valve is controlled by the PH–PL solenoid. The inhibitor solenoid controls the reverse inhibitor valve by cycling on and off. The inhibitor solenoid also controls high control pressure by applying reverse inhibitor pressure to the PH control valve.

The start clutch control valve regulates start clutch engagement according to throttle opening. The start clutch control valve is controlled by the start clutch solenoid, which is controlled by the TCM. If there is an electronic failure, the start clutch control circuit switches to hydraulic controls based on engine speed.

The shift valve body contains the shift valve and the shift control valve. The shift valve is controlled by shift valve pressure from the shift control valve. The shift valve distributes high and low pressure to pulleys to change the pulley ratios. The shift control valve controls the shift valve according to vehicle speed and throttle opening. The action of the shift control valve is controlled by the shift control solenoid. When there is an electronic failure, the shift control valve switches the shift inhibitor valve to uncover the port leading to the pilot regulator pressure to the start clutch; this allows the transaxle to work in spite of the electrical problems.

## Summary

- Transmissions are designed and built for particular applications. Things that must be considered in the design of a transmission is the power output of the engine, the weight of the vehicle, and the typical workload of the vehicle.
- Most manufacturers use an alphanumeric code that indicates the features of the transmission model.
- Most automatic transmissions use compound planetary gear sets to achieve the different forward gear ratios and a reverse gear.
- A good number of transmissions and transaxles are based on a Simpson gear train.
- A variable force motor or electronic pressure control solenoid may be used to change line pressure in response to engine speed and vehicle load.
- Torque converter clutch action may be controlled by a pulse width modulated solenoid.
- Through the years, Ford has probably used the Ravigneaux gear set more than other manufacturers.
- Many transmissions are now fitted with two or three simple planetary gearsets connected in tandem.
- Most transaxles used in Honda and Acura cars and Saturn vehicles use constant-mesh helical and square-cut gears, similar to that of a manual transmission.
- Constantly variable transmissions have no fixed forward speeds. Most of these units are equipped with two pulleys and a steel belt to automatically provide ratio changes.

## Review Questions

1. The coding used to identify the transmission model typically includes what kind of information?
2. Some transmission and transaxle control systems receive information through multiplexing. How does this work?
3. What is a VFM and how is it used?
4. When a Simpson gear based transmission is shifted into first gear, the input shaft is locked to the ________ planetary ring gear. The ________ ________ gear drives the front planet gears, which drive the ________ gear. The rear planet carrier is locked; therefore the sun gear spins the ________ ________ gears. The planet gears drive ________ ________ gear, which is locked to the output shaft, in a clockwise direction. The result is a forward gear reduction.
5. In some transmissions, an EPC solenoid replaces the conventional ________ setup to provide changes in pressure in response to ________.
6. During reverse gear in a Ravigneaux gear set, input is received at the ________ ________ gear and the planetary gear carrier is held. The clockwise rotation of

the __________ __________ gear drives the long pinion gears in a __________ direction. The long pinions then drive the ring gear and output shaft in a __________ direction with a gear reduction.

7. For low pulley ratios in a CVT, an increase of hydraulic pressure at the __________ pulley is proportional to the decrease of pressure at the __________ pulley. The opposite is true for high pulley ratios. __________ hydraulic pressure causes the driven pulley to __________ in size, while __________ pressure __________ the size of the drive pulley.
8. When discussing valve body assemblies in late-model transmissions, Technician A says that the valve body is no longer needed in some electronically controlled transmissions. Technician B says that most shift solenoid assemblies are mounted directly to the valve body. Who is correct?
   A. Technician A only
   B. Technician B only
   C. Both Technician A and Technician B
   D. Neither Technician A nor Technician B
9. When discussing current manufacturer transmission coding, Technician A says that Ford model numbers do not indicate the duty rating of the transmission. Technician B says that Chrysler model numbers indicate the number of planetary units used in the transmission. Who is correct?
   A. Technician A only
   B. Technician B only
   C. Both Technician A and Technician B
   D. Neither Technician A nor Technician B
10. When discussing Chrysler transaxles, Technician A says that if the controller loses source voltage, the transaxle will enter into limp-in mode. Technician B says that the transaxle will enter into limp-in if the controller senses a transmission failure. Who is correct?
    A. Technician A only
    B. Technician B only
    C. Both Technician A and Technician B
    D. Neither Technician A nor Technician B

# Section 4

## Diagnostics and In-Vehicle Service



## SECTION OBJECTIVES

After you have read, studied, and practiced the contents of this section, you should be able to:

- Identify and interpret transmission concern; ensure proper engine operation and determine necessary action.
- Diagnose transmission/transaxle gear reduction/multiplication concerns using driving, driven, and held member (power flow) principles.
- Diagnose unusual fluid usage, level, and condition problems.
- Diagnose noise and vibration problems.
- Diagnose electronic, mechanical, and vacuum control systems.
- Conduct preliminary checks on EAT systems and determine needed repairs or service.
- Retrieve trouble codes from common electronically controlled automatic transmissions and determine needed repairs or service.
- Inspect, test, and replace electrical/electronic switches, sensors, and solenoids.
- Replace automatic transmission fluid and filters.
- Inspect, replace, and align power train mounts.
- Inspect, adjust, and replace manual valve shift and TV linkages.
- Inspect and replace external seals and gaskets when the transmission is in the vehicle.
- Inspect and replace parking pawl, shaft, spring, and retainer when the transmission is in the vehicle.
- Adjust bands.
- Inspect, leak test, flush, and replace cooler, lines, and fittings.
- Diagnose hydraulically and electrically controlled torque converter clutches.

*Oldsmobile introduced the Hydramatic on 1940; this was the first fully automatic transmission in a mass-produced U.S. car. Two planetary gear sets were used and controlled by opposing throttle and governor pressures, to provide four forward speeds. Cadillac offered the Hydramatic in 1941 and Pontiac offered it in 1948.*

# Chapter 14 Diagnosing Automatic Transmissions

## *Introduction*

The major components of a transaxle are the same as those in a transmission, except the transaxle assembly includes the final drive and differential gears. Because of the similarities between a transmission and a transaxle, most of the diagnostic and service procedures are also similar. Therefore, all references to a transmission apply equally to a transaxle unless otherwise noted.

**You Should Know** *Whenever you are diagnosing or repairing a transaxle or transmission, make sure you refer first to the service manual for the specific transmission before you begin.*

## DIAGNOSTICS

The complexities involved in the operation of an automatic transmission **(Figure 1)** can make diagnosis quite complicated. This is especially true if a technician does not have a thorough understanding of the operation of a normally working transmission.

Automatic transmission problems are usually caused by one or more of the following conditions:

- Poor engine performance
- Poor maintenance
- Problems in the hydraulic system
- Abuse resulting in overheating
- Mechanical malfunctions
- Electronic failures
- Improper adjustments

Diagnosis should begin by checking the condition of the fluid and its level, conducting a thorough visual inspection, and by checking the various control systems. On late-model vehicles, the initial inspection should include a check of the electronic control system. These control systems vary with the type of transmission and the manufacturer and require specific diagnostic and service procedures. Therefore, separate chapters for these procedures are included in this book.

Poor engine performance can have a drastic effect on transmission operation. Low engine vacuum will cause a vacuum modulator to sense a load condition when it actually is not present. This will cause delayed and harsh shifts. Delayed shifts can also result from the action of the TV. assembly, if the engine runs so badly that the throttle pedal must be frequently pushed to the floor to keep it running or to get it going.

Engine performance also can affect torque converter clutch operation. If the engine is running too poorly to maintain a constant speed, the converter clutch will engage and disengage at higher speeds. The customer complaint may be that the converter chatters; however, the problem may be the result of engine misses.

If the vehicle has an engine performance problem, the cause should be found and corrected before any conclusions on the transmission are made. A quick way to identify if the engine is causing shifting problems is to connect a vacuum gauge to the engine and take a reading while the engine is running. The gauge should be connected to intake manifold vacuum, anywhere below the throttle plates. A normal vacuum gauge reading is steady and at 17 in. Hg. The rougher the engine runs, the more the gauge readings will fluctuate. The lower the vacuum readings, the more severe the problem.

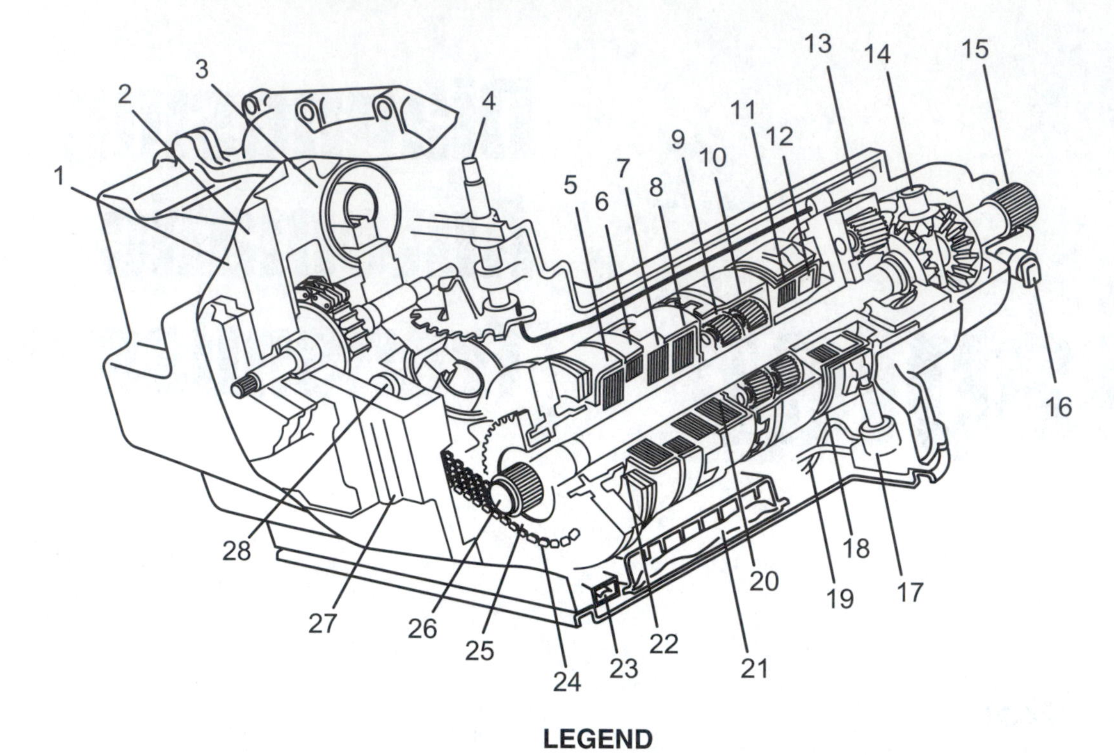


**LEGEND**

1. TFP manual valve position switch assembly
2. Control valve body assembly
3. Torque converter assembly
4. Manual shaft and detent lever assembly
5. 2nd Clutch
6. Reverse clutch
7. Coast clutch
8. Direct clutch
9. Reaction planetary gear set
10. Input planetary gear set
11. Forward clutch
12. Lo roller clutch
13. Parking lock actuator assembly
14. Differential and final drive assembly
15. Output stub shaft
16. Output speed sensor
17. Lo/Reverse servo assembly
18. Lo/Reverse band
19. Oil feed tube assembly
20. Input sprag clutch assembly
21. Oil filter assembly
22. 2nd roller clutch
23. Oil level control
24. Drive link assembly
25. Driven sprocket
26. Output shaft
27. Channel plate assembly
28. Input speed sensor

**Figure 1.** Basic components of a 4T65-E transaxle. Note the location of planetary gear set and apply and holding devices.

> **You Should Know** *At higher elevations, the atmospheric pressure is lower and so will be engine vacuum. Normally, measured vacuum will decrease one in. Hg for each 1,000 feet of elevation.*

Diagnosing a problem should follow a systematic procedure to eliminate every possible cause that can be corrected without removing the transmission. In order to properly diagnose a problem, you must totally understand the customer's concern or complaint. Make sure you know all you can about the conditions that exist when the problem occurs, such as:

- Cold or hot vehicle operating temperatures
- Cold or hot outside temperatures
- Loaded or unloaded vehicle
- City or highway driving
- Type of terrain
- Upshifting
- Downshifting
- Particular gear ranges
- Coasting
- Braking

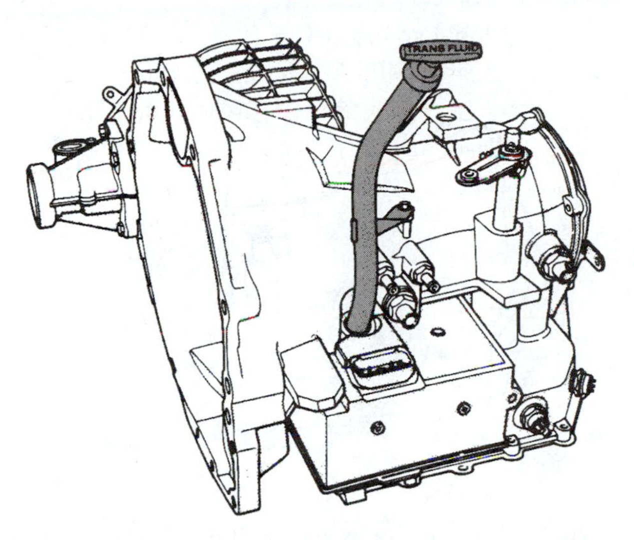


Figure 2. The dipstick and tube for an automatic transaxle.

## Fluid Check

Your diagnosis should begin with a fluid level check. To check the ATF level, make sure the vehicle is on a level surface. Before removing the dipstick, wipe all dirt off the protective disc and the dipstick handle **(Figure 2)**. On most automobiles, the ATF level can be checked accurately only when the transmission is at operating temperature and the engine is running. Remove the dipstick and wipe it clean with a lint-free white cloth or paper towel. Reinsert the dipstick, remove it again, and note the reading. Markings on a dipstick indicate ADD levels, and on some models, FULL levels for cool, warm, or hot fluid **(Figures 3 and 4)**. If the dipstick has a HOT level marked on it, this mark should be used only when the dipstick is hot to the touch.

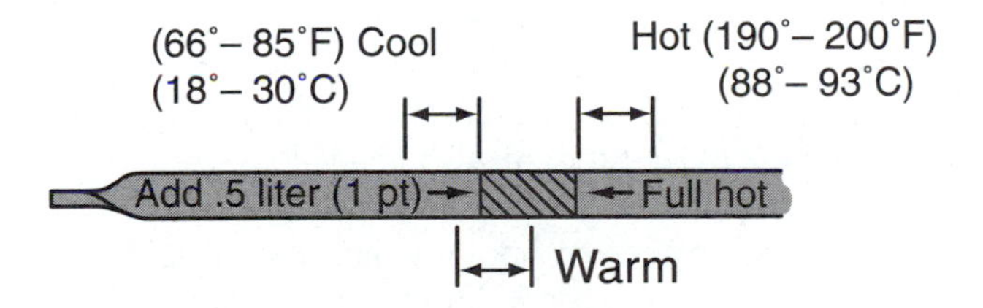


Figure 3. Typical dipstick markings for an automatic transmission.

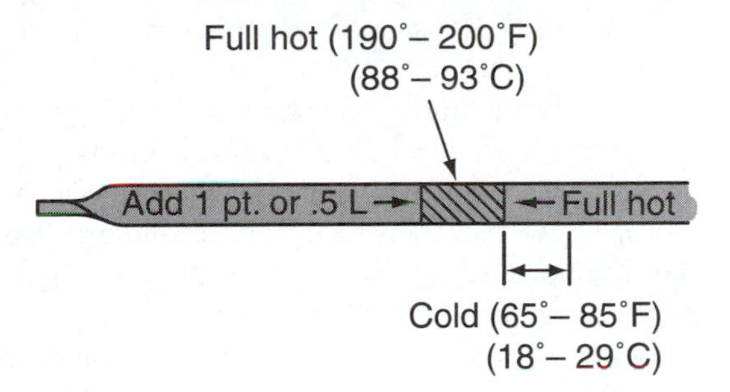


Figure 4. Typical dipstick markings for an automatic transaxle.

**You Should Know** *The temperature of the transmission or transaxle is a vital factor when checking ATF levels. Fluid level checking can be misleading because some transmissions have bimetallic elements that block fluid flow to the transmission cooler until a predetermined temperature is reached. The bimetal element keeps the fluid inside the transmission hot.*

If the fluid level is low and/or off the crosshatch section of the dipstick, the problem could be external fluid leaks. Check the transmission case, oil pan, and cooler lines for evidence of leaks.

Low fluid levels can cause a variety of problems. Air can be drawn into the oil pump's inlet circuit and mix with the fluid. This will result in aerated fluid, which causes slow pressure buildup, and low pressures, which will cause slippage between shifts. Air in the pressure regulator valve will cause a buzzing noise when the valve tries to regulate pump pressure.

Excessively high fluid levels can also cause **aeration**. As the planetary gears rotate in high fluid levels, air can be forced into the fluid. Aerated fluid can foam, overheat, and oxidize. All of these problems can interfere with normal valve, clutch, and servo operation. Foaming may be evident by fluid leakage from the transmission's vent.

The condition of the fluid should be checked while checking the fluid level. Examine the fluid carefully. The normal color of ATF is pink or red. If the fluid has a dark brownish or blackish color and/or a burned odor, the fluid has been overheated. A milky color indicates that engine coolant has been leaking into the transmission's cooler in the radiator. If there is any question about the condition of the fluid, drain out a sample for closer inspection.

After checking the ATF level and color, wipe the dipstick on absorbent white paper and look at the stain left by the fluid. Dark particles are normally band and/or clutch material, while silvery metal particles are normally caused by the wearing of the transmission's metal parts. If the dipstick cannot be wiped clean, it is probably covered with varnish, which results from fluid oxidation. Varnish will cause the spool valves to stick, causing improper shifting speeds. Varnish or other heavy deposits indicate the need to change the transmission's fluid and filter.

Temperature fluctuations from summer to winter can cause a thermal breakdown of AFT. Even high quality fluids can experience breakdown as a result of these frequent and extreme temperature changes. It is said that nearly 90 percent of all transmission failures are caused fluid breakdown or **oxidation**.

You Should Know *Contaminated fluid can sometimes be felt better than it can be seen. Place a few drops of fluid between two fingers and rub them together. If the fluid feels dirty or gritty, it is contaminated with burned frictional material.*

## FLUID LEAKS

Continue your diagnostics by conducting a quick and careful visual inspection. Check all drive train parts for looseness and leaks. If the transmission fluid was low or there was no fluid, raise the vehicle and carefully inspect the transmission for signs of leakage. Leaks are often caused by defective gaskets or seals. The housing itself may have a porosity problem, allowing fluid to seep through the metal. Case porosity is caused by tiny holes, which are formed by trapped air bubbles during the casting process. This problem may be corrected by using an epoxy-type sealer.

Common sources of leaks in a transaxle are the oil pan seal, rear and final drive covers, axle shaft seals, speedometer drive gear assembly, and electrical switches mounted into the housing **(Figure 5)**. Common sources of leaks in a transmission are the oil pan seal, housing vent, extension housing, speedometer drive gear assembly, and electrical switches mounted into the housing.

### Oil Pan

A common cause of fluid leakage is the seal of the oil pan to the transmission housing. If there are signs of leakage around the rim of the pan, retorquing the pan bolts may correct the problem. If tightening the pan does not correct the problem, the pan must be removed and a new gasket installed. Make sure the sealing surface of the pan's rim is flat and capable of providing a seal before reinstalling it. Keep in mind, some transmissions do not use a gasket, rather the pan is sealed with **RTV** sealer.

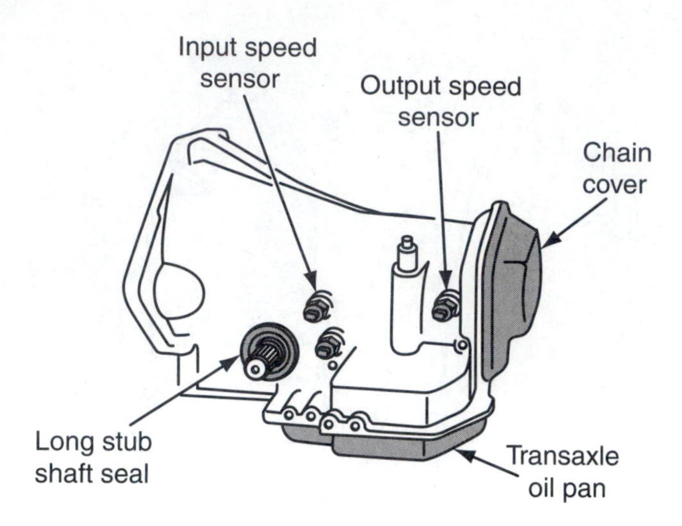


**Figure 5.** Possible sources of fluid leaks on this transaxle.

### Torque Converter

Torque converter problems can be caused by a leaking converter **(Figure 6)**. This type of problem may be the cause of customer complaints of slippage and a lack of power. To check the converter for leaks, remove the converter access cover and examine the area around the torque converter shell. An engine oil leak may be falsely diagnosed as a converter leak. The color of engine oil is different from transmission fluid and may help identify the true source of the leak. However, if the oil or fluid has absorbed much dirt, both will look the same. An engine leak typically leaves an oil film on the front of the converter shell, whereas a converter leak will cause the entire shell to be wet. If the transmission's oil pump seal is leaking, only the backside of the shell will be wet. If the converter is leaking or damaged, it should be replaced.

### Extension Housing

An oil leak stemming from the mating surfaces of the extension housing and the transmission case may be caused by loose bolts. To correct this problem, tighten the bolts to the specified torque. Also, check for signs of leakage at the rear of the extension housing. Fluid leaks from the seal of the extension housing can be corrected with the transmission in the car. Often, the cause for the leakage is a worn extension housing bushing, which supports the sliding yoke of the drive shaft. When the drive shaft is installed, the clearance between the sliding yoke and the bushing should be minimal. If the clearance is satisfactory, a new oil seal will correct the leak. If the clearance is excessive, the repair requires that a new seal and a new bushing be installed. If the seal is faulty, the transmission vent should be checked for blockage.

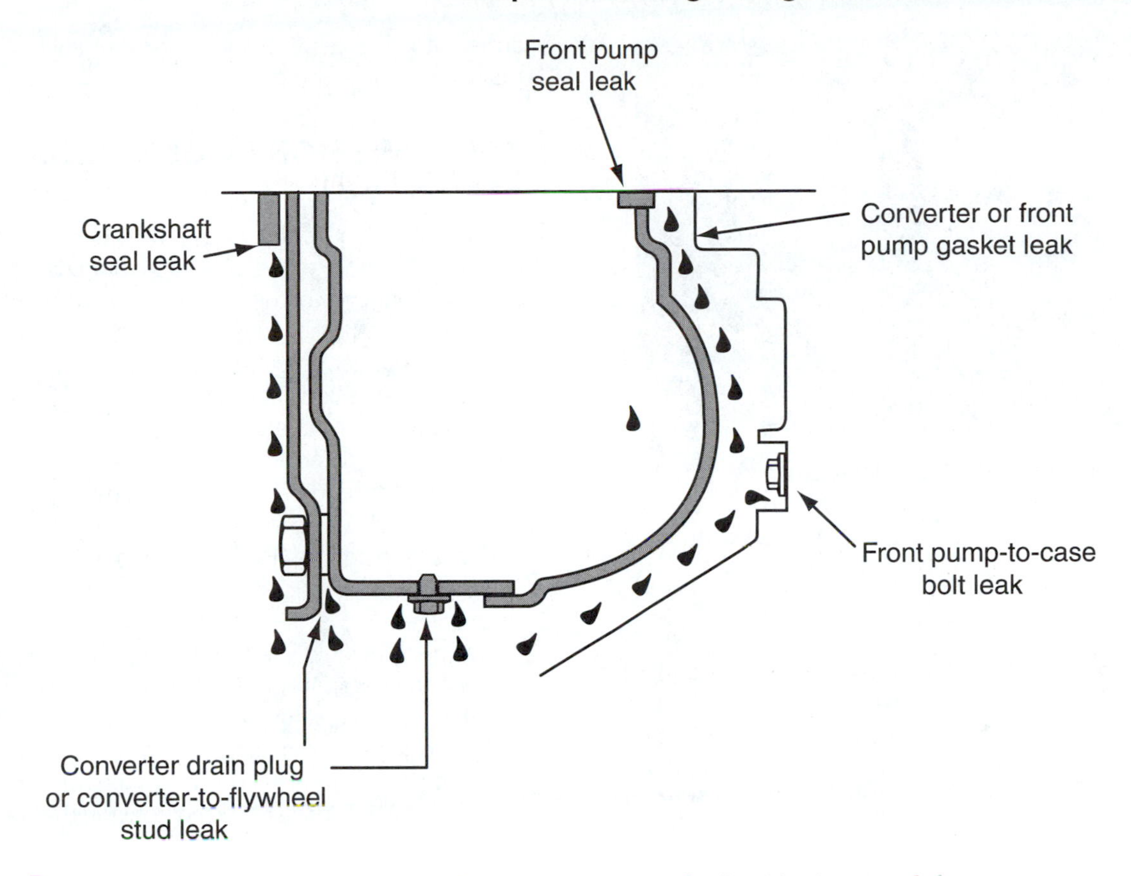


**Figure 6.** By determining the direction of fluid travel, the cause of a fluid leak around the torque converter can be identified.

## Speedometer Drive

The vehicle's speedometer can be purely electronic, which requires no mechanical hookup to the transmission, or it can be driven off, via a cable, the output shaft. If the transmission is equipped with a vehicle speed sensor, the bore and sensor can be a source of leaks. The sensor is retained in the bore with a retaining nut or bolt **(Figure 7)**. An oil leak at the speedometer cable or VSS can be corrected by replacing the O-ring seal.

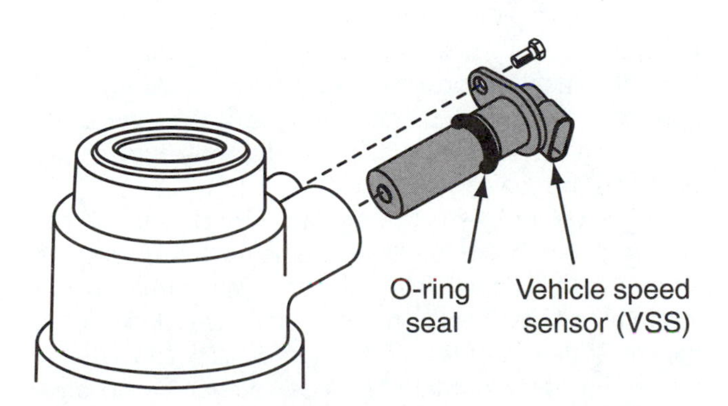


**Figure 7.** A VSS with its seal and retaining bolt.

## Electrical Connections

A visual inspection of the transmission should include a careful check of all electrical wires, connectors, and components. This inspection is especially important for electronically controlled transmissions and for transmissions that have a torque converter clutch. All faulty connectors, wires, and components should be repaired or replaced before continuing your diagnosis of the transmission.

Check all electrical connections to the transmission. Faulty connectors or wires can cause harsh or delayed and missed shifts. On transaxles, the connectors can normally be inspected through the engine compartment, whereas they can only be seen from under the vehicle on longitudinally mounted transmissions. To check the connectors, release the locking tabs and disconnect them, one at a time, from the transmission. Carefully examine them for signs of corrosion, distortion, moisture, and transmission fluid. Even the slightest amount of corrosion can affect the output of a sensor. Increased resistance will always change the voltage signal of a circuit. Also, check the connector at the transmission **(Figure 8)**. Using a small mirror and flashlight may help you get a good look at the inside of the connectors. Inspect the entire transmission wiring harness for tears and other damage. Road

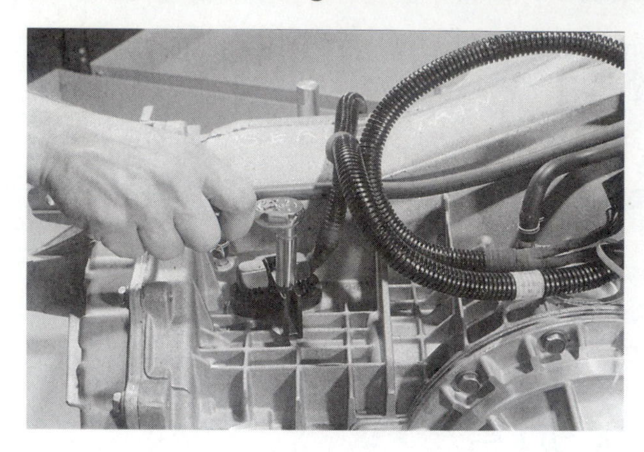

**Figure 8.** Carefully check all electrical wires and terminals connected to the transmission. Make sure they are not damaged or dirty and the connectors are tight.

debris can damage the wiring and connectors mounted underneath the vehicle.

Because the operation of the engine and transmission are integrated through the control computer, a faulty engine sensor or connector may affect the operation of both, the engine and the transmission. The various sensors and their locations can be identified by referring to the appropriate service manual.

## Checking the Transaxle Mounts

The engine and transmission mounts on FWD cars are important to the operation of the transaxle. Any engine movement may change the effective length of the shift and throttle cables and therefore may affect the engagement of the gears. Delayed or missed shifts may result from linkage changes as the engine pivots on its mounts. Problems with transmission mounts also may affect the operation of a transmission on a RWD vehicle, but this type of problem will be less detrimental than the same type of problem on FWD vehicles.

Many shifting and vibration problems can be caused by worn, loose, or broken engine and transmission mounts. Visually inspect the mounts for looseness and cracks. To get a better look at the condition of the mounts, pull up and push down on the transaxle case while watching the mount. If the mount's rubber separates from the metal plate or if the case moves up but not down, replace the mount. If there is movement between the metal plate and its attaching point on the frame, tighten the attaching bolts to an appropriate torque.

Then, from the driver's seat, apply the foot brake, set the parking brake, and start the engine. Put the transmission into a gear and gradually increase the engine speed to about 1,500–2,000 rpm. Watch the torque reaction of the engine on its mounts. If the engine's reaction to the torque appears to be excessive, broken or worn drive train mounts may be the cause.

## TRANSMISSION COOLER AND LINE INSPECTION

Transmission coolers are a possible source of fluid leaks. The efficiency of the coolers is also critical to the operation and longevity of the transmission. The vehicle may be equipped with the standard type cooler or also may have an auxiliary cooler. Both of these should be and are visually inspected and tested in the same way.

Follow these steps when inspecting the transmission cooler and associated lines and fittings:

1. Check the engine cooling system. The transmission cooler cannot be efficient if the engine's cooling system is defective. Repair all engine cooling system problems before continuing to check the transmission cooler.
2. Inspect the fluid lines and fittings between the cooler and the transmission **(Figure 9)**. Check these for looseness, damage, signs of leakage, and wear. Replace any damaged lines. Replace or tighten any leaking fittings.
3. Inspect the engine's coolant for traces of transmission fluid. If ATF is present in the coolant, the transmission cooler leaks.

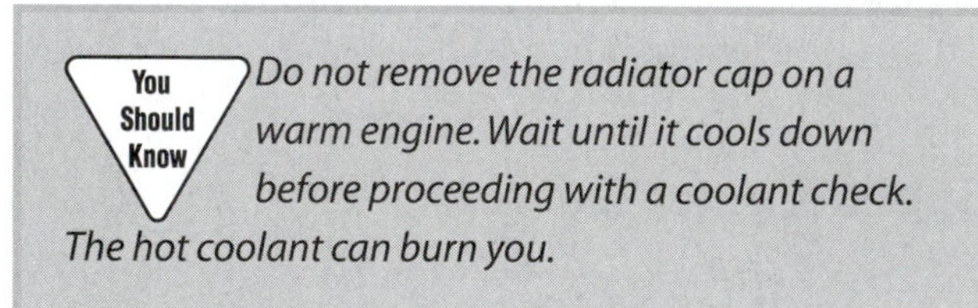


*Do not remove the radiator cap on a warm engine. Wait until it cools down before proceeding with a coolant check. The hot coolant can burn you.*

4. Check the transmission's fluid for signs of engine coolant. Water or coolant will cause the fluid to appear milky with a pink tint. This is also an indication that the transmission cooler leaks and is allowing engine coolant to enter into the transmission fluid.

The cooler can be checked for leaks by disconnecting and plugging the transmission to cooler lines at the radiator. Then, remove the radiator cap to relieve any pressure in the system. Tightly plug one of the ATF line fittings at the radiator. Using the shop air supply with a pressure regulator, apply 50 to 70 psi of air pressure into the cooler at the other cooler line fitting. Look into the radiator. If bubbles are observed, the cooler leaks.

The cooler system also should be checked for proper flow. To do this, make sure there are no fluid leaks. Start the engine and allow the transmission to reach normal operating temperature. Then, turn the engine off and check the fluid level. Remove the transmission dipstick and install a funnel into the dipstick tube. Raise the vehicle up on a hoist. Then, disconnect the cooler return line at the transmission. Securely fasten a rubber hose to the end of the

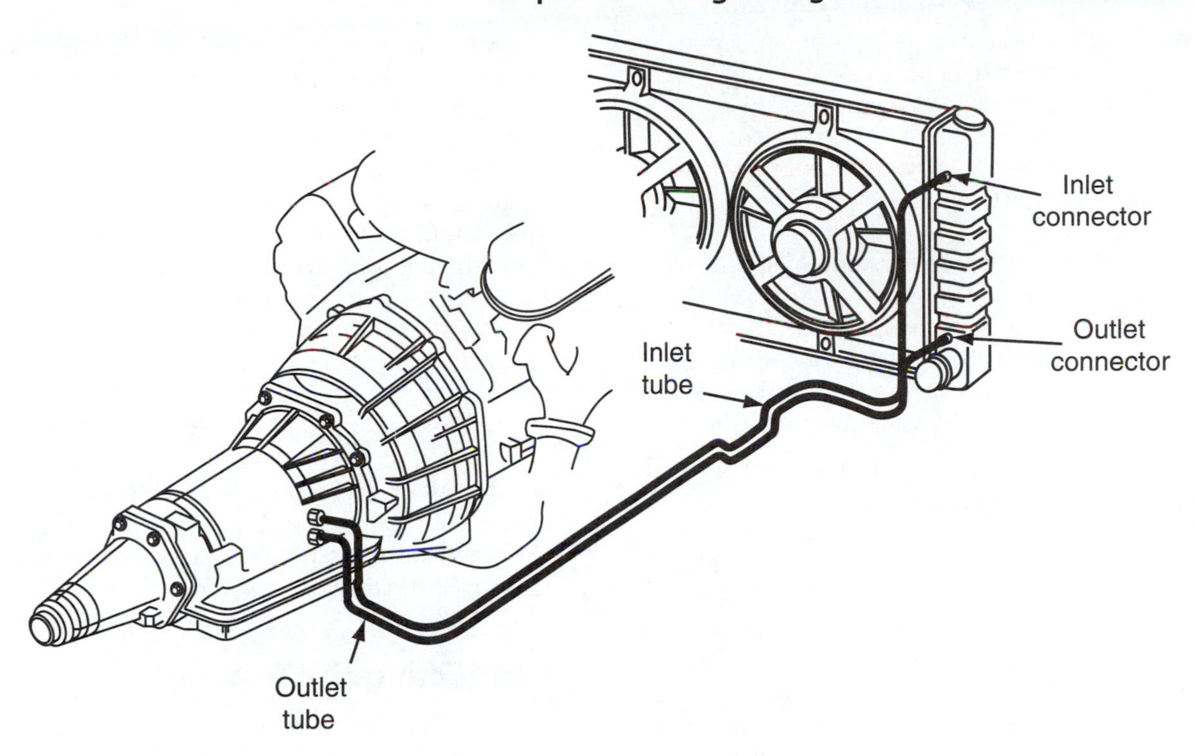


**Figure 9.** Inspect the fluid lines and fittings between the cooler and the transmission.

cooler line and place the other end of the hose into the funnel. Now, lower the vehicle and start the engine. Allow the engine to run at a fast idle and observe the flow of fluid into the funnel.

Another way to do this test is with an assistant. Run the engine and allow the fluid from the end of the return cooler line to enter into a container where you can easily observe how much is flowing **(Figure 10)**. Have your assistant pour the same amount of fluid into the filler tube as you are getting to flow into the container. If you get one pint, have the assistant pour in one pint. By replenishing the fluid, your assistant is ensuring that you will not run out

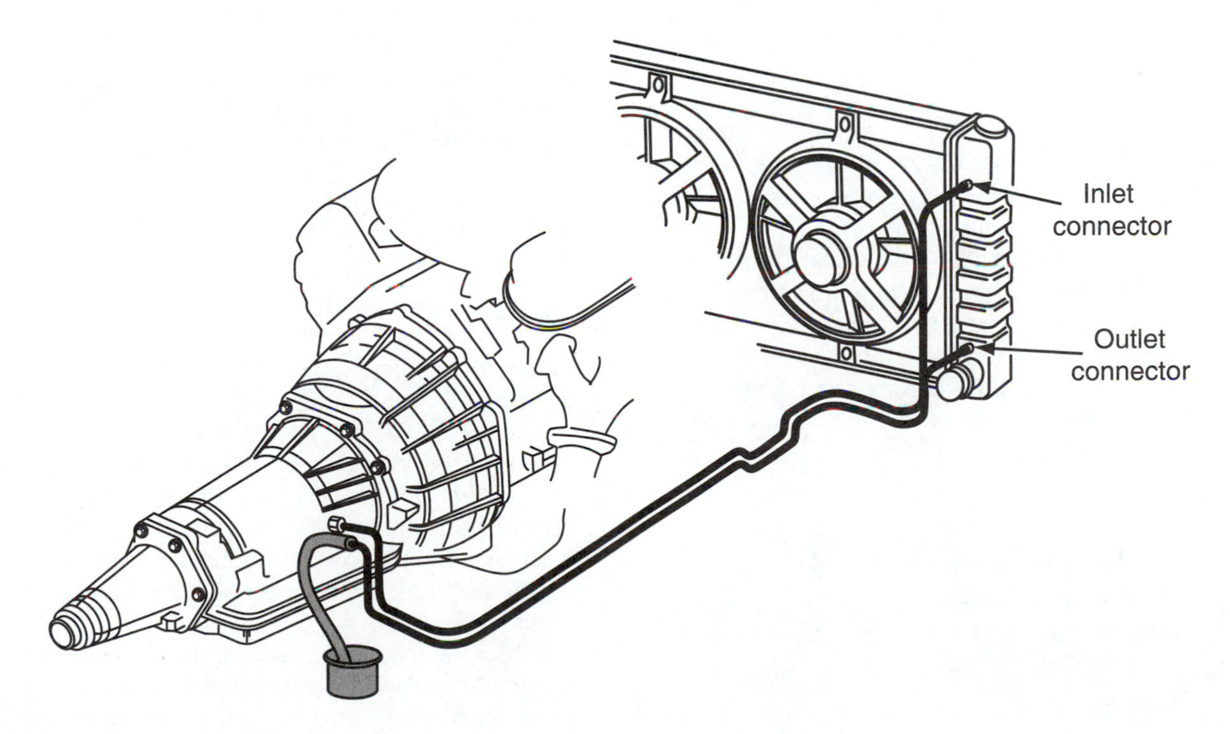


**Figure 10.** Setup for checking fluid flow.

fluid. Make sure the assistant keeps track of what is being poured in. Continue the test for 30 seconds.

The preferred way to check flow is with a flow meter. A flow meter is preferred because it has restrictions on the return side, which is closer to reality than an open hose. To use a flow meter, connect it into the cooler lines, start the engine, and observe.

Most manufacturers list specifications for transmission cooler flow rates. This is normally expressed in pints per second (such as 1 pint in 15–20 seconds). You should observe a steady flow of fluid. If the flow is erratic or weak, the cooler may be restricted, the transmission's pump may be weak, or there is an internal leak in the transmission.

## ROAD TESTING THE VEHICLE

All transmission complaints should be verified by road testing the vehicle and attempting to duplicate the customer's complaint. Knowledge of the exact conditions that cause the symptom and a thorough understanding of transmissions will allow you to accurately diagnose problems. Many problems that appear to be transmission problems may be caused by problems in the engine, drive shaft, U-joint or CV joint, wheel bearings, wheel/tire imbalance, or other conditions.

Make sure these are not the cause of the problem before you begin to diagnose and repair a transmission. Diagnosis becomes easy if you think about what is happening in the transmission when the problem occurs. If there is a shifting problem, think about the parts that are being engaged and what these parts are attempting to do.

### Test Preparation

Road testing a vehicle to duplicate the symptoms exactly is essential to proper diagnosis. During the road test, the transmission should be operated in all possible modes and its operation noted. The customer's complaint should be verified and any other transmission malfunctions should be noted. Your observations during the road test, your understanding of the operation of a transmission, and the information given in service manuals will help you identify the cause of any transmission problem. If conducted properly, road testing is diagnosis through a process of elimination.

**You Should Know** *Always refer to your service manual to identify the particulars of the transmission you are diagnosing. It is also helpful to check for any Technical Service Bulletins that may be related to the customer's complaint.*

Before beginning your road test, find and duplicate, from a service manual, the chart **(Figure 11)** that shows the band and clutch application for different gear selector positions. Using these charts will greatly simplify your diagnosis of automatic transmission problems. It also is wise to have a notebook or piece of paper to jot down notes about the operation of the transmission. If the transmission is electronically controlled, take a scan tool on the road test with you. It will be an added convenience if the scan tool has memory or a printer. Some manufacturers recommend the use of a symptoms chart or checklist that is provided in their service manuals **(Figure 12)**.

Prior to road testing a vehicle, always check the transmission fluid level and condition of oil and correct any engine performance problems. Inspect the transmission for signs of fluid leakage. If leakage is evident, wipe off the leaking oil and dust. Then, note the location of the leaks in your notebook or on the symptoms chart.

### Conducting the Test

Begin the road test with a drive at normal speeds to warm the engine and transmission. If a problem appears only when starting and/or when the engine and transmission are cold, record this symptom on the chart or in your notebook.

After the engine and transmission are warmed up, place the shift selector into the DRIVE position and allow the transmission to shift through all of its normal shifts. Any questionable transmission behavior should be noted.

During a road test, check for proper gear engagement as you move the selector lever to each gear position, including park. There should be no hesitation or roughness as the gears are engaging. Check for proper operation in all forward ranges, especially the 1–2, 2–3, 3–4 upshifts and converter clutch engagement during light throttle operation. These shifts should be smooth and positive and occur at the correct speeds. These same shifts should feel firmer under medium to heavy throttle pressures. Transmissions equipped with a torque converter clutch should be brought to the specified apply speed and their engagement noted. Again, record the operation of the transmission in these different modes in your notebook or on the diagnostic chart.

Force the transmission to kick down and record the quality of this shift and the speed at which downshifts. These will be later compared to the specifications. Manual downshifts also should be made at a variety of speeds. The reaction of the transmission should be noted, as should all abnormal noises, and the gears and speeds at which they occur.

After the road test, check the transmission for signs of leakage. Any new leaks and their probable cause should be noted. Then, compare your written notes from the road test to the information given in the service manual to identify

Range Reference

| Range | Park/ Neutral | Reverse | D | | | | 3 | | | 2 | | | 1 | |
|---|---|---|---|---|---|---|---|---|---|---|---|---|---|---|
| Gear | N | R | 1st | 2nd | 3rd | 4th | 1st | 2nd | 3rd | 1st | 2nd | 3rd** | 1st | 2nd*** |
| 1-2 Shift Solenoid | ON | ON | ON | OFF | OFF | ON | ON | OFF | OFF | ON | OFF | OFF | ON | OFF |
| 2-3 Shift Solenoid | OFF | OFF | OFF | OFF | ON | ON | OFF | OFF | ON | OFF | OFF | ON | OFF | OFF |
| 2nd Clutch | – | – | – | A | A* | A* | – | A | A* | – | A | A* | – | A |
| 2nd Roller Clutch | – | – | – | H | O | – | – | H | O | – | H | O | – | H |
| Int./4th Band | – | – | – | – | – | A | – | – | – | – | A | – | – | A |
| Reverse Clutch | – | A | – | – | – | – | – | – | – | – | – | – | – | – |
| Coast Clutch | – | – | – | – | – | A | A | A | A | A | A | A | A | A |
| Input Sprag | – | – | H | H | H | O | H | H | H | H | H | H | H | H |
| Direct Clutch | – | – | – | – | A | A | – | – | A | – | A | – | – | – |
| Forward Clutch | – | – | A | A | A | A* | A | A | A | A | A | A | A | A |
| Lo/Rev Band | A | A | – | – | – | – | – | – | – | – | – | – | A | – |
| Lo Roller Clutch | – | – | H | O | O | O | H | O | O | H | O | O | H | O |

A=Applied H=Holding O=Overrunning ON=The solenoid is energized. OFF= The solenoid is de-energized.
*=Applied with no load **Manual SECOND/THIRD gear is only available above approximately 100 km/k (62 mph).
***=Manual FIRST/SECOND gear is only available above approximately 60 km/h (37 mph). Manual FIRST/THIRD gear is also possible at high vehicle speed as safety feature.

**Figure 11.** Typical application chart for the brakes, clutches, and solenoids of a transmission. This chart should be referred to during road test and when determining the cause of any shifting problems.

the cause of the malfunction. The service manual usually has a diagnostic chart to aid you in this process.

The following problems and their causes are given as examples. The actual causes of these types of problems will vary with the different models of transmissions. Always refer to the appropriate band and clutch application chart when diagnosing shifting problems. Using these will allow you to identify the cause of the shifting problems through the process of elimination.

These are typical causes of common problems:

- If the shift for all forward gears is delayed, a slipping front or forward clutch is indicated.

Date ___/___/___

TRANSMISSION/TRANSAXLE-CONCERN CHECK SHEET
(*information required for technical assistance)

*VIN ______________ * Mileage________ R.O.# ____

*Model year ____ *Vehicle Model______ *Engine ____

* Trans. Model ______ *Trans. Serial __________

SERVICE ADVISOR

CUSTOMER'S CONCERN

Check the items that describe the concern:

| WHAT: | WHEN: | OCCURS: | USUALLY NOTICED: |
|---|---|---|---|
| _no power | _vehicle warm | _always | _idling |
| _shifting | _vehicle cold | _intermittent | _accelerating |
| _slips | _always | _seldom | _coasting |
| _noise | _not sure | _first time | _braking |
| _shudder | | | at ____MPH |

TECHNICIAN

Preliminary Check Procedures — NOTE FINDINGS

Inspect
*fluid level and condition
*engine performance-vacuum & EMC codes
*TV cable and/or modulator vacuum
*manual linkage adjustment
*road test to verify concern*

*Simulate the conditions under which the customer's concern was observed

* PROPOSED OR COMPLETED REPAIRS

TRANS. TEMPERATURE _____HOT _____COLD

*VACUUM READINGS AT MODULATOR
(180C, 250C, 350C, 400, 410-T4)
Readings at Modulator_______IN. HG.(Engine at Hot Idle, Trans. in Drive)
Check for Vacuum Response During Accelerator Movement

PRESSURE TEST

| PRN | MINIMUM SPEC. (from manual) | MINIMUM ACTUAL | MAXIMUM SPEC. (from manual) | MAXIMUM ACTUAL |
|---|---|---|---|---|
| P | | | | |
| R | | | | |
| N | | | | |
| D | | | | |
| O | | | | |
| 2 | | | | |
| 1 | | | | |

ROAD TEST

*FINDINGS BASED ON ROAD TEST

Check items on road test that describe customer's comments about:

| Garage Shift Feel | Upshifts | Downshifts | Torque Converter Clutch |
|---|---|---|---|
| _engine stops | _early | _busyness | _busyness |
| _harsh | _harsh | _harsh | _harsh |
| _delayed | _delayed | _delayed | _no release |
| _no drive | _slips | _slips | _shudder |
| | _no upshift | _no downshift | _early apply |
| | | | _no apply |
| | | | _late apply |

Concerns Occur When/During:

| Check | Gear | Range |
|---|---|---|
| | | P-N |
| | 1st | D (boxed) |
| | 2nd | |
| | 3rd | |
| | 4th | |
| | 1st | D |
| | 2nd | |
| | 3rd | |
| | 1st | 2 |
| | 2nd | |
| | 1st | 1 |
| | Reverse | R |

During
_1-2 upshift
_2-3 upshift
_3-4 upshift
_4-3 downshift
_3-2 downshift
_2-1 downshift

Throttle Position
_light
_medium
_heavy
_W.O.T.

Noise

| Type | When Noticed |
|---|---|
| _buzz | _always |
| _whine | _load sensitive |
| _clunk | _steering sensitive |
| | at ________MPH |
| | in ________gear |

| Pitch | Level |
|---|---|
| _low | _light |
| _medium | _medium |
| _high | _heavy |

**Figure 12.** Typical road test check list recommended by manufacturer to identify the cause of shifting problems.

- If there is a delay or slip when the transmission shifts into any two forward gears, a slipping rear clutch is indicated.
- If there is a shift delay while the transmission was in DRIVE and it shifted into third gear, either the front or rear clutch may be slipping. To determine which clutch is defective, look at the behavior of the transmission when one of the two clutches are not applied. For example, if the rear clutch is not engaged while the transmission is in reverse and there was no slippage in reverse, the problem of the slippage in third gear is caused by the rear clutch. If there is a delay only during shifts into reverse and from second to third gear, the Reverse/High clutch is slipping.
- If there is delayed shifting from first to second gear, there are intermediate clutch or band problems.
- If there is slippage in first gear when the gear selector is in the DRIVE position, but not when first gear is manually selected, the one-way or overrunning clutch may be the cause.

It is important to remember that delayed shifts or slippage also may be caused by leaking hydraulic circuits or sticking spool valves in the valve body. Because the application of bands and clutches are controlled by the hydraulic system, improper pressures will cause shifting problems. Other components of the transmission also can contribute to shifting problems. For example, on transmissions equipped with a vacuum modulator, if upshifts do not occur at the specified speeds or do not occur at all, either the modulator is faulty or the vacuum supply line is leaking.

## Noises and Vibrations

Abnormal transmission noises and vibrations can be caused by faulty bearings, damaged gears, worn or damaged clutches and bands, or a bad oil pump, as well as contaminated fluid or improper fluid levels. Vibrations also can be caused by torque converter problems. All noises and vibrations that occur during the road test should be noted, as well as, when they occur.

Often a customer will complain of a transmission noise, which in reality is caused by something else in the driveline and not the transmission or torque converter. Bad CV or U-joints, wheel bearings, brake calipers, and dragging brake pads can generate noises that customers and, unfortunately some technicians, mistakenly blame the transmission or torque converter. The entire driveline should be checked before assuming the noise is transmission related.

Transaxles have unique noises associated with them because of their construction. These noises can result from problems in the differential or drive axles. Use the following as a guide to determine if a noise is caused by the transaxle or by other parts in the drive train.

A knock at low speeds:

1. Worn drive axle CV joints
2. Worn side gear hub counterbore

Noise most pronounced on turns:

1. Differential gear noise

Clunk on acceleration or deceleration:

1. Loose engine mounts
2. Worn differential pinion shaft or side gear hub counterbore in the case
3. Worn or damaged drive axle inboard CV joints

Clicking noise in turns:

1. Worn or damaged outboard CV joint

Most vibration problems are caused by an unbalanced torque converter assembly, a poorly mounted torque converter, or a faulty output shaft. The key to determining the cause of the vibration is to pay particular attention to the vibration in relationship to engine and vehicle speed. If the vibration changes with a change in engine speed, the cause of the problem is most probably the torque converter. If the vibration changes with vehicle speed, the cause is probably the output shaft or the driveline connected to it. The later type of problem can be a bad extension housing bushing or universal joint, which would become worse at higher speeds.

Begin your diagnosis by determining if the cause of the problem is the drive line or the transmission. To do this, put the transmission in gear and apply the foot brakes. If the noise is no longer evident, the problem must be in the drive line or the output of the transmission. If the noise is still present, the problem must be in the transmission or torque converter.

Noise problems are also best diagnosed by paying a great deal of attention to the speed and the conditions at which the noise occurs. The conditions to pay most attention to the operating gear and the load on the drive line. If the noise is engine-speed related and is present in all gears, including Park and Neutral, the most probable source of the noise is the oil pump because it rotates whenever the engine is running. However, if the noise is engine related and is present in all gears except Park and Neutral, the most probable sources of the noise are those parts that rotate in all gears, such as the drive chain, the input shaft, and torque converter.

Noises that only occur when a particular gear is operating must be related to those components responsible for providing that gear, such as a band or clutch. On some RWD vehicles, noise only in first gear can be caused by a defective rear transmission mount. If the noise is vehicle related, the most probable causes are the output shaft and final drive assembly. Often the exact cause of noise and vibration problems can only be identified through a careful inspection of a disassembled transmission.

Whatever the exact fault in the transmission is the cause of a vibration or noise, the transmission will need to come out and the entire unit checked and carefully inspected. Proper diagnosis prior to disassembling the transmission will identify the specific areas that should be carefully looked at and can prevent unnecessary transmission removal and teardown.

## Summary

- Whenever you are diagnosing or repairing a transaxle or transmission, make sure you refer first to the service manual for the specific transmission before you begin.
- Diagnosis should begin by checking the condition of the fluid and its level, conducting a thorough visual inspection, and by checking the various control systems.
- If the fluid level is low and/or off the crosshatch section of the dipstick, the problem could be external fluid leaks. Check the transmission case, oil pan, and cooler lines for evidence of leaks.
- A visual inspection of the transmission should include a careful check of all electrical wires, connectors, and components.
- Many shifting and vibration problems can be caused by worn, loose, or broken engine and transmission mounts.
- Transmission coolers are a possible source of fluid leaks and the efficiency of the cooler is critical to the operation and longevity of the transmission.
- The cooler system should be checked for signs or internal and external leaks and for proper flow within the system.

- All transmission complaints should be verified by road testing the vehicle and attempting to duplicate the customer's complaint.
- During the road test, the transmission should be operated in all possible modes and its operation noted. All abnormal operations should be noted, along with the conditions that existed. The potential causes of the problem can be identified by a study of the transmission's band and clutch application chart.
- Abnormal transmission noises and vibrations can be caused by faulty bearings, damaged gears, worn or damaged clutches and bands, or a bad oil pump, as well as contaminated fluid or improper fluid levels.

## Review Questions

1. When diagnosing noises apparently from a transaxle assembly, Technician A says that a knocking sound at low speeds is probably caused by worn CV joints. Technician B says that a clicking noise heard when the vehicle is turning is probably caused by a worn or damaged outboard CV joint. Who is correct?
   A. Technician A only
   B. Technician B only
   C. Both Technician A and Technician B
   D. Neither Technician A nor Technician B
2. Technician A says that if the shift for all forward gears is delayed, a slipping front or forward clutch is normally indicated. Technician B says that a slipping rear clutch is indicated when there is a delay or slip when the transmission shifts into any forward gear. Who is correct?
   A. Technician A only
   B. Technician B only
   C. Both Technician A and Technician B
   D. Neither Technician A nor Technician B
3. Technician A says that delayed shifting can be caused worn planetary gear set members. Technician B says that delayed shifts or slippage may be caused by leaking hydraulic circuits or sticking spool valves in the valve body. Who is correct?
   A. Technician A only
   B. Technician B only
   C. Both Technician A and Technician B
   D. Neither Technician A nor Technician B
4. When checking the condition of a car's ATF, Technician A says that if the fluid has a dark brownish or blackish color and/or a burned odor, the fluid has been overheated. Technician B says that if the fluid has a milky color, this indicates that engine coolant has been leaking into the transmission's cooler. Who is correct?
   A. Technician A only
   B. Technician B only
   C. Both Technician A and Technician B
   D. Neither Technician A nor Technician B
5. When checking the engine and transmission mounts on a FWD car, Technician A says that any engine movement may change the effective length of the shift and throttle cables and therefore may affect the engagement of the gears. Technician B says that delayed or missed shifts are caused by hydraulic problems, not linkage problems. Who is correct?
   A. Technician A only
   B. Technician B only
   C. Both Technician A and Technician B
   D. Neither Technician A nor Technician B

# Basic EAT Diagnosis

## Introduction

One of the first tasks during diagnosis of an electronically controlled transmission is to determine if the problem is caused by the transmission or by electronics. To determine this, the transmission must be observed to see if it responds to commands given by the computer. Identifying whether the problem is the transmission or electrical will determine what steps need to be followed to diagnose the cause of the problem.

An electronic automatic transmission (EAT) will work only as well as the commands it receives from the computer **(Figure 1)**, even if the hydraulic and mechanical parts of the transmission are fine. All diagnostics should begin with a scan tool to check for trouble codes in the system's computer. After the received codes are addressed, you can begin a more detailed diagnosis of the system and transmission. Your next step may be manually activating the shift solenoids by connecting a jumper wire to them or by using a transmission tester that allows you to manually activate the solenoids. Prior to doing this, the wiring to the solenoids should be studied to determine if the computer activates them by supplying voltage to them or by completing the ground circuit. Also, you need to know what gear certain solenoids are activated in. This information can be found in the service manual.

### GUIDELINES FOR DIAGNOSING EATS

- Check all fuses and identify the cause of any blown fuses.
- Compare the wiring to all suspected components against the colors given in a service manual.
- When testing electronic circuits, always use a high impedance test light or DMM.
- If an output device is not working properly, check the power circuit to it.
- If an input device is not sending the correct signal back to the computer, check the reference voltage it is receiving and the voltage it is sending back to the computer.
- Before replacing a computer, check the solenoid isolation diodes according to the procedures outlined in the service manual.
- Make sure computer wiring harnesses do not run parallel with any high current wires or harnesses. The magnetic field created by the high current may induce a voltage in the computer harness. You also should be aware that antenna cables and CB radios could cause interference.
- When checking individual components, always check the voltage drop of the ground circuits. This becomes more and more important as cars are made of less material that conducts electricity well.
- Make sure the ignition is off whenever you disconnect or connect an electronic device in a circuit.
- All sensors should be checked in cold and hot conditions.
- All wire terminals and connections should be checked for tightness and cleanliness.
- Use TV-tuner cleaning spray to clean all connectors and terminals.
- Use dielectric grease at all connections to prevent future corrosion.
- If you must break through the insulation of a wire to take an electrical measurement, make sure you tightly tape over the area after you are finished testing.

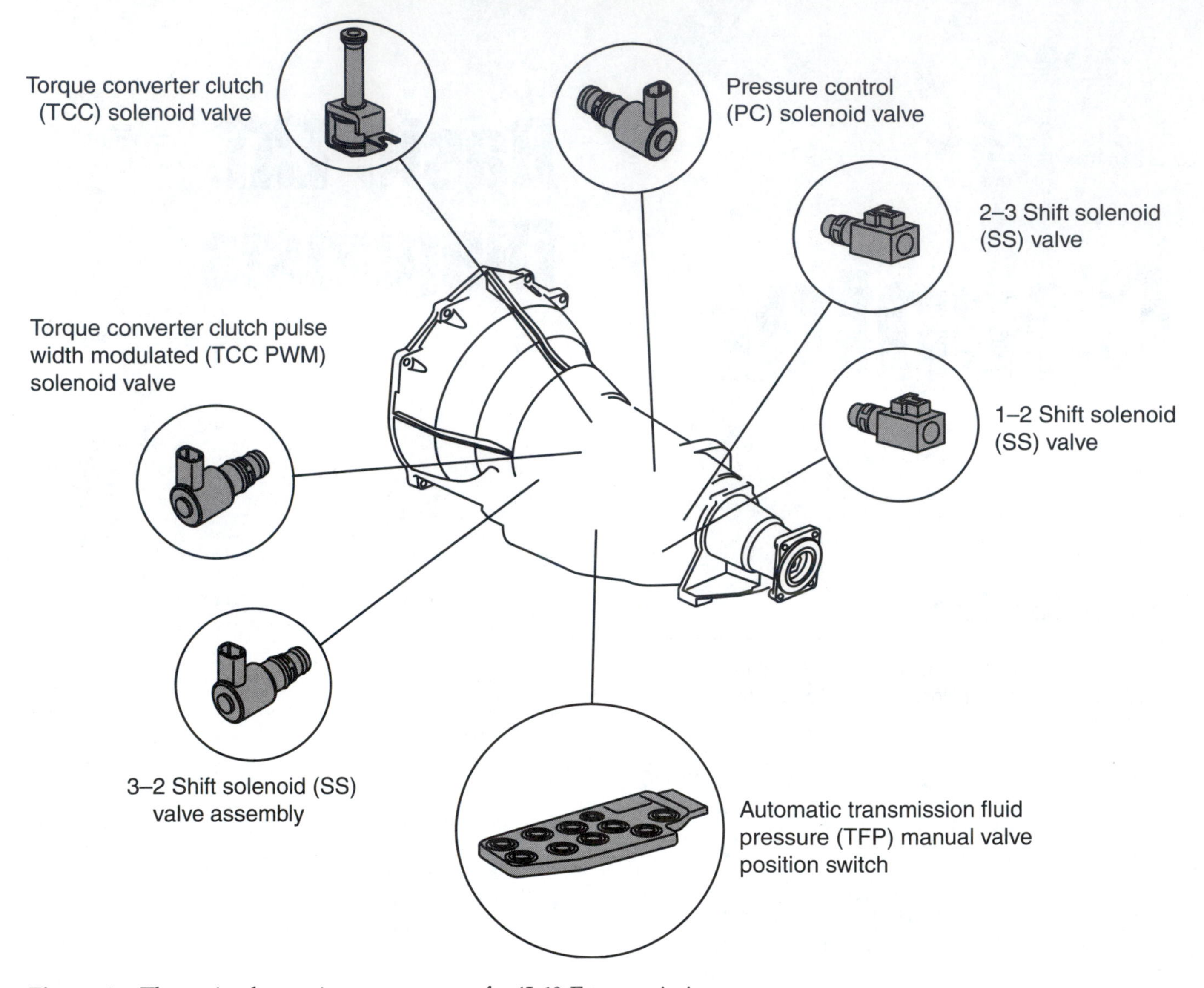


**Figure 1.** The main electronic components of a 4L60-E transmission.

## Electrostatic Discharge

Some manufacturers mark certain components and circuits with a code or symbol to warn technicians that they are sensitive to electrostatic discharge **(Figure 2)**. Static electricity can destroy or render a component useless.

When handling any electronic part, especially those that are static sensitive, follow the guidelines below to reduce the possibility of electrostatic buildup on your body and the inadvertent discharge to the electronic part. If you are not sure if a part is sensitive to static, treat it as if it is.

1. Always touch a known good ground before handling the part. This should be repeated while handling the part and more frequently after sliding across a seat, sitting down from a standing position, or walking a distance.

**Figure 2.** A manufacturer's warning symbol to show a circuit or component is sensitive to electrostatic discharge (ESD).

2. Avoid touching the electrical terminals of the part, unless you are instructed to do so in the written service procedures. It is good practice to keep your fingers off all electrical terminals as the oil from your skin can cause corrosion. Use an antistatic strap, if one is available.
3. When you are using a voltmeter, always connect the negative meter lead first.
4. Do not remove a part from its protective package until it is time to install the part.
5. Before removing the part from its package, ground yourself and the package to a known good ground on the vehicle.
6. When replacing a Programmable Read Only Memory (PROM) unit, ground your body by putting a metal wire around your wrist and connect the wire to a known good ground.

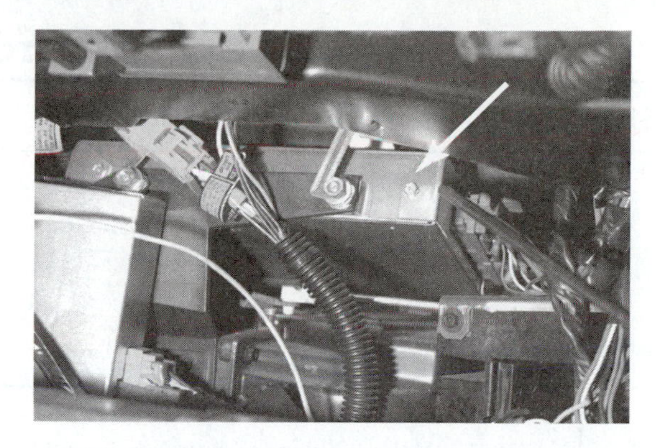

**Figure 3.** A typical PCM mounted under the dash.

## BASIC EAT TESTING

The best way to diagnose an electronically controlled transmission is to approach solving the problem in a logical way. To do this, you should follow these seven steps in order:

1. Verify the customer's complaint.
2. Conduct preliminary inspections and checks.
3. Check all service information for information about the complaint and the system, including service bulletins and recall notices.
4. Follow the diagnostic procedures outlined in the service manual for the specific complaint.
5. Interpret and respond to all diagnostic codes.
6. Define and isolate the cause of the complaint or problem.
7. Fix the problem and verify the repair.

It is important that you totally understand what the complaint or problem is before you venture in and try to find the cause. This is the purpose of the first four steps. This may include an interview or road test with the customer to thoroughly define the complaint and to identify when and where the problem occurs.

Because many EAT problems are caused by the basics, it is wise to conduct all of the preliminary checks required for a nonelectronically controlled transmission. Plus, you should conduct a thorough inspection of the electronic system. This inspection should include the retrieval of diagnostic codes. These codes will not only allow you to see what the PCM **(Figure 3)** sees as a transmission, problem but also will allow you to check for any engine problems. Whenever diagnosing a transmission, remember that an engine problem can and will cause the transmission to act abnormally.

Often accurately defining the problem and locating related information in TSBs and other materials can identify the cause of the problem. No manufacturer makes a perfect vehicle or transmission and, as the manufacturer recognizes common occurrences of a problem, they will issue a statement regarding the fix of the problem. Also, for many DTCs and symptoms, service manuals will give a simple diagnostic chart or path for identifying the cause of the problem **(Figure 4)**. These are designed to be followed step-by-step and will lead to a conclusion if you follow the path matched exactly to the symptom. Check all available information before moving on in your diagnostics.

Sometimes the symptom will not match any of those described in the service manual. This does not mean it is time to guess. It means it is time to identify clearly what is working right. By eliminating those circuits and components that are working correctly from the list of possible causes of the problem, you can identify what may be causing the problem and what should be tested further.

Although the first steps in troubleshooting include retrieving diagnostic codes, there are problems that will not be evident by a code. These problems may be solved with the diagnostic charts or pure logic. This logic must be based on a thorough understanding of the transmission and its controls.

Diagnostic codes should be used to recognize and locate problems. Codes that relate to transmission faults can be caused or detected by engine input or transmission input and/or output devices. It is also important to remember that codes can be set by out-of-range signals. This does not mean the sensor or sensor circuit is bad. It could mean the sensor is working properly but there is a mechanical or hydraulic problem causing the abnormal signals. Not only can internal transmission problems cause codes to be set, so can basic electrical problems. Problems such as loose connections, broken wires, corrosion, and poor grounds will affect the signals in that circuit.

To pinpoint the exact cause of a transmission problem, you will need to use basic electrical troubleshooting equipment, such as wiring diagrams, diagnostic charts, DMMs, lab scopes, special transmission testers, as well as scan tools. There are many varieties of electronically

**DTC P0742 TCC SYSTEM STUCK ON**

| Step | Action | Value(s) | Yes | No |
|---|---|---|---|---|
| 1 | Did you perform the Powertrain Diagnostic System Check? | — | *Go to Step 2.* | Go to *A Powertrain On Board Diagnostic (OBD) System Check* in Engine Controls. |
| 2 | 1. Install a *Scan Tool.*<br>2. Turn ON the ignition, with the engine OFF.<br>**Important:** Before clearing the DTC, use the *Scan Tool* in order to record the Freeze Frame and Failure Records. Using the Clear Info function erases the Freeze Frame and Failure Records from the PCM.<br>3. Record the DTC Freeze Frame and Failure records.<br>4. Clear the DTC.<br>5. Drive the vehicle in the D4 drive range in second, third or fourth gear under steady acceleration, with a TP angle at 20%.<br>While the *Scan Tool* TCC Enable status is No, does the *Scan Tool* display a TCC Slip Speed within the specified range? | –20 to +30 RPM | *Go to Step 3.* | Go to Diagnostic Aids. |
| 3 | The TCC is hydraulically stuck ON. Inspect for the following conditions:<br>• A clogged exhaust orifice in the TCC solenoid valve.<br>• The converter clutch apply valve is stuck in the apply position.<br>• A misaligned or damaged valve body gasket.<br>• A restricted release passage.<br>• A restricted transmission cooler line.<br>Did you find and correct the condition? | — | *Go to Step 4.* | — |
| 4 | Perform the following procedure in order to verify the repair:<br>1. Select DTC.<br>2. Select Clear Info.<br>3. Drive the vehicle in D4 under the following conditions:<br>Hold the throttle at 25 percent and accelerate to 88 km/h (55 mph). Ensure that the *Scan Tool* TCC Slip Speed is 130 to 2000 RPM for 4 seconds, with the TCC OFF.<br>4. Select Specific DTC.<br>5. Enter DTC P0742.<br>Has the test run and passed? | — | System OK. | *Go to Step 1.* |

**Figure 4.** An example of a manufacturer's diagnostic chart for a DTC.

controlled transmissions but all work in similar ways. Therefore, troubleshooting should always be approached in the same way.

## PRELIMINARY CHECKS

Troubleshooting a transmission's electronic control system is like any other electronic system; you need to make preliminary checks of the system before moving into to specific checks. The first step in a preliminary check is verifying system voltage. With a DMM, check the open-circuit voltage of the battery. Minimum voltage at the battery should be 12.6 volts. If the battery is below this, recharge the battery. If the voltage is still low after recharging, replace it.

Continue by checking the condition of the battery cables. Conduct a voltage drop test across each cable. Make sure the system is activated when doing a voltage drop

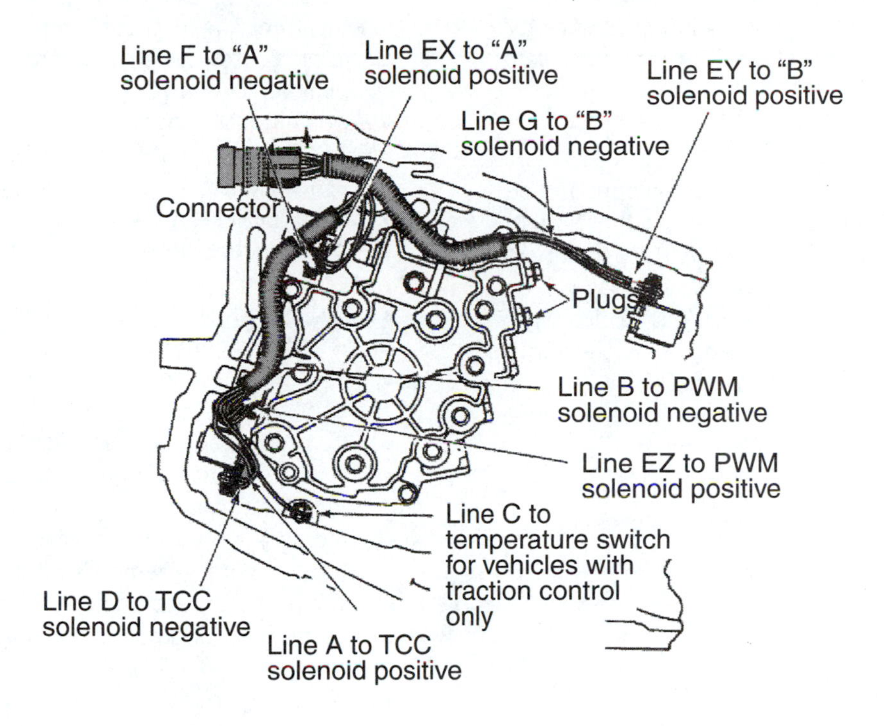


**Figure 5.** The position and condition of the transmission wiring harness and its connectors should be carefully checked.

test. There should be no more than 0.1 volt dropped across the positive or the negative cable. If the voltage drop exceeds this amount, clean or replace the cables.

A visual inspection of the circuit should follow. Carefully look at the entire system and check for damaged or corroded wires, loose connections, and damaged connectors **(Figure 5)**. Pay particular attention to the connectors. Make sure there are no bent or broken terminals. If any are present, replace the connector. Also, look inside each connector for signs of corrosion. If any is found, clean the wires and terminals. If the corrosion cannot be cleaned, replace the wires and/or the connector.

Continue your basic inspection with checking the fuse or fuses to the control module. To accurately check a fuse, either test it for continuity with an ohmmeter or check each side of the fuse for power when the circuit is activated.

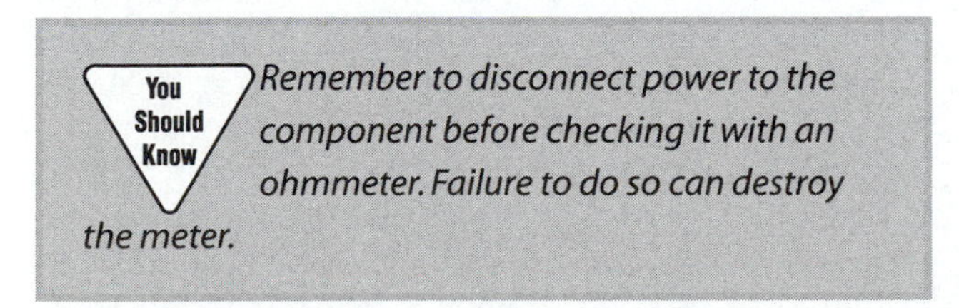


*Remember to disconnect power to the component before checking it with an ohmmeter. Failure to do so can destroy the meter.*

If the system has no apparent problems, continue testing. The diagnostic procedure for most EAT systems includes checking the system with a scan tool. The scan tool will display any trouble codes in the system. More important to the technician, most scan tools will display serial data. The serial data stream allows you to monitor system sensor and actuator activity during operation. Comparing the test values to the manufacturer's specifications will greatly help in diagnostics.

It is possible that the data displayed by a scan tool is not the actual value. Most computer systems will disregard inputs that are well out of range and rely on a default value held in its memory. These default values are hard to recognize and do little for diagnostics; this is why detailed testing with a DMM or lab scope is preferred by many technicians. These test instruments are also used to further test the system after a scan tool has identified a problem.

## Electronic Defaults

Often computer-controlled transmissions will start off in the wrong gear. This can happen for several reasons, either internal transmission problems or external control system problems. Internal transmission problems can be faulty solenoids or stuck valves. External problems can be the result of a complete loss of power or ground to the control circuit or a fail-safe protection strategy initiated by the computer to protect itself or the transmission from an observed problem. Typically, the default gear is simply the gear that is applied when the shift solenoids are off. When diagnosing a problem in an electronically controlled transmission, always refer to the appropriate service manual to identify the normal "default" operation of the transmission. By not recognizing that the transmission is operating in default, you could spend time tracing the wrong problem.

## ON-BOARD DIAGNOSTICS

Nearly all EATs have the ability to check input and output circuits during operation. When one of these circuits operates outside of the acceptable range, a diagnostic code is set. The setting of many of these codes will also cause the malfunction indicator lamp (MIL) to illuminate. This lamp informs the driver that a problem is present. It is important to remember that not all codes will cause the MIL to illuminate; therefore, an unlit MIL does not mean there are no DTCs in the computer's memory.

### Diagnostic Trouble Codes

Diagnostic trouble codes are what they are called. They are designed to help technicians identify and locate problems in the transmission's control system. DTCs from pre–OBD II systems can indicate a variety of things. Each manufacturer (and sometimes each vehicle model) used different codes to identify detected problems. It is very important that you refer to the service manual when interpreting DTCs.

Some OBD II codes are universal in that they are the same for all manufacturers and vehicle models. Other codes are manufacturer specific and must be interpreted through reference to the appropriate service manual.

DTCs are set by the computer when a signal from a component or circuit is not within the normal operating range for the operating conditions. A code also will be set when there is no signal from an input circuit or to an output device. EATs also have the ability to conduct a self-test. During this test, the sensor circuits are checked and if a problem is found, a DTC is set.

There are basically two types of DTCs. A hard code is a DTC that represents a problem that is present at the time of retrieval. These are the codes that should be responded to first during diagnostics. Soft codes are those DTCs that are not currently present. These codes can be retrieved and represent an intermittent problem or a problem that existed but is no longer present.

## BASIC DIAGNOSTICS OF OBD II SYSTEMS

OBD II systems were developed in response to the federal government's and the state of California's emission control system monitoring standards for all automotive manufacturers. The main goal of OBD II was to detect when engine or system wear or component failure causes exhaust emissions to increase by 50 percent or more. OBD II also called for standard service procedures without the use of dedicated special tools. To accomplish these goals, manufacturers needed to change many aspects of their electronic control systems. According to the guidelines of OBD II, all vehicles have:

- A universal diagnostic test connector, known as the Data Link Connector (DLC), with dedicated pin assignments.
- A standard location for the DLC. It must be under the dash on the driver's side of the vehicle and easily accessible.
- A standard list of Diagnostic Trouble Codes (DTCs).
- A standard communication protocol.
- The use of common scan tools on all vehicle makes and models.
- Common diagnostic test modes.
- Vehicle identification must be automatically transmitted to the scan tool.
- Stored trouble codes must be able to be cleared from the computer's memory with the scan tool.
- The ability to record and store in memory a snapshot of the operating conditions that existed when a fault occurred.
- The ability to store a code whenever something goes wrong and affects exhaust quality.
- A standard glossary of terms, acronyms, and definitions must be used for all emission-related components in the electronic control systems.

OBD II systems must illuminate the MIL **(Figure 6)** if the vehicle conditions would allow emissions to exceed 1.5 times the allowable standard for that model year based on a Federal Test Procedure (FTP). When a component or strategy failure allows emissions to exceed this level, the MIL is illuminated to inform the driver of a problem and a diagnostic trouble code is stored in the PCM.

Besides enhancements to the computer's capacities, some additional hardware is required to monitor the emissions performance closely enough to fulfill the tighter constraints and beyond merely keeping track of component failures. In most cases, this hardware consists of additional and improved sensors, upgrading specific connectors, and components to last the mandated 100,000 miles or 10 years, and a new standardized 16-pin DLC.

Instead of a fixed, unalterable PROM, the PCMs in most OBD II systems have an electronically erasable PROM (EEPROM) to store a large amount of information. The EEPROM is soldered into the PCM and is not replaceable. The EEPROM is an integrated circuit that contains the program used by the PCM to provide power train control. It is possible to erase and reprogram the EEPROM without removing this chip from the computer. When a modification to the PCM operating strategy is required, the EEPROM may be reprogrammed through the DLC using the special equipment.

For example, if the vehicle calibrations are updated for a specific car model, the recalibrating or reprogramming

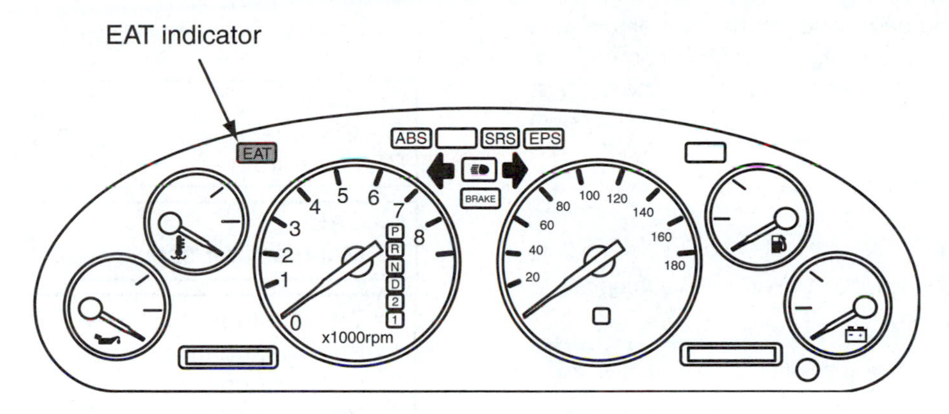


**Figure 6.** An example of a MIL that illuminates to warn the driver that transmission control has been switched to the default mode.

equipment can be to erase the EEPROM. After the erasing procedure, the EEPROM is reprogrammed with the updated information. The reprogramming information is provided by the manufacturer and is typically directed by a service bulletin or recall letter.

Computer systems without OBD II have the ability to detect component and system failure. Computer systems with OBD II are also capable of monitoring the ability of systems and components to maintain low emission levels.

## Data Link Connector

Standards require the DLC for OBD II systems to be mounted in the passenger compartment out of sight of vehicle passengers. The standard DLC **(Figure 7)** is a 16-pin connector. The same seven pins are used for the same information, regardless of the vehicle's make, model, and year. The connector is "D" shaped and has guide keys that allow the scan tool to be only installed one way. Use of standard

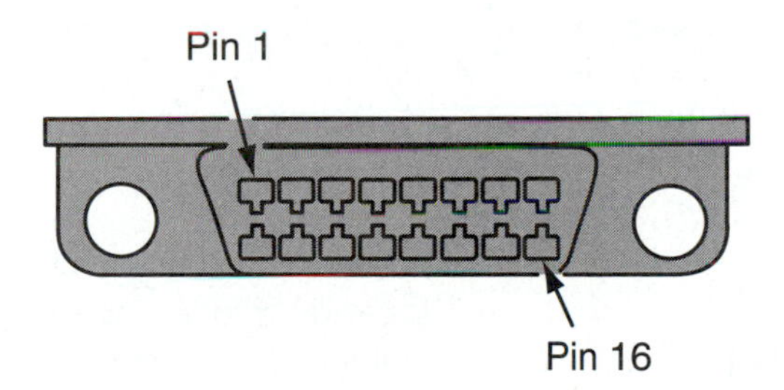


| | |
|---|---|
| Pin 1 | Secondary UART 8192 baud serial data<br>Class B 160 baud serial data |
| Pin 2 | J1850 Bus + line on 2 wire systems, or single wire (class 2) |
| Pin 3 | Ride control diagnostic enable |
| Pin 4 | Chassis ground pin |
| Pin 5 | Signal ground pin |
| Pin 6 | PCM/VCM Diagnostic enable |
| Pin 7 | K line for International Standards Organization (ISO) application |
| Pin 8 | Keyless entry enable, or MRD theft diagnostic enable |
| Pin 9 | Primary UART |
| Pin 10 | J1850 Bus-line for J1850-2 wire applications |
| Pin 11 | Electronic Variable Orifice (EVO) steering |
| Pin 12 | ABS diagnostic, or CCM diagnostic enable |
| Pin 13 | SIR diagnostic enable |
| Pin 14 | E&C bus |
| Pin 15 | L line for International Standards Organization (ISO) application |
| Pin 16 | Battery power from vehicle unswitched (4 AMP MAX) |

**Figure 7.** The purpose of each of the sixteen pins in a typical DLC for an OBD II system.

connector design and designation of the pins allows data retrieval with any scan tool designed for OBD II. Some European vehicles meet OBD II standards by providing the designated DLC along with their own connector for their own scan tool.

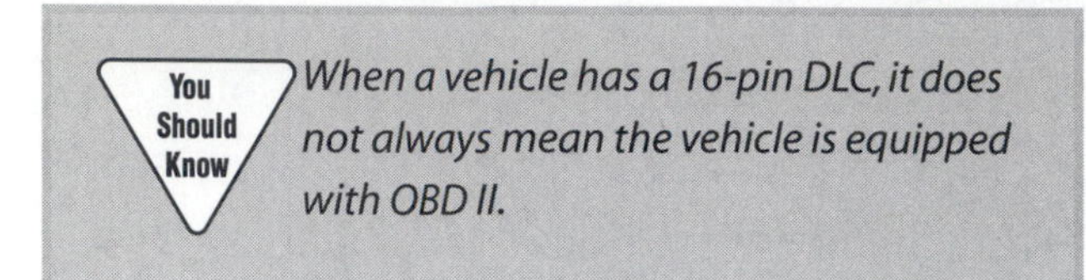


*When a vehicle has a 16-pin DLC, it does not always mean the vehicle is equipped with OBD II.*

The DLC is designed only for scan tool use. You cannot jump across any of the terminals to display codes on an instrument panel or other indicator lamp. Any generic scan tool can be connected to the DLC and can access the diagnostic data stream. The connector pins are arranged in two rows and are numbered consecutively. Seven of the sixteen pins have been assigned by the OBD II standard. The remaining nine pins can be used by the individual manufacturers to meet their needs and desires.

## Malfunction Indicator Lamp Operation

The MIL is only used to inform the driver that the vehicle should be serviced soon. It also informs a technician that the computer has set a trouble code. OBD II systems continuously monitor the entire system, switch on a MIL if something goes wrong, and store a fault code in the PCM when a problem is detected.

## Diagnostic Trouble Codes

OBD II DTCs are well defined and can lead a technician to the cause of a problem. A DTC is a five-character code with both letters and numbers **(Figure 8)**. This is called the alphanumeric system. The first character of the code is a letter. This defines the system where the code was set. The second character is a number. This defines the code as being a mandated code or a special manufacturer code. The third through fifth characters are numbers. These describe the fault. The third character tells you where the fault occurred. The remaining two characters describe the exact condition that set the code.

OBD II codes are quite definitive and can lead you to the cause of the problem. For example, there are five codes that relate to the transmission fluid temperature circuit. Code P0710 indicates there is a fault in the sensor's circuit, P0711 indicates the circuit is operating out of the acceptable range, P0712 indicates the circuit lower than normal input signals, P0713 indicates the input signals are higher than normal, and P0714 indicates the circuit appears to have an intermittent problem.

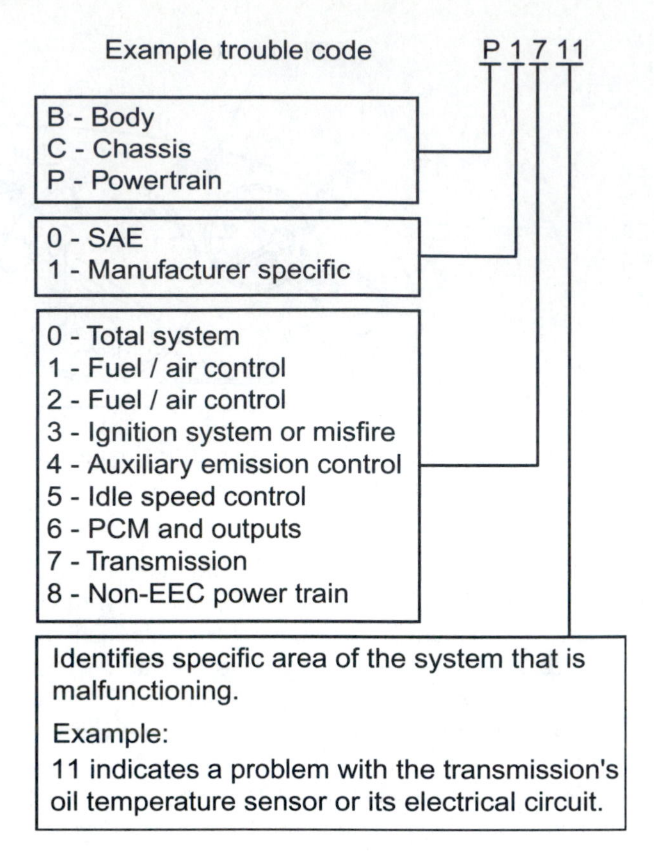


**Figure 8.** An explanation of the five digits in an OBD II trouble code.

## Test Modes

All OBD II systems have the same basic test modes. These test modes must be accessible with an OBD II scan tool. Mode 1 is the Parameter Identification (PID) mode. It allows access to certain data values, analog and digital inputs and outputs, calculated values, and system status information. Some of the PID values will be manufacturer specific, others are common to all vehicles.

Mode 2 is the Freeze Frame Data Access mode. This mode permits access to emission-related data values from specific generic PIDs. These values represent the operating conditions at the time the fault was recognized and logged into memory as a DTC. Once a DTC and a set of freeze frame data are stored in memory, they will stay in memory even if other emission-related DTCs are stored. The number of these sets of freeze frames that can be stored are limited.

There is one type of failure that is an exception to this rule—misfire. Fuel system misfires will overwrite any other type of data except for other fuel system misfire data. This data only can be removed with a scan tool. When a scan tool is used to erase a DTC, it automatically erases all freeze frame data associated with the events that lead to that DTC.

The basic advantage of the snapshot feature is the ability to look at the existing conditions when a code was set. This will be especially valuable for diagnosing intermittent problems. Whenever a code is set, a record of all related activities will be stored in memory. This allows you to look at the action of sensors and actuators when the code was set. This helps identify the cause of the problem.

Mode 3 permits scan tools to obtain stored DTCs. The information is transmitted from the PCM to the scan tool following an OBD II Mode 3 request. Either the DTC, its descriptive text, or both will be displayed on the scan tool.

The PCM reset mode (Mode 4) allows the scan tool to clear all emission-related diagnostic information from its memory. Once the PCM has been reset, the PCM stores an inspection maintenance readiness code until all OBD II system monitors or components have been tested to satisfy an OBD trip cycle without any other faults occurring. Specific conditions must be met before the requirements for a trip are satisfied.

Mode 5 is the oxygen sensor monitoring test. This mode gives the oxygen sensor fault limits and the actual oxygen sensor outputs during the test cycle. The test cycle includes specific operating conditions that must be met to complete the test. This information helps determine the effectiveness of the catalytic converter.

Mode 6 is the output state mode (OTM), which allows a technician to activate or deactivate the system's actuators through the scan tool. It is important to note that not all scan tools have this capability. When the OTM is engaged, the actuators can be controlled without affecting the radiator fans. The fans are controlled separately; this gives a pure look at the effectiveness and action of the outputs.

## OBD II Diagnostics

Testing of individual components is important in diagnosing OBD II systems. Although the DTCs in these systems give much more detail on the problems, the PCM does not know the exact cause of the problem. That is the technician's job.

OBD II systems note the deterioration of certain components before they fail, which allows owners to bring in their vehicles at their convenience and before it is too late. OBD II diagnosis is best done with a strategy-based procedure based on the flow charts and other information given in the service manuals. Check for any published TSBs relating to the exhibited symptoms. This should include videos, newsletters, and any electronically transmitted media. Do not depend solely on the diagnostic tests run by the PCM. A particular system may not be supported by one or more DTCs.

When the ignition is initially turned ON, the MIL will momentarily flash ON then OFF and remain ON until the engine is running or there are no DTCs stored in memory. Now connect the scan tool **(Figure 9)**. Enter the vehicle

**Figure 9.** A scan tool connected to a DLC.

identification information, and then retrieve the DTCs with the scan tool. The scan tool can be used to do more than simply retrieve codes. The information that is available, whether or not a DTC is present, can reduce diagnostic time. Some of the common uses of the scan tool are identifying stored DTCs, clearing DTCs, performing output control tests, and reading serial data **(Figure 10)**.

The scan tool must have the appropriate connector to fit the DLC on an OBD II system and the proper software for the vehicle being tested. Cable adapters are used to connect a standard scan tool to an OBD II system. The software inserts make the scan tool compatible with the PCM in the vehicle. Make sure the recommended insert for that make and model of vehicle is used in the scan tool. Always follow the instructions given in the scan tool's manual when testing the system.

If DTCs are stored in memory, the scan tool will display them. Record them. If there are multiple DTCs, they should be interpreted in the following order:

1. PCM error DTCs
2. System voltage DTCs
3. Component level DTCs
4. System level DTCs

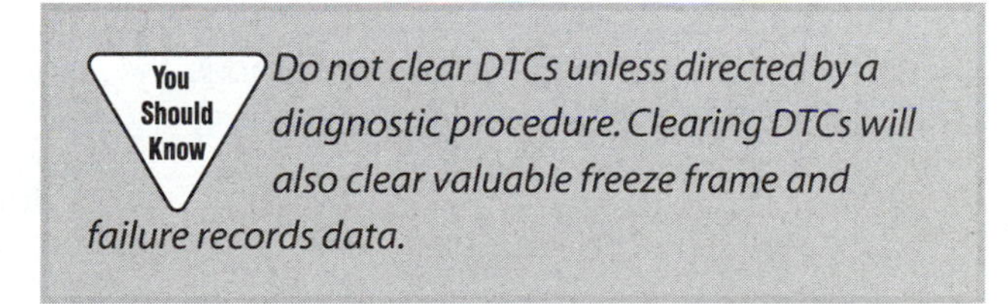


*Do not clear DTCs unless directed by a diagnostic procedure. Clearing DTCs will also clear valuable freeze frame and failure records data.*

After retrieving the codes, use the scan tool to command the MIL to turn OFF. If it turns OFF, turn it back ON and continue. If it stays ON, follow the manufacturer's diagnostic procedure for this problem.

The next step in diagnostics depends on the outcome of the DTC display. If there were DTCs stored in the PCM's

| SCAN TOOL DATA | | | |
|---|---|---|---|
| Engine Coolant Temperature Sensor (ECT) 0.46 V | Intake Air Temperature Sensor (IAT) 2.24 V | MAP Sensor 1.8 V<br>MAF Sensor 1.1 V | Throttle Position Sensor (TPS) 0.6 V |
| Engine Speed Sensor (RPM) 1500 rpm | Heated $O_2HO_2S$<br>Upstream 0.2–0.5 V<br>Downstream 0.1–0.3 V | Vehicle Speed Sensor (VSS) 0 mph | Battery Voltage (B+) 14.2 V |
| Idle Air Control Valve (IAC) 0 percent | Evaporative Emission Canister Solenoid (EVAP) OFF | Torque Converter Clutch Solenoid (TCC) OFF | EGR Valve Control Solenoid (EGR) 0 percent |
| Malfunction Indicator Lamp (MIL) OFF | Diagnostic Trouble Codes NONE | Open/Closed Loop CLOSED | Fuel Pump Relay (FP) ON |
| Fuel Level Sensor 3.5 V | Fuel Tank Pressure Sensor 2.0 V | Transmission Fluid Temperature Sensor 0.6 V | Transmission Turbine Shaft Speed Sensor 0 mph |
| Transmission Range Switch PARK | Transmission Pressure Control Solenoid 80% | Transmission Shift Solenoid 1 ON | Transmission Shift Solenoid 2 OFF |
| Measured Ignition Timing °BTDC | Base Timing: 10 | Actual Timing: 20 | |

**Figure 10.** An example of the information available with a scan tool.

memory, proceed by following the designated DTC diagnostic table to make an effective diagnosis and repair. If no DTCs were displayed, match the symptom to the symptoms listed in the manufacturer's symptom tables and follow the diagnostic paths or suggestions to complete the repair, or refer to the applicable component/system check. If there is not a matching symptom, analyze the complaint, then develop a plan for diagnostics. Utilize the wiring diagrams and theory of component and system operation.

It is possible that the customer's complaint is a normal operating condition for that vehicle. If this is suspected, compare the complaint to the driveability of a known good vehicle.

After the DTCs and the MIL have been checked, the OBD II data stream can be selected. In this mode, the scan tool displays all of the data from the inputs. The actual available data will depend on the vehicle and the scan tool; however, OBD II codes dictate a minimum amount of data. Always refer to the appropriate service manual and the scan tool's operating instructions for proper identification of normal or expected data.

## BASIC DIAGNOSTICS OF NON-OBD II SYSTEMS

Diagnosis on vehicles equipped with other than OBD II systems can be done by the self-diagnostics mode of the system and/or by connecting a scan tool to the DLC or ALDL (Assembly Line Data Link). Use the following discussions as examples only. Always refer to the correct service manual for the exact procedures that should be followed. The service manuals also give the correct interpretations for the trouble codes retrieved from the computer. Keep in mind, by law, all 1996 and newer vehicles must have an OBD II compliant system. Therefore, it is fair to say the following procedures apply only to vehicles produced in 1995 or before.

### Chrysler

Chrysler has basically two electronic control systems. One was first used in 1988 and has been used in a variety of models. The other system is a multiplexed system, which is used on later model minivans and cars. The original system is quite basic and similar to those used by other manufacturers. The control computer has self-diagnostic capabilities. The recommended way to retrieve trouble codes is through the use of Chrysler's scan tool, called a DRB-II.

The DRB-II is connected to the DLC, which is located by the left front shock tower. The red power supply lead is connected to the positive post of the battery. The scan tool has a variety of cartridges that contain the test parameters for the vehicle being tested. After the correct cartridge has been installed in the tester and the scan tool is connected to the vehicle, the tester's display will show a test pattern. After a few seconds, the display will change to read out the copyright date and revision level of the cartridge. After a few seconds, the display will ask for the vehicle's model year. The model year is selected by pressing either right or left arrow keys. Once the display shows the correct year, press the down arrow key.

The display will then ask for the system that will be tested. The system is selected by pressing either left or right arrow keys. Once the display indicates powertrain, press down arrow key. Other typical system choices are engine or EFI/Turbo.

The display will then ask for the desired test selection. The desired test is selected by pressing either the left or right arrow keys. The diagnostic test mode is used to see if there are any fault codes stored in the on-board diagnostic system memory of the computer. The circuit Actuation Test Mode (ATM) is used to make the computer cycle a solenoid on and off. The sensor test mode is used to see output signals of certain sensors as the computer receives them when the engine is not running. The engine running test mode is used to see sensor output signals as received by the computer while the engine is running.

To enter into the diagnostic mode, set the DRB-II to the engine off test. Turn the ignition switch to "ON–OFF, ON–OFF, ON" within five seconds. Then record all codes displayed on the tester. By depressing the "READ/HOLD" key until "HOLD" is displayed, you can stop the display of codes. To continue the display, press the "READ/HOLD" key.

To erase the fault codes put the system into ATM test mode 10. Then press the "READ/HOLD" key until "HOLD" is displayed. The display will then flash alternating "O"s. When the flashing "O"s stop and the display shows "00," all fault codes have been erased.

The other transmission control system is the Chrysler Collision Detection system, which is a multiplex serial data bus or network. The network consists of a pair of twisted wires, which interconnect the CCD modules. Each CCD module uses this network to communicate and exchange data with other modules on the network. The number of modules varies with the model of vehicle.

Diagnosis of this transaxle system is relatively simple. Nearly all of the important information is available through the use of a scan tool and the transaxle is equipped with many pressure taps. Both of these features give you an accurate look at the operation of the computer.

Remember, when the transaxle controller detects a serious problem, it will automatically go into the "limp-in" mode of operation. This mode defaults the transaxle to second gear only with second gear starts, reverse, and park operations. To determine why the computer has defaulted to this mode, connect a scan tool to the blue six-pin CCD connector under the dash. Retrieve and record the trouble codes that are displayed. Then, clear the memory and drive the vehicle until the complaint occurs again. Retrieve and write down the codes again. These codes are the most current and should be compared to the original codes.

If the scan tool is unable to display a code or is unable to recognize the computer or its circuits, the problem may be a faulty or dead bus. A dead bus is commonly caused by a body computer, transaxle controller, or any other module on the CCD bus. The exact bad module can be identified by disconnecting each module, one at a time, from the bus.

After you have one module disconnected, rerun the startup of the scan tool. If the bus becomes active, the disconnected control module was the cause of the previous dead bus. Repeat this process, by reconnecting the disconnected module and unplugging another. The module that is disconnected when the bus becomes active is the dead module.

If the cause of the dead bus was not identified, voltage checks should be made. Measure the voltage drop across the groundside of the diagnostic connector by inserting the positive voltmeter lead to the ground lead of the connector and the negative lead to the negative terminal of the battery. The maximum allowable voltage droop across the ground is 0.1 volts. The source voltage to the computer also should be checked. If battery voltage is not present at the computer, check the voltage of the battery. If it is okay, check the voltage drop in the supply circuit. If there is more than 0.1 volts dropped, identify the source of the resistance and repair the problem. Also, check the bus bias voltage by connecting the voltmeter's positive lead to either bias terminal in the diagnostic connector and the negative lead to the ground terminal in the connector. The bias voltage should be between 2.1 to 2.6 volts. Any incorrect voltages could be the cause for a dead bus and the cause of the problems should be corrected. After the electrical problems are repaired, the scan tool test should be conducted.

Using the scan tool, the trouble codes are retrieved and compared to the trouble code chart given in the service manual. These codes will identify problems circuits. The service manual also gives step-by-step procedures for pinpoint diagnostics of each trouble code. Always follow these, they are efficient and exact.

## Ford

Most early Ford EAT systems are based on their EEC IV or EEC V systems. Although the exact make-up of the systems varies according the transmission used and the model vehicle being tested, all follow similar procedures. The following equipment is recommended to diagnose and test an EEC system.

- Self-Test Automatic Read-out (STAR) Diagnostic Tester **(Figure 11)**—This tester is recommended but not required. The tester was designed for EEC systems and is used to display the service codes. There are also many aftermarket testers available for testing these

**Figure 11.** A STAR tester.

**Figure 12.** A breakout box with a transmission template.

systems. An Analog Volt/Ohmmeter with a 0–20 V DC range can be used as an alternate to a diagnostic tester.

- DMM—This multimeter must have a minimum impedance of 10 megohms.
- Breakout box—This is a jumper wire assembly that connects between the vehicle harness and the PCM. The breakout box **(Figure 12)** is required to perform certain tests on the system. During individual circuit tests, the procedures will call for probing particular pin numbers, these pin numbers relate to the pins of the breakout box.
- Vacuum gauge and vacuum pump—This can be one assembly or separate units.
- Tachometer—Must have a 0–6000 RPM range defined in 20 rpm increments.
- Spark tester—A modified side electrode spark plug with the electrode removed and an alligator clip attached may be used in place of the spark tester.
- MAP/BP tester—This tester plugs into the MAP/BP sensor circuit and the DMM. It is used to check input and output signals from the sensor, which produces a frequency signal. This signal may also be monitored on a scope set to a milli-volt scale.
- Other equipment—Timing light, fuel injection pressure gauge, nonpowered 12-volt test light, and a jumper wire, about 15 inches long.

The EEC system offers a variety of test sequences. These steps are called the Quicktest and should be carefully followed in sequence to get accurate readings. Failure to do so may result in misdiagnosis or the replacement of nonfaulty components. The steps are: test preparation and equipment hookup, Key On Engine Off (KOEO) self-test, Key On Engine Running (KOER) self-test, and continuous self-test.

After all tests, servicing, or repairs have been completed, the Quicktest should be repeated to ensure that all systems are working properly. The KOEO and KOER self-tests are designed to identify faults present at the time of the testing and not intermittent faults. Intermittent faults are detected by the Continuous self-test mode.

Before hooking up any equipment to diagnose the EEC system, the following checks should be made in preparation for further testing:

1. Verify the condition of the air cleaner and air ducting.
2. Check all vacuum hoses for leaks, restrictions, and proper routing.
3. Check the electrical connectors of the EEC system for corrosion, bent or broken pins, loose wires or terminals, and proper routing.
4. Check the PCM, sensors, and actuators for physical damage.
5. Set the parking brake and place shift lever in "P." DO NOT move the shift lever during the test unless the procedures call for it.
6. Turn off all lights and accessories. Make sure the vehicle's doors are closed when making voltage or resistance readings.
7. Check the level of the engine's coolant.
8. Start the engine and allow it to idle until the upper radiator hose is hot and pressurized.
9. Check for leaks around the exhaust manifold, exhaust gas oxygen sensor, and vacuum hose connections.

To retrieve codes with the analog VOM, turn the ignition key off. Then connect the jumper wire from Self-Test Input (STI) pigtail to pin #2 at the SELF-TEST connector. Set the VOM to the correct DC voltage scale. Connect the positive lead of VOM to the positive battery terminal. The negative lead of the meter is connected to the #4 pin of the SELF-TEST connector. Connect the timing light and proceed to the KOEO SELF-TEST. Codes are shown as voltage pulses (needle sweeps), pay attention to the length of the pauses in order to read the codes correctly.

To retrieve codes with the STAR tester, turn the ignition off. Then connect the color-coded adapter cable leads to the diagnostic tester. Connect the adapter cable's two service connectors to the vehicle's SELF-TEST connectors. Then, connect the timing light and proceed to the KOEO SELF-TEST.

> **You Should Know** *The SELF-TEST connector is found in different locations, depending upon the model of the vehicle. On most FWD models, the SELF-TEST connector is located in the right rear of engine compartment, near the firewall.*

All service codes are two or three digit numbers that are generated one digit at a time. There will be a two-second pause between each DIGIT in a code. There will be a four-second pause between each CODE. The continuous memory codes are separated from the functional test service codes by a six-second delay, a single one-half-second sweep, and another six-second delay. Always record the codes in the order received. If a diagnostic tester is used, it will count the pulses and display them as a digital code. If the "CHECK ENGINE" light is used, numeric service codes will be displayed at the light.

The first set of codes that will be displayed are the KOEO self-test codes. These codes will be repeated twice. They are followed by the Separator Pulse signal and the continuous memory codes. These codes will also be repeated twice. A KOEO code 11 (or 111) indicates the system has passed the self-test. Whenever a repair is made, repeat the Quicktest. To clear codes, remove the jumper wire while the codes are being displayed and before turning off the ignition. Record all displayed codes.

## General Motors

The 4T60-E and other GM transaxles use two electric solenoids to control transaxle upshifts and downshifts. The 4L80-E transmission functions in much the same way, however it is also fitted with a force motor, which controls hydraulic line pressure and a TCC solenoid. Some models have a TCM in addition to the PCM. In all systems, the PCM has self-diagnostic capabilities, which help identify which parts or circuits may need further testing.

The PCM constantly monitors all electrical circuits. If the PCM detects a problem or an out-of-range sensor input, it records a trouble code in its memory. If the problem continues for some time, the MIL will glow. It is possible for the PCM to have detected a trouble and not light the MIL. However, the appropriate code may be stored in its memory.

Many methods are commonly used to retrieve the trouble codes from the PCM's memory. The simplest method is to use the MIL. Other methods include the use of special scan tools.

To display the trouble codes with the MIL, locate the ALDL connector. Then, with the ignition on and engine off, jump across the A and B terminals of the connector. The MIL should begin to flash codes. Each code will be repeated three times. The first series of flashes is the first digit of the code and the second series is the second digit. Trouble codes are displayed starting with the lowest numbered code. Codes will continue until the jumper is disconnected from the ALDL.

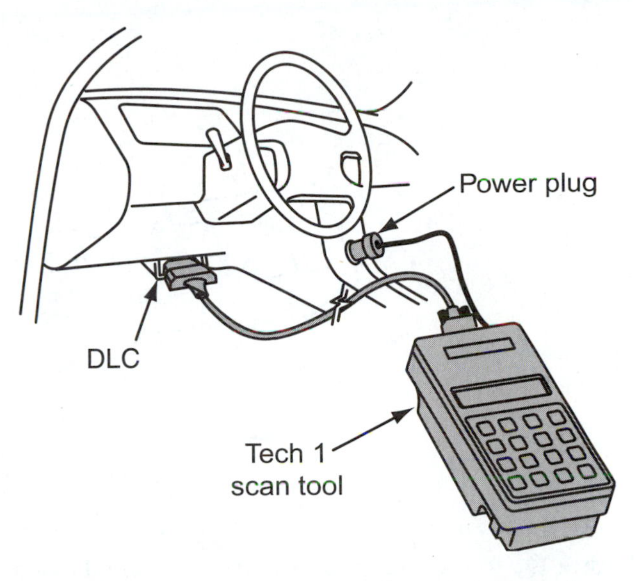


**Figure 13.** A transmission test box and Tech 1 scan tool used to diagnose early GM EATs.

A scan tool connected into the ALDL **(Figure 13)** can provide access to circuit voltage information, as well as trouble codes. By observing the various input voltages, you can identify out of specification input signals. Out of specification signals will not set a code unless they are out of the normal operating range. However, these signals will cause a driveability problem. To retrieve trouble codes with a scan tool, simply plug the tester lead into the ALDL. Then, program the scan tool for that particular vehicle according to the scan tools' instructions. Trouble codes will appear, digitally, on the tester.

To erase the codes after repairs have been made, turn the ignition off. With the jumper wire still connected, remove the control module fuse for 10 seconds. If the fuse cannot be located, disconnect the negative battery lead for 10 seconds. After the power has been disconnected from the PCM, the codes will be removed. So will the operating instructions on some models. For this reason, it is important that you follow the relearn procedures for the transmission.

## Honda

Honda and Acura transaxle trouble codes are communicated by either flashing an LED on a side of the transaxle controller, or by flashing the S or D4 shifter status light on the instrument panel after connecting a service connector **(Figure 14)**. Always refer to the correct service manual when attempting to retrieve codes.

The connector is a two pin, two-wire connector that is not connected to anything. Using a jumper wire, hook the

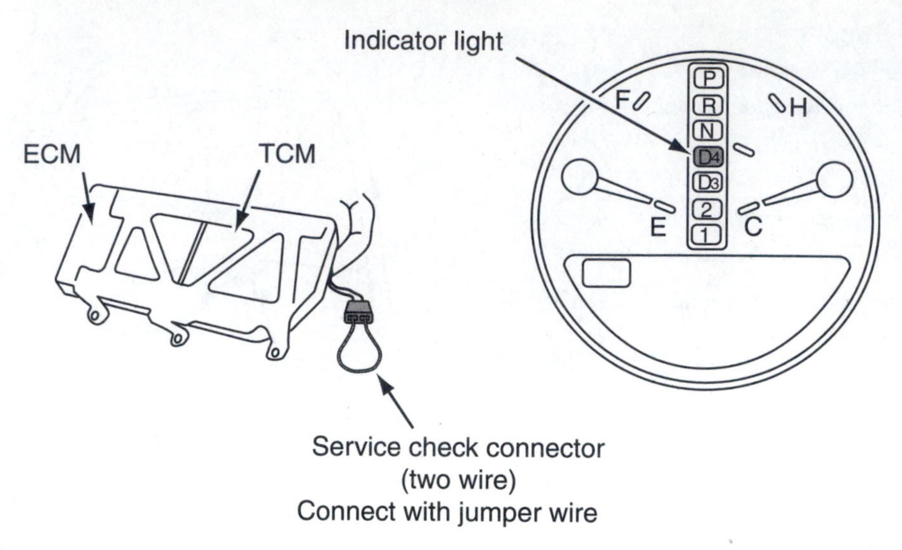


**Figure 14.** To retrieve trouble codes on some Hondas, connect two terminals of the Service Check Connector and observe the D4 indicator.

two pins together, turn the ignition switch ON, and watch either the S or D4 light on the instrument panel. The three possible locations for the connector are under the right side of the dash, behind the right side of the center console, or behind the left front edge of the passenger's carpet up against the firewall. If a trouble code is present, the light will either give short or long and short flashes. Short flashes are either a one- or two-digit trouble code. Long flashes are the tens digit of a two-digit code. Short flashes are about a half second long and long flashes are about a second and a half long. There is about a one-second pause between each code. All codes will be shown in sequence, lowest number first, up to the highest code. The sequence will repeat as long as the key is on and the connector is jumped across.

Some models have the diagnostic LED on the side of the transaxle controller. Read the flashes of the LED to determine the trouble code. The LED blinks all short flashes. Whenever the key is on and a problem exists, the LED will blink a series of short flashes, pauses, and then either repeats the same number of flashes (if there is only one code), or advances to the next number of blinks. Just count each series of flashes, and you have the code. The computer for Honda cars is located either under the front edge of the front passenger's carpet, behind the center console, or under the driver's seat. Acuras locate the computer behind the left side of the instrument panel, next to the inner fender panel, or under the right or left front seat with the LED facing the rear of the car.

The single and double-digit codes should be compared to the appropriate trouble code chart provided in the service manual. Like other computer self-diagnostic systems, these systems do not determine the exact fault; rather, they identify the problem area. After repairs have been made, the codes should be erased.

To erase the codes, remove the appropriate fuse for at least one minute or pull the negative cable of the battery for a minute. Because Honda and Acura systems use different fuses for different models to keep the memory alive, make sure you check the service manual to identify the proper fuse.

## Mazda

Trouble codes in Mazda EAT systems are displayed by either the HOLD light or the CHECK ENGINE light. If no problem exists, the HOLD or CHECK ENGINE light will come on for three seconds after turning the key on, and then it will go out. If a trouble does exist, the light will flash in a regular pattern until the trouble is no longer there. This is not a trouble code; rather it is an indication that a problem exists.

To retrieve the codes, locate the Diagnostic Request lead. This lead may be a single pin and wire blue connector, a single pin and wire green connector, or one of the pins in an integrated Diagnostic connector. Connect the lead to a good ground with a jumper wire. When the diagnostic pin is grounded, the HOLD or CHECK ENGINE light will display the code in long and short flashes. Long flashes, 1.2 seconds long, are the tens digit of a two digit trouble code, short flashes, 0.4 seconds long, are the ones digit of a one- or two-digit code. The lowest number codes will be shown first, then the higher number codes. There is a four-second pause between codes. The series of codes will repeat as long as the Diagnostic Request pin is grounded. If no codes are present when you ground this pin, the indicator light will remain off.

## Nissan

Transmission trouble code retrieval on Nissan computer shifted transmissions and transaxles are easy and uniform throughout the different models of vehicles. Simply turn the ignition switch to the ON position and move the shifter and the O/D OFF switch. Then an instrument panel light will flash codes. The codes are displayed by the O/D OFF light, POWER, or A/T CHECK light.

To begin the procedure, make sure the MODE switch is in the AUTO position. Then run the engine until it has reached normal operating temperature. Turn the key off, set the parking brake and do not touch the brake or throttle pedals until the procedure calls for you doing this. Move the shifter to D and turn the O/D switch OFF. Turn the ignition switch back to the ON position. After waiting at least two seconds, move the shifter to "2" and turn the O/D switch ON. Then, move the shifter to "1" and turn the O/D switch OFF. Now, slowly push the throttle pedal to the floor and slowly release it. The codes will now appear on the instrument panel.

The codes will appear in a sequence of eleven flashes. This sequence always begins with a long, two-second, starter flash, followed by ten shorter flashes. If a problem does not exist, the ten following flashes will be very short flashes, 0.2 seconds. If a problem exists, one of the ten short flashes will be a little longer, 0.8 seconds, than the rest in that sequence. To identify the code, count the short flashes and determine which one was the longer one. Begin counting after the longer initial flash. If the first flash was the longest short flash, the code is 1. After each sequence of 11 flashes, there is a 2.5 second pause, and then either the code or codes will repeat. All codes will automatically clear if the problem is fixed and the engine has been started two times after the system has been repaired.

## Saturn

Saturn EATs use five electronic valves that are controlled by the PCM to control shift feel and timing. The PCM has a self-diagnostic function. If an electrical problem occurs, the SES or "Shift to D2" lamp will come on or will start flashing. The PCM stores trouble codes whenever it has detected a fault in the engine or transaxle circuit. Problems are stored as hard or intermittent problems. The PCM will also store other problems but may not turn on the SES light in response to the problem. These codes are stored in the memory to aid in diagnostics.

The PCM recognizes three different types of faults: hard, intermittent, and malfunction history/information flags. Hard codes cause the SES light to glow and remain on until the problem is repaired. These codes can be interpreted by looking up the code in the appropriate trouble code chart. Intermittent codes cause the SES to flicker or glow and go out approximately 10 seconds after the fault disappears. The corresponding trouble code will remain, however, in memory to be retrieved by a technician. If the intermittent fault does not occur for 50 engine starts, the code will be erased from the computer's memory. Engine information flags will not cause the SES light to glow. Unlike the other types of codes, information flags and malfunction history will not be erased from the PCM memory. These codes can only be retrieved and erased from memory by using a Saturn PDT.

Trouble codes are read by counting the flashes of the SES lamp. The first series of flashes represents the first digit of the trouble code and the second series represents the second digit. The first code will always be code "12," followed by all other stored codes. Each code is repeated three times. When code "11" appears, that signals the end of the engine's self-diagnostic test and the beginning of the transaxle self-test. The "Shift to D2" lamp will start flashing any transaxle related codes stored in the transaxle controller. If the SES does not flash a code "11," there are no transaxle codes stored in memory.

To set the PCM into the self-diagnostic mode, turn the ignition on with the engine off. Connect a jumper wire from terminal B to terminal A at the PCM diagnostic connector. The SES should begin to flash codes. To end the self-diagnostic mode, turn the ignition off and remove the jumper wire.

To erase the trouble codes, turn the ignition on and make contact three times within five seconds with a jumper wire between terminals A and B at the diagnostic connector.

The PCM has a relearn function and the proper learn process should be followed anytime the battery has been disconnected.

## Toyota

A self-diagnostic feature is built into all Toyota EATs. A warning of a fault is indicated by the overdrive OFF indicator lamp. If a malfunction occurs in the speed sensor or solenoid circuits, the overdrive OFF lamp will blink to warn the driver of the fault. On some models, the diagnostic codes can be read by the number of blinks on the overdrive OFF lamp when terminals ECT and E2, in the diagnostic connector, are shorted together. Other models require that terminals TE and E be shorted together **(Figure 15)**. Make sure you refer to the appropriate service manual to determine which terminals should be shorted. Once the computer is set into its diagnostic mode, codes will begin to be displayed.

If a malfunction is in memory, the overdrive OFF lamp will blink once every 0.5 second. The first number of blinks equals the first digit of a two-digit code. After a 1.5 second pause, the second number of blinks will equal the second digit. If there are two or more codes, there will be a 2.5 second pause between each code. All codes should be interpreted by using the trouble code chart provided in the service manual.

The diagnostic codes can be erased from the computer's memory removing the DOME, ECU B+, or EFI fuse for at least 10 seconds with the key off.

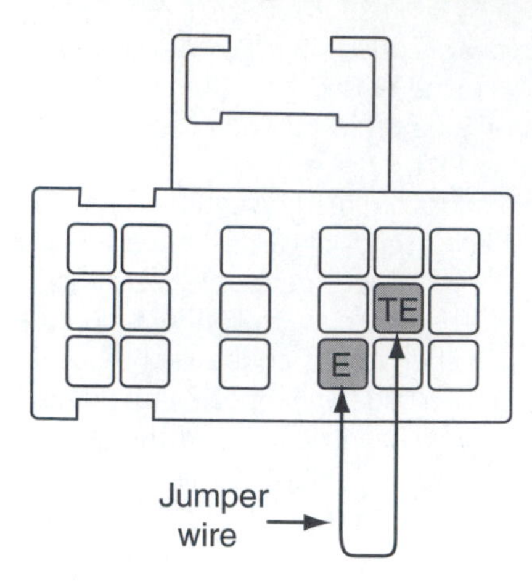


**Figure 15.** A typical Toyota diagnostic connector with a jumper wire inserted to activate the self-test mode.

## ROAD TEST

Critical to proper diagnosis of EAT and TCC control systems is a road test. The road test should be conducted in the same way as one for a nonelectronic transmission except that a scan tool also should be connected to the circuit to monitor engine and transmission operation.

**You Should Know** *Vehicles equipped with control computers may require a relearn procedure after the battery is disconnected. Many computers memorize and store vehicle operation patterns for optimum driveability and performance. When the vehicle's battery is disconnected, this memory is lost. The computer will use default data until new data from each key start is stored. Customers often complain of driveability problems during the relearn stage because the vehicle acts differently then before it was serviced. To reduce the possibility of complaints, the vehicle should be road tested and the correct relearn procedure followed.*

During the road test, the vehicle should be driven in the normal manner. All pressure and gear changes should be noted. Also, the various computer inputs should be monitored and the readings recorded for future reference. Some scan tools have the capability of printing out a report of the test drive. Critical information from the inputs includes engine speed, vehicle speed, manifold vacuum, operating gear, and the time it took to shift gears. If the scan tool does not have the ability to give a summary of the road test, you should record this same information after each gear or operating condition change.

**You Should Know** *If the transmission will not upshift after you have serviced a Hyundai with a Mitsubishi KM-series transmission, check the connectors. It is possible to mistakenly connect the output connector to the input sensor connector and vice versa. The connectors are identical, only the wire colors are different, and these should be used to verify a correct connection.*

## Basic Transmission Circuits

To summarize how the circuitry of an electronically controlled transmission interrelates to a computerized engine control system, a summary of the transmission controls for a Ford E4OD follows. This system is similar to the other systems used by Ford, as well as the other manufacturers. The transmission is controlled by the PCM of the engine control system. The computer receives inputs from both the engine and the transmission. Based on this information, the PCM can send signals to operate components in, both, the engine and transmission. There are many inputs to the PCM from the transmission. The Transmission Range (TR) Sensor that informs the PCM what gear has been selected. The Vehicle Speed Sensor, located in the same place as the governor would be, inputs the mph at which the vehicle is traveling. The Turbine Speed Sensor (not found on all models) sends signals to inform the PCM of the speed of the input to the transaxle. The Transmission Oil Temperature sensor, which monitors the temperature of the ATF. This signal determines whether the PCM should go to a cold-start shift cycle or shift according to its schedule.

There are four nontransmission inputs: the TP sensor, MAF, PIP, and brake on switch. The system uses five solenoids, which are located in the transmission: the electronic pressure control solenoid that regulates transmission operating pressure; a modulated solenoid that controls the converter clutch; and three shift solenoids that control fluid flow to the various holding and clutching devices **(Figure 16)**.

Because the engine and transmission control systems share information, a common component may affect both systems. Because the computer controls the outputs to the engine and the transmission based on input sensors common to both, it is very important that you perform basic engine checks before spending time diagnosing the transmission. This was important when diagnosing nonelectronically controlled transmissions, and it is even more important when diagnosing an EAT.

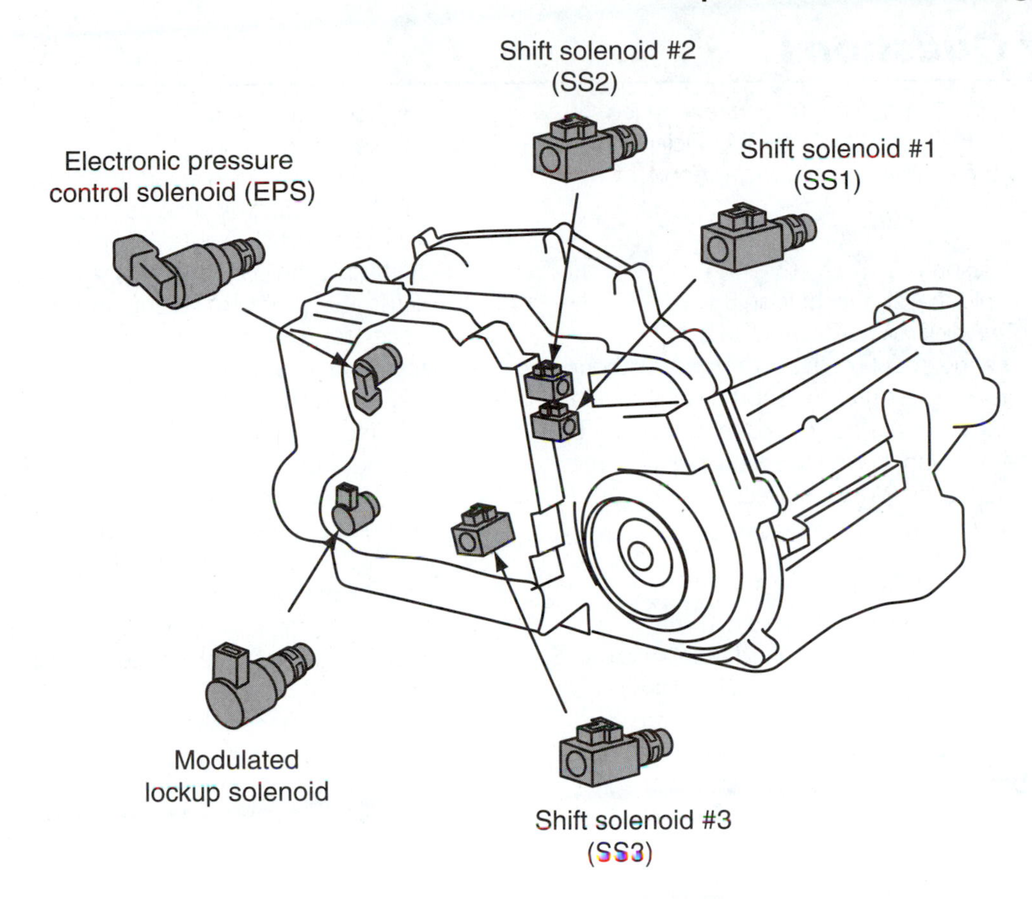


**Figure 16.** Location of the outputs on an E4OD transaxle.

# Summary

- One of the first tasks during diagnosis of an electronically controlled transmission is to determine if the problem is caused by the transmission or by electronics.
- All diagnostics should begin with a scan tool to check for trouble codes in the system's computer.
- When handling any electronic part, especially those that are static sensitive, try your best to reduce the possibility of electrostatic buildup on your body and the inadvertent discharge to the electronic part.
- The best way to diagnose an electronically controlled transmission is to approach solving the problem in a logical way.
- Diagnostic trouble codes are designed to help technicians identify and locate problems in the transmission's control system.
- On-Board Diagnostics II (OBD II) systems were developed with goal of detecting when engine or system wear or component failure causes exhaust emissions to increase by 50 percent or more.
- OBD II DTCs are well defined and can lead a technician to the cause of a problem.
- Testing of individual components is important in diagnosing OBD II systems. Although the DTCs in these systems give much more detail on the problems, the PCM does not know the exact cause of the problem. That is the technician's job.
- Some of the common uses of the scan tool are identifying stored DTCs, clearing DTCs, performing output control tests, and reading serial data.
- Diagnosis on vehicles equipped with other than OBD II systems can be done by the self-diagnostics mode of the system and/or by connecting a scan tool to the DLC or ALDL.
- Critical to proper diagnosis of EAT and TCC control systems is a road test. The road test should be conducted in the same way as one for a nonelectronic transmission except that a scan tool also should be connected to the circuit to monitor engine and transmission operation.

## Review Questions

1. Technician A says that scan tools are needed to retrieve codes on all models of cars. Technician B says that all scan tools provide a historical report of the computer system. Who is correct?
   A. Technician A only
   B. Technician B only
   C. Both Technician A and Technician B
   D. Neither Technician A nor Technician B
2. When checking the trouble codes on a Honda without OBD II, Technician A says that the short flashes should be counted, as this is the trouble code. Technician B says that there is about a one-second delay in the flashing between codes. Who is correct?
   A. Technician A only
   B. Technician B only
   C. Both Technician A and Technician B
   D. Neither Technician A nor Technician B
3. When checking the codes on a Ford AX4S with an OBD II compliant computer system, Technician A says that during the KOEO self-test, the codes are repeated only twice. Technician B says that the continuous memory codes appear after the separator pulse signal. Who is correct?
   A. Technician A only
   B. Technician B only
   C. Both Technician A and Technician B
   D. Neither Technician A nor Technician B
4. An electronically controlled transmission has erratic shifting. Technician A says that a poor PCM ground will cause this problem. Technician B says that a bad AC generator-to-battery circuit will cause erratic performance of a transmission or transaxle. Who is correct?
   A. Technician A only
   B. Technician B only
   C. Both Technician A and Technician B
   D. Neither Technician A nor Technician B
5. Technician A says that the second character code in an OBD II DTC is a designated code for future use and has little value to a technician. Technician B says that if the third character of a powertrain code is a 7 or 8, the problem is directly related to the transmission. Who is correct?
   A. Technician A only
   B. Technician B only
   C. Both Technician A and Technician B
   D. Neither Technician A nor Technician B

# Chapter 16 Detailed EAT Testing

## Introduction

The initial steps for troubleshooting an electronically controlled transmission include retrieving diagnostic codes. Diagnostic codes are used to recognize and locate problems. Remember, transmission related DTCs can be caused or detected by engine input or transmission input and/or output devices. Although the codes may appear to be caused by faulty input sensors or outputs, they may actually be caused by internal transmission problems. To pinpoint the exact cause of a transmission problem, you will need to use logic and basic electrical troubleshooting equipment, such as wiring diagrams, diagnostic charts, DMMs, lab scopes, special transmission testers, as well as scan tools.

Test equipment also may be required when the problem does not set a DTC or when verifying that a component or circuit is faulty. This chapter discusses the use of common diagnostic tools to test individual components. Before moving into this discussion, lab scopes and their general use are explained first as they are extremely valuable when diagnosing electronic systems.

### LAB SCOPES

A lab scope is the name given to a small oscilloscope **(Figure 1)**. An oscilloscope is a visual voltmeter. Lab scopes have become the diagnostic tool of choice for many good technicians. A scope allows you to see voltage changes over time. Voltage is displayed across the screen of the scope as a waveform. By displaying the waveform, the scope shows the slightest changes in voltage. This is a valuable feature for a diagnostic tool. With a scope, precise measurement of voltage is possible. When measuring voltage with an analog voltmeter, the meter only displays the average values at the point being probed. Digital voltmeters simply sample the voltage several times each second and update the meter's reading at a particular rate. If the voltage is constant, good measurements can be made with both types of voltmeters. A scope will display any change in voltage as it occurs. This is especially important for diagnosing intermittent problems.

The screen of a lab scope is divided into small divisions of time and voltage **(Figure 2)**. These divisions set up a grid pattern on the screen. The horizontal movement of the waveform represents time. Voltage is measured with the vertical position of the waveform. Because the scope displays voltage over time, the waveform moves from the left (the beginning of measured time) to the right (the end of measured time). The value of the divisions can be adjusted to improve the view of the voltage waveform. For example,

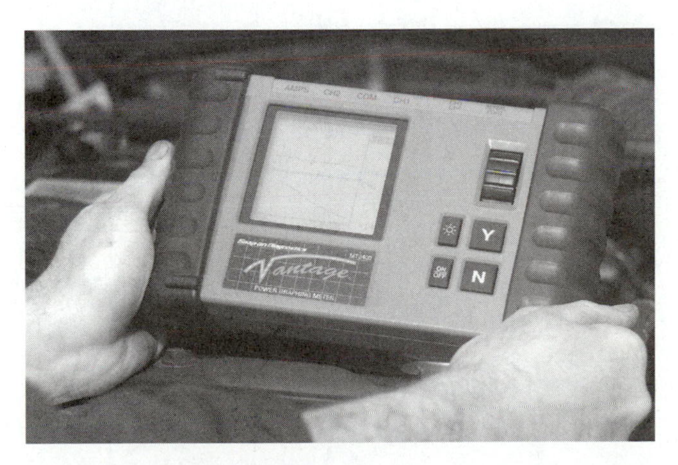


**Figure 1.** A hand-held dual-trace lab scope. (Courtesy of OTC Tool and Equipment, Division of SPX Corporation)

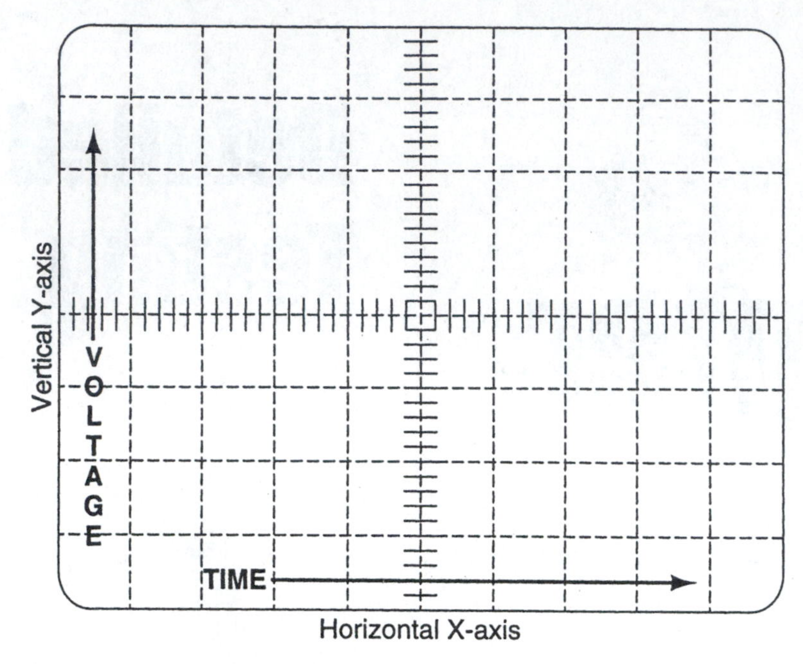


**Figure 2.** Grids on a scope screen serve as time and voltage references.

the vertical scale can be adjusted so that each division represents 0.5 volt and the horizontal scale can be adjusted so that each division equals 0.005 second (5 milliseconds). This allows you to view small changes in voltage that occur in a very short period of time. The grid serves as a reference for measurements.

Because a scope displays actual voltage, it will display all electrical noises or disturbances that accompany the voltage signal **(Figure 3)**. Noise is primarily caused by **radio frequency interference (RFI)**, which may come from the ignition system. RFI is an unwanted voltage signal that rides on a signal. This noise can cause intermittent problems with

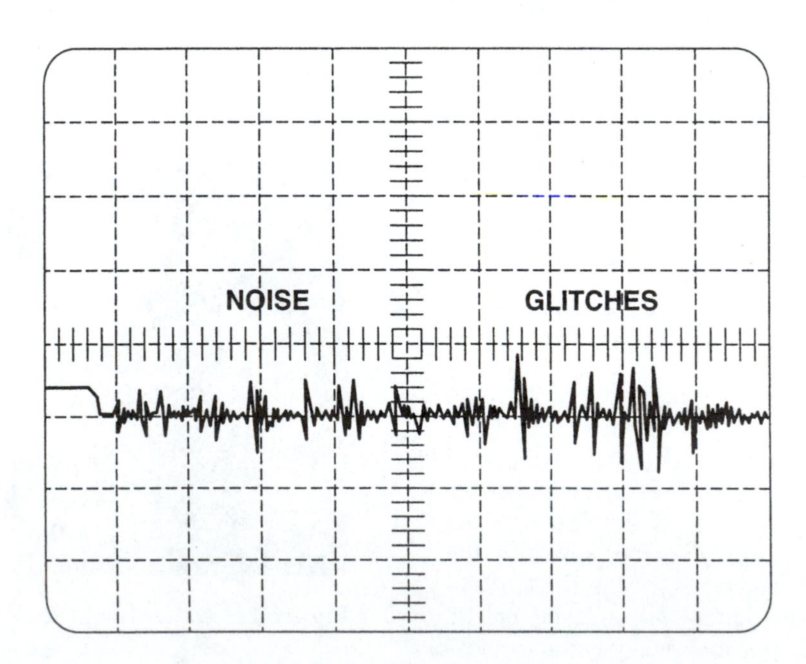


**Figure 3.** RFI noise and glitches may appear on the voltage signal.

unpredictable results. The noise causes slight increases and decreases in the voltage. When a computer receives a voltage signal with noise, it will try to react to the minute changes. As a result, the computer responds to the noise rather than the voltage signal.

Electrical disturbances or **glitches** are momentary changes in the signal. These can be caused by intermittent shorts to ground, shorts to power, or opens in the circuit. These problems can occur for only a moment or may last for some time. A lab scope is handy for finding these and other causes of intermittent problems. By observing the voltage signal and wiggling or pulling a wiring harness, any looseness can be detected by a change in the voltage signal.

## Analog versus Digital Scopes

Analog scopes show the actual activity of a circuit and are called real-time scopes. This simply means that what is taking place at the point being measured or probed is what you see on the screen. Analog scopes have a fast update rate, which allows for the display of activity without delay. The update rate is the time it takes the trace to move from the end of the screen on the right back to the left of the screen.

A digital scope, commonly called a **Digital Storage Oscilloscope (DSO)**, converts the voltage signal into digital information and stores it into its memory. Some DSOs send the signal directly to a computer or a printer, or save it to a disk. To help in diagnostics, a technician can "freeze" the captured signal for close analysis. DSOs also have the ability to capture low-frequency signals. Low-frequency signals tend to flicker when displayed on an analog screen. To have a clean waveform on an analog scope, the signal must be repetitive and occurring in real time. The signal on a DSO is not quite real time; rather, it displays the signal as it occurred a short time before.

This slight delay is actually very slight. Most DSOs have a sampling rate of one million samples per second. This is quick enough to serve as an excellent diagnostic tool. This fast sampling rate allows slight changes in voltage to be observed. Slight and quick voltage changes cannot be observed on an analog scope.

A DSO uses an A/D converter to digitize the input signal. Because digital signals are based on binary numbers, the trace appears slightly choppy when compared to an analog trace. However, the voltage signal is sampled more often, which results in a more accurate waveform. The waveform is constantly being refreshed as the signal is pulled from the scope's memory. Remember, the sampling rate of a DSO can be as high as a million times per second.

Both an analog and a digital scope can be dual trace scopes **(Figure 4)**. This means they both have the capability of displaying two traces at one time. By watching two traces simultaneously, you can watch the cause and effect of a sensor, as well as compare a good or normal waveform to the one being displayed.

## Waveforms

A waveform represents voltage over time. Any change in the **amplitude** or height of the trace indicates a change in the voltage. When the trace is a straight horizontal line, the voltage is constant. A diagonal line up or

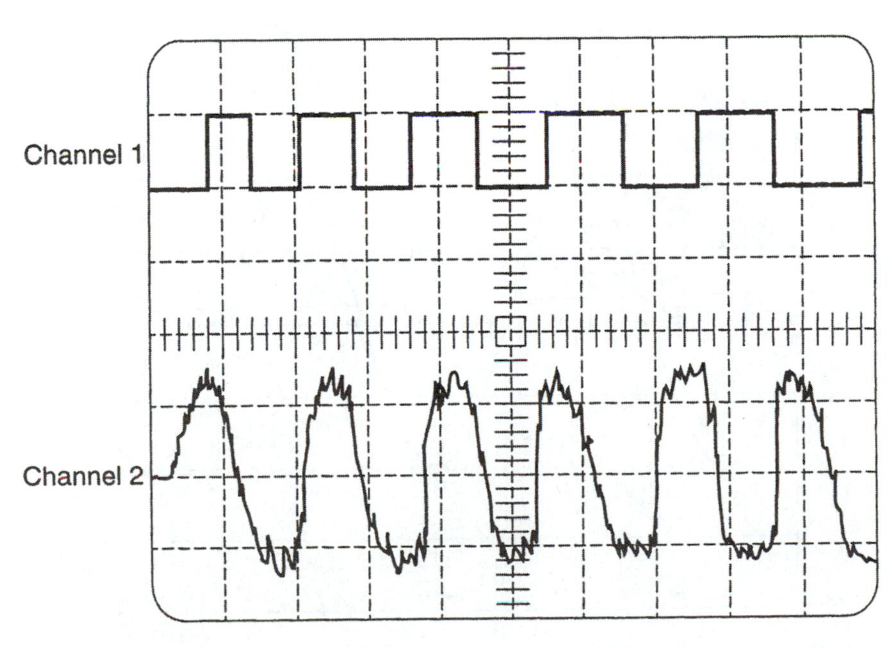


**Figure 4.** Some scopes can display two traces at one time; these are called dual trace scopes.

down represents a gradual increase or decrease in voltage. A sudden rise or fall in the trace indicates a sudden change in voltage.

Scopes can display AC and DC voltage; one at a time or both as in the case of noise caused by RFI. Noise results from AC voltage riding on a DC voltage signal. The consistent change of polarity and amplitude of the AC signal causes slight changes in the DC voltage signal. A normal AC signal changes its polarity and amplitude over a period of time. The waveform created by AC voltage is typically called a sine wave **(Figure 5)**. One complete sine wave shows the voltage moving from zero to its positive peak then moving down through zero to its negative peak and returning to zero. If the rise and fall from positive and negative are the same, the wave is said to be **sinusoidal**. If the rise and fall are not the same, the wave is non-sinusoidal. Therefore, it can be said that not all AC voltage waveforms are sine waves.

One complete sine wave is a **cycle**. The number of cycles that occur per second is the frequency of the signal. Checking frequency or cycle time is one way of checking the operation of some electrical components. Input sensors are the most common components that produce AC voltage. Permanent magnet voltage generators produce an AC voltage that can be checked on a scope **(Figure 6)**. AC voltage waveforms also should be checked for noise and glitches. These may send false information to the computer.

DC voltage waveforms may appear as a straight line or line showing a change in voltage. Sometimes a DC voltage waveform will appear as square wave or digital signal, which shows voltage making an immediate change **(Figure 7)**. Square waves are identified by having straight vertical sides and a flat type. This type of wave represents voltage being applied (circuit being turned on), voltage being maintained (circuit remaining on), and no voltage applied (circuit is turned off). Of course, a DC voltage waveform may also show gradual voltage changes.

## Scope Controls

Depending on the manufacturer and model of the scope, the type and number of its controls will vary. However, nearly all scopes have intensity, vertical (Y-axis) adjustments, horizontal (X-axis) adjustments, and trigger adjustments. The intensity control is used to adjust the brightness of the trace. This allows for clear viewing regardless of the light around the scope screen.

The vertical adjustment actually controls the voltage displayed. The voltage setting of the scope is the voltage that will be shown per division. If the scope is set at 0.5 volt (500 millivolts), this means a 5-volt signal will need 10 divisions. Likewise, if the scope is set to one volt, 5 volts will need only five divisions. When using a scope, it is important to set the vertical so that voltage can be accurately read. Setting the voltage too low may cause the waveform to move off the screen, whereas setting it too high may cause the trace to be flat and unreadable. The vertical position control allows the vertical position of the trace to be moved anywhere on the screen.

The horizontal position control allows the horizontal position of the trace to be set on the screen. The horizontal control is actually the time control of the trace. Setting the horizontal control is setting the time base of the scope's sweep rate. If the time per division is set too low, the complete trace may not show across the screen. Also, if the time

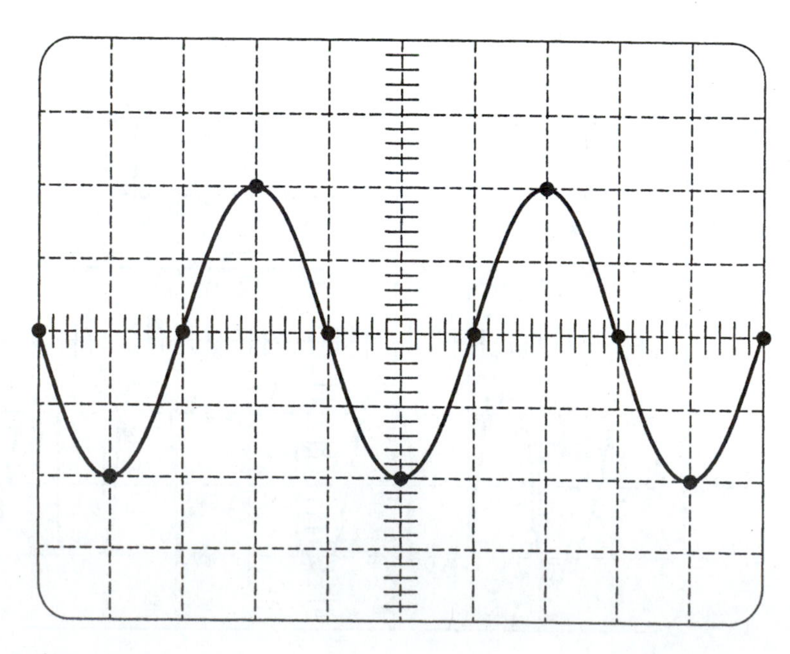

**Figure 5.** An AC voltage sine wave.

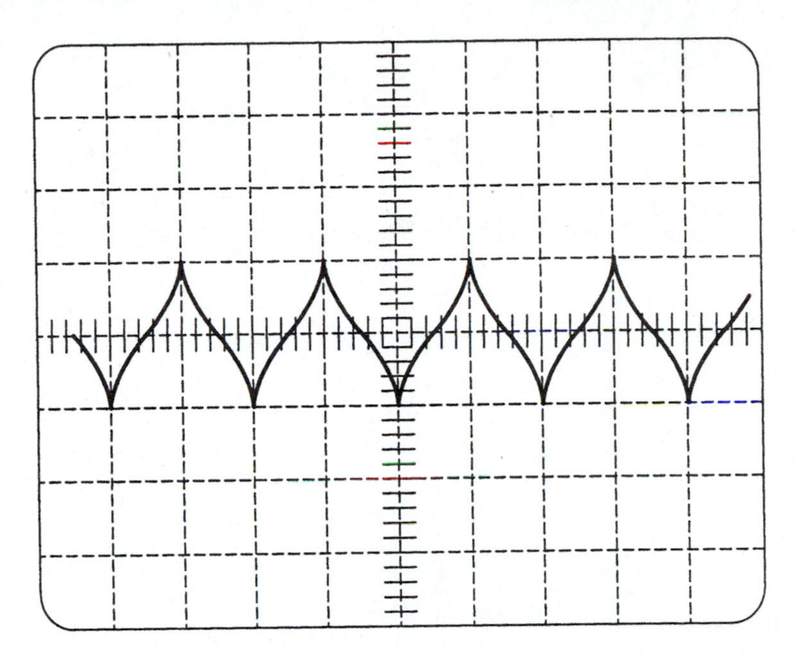

**Figure 6.** An AC voltage trace from a typical permanent magnet generator-type pickup or sensor.

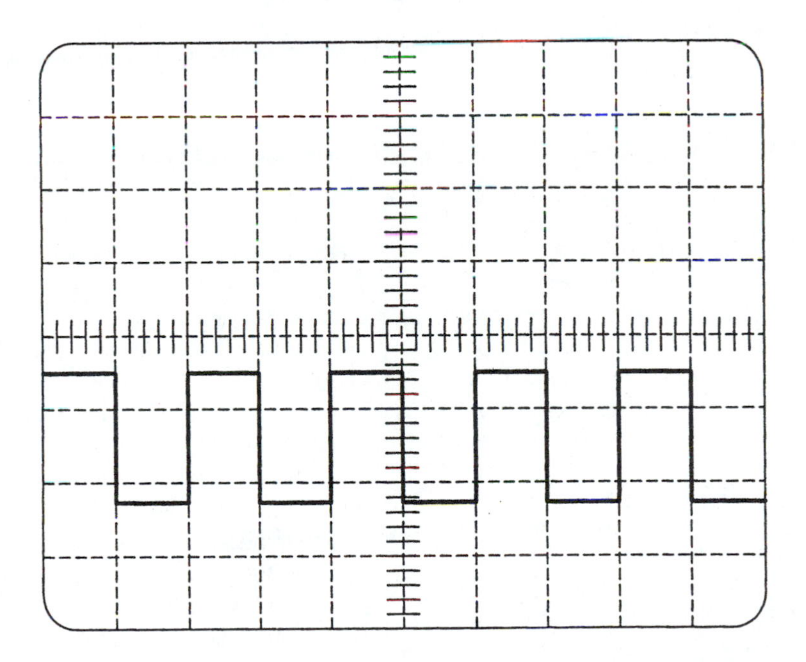

**Figure 7.** A typical square wave.

per division is set too high, the trace may be too crowded for detailed observation. The time per division (TIME/DIV) can be set from very short periods of time (millionths of a second) to full seconds.

Trigger controls tell the scope when to begin a trace across the screen. Setting the trigger is important when trying to observe the timing of something. Proper triggering will allow the trace to repeatedly begin and end at the same points on the screen. Typically, there are numerous trigger controls on a scope. The trigger mode selector has a NORM and AUTO position. In the NORM setting, no trace will appear on the screen until a voltage signal occurs within the set time base. The AUTO setting will display a trace regardless of the time base.

Slope and level controls are used to define the actual trigger voltage. The slope switch determines whether the trace will begin on a rising or falling of the voltage signal **(Figure 8)**. The level control determines where the time base will be triggered according to a certain point on the slope.

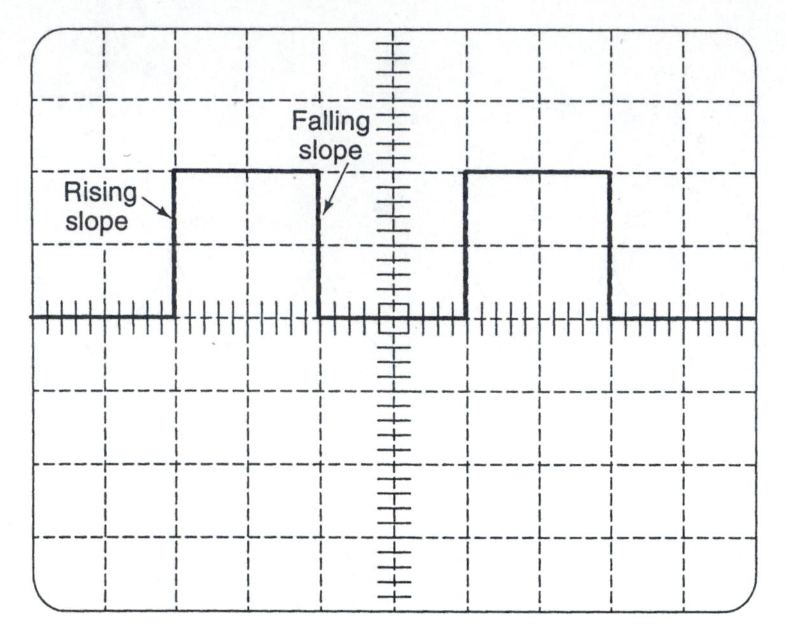


**Figure 8.** The trigger can be set to start the trace with a rise or fall of the voltage.

A trigger source switch tells the scope which input signal to trigger on. This can be Channel 1, Channel 2, line voltage, or an external signal. External signal triggering is very useful when desiring to observe a trace of a component that may be affected by the operation of another component. An example of this would be observing pressure control solenoid activity when changes in vehicle speed are made. The external trigger would be voltage change at the VSS. The displayed trace would be the cycling of the pressure control solenoid. Channel 1 and Channel 2 inputs are determined by the points of the circuit being probed. Some scopes have a switch that allows inputs from both channels to be observed at the same time or alternately.

## DETAILED SYSTEM TESTING

At times it may be difficult to determine if the problem is caused by something inside the transmission or if the cause is electronic. If you face this dilemma, take the vehicle on a road test and note all of the things that do not seem right. Then, remove the fuse for the transmission control unit. Doing this disables electronic control of the transmission and will cause it to operate in its default mode. Road test the vehicle again and observe the operation of the transmission. Keep in mind that during default operation, there will be no forward upshifts. If the problem or problems observed in the first road test still exist, the cause must be in the transmission.

### Computer Checks

Although there are DTCs that relate directly to the PCM or TCM, it may be difficult to identify a faulty computer unit. Because the computer reacts to the inputs it receives, abnormal input signals may result in abnormal computer operation. Also, most computers are designed to react to out-of-range inputs by ignoring the signals and using a fixed value held in its memory. This results in less than efficient operation.

Basically, if the computer system's inputs and outputs and the engine and hydraulic and mechanical functions of the transmission are all good, but a problem exists, the computer or its circuit is bad. A check of the computer circuit must always follow the suspicion that the computer is faulty.

To check the circuit, check for battery voltage to the computer. Use the wiring diagram in the service manual to identify the power feed for the computer. If less than battery voltage is available, conduct a voltage drop test to determine the location of the unwanted resistance.

If the computer is receiving battery voltage, check the computer's ground. The best way of doing this is to place one probe of a voltmeter on the case or ground connection of the computer and the other at a known good ground **(Figure 9)**. With the system powered, the meter should read less than 0.1-volt. If the reading is higher than that, conduct voltage drop tests on the rest of the ground circuit.

Most scan tools give you the capability of controlling shift points by manually turning on shift solenoids. If your scan tool does not have this feature, special testers are available to do this. This check operates the solenoids on command thereby bypassing the computer. Doing this, you can observe the operation of the solenoids and the transmission. If the customer's complaint is still evident when the computer is bypassed, you know the problem is not the

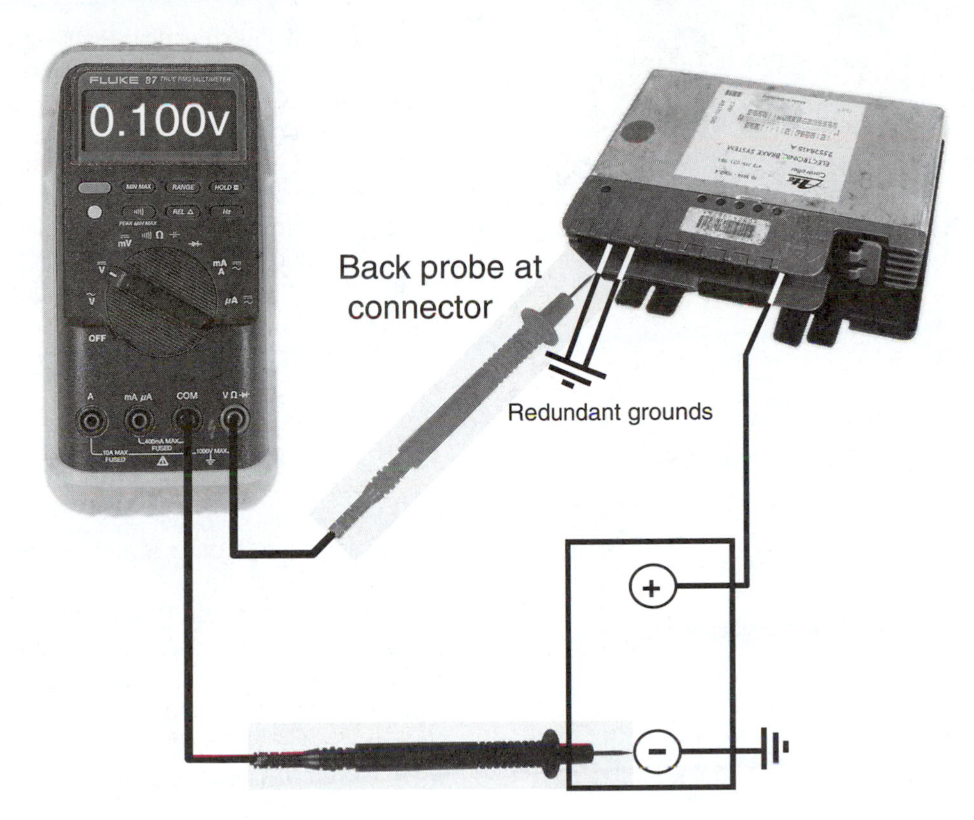


**Figure 9.** Check the power feed and the ground circuit before condemning a computer.

computer. If the problem is not evident, you know the problem is the computer and/or computer circuit.

## TCC Functional Test

On most vehicles, the engagement of the torque converter clutch can be checked during a road test. Drive the vehicle at a steady speed of 40 to 45 mph. Note the engine's speed on the tachometer. Then lightly tap the brake pedal, while maintaining road speed. Note the tachometer reading. If the brake switch caused the converter clutch to disengage, the engine's speed should increase. Shortly after disengagement, the clutch should reengage and engine speed should drop. If you observed no evidence of clutch engagement or disengagement, the TCC system **(Figure 10)** needs further testing.

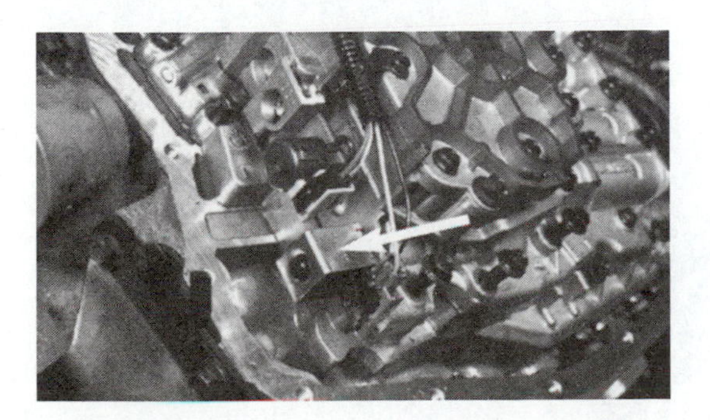

**Figure 10.** A torque converter clutch solenoid.

# DETAILED TESTING OF INPUTS

Depending on the operation of the system monitored, there are many different designs of sensors. Some sensors are nothing more than a switch that completes a circuit. Others are complex devices that react to chemical reactions and generate their own voltage during certain conditions. If the self-test sequence pointed to a problem in an input circuit, the input circuit should be tested to determine the exact malfunction. Often the manufacturers list specific procedures for specific sensors, always follow them.

## Testing Switches

Many different switches are used as inputs or control devices for EATs. Most of the switches are either mechanically or hydraulically controlled. The operation of these switches can be easily checked with an ohmmeter **(Figure 11)**. With the meter connected across the switch's leads, there should be continuity or low resistance when the switch is closed and there should be infinite resistance across the switch when it is open. A test light also can be used. When the switch is closed, power should be present at both sides of the switch. When the switch is open, power should be present at only one side.

The neutral safety switch and the kickdown switch are two examples of the switches used on some transmissions. There are many more possible switches connected to a transmission and they can serve many different purposes.

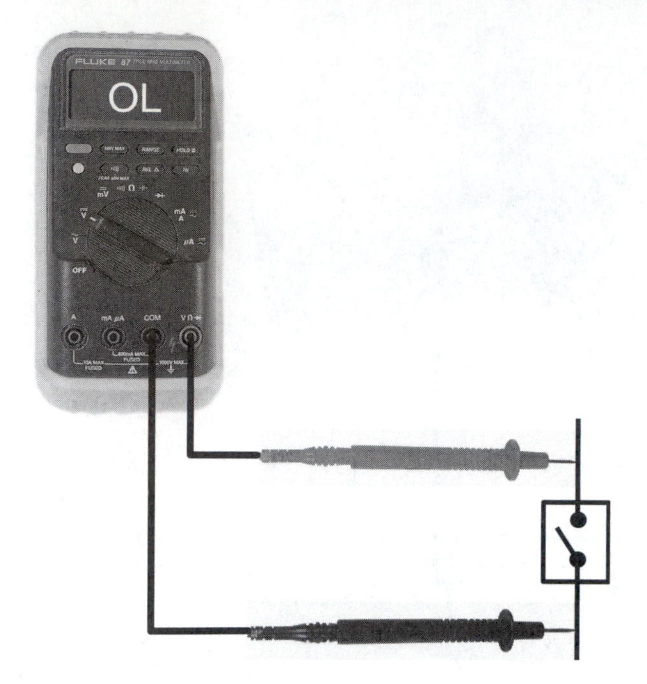


**Figure 11.** Checking a switch with an ohmmeter.

However, all of these switches either complete or open an electrical circuit. By completing a circuit, these switches either provide a ground for the circuit or connect two wires together. To open the circuit, switches either disconnect the ground from the circuit or disconnect two wires.

Prior to the use of electronic controls and their associated sensors, grounding switches were often used to control the engagement of torque converter clutches. These switches were controlled by governor pressure and supply the ground for the TCC solenoid when a particular speed is reached. Different TCC engagement speeds are possible through the use of different spring tensions on the switch. To engage the clutch earlier, a weaker spring is used and when the clutch engagement should occur later, a stronger spring is used. Grounding switches also are used to indicate or sense what gear the transmission is in. Other grounding type switches are used to control the back-up lamp circuit.

The Manual Lever Position or TR sensor **(Figure 12)** is a switch that provides information to the computer as to what operating range has been selected by the driver. Based on that information, the computer determines the proper shift strategy for the transmission. The following is a list of problems that may result from a faulty TR Sensor:

- No upshifting
- Slipping out of gear
- High line pressure in transmissions equipped with a pressure control solenoid
- Delayed gear engagements
- Engine starts in other lever positions besides park and neutral

Because this switch is open or closed, depending on position, it can be checked with an ohmmeter. By referring to a wiring diagram, you should be able to determine when the switch should be open. Doing this for some switch designs may be a little difficult as the switch assembly may be made up of more than one switch. Then connect the meter across the input and out of the switch. Move the lever into the desired position and measure the resistance. An infinite reading is expected when the switch is open. If there is any resistance, the switch should be replaced. The switch also should be replaced if there is some resistance when the switch is closed.

The pressure switches **(Figure 13)** used in today's transmissions either complete or open an electrical circuit. These switches are either normally open or normally closed. Normally open switches will have no continuity across the terminals until oil pressure is applied to it. Normally closed

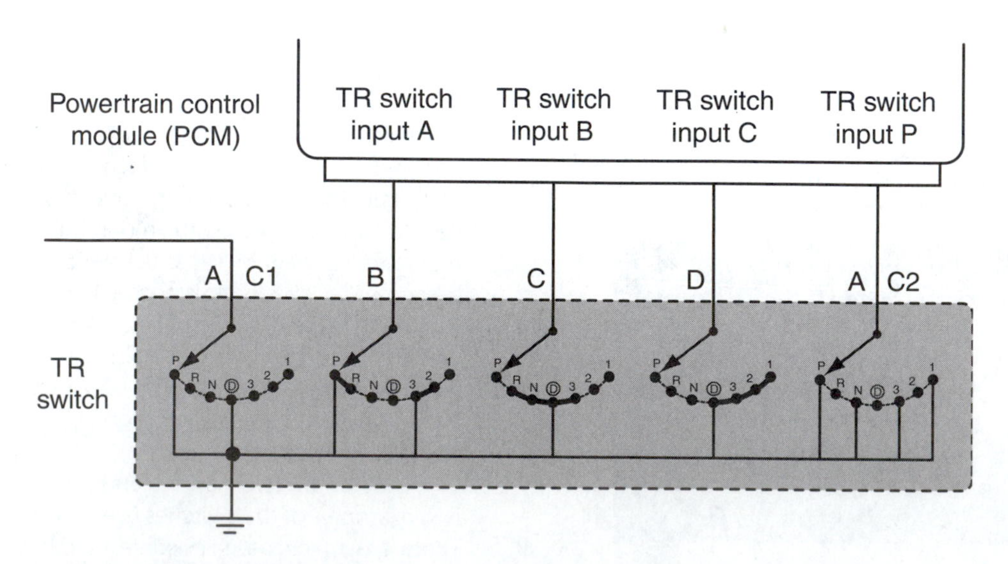


**Figure 12.** A Transmission Range (TR) switch.

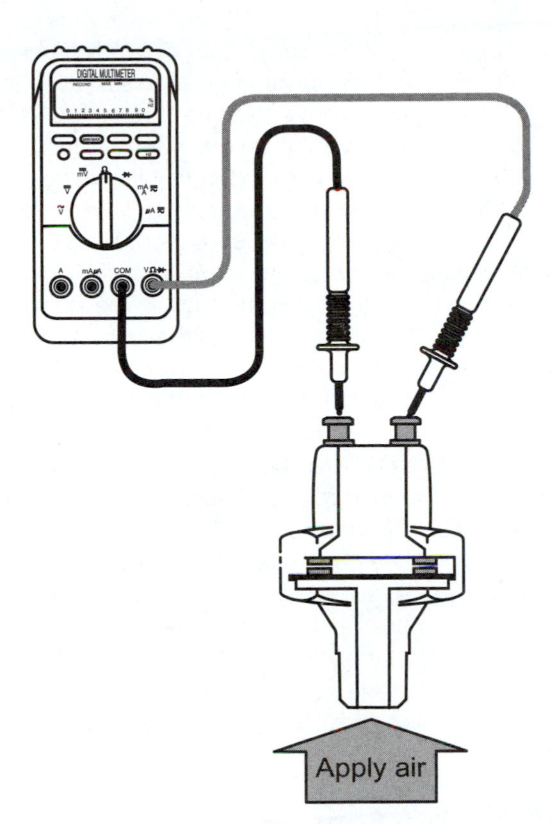


**Figure 13.** To check a normally open pressure switch, apply air to the bottom oil passage and check for continuity across the terminals.

switches will have continuity across the terminals until oil pressure is applied to it. Refer to the wiring diagram to determine the type of switch and test the switch with an ohmmeter. An ohmmeter can be used to identify the type of switch being used and can be used to test the operation of the switch. There should be continuity when the switch is closed and no continuity when the switch is open. Base your expected results on the type of switch you are testing.

Pressure switches can be checked by applying air pressure to the part that would normally be exposed to oil pressure. When applying air pressure to these switches, check them for leaks. Although a malfunctioning electrical switch will probably not cause a shifting problem, it will if it leaks; if the switch leaks off the applied pressure in a hydraulic circuit to a holding device, the holding member may not be able to function properly.

**You Should Know** *Always disconnect an electrical switch or component from the circuit or from a power source before testing it with an ohmmeter. Failure to do so will damage the meter.*

When possible, you should check pressure switches when they are installed and controlled by the vehicle. By watching an ohmmeter connected to a governor switch, you can check its spring rating. Connect the ohmmeter to the "A" terminal and ground. Bring the engine's speed up. When the ohmmeter reads about 25 ohms, the governor switch is closed. The speed at which this occurs is the speed required to close the grounding switch. Other grounding type switches can be checked in the same way. However, it is important that you identify if it is a normally open or closed switch before testing.

## Throttle Position Sensor

Another type of switch is a potentiometer **(Figure 14)**. Rather than open and close a circuit, a potentiometer controls the circuit by varying its resistance in response to something. A TP sensor is a potentiometer. It sends very low voltage back to the computer when the throttle plates are closed and increases the voltage as the throttle is opened.

A TP sensor sends information to the control computer as to what position the throttle is in. If this sensor fails, the following problems can result:

- No upshifts
- Quick upshifts
- Delayed shifts
- Line-pressure problems with transmissions that have a line pressure control solenoid
- Erratic converter clutch engagement

Most TP sensors receive a reference voltage of five volts. What the TP sensor sends back to the computer is determined by the position of the throttle. When the throttle is closed, it sends approximately 0.5 volt back to the computer. When the throttle begins to open, the voltage begins to increase. When the throttle reaches the wide-open position, approximately 4.5 volts are sent back to the computer. This increase in voltage should rise and fall smoothly with a change in throttle position. All changes in throttle position should result in a change in voltage.

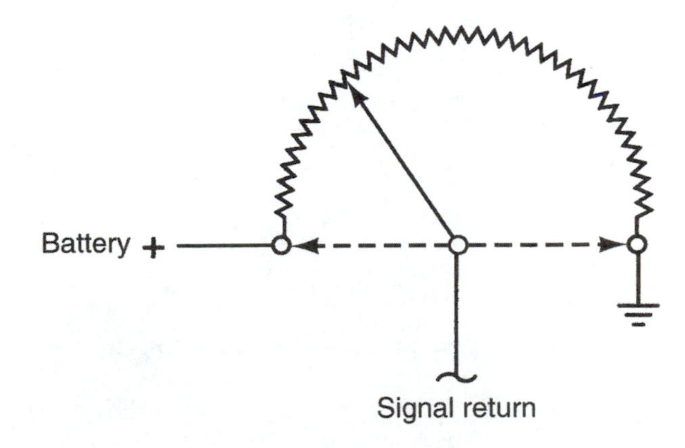


**Figure 14.** A potentiometer is used to send a signal voltage from the wiper of the switch.

A TP sensor can be checked with an ohmmeter or a voltmeter. If checked with an ohmmeter, you should be able to watch the resistance across the TP sensor change as the throttle is opened and closed. Often, there will be a resistance specification given in the service manual. Compare your reading to this.

With a voltmeter, you will be able to measure the reference voltage and the output voltage. Both of these should be within specified amounts. If the reference voltage is lower than normal, check the voltage drop across the reference voltage circuit from the computer to the TP sensor. If the TP sensor is found to be defective, it should be replaced.

Potentiometer testing with a lab scope is a good way to watch the sweep of the resistor. The waveform is a DC signal that moves up as the voltage increases **(Figure 15)**. Most potentiometers in computer systems are fed a reference voltage of five volts. Therefore, the voltage output of these sensors will range from 0.5 to 4.5 volts. The change in voltage should be smooth. Look for glitches or sudden changes in voltage **(Figure 16)**. These can be caused by changes in resistance or an intermittent open in the circuit or sensor.

## Mass-Airflow Sensor

A mass-airflow sensor is used to determine engine load by measuring the mass of the air being taken into the throttle body. The computer uses this information for fuel and air control. When this sensor fails or sends faulty signals, the engine runs roughly and tends to stall as soon as you put the transmission into gear. The mass airflow sensor is a wire, located in the intake air stream that receives a fixed voltage. The wire is designed so that it changes resistance in response to temperature changes. When the wire is hot, the resistance is high and less current flows through the wire. When the wire is cold, the resistance is low and larger amounts of current pass through it. The amount of air passing over the wire determines the amount of resistance the wire has. The computer monitors the amount of current flow and interprets the flow as the mass or volume of the air.

This sensor can typically be measured with a multimeter set to the Hz frequency range. Check the service manual for specific values. Normally at idle, 30 Hz is measured with the frequency increasing as the throttle opens. A scan tool can also be used to test this sensor; most scanners have a test mode that monitors mass-airflow sensors. The output of some MAFs can be observed with a DMM or lab scope, as their output is variable DC voltage. When diagnosing these systems, keep in mind that cold air is denser than warm air.

## Temperature Sensors

Temperature sensors are designed to change resistance with changes in temperature. A temperature sensor

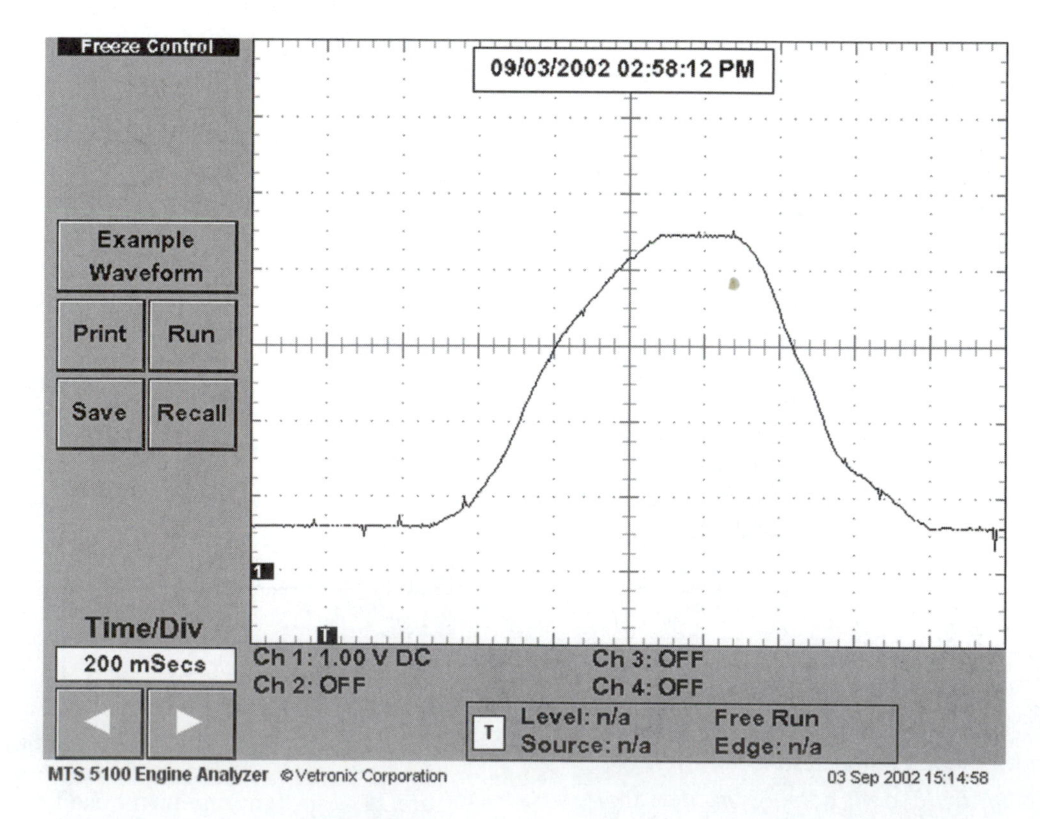


**Figure 15.** The voltage sweep of a good TP sensor.

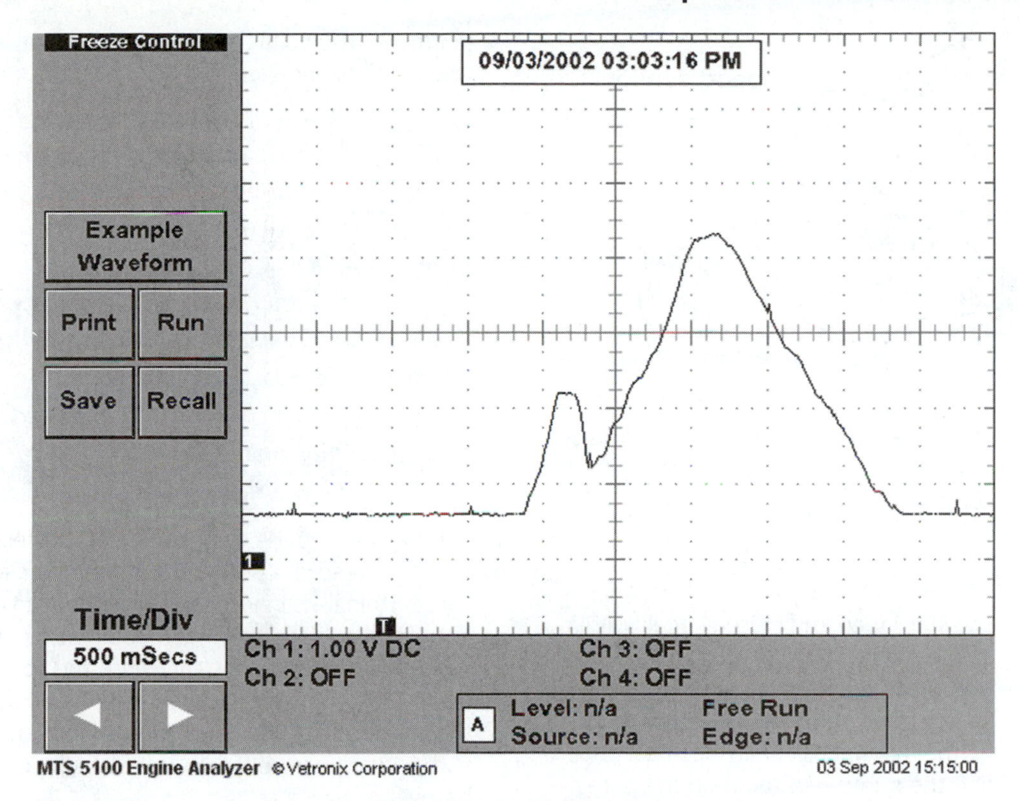


**Figure 16.** The voltage sweep of a faulty TP sensor or its circuit.

is based on a thermistor. Some thermistors increase resistance with an increase of temperature. Others decrease the resistance as temperature increases. Obviously, these sensors can be checked with an ohmmeter. To do so, remove the sensor. Then, determine the temperature of the sensor and measure the resistance across it. Compare your reading to the chart of normal resistances given in the service manual **(Figure 17)**.

Thermistors activity can be monitored with a lab scope. Connect the scope across to the output of the thermistor or temperature sensor. Run the engine and watch the waveform. As the temperature increases, there should be a smooth increase or decrease in voltage. Look for glitches in the signal. These can be caused by changes in resistance or an intermittent open.

| Transmission sensor Temperature to resistance | | | |
|---|---|---|---|
| C° | F° | (Ω) Minimum resistance | (Ω) Maximum resistance |
| 60° | 140° | 584 | 753 |
| 80° | 176° | 293 | 371 |
| 100° | 212° | 158 | 197 |
| 120° | 248° | 90 | 111 |
| 140° | 284° | 54 | 66 |
| 160° | 320° | 33 | 40 |
| 180° | 356° | 19 | 24 |
| 200° | 392° | 13 | 14 |

**Figure 17.** A chart showing the relationship between temperature and resistance for a temperature sensor.

## Speed Sensors

Electronically controlled transmissions rely on electrical signals from a speed sensor to control shift timing **(Figure 18)**. The use of this type sensor negates the need for hydraulic signals from a governor. When this sensor fails or sends faulty readings, it can cause complaints that are similar to those caused by a bad TP sensor. The most common complaints are no overdrive, no converter-clutch engagement, and no upshifts. Vehicle speed sensors provide road speed information to the computer.

There are basically two types of speed sensors, permanent magnetic (PM) generator sensors and reed style sensors. Both of these rely on magnetic principles. The reed style is simply a switch that closes every time a magnet passes by. The magnet is attached to a rotating shaft, typically the output shaft. The activity of the sensor can be checked with an ohmmeter. If the switch is disconnected and the output shaft rotated one complete revolution, the

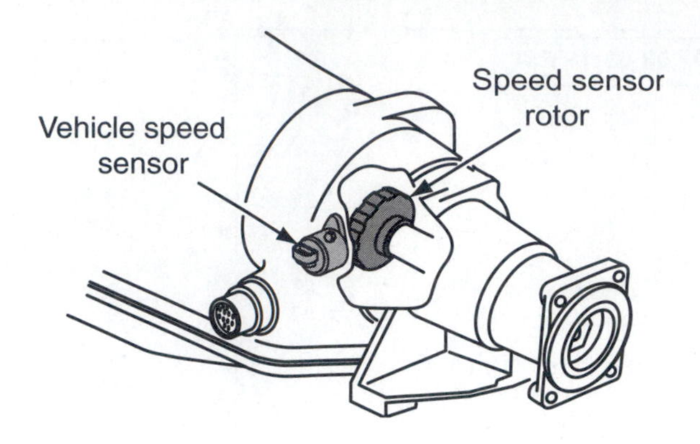


**Figure 18.** AC generators rely on a stationary magnet, the sensor assembly, and a rotor fitted with teeth.

ohmmeter should show an open and closed circuit within the one revolution of the shaft.

PM generators rely on a stationary magnet and a rotating shaft fitted with iron teeth **(Figure 19)**. Each time a tooth passes through the magnetic field an electrical pulse is present. By counting the number of teeth on the output shaft, you can determine how many pulses per one revolution you will measure with a voltmeter set to AC volts.

The operation of a PM generator can be tested with a DMM set to measure AC voltage. Raise the vehicle on a lift. Allow the wheels to be suspended and allowed to rotate freely. Connect the meter to the speed sensor. Start the engine and put the transmission in gear. Slowly increase the engine's speed until the vehicle is at approximately 20 mph, and then measure the voltage at the speed sensor. Slowly increase the engine's speed and observe the voltmeter. The voltage should increase smoothly and precisely with an increase in speed.

> **You Should Know** *When conducting this procedure, it is possible to damage the CV joints if the drive wheels are allowed to dangle in the air. Place safety stands under the front suspension arms to maintain proper drive axle operating angles. Also, when running the vehicle on a lift make sure you stay clear of the wheels and other rotating parts. If you are caught by something powered by the engine, serious injury can result.*

Magnetic pulse generators can be tested with a lab scope. Instead of connecting a voltmeter across the sensor's terminals, connect the lab scope leads. The expected pattern is an AC signal, which should be a perfect sine wave when the speed is constant. When the speed is changing, the AC signal should change in amplitude and frequency. If the readings are not steady and do not smoothly change with a change in speed, suspect a faulty connector, wiring harness, or sensor.

A speed sensor also can be tested with it out of the vehicle. Connect an ohmmeter across the sensor's terminals. The desired resistance readings across the sensor will vary with each and every individual sensor; however, you should expect to have continuity across the leads. If there is no continuity, the sensor is open and should be replaced. Reposition the leads of the meter so that one lead is on the sensor's case and other to a terminal. There should be no continuity in this position. If there is any measurable amount of resistance, the sensor is shorted.

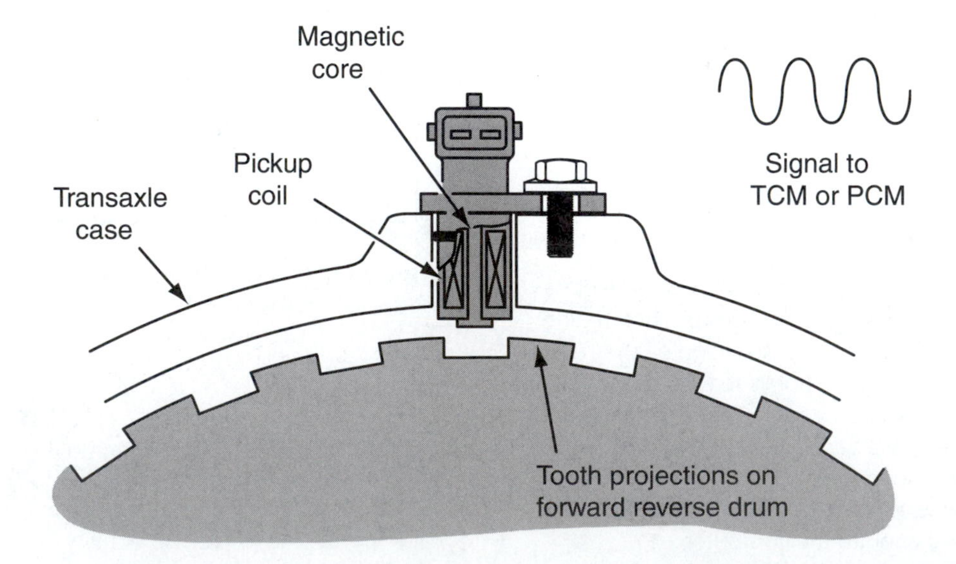


**Figure 19.** The vehicle speed sensor will display a pulse each time a tooth of the rotor passes by the sensor.

## DETAILED TESTING OF ACTUATORS

If you were unable to identify the cause of a transmission problem through the previous checks, you should continue your diagnostics with testing the solenoids. This will allow you to determine if the shifting problem is the solenoids or their control circuit, or if it is a hydraulic or mechanical problem in the transmission.

Before continuing, however, you must first determine if the solenoids are case grounded and fed voltage by the computer or if they always have power applied to them and the computer merely supplies the ground. When looking in the service manual to find this, also find the section that tells you which solenoids are on and which are off for each of the different gears.

To begin this test you should secure the tools and/or equipment necessary to manually activate the solenoids **(Figure 20)**. Switch panels are available that connect into the solenoid assembly and allow the technician to switch gears by depressing or flicking a switch. This check can be conducted on all transmission solenoids except those that have a duty cycle. Check the service manual to identify this type of solenoid.

To begin the test, disconnect the wiring harness that leads to the transmission's solenoids **(Figure 21)**. Make sure the ignition switch is off. If the solenoids are controlled by their ground, connect a jumper wire with an in line 20 amp fuse to the tester and a known good ground. Now complete the connection of the tester to the transmission and then program it according to the tester's instructions.

Start the engine and move the shift lever into Drive. With the tester, turn on solenoid #1; most transmissions will then shift into first gear **(Figure 22)**. Pay attention to how it shifted. Now increase your speed, and then turn on solenoid #2. The transmission should have immediately shifted into second gear. When third gear is selected, solenoid #1 will be turned off. Likewise when you want fourth gear, solenoid #2 will be turned off. After checking all gears under light, half, and full throttle, return to the shop and summarize your findings. Then disconnect the tester from the system and proceed to check solenoid circuits that seemed not to work correctly.

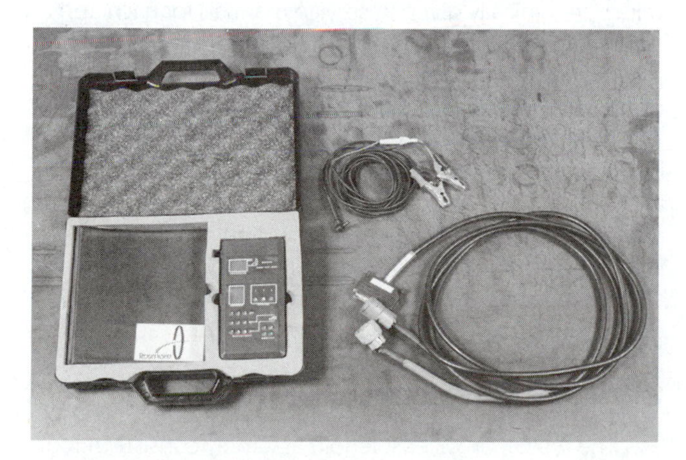

**Figure 20.** A solenoid tester.

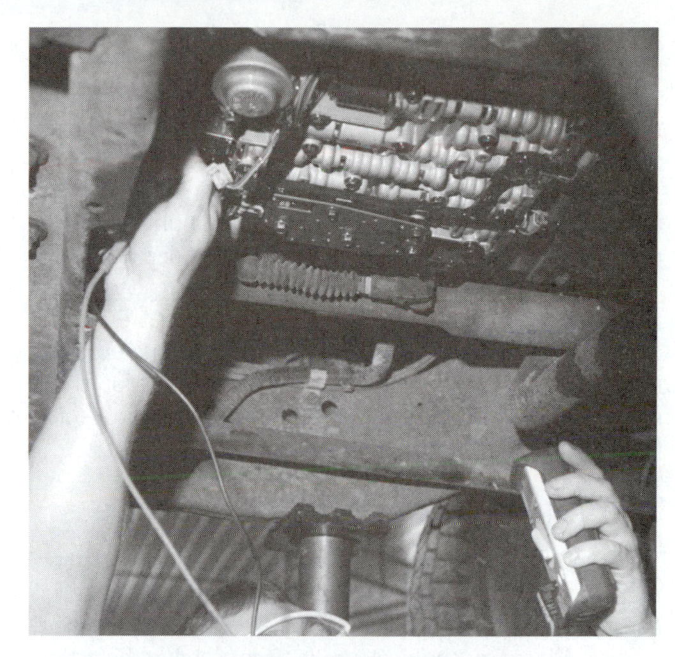

**Figure 21.** The solenoid tester being connected to the connector for the transmission's solenoids.

**Figure 22.** The solenoids are totally controlled by the tester.

**You Should Know** *This type of tester is easily made. Just secure a harness for the transmission you want to test. Connect the leads from the harness to simple switches. Follow the solenoid/gear pattern when doing this. To change gears, all you will need to do is turn off one switch and turn on the next.*

If the transmission shifted fine with the movement of the switches, you know that the transmission is fine. Any shifting problem must be caused by something electrical. If the transmission did not respond to the switch movements, the problem is probably in the transmission.

At times, a solenoid valve will work fine during light throttle operation but may not exhaust enough fluid when pressure increases. To verify that the valve is not exhausting, activate the solenoid then increase engine speed while pulling on the throttle cable. If the solenoid valve cannot exhaust, the transmission will downshift. Restricted solenoids should be suspect whenever the transmission shifts roughly or early under heavy loads or full throttle but shifts fine under light throttle.

## Testing Actuators with a Lab Scope

Actuators are electromechanical devices, meaning they are electrical devices that cause some mechanical action. When actuators are faulty, it is because they are electrically faulty or mechanically faulty. By observing the action of an actuator on a lab scope, you will be able to watch its electrical activity. Normally, if there is a mechanical fault, this will affect its electrical activity as well. Therefore, you get a good sense of the actuator's condition by watching it on a lab scope.

Some actuators are controlled pulse-width modulated signals **(Figure 23)**. These signals show a changing pulse width. These devices are controlled by varying the pulse width, signal frequency, and voltage levels. By watching the control signal, you can see the turning on and off of the solenoid **(Figure 24)**. The voltage spikes are caused by the discharge of the coil in the solenoid.

Both waveforms should be checked for amplitude, time, and shape. You also should observe changes to the pulse width as operating conditions change. A bad waveform will have noise, glitches, or rounded corners. You should be able to see evidence that the actuator immediately turns off and on according to the commands of the computer.

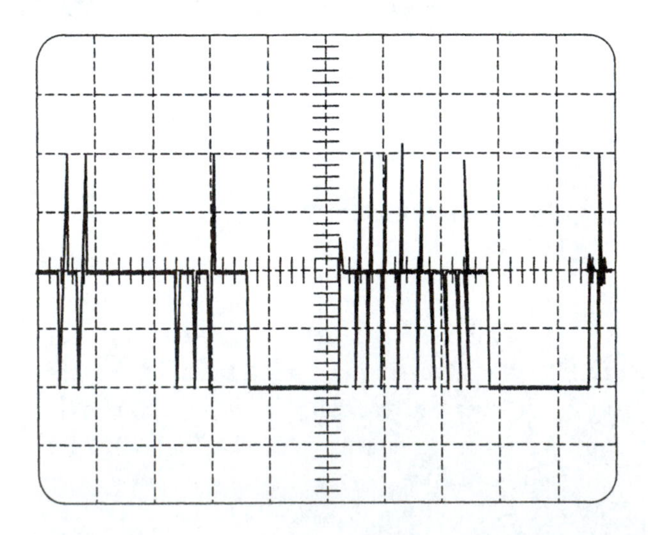

**Figure 23.** A typical control signal for a pulse width modulated solenoid.

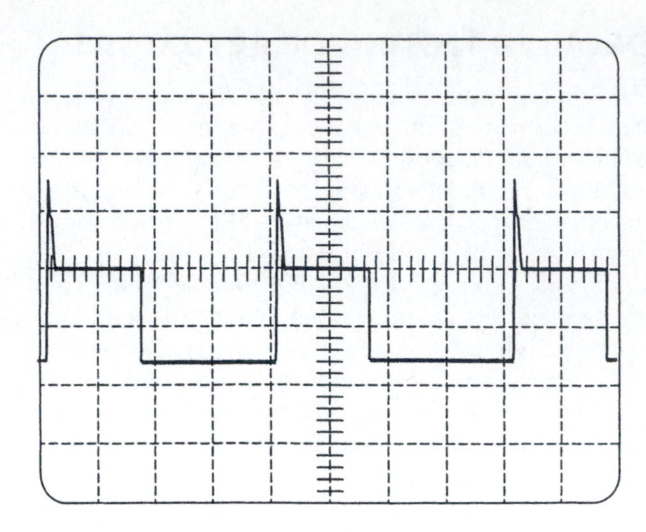

**Figure 24.** A typical control signal for a solenoid.

## Testing Actuators with an Ohmmeter

Solenoids can be checked for circuit resistance and shorts to ground. Typically, this can be done without removing the oil pan. The test can be conducted at the transmission case connector. By identifying the proper pins in the connector, individual solenoids can be checked with an ohmmeter. Remember, lower than normal resistance indicates a short, whereas higher than normal resistance indicates a problem of high resistance. If you get an infinite reading across the solenoid, the solenoid winding is open. The ohmmeter can also be used to check for shorts to ground. Just connect one lead of the ohmmeter to one end of the solenoid windings and the other lead to a good ground. The reading should be infinite. If there is any measurable resistance, the winding is shorted to ground.

Solenoids also can be tested on a bench. Resistance values are typically given in service manuals for each application **(Figure 25)**. A solenoid may be electrically fine but still may fail mechanically or hydraulically. A solenoid's check valve may fail to seat or the porting can be plugged. This is not an electrical problem; rather, it could be caused by the magnetic field collecting metal particles in the ATF and clogging the port or check valve. These would cause erratic shifting, no shift conditions, wrong gear starts, no or limited passing (kickdown) gear, or binding shifts. When a solenoid affected in this way is activated, it will make a slow dull thud. A good solenoid tends to snap when activated.

## Chrysler Solenoids

The typical Chrysler solenoid assembly consists of four solenoids, four ball valves, three pressure switches, and

| Components | Pass thru pins | Resistance at 20°C | Resistance at 100°C | Resistance to ground (case) |
|---|---|---|---|---|
| 1-2 shift solenoid valve | A, E | 19 - 24Ω | 24 - 31Ω | Greater than 250M Ω |
| 2-3 shift solenoid valve | B, E | 19 - 24Ω | 24 - 31Ω | Greater than 250M Ω |
| TCC solenoid valve | T, E | 21 - 24Ω | 26 - 33Ω | Greater than 250M Ω |
| TCC PWM solenoid valve | U, E | 10 - 11Ω | 13 - 15Ω | Greater than 250M Ω |
| 3-2 shift solenoid valve assembly | S, E | 20 - 24Ω | 29 - 32Ω | Greater than 250M Ω |
| Pressure control solenoid valve | C, D | 3 - 5Ω | 4 - 7Ω | Greater than 250M Ω |
| Transmission fluid temp. (TFT) sensor | M, L | 3088 - 3942Ω | 159 - 198Ω | Greater than 10M Ω |
| Vehicle speed sensor | A, B Vss conn | 1420Ω @ 25°C | 2140Ω @ 150°C | Greater than 10M Ω |

IMPORTANT: The resistance of this device is necessarily temperature dependent and will therefore vary far more than any other device. Refer to transmission fluid temp (TFT) sensor specifications.

**Figure 25.** Service manuals list the resistance values and test points for transmission solenoids and other components.

three resistors. The ball valves are operated by the solenoids and together they cause the transaxle to shift. The three pressure switches and resistors are used to signal the computer that a shift has occurred. The solenoids should be checked for proper operation. This can be done through the eight-pin connector at the assembly.

Using an ohmmeter with the harness disconnected from the assembly, check the resistance across pins 1 and 4, 2 and 4, and 3 and 4. The resistance should be between 270 and 330 ohms. With this test, you have checked the resistor for each of the switches. To check the action of these switches, apply air pressure to the feed holes for the overdrive, low-reverse, and overdrive pressure switches. With air applied, the resistance reading across each of the pin combinations should now be 0 ohm.

Each of the four solenoids also can be checked with an ohmmeter. Connect the meter across pins 4 and 5, 4 and 6, 4 and 7, and 4 and 8. The resistance across any of these combinations should be 1.5 ohms.

## Ford Solenoids

Ford's E4OD can be shifted without the computer by energizing the #1 and #2 shift solenoids in the correct sequence. Refer to **Figure 26** for the location of the pins in the E4OD harness connector. Connect a 12-volt power source to pin #1, then connect a ground lead to the following pins to shift the transmission: First gear—Pin #3; Second gear—Pin #3 and #2; Third gear—Pin #2; and Fourth gear—no pins grounded. If the transmission does not shift, check the solenoids.

Using an ohmmeter check the resistance of the #1 solenoid by measuring the resistance across pins #1 and #3. To check solenoid #2, measure across pins #1 and 2. And to check solenoid #3, measure the resistance across pins #1 and #4. The resistance across any of these should be 20 to 30 ohms.

To check the solenoids on an AX4S with an ohmmeter, disconnect the harness to the solenoid assembly. Measure

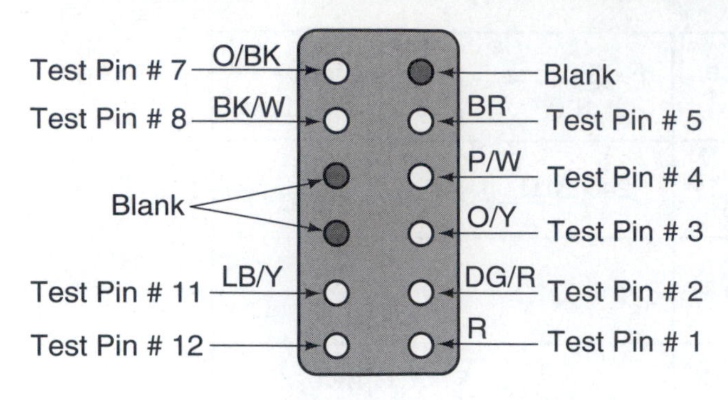


**Figure 26.** The transmission harness connector for an E4OD. The pin numbers are needed for testing of the solenoids.

the resistance across the appropriate pins and compare your readings to specifications. The shift solenoids should have a reading of 12-30 ohms; the EPC solenoid should have a reading of 2.5-6.5 ohms; the modulated converter clutch solenoid should have a reading of 0.75-22.0 ohms; and the regular converter clutch solenoid should have 16–40 ohms across its terminals.

## GM Solenoids

The solenoids of 4T60-E and 4L80-E transmissions can be checked with an ohmmeter. The terminals for these solenoids are accessible at the transmission harness **(Figure 27)**.

These, as are most, solenoids are simple on/off devices. Because of this, they can be checked through normal diagnostic methods. However, some transmissions, such as the GM 4L80-E, controls line pressure according to engine load signals. A simple on/off shift solenoid cannot regulate the flow of oil in a metered manner. Therefore, the 4L80-E, is equipped with a solenoid, which controls line pressure. This solenoid is a three-port, spool valve, electronic pressure regulator that controls pressure based on current flow through its coil winding. The amount of current flow through the solenoid is controlled by the PCM. As current through the pressure control solenoid increases, line pressure decreases.

Solenoids that are controlled by duty cycle or a constantly changing signal are more active than the simple on/off solenoids. Therefore, they are more likely to wear. Many transmission rebuilders recommend that this type solenoid be automatically replaced during a transmission overhaul. Duty cycle solenoids are less prone to clogging because they some are cycled full on every 10 seconds. This full on time helps to keep the regulator valve clean.

The action of a pressure control solenoid can be monitored on a DSO. Connect the DSO directly to the solenoid's circuit. The pattern should be compared to those given in the service manual. A PWM solenoid is used to apply and release the TCC in such a manner that the TCC will apply and release smoothly under all conditions. The PWM

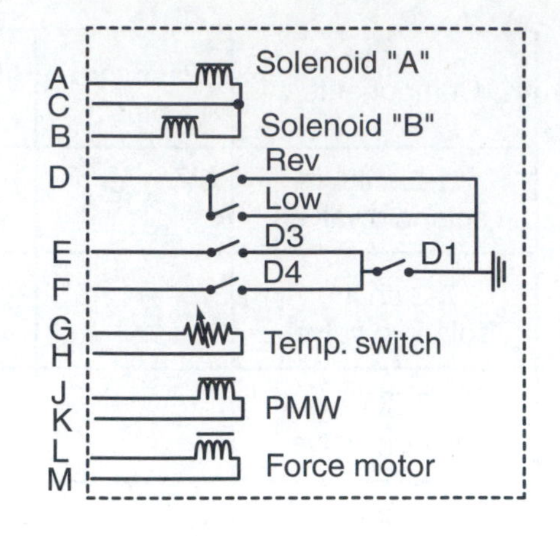


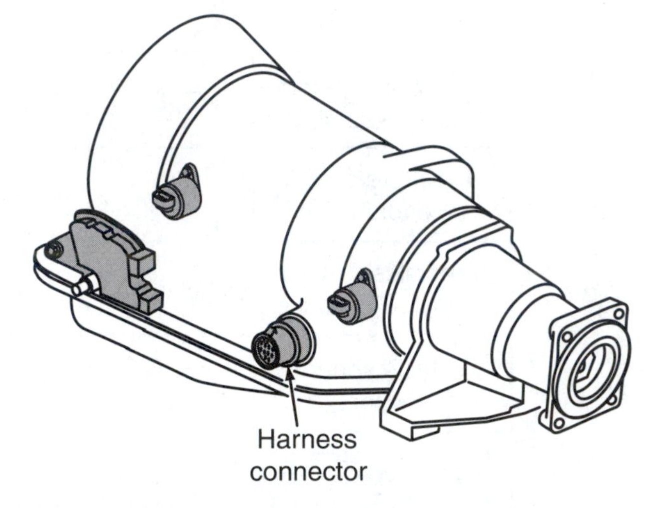


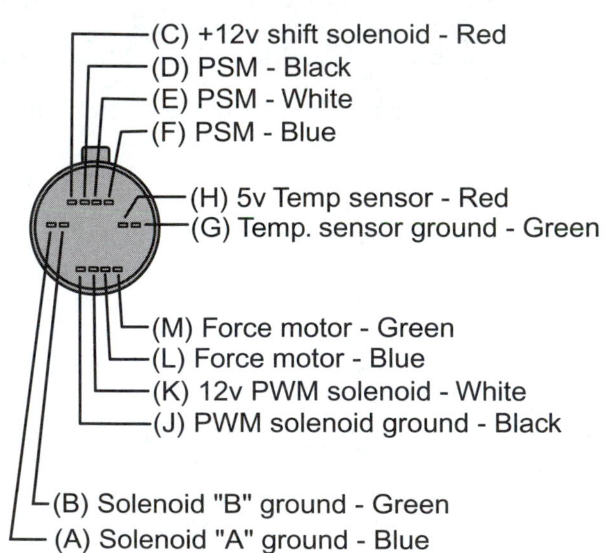


**Figure 27.** Testing of the electrical components in a 4L80-E can be done through the transmission harness connector.

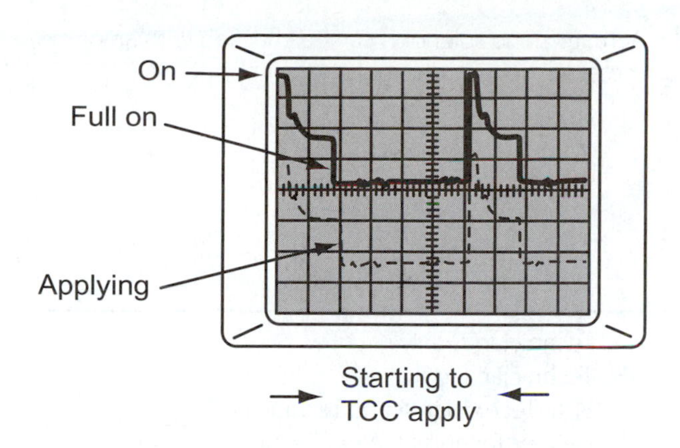


**Figure 28.** A normal scope pattern for a Pulse Width Modulated solenoid.

solenoid in the 4L80-E uses a negative duty cycle and its activity can also be monitored on a scope **(Figure 28)**.

## Honda Solenoids

Honda's shift solenoids are normally closed. When the computer sends battery voltage to them, they open and allow fluid to flow through them. When the fluid flows through the solenoid, it releases pressure that was preventing a shift valve from moving. When the pressure is exhausted, the valve is able to move and the transaxle shifts.

A Honda transmission can be shifted, without the computer, by jumping 12 volts to the shift solenoids. Apply voltage to solenoid B to operate in first gear, apply voltage to both shift solenoids for second gear, to solenoid A for third, and to none of them for fourth gear. If the transaxle shifts normally when the solenoids are energized, the solenoids sensors or the controller are at fault and the mechanics and hydraulics of the transaxle are normal.

To observe the action of the solenoids, connect a pressure gauge to the transaxle taps for the solenoids. Observe the pressure readings when the transaxle upshifts. If the pressure is greater than 3 psi when the solenoid valve is open, the valve has a restriction or the solenoid is not opening fully and not allowing a full release of pressure. If the valving is restricted, the solenoids should be replaced.

> **You Should Know** *There are small filters located in the valve body to protect the solenoids. If these filters are plugged, they will cause a restriction to oil flow. The filters are mounted in a preformed rubber gasket and can be cleaned; however, the rubber gaskets should be replaced on every rebuild.*

Use an ohmmeter to electrically check the solenoids. The resistance across a shift solenoid should be 12–24 ohms and 14–30 ohms across the lock-up solenoid. If the resistance of the solenoids is not within these figures, they should be replaced.

# Summary

- A lab scope is a visual voltmeter that shows voltage changes over time.
- Because a scope displays actual voltage, it will display all electrical noises or disturbances, such as RFI and other glitches that accompany the voltage signal.
- Analog scopes show the actual activity of a circuit and are called real-time scopes. A digital scope, commonly called a Digital Storage Oscilloscope (DSO), converts the voltage signal into digital information and stores it into its memory.
- Any change in the amplitude or height of the trace indicates a change in the voltage.
- On most vehicles, the engagement of the torque converter clutch can be checked during a road test.
- The operation of switches can be checked with an ohmmeter.
- A TP sensor is a potentiometer and sends information to the computer as to what position the throttle is in. If this sensor fails, there may be no upshifts, quick upshifts, delayed shifts, line-pressure problems with transmissions that have a line pressure control solenoid, and/or erratic converter clutch engagement.
- A faulty mass-airflow sensor or circuit may cause the engine to run roughly or stall as soon as you put the transmission into gear.
- Temperature sensors are designed to change resistance with changes in temperature and are critical to proper TCC engagement and shifting.
- Electronically controlled transmissions rely on electrical signals from a speed sensor to control shift timing.
- There are two types of speed sensors, permanent magnetic (PM) generator sensors and reed style sensors.

- The operation of a PM generator can be tested with a DMM.
- A check of the solenoids may lead to identifying if a shifting problem is caused by electronics or the transmission it self.
- Actuators and solenoids can be checked with an ohmmeter, lab scope, scan tool, and voltmeter.

# Review Questions

1. When checking a Transmission Range sensor, Technician A says that typically the switch should be open in all positions except Park and Neutral. Technician B says that a faulty TR sensor may allow the engine to start in other gear positions besides Park and Neutral. Who is correct?
   A. Technician A only
   B. Technician B only
   C. Both Technician A and Technician B
   D. Neither Technician A nor Technician B
2. Technician A says that a faulty TP sensor can cause delayed shifts. Technician B says that delayed shifts can be caused by an open shift solenoid. Who is correct?
   A. Technician A only
   B. Technician B only
   C. Both Technician A and Technician B
   D. Neither Technician A nor Technician B
3. Technician A says that some shift solenoids can be activated by providing a ground for the solenoid. Technician B says that some shift solenoids can be activated by applying hydraulic pressure to their valve. Who is correct?
   A. Technician A only
   B. Technician B only
   C. Both Technician A and Technician B
   D. Neither Technician A nor Technician B
4. A glitch appears in the waveform of a vehicle speed sensor. Which of the following is NOT a probable cause of the problem?
   A. A loose connector
   B. A damaged wire
   C. A poorly mounted sensor
   D. A damaged magnet in the sensor
5. A hot-wire MAF has low resistance across it when the wire is hot. Technician A says that this condition will cause the engine to stall as soon as you put the transmission into gear. Technician B says that this condition will cause the transmission to shift late. Who is correct?
   A. Technician A only
   B. Technician B only
   C. Both Technician A and Technician B
   D. Neither Technician A nor Technician B

Chapter 17

# Torque Converter Diagnosis

## *Introduction*

Many transmission problems are related to the torque converter. Normally, torque converter problems will cause abnormal noises, poor acceleration in all gears, normal acceleration but poor high speed performance, or transmission overheating.

To test the operation of the torque converter, many technicians perform a stall test. The stall test checks the holding capacity of the converter's stator overrunning clutch assembly, as well as the clutches and bands in the transmission.

However, torque converter problems often can be identified by the symptoms and therefore the need for conducting a stall test is minimized. If the vehicle lacks power when it is pulling away from a stop or when passing, it has a restricted exhaust or the torque converter's one-way stator clutch is slipping. To determine which of these problems is causing the power loss, test for a restricted exhaust first.

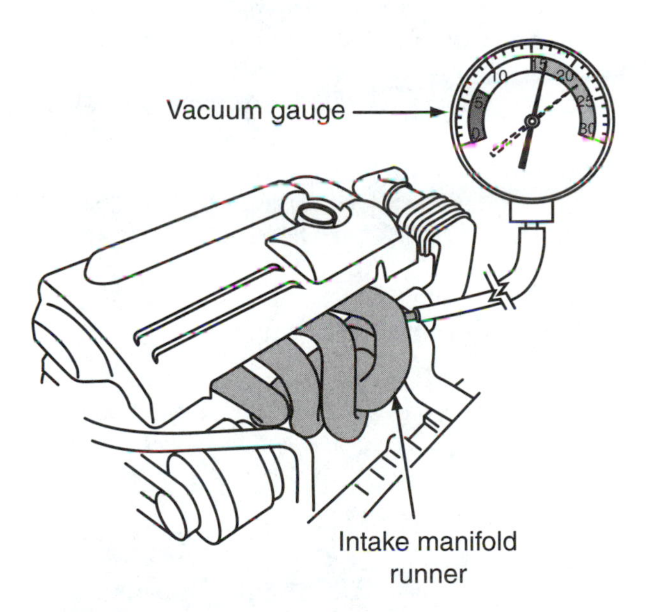


**Figure 1.** To check for an exhaust restriction, connect a vacuum gauge to the engine and observe the readings while the engine speed is raised and quickly lowered.

### EXHAUST RESTRICTION TEST

The easiest way to test for an exhaust restriction is to use a vacuum gauge. Connect the vacuum gauge to a source of engine manifold vacuum. Observe the vacuum reading with the engine at idle. Quickly open the throttle plates and observe the vacuum reading. Then quickly release the throttle plates to allow them to close. The vacuum reading should show an increase of about 5 in. Hg upon initial closing of the throttle plates **(Figure 1)**. If the vacuum did not increase with the closing of the throttle plates, a restricted exhaust is indicated. Common exhaust restrictions are a plugged catalytic converter or a collapsed exhaust pipe.

Another way to check for a restricted exhaust is to insert a pressure gauge in the exhaust manifold's bore for the oxygen sensor. With the gauge in place, start the engine. Bring the engine to 2000 rpm and keep it there while observing the gauge. If the exhaust is not restricted, the pressure reading will be less than 1.25 psi. An extremely restricted exhaust will give a reading of over 2.75 psi. If there is no evidence of a restricted exhaust, it can be assumed that the torque converter's stator clutch is slipping and not allowing any torque multiplication to take place in the converter. To repair this problem, the torque converter should be replaced.

If the engine's speed flares up when it is accelerated in DRIVE and does not have normal acceleration, the clutches or bands in the transmission are slipping. This symptom is similar to the slipping of a clutch in a manual transmission. This problem often is mistakenly blamed on the torque converter.

Technicians often blame the torque converter for problems simply based on the customer's complaint. Complaints of thumping or grinding noises are often blamed on the converter when they are really caused bad thrust washers or damaged gears and bearings in the transmission. This type noise also can be caused by nontransmission components such as bad CV joints and wheel bearings.

Also, many engine problems can cause a vehicle to act as if it has a torque converter problem. This is especially true of converter clutches, which may engage early or not at all. Engine problems can cause a vehicle to behave as if the torque converter is malfunctioning, or actually cause the converter clutch to lock up early or not at all. Clogged fuel injectors or bad spark-plug wires can seem like torque-converter complaints. Bad vacuum lines, EGR valves, or engine speed sensors can prevent the clutch from locking up at the proper time.

## STALL TESTING

Two methods are commonly used to check the operation of the stator's one-way clutch: the stall test and bench testing. To bench test a converter, it must be removed from the vehicle, and there is no need to do that unless we know the converter is at fault.

You Should Know *Stall testing is not recommended on many late-model transmissions. This test places extreme stress on the transmission and should only be conducted if recommended by the manufacturer.*

To conduct a stall test, connect a tachometer to the engine and position it so that it can be easily read from the driver's seat. Set the parking brake, raise the hood, and place blocks in front of the vehicle's nondriving tires **(Figure 2)**. Conduct the test outdoors if possible, especially if it is a cold day. If the test is conducted indoors, place a large fan in front of the vehicle to keep the engine cool. With the engine running, press and hold the brake pedal. Then move the gear selector to the DRIVE position and press the throttle pedal to the floor. Hold the throttle down for two seconds, then note the tachometer reading and immediately let off the throttle pedal and allow the engine to idle. Compare the measured stall speed to specifications **(Figure 3)**.

If the torque converter and transmission are functioning properly, the engine will reach a specific speed. If the tachometer indicates a speed above or below specifications, a possible problem exists in the transmission or torque converter. If a torque converter is suspected as being faulty, it should be removed and the one-way clutch checked on the bench.

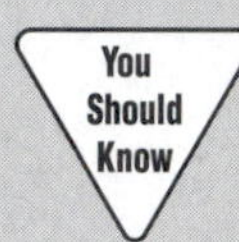

*If a stall test is not correctly conducted, the converter and/or transmission can be damaged.*

To prevent serious damage to the transmission, follow these guidelines when conducting a stall test:

1. Never conduct a stall test if there is an engine problem.
2. Check the fluid levels in the engine and transmission before conducting the test.
3. Engine should be at normal operating temperature during the test.
4. Never hold the throttle wide open for more than 5 seconds during the test.
5. Do not perform the test in more than two gear ranges without driving the vehicle a few miles to allow the engine and transmission to cool down.
6. After the test, allow the engine to idle for a few minutes to cool the transmission fluid before shutting off the ignition.

If the stall speed is below the specifications, a restricted exhaust or slipping stator clutch is indicated. If the stator's one-way clutch is not holding, ATF leaving the turbine of the converter works against the rotation of the impeller and slows down the engine. With both of these problems, the vehicle would exhibit poor acceleration. This is caused by a lack of power from the engine or no torque multiplication occurring in the converter. If the stall speed is only slightly below normal, the engine is probably not producing enough power and should be diagnosed and repaired.

If the stall speed is above specifications, the bands or clutches in the transmission may be slipping and not holding properly.

If the vehicle has poor acceleration but had good results from the stall test, suspect a seized one-way clutch. Excessively hot ATF in the transmission is a good indication that the clutch is seized. However, other problems can cause these same symptoms; therefore, be careful during your diagnosis.

A normal stall test will generate a lot of noise, most of which is normal. However, if you hear any metallic noises during the test, diagnose the source of these noises. On

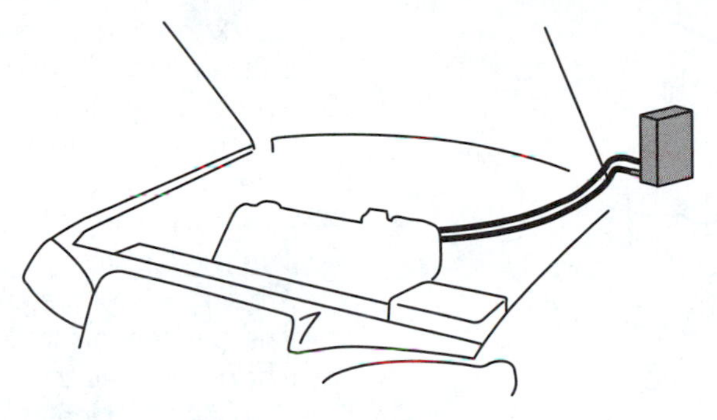

| Trouble | Probable cause |
|---|---|
| Stall rpm high in $D_4$, 2, 1 & R | Low fluid level or oil pump output<br>Clogged oil strainer<br>Pressure regulator valve stuck closed<br>Slipping clutch |
| Stall rpm high in R | Slippage of 4th clutch |
| Stall rpm high in 2 | Slippage of 2nd clutch |
| Stall rpm high in $D_4$ | Slippage of 1st clutch or 1st gear one-way clutch |
| Stall rpm low in $D_4$, 2, 1 & R | Engine output low<br>Torque converter one-way clutch slipping |

**Figure 2.** Before conducting a stall test, chock the wheels and place the tachometer in a position where it can be easily seen from the driver's seat.

**TORQUEFLITE TRANSMISSION STALL SPEED CHART**

| Engine liter | Transaxle type | Converter diameter | Stall rpm |
|---|---|---|---|
| 1.7 | A-404 | 9-1/2 inches (241 millimeters) | 2300-2500 |
| 2.2 | A-413 | 9-1/2 inches (241 millimeters) | 2200-2410 |
| 2.6 | A-470 | 9-1/2 inches (241 millimeters) | 2400-2630 |

**Figure 3.** A typical torque converter stall speed chart.

some vehicles, the vehicle can be run at low speeds on a hoist with the drive wheels free to rotate. If the noises are still present, the source of the noise is probable the torque converter. Before doing this, however, check the service manual. Some transaxles can be severely damaged if run with the wheels off the ground.

## Visual Inspection

If there was a noise coming from the torque converter during the stall test, visually inspect the converter before pulling the transmission out and testing or replacing the converter. Remove the torque converter access cover on the transmission **(Figures 4 and 5)** and rotate the engine with a remote starter button. While the torque converter is rotating with the engine, check to make sure the torque converter bolts are not loose and are not contacting the bell housing. Also, observe the action of the converter as it is spinning. If the converter wobbles, it may be because of a damaged flexplate or converter.

Check the converter's balance weights to make sure they are still firmly attached to the unit. Carefully inspect the flexplate for evidence of cracking or other damage. Also check the condition of the starter ring gear, the teeth of which should not be damaged. The gear should be firmly attached to the flexplate.

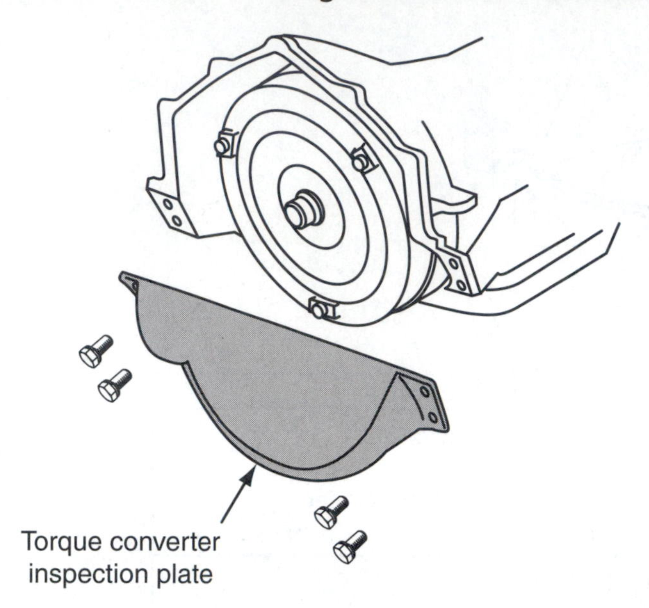


**Figure 4.** To inspect a torque converter when it is still in the vehicle, remove the access cover.

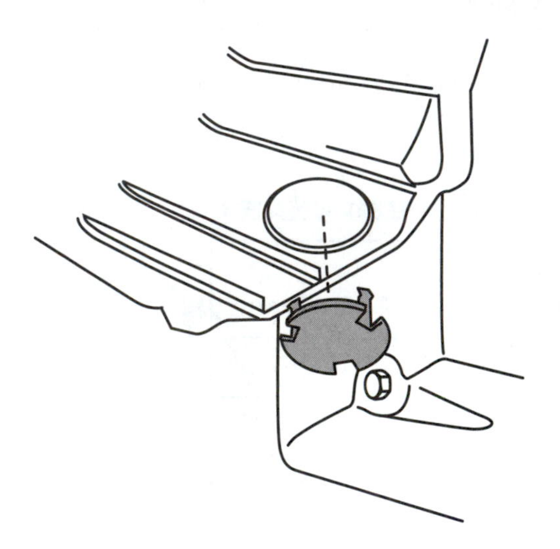

**Figure 5.** A torque converter access plug on a typical GM transmission.

*You Should Know* *Before rotating the engine with a remote starter button, make sure the engine's ignition or fuel injection system is disabled. Also, keep all parts of your body away from the rotating flexplate. Serious injury can result from being careless during this check.*

Check the torque converter for **ballooning**. If excessive pressure was able to build up inside the converter, the converter will expand, or balloon. Ballooning looks like the torque converter has been blown up like a balloon. A stuck converter check valve typically causes this. If the converter is ballooned, it should be replaced and the cause of the problem also repaired.

A ballooned torque converter also will damage other driveline components. If the converter is ballooned toward the rear of the unit, the transmission's oil pump is most likely bad. If it is ballooned toward the front, the crankshaft's thrust bearings are undoubtedly worn or damaged.

## TESTING CONVERTER CLUTCHES

Late-model transmissions are equipped with a torque converter clutch. Most converter clutches are controlled by the powertrain control module **(Figure 6)**. The computer turns on the converter clutch solenoid, which opens a valve and allows fluid pressure to engage the clutch. When the computer turns the solenoid off, the clutch disengages.

One of the trickiest parts of diagnosing a converter clutch problem is recognizing a normal acting converter clutch, as well as an abnormal acting clutch. You should pay attention to the action of all converter clutches, whether or not they are suspected of having a problem. By knowing what a normal clutch feels like, it is easier to feel abnormal clutch activity. A malfunctioning converter clutch can cause a wide variety of driveability problems. Normally, the application of the clutch should feel like a smooth engagement into another gear. It should not feel harsh, nor should there be any noises related to the application of the clutch.

To properly diagnose converter clutch problems, you must know when they should engage and disengage and understand the function of the various controls involved with the system. Although the actual controls for a converter clutch vary with the different manufacturers and models of transmissions, they all will have certain operating conditions that must be met before the clutch can be engaged.

Care should be taken during diagnostics because abnormal clutch action can be caused by engine, electrical, clutch, or torque converter problems.

Before the converter clutch is applied, the vehicle must be traveling at or above a certain speed. The vehicle speed sensor sends this speed information to the computer. Also, the converter clutch will not engage when the engine is cold; therefore, a coolant temperature (ECT) sensor provides the computer with information regarding engine temperature. During sudden deceleration or acceleration, the clutch should be disengaged. One of the sensors used to tell the computer when these driving modes are present is the TP sensor. Some transmissions use a third or fourth gear switch to inform the computer when the transmission is in those gears. The computer will then allow clutch

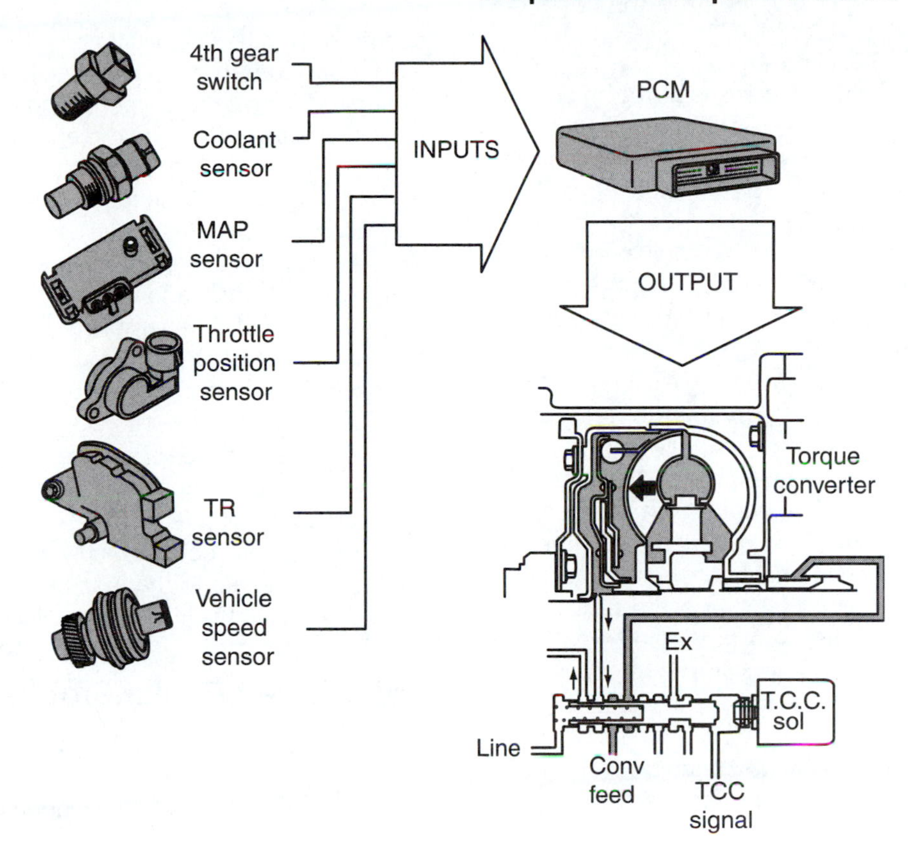


Figure 6. Typical electrical control circuitry for a lockup torque converter.

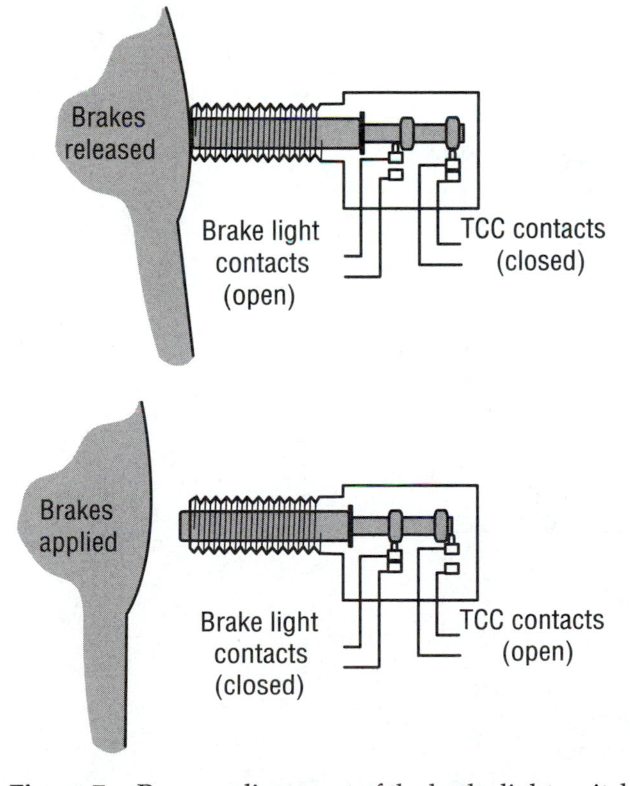


Figure 7. Proper adjustment of the brake light switch is essential for proper operation of a torque converter clutch.

engagement if other operating conditions are satisfactory. A brake switch is also used in some clutch circuits to disengage the clutch when the brakes are applied **(Figure 7)**. These key sensors, the VSS, ECT, TP, third/fourth gear switch, and brake switch, should be visually checked as part of your diagnosis of converter problems.

Diagnosis of a converter clutch circuit should be conducted in the same way as any other computer system. The computer will recognize problems within the system and store trouble codes that reflect the problem area of the circuit. The codes can be retrieved and displayed by an instrument panel light or a hand-held scan tool **(Figure 8)**.

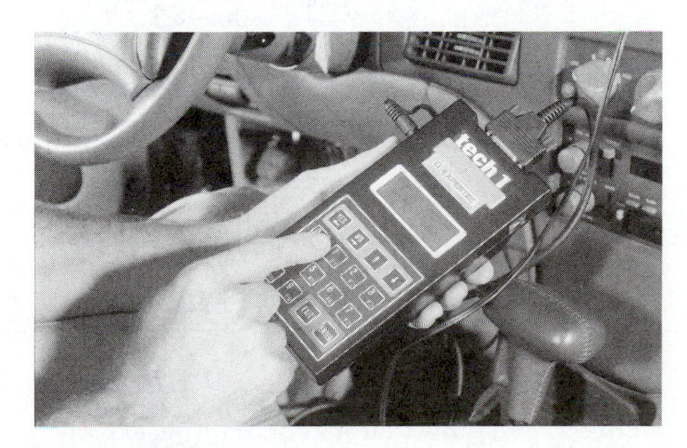


Figure 8. A scan tool.

## Engagement Quality

The engagement of the clutch should be smooth. If the clutch prematurely engages or is not being applied by full pressure, a shudder or vibration results from the rapid grabbing and slipping of the clutch. The clutch begins to engage, then slips, because it cannot hold the engine's torque and become fully engaged. The torque capacity of the clutch is determined by the oil pressure applied to the clutch and the condition of the frictional surfaces of the clutch assembly **(Figure 9)**.

If the shudder, sometimes called a stick-slip condition, is noticeable only during the engagement of the clutch, the problem is typically in the converter. When the shudder is only evident after the engagement of the clutch, the cause of the shudder is the engine, transmission, or another component of the driveline. Often the problems will not only affect TCC engagement but also will affect the operation of the transmission. Examples of these problems are worn valves or valve bores in the valve body, worn shafts or bearings, and worn sealing rings. If the shudder is caused by the clutch, the converter must be replaced to correct the problem.

When clutch apply pressure is low and the clutch cannot firmly engage fully, shudder will occur. A faulty clutch solenoid or its return spring may cause this. The valve controlled by the solenoid is normally held in position by a coil-type return spring. If the spring loses tension, the clutch will be able to prematurely engage. Because insufficient pressure is available to hold the clutch, shudder occurs as the clutch begins to grab and then slips. If the solenoid valve and/or return spring are faulty, they should be replaced, as should the torque converter.

An out-of-round torque converter prevents full clutch engagement, which also will cause shudder, as will broken or worn clutch dampener springs and contaminated clutch frictional material. The frictional material can become contaminated by metal particles circulating through the torque converter and collecting on the clutch.

In addition to replacing the torque converter and/or replacing other components of the system, one possible correction for shudder is the use of an ATF with friction modifiers. Some rebuilders may recommend that an oil additive be added to the ATF. The additive is designed to improve or alter the friction capabilities of regular ATF.

## TC-Related Cooler Problems

Vehicles equipped with a converter clutch may stall when the transmission is shifted into reverse gear. The cause of this problem may be plugged transmission cooler lines or cooler itself may be plugged. Fluid normally flows

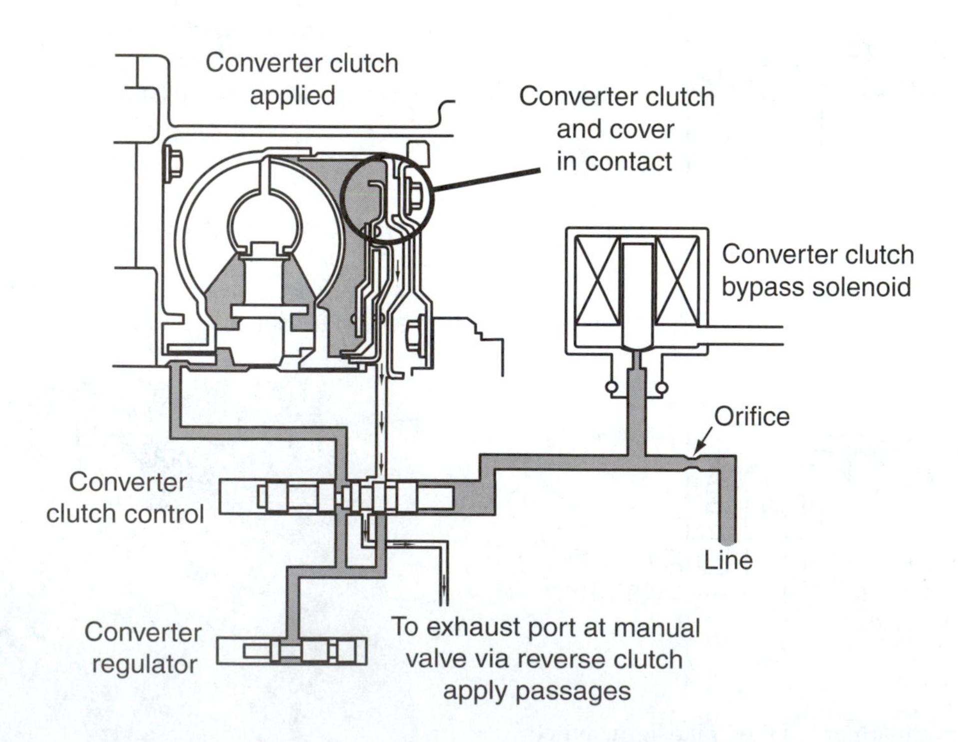


**Figure 9.** Fluid flow through the clutch circuit when the clutch is applied.

from the torque converter through the transmission cooler. If the cooler passages are blocked, fluid is unable to exhaust from the torque converter and the converter clutch piston remains engaged. When the clutch is engaged, there is no vortex flow in the converter and therefore little torque multiplication is taking place in the converter.

To verify that the transmission cooler is plugged, disconnect the cooler return line from the radiator or cooler. Connect a short piece of hose to the outlet of the cooler and allow the other end of the hose to rest inside an empty container. Start the engine and measure the amount of fluid that flows into the container after 20 seconds. Normally, one quart of fluid should flow into the container. If less than that filled the container, a plugged cooler is indicated.

To correct a plugged transmission cooler, disconnect the cooler lines at the transmission and the radiator. Blow air through the cooler, one end at a time, then through the cooler lines. The air will clear large pieces of debris from the transmission cooler. Always use low air pressure, no more than 50 psi. Higher pressures may damage the cooler. If there is little air flow through the cooler, the radiator or external cooler must be removed and flushed or replaced.

## GENERAL CONVERTER CONTROL DIAGNOSTICS

All testing of TCC controls should begin with a basic inspection of the engine and transmission. Too often technicians skip this basic inspection and become frustrated during diagnostics because of conflicting test results. Apparent transmission and torque converter problems are often caused by engine mechanical problems, broken or incorrectly connected vacuum hoses, incorrect engine timing, or incorrect idle speed adjustments. The basic inspection should include the following:

1. A road test to verify the complaint and further define the problem.
2. Careful inspection of the engine and transmission.
3. Check the PCM for codes. Then do a check of the mechanical condition of the engine, the output of ignition system, and the efficiency of the fuel system.
4. An idle speed and ignition timing check. If the timing is nonadjustable, check the operation of the electronic spark control system.
5. A check the entire intake system for vacuum leaks.

When inspecting wires and hoses, look for burnt spots, bare wires, and damaged or pinched wires. Make sure the harness to the electronic control unit has a tight and clean connection. Also, check the source voltage at the battery before beginning any detailed tests on an electronic control system. If the voltage is too low or too high, the electronic system cannot function properly.

On early TCC equipped vehicles, clutch engagement was controlled hydraulically. A switch valve was controlled by two other valves, the lockup and fail-safe valves, in the clutch control assembly **(Figure 10)**. The lockup valve responds to governor pressure and prevents lockup at speeds below 40 mph. The fail-safe valve responds to throttle pressure and permitted clutch engagement in high gear only. Problems with this system are diagnosed in the same way as other hydraulic circuits.

The clutch in most hydraulic TCC systems is applied when oil flow through the torque converter is reversed. This

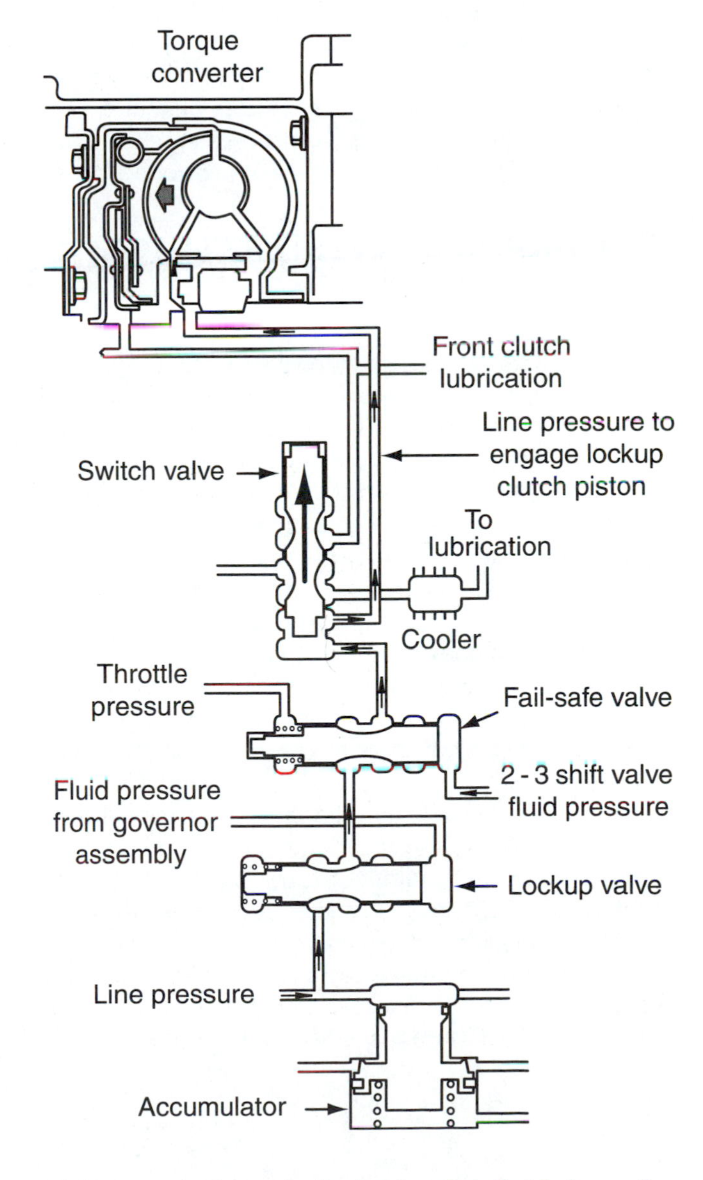


**Figure 10.** Action of the hydraulic fluid from the lockup valve to the torque converter.

change can be observed with a pressure gauge. Using the pressure gauge is also a good way to diagnose the clutch's hydraulic system. Connect a pressure gauge to the hydraulic line, with a "tee" fitting, from the transmission to the cooler. Position the gauge so that is easily seen from the driver's seat. Then raise the vehicle on a hoist with the drive wheels off the ground and able to spin freely. Operate the vehicle until the transmission shifts into high gear. Then maintain a speed of approximately 55 mph. Once the speed is maintained, watch the pressure gauge.

If the pressure decreases 5 to 10 psi, the converter clutch was applied. With this action, you should feel the engagement of the clutch, as well as a drop in engine speed. If the pressure changed but the clutch did not engage, the problem may be inside the converter or at the end of the input shaft. If the input shaft end is worn or the O-ring at the end is cut or worn, there will be a pressure loss at the converter clutch. This loss in pressure will prevent full engagement of the clutch. If the pressure did not change and the clutch did not engage, suspect a faulty clutch valve or control solenoid, or a fault in the solenoid control circuit.

## ELECTRONIC CONVERTER CLUTCH CONTROL DIAGNOSTICS

Each automobile manufacturer has specific test sequences that should be followed to test the system. However, there are certain checks that can be made to all systems. The first check is with the scan tool. After entering the appropriate information, retrieve the DTCs. If there are no trouble codes, check to make sure the converter clutch circuit has power. Check the circuit's fuse; make sure it is not blown. Then visually inspect all of the wires and connectors in the circuit. Make sure they are not loose, disconnected, or corroded.

If the clutch does not engage, check for power to the solenoid. If power is available, make sure the ground of the circuit is good. If there is power available and the ground is good, check the voltage drop across the solenoid. The solenoids should drop very close to source voltage. If less than that is measured, check the voltage drop across the power and ground sides of the circuit. If the voltage drop testing results in good results, remove the solenoid and test it with an ohmmeter. If the solenoid checks out fine with the ohmmeter, suspect clutch material, dirt, or other material plugging up the solenoid valve passages. If blockage is found, attempt to flush the valve with clean ATF. If the solenoid has a filter assembly **(Figure 11)**, replace the filter after cleaning the fluid passages. If the blockage cannot be removed, replace the solenoid.

If the clutch engages at the wrong time, a sensor or switch in the circuit is probably the cause. If clutch engagement occurs at the wrong speed, check all speed related sensors. A faulty temperature sensor may cause the clutch not to engage. If the sensor is not reading the correct temperature, the PCM may never realize the temperature is suitable for engagement. Checking the appropriate sensors can be done with a scan tool, DMM, and/or lab scope. A check of the sensors is normally part of the manufacturer's system check.

The method used to diagnose computer-controlled TCC systems varies with each manufacturer **(Figure 12)**. Although many of the procedural steps are quite different, all basically follow the same scheme and the same routine as followed for diagnosing electronically controlled transmissions.

When you have an intermittent problem, using soft codes may allow you to isolate the faulty circuit. The customer's explanation of the problem is also of great help during diagnostics of these problems. Finding out when and where the problem occurs can lead you to properly identifying the problem area. This information also will give you the operating conditions in which the problem exists, allowing you to duplicate the conditions to verify the complaint.

To verify that a component is faulty, test it with the appropriate meter. All faulty wires and connectors should be repaired or replaced. TCC solenoids are not rebuildable and are replaced when they are faulty.

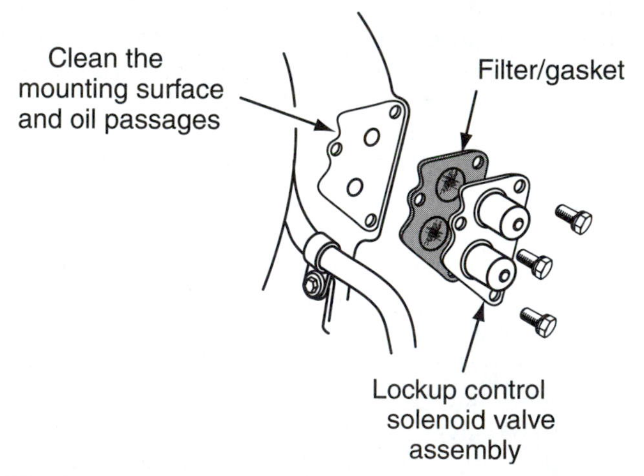


**Figure 11.** A torque converter clutch solenoid with a replaceable filter.

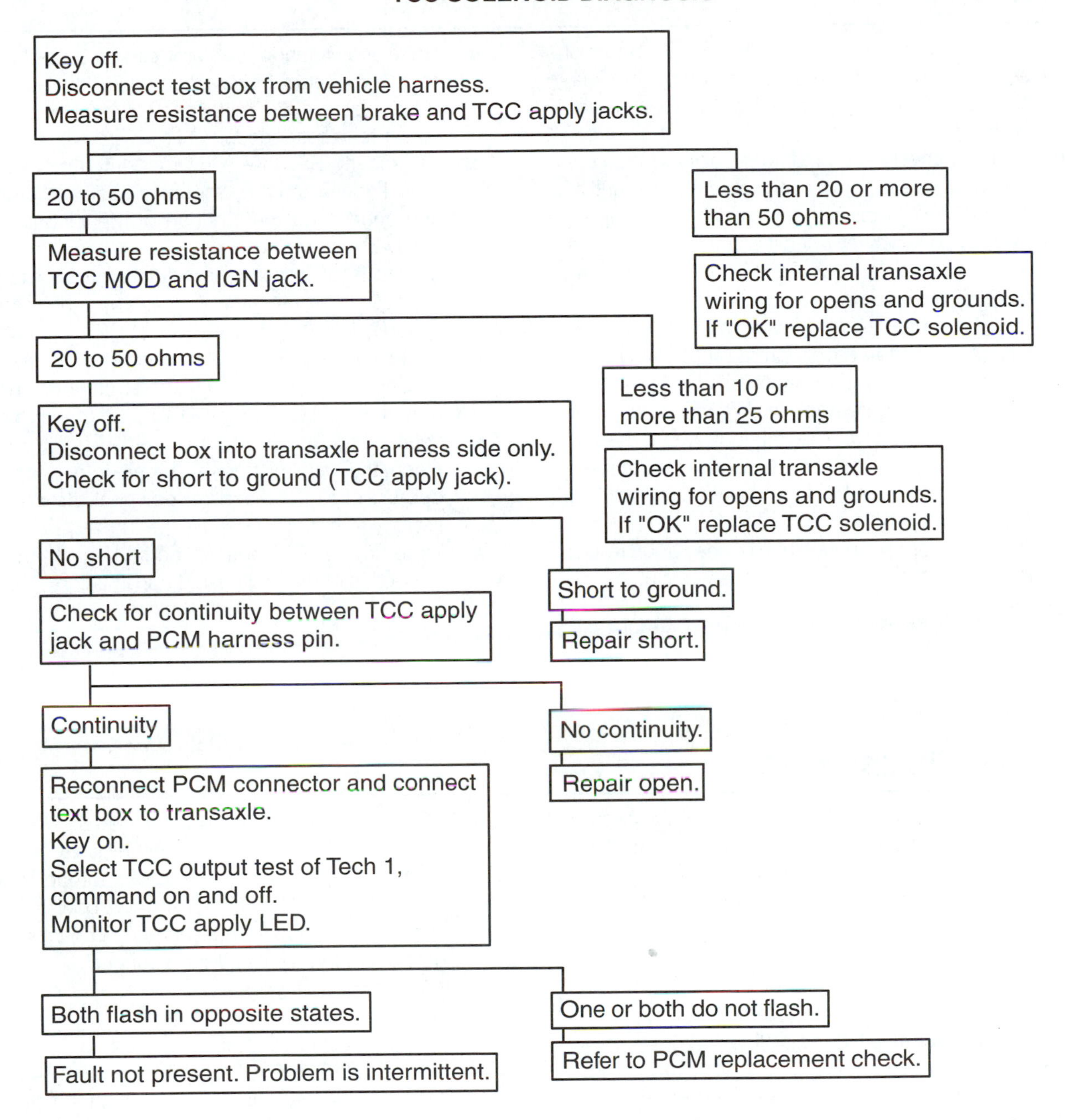


**Figure 12.** A typical TCC solenoid diagnostic chart. (Courtesy of General Motors Corporation, Service Operations)

## Summary


- Torque converter problems will cause abnormal noises, poor acceleration in all gears, normal acceleration but poor high speed performance, or transmission overheating.
- The stall test checks the holding capacity of the converter's stator overrunning clutch assembly, as well as the clutches and bands in the transmission.
- If the vehicle lacks power when it is pulling away from a stop or when passing, it has a restricted exhaust or the torque converter's one-way stator clutch is slipping.
- If the engine's speed flares up when it is accelerated in DRIVE and does not have normal acceleration, the clutches or bands in the transmission are slipping.
- If the stall speed is below the specifications, a restricted exhaust or slipping stator clutch is indicated. If the stall speed is above specifications, the bands or clutches in the transmission may be slipping and not holding properly.
- If the vehicle has poor acceleration but had good results from the stall test, suspect a seized one-way clutch.
- Noise and vibration problems can be caused by loose torque converter bolts, a damaged flexplate or converter, loose or missing balance weights, or the torque converter is ballooned.
- To properly diagnose converter clutch problems, you must know when they should engage and disengage and understand the function of the various controls involved with the system.
- Diagnosis of a converter clutch circuit should be conducted in the same way as any other computer system.
- If the clutch prematurely engages or is not being applied by full pressure, a shudder or vibration results from the rapid grabbing and slipping of the clutch.
- Vehicles equipped with a converter clutch may stall when the transmission is shifted into reverse gear because of plugged transmission cooler lines or cooler itself.
- All testing of TCC controls should begin with a basic inspection of the engine and transmission.


## Review Questions

1. Technician A says that a ballooned torque converter can cause damage to the oil pump. Technician B says that a ballooned torque converter is caused by excessive pressure in the torque converter. Who is correct?
   A. Technician A only
   B. Technician B only
   C. Both Technician A and Technician B
   D. Neither Technician A nor Technician B
2. Technician A says that if the stall speed of a torque converter is below specifications, a restricted exhaust or slipping stator clutch is indicated. Technician B says that if the stall speed is above specifications, the bands or clutches in the transmission may not be holding properly. Who is correct?
   A. Technician A only
   B. Technician B only
   C. Both Technician A and Technician B
   D. Neither Technician A nor Technician B
3. Technician A says that torque converter clutch control problems are always caused by electrical malfunctions. Technician B says that nearly all converter clutches are engaged through the application of hydraulic pressure on the clutch. Who is correct?
   A. Technician A only
   B. Technician B only
   C. Both Technician A and Technician B
   D. Neither Technician A nor Technician B
4. When checking the stall speed of a torque converter, Technician A maintains full-throttle power until the engine stalls. Technician B manually engages the TCC before running the test. Who is correct?
   A. Technician A only
   B. Technician B only
   C. Both Technician A and Technician B
   D. Neither Technician A nor Technician B
5. Technician A says that a lower than specified stall speed may be caused by a faulty stator one-way clutch. Technician B says that a higher than specified stall speed may be caused by faulty clutch packs. Who is correct?
   A. Technician A only
   B. Technician B only
   C. Both Technician A and Technician B
   D. Neither Technician A nor Technician B
6. Technician A says that a shudder after the converter clutch engages could be caused by a damaged or missing clutch check ball. Technician B says that driveline or converter shudder can be isolated by disconnecting the converter's clutch solenoid. Who is correct?

A. Technician A only
B. Technician B only
C. Both Technician A and Technician B
D. Neither Technician A nor Technician B

7. A customer complains that the engine seems to surge when driving at about 48 mph. Technician A says that there may be a problem with the torque converter clutch. Technician B says that a faulty vehicle speed sensor may cause this problem. Who is correct?
   A. Technician A only
   B. Technician B only
   C. Both Technician A and Technician B
   D. Neither Technician A nor Technician B

8. Technician A says that a plugged transmission cooler or cooler lines may cause the vehicle to stall when the transmission is shifted into a forward gear. Technician B says that a plugged transmission cooler or cooler lines may overheat the converter. Who is correct?
   A. Technician A only
   B. Technician B only
   C. Both Technician A and Technician B
   D. Neither Technician A nor Technician B

# Chapter 18 In-Vehicle Services

## Introduction

Transmissions are strong and trouble-free units that require little maintenance. Maintenance normally only includes periodic fluid and filter changes. Also, when there is a transmission problem, few service operations can be done without removing the unit from the vehicle. Those services and repairs, including changing the fluid and filter, are covered in this chapter. Linkage adjustments and other adjustments also can be made with the transmission in the vehicle. These adjustments are covered in the next chapter.

### FLUID CHANGES

The transmission's fluid and filter should be changed whenever there is an indication of oxidation or contamination. Periodic fluid and filter changes also are part of the preventative maintenance program for most vehicles. The frequency of this service depends on the conditions under which the transmission normally operates. Severe usage requires that the fluid and filter be changed every 15,000 miles. Severe usage is defined as:

a) More than 50 percent operation in heavy city traffic during hot weather (above 90°F).
b) Police, taxi, commercial-type operation, and trailer towing.

The mileage interval that a manufacturer will recommend for a fluid and filter change also will depend on the type of transmission. For example, some General Motors transmissions use aluminum valves in their valve body. Because aluminum is softer than steel, aluminum valves are less tolerant of abrasives and dirt in the fluid; therefore, to prolong the life of the valves and their bores, more frequent fluid changes are recommended by the GM.

**You Should Know** *Older transmissions did not require as frequent fluid and filter changes as do many later model vehicles. Customers should be made aware of the recommended frequency and the reason for this change. In older vehicles, both the engine and transmission ran cooler, which extended the life of transmission fluid. These transmissions operated at 175°F and the transmission fluid lasted about 100,000 miles before oxidizing. However, fluid life is halved with every 20° that its temperature rises above 175°. For example, fluid life expectancy drops to 50,000 miles at 195°F and just 25,000 miles at 215°F, which is the temperature at which most new transmissions run.*

Change the fluid only when the engine and transmission are at normal operating temperatures. On most transmissions, you must raise the vehicle and remove the oil pan to drain the fluid. Before removing the pan, remove any part that may interfere with the removal of the pan **(Figure 1)**. Some transmission pans on recent vehicles include a drain plug. A filter or screen is normally attached to the bottom of the valve body. Filters are made of paper or fabric and held in place by screws, clips, or bolts **(Figure 2)**. Filters should be replaced, not cleaned.

To drain and refill a typical transmission, the vehicle must be raised on a hoist. After the vehicle is safely in

**Figure 1.** Remove any parts that may interfere with the removal of the oil pan before attempting to remove it.

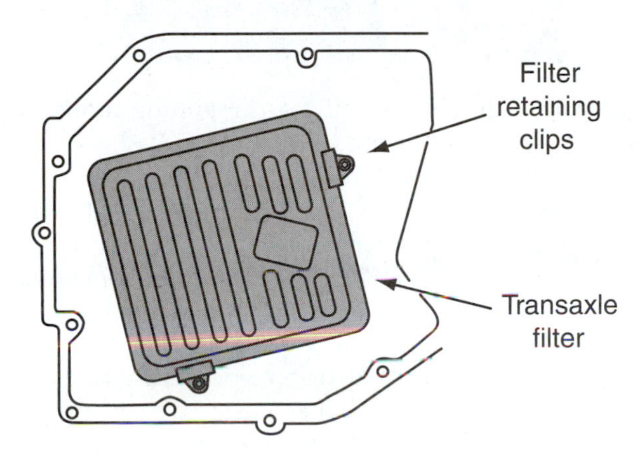


**Figure 2.** Transmission fluid filters are attached to the transmission case by screws, bolts, and/or retaining clips.

position, place a drain pan with a large opening under the transmission's oil pan **(Figure 3)**. Then loosen the pan bolts and remove all but three at one end of the pan. This will allow the fluid to flow out of the pan into the drain pan. It may be necessary to tap the pan at one corner to break it loose. Fluid will begin to drain from around the pan.

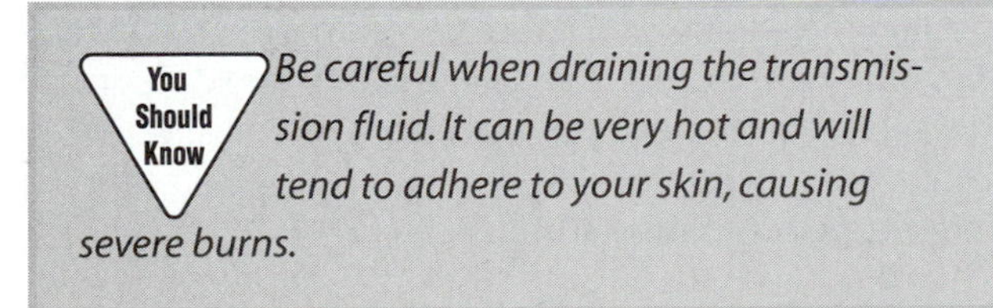


*Be careful when draining the transmission fluid. It can be very hot and will tend to adhere to your skin, causing severe burns.*

After draining, carefully remove the pan. There will be some fluid left in the pan, be prepared to dump it into the drain pan. Check the bottom of the pan for deposits and metal particles. Slight contamination—blackish deposits

**Figure 3.** Position a large drain pan under the transmission's oil pan.

**Figure 4.** Remove the filter and carefully inspect the residue in the filter.

from clutches and bands—is normal. Other contaminants should be of concern. Clean the oil pan and its magnet.

Remove the filter **(Figure 4)**, cut it open, and inspect it. Use a magnet to determine if metal particles are steel or aluminum. Steel particles indicate severe internal transmission wear or damage. If the metal particles are aluminum, they may be part of the torque-converter stator. Some torque converters use phenolic plastic stators; therefore, metal particles found in these transmissions must be from the transmission itself. Filters are always replaced, whereas screens are cleaned. Screens are removed in the same way as filters. Clean a screen with fresh solvent and a stiff brush.

Remove any traces of the old pan gasket on the case housing. Then, install a new filter and gasket on the bottom of the valve body and tighten the retaining bolts to the specified torque. If the filter is sealed with an O-ring, make sure it is properly installed.

Remove any traces of the old pan gasket from the oil pan and transmission housing. Make sure the mounting flange of the oil pan is not distorted and bent. Oil pans are typically made of stamped steel **(Figure 5)**. The thin steel tends to become distorted around the attaching boltholes. These distortions can prevent the pan from fitting tightly against the transmission case. To flatten the pan, place the

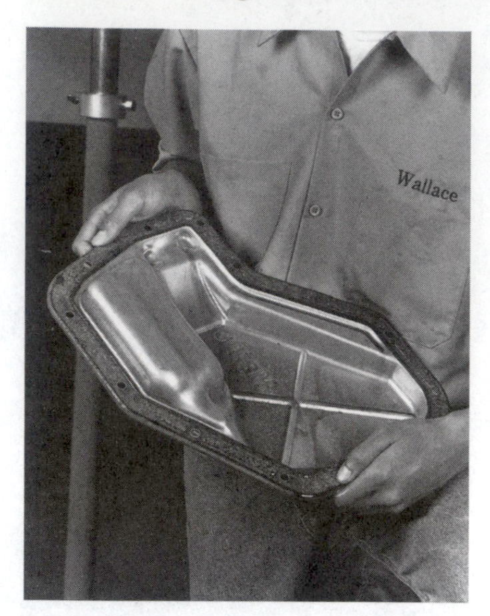


Figure 5. A typical oil pan.

Figure 6. Tighten the oil pan retaining bolts to specifications.

mounting flange of the pan on a block of wood and flatten one area at a time with a ball peen hammer.

Verify that the new gasket is correct by comparing the old with the new. Then, reinstall the pan using the gasket or sealant recommended by the manufacturer. Tighten the pan retaining bolts to the specified torque **(Figure 6)**; often this specification is given in inch-pounds rather than foot-pounds.

Transmission gaskets should not be installed with any type of liquid adhesive or sealant, unless specifically noted by the manufacturer. If any sealer gets into the valve body, severe damage can result. Also, sealant can clog the oil filter. If a gasket is difficult to install, a thin coating of transmission assembly lube or petroleum jelly can be used to hold the gasket in place when assembling the parts.

One type of gasket that presents unique installation problems is the cork-type gasket. These gaskets tend to change shape and size with changes in humidity. If a cork gasket is slightly larger than it should be, soak it in water and lay it flat on a warm surface. Allow it to dry before installing it. If the gasket is slightly smaller than required, soak it in warm water prior to installation. Another way to make a cork gasket grow is to lay in on a flat clean and hard surface. Then, with a hammer, strike it all the way around until it is the correct size.

Whenever you are using a cork gasket to create a seal between two parts, make sure you properly tighten the two surfaces together. Tighten the attaching bolts or nuts, in a staggered pattern, to the specified torque so that the gasket material is evenly squeezed between the two surfaces. If too much torque is applied, the gasket may split.

With the pan securely in place, lower the vehicle. Now, pour a little less than the required amount of fluid into the transmission through the dipstick tube. Always use the recommended type of ATF. The wrong fluid will alter the shifting characteristics of the transmission. For example, if type F fluid is used in a transmission designed for Dexron type fluid, the shifting will be harsher.

Start the engine and allow it to idle for at least one minute. Then, with the parking and service brakes applied, move the gear selector lever momentarily to each position, ending in the park. Recheck the fluid level and add a sufficient amount of fluid to bring the level to about one-eighth inch below the ADD mark.

**You Should Know** *On transmissions with a governor, after starting the engine, quickly move the gear selector to reverse. Many transmissions do not feed fluid to the governor when they are in reverse. By quickly placing the transmission in reverse, the chance of dirt entering the governor is greatly decreased.*

Run the engine until it reaches normal operating temperature. Then recheck the fluid level, it should be in the HOT region on the dipstick. Make sure the dipstick is fully seated into the dipstick tube opening. This will prevent dirt from entering into the transmission.

## Parking Pawl

Anytime you have the oil pan off, you should inspect the transmission parts that are exposed. This is especially true of the parking pawl assembly **(Figure 7)**. This component is typically not hydraulically activated; rather, the

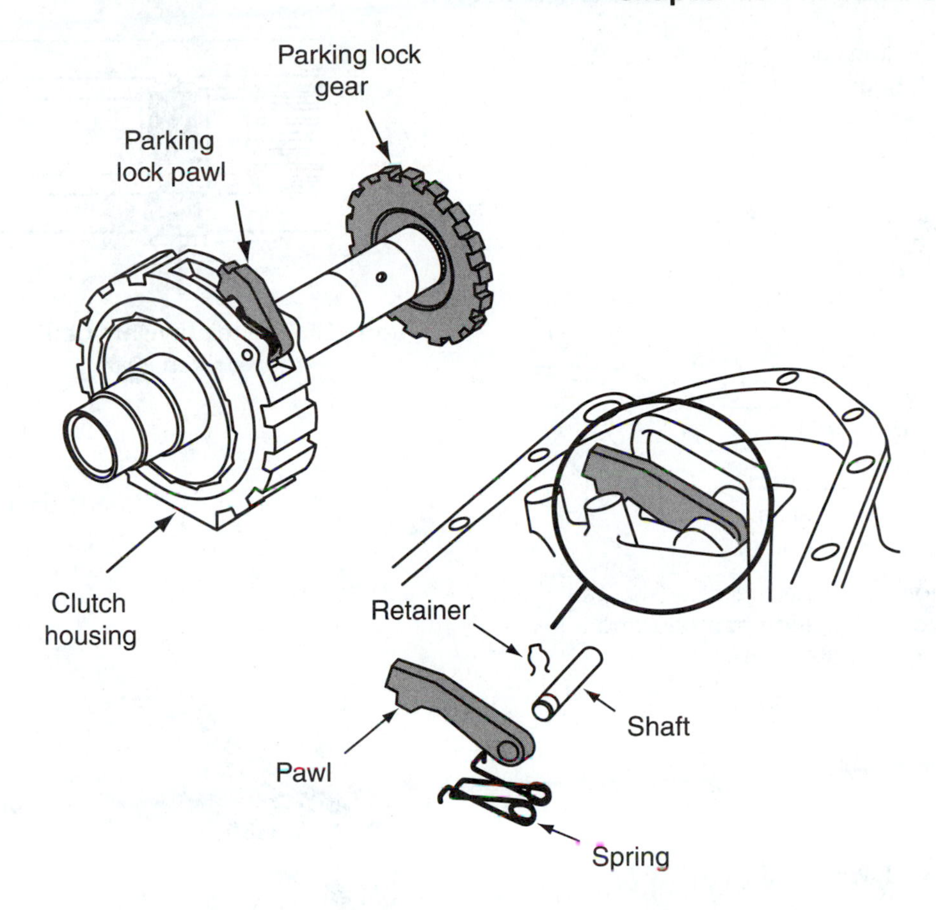


**Figure 7.** The entire parking pawl and gear assembly should be carefully inspected for wear or defects.

gearshift linkage moves the pawl into position to lock the output shaft of the transmission. Unless the customer's complaint indicates a problem with the parking mechanism, no test will detect a problem here.

Check the pawl assembly for excessive wear and other damage. Also, check to see how firmly the pawl is in place when the gear selector is shifted into the PARK mode. If the pawl can be easily moved out, it should be repaired or replaced.

## REAR OIL SEAL AND BUSHING REPLACEMENT

An oil leak stemming from the mating surfaces of the extension housing and the transmission case may be caused by loose bolts. To correct this problem, tighten the bolts to the specified torque. Also, check for signs of leakage at the rear of the extension housing. Often, the cause for the leakage is a worn extension housing bushing, which supports the sliding yoke of the drive shaft. When the drive shaft is installed, the clearance between the sliding yoke and the bushing should be minimal. If the clearance is satisfactory, a new oil seal will correct the leak. If the clearance is excessive, the repair requires that a new seal and a new bushing be installed. If the seal is faulty, the transmission vent should be checked for blockage.

Procedures for the replacement of the rear oil seal and bushing vary little with each car model. General procedures for the replacement of the oil seals and bushings follow.

To replace the rear seal:

1. Remove the drive shaft.
2. Remove the old seal from the extension housing **(Figure 8)**.
3. Lubricate the lip of the seal, and then install the new seal in the extension housing.
4. Install the drive shaft.

To replace the rear bushing and seal:

1. Remove the drive shaft from the car.
2. Insert the appropriate puller tool into the extension housing until it grips the front side of the bushing.
3. Pull the seal and bushing from the housing.
4. Drive a new bushing into the extension housing.
5. Install a new seal in the housing **(Figure 9)**.
6. Install the drive shaft.

### Speedometer Drive Service

An oil leak at the speedometer cable or VSS can be corrected by replacing the O-ring seal at the speedometer or

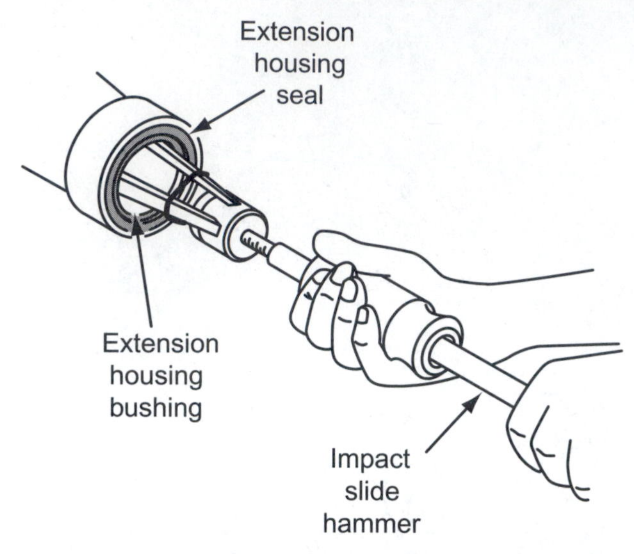


**Figure 8.** Extension housing rear seals can be removed with a slide hammer and a special removal tool.

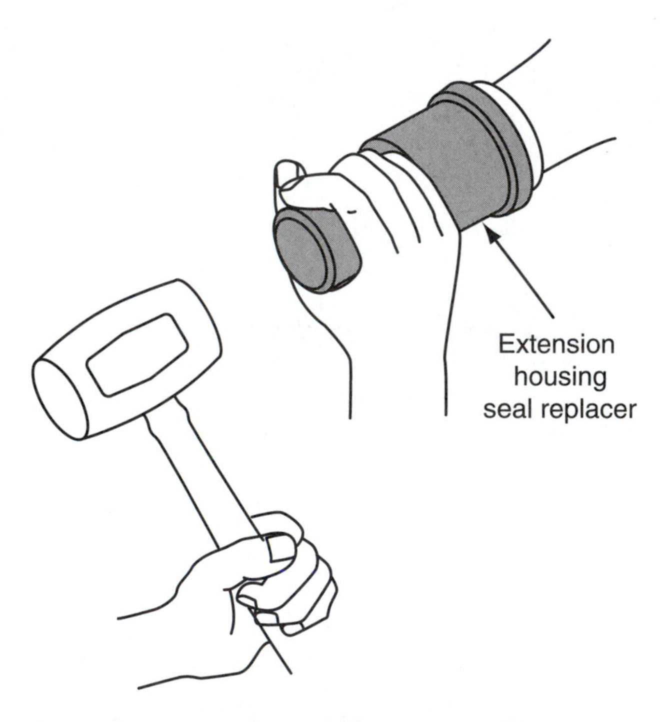


**Figure 9.** Rear seals can be easily installed into the extension housing with the proper driver and a hammer.

VSS drive. Also, speedometer problems may be caused by the speedometer or VSS drive gear. This problem also may be caused by a faulty cable, driven gear, or speedometer. A damaged drive gear can cause the driven gear to fail; therefore, both should be carefully inspected **(Figure 10)**. On some transmissions, the speedometer drive gear is a set

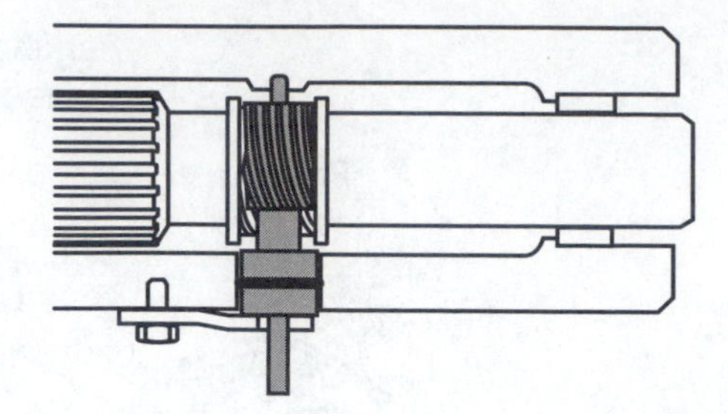

**Figure 10.** The relationship of the speedometer drive and driven gears.

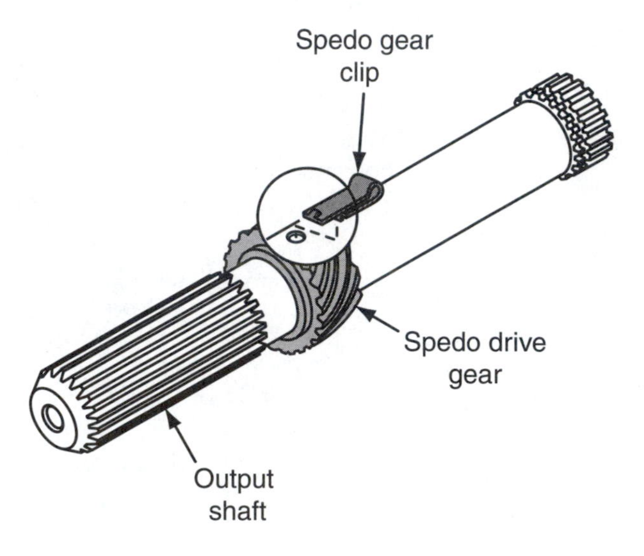


**Figure 11.** The speedometer drive gear is often retained to the output shaft by a clip.

of gear teeth machined into the output shaft. Inspect this gear. If the teeth are slightly rough, they can be cleaned up and smoothened with a file. If the gear is severely damaged, the entire output shaft must be replaced. Other transmissions have a drive gear which is splined to the output shaft, held in place by a clip **(Figure 11)**, or driven and retained by a ball which fits into a depression in the shaft. If a clip is used, it should be carefully inspected for cracks or other damage.

The drive gear can be removed and replaced, if necessary. The driven gear or speedometer pinion gear is normally attached to the transmission end of the speedometer cable **(Figure 12)**. If the drive gear is damaged, it is very likely that the driven gear also will be damaged. Most driven gears are made of plastic to reduce speedometer noise; therefore, they are weaker than most drive gears. Also, check the retainer of the driven gear on the speedometer cable.

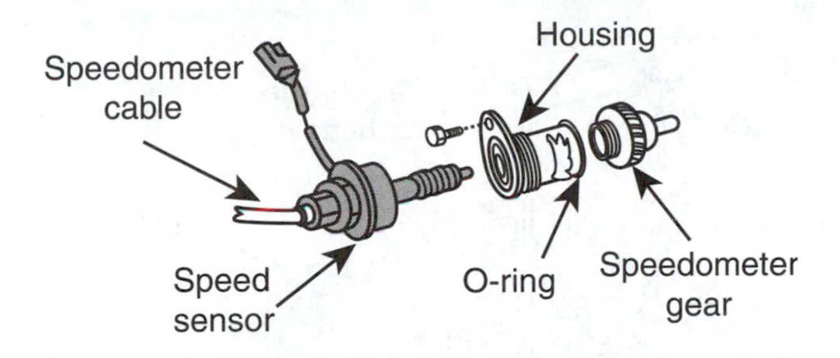


**Figure 12.** The speedometer cable and gear normally fit into a bore in the transmission case or extension housing.

## TRANSAXLE MOUNTS

If it is necessary to replace the transaxle mount, make sure you follow the manufacturer's recommendations for maintaining the alignment of the driveline. Failure to this may result in poor gear shifting, vibrations, and/or broken cables. Some manufacturers recommend that a holding fixture or special bolt be used to keep the unit in its proper location.

When removing the transaxle mount, begin by disconnecting the battery's negative cable. Disconnect any electrical connectors that may be located around the mount. It may be necessary to move some accessories, such as the horn, in order to service the mount without damaging some other assembly. Be sure to label any wires you remove to facilitate reassembly.

Install the engine support fixture **(Figure 13)** and attach it to an engine hoist. Lift the engine just enough to take the pressure off the mounts. Remove the bolts attaching the transaxle mount to the frame and the mounting bracket, and then remove the mount.

To install the new mount, position the transaxle mount in its correct location on the frame and tighten its attaching bolts to the proper torque. Install the bolts that attach the mount to the transaxle bracket. Some manufacturers require that the mount bolts be discarded and new ones installed during reassembly. Prior to tightening these bolts, check the alignment of the mount. Once you have confirmed that the alignment is correct, tighten all loosened bolts to their specified torque. Remove the engine hoist fixture from the engine and reinstall all accessories and wires that may have been removed earlier.

> **You Should Know** *In the procedure for removing the mount, some manufacturers recommend the use of a special alignment bolt, which is installed in an engine mount. This bolt serves as an indicator of power train alignment. If excessive effort is required to remove the alignment bolt, the power train must be shifted to allow for proper alignment.*

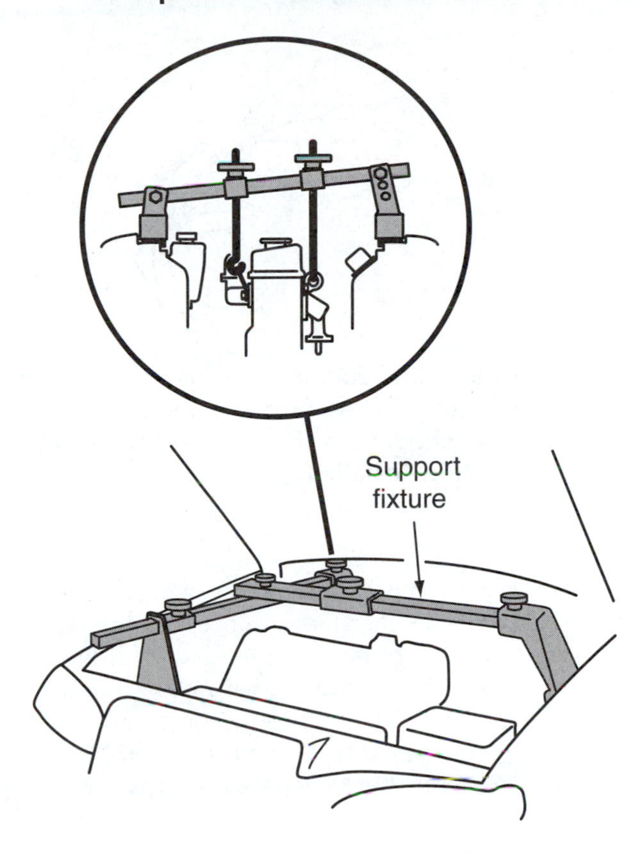


**Figure 13.** A typical engine support fixture in place to remove and/or install engine and transaxle mounts.

## TRANSMISSION COOLER SERVICE

Vehicles equipped with an automatic transmission can have an internal or external transmission fluid cooler, or both. The basic operation of either type of cooler is that of a **heat exchanger**. Heat from the fluid is transferred to something else, such as a liquid or air. Hot ATF is sent from the transmission to the cooler, where it has some of its heat removed, and then the cooled ATF returns to the transmission.

The engine's cooling system is the key to efficient transmission fluid cooling. If anything affects engine cooling, it also will affect ATF cooling. The engine's cooling system should be carefully inspected whenever there is evidence of ATF overheating or a transmission cooling problem.

If the problem is the transmission cooler, examine it for signs of leakage. Check the dipstick for evidence of coolant mixing with the ATF. Milky fluid indicates that engine coolant is leaking into and mixing with the ATF because of

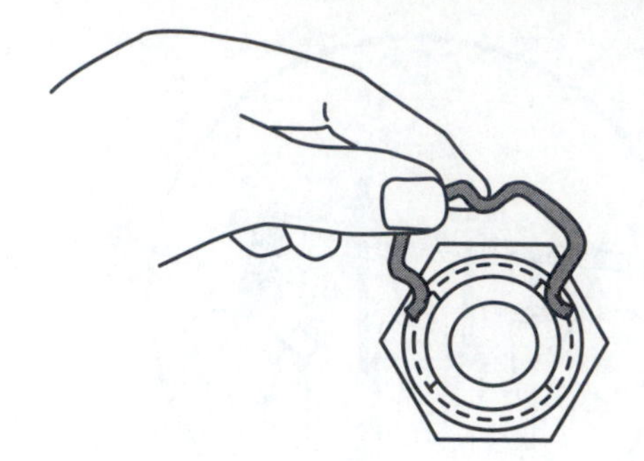

**Figure 14.** Cooler line fittings may use a retaining ring to lock the fitting to the line. These rings should never be reused and must be properly installed.

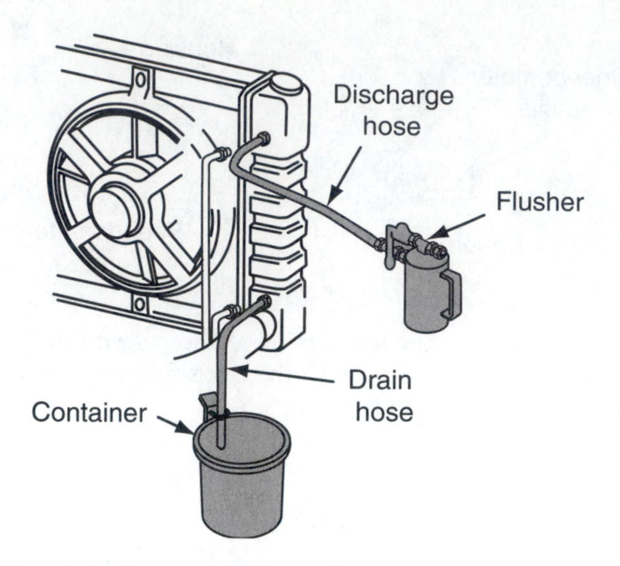


**Figure 15.** Setup for flushing a transmission's cooling system.

a leak in the cooler. At times, the presence of ATF in the radiator will be noticeable when the radiator cap is removed, as ATF will tend to float to the top of the coolant. A leaking transmission cooler core can be verified with a leak test.

If there is an indication of a restriction or build up of dirt in the cooler, it should be cleaned and/or flushed. The cooler and lines also should be flushed if the fluid was found to be contaminated. Before doing this, disconnect the cooler lines at the transmission **(Figure 14)**.

Identify the design of the cooler. A tube-type transmission cooler can be cleaned by using cleaning solvent or mineral spirits and compressed air. A fin-type cooler, however, cannot be cleaned in this same way. Therefore, normal procedure includes replacing the radiator (which includes the cooler). In both cases, the cooler lines should also be flushed to remove any debris.

Properly flushing a cooler will remove nearly all of the debris, leaving a clean, dry system. To do this, a transmission cooler flusher **(Figure 15)** is required. The flusher's container is filled with flushing fluid. Compressed air pushes the fluid through the cooler and lines. The system is then cleaned with forced hot water. When the discharge from the system is clear, compressed air is released into the cooler until no moisture is evident in the discharge. With a hand pump, force one quart of clean ATF into the cooler. Now, the lines can be reconnected and the engine started. Check and correct the fluid level.

If the cooler is plugged, apply compressed air through one port of the cooler then to the other port. The air should blow any debris out of the cooler. Always use low air pressure, no more than 50 psi. Higher pressures may damage the cooler. After the air pressure has removed the restrictions, flush the cooler. If little air was able to pass through the cooler, the cooler is severely plugged and may need to be replaced.

If the vehicle has two coolers, pull the inlet and outlet lines from the side of the transmission and using a hand pump, pump mineral spirits into the inlet hose until clear fluid pours out of the outlet hose. To remove the mineral spirits from the cooler, release some compressed air into the inlet port, then pump one quart of ATF into the cooler. Mineral spirits is an alcohol-based liquid that leaves no surface residue after it dries.

> **You Should Know** *A transmission cooler is a good hiding place for debris when there has been a major transmission failure. Blowing out the cooler and lines will not always remove all of the dirt. Install a temporary auxiliary oil filter in the cooler lines to catch any leftover debris immediately after rebuilding a transmission.*

Check the condition of the cooler lines from their beginning to their ends. A line that has been damaged will reduce oil flow through the cooler and will shorten the life of the transmission. If the steel cooler lines need to be replaced, use only double-wrapped and brazed steel tubing. Never use copper or aluminum tubing to replace steel tubing. The steel tubing should be double-flared and installed with the correct fittings.

## Summary

- A transmission's fluid and filter should be changed whenever there is an indication of oxidation or contamination and when the manufacturer recommends it as part of the preventative maintenance program.
- Severe duty use means the fluid and filter should be changed more frequently than the time specified in the preventative maintenance schedule.
- After the fluid is drained and the filter removed, they should be checked for contaminants, as these will help identify many transmission problems.
- Anytime you have the oil pan off, you should inspect the parking pawl assembly.
- The extension housing bushing and seal can be replaced when the transmission is still in the vehicle.
- An oil leak at the speedometer cable or VSS can be corrected by replacing the O-ring seal at the speedometer or VSS drive.
- Bad engine or transaxle mounts can cause poor gear shifting, vibrations, and/or broken shift cables.
- The transmission cooler should be flushed whenever the fluid is contaminated or there is evidence of a restriction in the cooler.

## Review Questions

1. What is indicated by aluminum particles in a transmission's fluid or filter?
2. When discussing the cause of an oil leak at the rear of the extension housing, Technician A says that the problem may be a bad seal. Technician B says that the problem may be a worn extension housing bushing. Who is correct?
   A. Technician A only
   B. Technician B only
   C. Both Technician A and Technician B
   D. Neither Technician A nor Technician B
3. Why is it important to maintain the correct alignment of the driveline when replacing an engine or transmission mount?
4. What can cause an inoperative speedometer?
5. Technician A says that a fin-type cooler can be cleaned by using cleaning solvent or mineral spirits and compressed air. Technician B says that a tube-type transmission cooler can be cleaned by using cleaning solvent or mineral spirits and compressed air. Who is correct?
   A. Technician A only
   B. Technician B only
   C. Both Technician A and Technician B
   D. Neither Technician A nor Technician B

# Chapter 19 External Adjustments and Service

## Introduction

Many transmission problems are caused by improper adjustment of the linkages. All transmissions have either a cable or rod type gear selector linkage and some sort of neutral safety switch. Also, some transmissions have a throttle valve linkage and/or a kickdown switch. All of these may need adjustment to correct problems or as part of a preventative maintenance program. Another adjustment that is done periodically on a few transmissions is the bands.

### GEAR SELECTOR LINKAGE

A worn or misadjusted gear selection linkage (often called the manual linkage because it controls the manual shift valve) will affect transmission operation. The transmission's manual shift valve must be in the correct position in order to completely engage the selected gear **(Figure 1)**. Partial manual shift valve engagement will not allow the proper amount of fluid pressure to reach the rest of the valve body. If the linkage is misadjusted, poor gear engagement, slipping, and excessive wear can result. The gear selector linkage should be adjusted so the manual shift valve detent position in the transmission matches the selector level detent and position indicator.

To check the adjustment of the linkage, move the shift lever from the PARK position to the lowest DRIVE gear. Detents should be felt at each of these positions. If the detent cannot be felt in either of these positions, the linkage needs to be adjusted. When moving the shift lever, pay attention to the gear position indicator. Although the indicator will move with an adjustment of the linkage, the pointer may need to be adjusted so that it shows the exact gear after the linkage has been adjusted.

**You Should Know** *Always set parking brake before moving the gear selector through its positions. This will prevent the vehicle from rolling over you or someone else.*

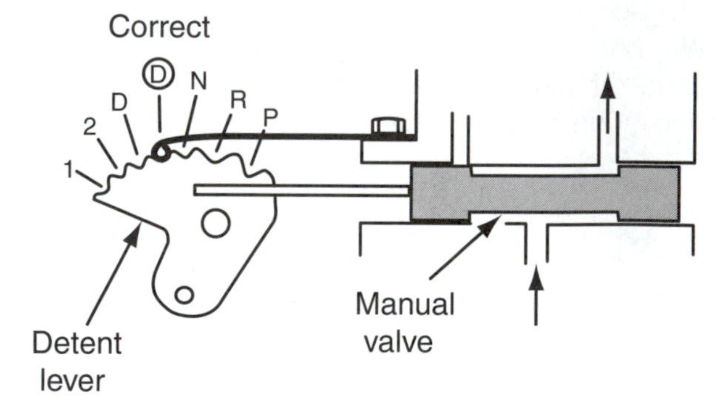


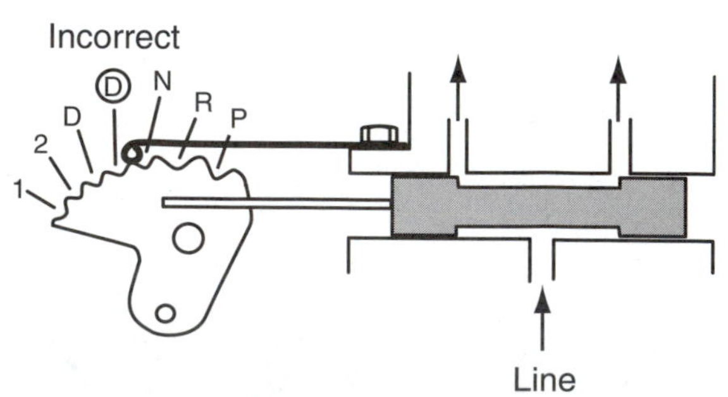


**Figure 1.** Incorrect linkage adjustments may cause the manual shift valve to be positioned improperly in its bore and cause slipping during gear changes.

Normal operation of a **neutral safety switch**, also called an inhibitor switch, provides a quick check for the adjustment of the gear selector linkage. To do this, move the selector lever slowly until it clicks into the PARK position. Turn the ignition key to the "Start" position. If the starter operates, the PARK position is correct. After checking the PARK position, move the lever slowly toward the NEUTRAL position until the lever drops at the end of the N stop in the selector gate. If the starter also operates at this point, the gearshift linkage is properly adjusted. This quick test also tests the adjustment of the neutral safety switch. If the engine does not start in either or both of these positions, the neutral safety switch or the gear selector linkage needs adjustment or repair.

On late-model transmissions with a TR sensor, the selected gear should be illuminated on the instrument panel when the gear selector is placed in the proper detent for a gear. If this does not happen, either the TR sensor or the linkage needs to be adjusted.

## Linkage Adjustment

**You Should Know** *Because you must work under the car to adjust most shift linkages, make sure you properly raise and support the car before working under it. Also, wear safety glasses or goggles when working under the car.*

To adjust a typical floor mounted gear selector linkage:

1. Place the shift lever into the DRIVE position.
2. Loosen the locknuts and move the shift lever until DRIVE is properly aligned and the vehicle is in the "D" range.
3. Tighten the locknut **(Figure 2)**.

To adjust a typical cable-type linkage:

1. Place the shift lever into the PARK position.
2. Loosen the clamp bolt on the shift cable bracket.
3. Make sure the preload adjustment spring engages the fork on the transaxle bracket.
4. By hand, pull the shift lever to the front detent position (PARK), and then tighten the clamp bolt. The shift linkage should now be properly adjusted.
5. If the cable assembly is equipped with an adjuster **(Figure 3)**, move shift lever into the PARK position and adjust the cable by rotating the adjuster into the lock position. Typically, the adjuster will click when the lock is fully adjusted.

To adjust a typical rod-type linkage, loosen or disconnect the shift rod at the shift lever bracket. Then, place the gear selector into PARK and the manual shift valve lever into the PARK detent position. With both levers in position, tighten the clamp on the sliding adjustment to maintain their relationship. On some vehicles, you may need to adjust the neutral safety switch after resetting the linkage.

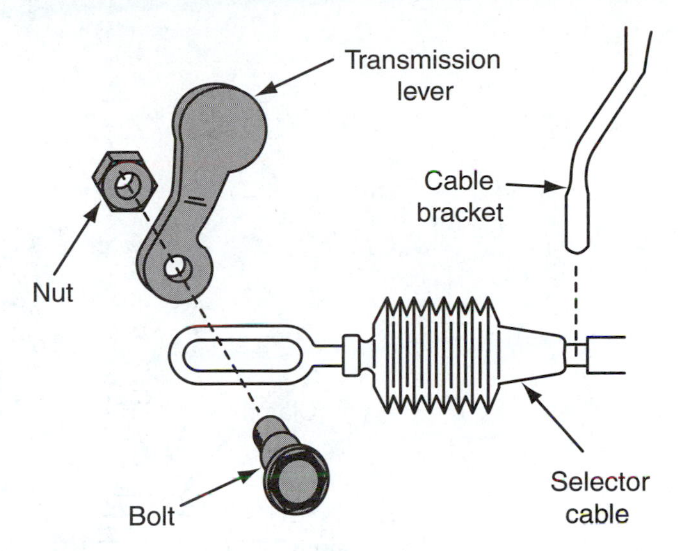


**Figure 2.** Gearshift linkages normally attach directly to the shift lever at the transmission and are retained by a locknut.

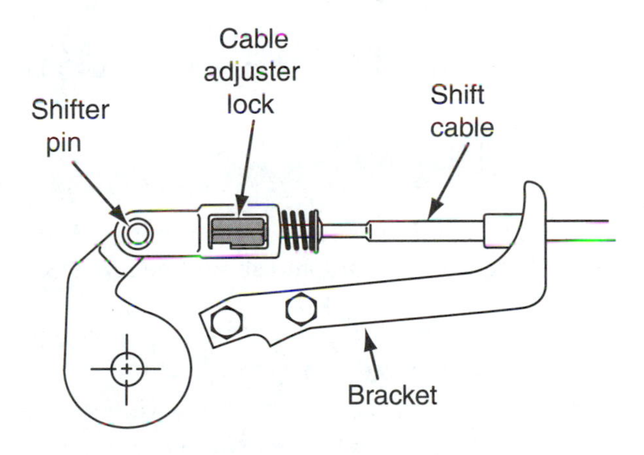


**Figure 3.** A shift cable adjuster lock assembly.

After adjusting any type of shift linkage, recheck it for detents throughout its range. Make sure a positive detent is felt when the shift lever is placed into the PARK position, as a safety measure. If you are unable to make an adjustment, the levers' grommets may be badly worn or damaged and should be replaced **(Figure 4)**. When it is necessary to disassemble the linkage from the levers, the plastic grommets used to retain the cable or rod should be replaced. Use a prying tool to force the cable or rod from the grommet, and then cut out the old grommet. Pliers can be used to snap the new grommets into the levers and the cable or rod into the levers.

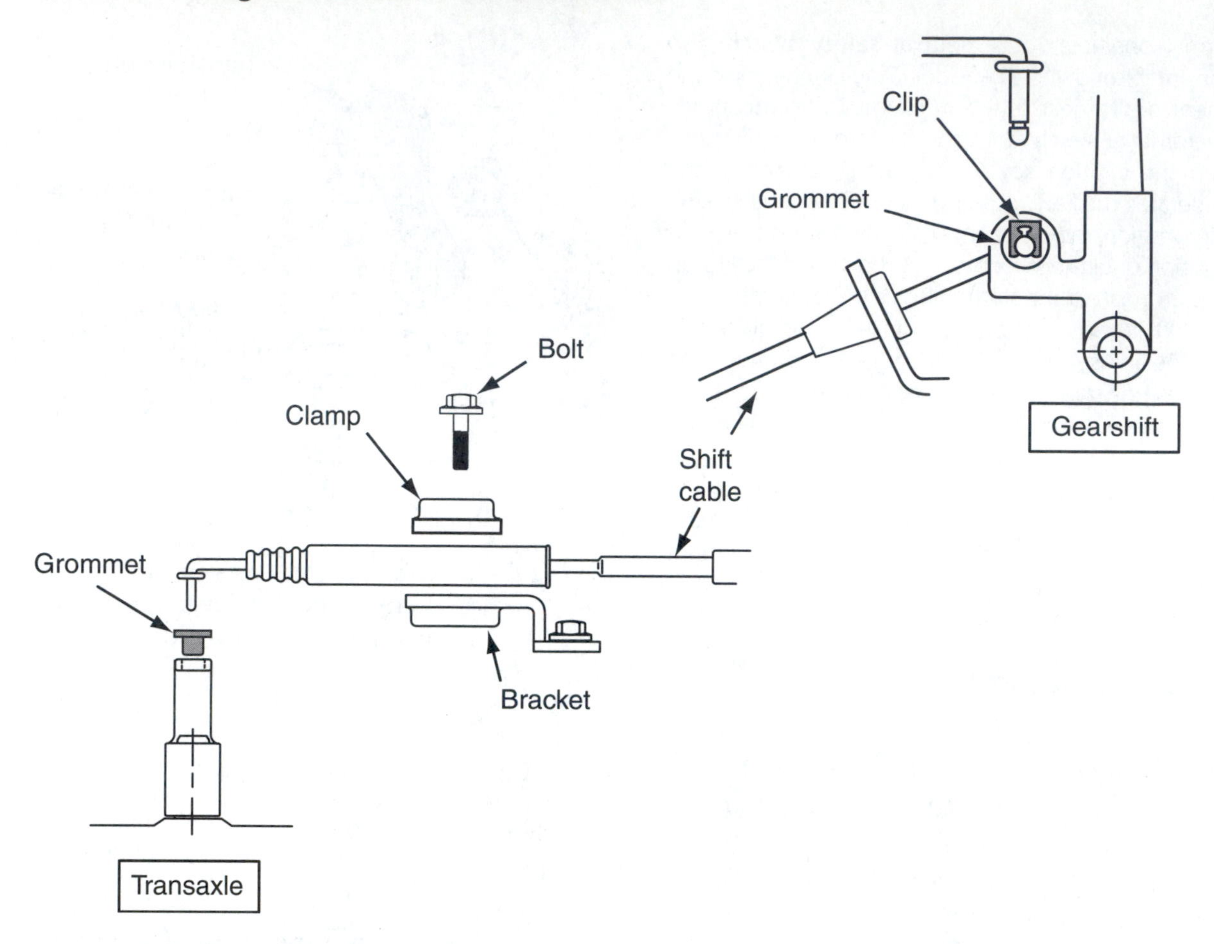


**Figure 4.** If the shift cable cannot be properly adjusted, the cable is distorted or some of the various grommets or brackets are worn and should be replaced.

**You Should Know** *If the proper prying tool is not used, the shift lever in some transmissions can easily break. If this tool is not available, apply some heat to the grommet. This should allow for easy separation of the cable or rod from the lever.*

## NEUTRAL SAFETY SWITCH

Most neutral safety switches, or TR sensors, are combinations of a neutral safety switch and a backup lamp switch. Others have separate units for these two distinct functions. When the reverse light switch is not part of the neutral safety switch it seldom needs adjustment as it is directly activated by the transmission **(Figure 5)** or the gear selector linkage. A neutral safety switch allows current to pass from the starter switch to the starter when the lever is placed in the "P" or "N" positions. Some neutral safety switches are nonadjustable. If these prevent starting in P and/or N and the gear selector linkage is correctly adjusted, they should be replaced.

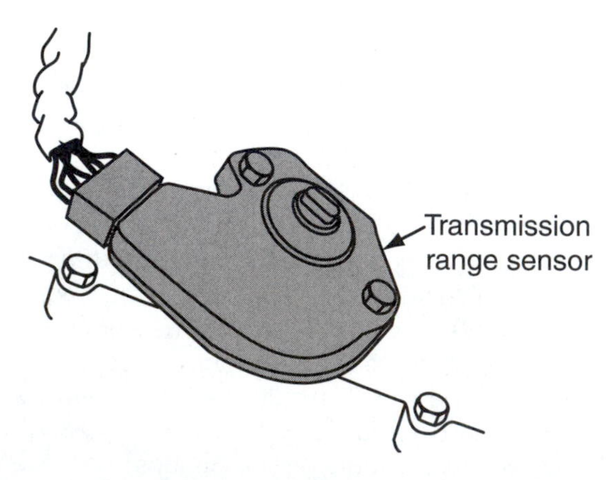


**Figure 5.** A digital transmission range (TR) sensor that serves as a neutral start switch, backup lamp switch, and as an input to the computer. It is located on the side of the transmission at the manual lever.

A voltmeter can be used to check the switch for voltage when the ignition key is turned to the START position with the shift lever in P or N. If there is no voltage, the switch should be adjusted or replaced.

To adjust a typical neutral safety switch, place the shift lever in Neutral. Loosen the attaching bolts for the switch. Move the switch until the alignment pin falls into the hole in its rotor. Then, tighten the attaching bolts. Recheck the switch for continuity. If no voltage is present, replace the switch.

## THROTTLE VALVE LINKAGES

The throttle valve cable connects the movement of the throttle pedal movement to the throttle valve in the transmission's valve body. On some transmissions, the TV linkage may control both the downshift valve and the throttle valve. Others use a vacuum modulator to control the throttle valve and a throttle linkage to control the downshift valve. Late-model transmissions may not have a throttle cable. Instead, they rely on electronic sensors and switches to monitor engine load and throttle plate opening **(Figure 6)**. The action of the throttle valve produces throttle pressure **(Figure 7)**. Throttle pressure is used as an indication of engine load and influences the speed at which automatic shifts will take place.

A misadjusted TV linkage also may result in a throttle pressure that is too low in relation to the amount that the throttle plates are open, causing early upshifts. Throttle pressure that is too high can cause harsh and delayed upshifts and part and wide-open throttle downshifts will occur earlier than normal. When adjusting the TV and downshift linkages, always follow the manufacturer's recommended procedures. An adjustment as small as a half turn can make a big difference in shift timing and feel.

Some GM transmissions use a pull-push plastic lock or a metal push-release tab on the TV cable instead of a threaded adjustment. Follow the procedure given in the service manual to adjust this type of cable. If the adjustment is off by a single click of the adjusting mechanism, shift timing will be wrong.

**Figure 6.** Throttle cable and solenoid connections to an EAT.

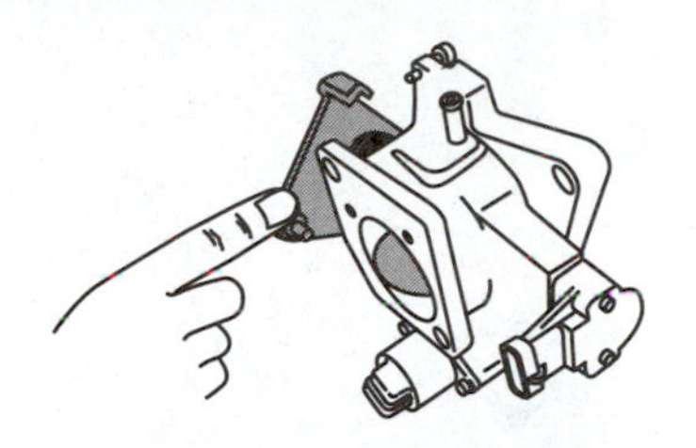


**Figure 7.** The movement of the throttle plate causes the TV valve to move.

> **You Should Know** *It is important that you check the service manual before proceeding to make adjustments to the throttle or downshift linkages. Some of these linkage systems are not adjustable and if you loosen them in an attempt to adjust them, you may break them.*

To adjust a typical throttle cable:

1. Run engine until it has reached normal operating temperature.
2. Loosen the cable mounting bracket or swivel lock screw.
3. Position the bracket so that both bracket alignment tabs are touching the transaxle, and then tighten the lock screw to the recommended torque.
4. Release the cross-lock (called the readjust tab on some transmissions) on the cable assembly.
5. Make sure the cable is free to slide all the way toward the engine, against its stop, after the cross-lock or locking tab is released **(Figures 8 and 9)**.
6. Move the throttle control lever clockwise against its internal stop, then press the cross-lock downward into its locked position.

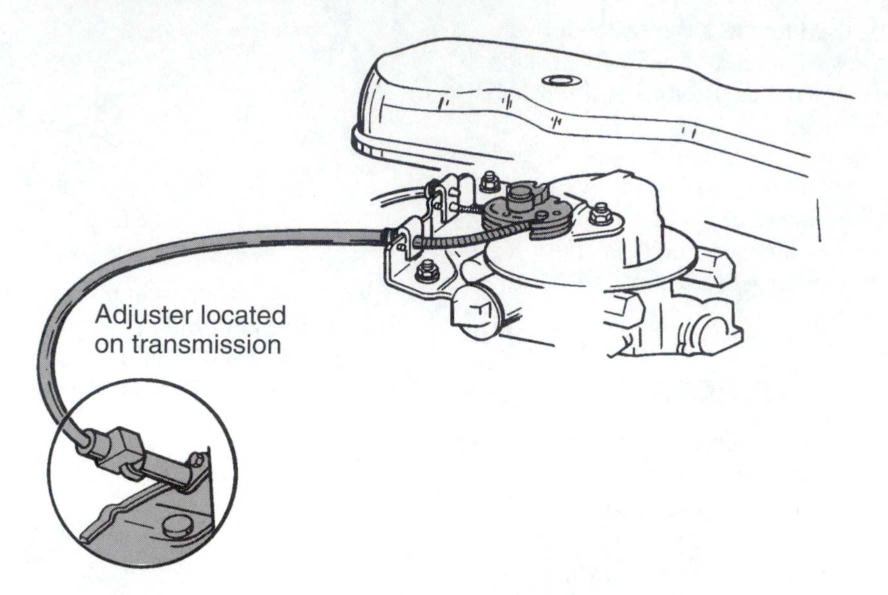


**Figure 8.** The cable adjuster may be located at the transmission or transaxle or at the throttle lever.

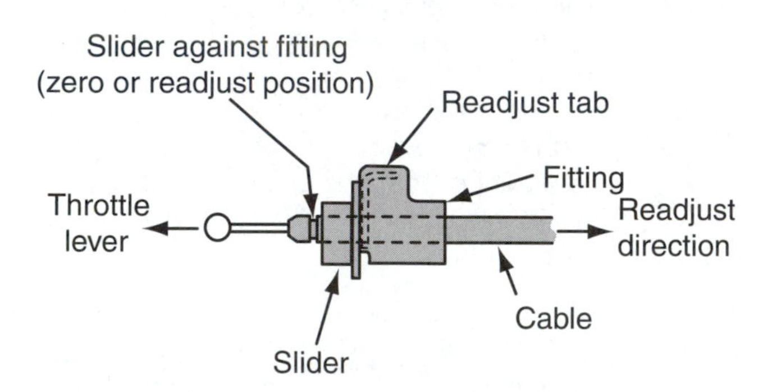


**Figure 9.** Typical adjusting mechanism for a TV cable.

7. At this point, the cable is adjusted and the backlash in the cable removed.
8. Check the cable for free movement by moving the throttle lever counterclockwise and slowly release it to confirm it will return clockwise.
9. No lubrication is required for any component of the throttle cable system.
10. Check the adjustment of the cable by conducting a system pressure test.

To adjust a typical downshift linkage:

1. Run the engine until it has reached normal operating temperature.
2. Put the transmission in neutral with the parking brake set and allow the engine to run at its normal idle speed.
3. Using the specified amount of pressure, press down on the downshift rod.
4. Rotate the adjustment screw to obtain the specified clearance between the screw and the throttle arm.

## KICKDOWN SWITCH ADJUSTMENT

Most late-model transmissions are not equipped with downshift linkages; rather, they may use a kickdown switch typically located at the upper post of the throttle pedal **(Figure 10)**. Some kickdown circuits are part of the computer control circuit, based on signals from a throttle position sensor. Movement of the throttle pedal to the wide-open position signals to the transmission that the driver desires a forced downshift.

To check the operation of the switch, fully depress the throttle pedal and listen for a click that should be heard just before the pedal reaches its travel stop. If the click is not heard, loosen the locknut and extend the switch until the pedal lever makes contact with the switch. If the pedal contacts the switch too early, the transmission may downshift during part-throttle operation.

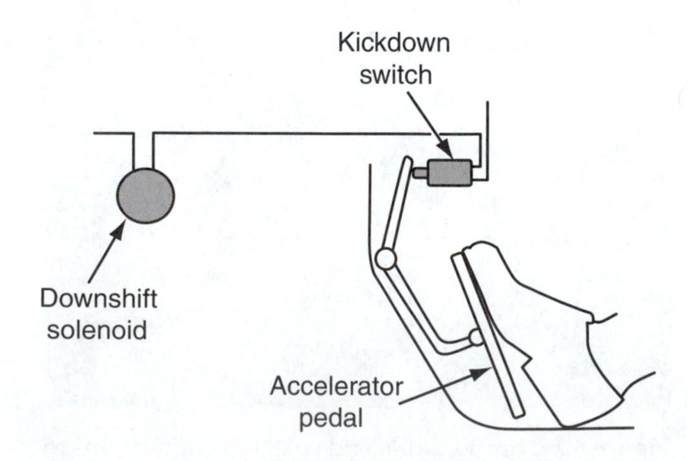


**Figure 10.** Action of a kickdown switch.

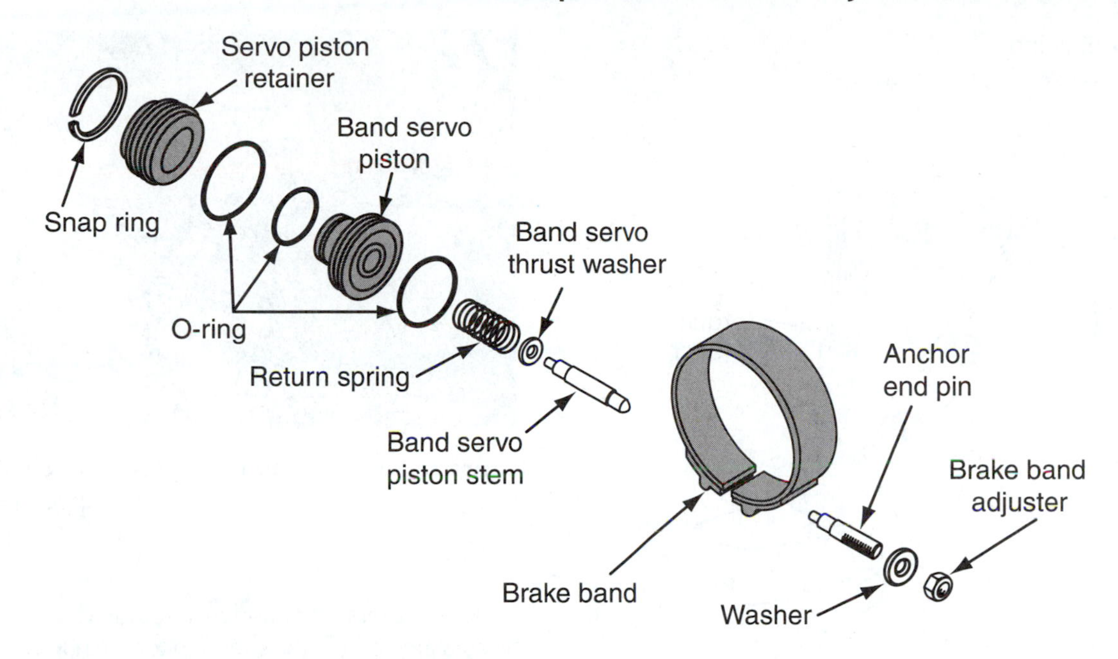


**Figure 11.** A band and servo assembly with an adjuster.

If you feel and hear the click of the switch but the transmission still does not kick down, use an ohmmeter to check the continuity of the switch when it is depressed. (Make sure the ignition switch is off before connecting the meter.) An open switch will prevent forced downshifting, whereas a shorted switch can cause upshift problems. Defective switches should be replaced.

## BAND ADJUSTMENT

If a transmission problem still exists after the shift linkage and throttle pressure cable and rod have been adjusted and all hydraulic circuits seem to be fine, the bands of the transmission may need adjustment **(Figure 11)**. To help identify if a band adjustment will correct the problem, refer to the written results of your road test. Compare your results with the Clutch and Band Application Chart in the service manual. If slippage occurs when there is a gear change that requires the holding by a band, the problem may be corrected by adjusting the band.

On some vehicles, the bands can be adjusted externally **(Figure 12)** with a torque wrench. On others, the transmission fluid must be drained and the oil pan removed. Still others can only be adjusted when the transmission is disassembled. These bands are adjusted with selective size pins.

Locate the band adjusting nut, then clean off all dirt on and around the nut. Now, loosen the band adjusting bolt locknut and back it off approximately five turns **(Figure 13)**.

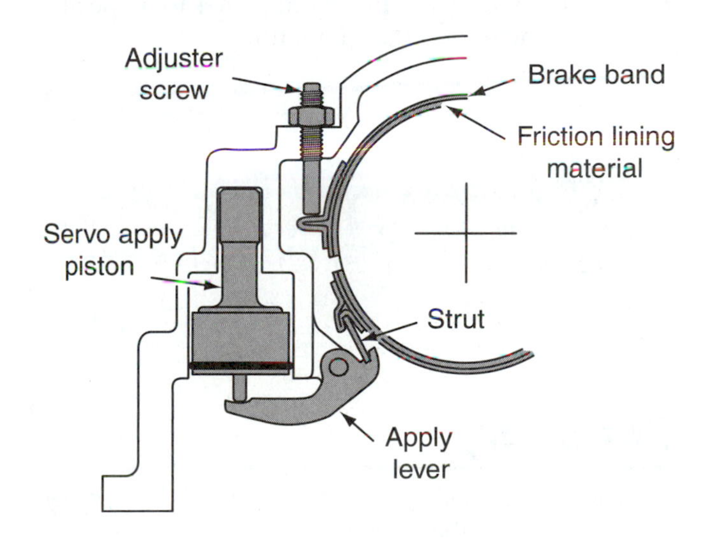


**Figure 12.** A band assembly with an external adjuster screw.

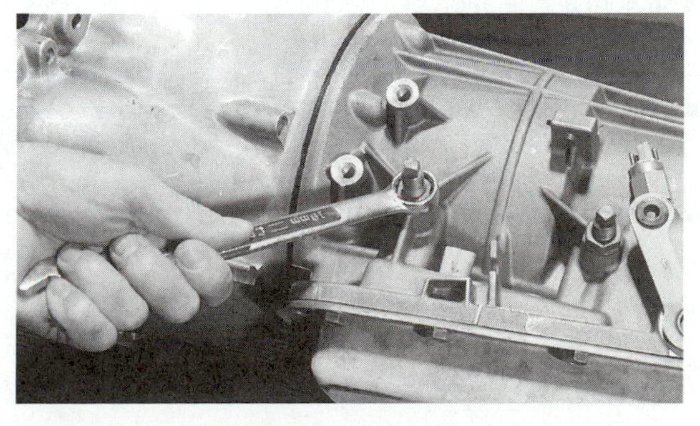

**Figure 13.** Loosen the adjusting screw locknut before attempting to adjust the band.

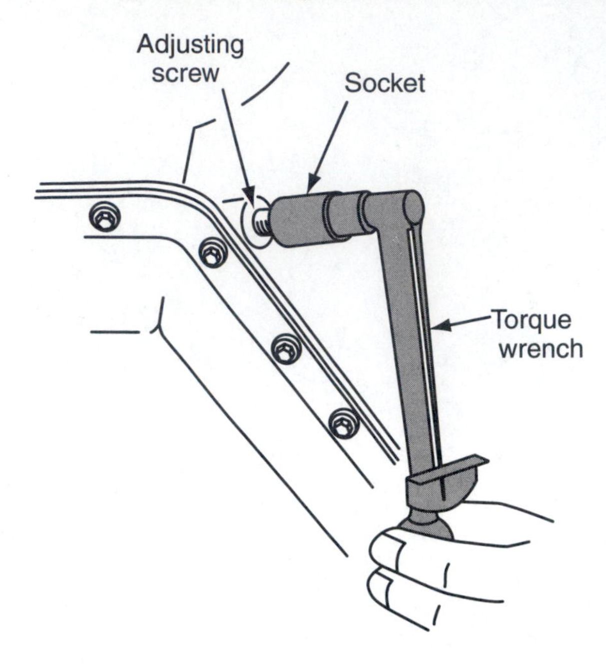


Figure 14. Bands are typically adjusted to a specific inch-pound torque setting.

> You Should Know *Do not excessively back off the adjusting stem as the anchor block may fall out of place and it will be necessary to remove and disassemble the transmission to fit it back in place.*

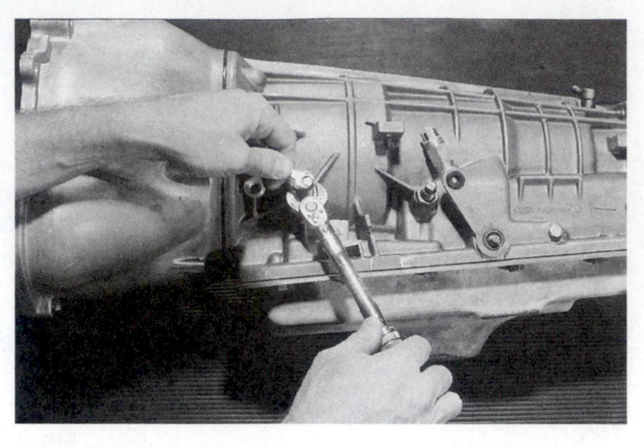

Figure 15. Hold the adjusting screw in position and tighten the locknut to the specified torque.

Use a calibrated pound-inch torque wrench to tighten the adjusting bolt to the specified torque **(Figure 14)**. Then, back off the adjusting screw the specified number of turns and tighten the adjusting bolt locknut to specifications, while holding the adjusting stem stationary **(Figure 15)**. Reinstall the oil pan with a new gasket and refill the transmission with fluid. If the transmission problem still exists, an oil pressure test or transmission teardown must be done.

## Summary

- Many transmission problems are caused by improper adjustment of the linkages.
- A worn or misadjusted gear selection linkage will affect transmission operation because it prevents the proper position of the manual shift valve.
- Normal operation of a neutral safety switch provides a quick check for the adjustment of the gear selector linkage.
- Most neutral safety switches, or TR sensors, are combinations of a neutral safety switch and a backup lamp switch.
- The TV cable connects the movement of the throttle pedal movement to the throttle valve in the transmission's valve body.
- Late-model transmissions may rely on electronic sensors and switches to monitor engine load and throttle plate opening.
- A misadjusted TV linkage may also cause early upshifts, harsh and delayed upshifts, or early part- and wide-open throttle downshifts.
- Most late-model transmissions use a kickdown switch or the TP sensor for automatic downshifting when the throttle is quickly opened.

# Review Questions

1. What can result from a misadjusted throttle linkage that allows for excessive throttle pressures?
2. Describe two quick ways the adjustment of the gear selector linkage can be checked.
3. When discussing neutral safety switches, Technician A says that most neutral safety switches and TR sensors are combinations of a neutral safety switch and a backup lamp switch. Technician B says that some neutral safety switches are nonadjustable. Who is correct?
   A. Technician A only
   B. Technician B only
   C. Both Technician A and Technician B
   D. Neither Technician A nor Technician B
4. When discussing proper band adjustment procedures, Technician A says that on some vehicles the bands can be adjusted externally with a torque wrench. Technician B says that a calibrated inch-pound torque wrench is normally used to tighten the band adjusting bolt to a specified torque. Who is correct?
   A. Technician A only
   B. Technician B only
   C. Both Technician A and Technician B
   D. Neither Technician A nor Technician B
5. When discussing the proper way to diagnose a kickdown switch, Technician A says that, when the throttle pedal is fully depressed, a click should be heard just before the pedal reaches its travel stop. If the click is not heard, the switch should be replaced. Technician B says that, if the transmission cannot be forced to automatically downshift, the kickdown switch is open and should be replaced. Who is correct?
   A. Technician A only
   B. Technician B only
   C. Both Technician A and Technician B
   D. Neither Technician A nor Technician B

## Hydraulic System and Controls



## SECTION OBJECTIVES

After you have read, studied, and practiced the contents of this section, you should be able to:

- Describe the design and operation of the hydraulic controls and valves used in modern transmissions and transaxles.
- Describe the various configurations of a spool valve and explain how it can be used to open and close various hydraulic circuits.
- Explain the role and operation of the following components of the transmission control system: pressure regulation valve, throttle valve, governor assembly, manual valve, shift valves, and kick-down assembly.
- Describe the operation of the main pressure regulator valve.
- Describe how fluid pressure can be controlled by a solenoid.
- Describe the different designs of a governor and explain the operation of each type.
- Explain why load-sensing devices are necessary for automatic transmission efficiency.
- Describe the purpose of a transmission's valve body.
- Name and explain the different designs of pumps used in modern transmissions.
- Trace through the oil circuit for a transmission and describe where the fluid flows in each transmission range.
- Perform pressure tests; determine necessary action.
- Inspect, adjust, or replace (as applicable) vacuum modulator; inspect and repair or replace lines and hoses.
- Inspect, repair, and replace governor assembly.

*The Chevrolet Powerglide was a popular transmission in the 1950s. For its time, it had a very complex hydraulic system. The valve body and accumulator assembly contained a total of seven valves, two of which were simple check valves.*

# Chapter 20 Basic Hydraulic Systems

## Introduction

An automatic transmission receives engine power through a torque converter, which is indirectly attached to the engine's crankshaft. Hydraulic pressure in the converter allows power to flow from the engine to the transmission's input shaft. The input shaft drives a planetary gear set, which provides the different forward gears, a neutral position, and one reverse gear. Power flow through the gears is controlled by multiple friction disc packs, one-way clutches, and/or friction bands. These hold or drive a member of the gear set when hydraulic pressure is applied to them. Hydraulic pressure is routed to the correct apply device by the transmission's valve body, which controls the pressure and direction of the hydraulic fluid **(Figure 1)**.

Although most discussions of an automatic transmission's are focused on the fluid flows and pressures used to provide for gear selection, it is important to remember that ATF is also used to lubricate the transmission. Part of the pump's output is always directed to lubricate the moving parts of the transmission. ATF flows through various circuits and passages in the case, valve body, shafts, and gear sets to reach these moving parts. Normally, the fluid flowing from the oil cooler is used as the lubricating oil.

**Figure 1.** A typical valve body.

### VALVE BODIES

Although the pump is the source of all fluid flow through the transmission, the valve body regulates and directs the fluid flow to provide for the gear changes. Many different valves and hydraulic passages **(Figure 2)** are used to regulate, direct, and control the movement of the fluid after it leaves the pump. The basic types of valves used in automatic transmissions are ball, poppet, or needle check valves and relief valves, orifices, and spool valves. The purpose of these valves is start, stop, or regulate and direct the flow of fluid throughout the transmission. Fluid flow through the passages and bores is controlled by a single valve or a series of valves. Most of the control valves operate automatically to direct the fluid as needed to release or apply the brakes and clutches in response to engine and vehicle load, as well as to meet the needs and desires of the driver.

A simple check valve is used to block a hydraulic passageway. It normally stops fluid flow in one direction while allowing flow in the opposite direction. A **relief valve** prevents or allows fluid flow until a particular pressure is reached, then it either opens or closes the passageway. Relief valves are used to control maximum pressures in a hydraulic circuit. **Orifices** are used to regulate and/or control fluid pressures. Spool valves are normally used as flow-directing valves and the most commonly used type of valve in an automatic transmission.

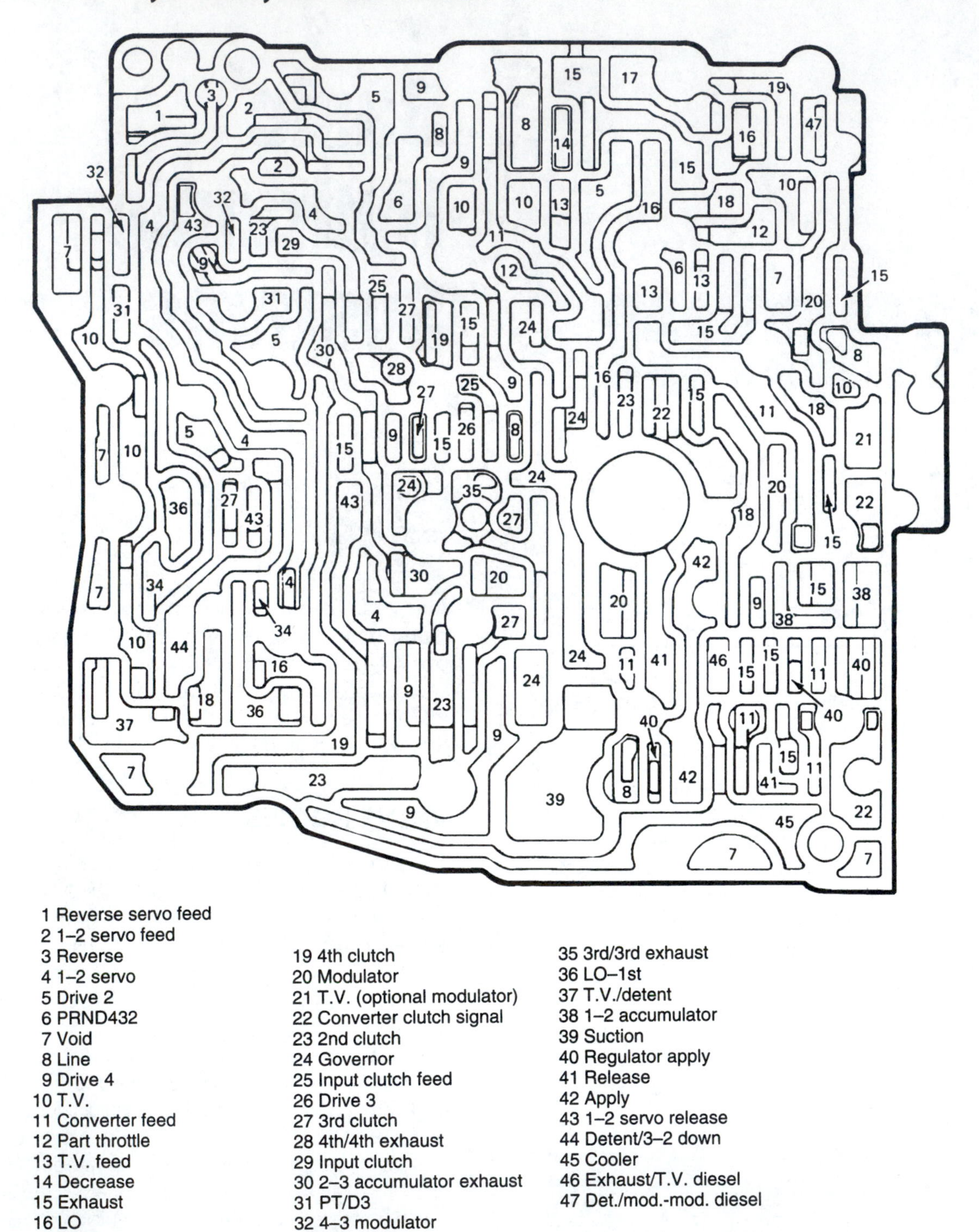


**Figure 2.** Passages of a typical valve body.

Fluid flow to and from the valve body is routed through passages and bores. Although many of these are located in the transmission case and pump housing, most will be found in the valve body. Most of the transmission's valves also are housed in the valve body, as well as many different check balls.

The separator and transfer plates are designed to seal off some of the passages and they contain some openings that help to control and direct fluid flow through specific passages. The three parts are typically bolted together and mounted as a single unit to the transmission housing. Other transmission designs may separate valve body

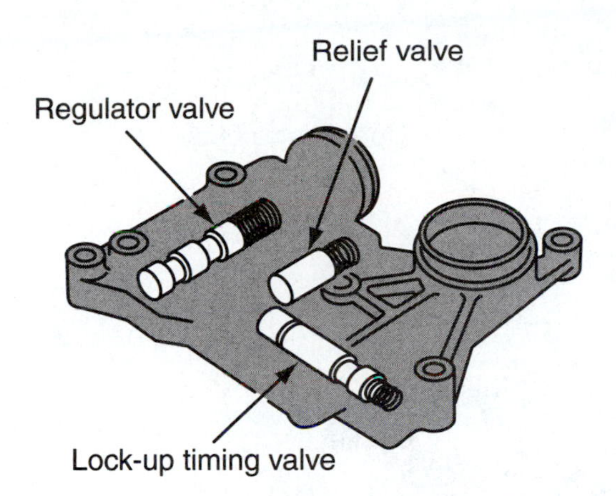


**Figure 3.** This pressure regulator valve body assembly is part of the main valve body assembly.

assemblies, each with a purpose. For example, the pressure regulator assembly shown in **Figure 3** is bolted to the main valve body in **Figure 4**. In other designs, the valve body is comprised on three major units **(Figure 5)**; each unit is separated from the other by a gasket and separator plate.

## Check Valves

Check valves are used to hold fluid in cylinders and to prevent fluid from returning to the reservoir. A check valve opens when fluid is flowing and closes when the flow stops. The valve also closes when fluid pressure is applied to the outlet side of the valve. Check valves can also serve as a one-way valve and the direction of the fluid flow controls and operates the check valve **(Figure 6)**. Ball-type check valves are commonly used to redirect fluid flow or to stop it from back flowing to the reservoir. The seats for the ball-type check valves are often holes cut into the valve body's **separator plate (Figure 7)**.

Needle-type and poppet-type check valves also are used to prevent back flow. **Needle check valves** are not used very frequently in automatic transmissions but operate in the same way as a ball-type valve.

A return spring is used with a poppet valve and allows the valve to totally stop back flow. The valve is forced closed by the spring as soon as spring pressure is greater than the pressure of the fluid. Ball-type check valves may also be fitted with a return spring to accomplish the same task as a poppet valve. A spring loaded check valve will open whenever fluid pressure on the inlet side of the valve exceeds the tension of the spring. The valve will remain open until the fluid's pressure drops below the spring's tension. Check valves without a return spring will not seat until there is a small amount of backflow.

A ball-type check valve without a return spring can be used as a **two-way check valve**. This type valve is used where hydraulic pressure from two different sources is to be sent to the same outlet port. When hydraulic pressure on one side of the valve is stronger than the pressure on the other side of the valve, the ball moves to the weaker side and closes that port. The ball will toggle between the ports in response to differing pressures. Both ports are open at the same time unless the pressures on both sides of the ball are equal. Normally, if the pressures were equal, the ball would be centered and would block off the outlet port.

## Relief Valves

A check valve fitted with a spring designed to prevent the valve from opening until a specific pressure is reached is called a pressure relief valve **(Figure 8)**. A pressure relief valve is used to protect the system from damage because of excessive pressure. When the pressure builds beyond the rating of the spring, the valve will open and reduce the pressure by exhausting some fluid back to the reservoir. After the pressure decreases, the valve again closes allowing normal fluid flow.

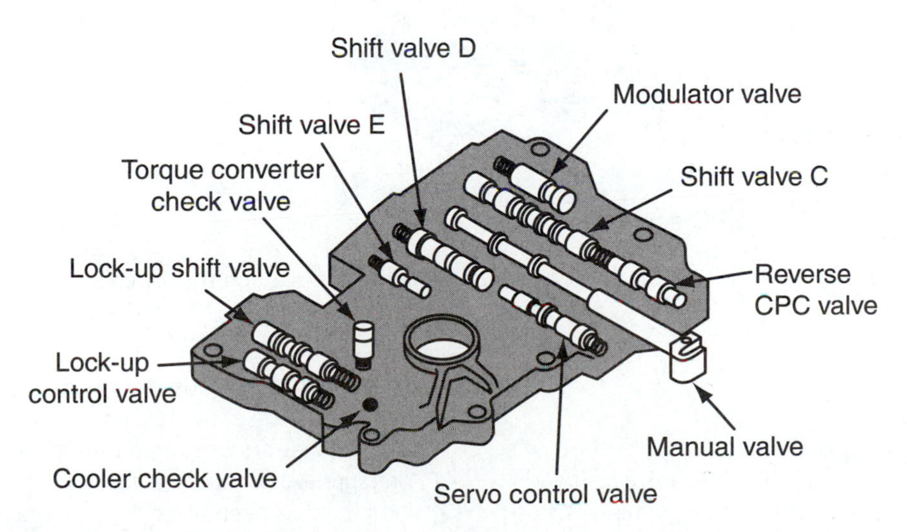


**Figure 4.** The pressure regulator assembly in Figure 3 bolts to the top of this valve body assembly.

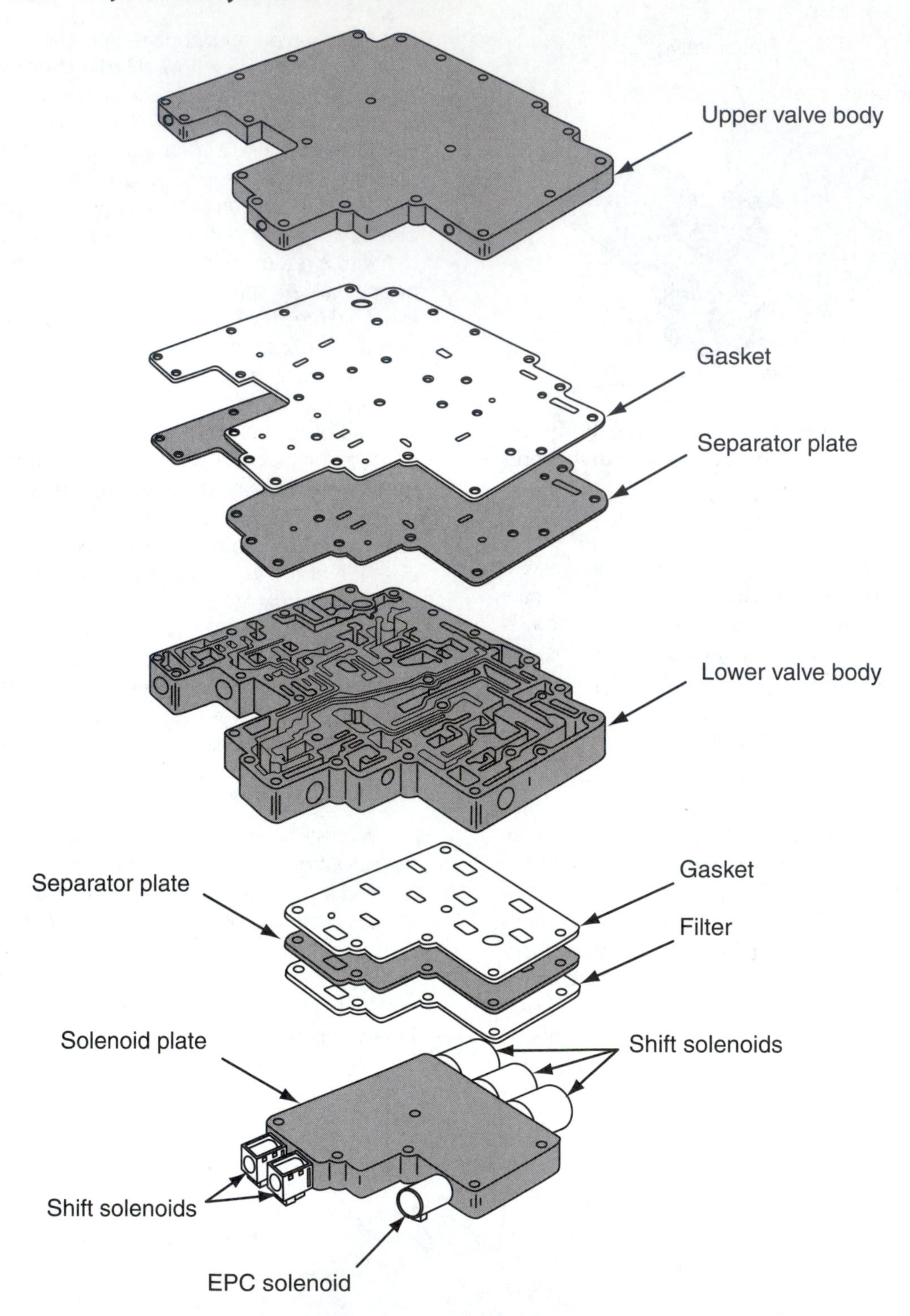


**Figure 5.** This valve body assembly is made up of three separate units, each mounted with gaskets and a separator plate.

## Orifices

The dynamic pressure of a fluid is affected by the fluid's movement from one point to another through a restriction. The hydraulic system in a transmission has many restrictions, such as connecting lines, small bores, and orifices. These restrictions cause the pressure of the fluid to increase at the inlet side of the restriction and decrease at the outlet side. Orifices are used in transmissions to control dynamic pressures. An orifice is a passage in a hydraulic system or item that has been placed in a hydraulic circuit that restricts and slows down fluid flow **(Figure 9)**.

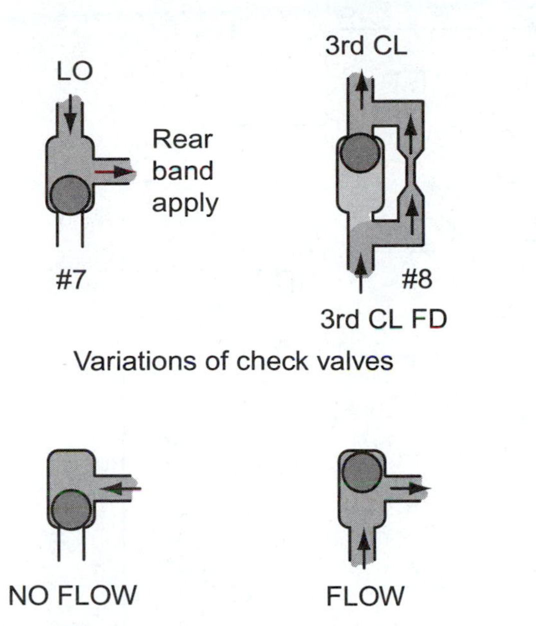


**Figure 6.** Action of a check ball valve.

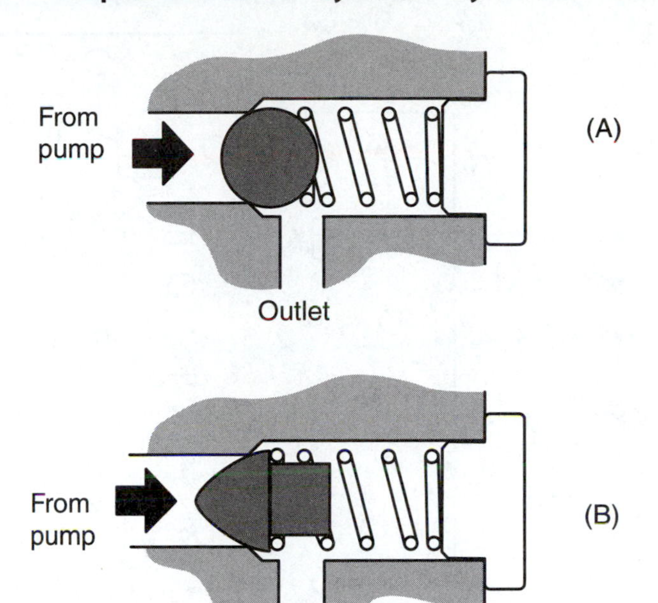


**Figure 8.** A relief valve is typically a (A) ball check valve or a (B) poppet valve with a spring holding it in position.

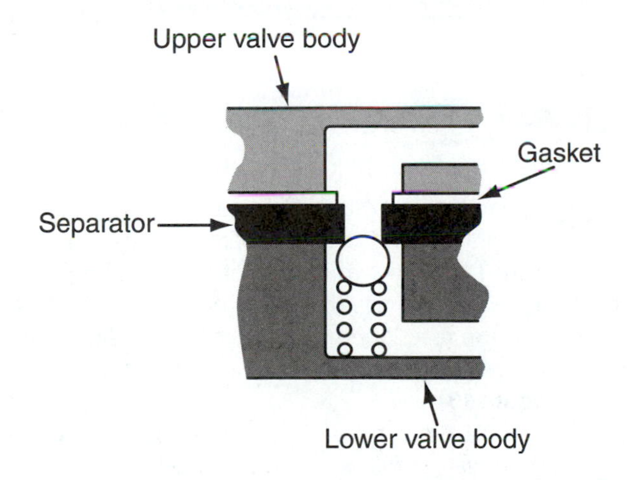


**Figure 7.** The seat for a check ball-type valve may be located in the valve body's separator plate.

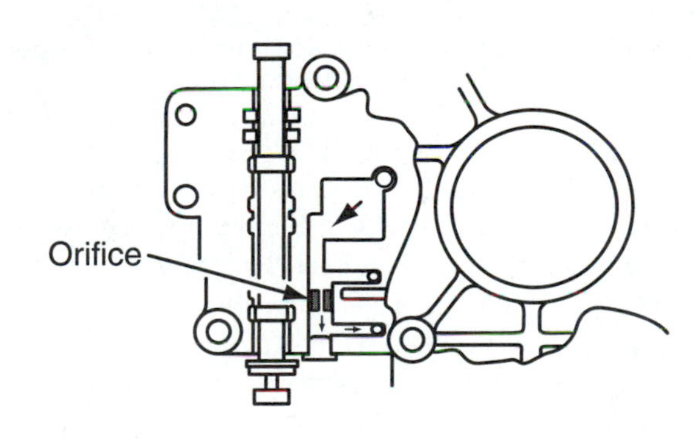


**Figure 9.** Location of a fixed orifice to restrict fluid flow.

The flow is slowed down, which results in higher pressure on the inlet side of the restriction than on the outlet side of the restriction. The opening or size of the orifice will determine the amount of pressure decrease. Pressure drops only when fluid flows through the orifice. When the flow stops, the pressure becomes static or equalized. Orifice sizes are specifically selected to meet the needs of the transmission **(Figure 10)**. Orifices can be cast as part of the hydraulic passages in the transmission's case or valve body, or they can be line plugs with precisely drilled holes.

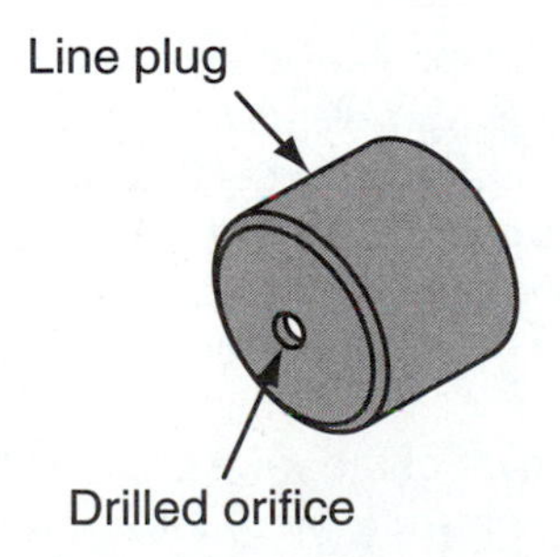


**Figure 10.** The size of the opening in an orifice determines its flow rate.

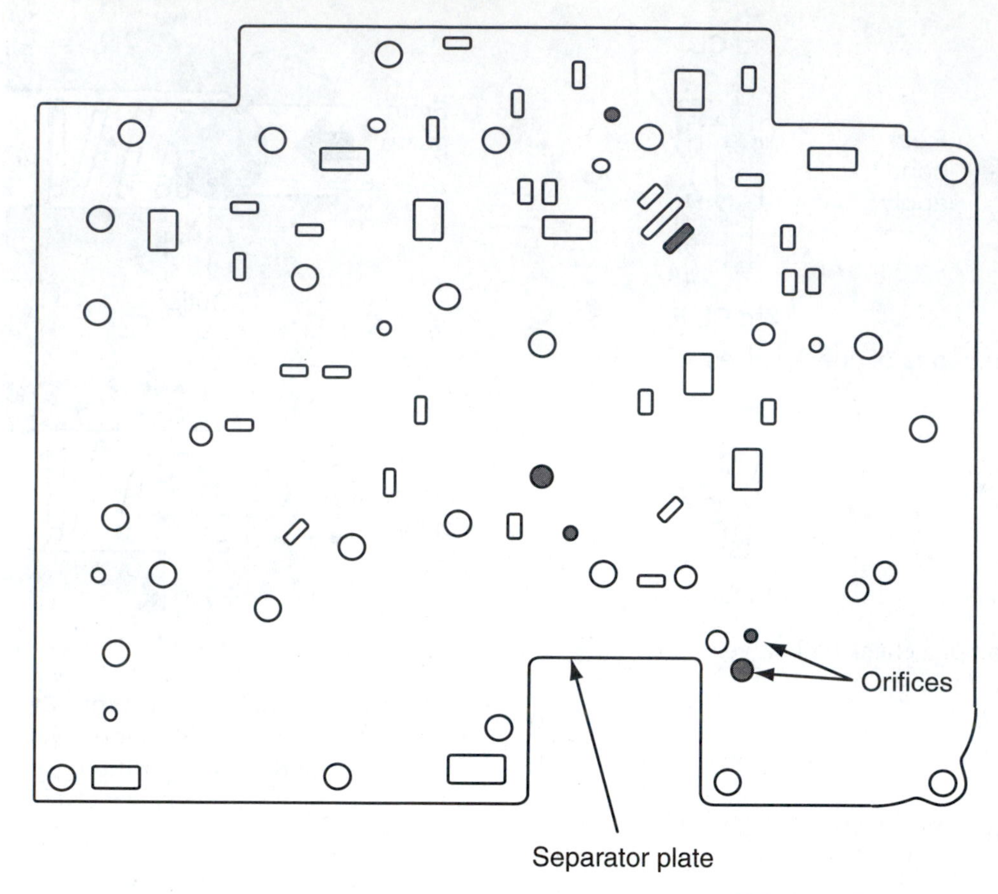


**Figure 11.** A separator plate with orifices and other bores for a valve body assembly.

Often the orifices are found in the valve body's spacer or separator plate and/or the plate to valve body gasket **(Figure 11)**. The other bores and openings are precisely drilled to match the passages in and out of the valve body. The orifices are made smaller than the connecting passages to cause a restriction to fluid flow.

A number of orifices are often placed in series to provide a cushioning effect on the hydraulic system. Any number of restrictions may be present in a line and these allow for a gradual activation of an apply device, which improves shift quality.

## Spool Valves

The lands of a spool valve are connected by the valve's stem. The stem has a smaller diameter than the lands but is not a precisely machined part of the valve. The valley of the valve is made smaller to allow fluid to pass between the lands and through the adjoining ports. The movement of the valve allows fluid to flow from various inlets to various outlets.

The movement of the valve is controlled by hydraulic pressure, spring pressure, mechanical linkage, or a combination of hydraulic and spring pressure. A spool valve will always move in the direction of the greater force. When the valve is controlled by a spring and hydraulic pressure, the tension of the spring keeps the valve at one end of the bore **(Figure 12)**. On the other end of the valve (opposite of the spring), fluid enters through a port. When the fluid pressure on the reaction end of the valve is greater than the tension of the spring, the valve moves against the spring **(Figure 13)**.

A spool valve can have different sized lands (sized in diameter). The larger diameter provides for a greater surface area for the hydraulic pressure to work against. If pressure is applied to the valley between two differently sized lands, the valve will move to the direction of the larger land, simply because there is greater pressure there.

When spool valves with different sized lands are fitted with a spring, the mechanical advantage of hydraulics dictates when the valve will move. If the spring is behind the smaller land, hydraulic pressure on the larger land effectively increases the spring tension on the valve. Hydraulic pressure on the other side of the valve must be now greater before it can overcome the tension of the spring and move the spool valve. If the spring is located behind the larger land, less hydraulic pressure will be needed to overcome the tension of the spring.

Hydraulic flow can move the valves against spring pressure or move the valve back and forth in its bore. The

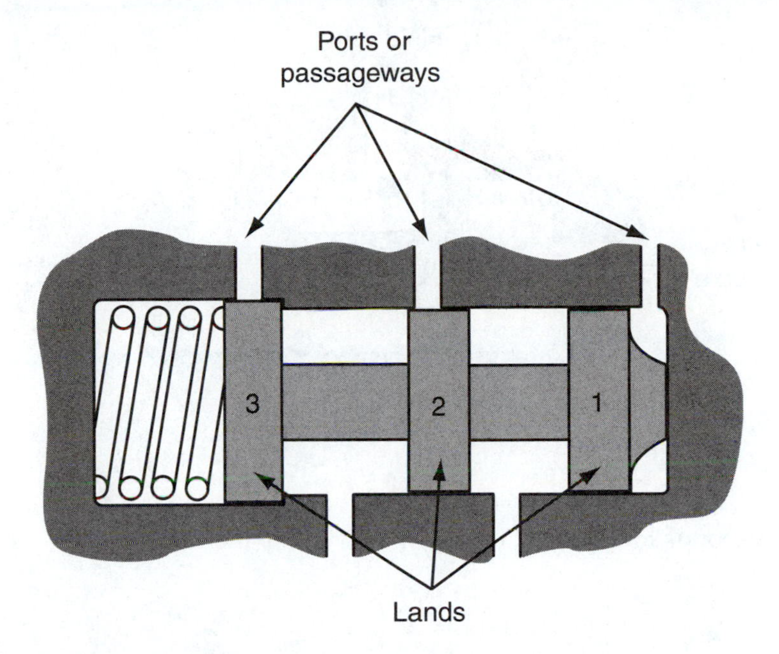


Figure 12. A spring loaded spool valve at rest.

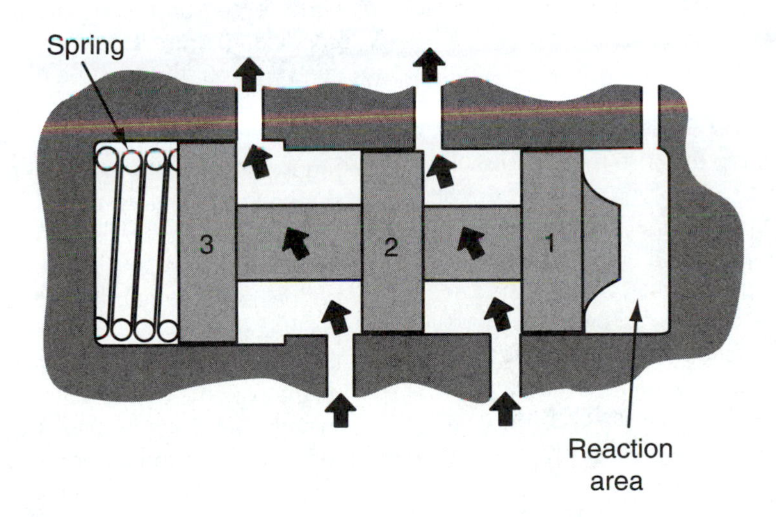


Figure 13. A spring loaded spool valve with sufficient pressure on the reaction area to overcome the tension of the valve's spring.

latter is the case for a regulating spool valve. If the pressure of the fluid increases to levels that exceed the tension of the spring, the valve will move to open an exhaust port, which allows some of the fluid to return to the reservoir. Once the pressure has been reduced to a pressure below the tension of the spring, the valve will move back to its original position. This toggling back and forth controls the fluid pressure at the valve.

## Relay Valves

A **relay valve** is a spool valve that can have several lands and reaction areas. It is used to control the direction of fluid flow and does not control pressure. A constant supply of fluid is fed to the valve and the valve's position directs the fluid to the appropriate hydraulic passageway. If a relay valve is fitted into a bore with one inlet and three separate outlets, the position of the valve will determine where the fluid will be directed.

A relay valve is held in position in a bore by spring tension, auxiliary fluid pressure, or a mechanical linkage. Forces from hydraulic pressure or mechanical linkages oppose spring tension and move the relay valve to a different position in its bore. In each position, the valve's valley may or may not be aligned outlet ports. Fluid flows from an inlet port across the valve's valley to an outlet port, if the port is

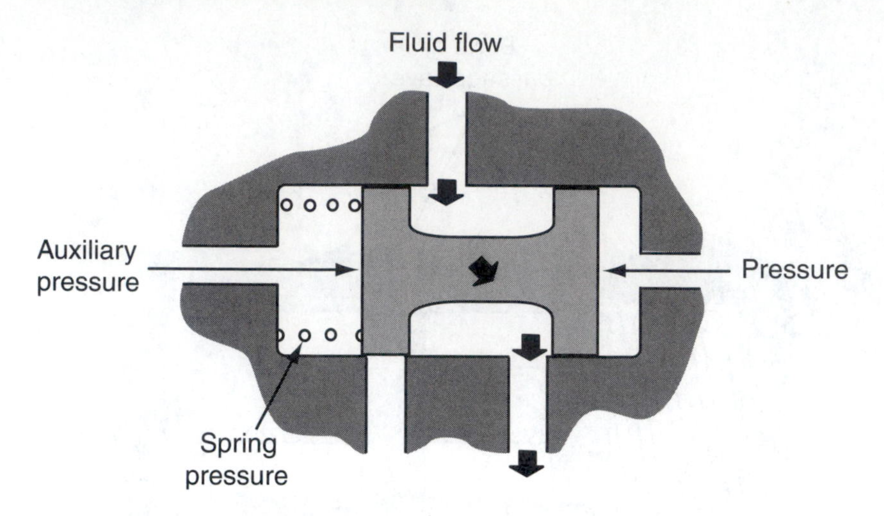


**Figure 14.** A relay valve responds to opposing forces on each end of the valve. A high pressure always moves to a lower pressure.

not blocked by a valve land **(Figure 14)**. If the land is blocking the port, fluid flow to that circuit is stopped.

## Regulator Valves

Regulator valves are valves that change the pressure of the fluid. The pressure change occurs as the position of the valve allows some of the fluid to exhaust. These valves are widely used in transmissions to control line pressure, as well as the pressures used to control or signal shift points.

Regulator valves have a spring that holds the valve in a position until the fluid pressure is great enough to compress the spring. At that time, the valve moves and allows some fluid to exhaust. Once the fluid begins to exhaust, the pressure on the valve decreases. The valve remains in the exhausted position until the fluid's pressure is not great enough to work against the tension of the spring.

## PRESSURE REGULATOR VALVES

Many automatic transmissions use a positive displacement pump. As the pump delivers fluid to the transmission, the pressure increases as the engine's speed increases. Enough pressure can be generated to stall the pump. The pressure regulator valve serves as a relief valve in the pump output circuit to develop, regulate, and maintain mainline pressure.

Pressure regulating valves use principles of both a pressure relief valve and a spool valve. As a relief valve, they respond and move to pressures that are either greater or less than the tension of the valve's spring. As a spool valve, pressure regulator valves toggle back and forth in their bore to open and close an exhaust port. Some pressure regulator valves incorporate a boost valve, which increases the regulated pressure.

The pressure regulator valve usually has three primary purposes and modes of operation: filling the torque converter, exhausting fluid flow, and establishing a balanced condition **(Figure 15)**. When the pump begins to turn and send fluid out through the transmission there is little resistance to fluid flow in the system; therefore, pressure does not build up. During this time, the springs hold the pressure regulator valve in the closed position and fluid flows throughout the transmission.

Although fluid is flowing, no fluid pressure is applied to the apply devices and the torque converter. This condition normally exists when the manual shift valve is in the PARK position **(Figure 16)**. When the manual shift valve is moved, fluid from the pump is sent to the pressure regulator valve. As the pressure on the valve increases beyond a predetermined level, normally about 60 psi, the regulator valve is moved against the tension of the springs. This movement opens an additional outlet port that sends fluid out to the torque converter and to the appropriate apply device, if appropriate **(Figure 17)**. The opening of this port also decreases the pressure in the system.

Fluid flows into the converter circuit and the converter will become pressurized. The pressure regulator valve has several lands, so, when different fluid pressures are applied, converter pressure can vary in proportion to torque requirements. Converter pressure is used to transmit torque and keep ATF circulating in and out of the torque converter. This action reduces the possibilities of forming air bubbles in the fluid and helps to keep the fluid cool. Converter pressures typically do not vary when the converter is in its clutch apply mode of operation.

Once the converter is pressurized, the pressures of the fluid can again begin to increase. Therefore, an additional exhaust port is needed to regulate the pressure from the pump. The resistance to fluid flow at the converter

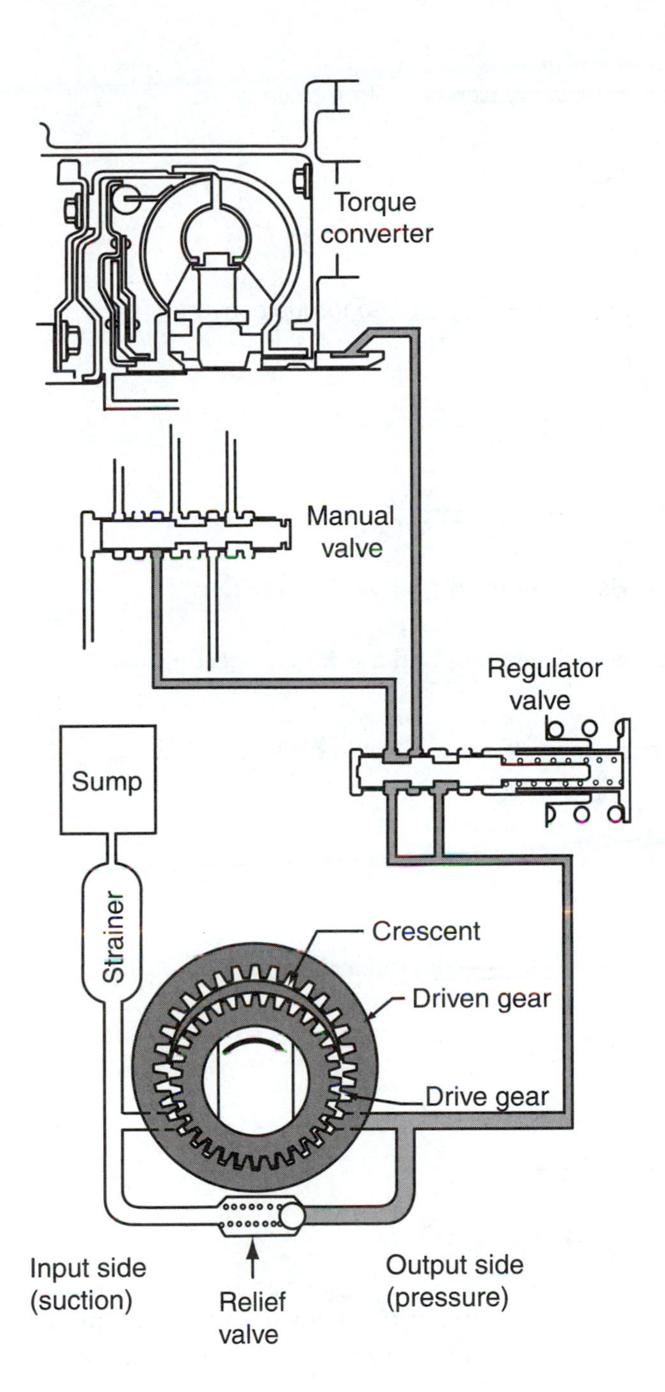


**Figure 15.** Oil pump output is regulated by the pressure regulator valve.

increases pressure, moves the pressure regulator valve, and uncovers the exhaust port. All fluid pressure not needed for transmission operation is exhausted back into the fluid reservoir.

With fluid pressure acting at one end of the valve and spring tension at the opposite end, the pressure regulator valve is in a balanced position. As soon as the fluid pressure is less than the tension of the spring, the valve will move and allow pressure to build up again. This process is ongoing when the transmission is operating. The constant toggling of the pressure regulating valve maintains a constant pressure in the system. The valve's spring controls the mainline pressure. If the pressure begins to decrease, the spring will move the valve and prevent some of the fluid flow to the reservoir to maintain the desired mainline pressure. If the pressure increases, the valve again moves and reopens the exhaust outlet to the reservoir.

## SHIFT VALVES

Shift valves **(Figure 18)** control the upshifting and downshifting of the transmission by controlling the flow of fluid to the apply devices that engage the different gears. Transmissions have several shift valves that provide for control all of possible gear changes. Most shift valves are spool valves operated by two different hydraulic pressure sources that oppose each other. As one pressure gains strength over the other, the valve moves in the direction of the lower pressure.

For example, when the governor sends a signal that would normally force an upshift, throttle pressure may delay the shift, which allows for continued operation in a lower gear when the vehicle still needs the gear reduction. When operating under a load, governor pressure must overcome the throttle pressure plus the spring tension on the shift valve before it can force an upshift.

### Manual Valve

The **manual valve** is a spool valve operated manually by the gear selector linkage. When the gear selector is placed in a forward or reverse gear, the manual valve directs mainline pressure to the circuits that apply the reaction members involved in that operating range. The gear selector linkage positions the manual valve in the valve body. If the driver selects the Drive range, the manual valve is moved into a position that allows fluid pressure to flow across the manual valve valley through its outlet port to activate the forward circuit. If the gear selector is moved to the reverse position, the manual valve is moved to open the reverse inlet and outlet ports to activate the reverse circuit.

### Kickdown Valve

A valve body is also fitted with a kickdown circuit to provide a downshift when additional torque is needed. When the throttle is quickly opened, throttle pressure increases rapidly and is directed to the kickdown valve. This moves the kickdown valve to open a port, which allows mainline pressure to flow against the shift valve. The spring tension on the shift valve, kickdown pressure, and throttle pressure push on the end of the shift valve causing it to move to the downshift position, forcing a quick downshift.

A transmission also can automatically downshift when operating under a load, such as climbing a hill. During this

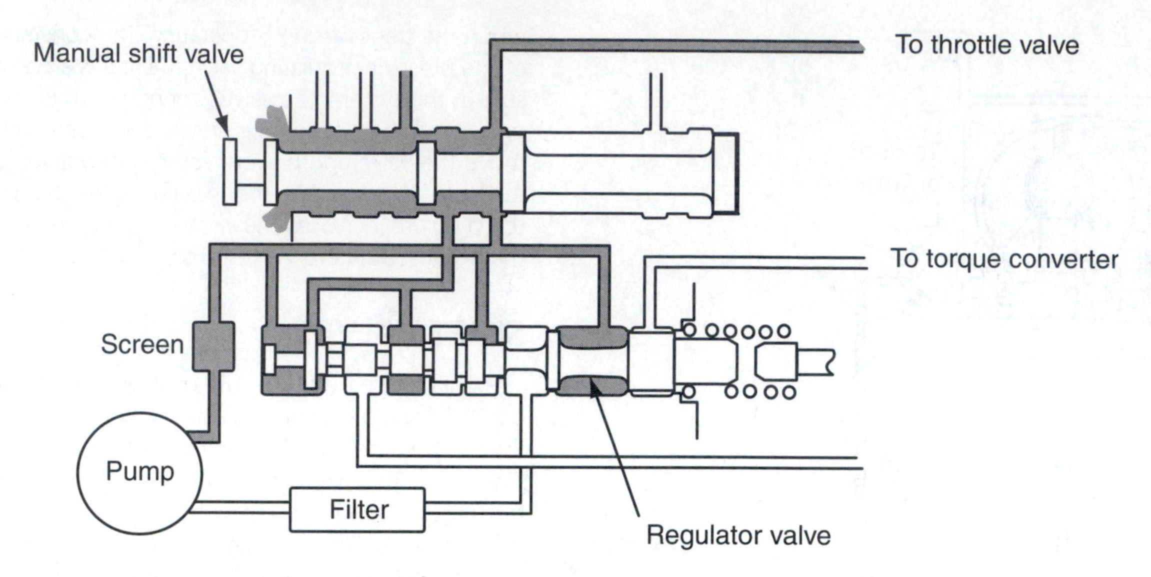


**Figure 16.** The path of fluid in a transmission when the transmission is in park and the engine speed is low.

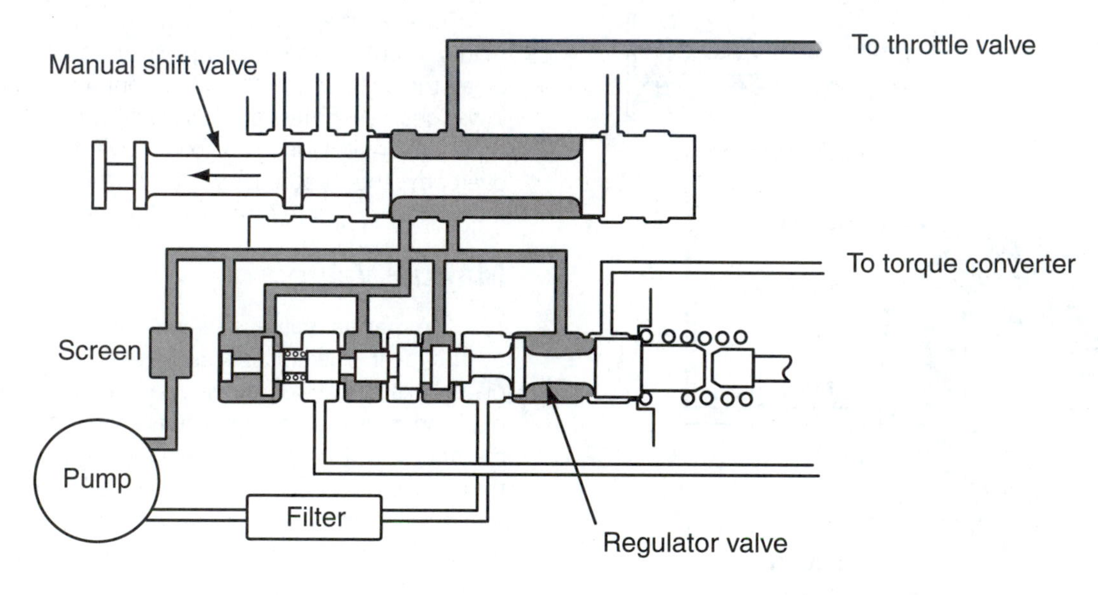


**Figure 17.** The path of fluid in a transmission when the transmission is not in park and power can be transmitted through the torque converter into the transmission.

time, throttle pressure exceeds governor pressure and forces a downshift.

Some transmissions use a solenoid kickdown valve operated electrically by a switch on the linkage of the throttle pedal. This switch senses when the throttle plate is wide-open and allows the kickdown valve to increase throttle pressure, which forces an immediate downshift.

## GOVERNORS

The governor valve controls transmission shift points based on vehicle speed. Fluid flows from the pump throughout the transmission and eventually into the governor circuit. When the flow of fluid reaches the shift valves, the forces behind the valves offer a resistance. This resistance to fluid flow causes a buildup of pressure in the governor circuit. Depending on the type of governor, it will exhaust fluid flow or restrict fluid flow from the circuit, thereby regulating pressure. The governor pressure increases with road speed.

Mechanical governors respond to vehicle speed through the action of movable weights that respond to centrifugal force. As the speed of the output shaft increases, centrifugal force moves the weights farther and farther from their rotating axis. The movement of the weights causes the governor valve to direct more fluid flow to the shift valve.

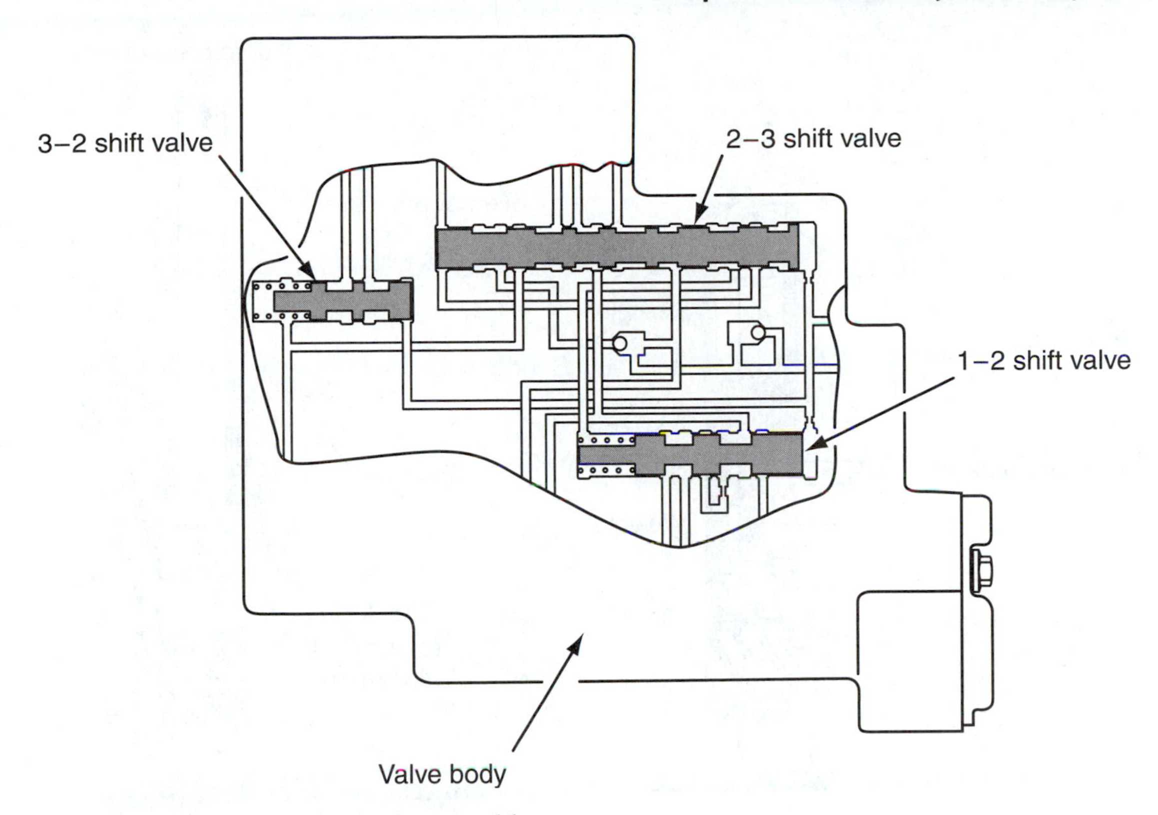


**Figure 18.** The shift valves in a valve body assembly.

As the pressure increases, the spring tension on the shift valve is eventually overcome and the valve moves to the upshifted position. In this position, the inlet and outlet ports of the valve bore are in the same valley of the shift valve. This results in an upshift as fluid pressure flows out of the outlet port and through the connecting worm tracks to engage the apply device for the next higher gear. A decrease in output shaft speed will result in a decrease in pressure and will cause a downshift.

Many different designs of governors have been used in automatic transmissions. All governors, regardless of the design, provide a variable pressure in relation to road speed. The governors in automatic transmissions either rotate with, or are turned by, the transmission's output shaft. Governors are mounted in the transmission case and driven off the output shaft by a worm gear or they are mounted directly on the output shaft. On many FWD vehicles, the governor is driven by a gear on the final drive unit.

The governor assembly normally consists of a small separate valve body with fluid passages for line pressure, governor pressure and an exhaust port to send fluid back to the reservoir. The governor's valve body has one or two valves controlled by the weights and spring tension. When the vehicle is stopped, fluid to the governor is blocked. As the vehicle begins to move, the governor rotates and centrifugal force will begin to move the valves and/or weights. This allows mainline fluid flow to enter the governor circuit, which sends fluid to the shift valves.

## Types of Governors

There are three common designs of governors: shaft-mounted, gear-driven with check balls, and gear-driven with a spool valve **(Figure 19)**. Most of these designs have primary and secondary weights, springs, and valves connected to the output shaft by a valve shaft. The heavier primary weights move out at low speeds and the lighter secondary springs move out at higher speeds.

The shaft-mounted governor has a spring-loaded spool valve attached to the output shaft of the transmission. The governor's weights are mounted so that the primary weight moves against spring tension while the secondary weight moves against mainline pressure. The combined force of the weights and springs on one side of the output shaft is much greater than the force of the valve on the other side. Therefore, as the output shaft rotates faster, the valve is pulled inward by the valve shaft. As speed increases, fluid is exhausted out through a port that feeds pressure to the shift valves.

As the output shaft's speed increases, the governor gradually opens the circuit and increases governor pressure, until the governor pressure equals mainline pressure. At this point, fluid is exhausted and the pressure decreases. After the fluid is exhausted, the valve once again is pulled inward by the rotating weights, allowing mainline pressure into the governor passage and the cycle starts again.

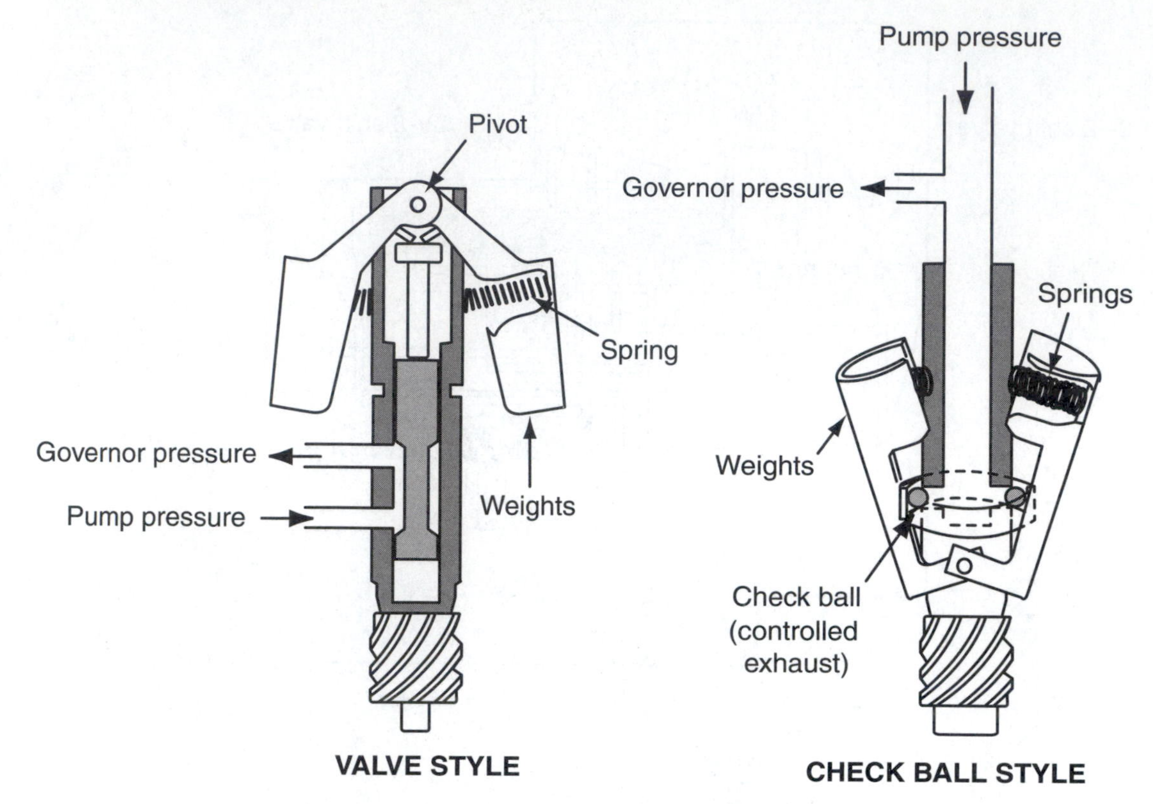


**Figure 19.** Gear driven, shaft-mounted governors with a spool valve and with check balls.

It is not unusual for a governor to still be regulating at 60 or 70 mph. Some may regulate at even higher speeds depending on the weights and the spring tension calibration. When the vehicle is stopped, the governor spring forces the valve closed, which blocks mainline pressure from the governor circuit.

Gear-driven governors with check balls mount to the transmission at a right angle to the output shaft. The centrifugal weights directly on the check balls that open or close two fluid passages. The check balls are forced open by mainline pressure to exhaust fluid. When the vehicle is stopped, the check balls unseat and fluid flow to the governor is allowed to bleed off. As vehicle speed increases, the weights apply force to seat the check balls and restrict the flow of escaping fluid. This increases governor pressure, up to the point where no fluid escapes and governor pressure equals mainline pressure.

The gear-driven governor with a spool valve is usually driven by, and mounted at a right angle to, the transmission output shaft. Hydraulically, this design operates much like the shaft-mounted governor; however, its mechanical operation is different. The weights move by centrifugal force, indirectly control the spool valve through levers attached to the weights. The weights also serve as levers. They act against a spool valve located in the governor shaft. As vehicle speed increases, centrifugal force moves the weights out causing the spool valve to move. This closes the exhaust port and directs fluid to the shift valves.

## Vehicle Speed Sensors

Electronically controlled transmissions do not rely on hydraulic signals from a governor to determine when to shift, nor are they fitted with a speedometer cable. Instead, they use permanent magnet (PM) generators to sense vehicle speeds.

A PM generator is an electronic device that utilizes a magnetic pickup sensor located on the transmission/transaxle housing and a trigger wheel mounted on the output shaft. The trigger wheel has a number of projections evenly spaced around it. As the wheel rotates, the projections move by the magnetic pickup. The passing of the projections produces a voltage in a coil inside the pickup assembly. This voltage is actually AC voltage, but is changed to a digital DC voltage by a converter before it is sent to the TCM. The frequency of the pulsations and the amount of voltage generated by the pickup assembly is translated into vehicle speed by the computer.

## LOAD SENSORS

Shift timing and quality should vary with engine load, as well as with vehicle speed. Throttle pressure increases the fluid pressure applied to the apply devices of the planetary units to hold them tightly to reduce the chance of slipping while the vehicle is operating under heavy load. Throttle pressure interacts with mainline pressure to coordinate shift points and shift quality with vehicle and engine load. On some transmissions, a vacuum modulator is used to control throttle pressure **(Figure 20)**. Transmissions that are not equipped with a vacuum modulator use a throttle cable **(Figure 21)** or electronic devices, such as a MAP sensor, to sense engine load and change fluid pressures. It is not safe to assume that not all electronic transmissions are equipped with a vacuum modulator, some are. For example, GM's 4T60-E uses a vacuum modulator, in addition to electronic inputs and shift solenoids, to control shift timing and quality.

The terms modulator and throttle are used to describe the same function and pressure in a transmission. Although a vacuum modulator is very different from an engine load sensor, they both generate a hydraulic engine torque signal based on engine load. As a result of this signal, the transmission has many different possible shift points. Each of those shift points is dictated by engine speed and load. This feature allows for a delay in shifting when load demands are high.

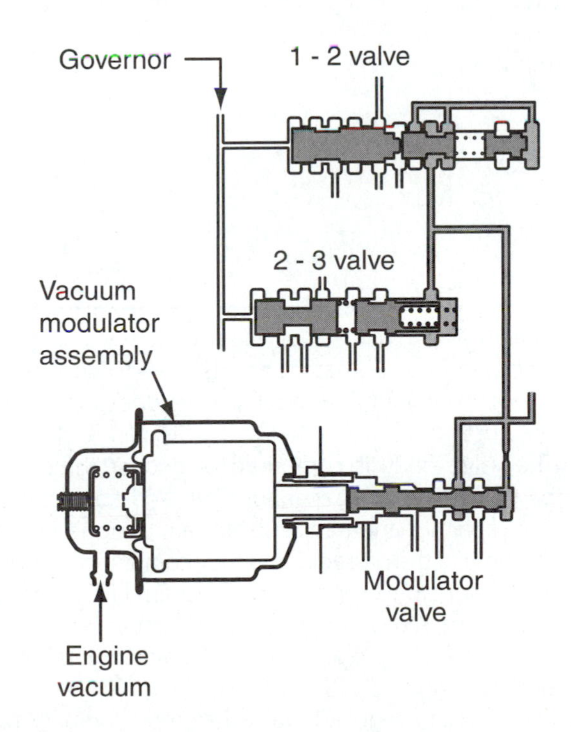


**Figure 20.** Action of a vacuum modulator and the modulated pressure it generates.

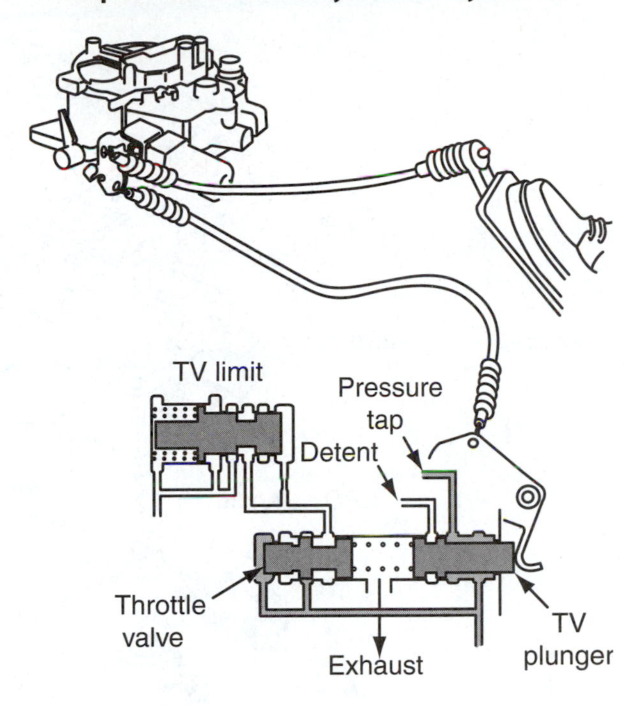


**Figure 21.** Typical cable-type throttle pressure control system.

## ELECTRONIC SHIFT-TIMING CONTROLS

Most current transmissions use electronics to control shift timing and quality. Shifting electronically allows for shifting at more optimum times than does shifting by hydraulics. The use of electronics also reduces the occurrence of many common transmission problems, such as dirty, seized, or worn springs, check balls, valves, and orifices.

Shifting is based on inputs to a control computer from various sensors, such as engine temperature, engine speed, engine load, vehicle speed, throttle position, and gear selector position. The computer may be the vehicle's engine control computer or one designated just for transmission control. The computer compares the information from the sensors against the shifting instructions programmed into it. The computer then controls the appropriate solenoid valves to provide for optimum shift timing.

Most late-model transmissions use an EPC solenoid **(Figure 22)** to regulate transmission pressure. These solenoids regulate pressure according to the current that flows through its windings. The resultant magnetic field moves the solenoid's valve, which in turn regulates the pressure. The current is provided by the PCM. The PCM relies on inputs from the electronic control system to determine the required and desired fluid pressure. The EPC solenoid typically takes the place of the throttle valve and/or vacuum modulator.

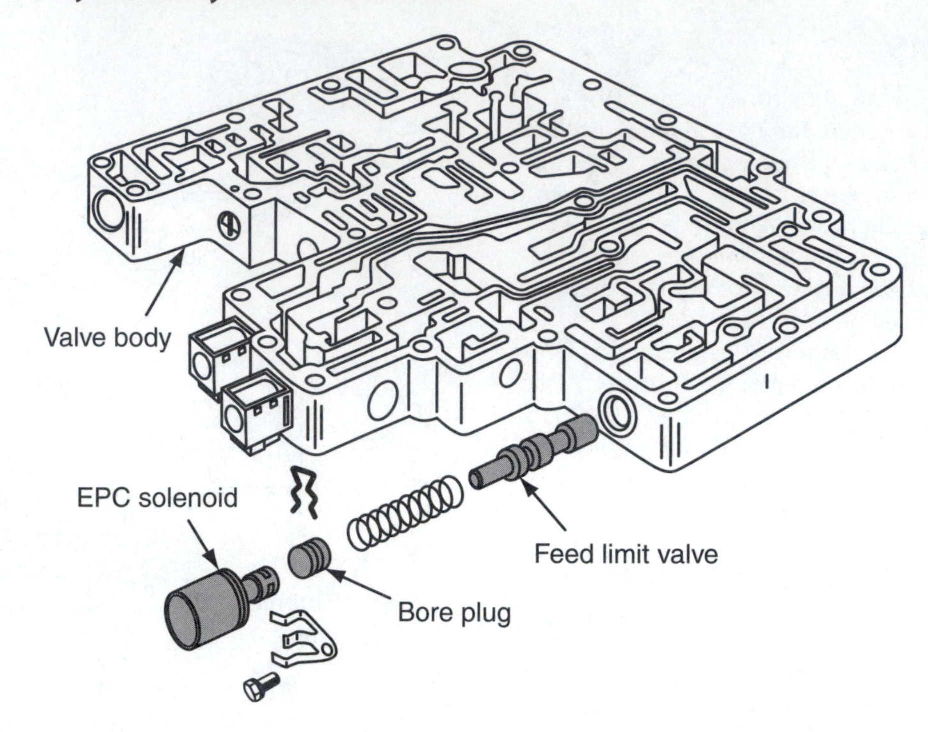


**Figure 22.** Location of an Electronic Pressure Control (EPC) solenoid in a valve body.

The PCM changes the amperage to the solenoid by altering the duty cycle of the solenoid. High amperage (long duty cycle) from the PCM causes minimum line pressure and low amperage (short duty cycle) results in high pressure.

## Solenoid Operated Valves

Most solenoid operated valves are ball-type valves, which open and close a hydraulic passage. They are designed to block fluid flow when voltage is applied to the solenoid and to allow fluid flow when voltage is not applied. Most transmissions have two solenoid valves, which control the shifting through all of the forward gears. By controlling which solenoid is energized, the computer controls the shift timing.

# *Summary*

- The valve body is the master flow control for an automatic transmission. It contains the passages and numerous bores fitted with many spool-type and other valves. Each passage, bore, and valve forms a specific oil circuit with a specific function.
- A relief valve prevents or allows fluid flow until a particular pressure is reached. Then, it either opens or closes the passageway. Relief valves are used to control maximum pressures in a hydraulic circuit.
- Orifices are used to regulate and/or control fluid volumes.
- Fluid passages in the bore are closed and opened as the lands of the spool cover and uncover the passage openings.
- The movement of a spool valve is controlled by either mechanical or hydraulic forces.
- A relay valve is a spool valve with several spools, lands, and reaction areas and is used to control the direction of fluid flow; a relay valve does not control pressure.
- The valve body is comprised of two or three main parts: a valve body, separator plate, and transfer plate. The separator and transfer plates are designed to seal off some of the passages and they contain some openings that help to control and direct fluid flow through specific passages. The three parts are typically bolted together and mounted as a single unit to the transmission housing.
- The pressure regulator valve has three primary purposes and modes of operation: filling the torque converter, exhausting fluid pressure, and establishing a balanced condition.

- Shift valves control the up and downshifting of the transmission by controlling the flow of fluid to the apply devices that engage the different gears.
- The manual valve is a spool valve operated manually by the gear selector linkage.
- There are three common designs of governors: shaft-mounted, gear-driven with check balls, and gear-driven with a spool valve.
- A PM generator is an electronic device that utilizes a magnetic pickup sensor located on the transmission/transaxle housing and a trigger wheel mounted on the output shaft.

## *Review Questions*

1. What is the purpose of the transfer and separator plates in a valve body assembly?
2. What is the purpose of the shift valves?
3. Technician A says that a relay valve is used to control the direction of fluid flow. Technician B says that a control valve is used as a flow-directing valve. Who is correct?
   A. Technician A only
   B. Technician B only
   C. Both Technician A and Technician B
   D. Neither Technician A nor Technician B
4. Which of the following valves does not regulate pressure?
   A. Manual valve
   B. Throttle valve
   C. Pressure regulator valve
   D. Governor valve
5. Technician A says that the movement of a spool valve is controlled by mechanical forces. Technician B says that the movement of a spool valve is controlled by hydraulic forces. Who is correct?
   A. Technician A only
   B. Technician B only
   C. Both Technician A and Technician B
   D. Neither Technician A nor Technician B

# Chapter 21 Transmission Pumps

## Introduction

A transmission's fluid pump is the source of all pressure in the system. It provides a flow of fluid, which transmits a force. The force, in turn, can cause movement; therefore, the fluid is able to do work. This work moves valves and the fluid moves through the valves to apply brakes and clutches.

### PUMPS

The torque converter and transmission rely on the pump **(Figure 1)** to provide the circulation of the ATF through the transmission. Although the pump is the source of all fluid flow through the transmission, the valve body regulates and directs the fluid flow to provide for the gear changes. The pump is driven by the hub of the torque converter or a shaft from the converter; therefore, it operates whenever the engine is running. Three types of pumps are commonly used: the **gear-type**, **rotor-type**, and **vane-type pumps**.

**Figure 1.** A typical transmission pump.

Gear and rotor-type pumps are considered **fixed displacement pumps**. Displacement is the volume of fluid moved or displaced during each cycle of a pump. Fixed displacement means the inside of the pump does not change. For each shaft rotation, there is a fixed amount of fluid that a fixed displacement pump causes to flow. The flow rate of these pumps depends on rotational speed. As the speed increases, so does the flow rate. If at a given speed, the flow rate requirements decrease; the pump cannot decrease the flow, thereby wasting energy and horsepower.

Modern transmissions commonly use vane-type pumps, because they are usually variable in displacement. If, at a given speed, the flow rate requirements decrease, the inside of the pump has the ability to decrease the displacement causing a decrease in fluid flow. This saves energy. With a variable displacement pump, decreasing fluid output can happen without a decrease in speed.

A transmission's pump is commonly called the front pump because of its location at the front of the housing. The pump is a separate unit that is sealed to the housing by an outer seal **(Figure 2)** and a gasket **(Figure 3)**. This prevents the fluid from leaking out and maintains the pressure built up by the fluid flow. Pumps also may be fitted with a plate installed between the backside of the pump and the housing **(Figure 4)**. This plate directs the fluid, both in and out of the pump, to the correct passages within the transmission.

### Pump Drives

Automatic transmission pumps are driven by the engine's crankshaft through the torque converter housing.

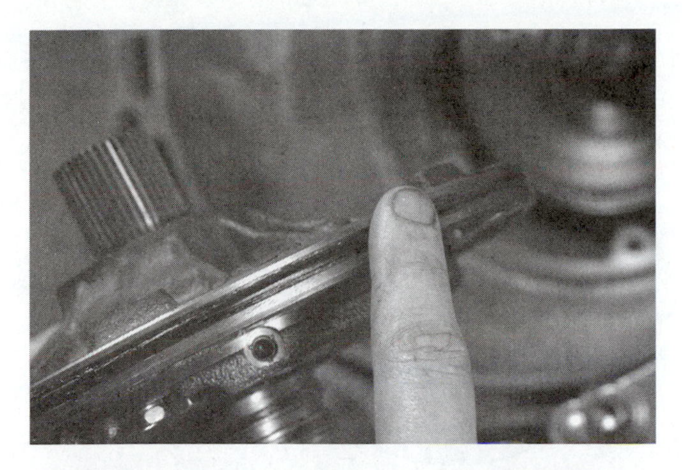

Figure 2. The outer seal on a transmission pump.

Figure 3. A pump and its gasket.

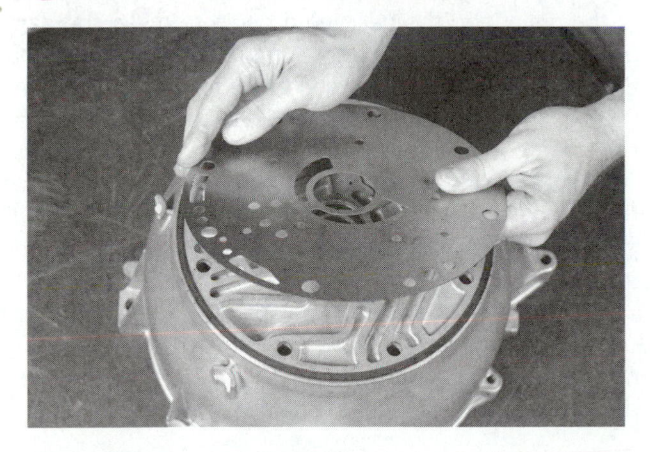

Figure 4. A typical mounting or separator plate that fits between the pump and the housing.

The pumps in all RWD and some FWD transmissions are driven externally by the torque converter's drive hub **(Figure 5)**. The hub has two slots or flats machined in it, which fit into the pump drive. As the converter rotates, the pump is driven directly by the hub. Many FWD transaxles use a hex-shaped or splined shaft fitted in the center of the converter cover to drive the pump **(Figure 6)**. This system is

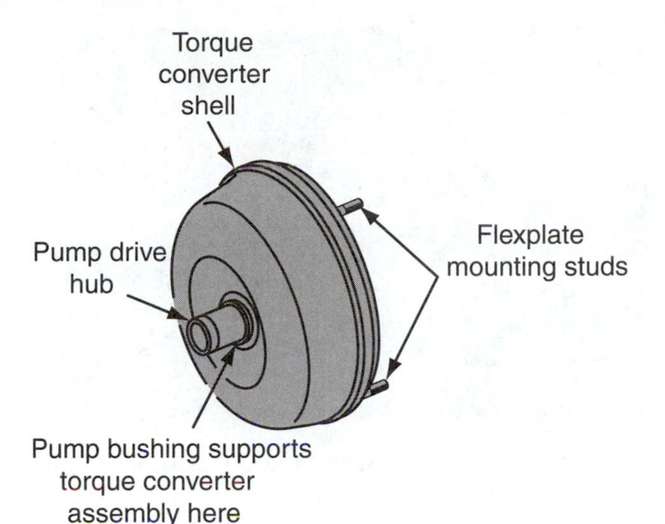


Figure 5. Pump drive on a typical torque converter.

Figure 6. A typical oil pump drive shaft used on some transmissions and transaxles.

called an internal drive. The pump shaft is sealed to prevent fluid from leaking past it **(Figure 7)**.

The Honda CVT uses a chain drive to drive the pump. Many older and a few late-model transmissions use a second pump mounted at the rear of the transmission case and driven by the output shaft. This second pump allows for pump operation whenever the output shaft is rotating, such as when the vehicle is being push-started.

## BASIC OPERATION

During pump operation, ATF moves from an area of high pressure to an area of low pressure. It begins by the pump creating a low pressure area at its inlet port. The oil

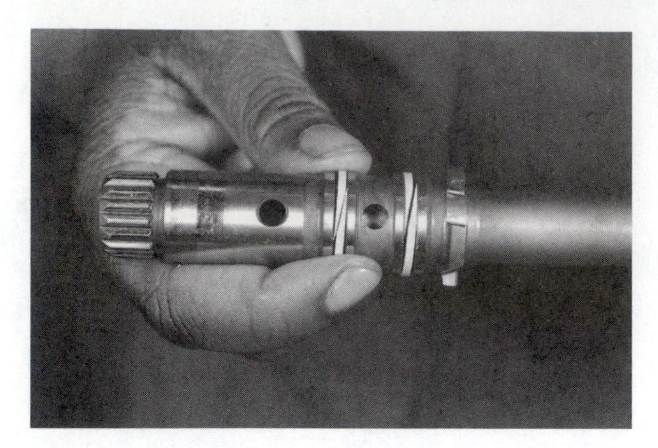

Figure 7. The pump drive shaft is sealed to the inside of the pump.

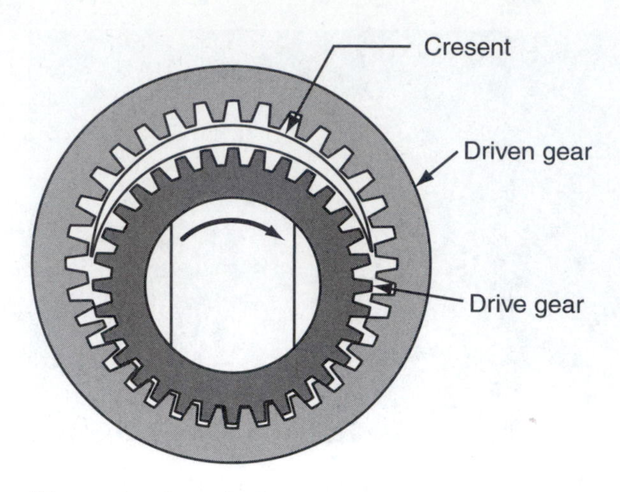


Figure 8. A typical crescent-type pump.

pan is vented to the atmosphere and atmospheric pressure forces the fluid into the inlet port of the pump. The fluid is now at atmospheric pressure and is moved by the pump through the outlet port, causing fluid flow. What happens next depends on the type of pump.

## Gear-Type Pumps

External tooth gear-type pumps consist of two gears in mesh to cause fluid flow. The gears rotate on their own shafts and in opposite directions. The gears are assembled in a housing that surrounds and totally encloses the gears. The shafts are sealed in the housing by bushings and the housing is normally sealed with a cover. In order for the pump to create low inlet pressures, the pump housing must be sealed. Some gear-type pumps use a wear plate instead of a pump cover to seal the gears in the housing.

One pump gear may be driven by the torque converter and drives the other gear. As the gears rotate, the gear teeth move in and out of mesh. As the teeth move out of mesh, inlet oil is trapped between the gear teeth and the walls of the housing. The trapped oil is carried around with the teeth until the gear again meshes. At this time, the meshing of the teeth forces the oil out. The continuous release of trapped oil provides for flow as the fluid is pushed out of the pump's outlet port. The meshing of the gears also forms a seal that stops the fluid from moving out the inlet port. Atmospheric pressure on the fluid ensures that the gap between the gear teeth will be refilled with fluid.

This type of pump is made with close tolerances. Excessive wear or play between the teeth of the gears or between the gears and the housing or pump cover will reduce the output and efficiency of the pump.

The most common type of gear pump is the **crescent** pump, which also uses two gears. One gear of this pump has internal teeth and the other has external teeth. The smaller gear with external teeth is in mesh with one part of the larger gear. In the gap where the teeth are not meshed is a crescent-shaped separator **(Figure 8)**. This crescent is a half-moon shaped part that isolates the inlet side of the pump from the outlet side of the pump. The small gear is driven by the torque converter and it drives the larger gear. The gears' teeth mesh tightly together. As the gears rotate, the teeth mesh then and separate. As they separate, a low pressure is created between the gear teeth. The fluid is pushed by atmospheric pressure into this void until it is full.

As the gears rotate, oil is trapped between the teeth and the crescent. The fluid is then carried around the pump housing toward the outlet port. As the gear teeth get close to the outlet and the gear teeth begin to mesh, the gap between the gear teeth begins to narrow. This narrowing of the gap continues until the gear teeth fully mesh. The narrowing of the gap squeezes the fluid and forces it through the outlet port as the gears move into mesh. A continuous flow of oil toward the outlet port pushes the fluid out into the transmission's hydraulic circuit. The crescent blocks the pressurized fluid and prevents it from leaking back toward the pump's inlet port.

Crescent pumps are made with close tolerances and can deliver the same amount of fluid each time the gear makes one complete revolution. The rate of fluid delivery does change with changes in engine speed.

## Rotor-Type Pump

A rotor-type or gerotor pump is a variation of the gear-type pump. However, instead of gears, an inner and outer rotor turns inside the housing **(Figure 9)**. A rotor utilizes rounded lobes instead of teeth. Crescent and rotor-type pumps are sometimes referred to as **IX pumps**; IX is an abbreviation that refers to the internal/external design of the gears or lobes used in these pumps.

The lobes of one rotor mesh with the recess area between the lobes of the other rotor. The torque converter

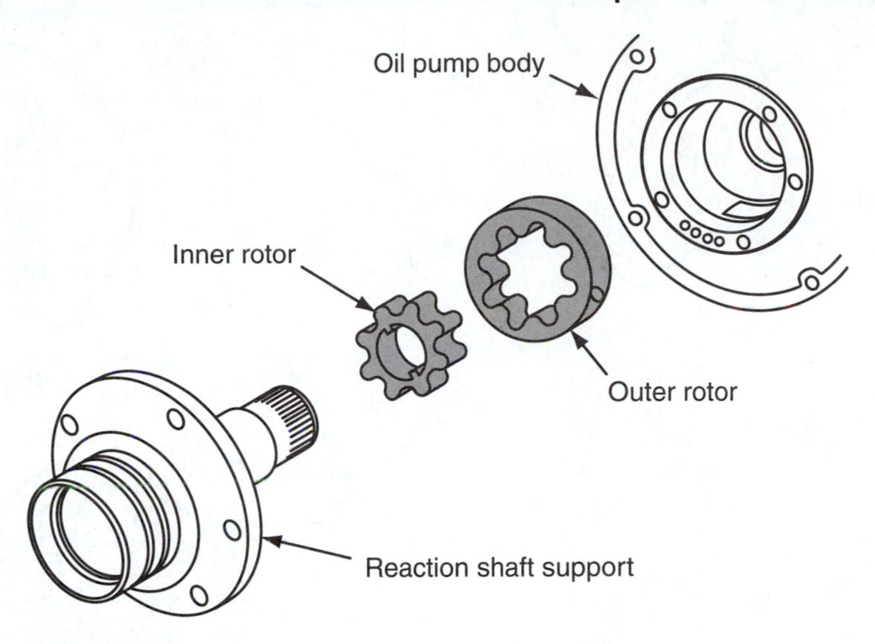


**Figure 9.** A typical rotor-type pump.

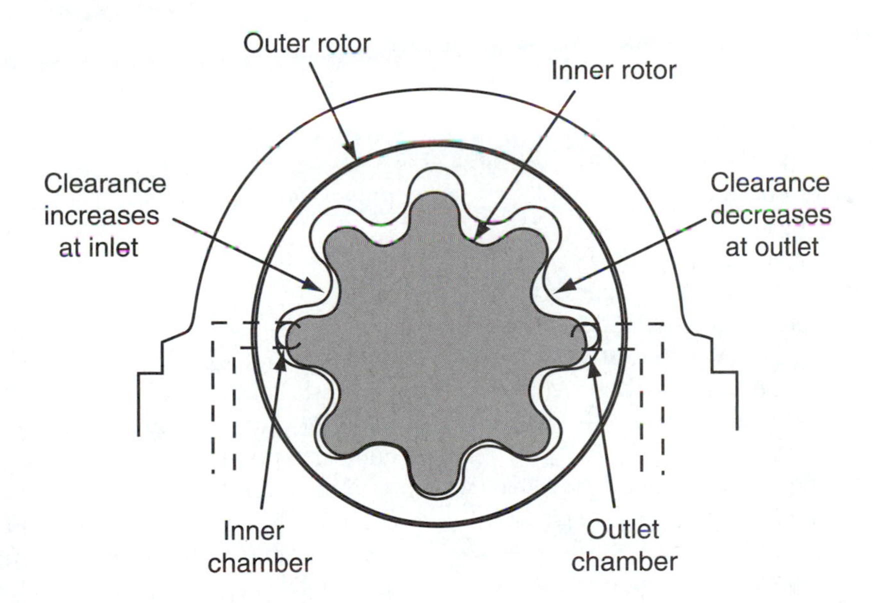


**Figure 10.** Action of rotors to move the fluid.

drives the inner rotor, which rotates inside the outer rotor or rotor ring. The inner rotor has one less lobe than the ring, so that only one lobe is engaged with the outer ring at any one time.

Fluid is carried in the recess between the lobes and squeezed out toward the outlet port as the lobe moves into the recess of the outer ring **(Figure 10)**. A constant supply of fluid being forced out of outlet port supplies fluid for the operation of the transmission. Fluid is prevented from backing up into the inlet port by the action of the lobes sliding over the lobes of the outer ring. The lobe-to-lobe contact causes them to seal against each other and creates small fluid chambers whose volumes increase as the lobes separate on the inlet side. This creates low pressure that allows fluid to be pushed into the pump inlet. As the lobes move into the recesses of the outer ring, oil is squeezed out.

## Vane-Type Variable Capacity Pumps

Many late-model automatic transmissions are equipped with vane-type pumps **(Figure 11)**. This type of

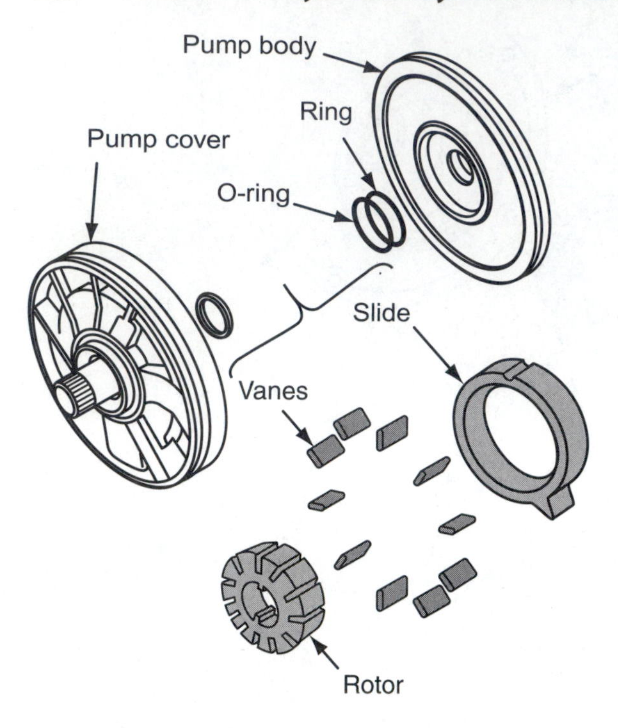


**Figure 11.** Typical vane-type pump.

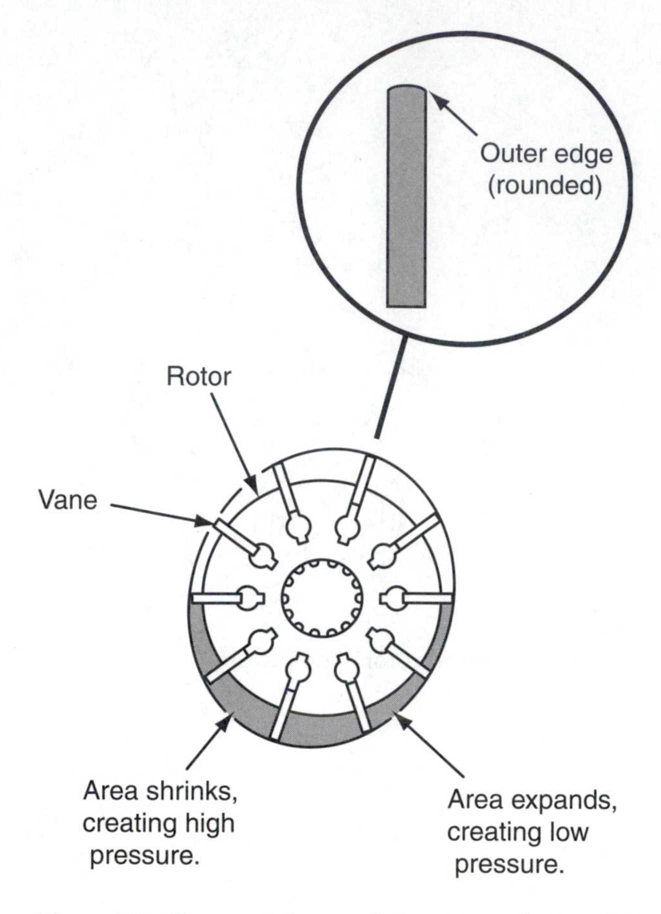


**Figure 12.** The position and location of the slides determines the output of the pump.

pump is a **variable-capacity pump** or variable displacement pump whose output can be reduced when high fluid pressures are not necessary. To monitor and control output flow, a sample of the output is applied to the backside of the pump's slide. The pressure moves the slide against spring pressure and changes its position in relation to the rotor. This action controls the size variations between the chamber at the inlet and the outlet ports, which in turn controls the output pressure of the pump.

The rotor and vanes of the pump are contained within the bore of a movable slide that is able to pivot on a pin. The position of the slides determines the output of the pump **(Figure 12)**. When the slide is in the fully extended position, the slide and rotor vanes are at maximum output. As the rotor and vanes rotate within the bore of the slide, the changes in the chamber size form low- and high-pressure areas. Fluid trapped between the vanes at the inlet port is moved to the outlet side. As the slide pivots toward the center or away from the fully extended position, a greater amount of fluid is allowed to move from the outlet side back to the inlet side. When the slide is centered, there is no output from the pump and a neutral or no-output condition exists. Because the slide moves in response to the sample of pressure sent to it, the slide can be at an infinite number of positions.

The output of a variable-capacity pump will vary according to the needs of the transmission not according to the engine's speed. Therefore, it does not waste energy as a positive displacement pump does. A variable-capacity pump can deliver a large volume of fluid when the need is great, especially at low pump speeds. At high pump speeds, the pressure requirements are normally low and the variable-capacity pump can reduce its output accordingly **(Figure 13)**. Once the needs of the transmission are met, the pump delivers only the amount of fluid that is needed to maintain the regulated pressure.

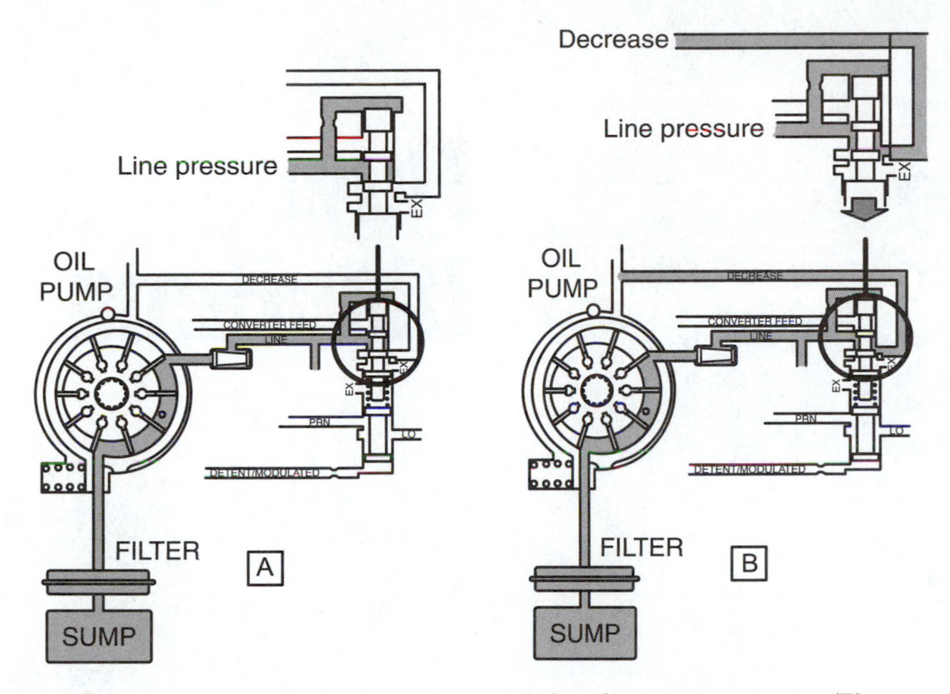


**Figure 13.** Action of vane-type pump, maximum output (A) and minimum output (B).

## Summary

- Three types of pumps are commonly used: the gear-type, rotor-type, and vane-type.
- The pumps in all RWD and some FWD transmissions are driven externally by the torque converter's drive hub. The hub has two slots or flats machined in it, which fit into the pump drive.
- Many FWD transaxles use a hex-shaped or splined shaft fitted in the center of the converter cover to drive the pump.
- When the pump is operating, the pump creates a low pressure area at its inlet port. The oil pan is vented to the atmosphere and atmospheric pressure forces the fluid into the inlet port of the pump. The fluid is now moved by the pump and forced out, at a higher pressure, through the outlet port.
- Gear-type pumps consist of two gears in mesh to move the fluid.
- Another common type of gear pump is the crescent pump, which also uses two gears. One gear of this pump has internal teeth and the other has external teeth. The smaller gear with external teeth is in mesh with one part of the larger gear.
- A rotor-type pump is a variation of the gear-type pump. However, instead of gears, an inner and outer rotor turns inside the housing.
- Vane-type pumps are variable-displacement pumps whose output can be reduced when high fluid pressures are not necessary.

## Review Questions

1. What is a fixed displacement pump?
2. What is an IX pump?
3. What is the most common type of pump used on today's vehicles, and why?
4. Basically describe how a variable vane-type pump regulates its output pressure.
5. Technician A says that in a vane-type pump, the vane ring contains several sliding vanes that seal against a slide mounted in the pump housing. Technician B says that vane-type pumps are variable-displacement pumps whose output can be reduced when high fluid pressures are not necessary. Who is correct?
   A. Technician A only
   B. Technician B only
   C. Both Technician A and Technician B
   D. Neither Technician A nor Technician B

# Flow Diagrams

## Introduction

The valve body is the master flow control for an automatic transmission. It contains the passages and numerous bores fitted with many spool-type and other type valves. Each passage, bore, and valve forms a specific **oil circuit** with a specific function. These oil circuits can be identified through the use of flow diagrams. These diagrams also can be used to clearly see how each valve functions to allow the transmission to operate in any particular gear range.

By tracing through an oil circuit, the fluid flow through passages, bores, and valves to the components that make the actual shift can be traced. An automatic transmission has a circuit for each of its functions.

To understand how a particular transmission works, locate a flow diagram for that transmission and trace the oil circuit for each gear. By carefully studying the oil circuits in this chapter, you will not only gain a better understanding of the operation of an automatic transmission but you also will be able to understand the oil circuits of other transmissions.

## FLOW DIAGRAM

We will discuss the fluid flow for a GM 4T60-E transmission **(Figure 1)**. This is a commonly used unit and many variants of it are also used. The 4T60-E is an electronically controlled four-speed automatic transaxle with a torque converter clutch. We also will use flow diagrams for the transaxle as we discuss the various gear selector positions and operating gear ranges. It may be wise to review the first diagram, become familiar with the legend, and then locate the major valves and components of this transmission on the diagram.

### Park or Neutral

When the gear selector is in Park or Neutral **(Figure 2)** and the engine running, fluid is pulled from the transmission's oil pan, through the filter, and into the pump. The pump supplies the line pressure, according to the calibration of the pressure regulator. Line pressure is directed to the following:

*Manual valve*—Controlled by the gear selector, directs line pressure into the PRN passage in Park, Reverse, and Neutral.

*Pressure modulator valve*—Regulates pump output in response to spring force on the pressure regulator valve, modulator pressure, and pressure at the reverse boost valve.

*Modulator valve*—Controlled by the vacuum modulator, it regulates modulator pressure inversely to engine vacuum. Modulator pressure is directed to the line boost valve, 1–2 accumulator valve, secondary 1–2 accumulator valve, 2–3 accumulator valve, and 3–4 accumulator valve.

*2–3 shift valve*—Line pressure passes through the 2–3 shift valve and feeds the input clutch feed passage. Line pressure then passes through the 3–4 shift valve, around the #3 check ball to apply the input clutch. Although the input clutch is applied and the input sprag clutch is holding, they are not effective because neither the forward band nor the reverse band is applied.

*Converter clutch valve*—Spring force and line pressure holds the valve against the converter clutch solenoid allowing converter feed pressure into the release passage and TCC accumulator passage. Release fluid is then directed to the spring side of the TCC

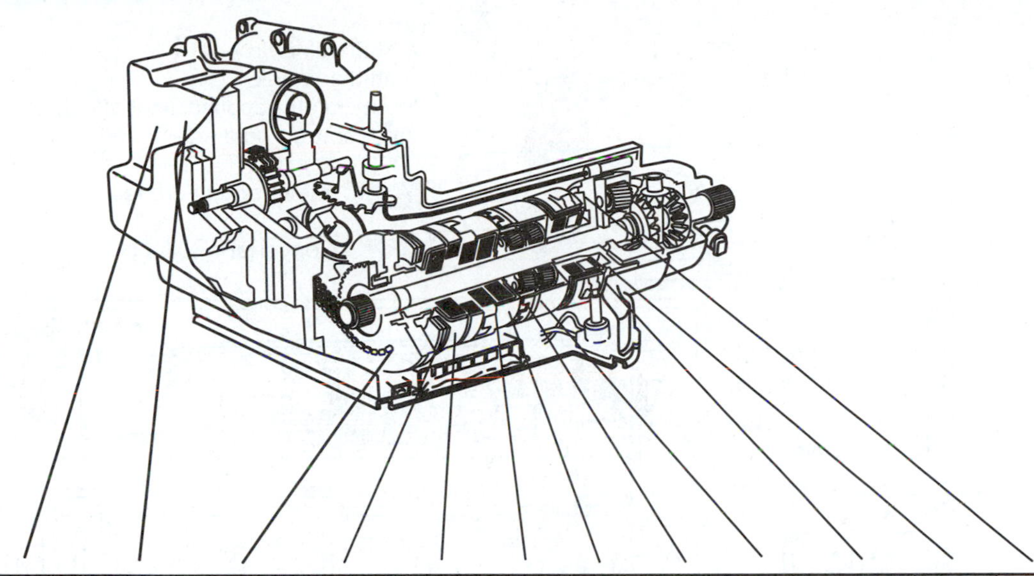

| Range | Gear | A solenoid | B solenoid | 4th clutch | Reverse band | 2nd clutch | 3rd clutch | 3rd roller clutch | Input clutch | Input sprag | Forward band | 1-2 roller clutch | 2-1 band |
|---|---|---|---|---|---|---|---|---|---|---|---|---|---|
| P - N | | On | On | | | | | | * | * | | | |
| D | 1st | On | On | | | | | | Apply | Hold | Apply | Hold | |
| | 2nd | Off | On | | | Apply | | | * | Orun | Apply | Hold | |
| | 3rd | Off | Off | | | Apply | Apply | Hold | | | Apply | | |
| | 4th | On | Off | Apply | | Apply | * | Orun | | | Apply | | |
| D | 3rd | @Off | @Off | | | Apply | Apply | Hold | Apply | Hold | Apply | | |
| | 2nd | @Off | @On | | | Apply | | | * | Orun | Apply | Hold | |
| | 1st | @On | @On | | | | | | Apply | Hold | Apply | Hold | |
| 2 | 2nd | @Off | @On | | | Apply | | | * | Orun | Apply | Hold | Apply |
| | 1st | @On | @On | | | | | | Apply | Hold | Apply | Hold | Apply |
| 1 | 1st | @On | @On | | | | Apply | Hold | Apply | Hold | Apply | Hold | Apply |
| R | Rev | On | On | | Apply | | | | Apply | Hold | | | |

* Applied by not effective
On - Solenoid energized
Off - Solenoid de-energized

@ The solenoid's state follows a shift pattern, which depends upon vehicle speed and throttle opening. It does not depend upon the selected gear.

**Figure 1.** Range reference chart for a 4T60-E transaxle.

accumulator piston and to the release side of the converter clutch plate.

*"A" Shift solenoid*—This is energized and line pressure moves the 1–2 shift valve against spring pressure and directs fluid into the solenoid's passage. "A" shift solenoid fluid is then directed to the 3–4 shift valve.

*"B" Shift solenoid*—This is energized and line pressure is directed into its passage. The fluid is then directed to the 4–3 manual downshift valve and moves the valve against spring pressure. The fluid is also directed to the 3–2 manual downshift valve.

## Reverse

When Reverse is selected by the gear selector, the manual valve moves and allows line pressure to move from the PRN passage into the reverse fluid passage **(Figure 3)**. Reverse fluid seats check ball #5 and directs fluid to the reverse servo boost valve and through the reverse servo feed orifice. After passing through the orifice, the fluid enters the reverse servo passage and forces the servo's piston to move against the servo's spring pressure. The movement of the piston applies the reverse band.

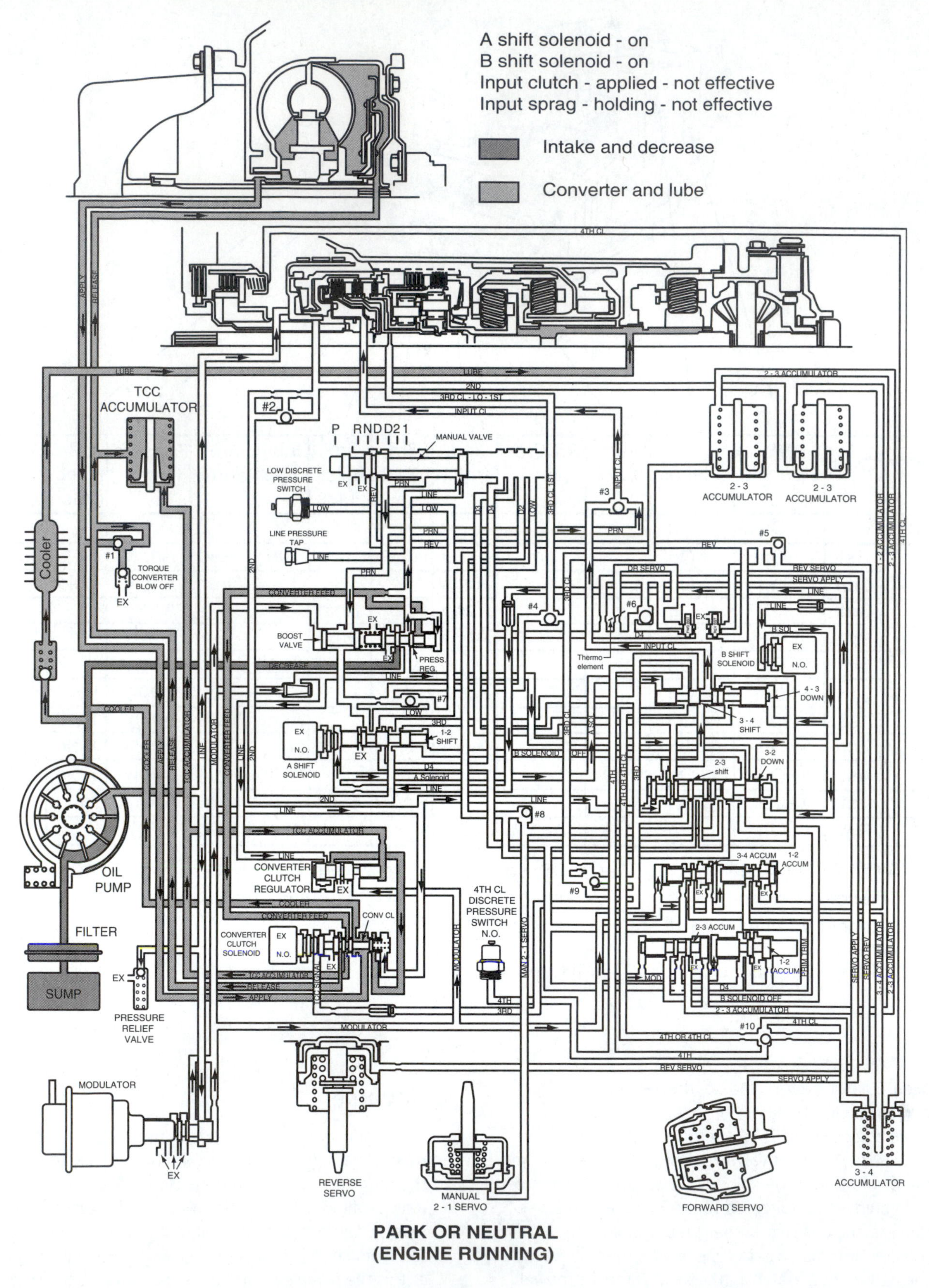


Figure 2. Oil circuit for a 4T60-E when in Park or Neutral.

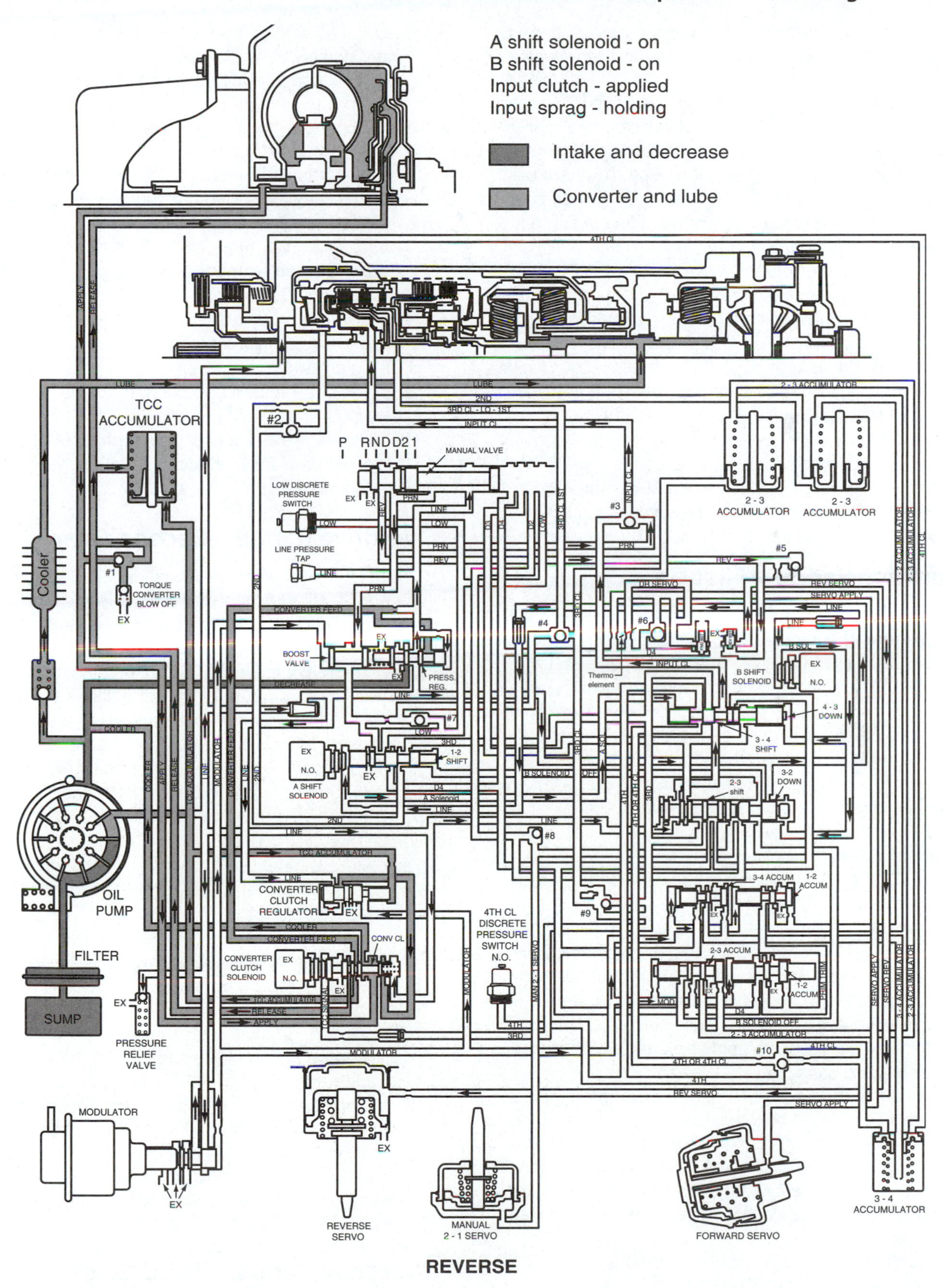


Figure 3. Oil circuit for a 4T60-E when in Reverse.

*Manual valve*—Controlled by the gear selector, directs line pressure into the PRN passage in Park, Reverse, and Neutral.

*#5 check ball*—Blocks the reverse servo feed passage forcing reverse fluid through an orifice in the spacer plate and into the reverse servo passage. When the manual valve is moved out of reverse, the check ball unseats allowing the reverse servo fluid to exhaust through the check ball's seat rather than the orifice.

*Reverse boost valve*—During hard acceleration, when in reverse, the reverse boost valve can allow the reverse fluid to bypass the reverse feed orifice and enter directly into the reverse servo passage. This allows for a quick fill of the servo passage and for a quick application of the reverse band. Both of these actions will prevent band slippage during abusive shifts from Park or Neutral into Reverse.

*Reverse servo assembly*—Applies the reverse band in response to reverse servo fluid pressure feeding into the servo cover side of the reverse servo piston.

*Reverse band*—Wraps around the second clutch housing and holds the input carrier allowing for reverse.

## Overdrive Range—First Gear

When the gear selector is moved to the Overdrive range **(Figure 4)**, the position of the manual valve allows line pressure to fill the D–4 passage. D–4 pressure seats the #6 check ball and sends fluid through the forward servo feed orifice into the servo apply passage. Servo apply fluid applies the forward band. The transmission is now operating in first gear. D–4 fluid also feeds the servo apply passage through the thermo element on the spacer plate. D–4 pressure is also directed, by the manual valve, to the 1–2 shift valve, 2–3 shift valve, 1–2 accumulator valve, 2–3 accumulator valve, and 3–4 accumulator valve.

*Manual valve*—Controlled by the gear selector, directs line pressure into the D–4 passage.

*#6 check ball*—Blocks the forward servo apply passage forcing D–4 pressure to the forward servo boost valve, the forward servo feed orifice, and thermo element. When the manual valve is moved from Drive to Park or Neutral, the check ball unseats and allows for a quick exhaust of servo apply fluid and the release of the forward band.

*Forward servo boost valve*—Opens under high line pressure allowing D–4 fluid to bypass the feed orifice and thermo element to enter the servo apply passage. This provides for a quick fill of the servo passage and for a quick apply of the forward band, which prevents slippage during abusive shifts from Park or Neutral to Drive.

*Forward servo assembly*—Applies and hold the forward band during all forward gears.

*Forward band*—Wraps around and holds the 1–2 support outer race.

*1–2 accumulator valve*—Fed by D–4 pressure, it regulates primary trim fluid pressure to the secondary 1–2 accumulator valve in proportion to changes in modulator pressure.

*Secondary 1–2 accumulator valve*—Fed by primary trim fluid pressure, it regulates 1–2 accumulator pressure in proportion to modulator pressure. When primary trim reaches a specified pressure, it is lowered by the secondary 1–2 accumulator valve.

*2–3 accumulator valve*—Fed by D–4 pressure, it regulates 2–3 accumulator pressure in proportion to modulator pressure.

*3–4 accumulator valve*—Fed by D–4 pressure, it regulates 3–4 accumulator pressure in proportion to modulator pressure.

*2–3 shift valve*—Allows D–4 pressure to enter the auxiliary input clutch feed passage.

*1–2 shift valve*—Held against spring pressure by "A" shift solenoid pressure, D–4 fluid stops at this valve until a 1–2 shift is made.

## Overdrive Range—Second Gear

When the PCM receives the correct information from its sensors, it will turn off the "A" shift solenoid. This allows the line pressure at the 1–2 shift valve to exhaust through the solenoid. The spring pressure on the 1–2 shift valve causes the valve to move once the line pressure is exhausted. The movement of the valve allows the fluid to enter the 2$^{nd}$ fluid passage. 2$^{nd}$ fluid feeds to the #2 check ball and seats it, forcing the fluid through a feed orifice before applying the 2$^{nd}$ clutch. At the same time, 2$^{nd}$ clutch fluid is fed to the 1–2 accumulator piston and moves the piston up against spring pressure and 1–2 accumulator fluid pressure, cushioning 2$^{nd}$ clutch apply. Second gear is engaged. 1–2 accumulator fluid is then forced through the 1–2 accumulator passage and exhausts at the 1–2 secondary accumulator valve.

*1–2 shift valve*—With the "A" shift solenoid off, spring pressure moves the valve allowing D–4 pressure to enter the 2$^{nd}$ passage directing 2$^{nd}$ fluid to the #2 check ball.

*#2 check ball*—2$^{nd}$ fluid seats the check ball and is then forced through an orifice into the 2$^{nd}$ clutch passage to apply the 2$^{nd}$ clutch.

*1–2 accumulator piston*—Cushions the apply of the 2$^{nd}$ clutch using 2$^{nd}$ fluid to force the piston against spring force and 1–2 accumulator fluid.

*Secondary 1–2 accumulator valve*—Regulates the exhaust rate of 1–2 accumulator fluid during the apply of the 2$^{nd}$ clutch.

## Overdrive Range—Third Gear

When ideal operating conditions are met, the PCM turns off the "B" shift solenoid allowing the fluid at the

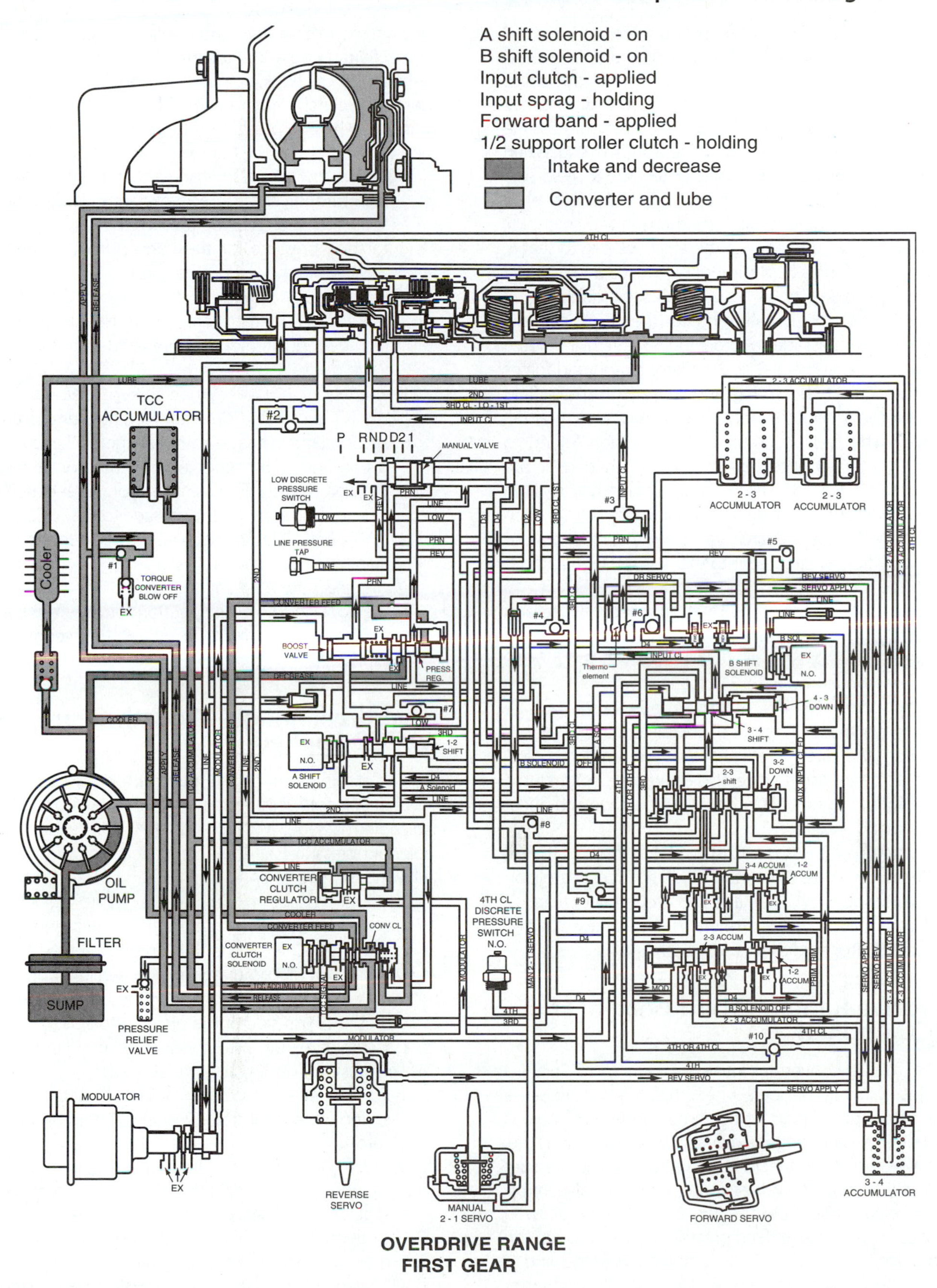


**Figure 4.** Oil circuit for a 4T60-E when in Overdrive range and operating in first gear.

solenoid to exhaust. This allows spring force to move the 4–3 manual downshift valve. Also line pressure at the end of the 2–3 shift valve moves the 3–2 manual downshift valve against spring pressure. Line pressure at the 3–2 downshift valve also enters the "B" shift solenoid passage and helps prevent the 1–2 shift valve from moving. When the 2–3 shift valve moves, D–4 fluid enters the 3rd fluid passage. The 3rd fluid passage feeds fluid to the converter clutch solenoid (where it exhausts through the solenoid), the 1–2 shift valve, and against the #9 check ball to seat it. This last action forces fluid through an orifice and into the 3rd clutch passage. The 3rd clutch passage directs fluid to seat the #4 check ball sending the fluid into the 3rd clutch/Lo–1st passage to apply the 3rd clutch. 3rd clutch fluid is also sent to the 2–3 accumulator piston and forces the piston against the spring and 2–3 accumulator fluid to cushion the 3rd clutch apply. Third gear is now engaged **(Figure 5)**. The 2–3 accumulator fluid is then forced to exhaust at the 2–3 accumulator valve. In third gear, the input clutch is released to allow the input clutch apply fluid to exhaust through the 3–4 shift valve into the input clutch feed passage. At the 2–3 shift valve, the exhausting input clutch apply fluid is directed into the D–3 passage and out through the manual valve.

*#9 check ball*—Forces fluid through a feed orifice and into the 3rd clutch passage.

*#4 check ball*—Directs 3rd clutch fluid into the 3rd/Lo–1st passage to apply the 3rd clutch.

*2–3 shift valve*—Allows D–4 fluid to enter the 3rd passage to move the 2–3 accumulator piston, apply the 3rd clutch, and direct fluid to the converter clutch solenoid. The 2–3 shift valve also allows the input clutch apply fluid to exhaust into the D–3 passage and out of the manual valve.

*2–3 accumulator piston*—Cushions the apply of the 3rd clutch by using 3rd clutch fluid to force the piston against spring pressure and 2–3 accumulator fluid.

*2–3 accumulator valve*—Regulates the exhaust of 2–3 accumulator fluid during 3rd clutch apply.

## Overdrive Range—Third Gear w/TCC Applied

The torque converter clutch can be applied in third gear. This will only happen when the PCM receives the appropriate input signals. At this time, the TCC solenoid is turned on and is no longer exhausting the TCC signal fluid. This moves the converter clutch valve against line pressure and the strength of the valve's spring. With the valve in this position, release fluid from the converter clutch is allowed to exhaust. The converter feed pressure passes through the clutch valve, enters the TCC accumulator passage, and is sent to the TCC accumulator piston and the converter clutch regulator valve. Regulated apply fluid enters the apply passage. The apply fluid seats the #1 check ball against the release fluid and applies the converter clutch. Apply fluid is also at the torque converter blowoff valve to exhaust excess fluid pressure. Converter feed fluid at the converter clutch valve is allowed to feed into the cooler passage where it passes through the cooler check valve and cooler, and into the transaxle's lubrication system.

*TCC accumulator assembly*—Absorbs converter feed fluid through the TCC accumulator passage and uses spring force to provide an increasing TCC accumulator bias pressure for the converter clutch regulator valve when the TCC's plate is applied.

*Converter clutch regulator valve*—Biased by modulator and TCC accumulator pressure, it uses line pressure to provide regulated apply fluid to the converter clutch valve.

*Converter clutch valve*—When shifted, release fluid from the TCC exhausts. Converter feed fluid enters the TCC accumulator passage to move the TCC accumulator piston and is also directed into the cooler passage. Regulated apply fluid from the converter clutch valve can now enter the apply passage to the TCC.

*#1 check ball*—Blocks release fluid while sending apply fluid to the TCC blow off valve.

*Cooler check valve*—Allows cooler fluid to pass through the cooler, provides lubrication for the transaxle, and prevents converter drain back when the engine is not running.

## Overdrive Range—Fourth Gear w/TCC Applied

To obtain fourth gear **(Figure 6)**, the PCM turns on the "A" shift solenoid. This directs line pressure into the "A" shift solenoid passage and to the 3–4 shift valve. "A" shift solenoid pressure shifts the valve against spring pressure allowing 3rd fluid to enter the 4th fluid passage and the 4th clutch fluid passage. 4th fluid is directed to the 4th clutch discrete pressure switch to close the switch and seat the #10 check ball to force the fluid through a feed orifice to move the 3–4 accumulator piston while applying the 4th clutch. Fourth gear is now engaged. 3–4 accumulator fluid on the spring side of the piston is forced through the 3–4 accumulator pin into the 3–4 accumulator passage and exhausts at the 3–4 accumulator valve.

*1–2 shift valve*—Is held against the "A" shift solenoid by spring pressure and the "B" shift solenoid by fluid pressure.

*3–4 shift valve*—When shifted against spring pressure, it allows 3rd fluid to enter the 4th and 4th clutch fluid passages. 4th fluid seats the #10 check ball and is forced through a feed orifice before moving the 3–4 accumulator piston and applying the 4th clutch.

*4th clutch discrete switch*—A normally open switch that closes when 4th fluid pressure is fed from the 3–4 shift valve. When the switch is closed, it completes the

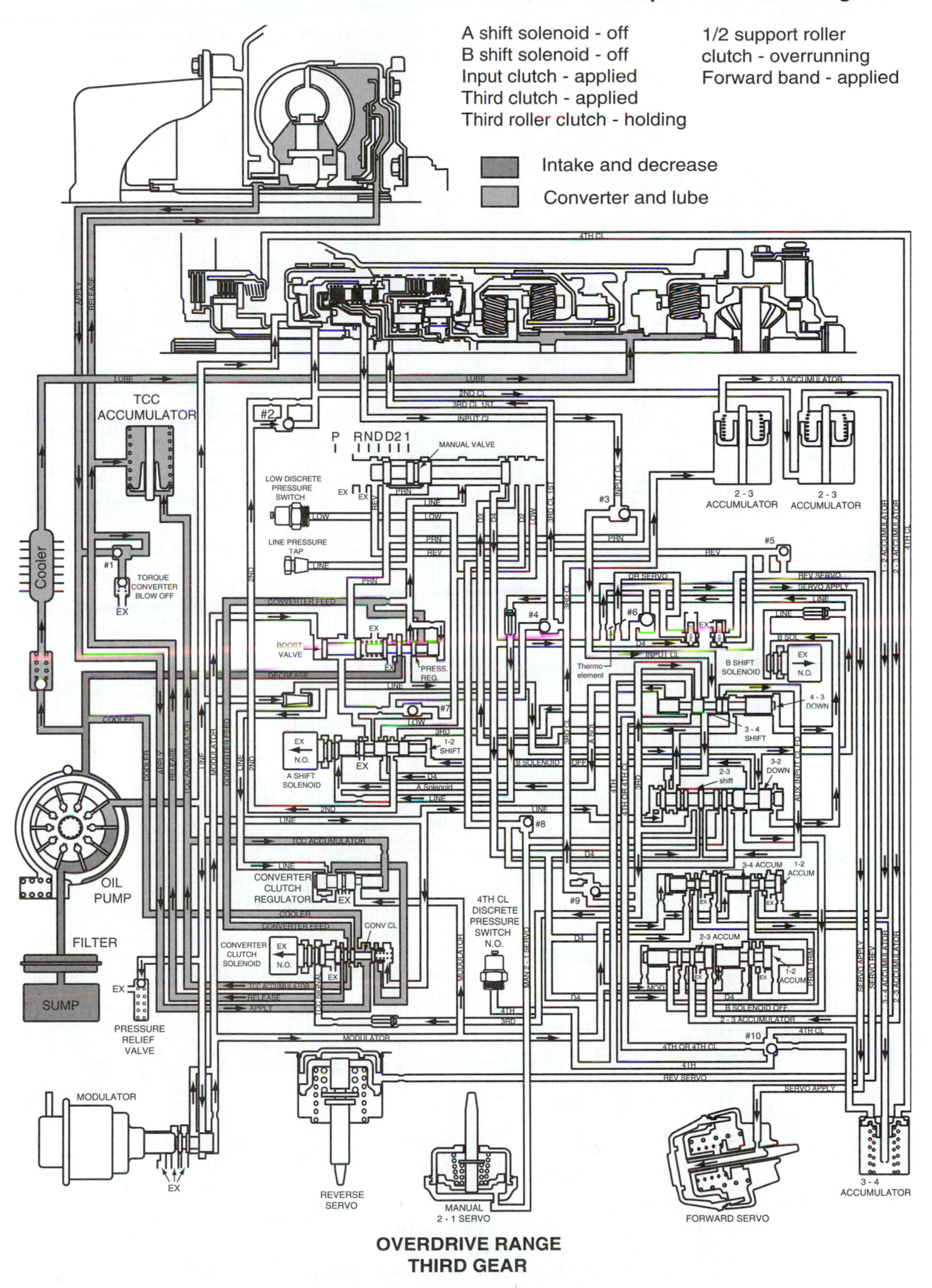


**Figure 5.** Oil circuit for a 4T60-E when in Overdrive range and operating in third gear.

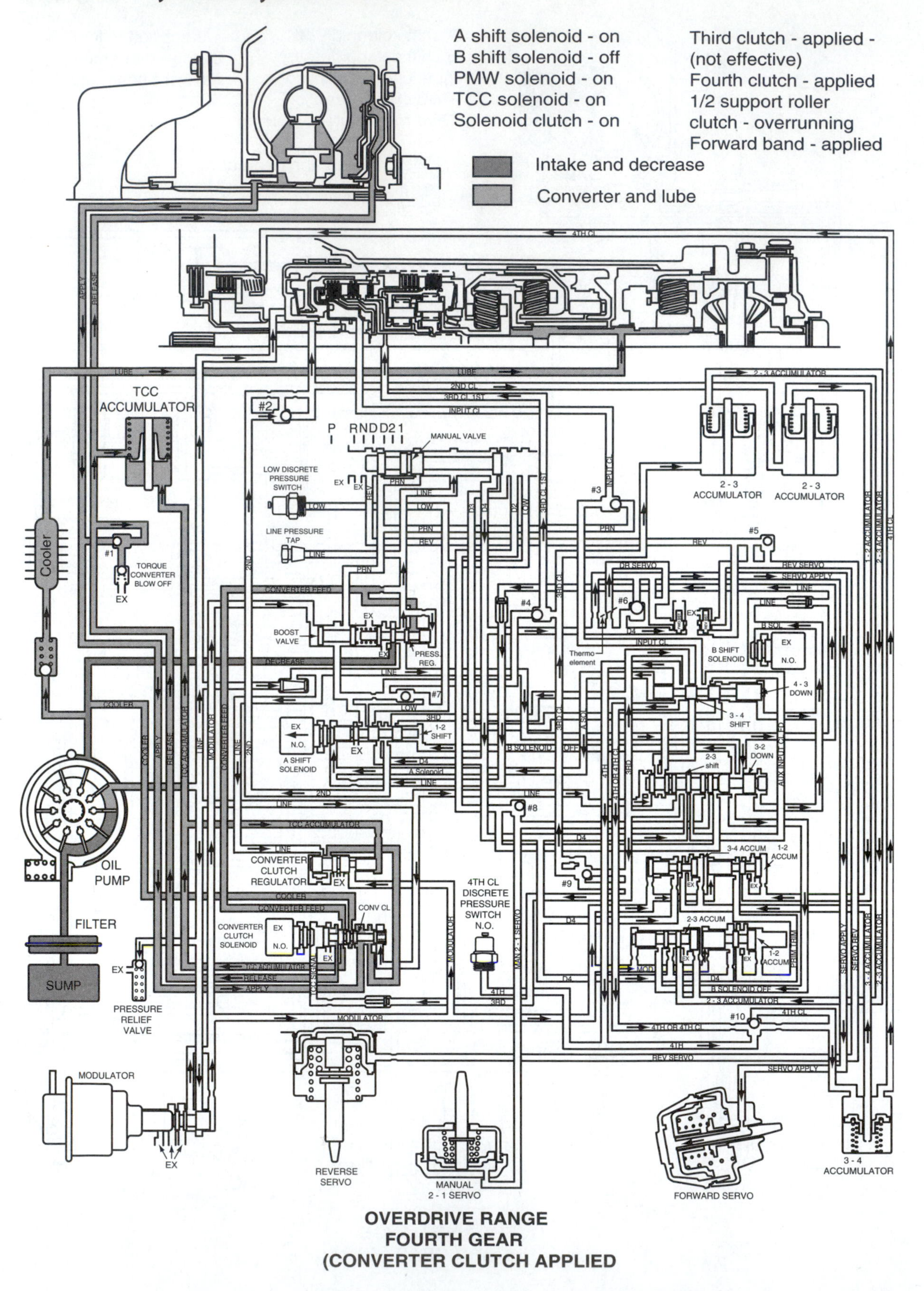


**Figure 6.** Oil circuit for a 4T60-E when in Overdrive range and operating in fourth gear.

circuit from the PCM to ground and informs the PCM that the transaxle is in fourth gear.

*#10 check ball*—Forces 4th fluid through a feed orifice into the 4th clutch passage to move the 3–4 accumulator piston.

*3–4 accumulator piston*—Cushions the apply of the 4th clutch by using 4th clutch fluid to move the piston against spring pressure and 3–4 accumulator fluid.

*3–4 accumulator valve*—Controls the rate of exhaust of 3–4 accumulator fluid during 4th clutch apply.

## Overdrive Range—4–3 Downshift w/TCC Released

During light acceleration or slight load changes, the PCM will turn off the torque converter clutch. Often, the PCM will order a 4–3 downshift during these same conditions **(Figure 7)**. When the TCC solenoid is turned off, line pressure and spring tension at the converter clutch valve move the valve to allow TCC signal fluid to exhaust through the converter clutch solenoid. Converter feed fluid enters the release passage; seats the #1 check ball against apply fluid and feed the release side of the torque converter clutch plate. Apply fluid from the torque converter is directed through the apply passage to the converter clutch valve to the cooler passage. TCC accumulator fluid is directed through the converter clutch valve and enters the release passage. It also is directed to the converter clutch regulator valve to control the release of the converter clutch plate.

When the PCM directs a downshift, it turns off the "A" shift solenoid. This allows the "A" shift solenoid fluid at the 3–4 shift valve to exhaust through the solenoid. Spring pressure at the 3–4 shift valve moves the valve, which allows 4th clutch apply fluid to be directed through the 3–4 accumulator assembly, seat the #10 check ball against 4th clutch apply fluid, and exhaust at the 3–4 shift valve. Third gear is engaged.

*TCC solenoid*—When deenergized it allows the converter clutch valve to move and exhaust TCC signal fluid through the solenoid.

*Converter clutch valve*—When the TCC solenoid is off, the converter feed fluid passes through the valve into the converter passage and converter apply fluid is directed into the cooler passage. TCC accumulator fluid also exhausts into the release passage and acts as a bias on the converter clutch regulator valve.

*Converter clutch regulator valve*—Uses filtered line fluid to supply regulated apply fluid to the converter clutch valve. Regulation is controlled by TCC accumulator and modulator fluid pressures.

*TCC accumulator*—Spring force and release fluid pressure returns the TCC accumulator piston to its off position.

*#1 check ball*—Converter release fluid seats the check ball against apply fluid to allow release of the TCC's plate.

*3–4 shift valve*—When shift solenoid "A" exhausts, spring force moves the valve allowing 4th clutch fluid to exhaust through the valve.

*3–4 accumulator*—Allows 4th clutch exhaust fluid to pass through the accumulator to #10 check ball.

*#10 check ball*—Forces 4th clutch fluid through an exhaust orifice to the 3–4 shift valve where it exhausts.

*4th clutch discrete switch*—Opens and sends a signal to the PCM that the transaxle is no longer operating in fourth gear.

## Overdrive Range—3–2 Downshift

During all full throttle downshifts, low engine vacuum allows the modulator valve to move and increase modulator pressure to the line boost valve. At the same time, pump output increases because of the higher engine speeds and forces a higher volume of fluid into the line circuit. The increased modulator pressure and increased fluid volume at the pressure regulator valve creates higher line pressures. A full throttle 3–2 downshift occurs when the "B" shift solenoid is turned on by the PCM. Line pressure at the solenoid is forced into the "B" solenoid passage and is directed to the 4–3 manual downshift valve and 3–2 downshift valve. "B" solenoid fluid forces the 3–2 manual downshift valve and the 2–3 shift valve to move. Line pressure at the 2–3 shift valve is then directed into the input clutch feed passage, through the 3–4 shift valve, into the input clutch passage. Input clutch fluid seats #3 check ball and applies the input clutch. The 3rd clutch is released by allowing 3rd clutch/Lo–1st fluid to exhaust to the #4 check ball and into the 3rd clutch passage. 3rd clutch fluid is directed to the #9 check ball forcing it through the exhaust orifice and into the 3rd fluid passage to the 2–3 shift valve where it exhausts.

*Modulator valve*—Moves in response to engine vacuum and increases modulator pressure to the line boost valve.

*Line boost valve*—Moves in response to modulator pressure to increase line pressure through the pressure regulator valve.

*4–3 manual downshift valve*—Shifted by the "B" solenoid fluid and prevents the 3–4 shift valve from moving.

*3–2 manual downshift valve*—Shifted by the "B" solenoid fluid to downshift the 2–3 shift valve from moving.

*2–3 shift valve*—When downshifted, it allows line pressure to enter into the input clutch feed passage and directs it to the 3–4 shift valve.

*3–4 shift valve*—Directs input clutch feed fluid into the input clutch passage and sends it to the #3 check ball.

*#3 check ball*—Seats against the PRN passage to allow input clutch fluid to apply the input clutch.

*#4 check ball*—Directs 3rd clutch/Lo–1st exhaust fluid into the 3rd clutch passage.

*#9 check ball*—Directs 3rd clutch exhaust fluid into the 3rd fluid passage. 3rd fluid is sent to the 2–3 shift valve where it exhausts.

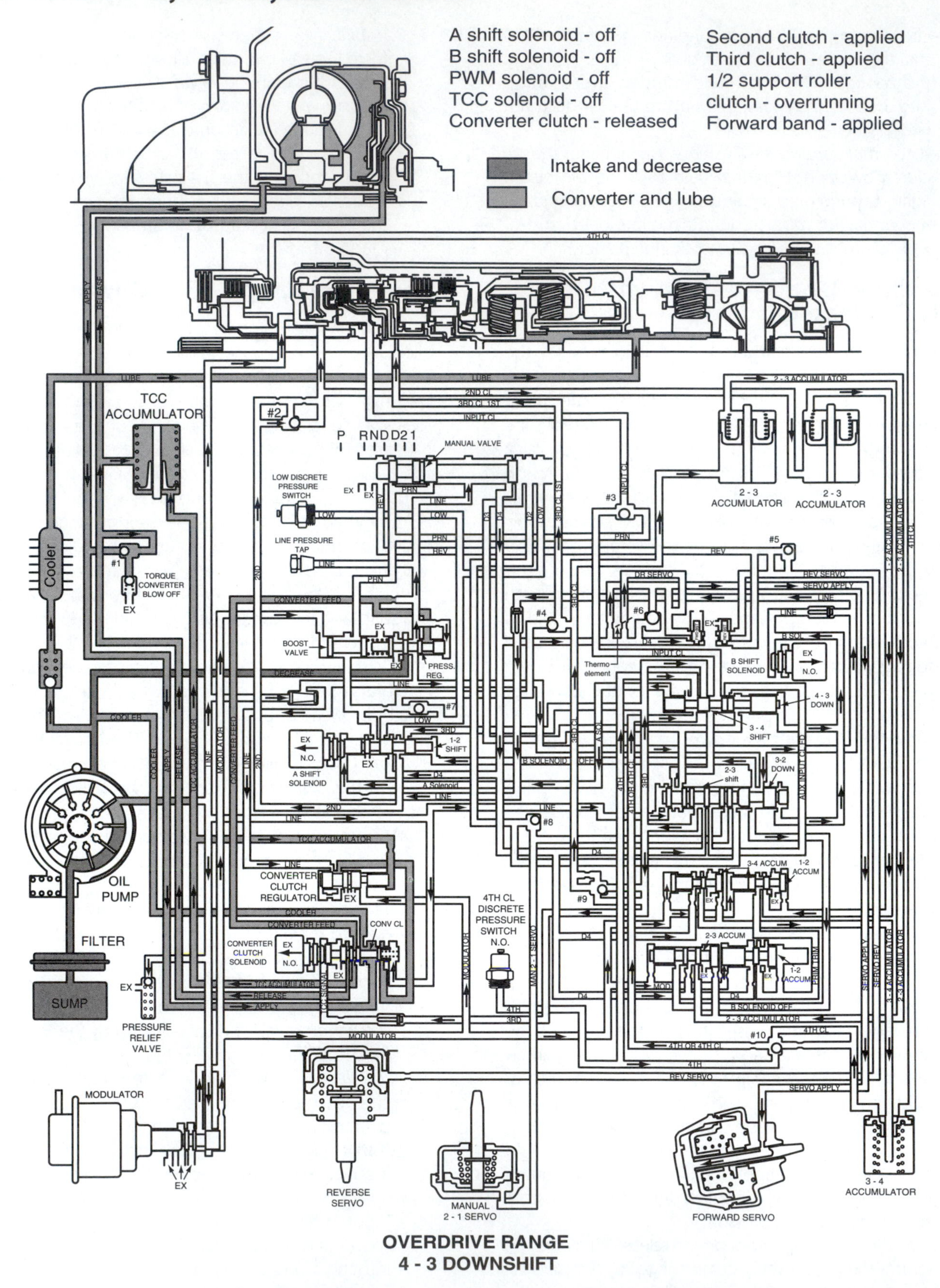


**Figure 7.** Oil circuit for a 4T60-E when in Overdrive range and downshifting from fourth to third gear.

## Manual Gear Selections

The gear selector allows the driver to select a particular gear for downshifting or eliminating the chance of the transmission upshifting. The manual gear selections that are available will vary by application. Some transmissions allows for shifting into all forward gears and keeping them in the chosen gear. Others allow for manual downshifting, but the choice of the gear is left to the PCM and transmission. Still others allow the driver to select a gear and hold it until speed and other conditions dictate that an upshift should take place. By moving the gear selector into a lower operating gear, the manual valve attempts hydraulically to override the electronic control system for manual downshifting.

The following discussions give a brief discussion of what takes place as the gear selector is moved to manually downshift one gear at a time.

**Manual third.** When the gear selector is moved into the Drive range from Overdrive, the manual valve allows line pressure to enter the D–3 passage. D–3 fluid is then directed to the 3–4 shift valve and moves the valve against "A" shift solenoid pressure. D–3 fluid is also sent to the 2–3 shift valve where it enters the input clutch feed passage. Input clutch feed fluid passes through the 3–4 shift valve into the input clutch passage, seats #3 check ball against the PRN passage and applies the input clutch. 4th clutch apply fluid at the 4th clutch is now directed through the 3–4 accumulator to the #10 check ball. Exhausting 4th clutch fluid seats #10 against the 4th passage directing fluid through an exhaust orifice up to the 3–4 shift valve where it exhausts. 4th fluid pressure at the 4th discrete switch also exhausts at the 3–4 shift valve allowing the switch to open and signal to the PCM that the transaxle is no longer operating in 4th gear.

Because D–3 fluid overrides "A" solenoid fluid at the 3–4 shift valve, a manual 4–3 downshift will result even if the solenoid is turned on. Once the PCM turns off the solenoid, fluid at the solenoid will exhaust through the solenoid.

**Manual second.** When the gear selector is moved into second gear, the manual valve allows line pressure to enter the D–2 passage. D–2 fluid seats the #8 check ball and is directed to the 2–3 shift valve and moves it. This allows D–2 fluid to enter the manual 2–1 servo feed passage and applies the 2–1 band. The 3rd clutch is released by allowing 3rd clutch/Lo–1st apply fluid to exhaust by seating the #4 check ball against the Lo–1st passage and into the 3rd clutch passage. 3rd clutch fluid is then directed to the #9 check ball and seats it against the 3rd fluid passage forcing the fluid through an exhaust orifice and into the 3rd fluid passage. 3rd fluid is then sent to the 2–3 shift valve where it exhausts. Because D–2 fluid overrides line pressure at the 2–3 shift valve, a manual 3–2 downshift results even if shift solenoid "B" is on. Once the PCM receives the correct inputs, it shut turn off the solenoid. The fluid at the solenoid is then directed to the 4–3 manual downshift valve and the 3–2 manual downshift valve.

**Manual first.** When the gear selector is moved into first gear, the manual valve allows line pressure to enter the Lo passage. Lo fluid pressure is sent to the discrete switch and closes it completing the circuit from the PCM to ground. Lo fluid pressure is also sent to the #7 check ball which seats against the Lo–1st fluid passage. The PCM turns on the "A" shift solenoid. The fluid from the solenoid holds the 1–2 shift valve and is also directed to the 3–4 shift valve.

The 2nd clutch is released by 2nd clutch apply fluid exhausting to the #2 check ball and through the exhaust orifice into the 2nd fluid passage. 2nd fluid is then directed through the 1–2 shift valve and into the 3rd fluid passage where it exhausts at the 2–3 shift valve. At the same time, Lo fluid is directed through the 1–2 shift valve and into the Lo–1st fluid passage. Lo–1st fluid is sent to the pressure regulator valve to boost line pressure and to the Lo blow off valve. Lo–1st fluid also seats the #4 check ball against the 3rd clutch passage and directs the fluid into the 3rd clutch/Lo–1st passage to apply the 3rd clutch.

## GUIDELINES

The use of flow diagrams is invaluable to identifying the cause of a problem. The results of the road test and input from the customer will help you define the problem. Once you know these, you study the flow diagram to eliminate causes for the problem and perhaps identify the true cause. The following should be followed to make sense out of any flow diagram you look at:

1. Identify the operating gear(s) with the problem.
2. If the problem exists only during some conditions, think about what is happening then, and about what that will affect.
3. Trace through one circuit (one operating gear) at a time. Flow diagrams are available for each of the possible gear ranges plus the following circuits: supply, main control pressure, converter and cooler, governor, throttle valve, boosted throttle, modulated throttle, accumulator, and converter clutch.
4. Begin at the applied member or devices for a particular gear and trace backwards.
5. Determine what type and the amount of pressure the manual shift valve circuit and the modulator valve circuit is receiving.
6. Trace the main path of the circuit first, and then trace the effect of the alternate circuits.
7. Because all diagrams do not show the actual position of the valves, follow the flow based on your knowledge.
8. Flow diagrams do not show the exact location of the components, so do not be misled by them. They are simply a summary of the flow.

9. Trace through the fluid passages as they are drawn.
10. Pay particular attention to the direction of the fluid flow and the placement of the orifices and check valves.

If the flow diagram is not colored, follow the fluid flow through with a colored pencil. Use one color for line pressure, another for throttle pressure, another for exhaust, and so on.

## Summary


- Each passage, bore, and valve in the valve body forms a specific oil circuit with a specific function.
- Flow diagrams can be used to clearly see how each valve functions to allow the transmission to operate in any particular gear range.
- When the gear selector is in Park or Neutral, line pressure is directed to the manual valve, pressure modulator valve, modulator valve, 2–3 shift valve, converter clutch valve, "A" shift solenoid, and "B" shift solenoid.
- When Reverse is selected, line pressure moves from the PRN passage to the reverse servo boost valve and through the reverse servo feed orifice. After passing through the orifice, the fluid enters the reverse servo passage and the reverse band is applied.
- When the gear selector is moved to the overdrive range, line pressure is directed through the forward servo feed orifice into the servo apply passage to apply the forward band and first gear is obtained.
- In the overdrive range and the operational gear is first gear and a specified speed and load is reached, the "A" shift solenoid is turned off the fluid at the 1–2 shift valve exhausts and the valve moves to direct fluid to the 2nd clutch and 2nd gear is engaged.
- While in second gear and ideal operating conditions are met, the "B" shift solenoid is turned off. Fluid is directed to apply the 3rd clutch and 3rd gear is engaged.
- When operating conditions are within a specified range, the TCC can be applied in third gear.
- To obtain fourth gear while in overdrive, "A" shift solenoid is turned on directing fluid to the 4th clutch and fourth gear is engaged.
- Automatic downshifts occur during acceleration or load increases.
- The gear selector allows the driver to select a particular gear for downshifting or eliminating the chance of the transmission upshifting. The manual gear selections that are available will vary by application.
- The use of flow diagrams is invaluable to identifying the cause of a problem identified during a road test.


## Review Questions

1. Describe the path of line pressure when a 4T60-E is operating in first gear and the gear selector is in overdrive.
2. When a 4T60-E is in Park or Neutral and the engine is running, why are the input clutch and input sprag not effective?
3. List five things you should do to use a flow diagram to identify the cause of a problem.
4. What will most likely occur if the #6 check ball cannot unseat itself because of dirt in the passage?
5. What influences the pressure regulated by the pressure modulator valve?

# Diagnosing Hydraulic Circuits

## Introduction

The best way to identify the exact cause of the problem is to use the results of the road test, logic, and the flow diagrams for the transmission being worked on. Before doing this however, you should always check all sources for information about the symptom first. Also, make sure you check the basics: trouble codes in the computer, fluid level and condition, leaks, and mechanical and electrical connections. Using the flow diagrams, you can trace problems to specific valves, servos, clutches, and bands.

The basic oil flow is the same for all transmissions. The oil pump supplies the fluid flow that is used throughout the transmission. Fluid from the pump always goes to the pressure-regulating valve. From there, the fluid is directed to the manual shift valve. When the gear selector is moved, the manual valve directs the fluid to other valves and to the apply devices. By following the flow of the fluid with the flow diagram, you can identify which valves and apply devices should be operating in each particular gear selector position. Through a process of elimination, you can identify the most probable cause of the problem **(Figure 1)**. Keep in mind, if a component is involved in the shifting of two different gear ranges but only one of those ranges appears to have a problem, that component is not a likely cause of the problem. Look for causes that would affect only the problem gear.

### DIAGNOSE HYDRAULIC AND VACUUM CONTROL SYSTEMS

In most cases, the transmission or transaxle will need to be removed to repair or replace the items causing the problem. However, some transmissions allow for a limited amount of service to the apply devices and control valves.

**You Should Know** *Whenever there is a shifting problem that is not readily diagnosed, check the service manual and technical service bulletins before making assumptions about the problem or getting frustrated. Sometimes the cause of a problem is something that you would least expect. For example: A single, and hard to diagnose, problem in a 4L30 transmission can cause it to have no movement, forward movement in neutral, no forward movement, or no reverse gear. All of these complaints may be caused by a defective center support. The center support is made of two pieces, the aluminum housing and the steel sleeve, which is pressed into the center of the support. Grooves machined into the sleeve provide channels for several oil paths through the support. If the sleeve rotates, the potential for several problems exists. The most common complaint is a bindup in reverse. It is not recommended that the sleeve be removed and a new one installed, rather the entire center support assembly should be replaced.*

Mechanical and/or vacuum controls also can contribute to shifting problems. The condition and adjustment of the various linkages and cables should be checked whenever there is a shifting problem. If all checks indicate

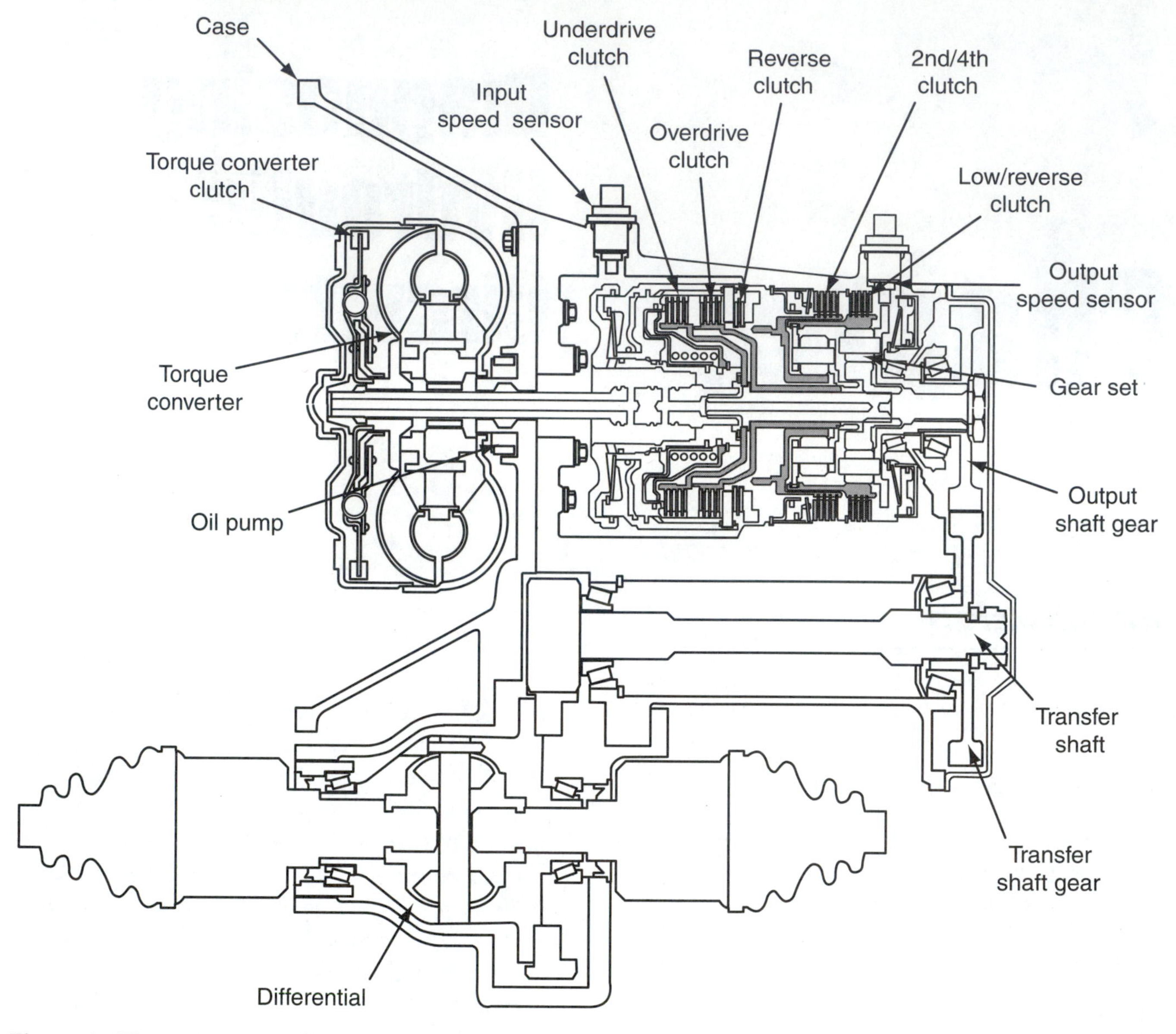


**Figure 1.** The major parts of a transaxle.

that the problem is either an apply device or the valves or passages, an air pressure test can help identify the exact problem. Air pressure tests also are performed during disassembly to locate leaking seals and during reassembly to check the operation of the clutches and servos **(Figure 2)**.

## Air Pressure Test

An air pressure test is conducted by applying clean, moisture-free air, at approximately 40 psi, through a rubber tipped air nozzle. With the valve body removed, direct air pressure to the case holes that lead to the servo, accumulator, and clutch apply passages **(Figure 3)**. Cover the vent hole of the circuit being tested with a clean, lint-free shop towel; this will catch any fluid that may spray out. You should clearly

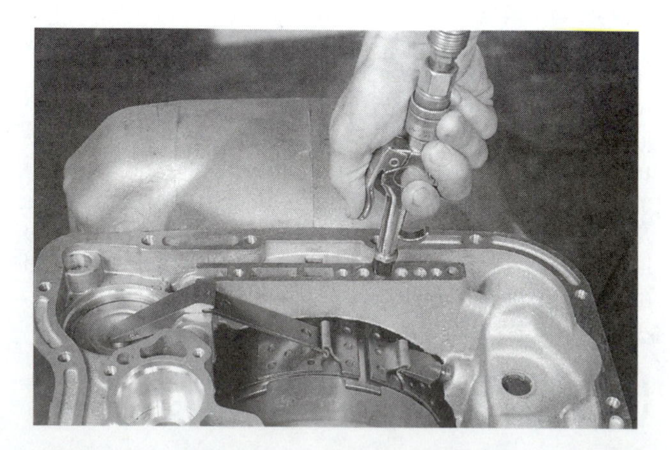

**Figure 2.** Air pressure test is being done after a transmission has been rebuilt.

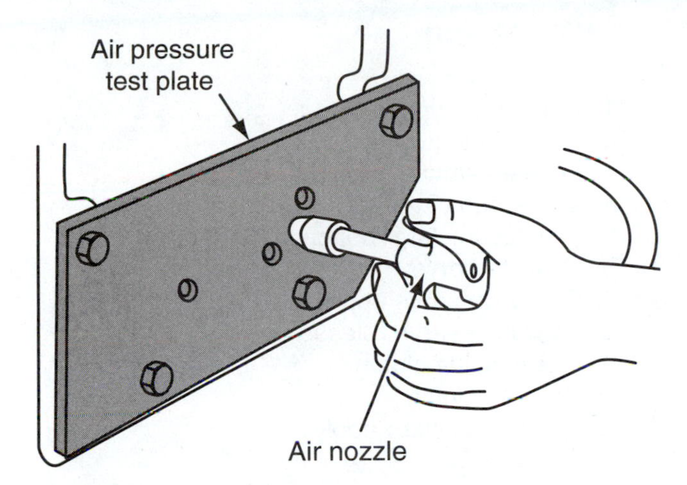


**Figure 3.** Using air pressure through a test plate to test hydraulic circuits.

hear or feel a dull thud that indicates the action of a holding device. If a hissing noise is heard, a seal is probably leaking in that circuit **(Figure 4)**. If you cannot hear the action of the servo or clutch, apply air pressure again and watch the assembly to see if it is reacting to the air. A servo or servo should react immediately to the release of the apply air. If it does not, something is making it stick. Repair or replace the apply devices if they do not operate normally.

Air pressure can normally be directed to the following circuits, through the appropriate hole for each: front clutch, rear clutch, kick down servo, low servo, and reverse servo. Some manufacturers recommend the use of a specially drilled plate and gasket, which is bolted to the transmission case **(Figure 5)**. This plate not only clearly identifies which passages to test but also seals off the other passages. Air is applied directly through the holes in the plate.

## PRESSURE TESTS

If you cannot identify the cause of a transmission problem from your inspection or road test, a pressure test should be conducted. This test measures the fluid pressure of the different transmission circuits during the various operating gears and gear selector positions. Refer to the illustrations shown **(Figures 6, 7, 8, and 9)** for a quick review of typical power flows in the various gears. The number of hydraulic circuits that can be tested varies with the different makes and models of transmissions. However, most transmissions are equipped with pressure taps, which allow the pressure test equipment to be connected to the transmission's hydraulic circuits **(Figures 10 and 11)**. Pressure testing checks the operation of the oil pump, pressure regulator valve, throttle valve, and vacuum modulator system (if the vehicle is equipped with one), plus the governor assembly. Pressure tests can be conducted on all transmissions whether their pressures are regulated by a variable force motor, vacuum modulator, or through conventional valving.

A pressure test has its greatest value when the transmission shifts roughly or when the shift timing is wrong. Both of these problems may be caused by excessive line pressure, which can be verified by a pressure test.

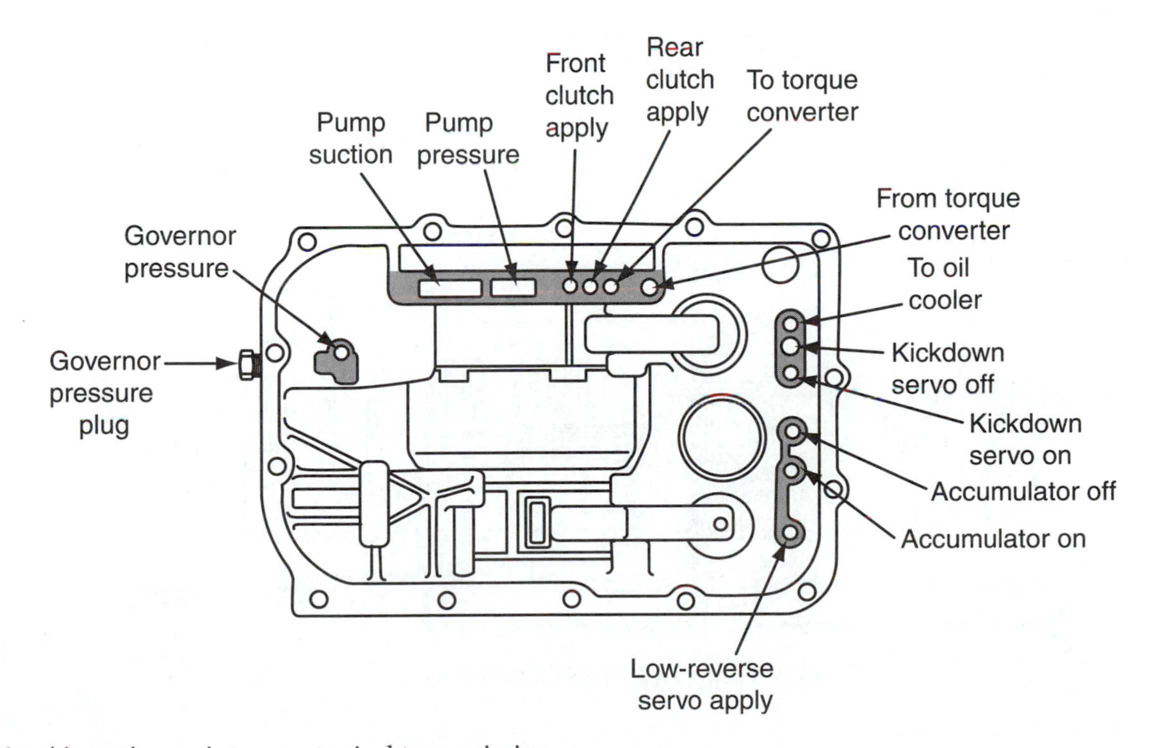


**Figure 4.** Air testing points on a typical transmission.

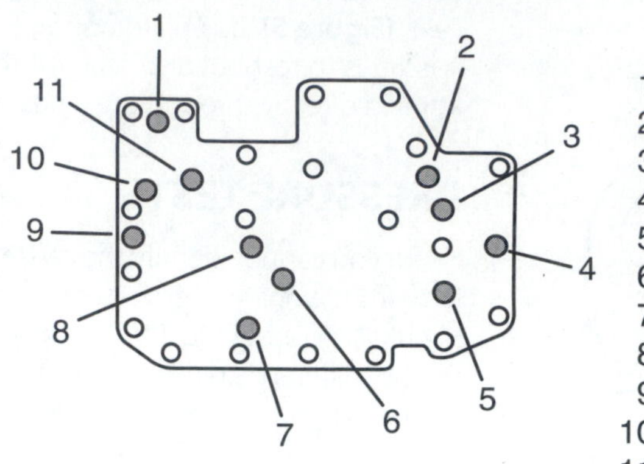


**LEGEND**

1. Converter bypass
2. Direct clutch
3. Forward clutch
4. 2–3 accumulator top
5. 2–3 accumulator bottom
6. Reverse servo
7. Overdrive servo apply
8. Overdrive servo release
9. Intermediate clutch
10. Reverse clutch
11. 1–2 accumulator apply

**Figure 5.** An example of a transmission air test plate.

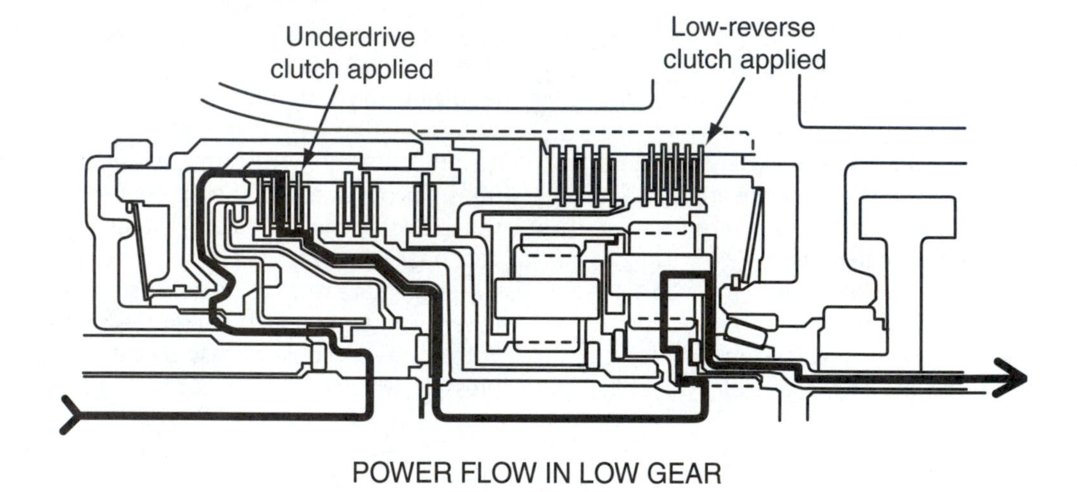


**Figure 6.** Power flow in low gear.

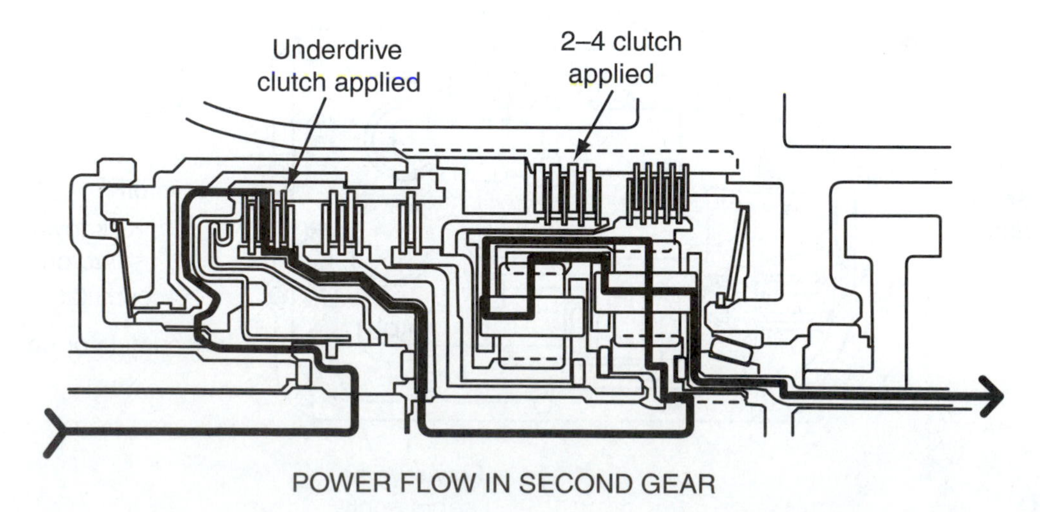


**Figure 7.** Power flow in second gear.

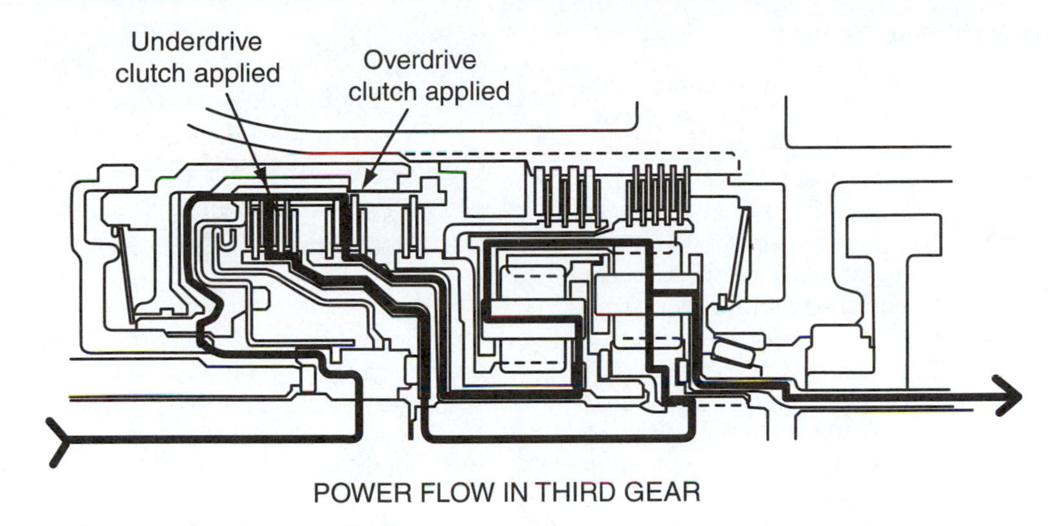


**Figure 8.** Power flow in third gear.

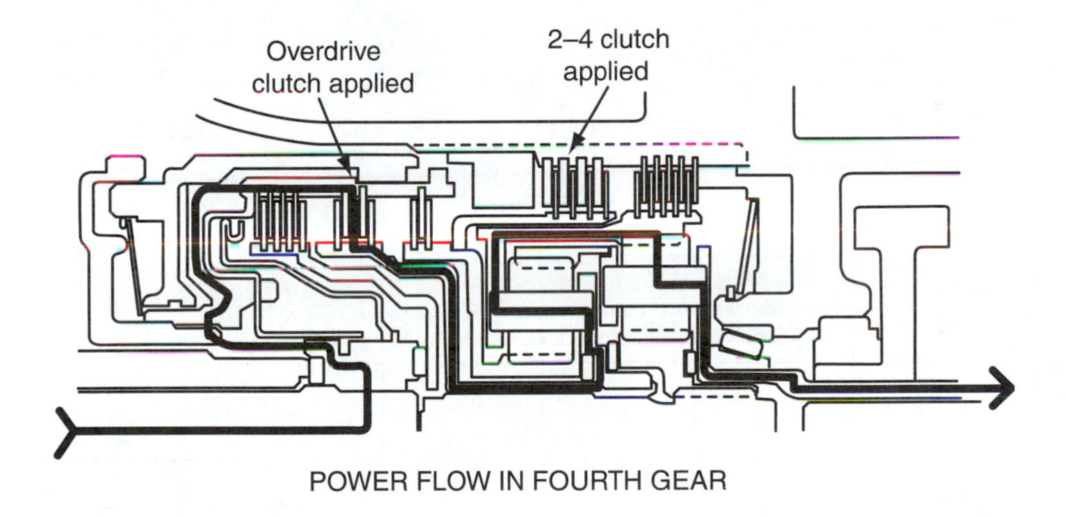


**Figure 9.** Power flow in fourth gear.

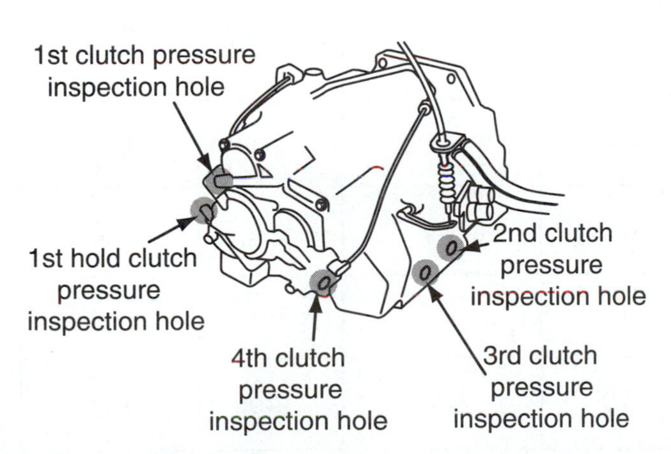


**Figure 10.** Pressure taps on the outside of a typical Honda transaxle case.

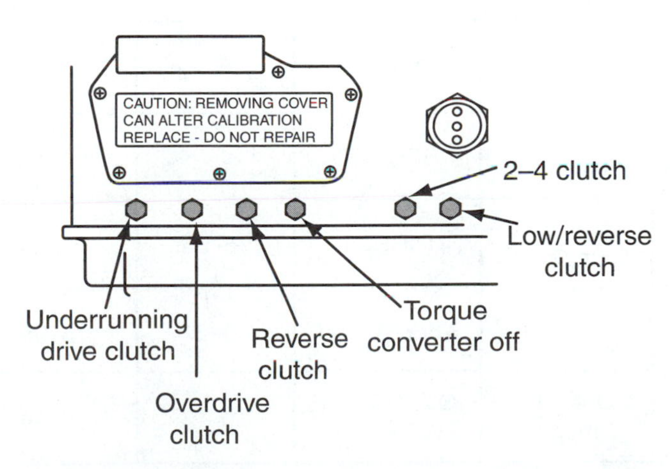


**Figure 11.** Pressure taps on the outside of a typical Chrysler transaxle.

## Conducting a Pressure Test

Before conducting a pressure test on an electronic automatic transmission, check and correct all trouble codes retrieved from the system. Also make sure the transmission fluid level and condition is okay and that the shift linkage is in good order and properly adjusted. To conduct a pressure test, use a tachometer, two pressure gauges (electronic gauges if available), the correct manufacturer specifications **(Figure 12)**, and (for vehicles equipped with a vacuum modulator) a vacuum gauge and a hand-held vacuum pump.

The pressure is best conducted with three pressure gauges, but two will work. Two of the gauges should read up to 400 psi and the other to 100 psi. The two 400 psi gauges are usually used to check mainline and an individual circuit, such as mainline and direct or forward circuits. If a circuit is 10 psi lower than mainline pressure when they are both tested at exactly the same time, a leak is indicated. This is why it is best to use two 400 psi gauges. A 100 psi gauge may be used be used on TV and governor circuits. This insures an observation of the critical pressures from these two circuits. Next to the scan tool, a pressure gauge is the most valuable tool for automatic transmission diagnostics.

**You Should Know** *When using two or more pressure gauges, be certain to put them on a "Tee" fitting at the same time to calibrate them. The gauges should be calibrated to read the same at 50, 75, 100, 125, and 150 psi.*

The pressure gauges are connected to the pressure taps **(Figure 13)** in the transmission housing and routed so that the gauges can be seen by the driver. The vehicle is then road tested and the gauge readings observed during the following operational modes: slow idle, fast idle, and wide-open throttle (WOT).

**You Should Know** *Always stop the engine when connecting and disconnecting the pressure gauges at the transmission. The fluid is under pressure and can easily spray at you and get in your eyes, if a hose, fitting, or gauge leaks.*

| Gear Selector Position | Actual Gear | PRESSURE TAPS | | | | | |
|---|---|---|---|---|---|---|---|
| | | Underdrive Clutch | Overdrive Clutch | Reverse Clutch | Torque Converter Clutch Off | 2/4 Clutch | Low/Reverse Clutch |
| PARK* 0 mph | PARK | 0–2 | 0–5 | 0–2 | 60–110 | 0–2 | 115–145 |
| REVERSE* 0 mph | REVERSE | 0–2 | 0–7 | 165–235 | 50–100 | 0–2 | 165–235 |
| NEUTRAL* 0 mph | NEUTRAL | 0–2 | 0–5 | 0–2 | 60–110 | 0–2 | 115–145 |
| L # 20 mph | FIRST | 110–145 | 0–5 | 0–2 | 60–110 | 0–2 | 115–145 |
| 3 # 30 mph | SECOND | 110–145 | 0–5 | 0–2 | 60–110 | 115–145 | 0–2 |
| 3 # 45 mph | DIRECT | 75–95 | 75–95 | 0–2 | 60–90 | 0–2 | 0–2 |
| OD # 30 mph | OVERDRIVE | 0–2 | 75–95 | 0–2 | 60–90 | 75–95 | 0–2 |
| OD # 50 mph | OVERDRIVE WITH TCC | 0–2 | 75–95 | 0–2 | 0–5 | 75–95 | 0–2 |

* Engine speed at 1500 rpm.
# CAUTION: Both front wheels must be turning at the same speed.

**Figure 12.** A pressure chart. These pressures are for an engine speed of 1500 rpm.

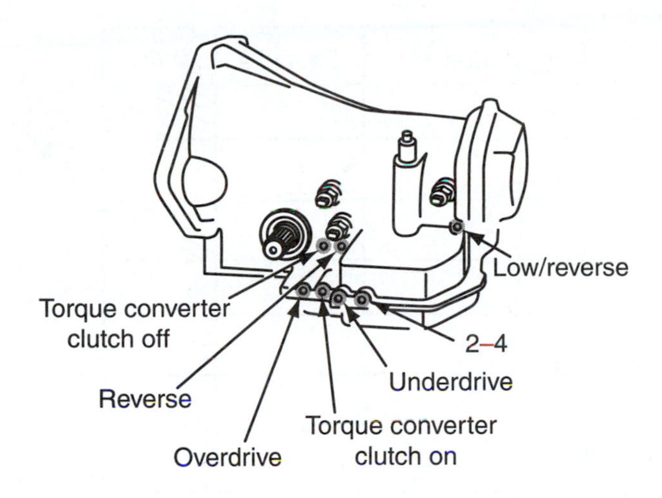


**Figure 13.** The pressure taps on a typical transaxle.

During the road test, observe the starting pressures and the steadiness of the increases that should occur with slight increases in load. The amount the pressure drops as the transmission shifts from one gear to another should also be noted. The pressure should not drop more than 15 psi between shifts.

Any pressure reading not within the specifications indicates a problem **(Figure 14)**. Typically, when the fluid pressures are low, there is an internal leak, clogged filter, low oil pump output, or faulty pressure regulator valve. If the fluid pressure increased at the wrong time or the pressure was not high enough, sticking valves or leaking seals are indicated. If the pressure drop between shifts was greater than approximately 15 psi, an internal leak at a servo or clutch seal is indicated. Always check the manufacturer's specifications for maximum drop off before jumping to any conclusions.

To maximize the usefulness of a pressure test and to be better able to identify specific problems, begin the test

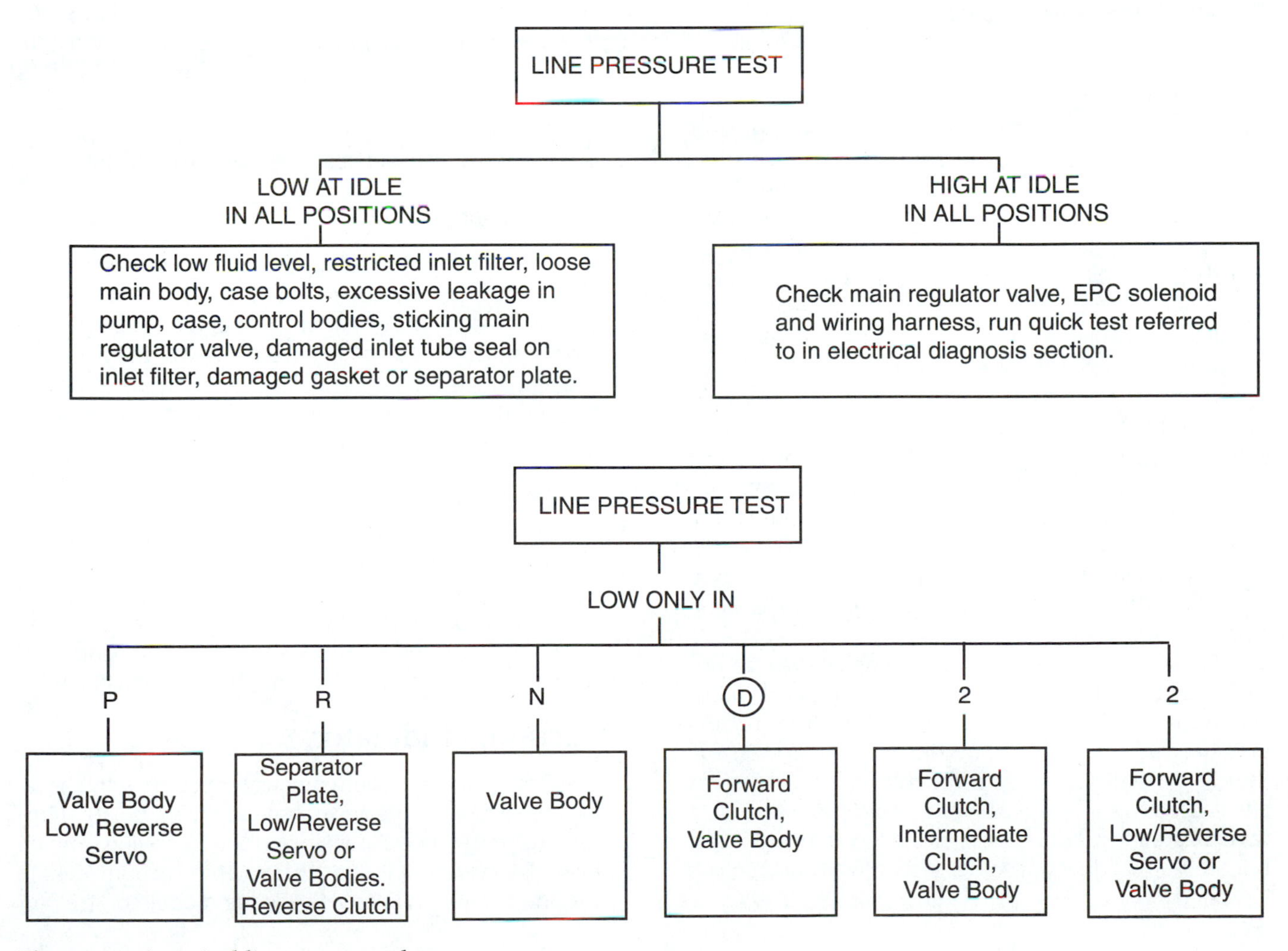


**Figure 14.** A typical line pressure chart.

by measuring line pressure. Mainline pressure should be checked in all gear ranges and at the three basic engine speeds. It is very helpful for you to make a quick chart such as the one shown in **Figure 15** to record the pressures during the test.

If the pressure in all operating gears is within specifications at slow idle, the pump and pressure regulator are working fine. If all pressures are low at slow idle, it is likely that there is a problem in the pump, a stuck open pressure regulator, clogged filter, improper fluid level, or there is an internal pressure leak. To further identify the cause of the problem, check the pressure in the various gears when the engine is at a fast idle.

If the pressures at fast idle are within specifications, the cause of the problem is normally a worn oil pump; however, the problem may be an internal leak or the pressure regulator valve is stuck open. Internal leaks typically are more evident in a particular gear range because that is when ATF is being sent to a particular device through a particular set of valves and passages. If any of these leak, the pressure will drop when that gear is selected or when the transmission is operating in that gear.

Further diagnostics can be made by observing the pressure change when the engine is operated at WOT in each gear range. A clogged oil filter will normally cause a gradual pressure drop at higher engine speeds because the fluid cannot pass through the filter fast enough to meet the needs of the transmission and the faster turning oil pump. If the fluid pressure changed with the increase in engine speed, a stuck pressure regulator is the most probable cause of the problem. A stuck open pressure regulator may still allow the pressure to build with increases in engine speed, but it will not provide the necessary boost pressures.

If the pressures are high at slow idle, a stuck closed pressure regulator, a misassembled or stuck open boost valve, a poorly adjusted or damaged shift cable, or a stuck open throttle valve is indicated. If all of the pressures are low at WOT, pull on the TV cable or disconnect the vacuum hose to the vacuum modulator. If this causes the pressures to be in the normal range, the low pressure is caused by a faulty cable or there is a problem in the vacuum modulator or vacuum lines. If the pressures stayed below specifications, the most likely causes of the problem are the pump or the control system.

If all pressures are high at WOT, compare the readings to those taken at slow idle. If they were high at slow idle and WOT, a faulty pressure regulator is indicated. If the pressures were normal at slow idle and high at WOT, the throttle system is faulty. To verify that the low pressures are caused by a weak or worn oil pump, conduct a **reverse stall test**. If the pressures are low during this test but are normal during all other tests, a weak oil pump is indicated.

| | Slow idle | Fast idle | WOT |
|---|---|---|---|
| P | | | |
| R | | | |
| N | | | |
| D | | | |
| 3 | | | |
| 2 | | | |
| 1 | | | |

**Figure 15.** A template for a pressure chart to be filled out during testing to aid in diagnostics.

On transmissions equipped with an electronic pressure control (EPC) solenoid, if the line pressure is not within specifications, the EPC pressure needs to be checked. To do this, connect the pressure gauge to the EPC tap **(Figure 16)**. Start the engine and check EPC pressure. Use the appropriate pressure chart for specifications. If the EPC pressure is not within specifications **(Figure 17)**, follow the recommended procedures for testing the EPC. If the pressure is okay, be concerned with mainline pressure problems.

## Governors

If the pressure tests suggested that there was a governor problem, it should be removed, disassembled, cleaned, and inspected. Some governors are mounted internally and the transmission must be removed to service the governor, whereas others can be serviced by removing the extension housing or oil pan, or by detaching an external retaining clamp and then removing the unit.

Improper shift points are typically caused by a faulty governor or governor drive system. However, some electronically controlled transmissions do not rely on the hydraulic signals from a governor; rather, they rely on the electrical signals from these sensors. Sensors, such as speed and load sensors, signal to the transmission's computer when gears should be shifted. Faulty electrical components and/or loose connections also can cause improper shift points.

## Vacuum Modulator

Diagnosing a vacuum modulator begins with checking the vacuum at the line or hose to the modulator **(Figure 18)**. The modulator should be receiving engine manifold vacuum. If it does, there are no vacuum leaks in the line to the modulator. Check the modulator itself for

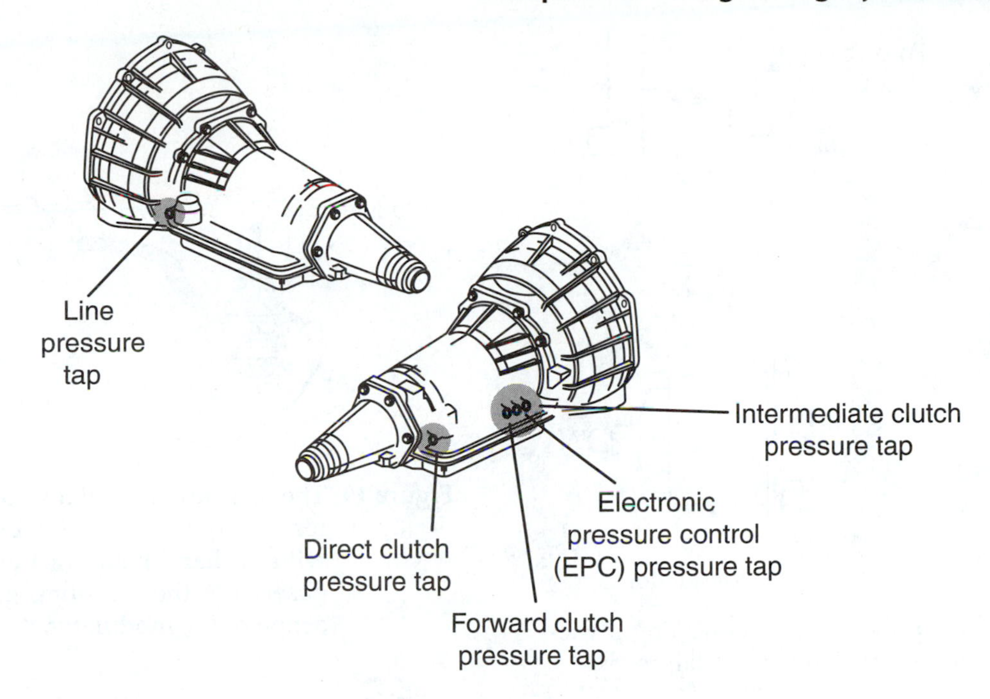


**Figure 16.** Location of EPC pressure tap.

| Transmission Pressure with TP at 1.5 Volts and Vehicle Speed above 8 Km/h (5 mph) | | | | | |
|---|---|---|---|---|---|
| Gear | EPC Tap | Line Pressure Tap | Forward Clutch Tap | Intermediate Clutch Tap | Direct Clutch Tap |
| 1 | 276–345 kPa (40–50 Psi) | 689–814 kPa (100–118 Psi) | 620–745 kPa (90–108 Psi) | 641–779 kPa (93–113 Psi) | 0–34 kPa (0–5 Psi) |
| 2 | 310–345 kPa (45–50 Psi) | 731–869 kPa (106–126 Psi) | 662–800 kPa (96–116 Psi) | 689–827 kPa (100–120 Psi) | 655–800 kPa (95–116 Psi) |
| 3 | 341–310 kPa (35–45 Psi) | 620–758 kPa (90–110 Psi) | 0–34 kPa (0–5 Psi) | 586–724 kPa (85–105 Psi) | 551–689 kPa (80–100 Psi) |

**Figure 17.** A pressure chart for a transmission equipped with an EPC.

leaks with a hand-held vacuum pump **(Figure 19)**. The modulator should be able to hold approximately 18 in. Hg. If transmission fluid is found when you disconnect the line at the modulator, the vacuum diaphragm in the modulator is leaking and the modulator should be replaced. If the vacuum source, vacuum lines, and vacuum modulator are in good condition but shift characteristics indicate a vacuum modulator problem, the modulator may need adjustment.

If no engine vacuum was found at the modulator, check the vacuum line from the engine to the modulator. If the engine is running, it is very unlikely that it would have zero vacuum so some vacuum should be present unless the line or fittings leak.

If the vacuum to modulator was low, check engine vacuum. Run the normal engine vacuum tests to determine where the problem is. A common problem for low vacuum is a restricted exhaust.

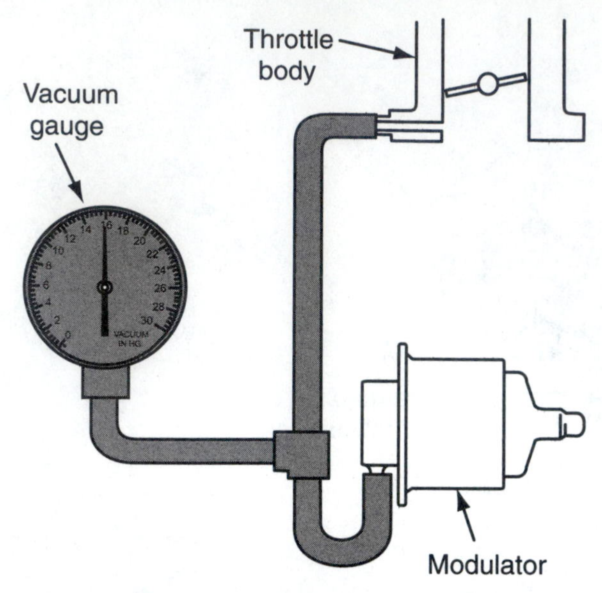


Figure 18. The correct hookup for connecting a vacuum gauge to vacuum modulators.

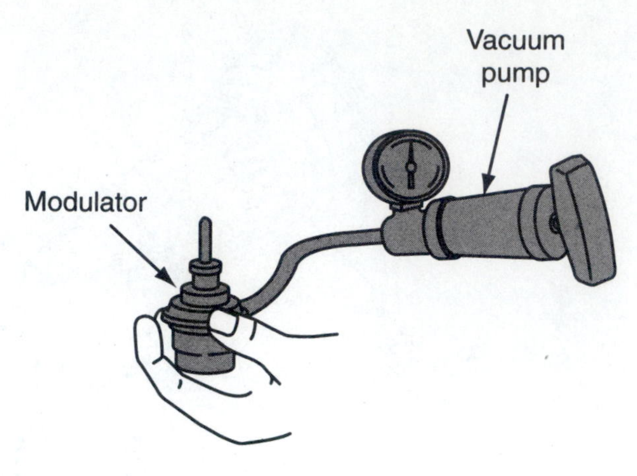


Figure 19. The vacuum modulator can be checked for leakage and action by activating it with a hand-held vacuum pump and observing the vacuum gauge and the action of the modulator.

A vacuum drop test is a very important test. It checks the soundness of the vacuum sources and connecting lines and hoses. With the engine running, foot on the brake pedal, and the transmission in gear, quickly press the gas pedal to the floor then release it. The reading on the vacuum gauge should fall to zero and return to about 17 in. Hg immediately. If it falls slowly or goes down only to about 5 in. Hg, there is a restriction. A restriction is also noted by a slow return to 17 in. Hg.

Most modulators must be removed to be adjusted, if it is adjustable. However, there are some that have an external adjustment. This adjustment allows for fine-tuning of modulator action. To remove a vacuum modulator from the transmission, loosen the retaining clamp and bolt. Some units are screwed into the transmission case. When pulling the modulator out of the housing, be careful not to lose the modulator actuating pin, which may fall out as the modulator is removed. Use a hand-held vacuum pump with a vacuum gauge and the recommended gauge pins to adjust the modulator according to specifications.

## Summary

- An air pressure test is conducted to check the operation of servos and clutches and to check for leaks.
- A pressure test measures the fluid pressure of the different transmission circuits during the various operating gears and gear selector positions.
- A pressure test has its greatest value when the transmission shifts roughly or when the shift timing is wrong.
- The pressure gauges are connected to the pressure taps in the transmission housing and the gauges typically are read during a road test.
- When measured pressures are low, there is an internal leak, clogged filter, low oil pump output, or faulty pressure regulator valve. If the fluid pressure increased at the wrong time or the pressure was not high enough, sticking valves or leaking seals are indicated. If the pressure drop between shifts was greater than approximately 15 psi, an internal leak at a servo or clutch seal is indicated.
- The operation of a vacuum modulator is checked by verifying that vacuum is supplied to it then checked for leaks.

## Review Questions

1. What are the most likely causes for low pressure readings in all gear ranges when the engine is at low speeds?
2. When discussing the results of an oil pressure test, Technician A says that when the fluid pressures are high; internal leaks, a clogged filter, low oil pump output, or a

faulty pressure regulator valve are indicated. Technician B says that if the fluid pressure increased at the wrong time, an internal leak at the servo or clutch seal is indicated. Who is correct?
   A. Technician A only
   B. Technician B only
   C. Both Technician A and Technician B
   D. Neither Technician A nor Technician B

3. Technician A says that low engine vacuum will cause a vacuum modulator to sense a load condition when it actually is not present, as this will cause delayed and harsh shifts. Technician B says poor engine performance can cause delayed shifts through the action of the TV assembly. Who is correct?
   A. Technician A only
   B. Technician B only
   C. Both Technician A and Technician B
   D. Neither Technician A nor Technician B

4. When discussing a pressure test, Technician A says that this test is the most valuable diagnostic check for slippage in one gear. Technician B says that the test can identify the cause of late or harsh shifting. Who is correct?
   A. Technician A only
   B. Technician B only
   C. Both Technician A and Technician B
   D. Neither Technician A nor Technician B

5. Pressure testing revealed low pressure in all gears. Technician A says that using a reverse stall test could isolate the faulty component. Technician B says that a reverse stall test will determine whether the low-reverse servo is the malfunctioning component. Who is correct?
   A. Technician A only
   B. Technician B only
   C. Both Technician A and Technician B
   D. Neither Technician A nor Technician B

# Section 6

## Internal Parts



## SECTION OBJECTIVES

After you have read, studied, and practiced the contents of this section, you should be able to:

- Describe the construction and operation of typcal Simpson gear-based transmissions.
- Describe the construction and operation of Ravigneaux gear-based transmissions.
- Describe the construction and operation of transmissions that use planetary gearsets in tandem.
- Explain how a brake band works and what its purpose is.
- Identify the basic components in a hydraulic servo and describe their functions.
- Describe the different types of one-way clutches used in automatic transmissions.
- Explain how a roller-type or sprag-type one-way clutch works.
- Identify the components in a hydraulic multiple disc clutch and describe their functions.
- Explain the purpose and operation of an accumulator and modulator valve.
- Explain the purposes of seals and gaskets that are found in an automatic transmission/transaxle.
- Describe the purpose of a differential.
- Identify the major components of a differential and explain their purpose.
- Explain the operation of a FWD differential and its drive axles.
- Describe the different designs of four-wheel-drive systems and their applications.

*The idea for using a band as a brake or holding device was first applied to the vehicle's brakes. Before cars featured hydraulic internal brakes, a band was fitted around the outside of a brake drum and was tightened mechanically through a series of levers.*

# Chapter 24 Planetary Gearsets

## *Introduction*

The operation of an automatic transmission allows the engine to move heavy loads with little effort. As the heavy load decreases or the vehicle begins to move, higher gear ratios are required to keep the vehicle moving. By providing different gear ratios, a transmission provides for performance and economy over the entire driving range.

Most automatic transmissions use planetary gearsets to provide for the different gear ratios. The gear ratios are selected manually by the driver or automatically by the hydraulic control system, which engages and disengages the clutches and brakes used to shift gears. Planetary gears **(Figure 1)** are always in constant mesh; therefore, they allow quick, smooth, and precise gear changes without the worry of clashing or partial engagement.

## PLANETARY GEARS

In a simple planetary gearset, the sun gear **(Figure 2)** is located in the center and meshes with the teeth of the planetary pinion gears. Planetary pinion gears are fitted into a framework called the planetary carrier. The planetary carrier can be made of cast iron, aluminum, or steel plate and is designed with a shaft for each of the planetary pinion gears.

A planetary gearset normally has three or four planet pinion gears; each pinion gear spins on its own separate shaft. For heavy loads, the number of planet gears is increased to spread the workload over more gear teeth. Planetary pinion gears rotate on needle bearings. The carrier and pinions are considered one unit—the mid-size gear member **(Figure 3)**.

The planetary pinions are surrounded by the annulus or ring gear **(Figure 4)**, which is the largest part of the

**Figure 1.** A typical planetary gearset.

**Figure 2.** A sun gear with its thrust bearing.

**Figure 3.** A planetary carrier assembly on the left and a ring gear on the right.

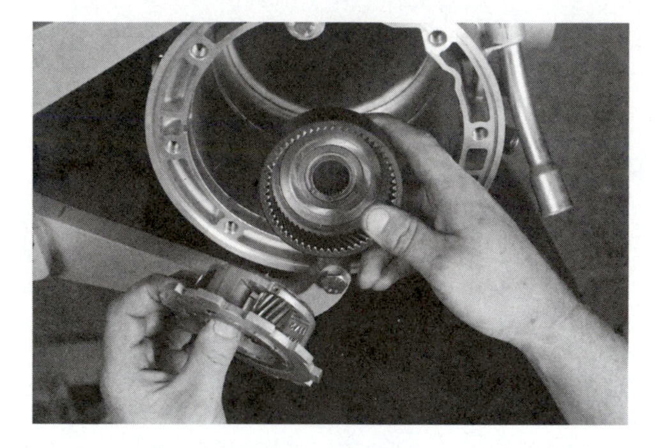

**Figure 4.** A ring gear and carrier.

simple gear set. The ring gear holds the entire gear set together and provides great strength to the unit.

The planet pinion gears are in constant mesh with the sun and ring gears. Each member of a gear set can spin (revolve) or be held at rest. Power transfer through a planetary gearset is only possible when one of the members is held or if two of the members are locked together.

Any one of the three members can be used as the driving or input member. At the same time, another member might be kept from rotating and thus becomes the held or stationary member. The third member then becomes the driven or output member. Depending on which member is the driver, which is held, and which is driven, either a torque increase or a speed increase is produced by the planetary gearset. Output direction also can be reversed through various combinations.

Planetary gears are used because:

- They are compact yet strong enough to handle great amounts of torque.
- All members of a gear set share a common axis, which provides for a very compact unit.
- The gears are typically helically cut gears, which offer quiet operation.
- The gears are always in constant mesh. This reduces the risk of gear damage and allows for smooth and quick gear ratio changes.
- Holding a gear set member or locking them together is easy because of the location of the different members of the gear set.
- The gears within the gear set can function independently of each other and can turn on their own centers when revolving around the sun gear.

## Reaction Members

Reaction members are those parts of a planetary gearset that are held in order to produce an output motion. Other members of the gear set react against the held member. Devices, such as multiple friction disc packs, brakes, and one-way overrunning clutches are used in automatic transmissions to hold or drive members of the planetary gearset in order to provide for the various gear ratios and directions. One-way overrunning clutches are purely mechanical devices, whereas clutches and brakes are hydraulically controlled mechanical devices. Most automatic transmissions use more than one type of these devices, some use all three.

The bulk of a transmission's interior is comprised of these devices. Each member of the compound planetary gear unit is attached to some sort of a holding or driving device. This results in many apply or brake devices attached to a simple compound planetary unit because each member of the gear set can be used as the input, output, or reaction member.

Brake bands work by holding a drum, attached to a planetary member, stationary. Multiple friction discs can ground or lock a planetary member to the transmission housing. One-way clutches prevent a planetary member from rotating in one direction.

# LAWS OF PLANETARY GEAR ACTION

Changing gears in an automatic transmission is all about planetary gear control. The output of a gear set is always the same, regardless of the size of the gears. The output depends only on what member is the input, the output, and what is held **(Figure 5)**.

When the ring gear and pinion gears are able to rotate at the same time, the pinions always rotate in the same direction as the ring gear. The sun gear, however, always rotates in the opposite direction as the pinion gears. The ring gear has internal gear teeth and the planet pinions and sun gears have external teeth. When an external gear drives another external gear, the output gear always rotates in the opposite direction as the input gear. However, when an external gear is in mesh with an internal gear, the two gears rotate in the same direction.

| SUN | CARRIER | RING | SPEED | TORQUE | DIRECTION |
|---|---|---|---|---|---|
| Input | Output | Held | Maximum reduction | Maximum increase | Same as input |
| Held | Output | Input | Minimum reduction | Minimum increase | Same as input |
| Output | Input | Held | Maximum increase | Maximum reduction | Same as input |
| Held | Input | Output | Minimum increase | Minimum reduction | Same as input |
| Input | Held | Output | Reduction | Increase | Opposite of input |
| Output | Held | Input | Increase | Reduction | Opposite of input |

**Figure 5.** The basic laws of simple planetary gear action.

When the planet carrier is the output member, it always goes in the same direction as the input gear. Likewise, when the planet carrier is in the input, the output gear member will always rotate in the same direction as the carrier.

Remember, whenever a small gear drives a larger gear, output torque is increased and output speed is decreased. Also, whenever a large gear drives a smaller gear, output torque is decreased and output speed is increased.

## Neutral

When there is an input to the gear set but no reaction member, the gear set is in neutral and there will be no output. This is the neutral position of a planetary gearset.

Consider the planetary gearset shown in **Figure 6**, the input gear is the sun gear and it rotates clockwise. The output is the planet carrier, which is held stationary by the vehicle's weight on the drive wheels. This allows the pinions to rotate on their own shafts and drive the ring gear in a counterclockwise direction. The planet pinions and the ring gear simply rotate or freewheel around the sun gear.

## Gear Reduction

When the output member turns at a slower speed than the input member does, gear reduction has occurred and this results in increased torque on the output member. This can be accomplished by holding the sun gear. If power is applied to the ring gear, the planet gears will spin on their shafts in the carrier **(Figure 7)**. Because the sun gear is being held, the spinning planet gears will walk

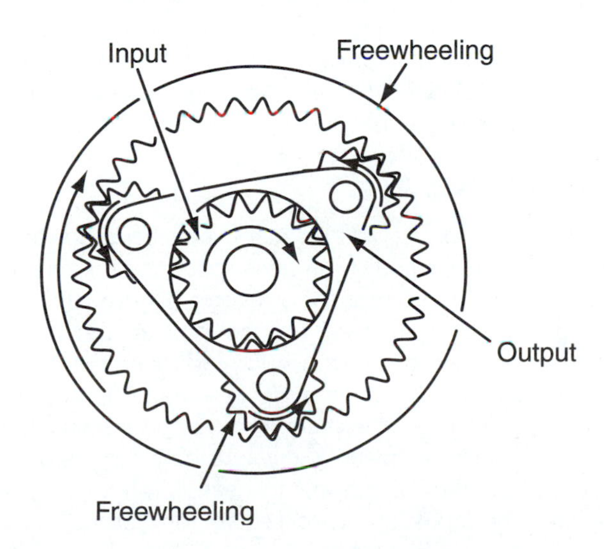


**Figure 6.** A planetary gearset in its neutral position.

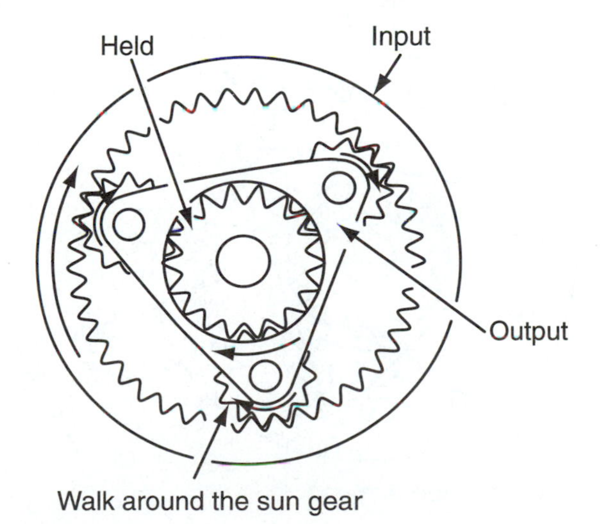


**Figure 7.** Gear reduction with the ring gear as the input.

around the sun gear and carry the carrier with them. This causes a gear reduction because one complete revolution of the ring gear will not cause one complete revolution of the carrier. The carrier is turning at a slower speed and therefore is increasing the torque output. By holding the sun gear, the carrier moves in the same direction as the ring gear.

Gear reduction also can occur if the ring gear is held and the sun gear is the input gear **(Figure 8)**. The planet pinions will rotate on their own shafts and walk around the inside of the ring gear but in the opposite direction as the sun gear. This causes the carrier to rotate in the same direction as the sun gear. The carrier will rotate slower than it did when the ring gear drove it. This results in greater gear reduction and therefore greater torque.

To have gear reduction, a planetary member must be held stationary. To do this a brake band may be used to clamp around a rotating drum and stop it from rotating. When the band is applied, the member (either the sun or ring gear) in the drum becomes the reaction member.

Multiple disc packs are also used to hold a planetary member stationary. The outer edges of the steel clutch discs are splined and fit into internal splines in the transmission housing. Either the sun or the ring gear is attached to the friction discs. When the assembly is applied, the friction discs squeeze against the steel discs and the sun or ring gear is locked to the case.

One-way clutches are also used to hold members. If the sun gear is connected to a one-way clutch, it will only be able to rotate freely in one direction. If the input member is the ring gear, the planet carrier will rotate with the ring gear. This would cause the sun gear, if not held by another brake, to attempt to rotate in the opposite direction. This tendency

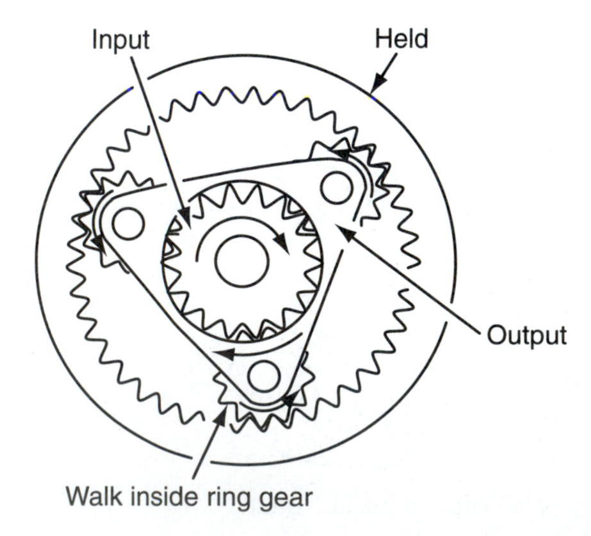


**Figure 8.** Gear reduction with the sun gear as the input.

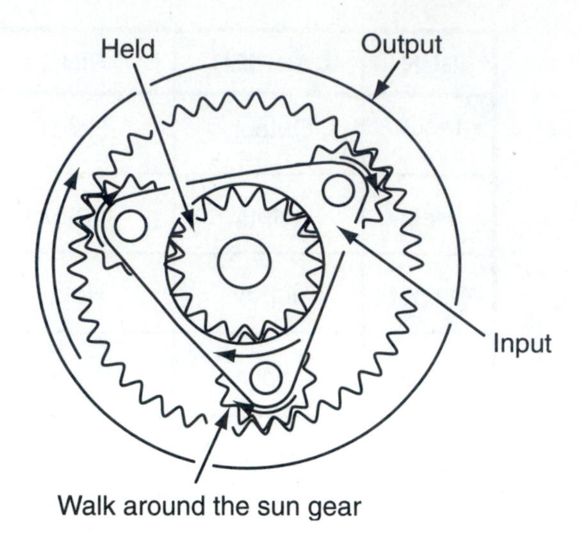


**Figure 9.** Overdrive with the sun gear held.

will cause the one-way clutch to lock and stop the sun gear from turning.

## Overdrive

When the planet carrier is the driving member of the gear set and the sun gear is held, overdrive occurs **(Figure 9)**. When the planet carrier is the input member of the gear set and the sun gear is held, overdrive occurs. As the planet carrier rotates, the pinion gears are forced to walk around the held sun gear, which drives the ring gear faster. One complete rotation of the planet carrier causes the ring gear to rotate more than one complete revolution in the same direction. This provides more output speed but less torque or overdrive.

A higher speed overdrive is possible by holding the ring gear stationary. With input on the planet carrier, the pinion gears are forced to walk around the inside of the ring gear, driving the sun gear clockwise. The planetary carrier rotates much less than one turn to rotate the sun gear one complete revolution. The result is a great reduction in torque output and a maximum increase in output speed.

## Direct Drive

If any two of the gear set members receive power in the same direction and at the same speed, the third member is forced to move with the other two **(Figure 10)**. If the ring gear and the sun gear are the input members, the internal teeth of the ring gear will try to rotate the planetary pinions in one direction, while the external teeth of the sun gear will try to drive them in the opposite direction. This action locks the planetary pinions between the other members and the entire planetary gear set rotates as a single unit. The input is now locked to the output, which results in direct drive. One input revolution equals one output revolution for a gear ratio of 1:1.

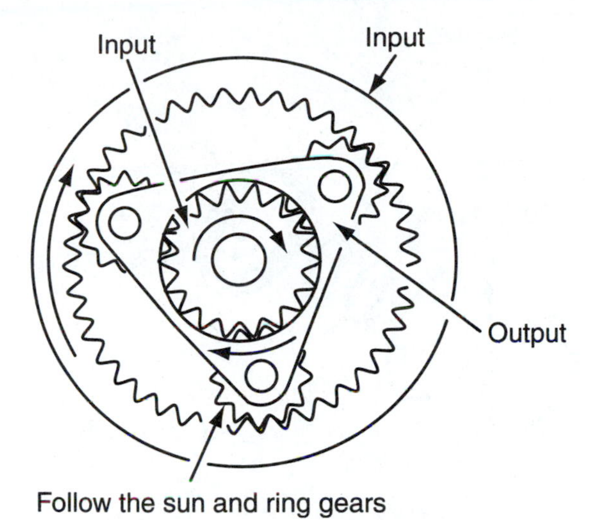


Figure 10. Direct drive with input to the ring and sun gears.

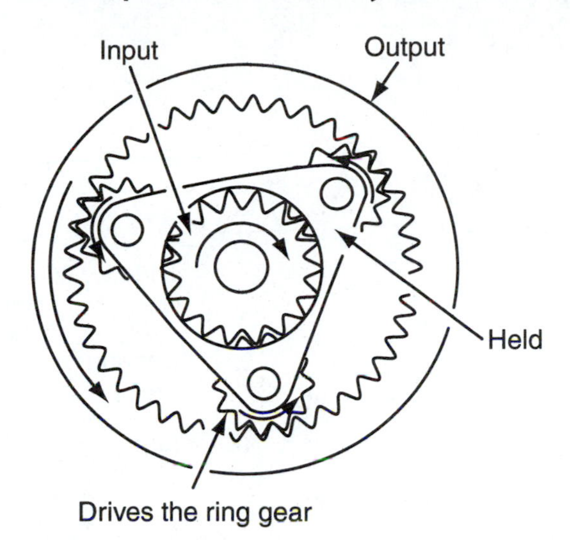


Figure 11. Reverse gear with the planet carrier held and the ring gear as the output.

## Reverse

If the planet carrier is held and the sun gear is rotated in a clockwise rotation **(Figure 11)**, the ring gear will rotate in a counterclockwise direction. The planet gears cannot travel around the teeth of an enmeshed gear; rather, the carrier is held in place and the planet gears spin on their shafts. The sun gear spins the planet gears, which drive the ring gear in the opposite direction but at a slower speed. Therefore, the gear set is providing reverse gear with reduction.

Reverse with overdrive is possible by applying the input to the ring gear and using the sun gear as the output member. A reversal of direction is obtained whenever the planet carrier is stopped from turning and power is applied to either the sun gear or the ring gear. This causes the pinion gears to act as idler gears, thus driving the output member in the opposite direction as the input, with a great amount of overdrive.

# COMPOUND PLANETARY GEARSETS

To provide more gear reduction gear ratios, transmissions are equipped with more than one planetary gearset. The additional gear sets are arranged in a number of different ways and this arrangement varies with the transmission models. When two planetary units share a member, they are called compound planetary units.

## Simpson Gear Set

One half of the Simpson gear set or one planetary unit is referred to as the front planetary and the other planetary unit is the rear planetary. The two planetary units do not need to be the same size or have the same number of teeth on their gears. The size and number of gear teeth determine the actual gear ratios obtained by the compound planetary gear assembly.

Gear ratios and direction of rotation are the result of applying torque to one member of either planetary unit, holding at least one member of the gear set, and using another member as the output **(Figure 12)**. For the most part, each automobile manufacturer uses different parts of the planetary assemblies as an input, output, and reaction members; this also varies with the different transmission models from the same manufacturer. There are also many different apply devices used in the various transmission designs.

## Ravigneaux Gear Set

The Ravigneaux gear train, like the Simpson gear train provides forward gears with a reduction, direct drive, overdrive, and a reverse operating range. The Ravigneaux offers some advantages over a Simpson gear train. It is very compact. It can carry large amounts of torque because of the great amount of tooth contact. It also can have three different output members. However, it has a disadvantage to students and technicians, it is more complex and therefore its actions are more difficult to understand.

The Ravigneaux gear train is designed to use two sun gears, one small and one large. They also have two sets of planetary pinion gears, three long pinions and three short pinions. The planetary pinion gears rotate on their own shafts that are fastened to a common planetary carrier. A single ring gear surrounds the complete assembly.

The small sun gear is meshed with the short planetary pinion gears. These short pinions act as idler gears to drive the long planetary pinion gears. The long planetary pinion gears mesh with the large sun gear and the ring gear.

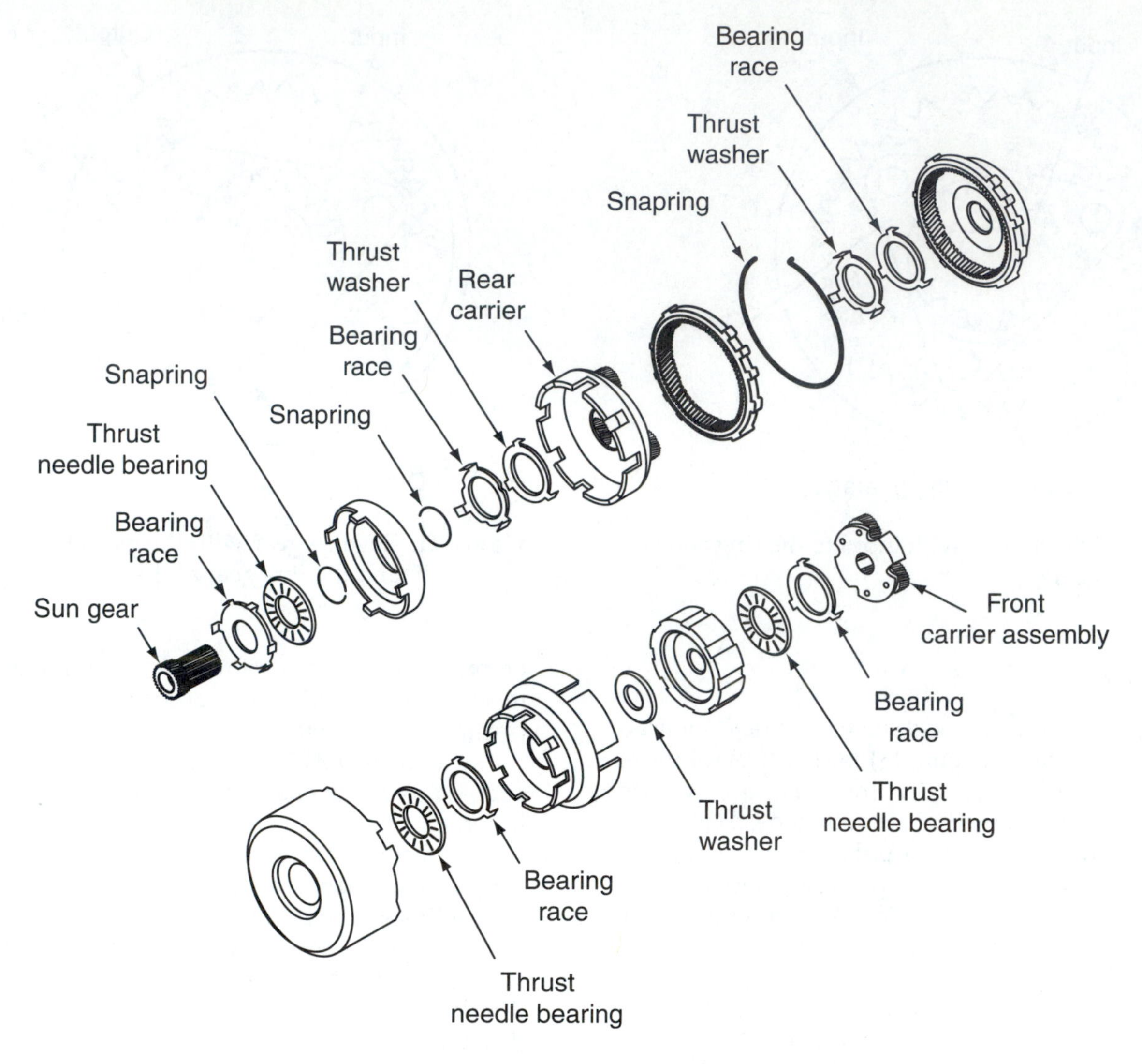


**Figure 12.** A Simpson planetary gearset.

## Tandem Gear Sets

Rather than rely on a compound gear set, some automatic transmissions use two or three simple planetary units in series. In this type of arrangement, gear set members are not shared instead the holding devices are used to lock different members of the planetary units together.

The combination of the planetary units functions much like a compound unit. The tandem units do not share a common member, rather certain members are locked together or are integral with each other. The front planetary carrier is locked to the rear ring gear and the front ring gear is locked to the rear planetary carrier.

Rather than relying on the use of a compound gear set, some automatic transmissions use two simple planetary units in series. In this type of arrangement, gear set members are not shared; rather, certain members are locked together or are integral with each other. For example, the front planetary carrier may be locked to the rear ring gear and the front ring gear locked to the rear planetary carrier **(Figure 13)**. Through such an arrangement, the output of one gear set can become the input for the next.

## Nonplanetary Gear Sets

Honda and Saturn transaxles are unique in that they do not use a planetary gearset to provide for the different gear ranges. Constant-mesh helical and square-cut gears **(Figure 14)** are used in a manner similar to that of a manual transmission.

These transaxles have a mainshaft and countershaft on which the gears ride. The two shafts run parallel to each and some later models have a third parallel shaft.

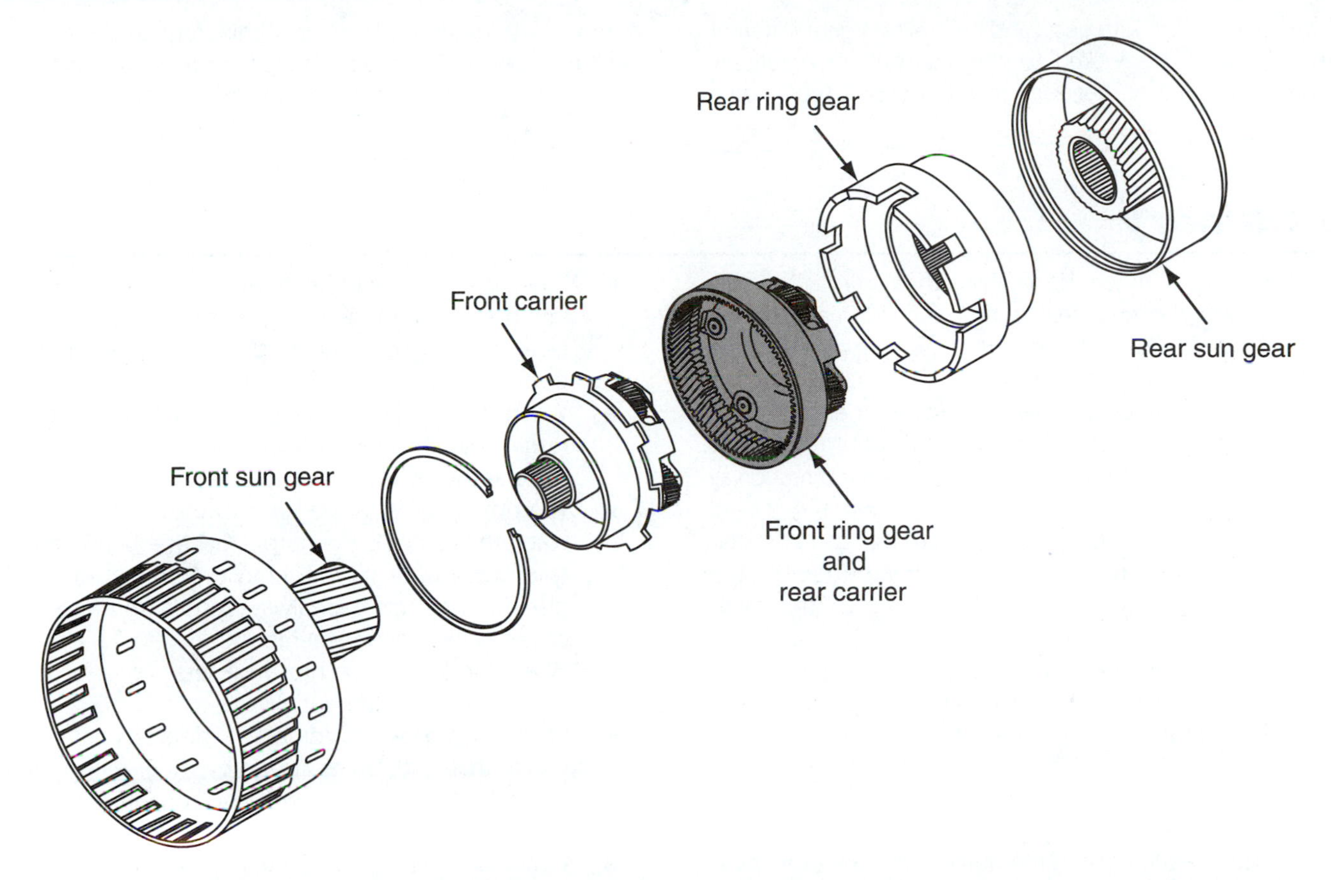


**Figure 13.** Two planetary units with the ring gear of one gear set connected to the planet carrier of the other.

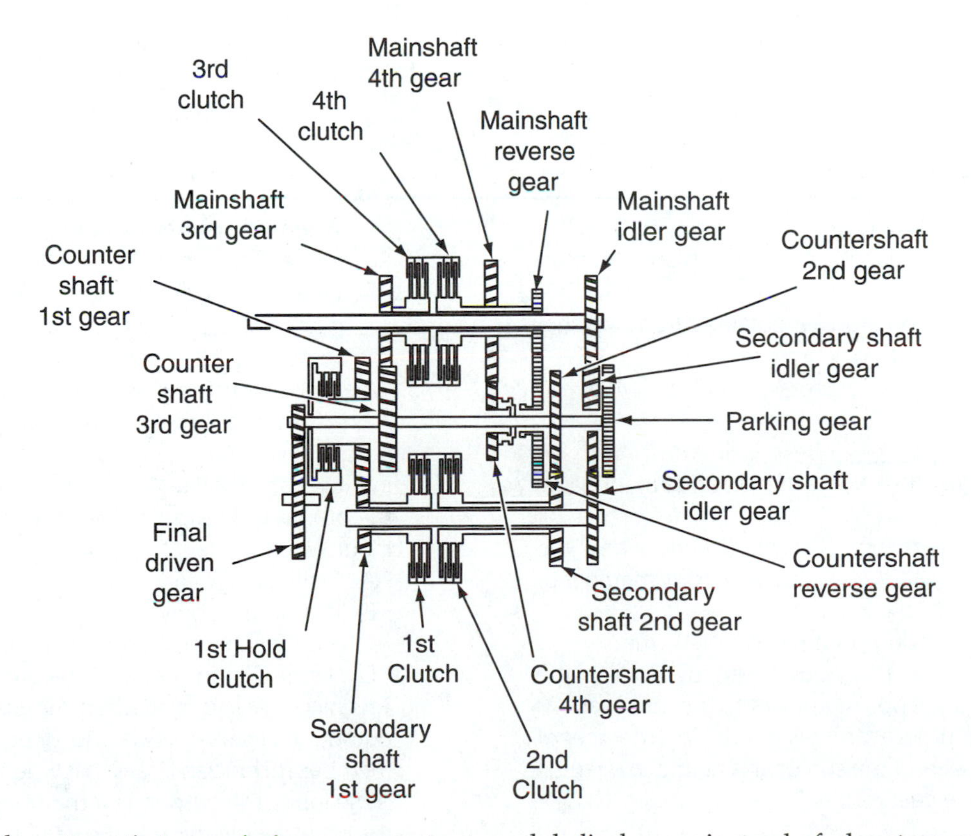


**Figure 14.** Honda automatic transmissions use constant-mesh helical gears instead of planetary gearsets.

To provide the forward gears and one reverse gear, different pairs of gears are locked to the shafts by hydraulically controlled clutches. Reverse gear is obtained, on some models, through the use of a shift fork that slides the reverse gear into position. The power flow through these transaxles is also similar to that of a manual transaxle.

## Summary

- When the sun gear is the input, the carrier is the output, and the ring gear is the reaction member, the output will have maximum reduction in the same direction as the input.
- When the sun gear is the reaction member, the carrier is the output, and the ring gear is the input, the output will have minimum reduction in the same direction as the input.
- When the sun gear is the output, the carrier is the input, and the ring gear is the reaction member, the output will have maximum overdrive in the same direction as the input.
- When the sun gear is the reaction member, the carrier is the input, and the ring gear is the output, the output will have minimum overdrive in the same direction as the input.
- When the sun gear is the input, the carrier is the reaction member, and the ring gear is the output, there will be gear reduction and the output will be in the opposite direction as the input.
- When the sun gear is the output, the carrier is the reaction member, and the ring gear is the input, there will be an overdrive condition and the output will be in the opposite direction as the input.
- When any two members are held together, direct drive results and the output speed and direction are the same as the input.
- The planet carrier is the key member of the gear set: when the carrier is the output, gear reduction always takes place, when it is the input, overdrive always takes place, and reverse gear always results from using the carrier as the reaction member.
- When there is input into a planetary gearset but a member is not held, neutral results.
- Compound gear sets combine simple planetary gearsets so that load can be spread over a greater number of teeth for strength and also to obtain the largest number of gear ratios possible in a compact area.
- Honda and Saturn nonplanetary based transaxles use constant-mesh helical and square-cut gears.

## Review Questions

1. How does direct drive result in a planetary gearset?
2. How is reverse gear accomplished in a planetary gearset?
3. How can a gear reduction be obtained from a planetary gearset?
4. What happens in a planetary gearset when no member is held?
5. When an ____________ gear is in mesh with an ___________ gear, the two gears will rotate in the same direction.
6. Forward gear reduction always occurs when the ____________ ____________ is the output member of the gear set.
7. Which of the following statements is NOT true?
   A. Gear reduction is accomplished by holding the ring gear and applying power to the sun gear.
   B. When the planet carrier is the driving member of the gear set and the sun gear is held, a low reduction reverse gear results.
   C. When the planet carrier is held, the output will rotate in the opposite direction as the input.
   D. When the planet carrier is the input, the output gear member always follows the direction of the carrier.
8. When discussing reverse gear in a planetary gearset, Technician A says that when the ring gear is held, the output will rotate in the opposite direction as the input. Technician B says that when the sun gear is held, the output will rotate in the opposite direction as the input. Who is correct?
   A. Technician A only
   B. Technician B only
   C. Both Technician A and Technician B
   D. Neither Technician A nor Technician B
9. Technician A says that when the planet carrier is the output, it always follows the direction of the input member. Technician B says that when the planet carrier is the input, the output gear member always moves in the opposite direction as the carrier. Who is correct?

A. Technician A only
B. Technician B only
C. Both Technician A and Technician B
D. Neither Technician A nor Technician B

10. Technician A says that when the ring gear and carrier pinions are free to rotate at the same time, the pinions will always follow the same direction as the ring gear. Technician B says that the sun gear always rotates opposite of the rotation of the pinion gears. Who is correct?
    A. Technician A only
    B. Technician B only
    C. Both Technician A and Technician B
    D. Neither Technician A nor Technician B

# Friction and Reaction Units

## Introduction

The reaction and friction devices in an automatic transmission are what make everything else work. The hydraulic system with its complexity does no more than activate these devices. Covered in this chapter are the various brakes and clutches used in transmissions **(Figure 1)**.

## BRAKE BANDS

A band is an externally contracting brake assembly that is positioned around the outside of a drum **(Figure 2)**. The drum is connected to a member of the planetary gear set. Bands are simply flexible metal strips lined with either a semi-metallic or organic friction material. Semi-metallic linings are very durable and can work well under

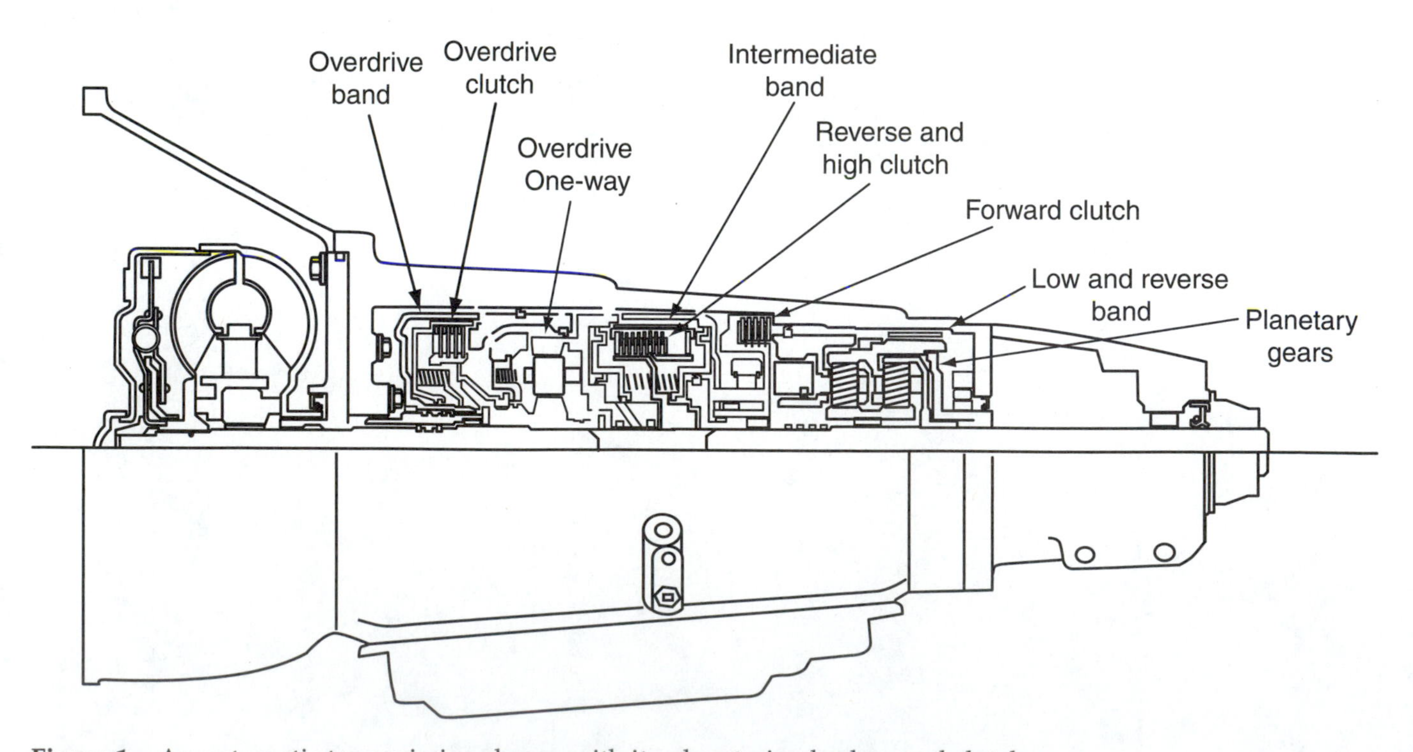


**Figure 1.** An automatic transmission shown with its planetaries, brakes, and clutches.

**Figure 2.** A typical band.

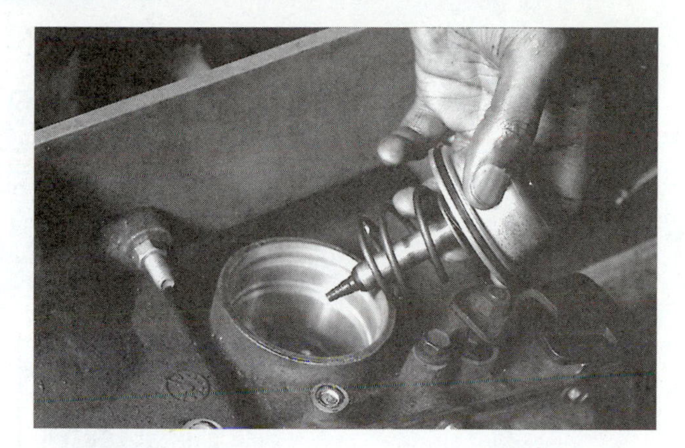

**Figure 3.** A servo unit.

high pressures. This type lining is typically used in reverse gear because they tend to wear the drum surface when they are tightened around a drum that is rotating with high torque. In reverse, the band does not need to stop the drum from rotating; rather, it keeps it from rotating.

Organic friction lining materials are commonly used on bands that stop a drum rotating with high torque. The lining material is typically a soft paper pulp-based or cellulose-based material. These materials do not wear away the surface of the drum and provide for a good clamp on the drum. The heat produced by the friction of the band is very high (up to 800°F). The fluid present on the drum or in the band's lining quickly evaporates in the presence of this heat, thus creating more frictional heat. The soft lining, however, is saturated with fluid. As the band tightens, the fluid in the lining is squeezed out. The fluid replaces the fluid lost by evaporation and cools the band. To aid in this cooling process, bands are made with grooves cut into the friction surface.

The construction of the drum is critical to band action. If the drum is made of soft iron, the squeezing action of the band can distort the drum. However, the rough surface of this type drum aids in the clamping effort of the band. The rough surface also holds fluid in its pores and this fluid can cause glazing on the drum and band after some use. Drums made with hard iron will not distort and will withstand high clamping forces. This type drum can be finished with a smooth finish that does not wear the band's surface.

A band is applied hydraulically by a servo assembly **(Figure 3)**. Hydraulic pressure moves the servo piston, which compresses the servo spring and directly applies the band or does so through a mechanical linkage **(Figure 4)**. To release the band, hydraulic pressure to the servo is diverted and the spring moves the piston back into its bore.

A band and its servo are always used as holding devices and never are used to drive a member of the gear set. A band is anchored at one end and force is applied against the other end. As this force is applied, the band contracts around the rotating drum and squeezes it to a stop. The amount of pressure that stops the drum from rotating is determined by the length and width of the band and by the amount of force applied against the band's unanchored end. Rods or struts, which may be placed between the anchor and the band or be at both sides of the band, are used to apply force against the band.

A band can be positioned to allow the applying pressure to move against the direction of drum rotation or with it. If the band is mounted so that the force is applied to same direction as drum rotation, the movement of the drum adds to the applying force and less hydraulic pressure is needed. When the band moves in the opposite direction of drum rotation, the drum opposes the band and more pressure is needed to stop the drum.

Although all transmission bands are made of flexible steel and their inside surfaces lined with frictional material, they differ in size and construction depending on the amount of work they are required to do. A band that is split with overlapping ends is called a double-wrap band. A one-piece band that is not split is called a single-wrap band.

Two types of single-wrap bands **(Figure 5)** are commonly used in transmissions. One is made of a light and flexible steel; the other type is made of a heavy and more rigid cast iron. The heavy bands typically have a metallic lining material that can withstand large gripping pressures. Light bands are lined with a less-abrasive material that helps limit drum wear.

Double-wrap bands **(Figure 6)** have a smoother and uniform grip and lend themselves more to self-energizing. A double-wrap band readily conforms to the shape of a drum; therefore, it can provide greater holding power for a given application force. A double-wrap band also requires less hydraulic pressure than a single-wrap band to produce the same amount of holding power.

To prevent harsh changing of gears, which would result from quickly stopping the movement of a gear set member, bands are designed to slip as they are being applied. The amount they slip increases as their lining

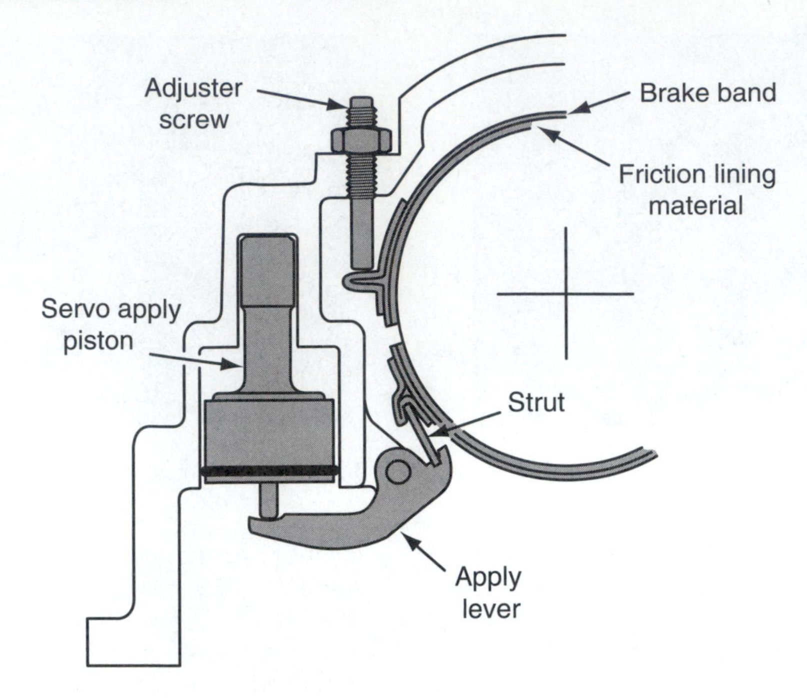


**Figure 4.** Band in place with its servo and apply lever.

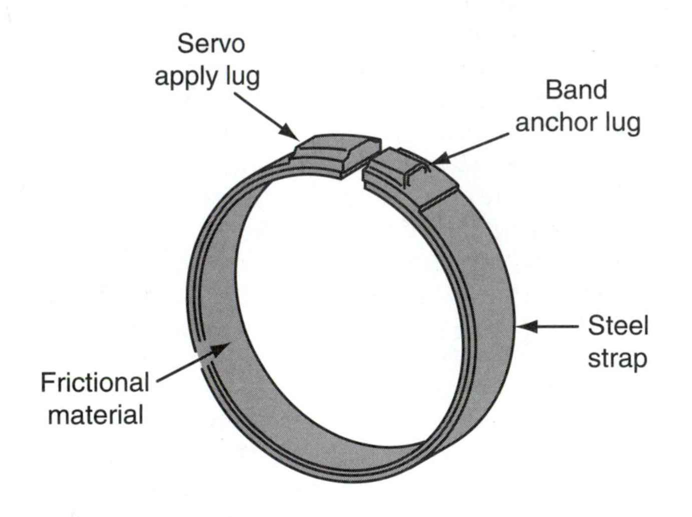


**Figure 5.** A single-wrap band.

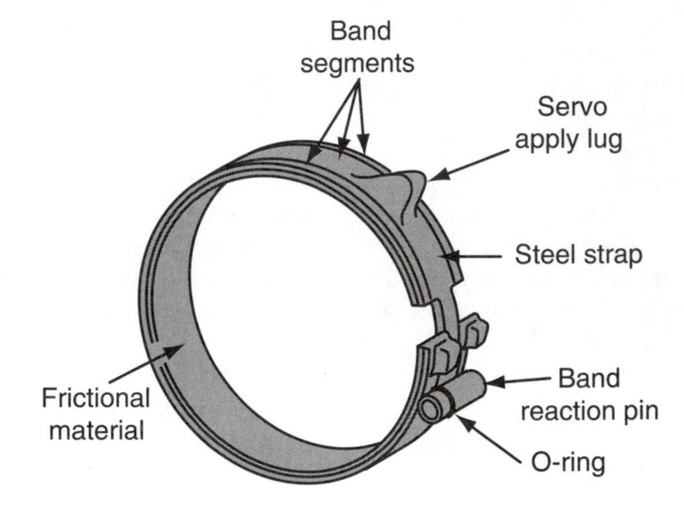


**Figure 6.** A double-wrap band.

wears. This wear increases the clearance between the band and the drum, and reduces the holding force of the band. Because of this, the bands in older transmissions need to be adjusted periodically. However, as band designs have improved, periodic band adjustment is not needed on most new transmissions. On units that require periodic adjustment of the bands, the clearance between the band and the drum is set with an adjustment screw that also serves as the anchor for the band. Excessive slippage will cause a band to burn or become glazed as well as cause poor power transfer through the transmission.

Several factors contribute to the effectiveness of a band:

- The type of frictional material used for the lining.
- The composition of the drum and its surface finish.
- Sufficient fluid flow to cool the band and drum.
- Condition and type of transmission fluid.
- Proper grooving of the lining material to aid in the cooling process.
- Proper adjustment of the band.
- The force used to apply the band.

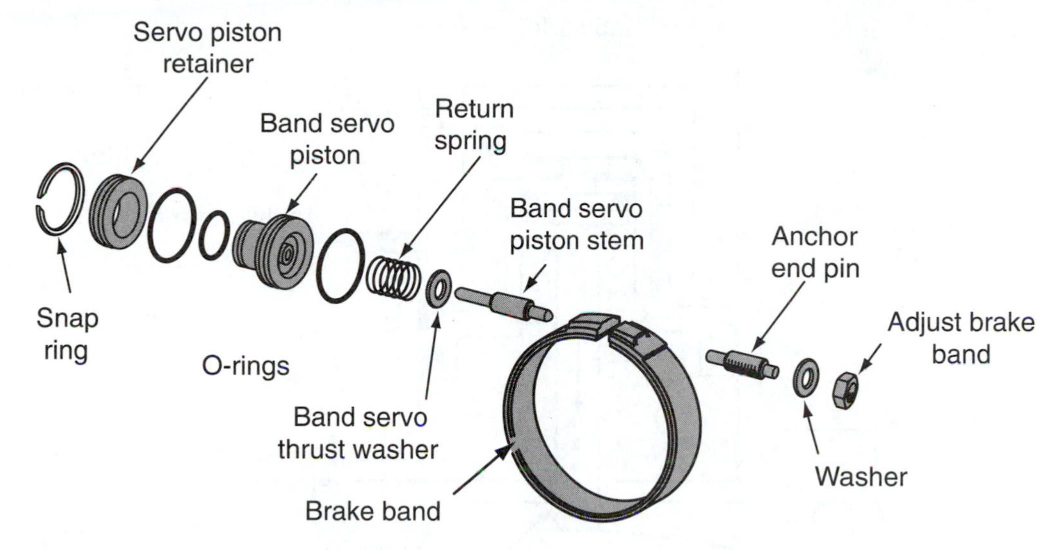


**Figure 7.** A typical band and servo assembly.

## Servos

The servo **(Figure 7)** contracts the band when hydraulic pressure pushes against the servo's piston and overcomes the tension of the servo's return spring. This action moves an operating rod toward the band, which squeezes the band around the drum. When hydraulic pressure to the servo apply port is stopped, the band is released and the return spring returns the piston to its original position. On some transmissions, the return of the piston is aided by hydraulic pressure sent to the release side of the servo. Once the piston has returned, this release pressure keeps the band from applying until apply pressure is again sent to the servo.

A servo unit consists of a piston in a cylinder and a piston return spring. The cylinder may be part of the transmission housing's casting or it may be a separate unit bolted to the case. The force from the servo's piston acts directly on the end of the band by an apply pin or through a lever arrangement that provides mechanical advantage or a multiplying force. The band squeezes a drum attached to a gear set member.

The servo unit must apply the band securely so that it can rigidly hold a member of the planetary gear set in order to cause forward or reverse gear reduction. To assist the hydraulic and mechanical apply forces, the servo and band anchor are positioned to take advantage of drum rotation. When the band is applied, it becomes self-energized and wraps itself around the drum in the same direction as drum rotation. This self-energizing effect reduces the force that the servo must produce to hold the band.

To release the servo apply action on the band, the servo apply oil is exhausted from the circuit or a servo release oil is introduced on the servo piston that opposes the apply oil. When servo release oil is introduced on a shift, the hydraulic mainline pressure acting on the top of the piston plus the servo return spring will overcome the servo apply oil and move the piston down.

The application force exerted on it by the servo determines the clamping pressure of a band. Because the area of the servo's piston is relatively large, the application force of the servo is considerably greater than the line pressure delivered to the servo. This increase in force is necessary to clamp the band tight enough around the spinning drum to bring it to a halt.

Servos are designed to react to varying pressures and apply different amounts of application forces to meet the needs of the transmission. The total force of a piston is equal to the pressure applied to it multiplied by the surface area of the piston. Therefore, if the servo piston has a surface area of 2.5 square inches and the normal pressure sent to the servo is 50 psi, the application force of the servo on the band will be 125 psi. When the pressure sent to the servo increases to 100 psi, such as during heavy loads, the application force on the operating rod would increase to 250 psi **(Figure 8)**. This increase in force allows for increased clamping pressures during heavy load conditions and at times when slipping will most likely take place.

The operating rod of a servo tightens a band through one of three basic linkage designs: straight, lever, or cantilever. The straight linkage design uses a straight rod or strut to transfer the piston's force to the free end of the band. This type is used only when the servo is placed where it can act directly on a band and when the servo's piston is large enough to hold the band when maximum torque is applied to the drum. Some transmissions that use a straight linkage have specially designed rods, which are graduated and designed to minimize the need for periodic band adjustment. Band adjustment is done by selecting the

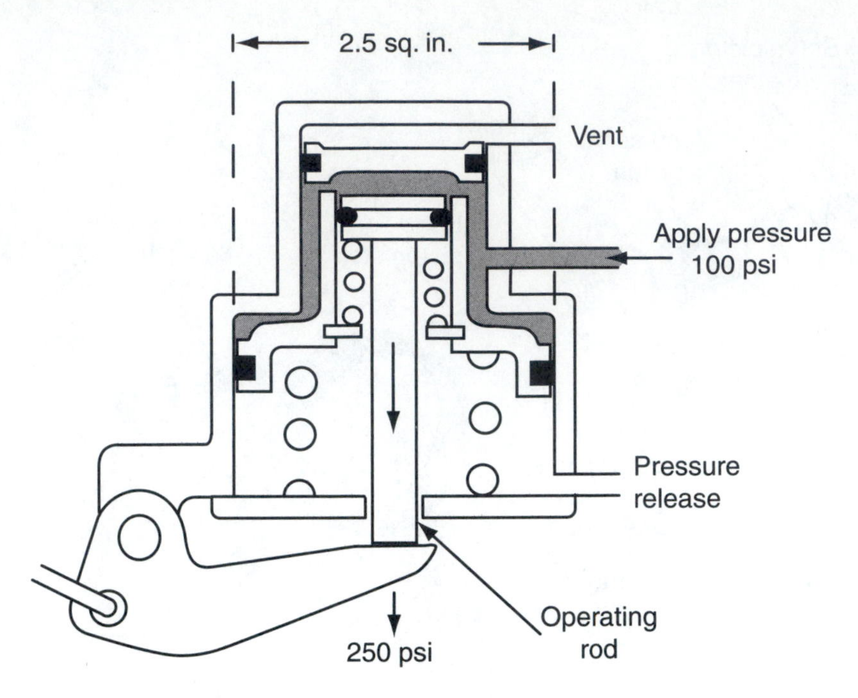


**Figure 8.** The surface area of a servo piston increases fluid pressure so more force is on the operating rod.

proper rod length. When the correct rod is installed in the servo, the piston will apply a specific amount of force against a band.

The lever-type linkage uses a lever to move the rod or strut that actually applies the band. This type of linkage increases the application force of the piston, because the **fulcrum** of the lever is closer to the band **(Figure 9)**.

A **cantilever**-type linkage uses a lever and a cantilever to act on both ends of an unanchored band. As the servo piston applies force to the operating rod, the rod moves the lever and applies force to one end of the band. The movement of the piston also pulls the cantilever toward its pivot pin, thereby clamping the ends of the band together and tightening the band around the drum. A cantilever linkage increases band application force and allows for smooth band application, because the band self-centers and contracts evenly around the drum.

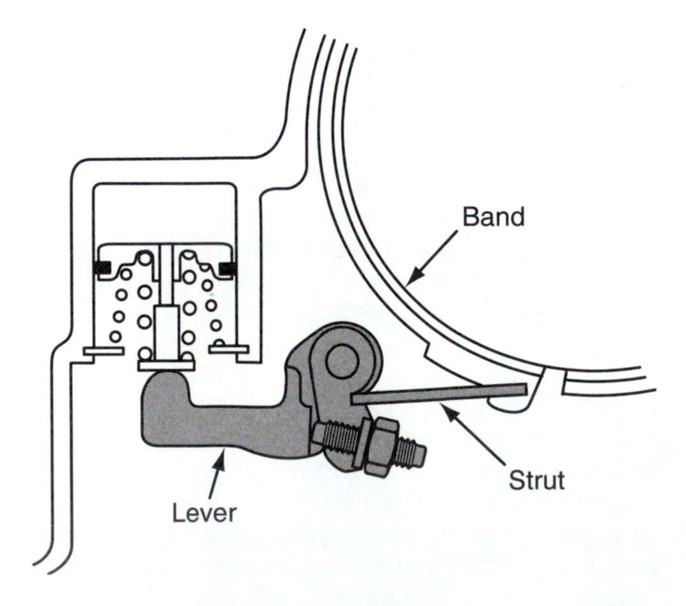


**Figure 9.** A lever-type servo linkage.

## ONE-WAY CLUTCHES AND BRAKES

One-way or overrunning clutches are holding or braking devices. They operate mechanically, not hydraulically, and are considered apply devices. The one-way clutch allows rotation in only one direction and operates at all times, whereas a disc pack or a band allows or stops rotation in either direction and operates only when hydraulic pressure is applied to them.

One-way clutches can freewheel or rotate without affecting the input or output of the planetary gear set. When one-way clutches are freewheeling, they are off or ineffective. Freewheeling normally takes place when the clutch is rotating in a counter clockwise direction.

Overrunning clutches can be either roller or sprag types. A roller clutch uses roller bearings held in place by springs to separate the inner and outer race of the clutch assembly. Around the inside of the outer race are several cam-shaped indentations. The rollers and springs are located in these pockets. Rotation of one race in one direction locks the rollers between the two races causing both to rotate together **(Figure 10)**. When a race is rotated in the opposite direction, the roller bearings move into the

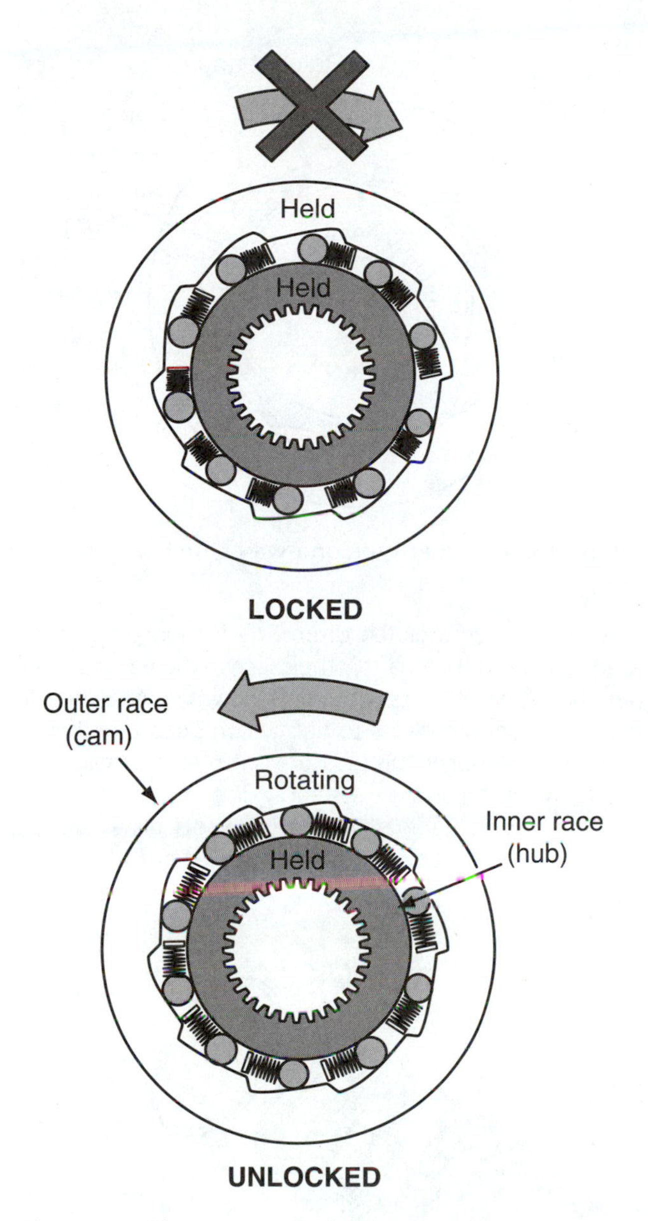


Figure 10. The action of a one-way roller clutch.

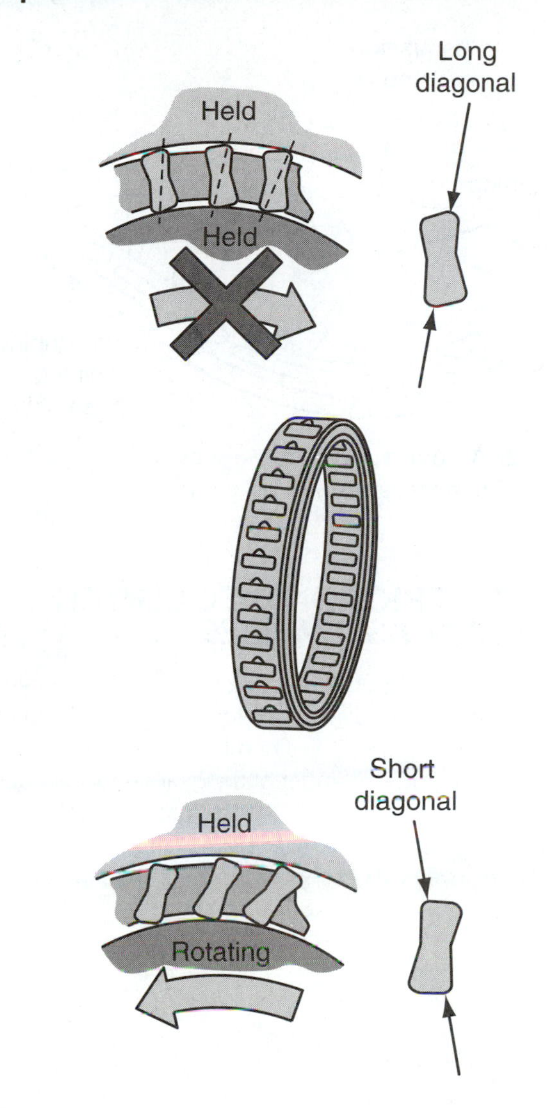


Figure 11. The action of a sprag-type one-way clutch.

pockets and are no longer locked between the races and the races can turn independently.

A one-way sprag clutch uses figure-eight-shaped metal pieces called sprags. The sprags lock between the races when a race is turned in one direction only. The sprags are longer than the distance between the two races of the clutch assembly. Cages keep the sprags equally spaced around the diameter of the races. Springs hold the sprags at the correct angle and maintain contact of the sprags with the races. When a race turns in one direction, the sprags tilt and allow the races to move independently **(Figure 11)**. When a race is moved in the opposite direction, the sprags straighten up and lock the two races together.

All types of one-way clutches apply and release quickly in response to the rotational direction of the races. This allows for smooth gear changes. Either type can be used to hold a member of the planetary gear set by locking the inner race to the transmission housing **(Figure 12)**. When the transmission is in a low gear and is coasting, the drive wheels rotate the transmission's output shaft with more power than present on the input shaft. This allows the sprags or rollers to begin freewheeling, which puts the gear set into neutral and disallows engine compression from helping in the slowing down or braking process. Engine braking is provided in many different ways by the various transmission designs.

Some aftermarket and a few manufactured (for example, in the 4R70W) one-way clutches are ratchet-types **(Figure 13)**. These work in the same way as a ratchet wrench.

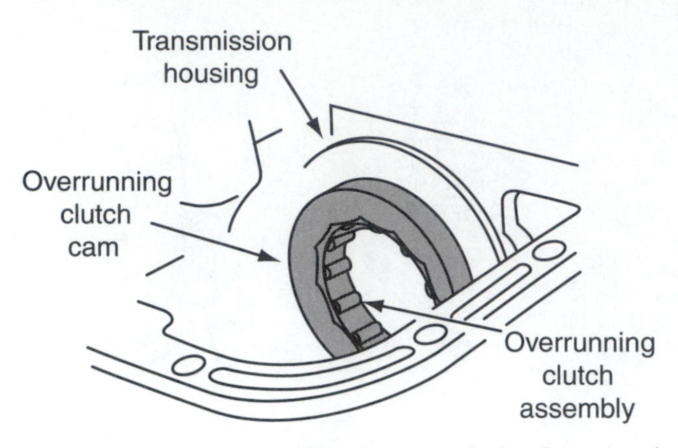


**Figure 12.** An overrunning (one-way) clutch secured in a transmission housing.

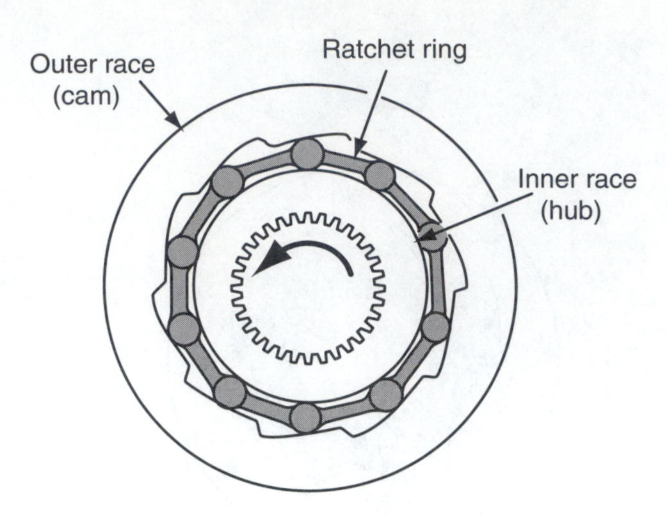


**Figure 13.** A ratchet-type one-way clutch assembly.

## MULTIPLE FRICTION DISC CLUTCH AND BRAKE ASSEMBLIES

A multiple friction disc pack has much more friction surface area than a band and therefore has greater holding capabilities. The multiple disc assembly **(Figure 14)** can be used to drive or hold a member of the gear set by connecting it to the transmission's case or to a clutch drum. A multiple disc pack consists of several plates lined with friction material and several **steel separator discs** that are placed alternately inside a clutch drum. The friction plates have rough frictional material on their faces, whereas the steel discs have smooth faces without friction material. Hydraulic pressure causes the discs to tightly compress together to apply the assembly. If this pressure is vented, return springs retract the piston, and the assembly disengages.

The friction discs are basically a steel plate with its sides lined with friction material. Metallic, semi-metallic, and paper-based materials are used as this frictional lining.

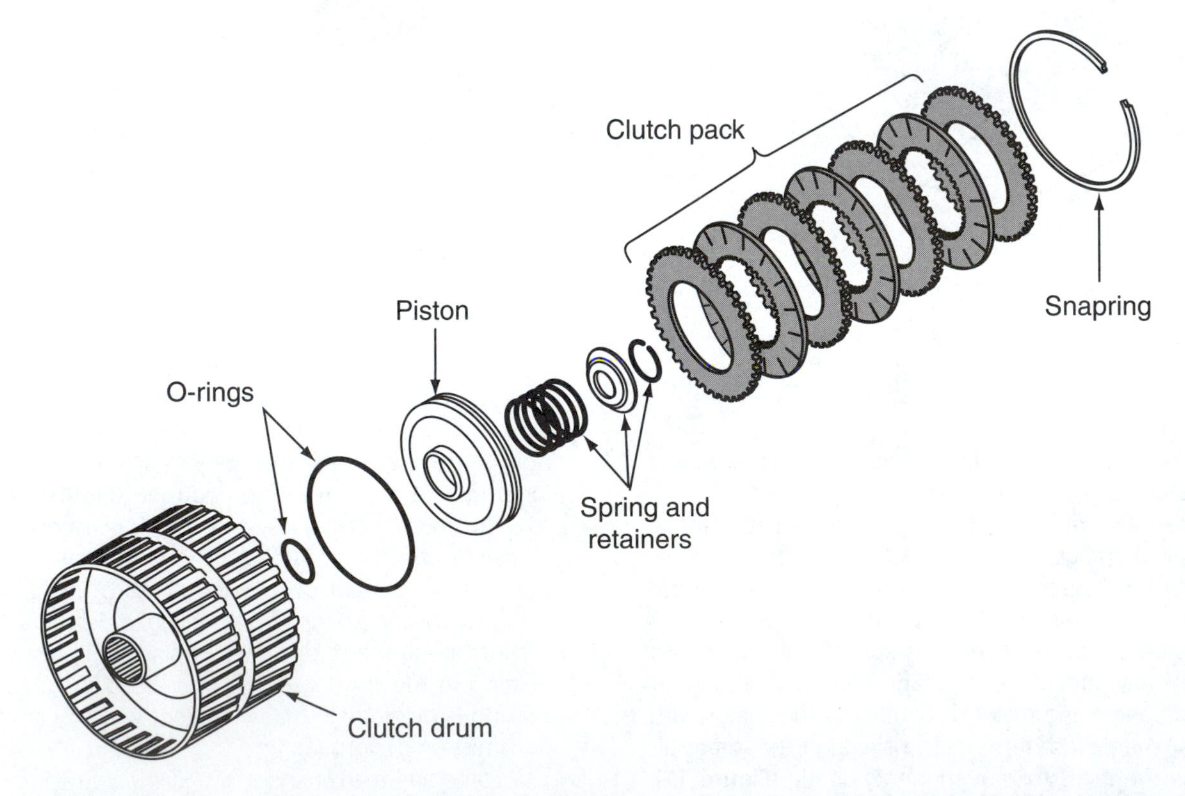


**Figure 14.** Different pairs of gears are locked or individual gear members are held by hydraulically controlled multiple-disc assemblies.

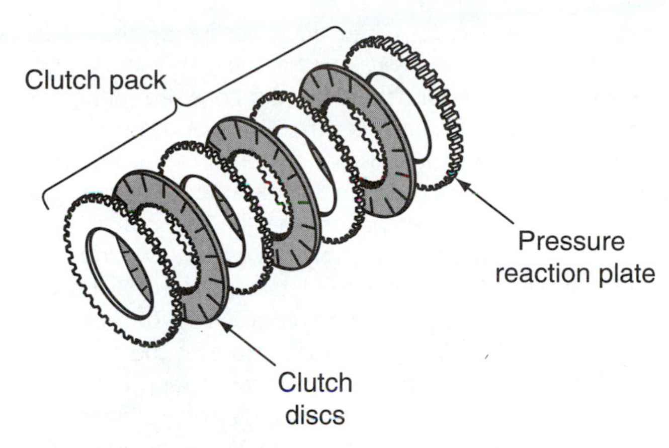


**Figure 15.** The clutch discs have internal splines, and the pressure plate and the clutch plates have external splines.

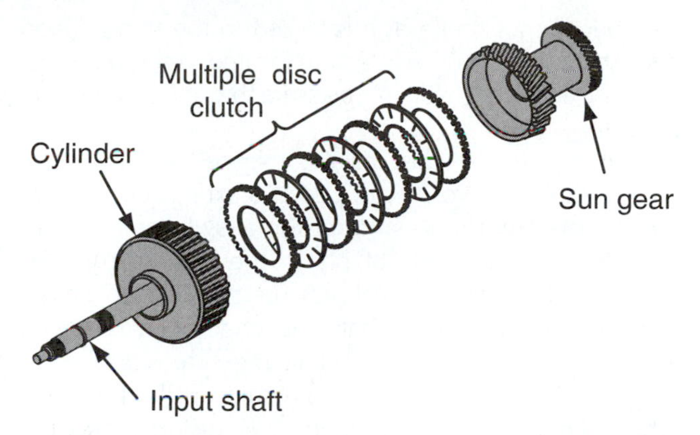


**Figure 16.** A multiple disc pack used as a driving clutch.

Paper cellulose is most commonly used because it offers good holding power without the high frictional wear of metallic materials. Friction plates often have grooves cut in them to help keep them cool, thereby increasing their effectiveness and durability.

A multiple-disc pack also contains one or more return springs, return spring retainers, seals, one or more pressure plates, and snaprings. The seals hold in the hydraulic pressure during application of the pack. A pressure plate **(Figure 15)** provides the clamping surface for the plates and is installed at one, or both ends of the pack. The snaprings are used to hold the parts in the pack together. In a typical pack, the apply piston at the rear of the drum is held in place by the return springs and a spring retainer secured by a snap ring. Hydraulic pressure moves the piston against return spring pressure and clamps the plates against the pressure plate. The friction between the plates locks them together, causing them to turn as a unit.

One set of plates has splines on its inner edges, whereas the other set is splined on its outer edges. The splines of each set fit into matching splines on either a shaft, a drum, a member of the gear set, or the transmission case. When the friction disc pack is applied, the components meshed with the splines of the plates are mechanically connected when the sets of plates are locked together. These packs can be holding or driving devices, depending on what they are splined to.

When multiple disc packs lock two members of a planetary gear set together, the pack is a driving device **(Figure 16)**. One set of discs is splined to one member of the planetary gear set and the other set is splined to another member of the gear set. The two members will rotate together at the same speed, when the pack is applied.

Driving clutches can have a set of friction plates splined to the transmission's input shaft. The set of steel plates is splined to the inside of the drum, which is connected to a member of the planetary gear set. When the pack is applied, the piston clamps the sets of the discs together. The drive discs then engage with the driven discs, which in turn rotates the drum and the attached gear set member at the same speed. Hydraulic pressure is applied to the assembly through a passage inside the input shaft. When the hydraulic pressure is exhausted, the drive discs rotate with the input shaft but are not engaged with the driven discs splined to the drum.

Driving clutches also can have a drum splined directly to the input shaft. Inside the drum and splined to it are the drive discs. The driven discs are splined to the outside of a hub splined to the output shaft or attached to member of the gear set. When the pack is applied, the plates mechanically lock the drum and hub together.

A clutch pack also may be used to hold one member of the gear set **(Figure 17)**. The friction discs are splined on their inner edges and are fit into matching splines on the outside of a drum. The steel discs are splined on their outer edges and fit into matching splines machined into the transmission case. When the pack is applied, the gear set

**Figure 17.** The splines in the housing allow a multiple-friction disc pack to lock a planetary gear-set member to the transmission case.

member cannot rotate, as it is locked to the transmission case. The apply piston is in a bore located in the transmission case and is fed hydraulic pressure through a passage in the case.

When a friction disc pack is released, the hydraulic pressure to the piston is stopped and exhausted. The piston return spring moves the piston allowing for a clearance between the sets of plates, thereby disengaging them. To relieve any residual pressure, the assembly has a vent port with a check ball. The check ball is forced against its seat when full hydraulic pressure is applied to the pack. This holds all of the pressure inside the drum. When the pressure is stopped, only residual pressure remains on the check ball. Centrifugal force pulls the ball from its seat and allows the fluid to escape from the drum through the vent port. If the residual pressure is not relieved, some pressure may remain on the piston, causing partial engagement of the pack. Not only would this cause the discs to wear prematurely, it also would adversely affect shift quality.

The check ball and vent port are located in the drum or the piston. Some transmissions are fitted with a metered orifice in the drum or the piston. A metered orifice controls the release of the hydraulic pressure and therefore serves the same purpose as a vent port and check ball.

Disc pack apply pistons are retracted by one large coil spring **(Figure 18)**, several small springs **(Figure 19)**, or a single Belleville spring. The return spring used is determined by the pressure needed to release the piston to prevent dragging. However, the amount of spring tension is limited to minimize the resistance to piston application. Pistons fitted with multiple springs have spring pockets machined into the assembly. Often there are fewer springs than there are pockets. This is an indication that the manufacturer uses the same assembly for different applications and springs are added or subtracted to meet the needs of those applications. The manufacturer also may make other changes to the basic assembly to change the load-carrying capacity of the pack. One of these changes may be to use a different number of plates in the assembly. This is done by using a different thickness of the piston or retainer plate, or by changing the location of the snap ring groove in the drum.

A Belleville spring **(Figure 20)** improves the clamping force of the assembly. As the piston moves to apply the pack, it moves the inner ends of the Belleville spring fingers into contact with the pressure plate to apply the assembly. The fingers act as levers against the pressure plate and increase

**Figure 19.** A piston return assembly with several small coil springs.

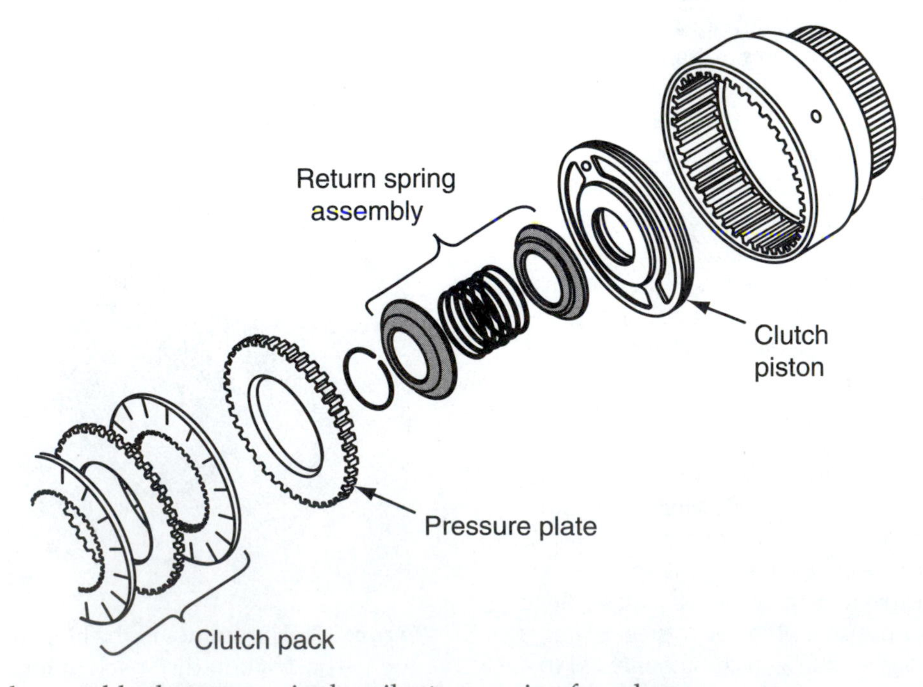


**Figure 18.** A clutch assembly that uses a single coil return spring for release.

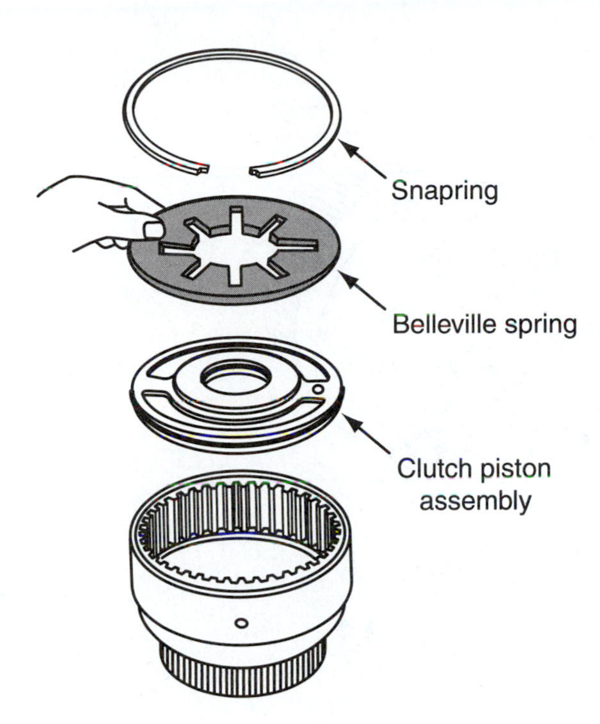


**Figure 20.** This clutch assembly uses a single Belleville-type return spring.

the application force of the pack. When hydraulic pressure to the piston stops, the spring relaxes and returns to its original shape forcing the piston back and releasing the pack.

Several factors contribute to the effectiveness of a clutch pack:

- The type of frictional material used on the friction discs.
- The composition of the steel plates and their surface finish.
- Sufficient fluid flow to cool the clutch pack.
- Condition and type of transmission fluid.
- Proper grooving of the lining on the friction plates to aid in the cooling process.
- Proper clutch plate clearances.
- The force used to apply the clutch.

## SHIFT FEEL

All transmissions are designed to change gears at the correct time. However, they are also designed to provide for positive change of gears without jarring the driver or passengers. If a brake or clutch is applied too quickly, a harsh shift will occur. "Shift feel" is controlled by the pressure at which each reaction member is applied or released, the rate at which each is pressurized or exhausted, and the relative timing of the apply and release of the members.

Shift feel also is affected by fluid type, the momentary engagement of a component in a different circuit, pulsed pressures, the clearance of the apply devices, and many more design features of the various transmission models.

To improve shift feel, a band is often released while a multiple-disc pack is applied. The timing of these two actions must be just right or both components will be released or applied at the same time. This would cause engine flareup or driveline shudder. Several other methods are used to smoothen gear changes and improve shift feel.

Shift feel can be improved by using a restricting orifice or an accumulator piston in the brake or clutch apply circuit. A restricting orifice in the passage to the apply piston restricts fluid flow and slows the pressure increase at the piston by limiting the quantity of fluid that can pass in a given time. An accumulator piston slows pressure buildup at the apply piston by diverting a portion of the pressure to a second piston in the same hydraulic circuit. This delays and smooths the application of a clutch or brake.

The use of hydraulic power to apply brakes and clutches is confined to a single device—a hydraulic piston. The pistons are housed in cylinder units known as servo and clutch assemblies. The piston converts the force in the fluid into a mechanical force capable of handling large loads. Hydraulic pressure applied to the piston, strokes the piston in the cylinder, and applies its load. During the power stroke, a mechanical spring or springs are compressed to provide a way to return the piston to its original position. The springs also determine when the apply pressure buildup will stroke the piston. This is critical to clutch/brake life and shift quality.

In rotating friction disc units, a problem can arise when the pack is not engaged. With the pack off, the drum still spins at an overspeed. The high-speed rotation can create sufficient centrifugal force in the residual or remaining fluid in the apply cylinder to partially engage the pack. This creates unwanted drag between the plates. To prevent this, a centrifugal check ball relief valve is incorporated in the drum or piston **(Figure 21)**.

The check ball operates in a cavity with a seat. A small hole or orifice is tapped from the seat for pressure relief. When the unit is applied, fluid pressure holds the ball on its seat and blocks off the orifice. In the released position, centrifugal force moves the ball off its seat, allowing the residual fluid behind the pack's piston to be released.

### Accumulators

Shift quality is dependent upon how quickly the apply device is engaged by hydraulic pressure and the pressure exerted on the piston. Some apply circuits use an accumulator to slow down application rates without decreasing the holding force of the apply device.

Most accumulators rely on the action of a piston or valve to delay the delivery of pressure to a clutch or band. An accumulator **(Figure 22)** works like a shock absorber and cushions the application of servos and disc packs. An accumulator temporarily diverts some of the apply fluid into a parallel circuit or chamber. This allows the pressure to

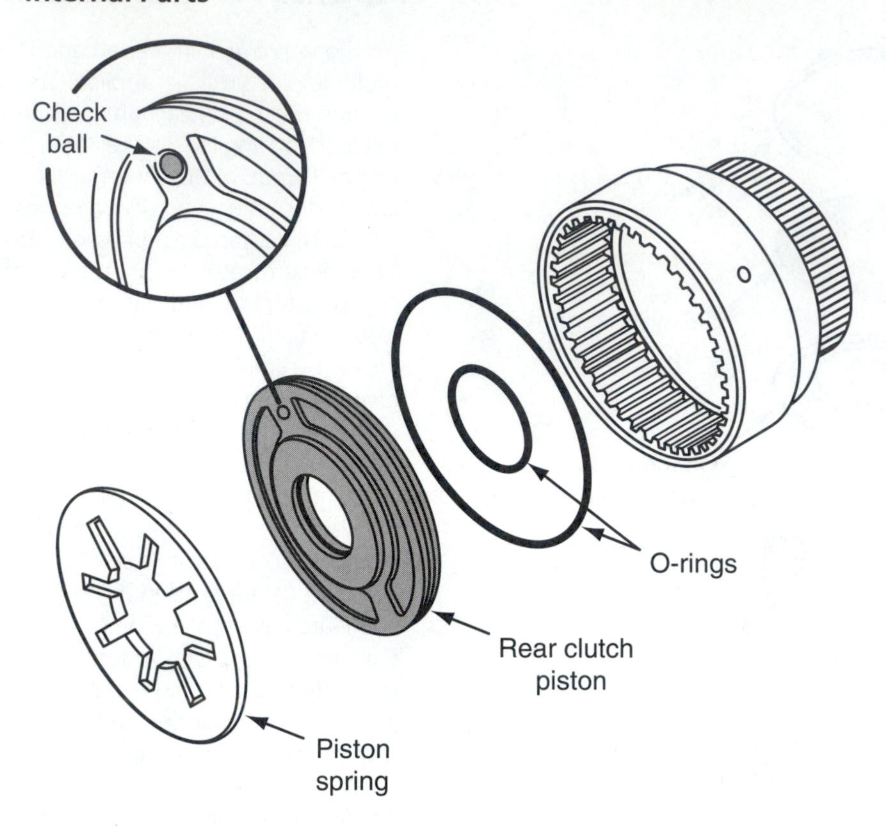


**Figure 21.** The typical location of the check ball in the piston for a clutch assembly.

increase gradually, and it provides for smooth engagement of a brake or clutch.

A piston-type accumulator is similar to a servo in that it consists of a piston and cylinder **(Figure 23)**. The pressure required to activate the piston in an accumulator is controlled by a spring or by fluid pressure. As the spring compresses, pressure at the servo or disc pack increases. This increase in pressure activates the accumulator, which delays the delivery of high pressure to the servo **(Figure 24)**. As a result, the shift takes slightly longer but is less harsh.

When a change of gears is ordered, the shift valve provides a rapid flow of fluid to the servo or disc pack. This causes

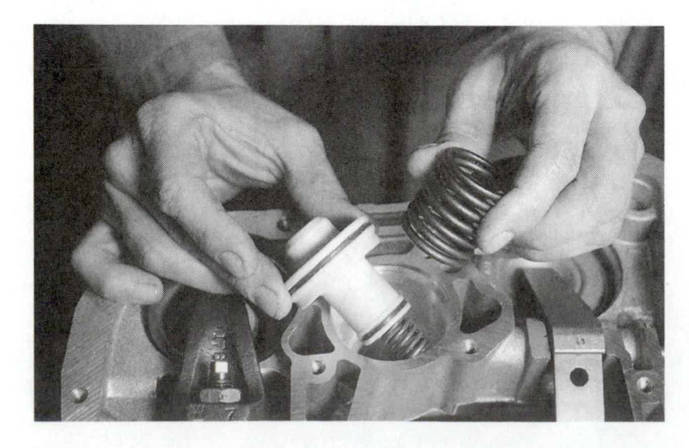

**Figure 22.** An accumulator assembly.

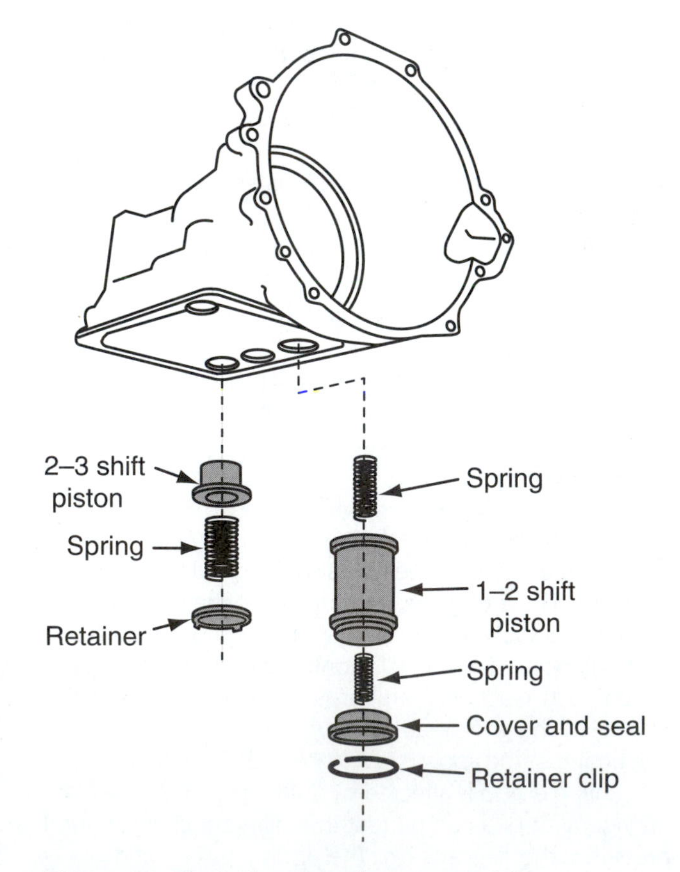


**Figure 23.** Typical accumulator assemblies.

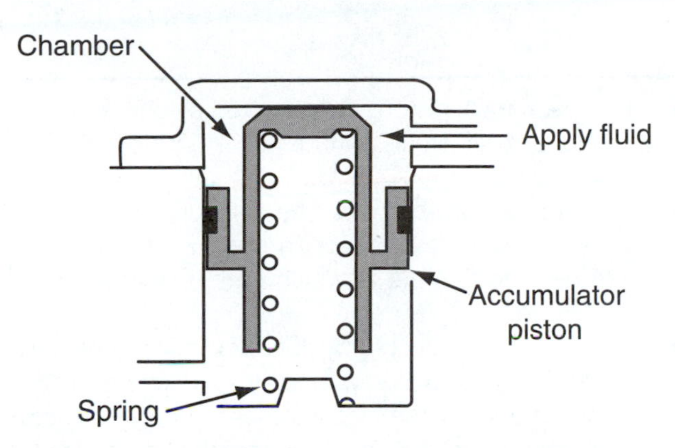


**Figure 24.** An accumulator works like a shock absorber to cushion the application of servos and clutches.

engagement of the brake or clutch. When a band is engaging, fluid pressure is used to clamp the band around its drum. After the band makes full contact with the drum, pressure in the circuit increases, because the servo has a harder time moving the band tighter. Likewise, as the multiple disc pack is engaging, the plates are squeezed together and pressure builds as the clearance between the plates is decreased.

With this rapid increase in pressure, the accumulator's piston begins to move against spring tension. The movement of the piston controls the amount of fluid sent to the servo or clutch assembly. Once pressure in the accumulator is great enough to overcome the tension of the spring, full pressure is released to the servo or clutch.

Valve-type accumulators are not as common as the piston-types. When a shift is ordered, these units allow only enough fluid to apply the brake or clutch. After initial engagement of the clutch or brake, the accumulator valve allows more pressure to the apply device and eventually allows full pressure. This delay is accomplished through an interaction of springs, valves, and an orifice. The orifice allows some pressure to reach the apply device initially. As the fluid passes through the orifice, pressure builds and begins to overcome the tension of the spring. At this time, more fluid pressure is sent to the brake or clutch. In a short time, pressure will overcome the tension on the spring and full pressure is applied to the brake or clutch.

The tension of the spring in an accumulator controls its action. Many accumulators also rely on throttle pressure and mainline pressure to control the action of the accumulator. Doing this allows for quick, firm shifts during heavy loads, and also for soft shifts during light loads.

Some designs do not rely on an accumulator for shift quality; rather, a restrictive orifice in the line to the servo or multiple disc pack's piston is used. This restriction decreases the amount of initial apply pressure but will eventually allow for full pressure to act on the piston.

Several transmissions use servo units that also work as an accumulator **(Figure 25)**. This servo/accumulator unit actually keeps the band applied during the initial engagement of the clutch pack. As the clutch becomes more engaged, the band is released. This action prevents the harsh engagement of the upshift.

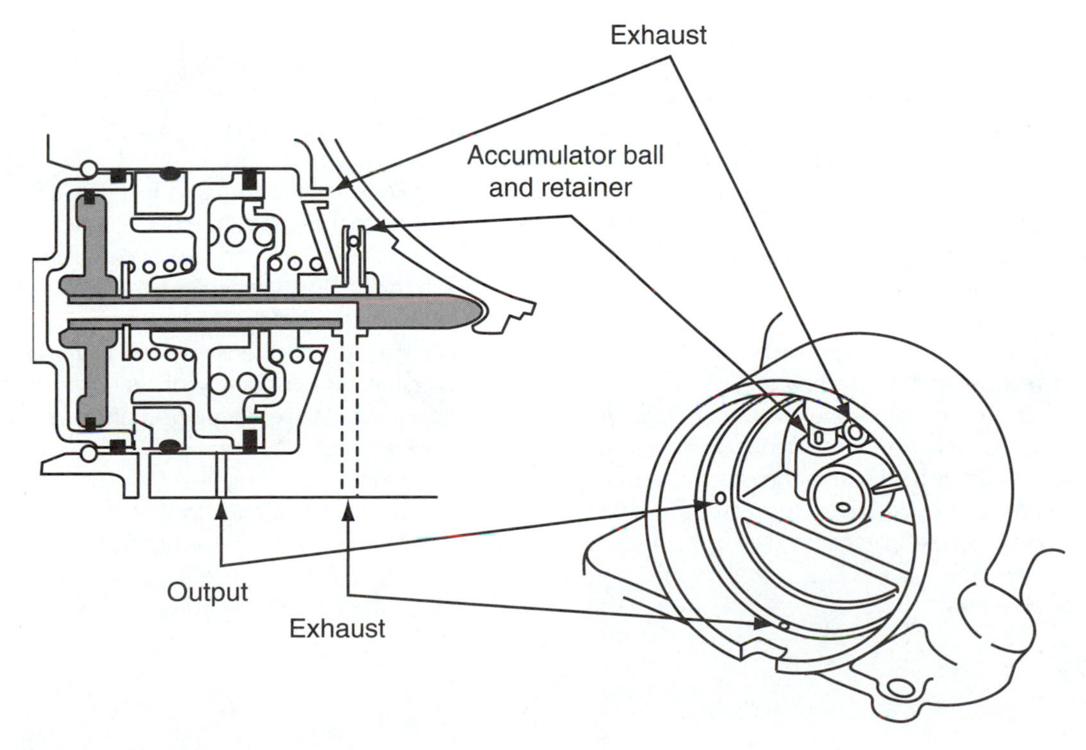


**Figure 25.** A servo unit with a built-in accumulator.

## Summary

- A band is an externally contracting brake assembly positioned around a drum. The drum is connected to a member of the planetary gear set.
- A servo is a hydraulically operated piston assembly used to apply a band.
- The operating rod of a servo tightens a band through one of three basic linkage designs: straight, lever, or cantilever.
- One-way clutches are pure mechanical devices and can be either roller or sprag types.
- A one-way clutch allows rotation in only one direction at all times, whereas a band and multiple disc pack allows or stops rotation in either direction and operates only when hydraulic pressure is applied to them.
- The multiple friction disc pack can be used to drive or hold a member of the planetary gear set by connecting it to the transmission's case or to a drum.
- When multiple-disc packs lock two members of a planetary gear set together, the pack is a driving device.
- Disc pack apply pistons are retracted by one large coil spring, several small springs, or a single Belleville spring.
- Shift feel is controlled by the pressure and relative timing at which each reaction member is applied or released and the rate at which each is pressurized or exhausted.
- Shift feel is also affected by fluid type, the momentary engagement of a component in a different circuit, pulsed pressures, the clearance of the apply devices, and many more design features of the various transmission models.
- Some apply circuits use an accumulator to slow down application rates without decreasing the holding force of the apply device.

## Review Questions

1. What is a reaction member and why is it called that?
2. Briefly describe how a multiple friction disc pack works.
3. How is a band applied?
4. In most cases, a band is used to hold the __________ or __________ gear.
5. When discussing the operation of a multiple disc assembly, Technician A says the seals of the pack hold in the hydraulic pressure during application of the pack. Technician B says that the pressure plate provides the clamping surface for the plates and is installed at one, or both, ends of the pack. Who is correct?
   A. Technician A only
   B. Technician B only
   C. Both Technician A and Technician B
   D. Neither Technician A nor Technician B
6. When discussing brake band operation, Technician A says that a servo is a hydraulically operated piston assembly used to apply the band. Technician B says that an accumulator is a hydraulic piston assembly that helps a servo to quickly apply the band. Who is correct?
   A. Technician A only
   B. Technician B only
   C. Both Technician A and Technician B
   D. Neither Technician A nor Technician B
7. Technician A says that one-way overrunning clutches can be either roller or sprag type. Technician B says that a roller clutch utilizes roller bearings held in place by sprags that separate the inner and outer race of the clutch assembly. Who is correct?
   A. Technician A only
   B. Technician B only
   C. Both Technician A and Technician B
   D. Neither Technician A nor Technician B
8. Technician A says that an accumulator relies on the action of a piston or a valve to delay the delivery of high pressure to a clutch or band. Technician B says that an accumulator cushions sudden increases in hydraulic pressure by temporarily diverting some of the apply fluid into a parallel circuit or chamber. Who is correct?
   A. Technician A only
   B. Technician B only
   C. Both Technician A and Technician B
   D. Neither Technician A nor Technician B
9. Technician A says that when a band is applied, it always wraps itself around the drum in the same direction as

drum rotation. Technician B says that the self-energizing effect of a band reduces the force that the servo must produce to hold the band. Who is correct?

A. Technician A only
B. Technician B only
C. Both Technician A and Technician B
D. Neither Technician A nor Technician B

10. Which of the following factors does NOT contribute to the effectiveness of a band?
    A. The composition of the drum and its surface finish.
    B. Condition and type of transmission fluid.
    C. The direction of rotation of the drum.
    D. The force used to apply the band.

# Bearings, Thrust Washers, and Seals

## *Introduction*

An automatic transmission has many parts that have little to do with shifting but are very important to its operation and reliability. Shafts, bearings, bushings, thrust washers, gaskets, and seals are quite vital to the efficiency of an automatic transmission.

### SHAFTS

The shafts inside a transmission have an important role. They transfer torque from one component to another. All transmissions have at least two shafts—an input and an output shaft. The input shaft **(Figure 1)** connects the output of the torque converter to the driving members inside the transmission. Each end of the input shaft is externally splined to fit into the internal splines of torque converter's turbine and the driving member in the transmission. Input can be received by any of the members of the gear set and, when in neutral, it may be delivered to two of them. In many transmissions there is a tube **(Figure 2)**, called the stator shaft, which surrounds the input shaft. The stator shaft is splined to the torque converter's stator and is a stationary shaft.

The output shaft connects the driven gear set members to the final drive gear set **(Figure 3)**. The rotational torque and speed of this shaft varies with input speed and the operating gear. The output shaft may be splined to any member of the planetary gear set, but only one. On some transaxles with a planetary gear final drive unit, the output shaft is connected to two different members of the gear set through two different apply devices.

Some transaxles have additional shafts. These shafts are actually a continuation of the input and output shafts.

**Figure 1.** An input shaft protruding from the front of a transmission housing.

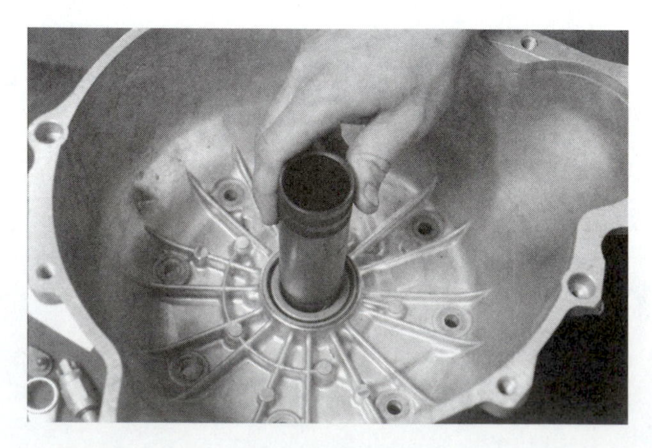

**Figure 2.** A stator shaft and support.

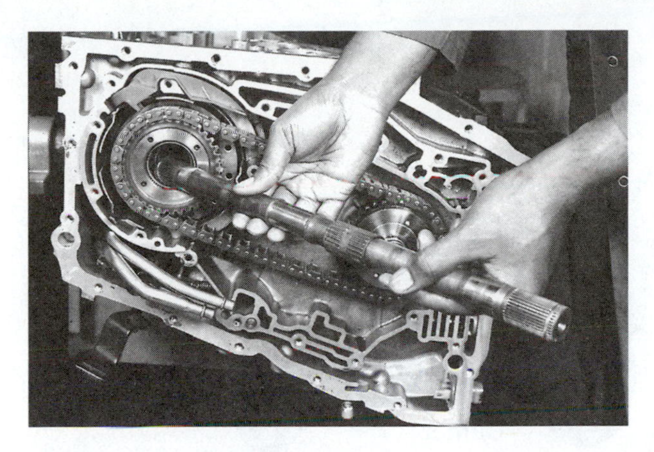

**Figure 3.** An output shaft.

**Figure 4.** A transmission has many small shafts, in addition to the input and output shafts. This is the final drive sun gear shaft from a transaxle.

They are placed in parallel where the rotating torque can be easily transferred from one shaft to another. The shafts are divided to keep the transaxle unit compact.

An additional shaft is also found in some transaxles. These shafts are typically referred to as transfer shafts and transfer the torque on the output shaft to the final drive unit.

There also are small shafts **(Figure 4)** throughout transmissions and transaxles that serve to connect the planetary units to each other or to apply devices.

## BEARINGS, BUSHINGS, AND THRUST WASHERS

A bearing is a device placed between two surfaces to reduce friction and wear. Most bearings have surfaces that either slide or roll against each other **(Figure 5)**. Sliding bearings are used when parts have low rotating speeds, very large bearing surfaces compared to the surfaces present, and/or low use. Rolling bearings are used in high-speed applications, high load with relatively small bearing surfaces, and high use.

Transmissions use sliding bearings that are composed of a relatively soft bronze alloy. Many are made from steel with the bearing surface bonded or fused to the steel. Those that take radial loads are called bushings and those that take axial loads are called thrust washers. The bearing's surface usually runs against a harder surface such as steel to produce minimum friction and heat wear characteristics.

Bushings are cylindrically shaped and are usually press fit. Because bushings are made of soft metal, they act like a bearing and support many of the transmission's rotating parts. They also are used to guide the movement of various valves in the transmission's valve body. Bushings also can be used to control fluid flow. Some restrict the flow from one part to another, whereas others are made to direct fluid flow to a particular point or part in the transmission.

Thrust washers, which often serve as both a bearing and a spacer, are made in various thicknesses **(Figure 6)**. They may have one or more tangs or slots on the inside or outside circumference that mate with the shaft bore to keep them from turning **(Figure 7)**. Some low-load thrust washers are made of nylon or Teflon. Others are fitted with rollers to reduce friction and wear.

Thrust washers normally control free axial movement or end play. Because some end play is necessary in all transmissions because of heat expansion, proper end play is often accomplished through selective thrust washers. These thrust washers are inserted between various parts of the transmission. Whenever end play is set, it must be set to manufacturer's specifications. Thrust washers work by filling the gap between two objects and become the primary wear item because they are made of softer materials than the parts they protect. Normally, thrust washers are made of copper faced soft steel, bronze, nylon, or plastic.

Torrington bearings, also called caged needle bearings **(Figure 8)**, are thrust washers fitted with roller bearings. These thrust bearings are used to limit end play while reducing the friction between two rotating parts. Most often, Torrington bearings are used in combination with flat thrust washers to control end play of a shaft or the gap between a gear and its drum.

The bearing surface is greatly reduced through the use of roller bearings. The simplest roller bearing design leaves enough clearance between the bearing surfaces of two sliding or rotating parts to accept some rollers. Each roller's two points of contact between the bearing surfaces are so small that friction is greatly reduced. The bearing surface is more like a line than an area.

If the roller length to diameter is about 5:1, or more, the roller is called a needle and such a bearing is called a needle bearing **(Figure 9)**. Sometimes the needles are loose or they can be held in place by a steel cylinder or by

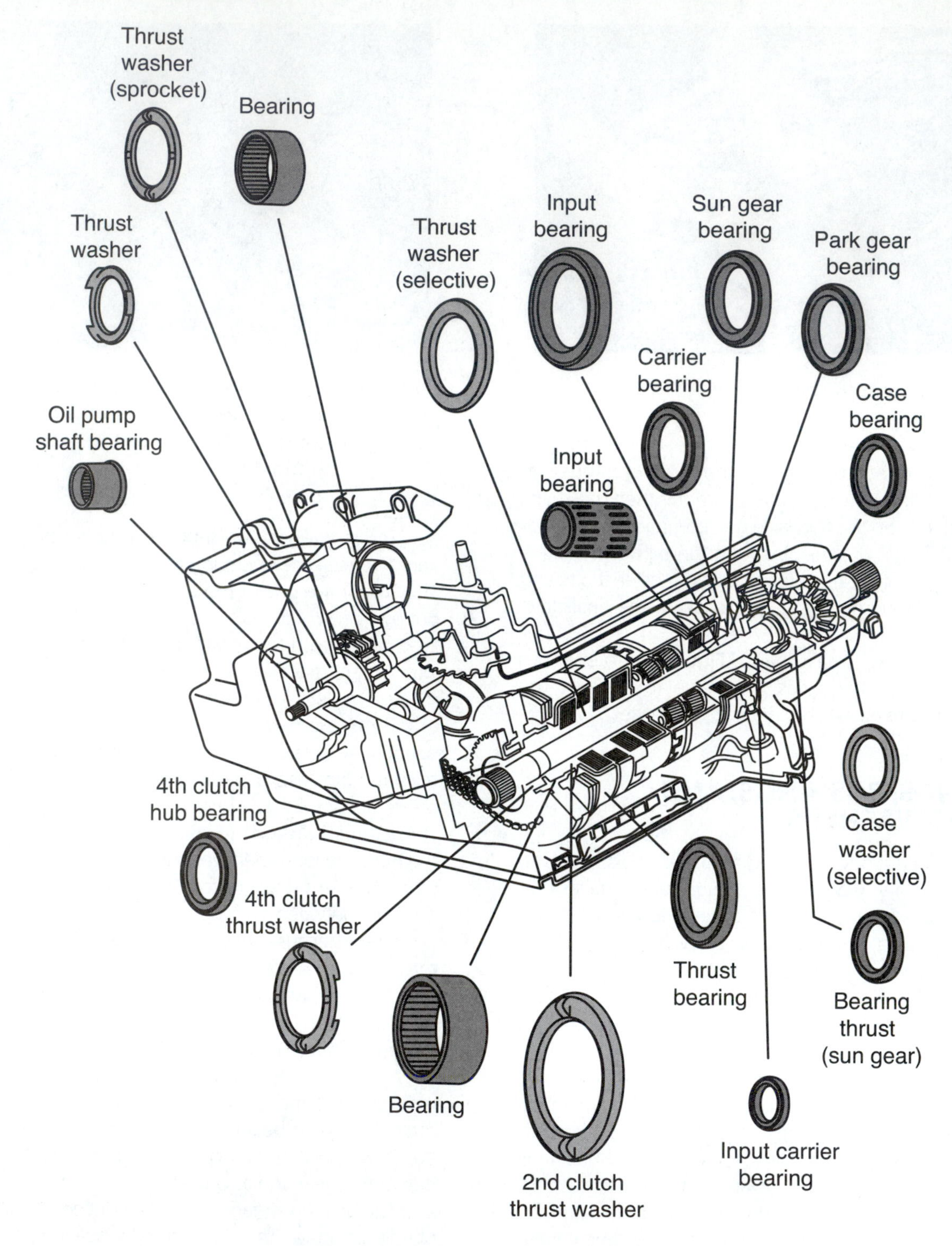


**Figure 5.** Locations of the various bushings, bearings, and thrust washers in a typical transmission.

rings at each end. Often the latter are drilled to accept pins at the ends of each needle that act as an axle. These small assemblies help save the agony of losing one or more loose needles and the delay caused by searching for them.

Many roller bearings are constructed as assemblies. The assemblies include an inner and outer race, the rollers, and a cage. Roller bearings are designed for radial loads. Tapered roller bearings are designed for both radial and

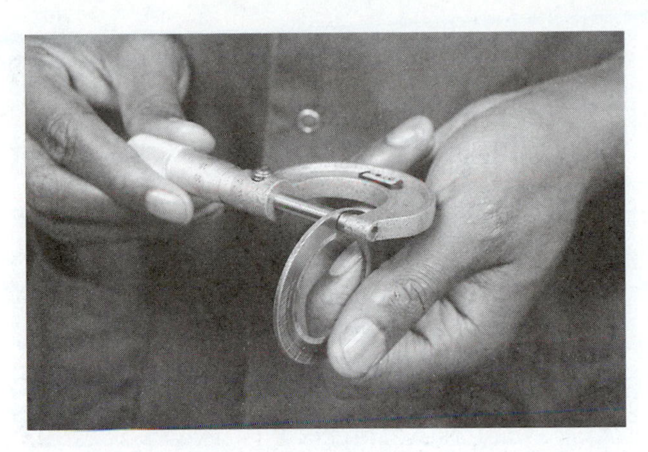

**Figure 6.** Thrust washers often are available in various thicknesses to provide for the correct endplay or clearance.

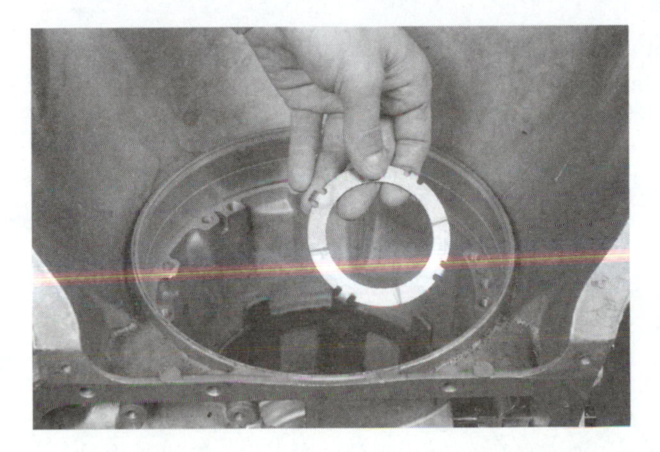

**Figure 7.** Thrust washers may have one or more tangs or slots on the inside or outside circumference to keep them from turning.

axial loads and are rarely used in automatic transmissions. Ball bearings are similar to roller bearings, except that their races are grooved to accept balls. Ball bearings can withstand heavy radial loads, as well as light axial loads.

## Snaprings

Many different sizes and types of snaprings are used in today's transmissions **(Figure 10)**. External and internal snaprings are used as retaining devices throughout the transmission. Internal snaprings are used to hold servo assemblies and clutch assemblies together. In fact, snaprings are also available in several thicknesses and may be used to adjust the clearance in multiple-disc clutches. Some snaprings for clutch packs are waved to smooth clutch application. External snaprings are used to hold gear and clutch assemblies to their shafts.

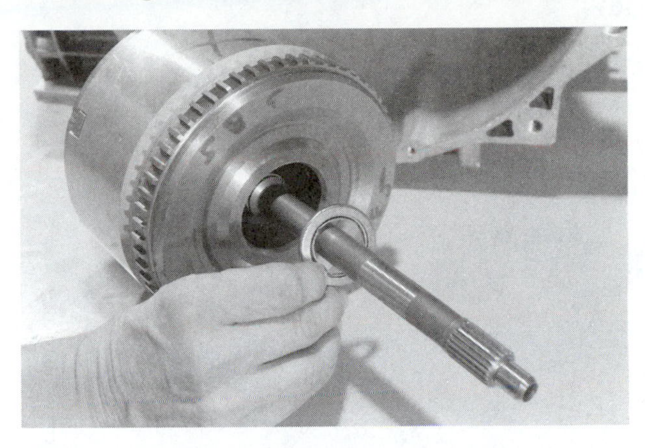

**Figure 8.** A needle bearing.

**Figure 9.** A caged needle bearing.

# GASKETS AND SEALS

The gaskets and seals of an automatic transmission help contain the fluid within the transmission and prevent it from leaking out of the various hydraulic circuits. Different types of seals are used; they can be made of rubber, metal, or Teflon **(Figure 11)**. Transmission gaskets are made of rubber, cork, paper, synthetic materials, or plastic.

## Gaskets

Gaskets are used to seal two parts together or to provide a passage for fluid flow from one part of the transmission to another. Gaskets can be divided into two separate groups, hard and soft, depending on their application. **Hard gaskets** are used whenever the surfaces to be sealed are smooth. This type of gasket is usually made of paper. A common application of a hard gasket is the gasket used to seal the valve body and oil pump against the transmission case. Hard gaskets also are often used to direct fluid flow or seal off some passages between the valve body and the separator plate **(Figure 12)**.

**Figure 10.** An example of a snapring used to hold a drum to a shaft.

**Figure 12.** A hard gasket is used to direct fluid flow or seal off some passages between the valve body and the separator plate.

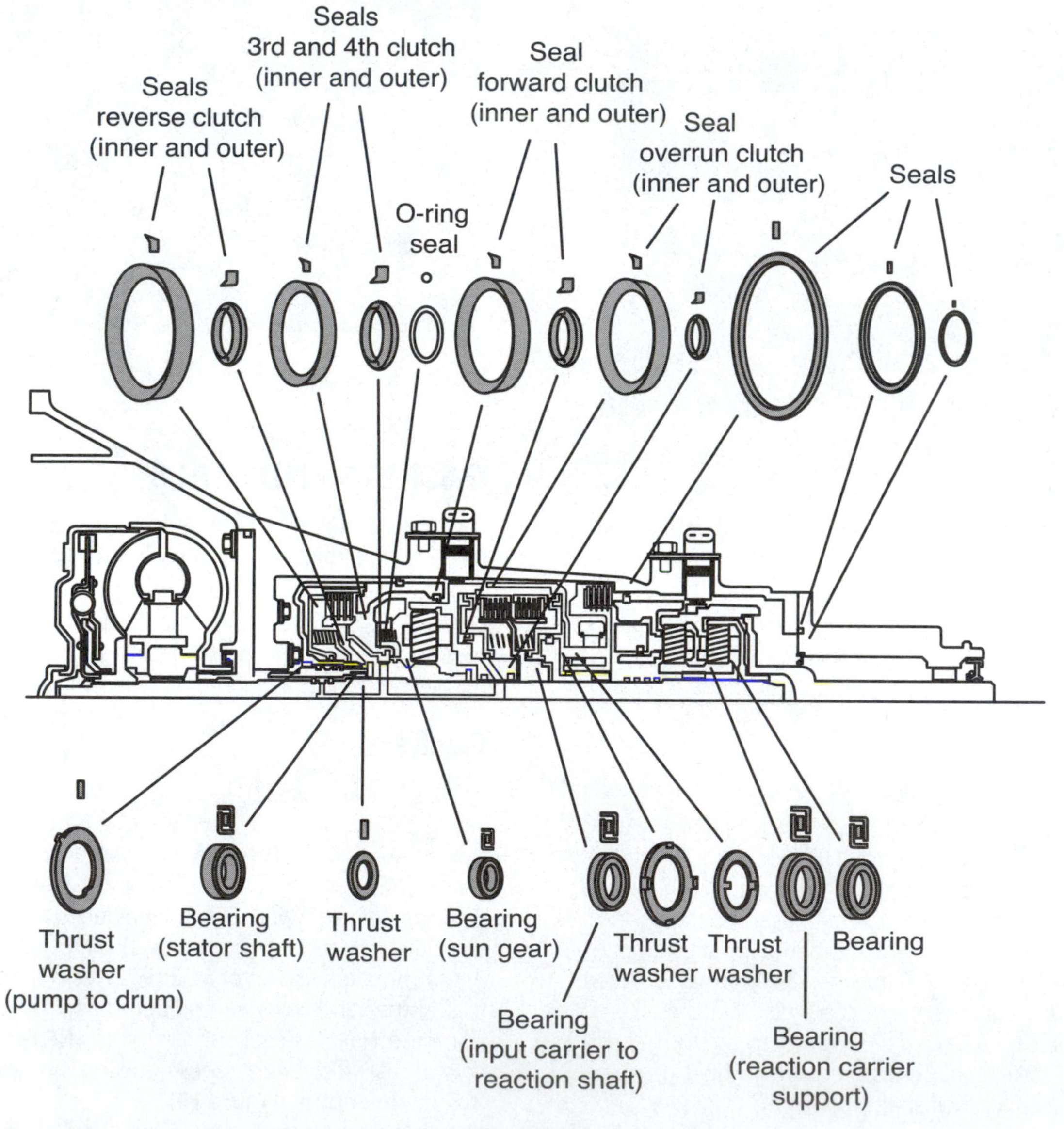


**Figure 11.** Locations of various seals and gaskets in a typical transmission.

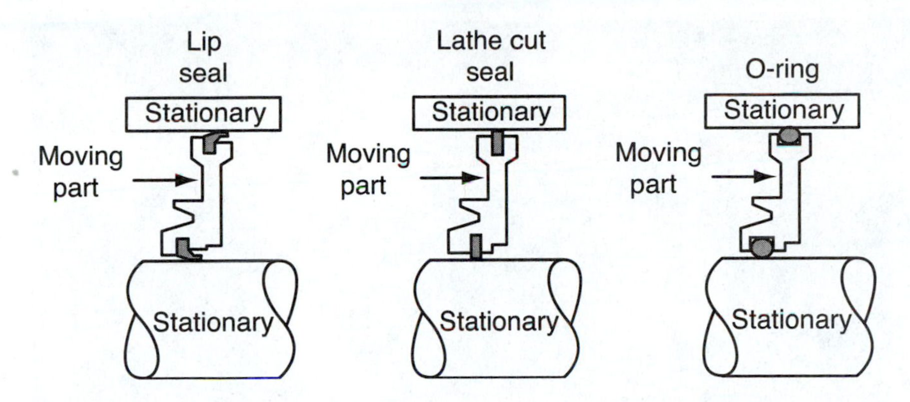


**Figure 13.** Three types of seals shown in their typical position and mountings.

Gaskets that are used when the sealing surfaces are irregular or in places where the surface may distort when the component is tightened into place are called **soft gaskets**. A typical location of a soft gasket is the oil pan gasket, which seals the oil pan to the transmission case. Oil pan gaskets are typically a **composition gasket** made with rubber and cork. However, some transmissions use an RTV sealant instead of a gasket to seal the oil pan.

## Seals

As the valves and shafts move within a transmission, it is essential that the fluid and pressure be contained within its bore. Any leakage would decrease the pressure and result in poor transmission operation. Seals are used to prevent leakage around valves, shafts, and other moving parts. Rubber, metal, or Teflon materials are used throughout a transmission to provide for **static** and **dynamic sealing**. Both static and dynamic seals can provide for positive and nonpositive sealing. A definition of each of the different basic classifications of seals follows:

*Static*—A seal used between two parts that do not move in relationship to each other.

*Dynamic*—A seal used between two parts that do move in relationship to each other. This movement is either a rotating or reciprocating (up and down) motion.

*Positive*—A seal that prevents all fluid leakage between two parts.

*Nonpositive*—A seal that allows a controlled amount of fluid leakage. This leakage is typically used to lubricate a moving part.

Three major types of rubber seals are used in automatic transmissions: the O-ring, the lip seal, and the **lathe-cut seal** or square-cut seal **(Figure 13)**. Rubber seals are made from synthetic rubber rather than natural rubber.

O-rings are round seals with a circular cross section. Normally, an O-ring is installed in a groove cut into the inside diameter of one of the parts to be sealed. When the other part is inserted into the bore and through the O-ring, the O-ring is compressed between the inner part and the groove. This pressure distorts the O-ring and forms a tight seal between the two parts **(Figure 14)**.

O-rings can be used as dynamic seals but are most commonly used as static seals. An O-ring can be used as a dynamic seal when the parts have relatively low amounts of axial movement. If there is a considerable amount of axial movement, the O-ring will quickly be damaged as it rolls within its groove. O-rings are never used to seal a shaft or part that has rotational movement.

Lip seals are used to seal parts that have axial or rotational movement. They are round to fit around a shaft and into a bore. The primary sealing part is a flexible lip **(Figure 15)**. The flexible lip is normally made of synthetic rubber and shaped so that it is flexed when it is installed to apply pressure at the sharp edge of the lip. Lip seals are used around input and output shafts to keep fluid in the housing and dirt out. Some seals are double-lipped.

When the lip is around the outside diameter of the seal, it is used as a piston seal **(Figure 16)**. Piston seals are designed to seal against high pressures and the seal is positioned so that the lip faces the source of the pressurized fluid. The lip is pressed firmly against the cylinder wall as the fluid pushes against the lip, this forms a tight seal. The lip

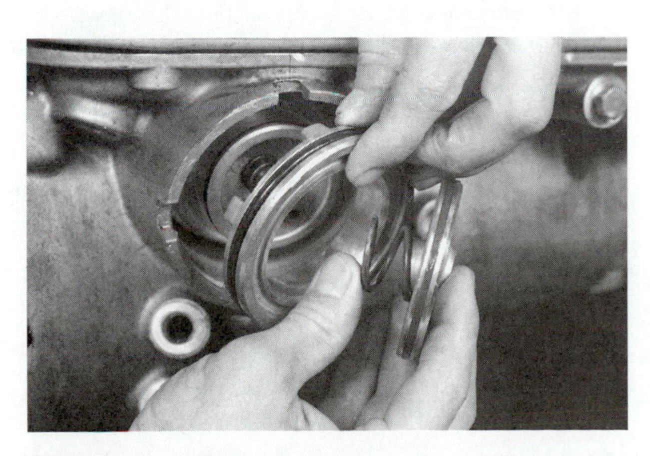

**Figure 14.** An O-ring seal on a servo cover.

**Figure 15.** A typical lip seal.

then relaxes its seal when the pressure on it is reduced or exhausted.

Lip seals also are commonly used as shaft seals. When used to seal a rotating shaft, the lip of the seal is around the inside diameter of the seal and the outer diameter is bonded to the inside of a metal housing. The outer metal housing is pressed into a bore. To help maintain good sealing, a garter spring is fitted behind the lip. This spring pushes on the lip to provide uniform contact on the shaft. Shaft seals are not designed to contain pressurized fluid; rather, they are designed to prevent fluid from leaking over the shaft and out of the housing. The tension of the spring and of the lip is designed to allow an oil film of about 0.0001 of an inch. This oil film serves as a lubricant for the lip. If the tolerances increase, fluid will be able to leak past the shaft. If the tolerances are too small, excessive shaft and seal wear will result.

A **square-cut seal** is similar to an O-ring; however, a square-cut seal can withstand more axial movement than an O-ring can. Square-cut seals are also round seals but have a rectangular or square cross section. This design prevents the seal from rolling in its groove when there is a large amount of axial movement. Added sealing comes from the distortion of the seal during axial movement. As the shaft moves, the outer edge of the seal moves more than the inner edge causing the diameter of the sealing edge to increase, which creates a tighter seal **(Figure 17)**.

## Metal Sealing Rings

There are some parts of the transmission that do not require a positive seal and where some leakage is acceptable. These components are sealed with ring seals, which fit into a groove on a shaft **(Figure 18)**. The outside diameter of the ring seals slide against the walls of the bore into which the shaft is inserted. Most ring seals are placed near pressurized fluid outlets on rotating shafts to help retain pressure. Ring seals are made of cast iron, nylon, or Teflon.

Three types of metal seals are used in automatic transmissions: butt-end seals, open-end seals, and hook-end seals. In appearance, butt-end and open-end seals are much the same; however, when an open-end seal is installed, there is a gap between the ends of the seal. When a butt-end seal is installed, the square cut ends of the seal touch or

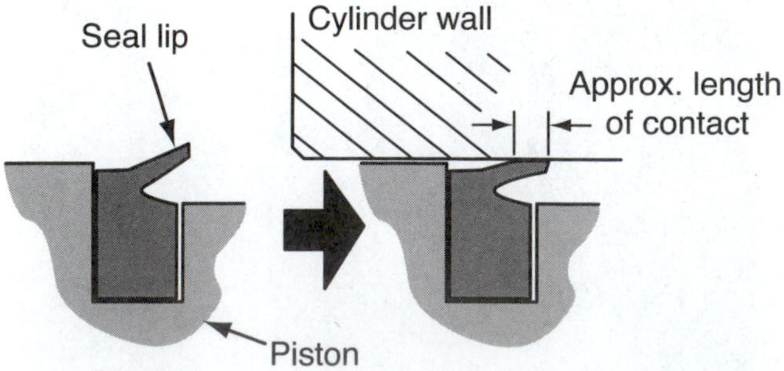

**Figure 16.** Sealing action of a lip seal.

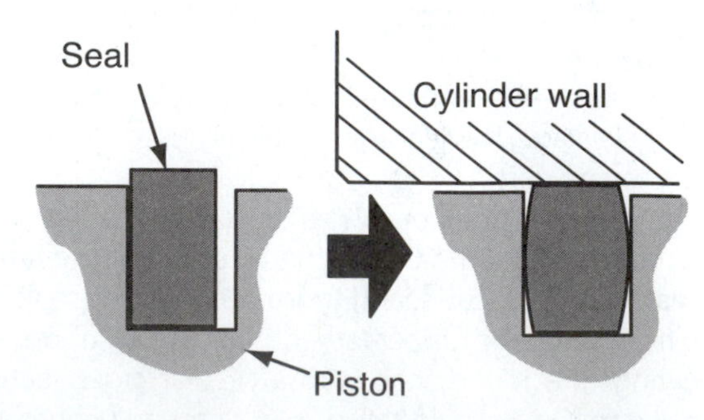


**Figure 17.** Sealing action of a square-cut seal as a piston moves in its bore.

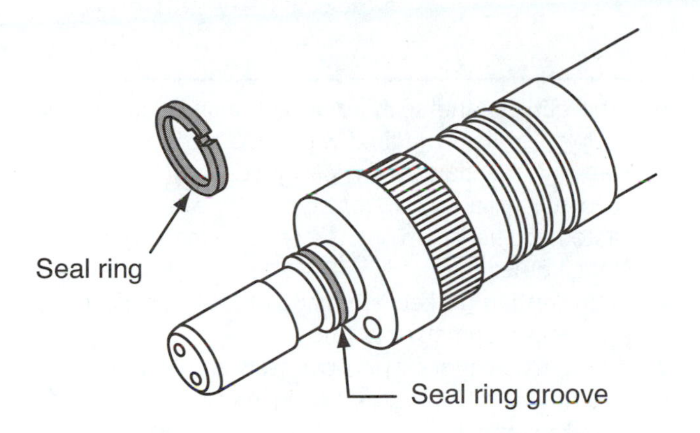


**Figure 18.** Metal sealing rings are fit into grooves on a shaft.

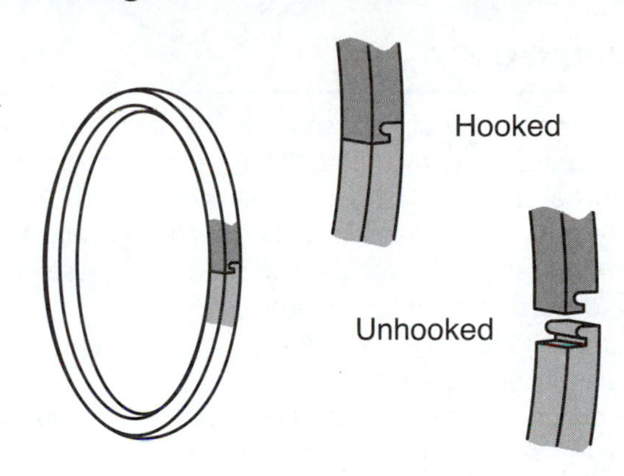


**Figure 19.** Hook-end sealing rings.

butt against each other. Hook-end seals **(Figure 19)** have small hooks at their ends, which are locked together during installation to provide better sealing than the open-end or butt-end seals.

## Teflon seals

Some transmissions use Teflon seals instead of metal seals. Teflon provides for a softer sealing surface that results in less wear on the surface that it rides on and, therefore, a longer-lasting seal. Teflon seals are similar in appearance to metal seals except for the hook-end type. The ends of locking-end Teflon seals are cut at an angle **(Figure 20)** and the locking hooks are staggered somewhat.

Many late-model transmissions are equipped with solid one-piece Teflon seals. Although the one-piece seal requires some special tools for installation, they provide for a near positive seal. These Teflon rings seal much better than other metal sealing rings.

GM uses a different type of synthetic seal on some late-model transmissions. The material used in these seals is Vespel, which is a flexible but highly durable plastic-like material. Vespel seals are found on 4T60-E and 4T80-E transaxles.

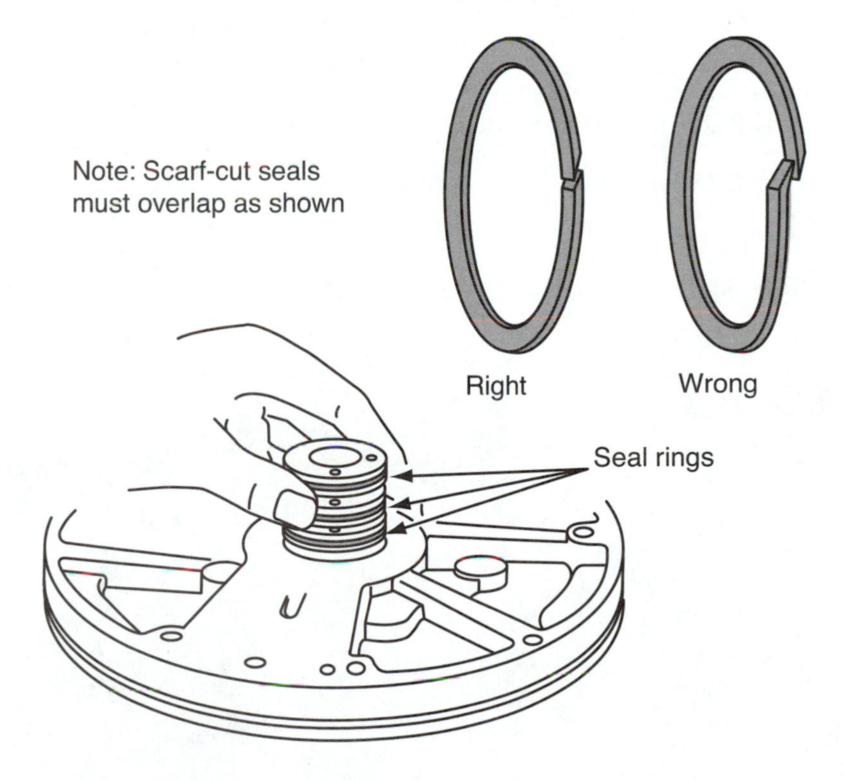


**Figure 20.** Scarf-cut seals. Notice that the ends of the seal are cut at opposing angles.

## Summary

- The input shaft of a transmission connects the output of the torque converter to the driving members inside the transmission. The output shaft connects the driven member of the gear sets to the final drive gear set.
- Internal snaprings are used to hold servo assemblies and clutch assemblies together.
- External snaprings are used to hold gear and clutch assemblies to their shafts.
- A bearing is a device placed between two bearing surfaces to reduce friction and wear.
- Bushings act like a bearing and support many of the transmission's rotating parts.
- Thrust washers normally control free axial movement or endplay.
- Torrington bearings are thrust washers fitted with roller bearings.
- The gaskets and seals of an automatic transmission help to contain the fluid within the transmission and prevent the fluid from leaking out of the various hydraulic circuits. Different types of seals are used in automatic transmissions; they can be made of rubber, metal, or Teflon.
- Transmission gaskets are made of rubber, cork, paper, synthetic materials, or plastic.
- Three major types of rubber seals are used in automatic transmissions: the O-ring, the lip seal, and the square-cut seal.
- Three types of metal seals are used in automatic transmissions: butt-end seals, open-end seals, and hook-end seals.

## Review Questions

1. What are the input and output shafts connected to in a typical transmission?
2. Most automatic transmissions use specially designed snaprings. What is so special about them?
3. Why are Torrington bearings used in automatic transmissions?
4. The three types of metal seals used in automatic transmissions are the ____________________, ____________________, and ____________________ seals.
5. Technician A says that a nonpositive static seal will allow some fluid leakage between two parts that do not move in relationship to each other. Technician B says that a positive dynamic seal allows a controlled amount of leakage between two parts that move in relationship to each other. Who is correct?
   A. Technician A only
   B. Technician B only
   C. Both Technician A and Technician B
   D. Neither Technician A nor Technician B

# Chapter 27 Transfer and Final Drives

## *Introduction*

The last set of gears in the drive train is the final drive. In most RWD cars, the final drive is located in the rear axle housing. On most FWD cars, the final drive is located within the transaxle. Some FWD cars with longitudinally mounted engines have the final drive in a separate case that bolts to the transmission. When the final drive is part of the transaxle, it is serviced with the transmission. Also, some 4WD vehicles use a modified transmission or transaxle to connect to the transfer case, which sends power to the front and rear wheels.

## FINAL DRIVES

RWD final drives normally use a hypoid final drive gear set that turns the power flow 90° from the drive shaft to the drive axles. On FWD cars with a transversely mounted engine, the power flow axis is naturally parallel to that of the drive axles; therefore, the power does not need to turn. A transaxle's final drive gears provide a way to transmit the transmission's output to the differential section of the transaxle. Helical gear **(Figure 1)**, planetary gear, and chain final drive arrangements are found with transversely mounted engines. Hypoid final drive gear assemblies are normally found in vehicles with a longitudinally placed engine. The hypoid assembly is basically the same unit as would be used on RWD vehicles and is mounted directly to the transmission.

The pinion and ring gears and differential assembly are normally located within the transaxle housing. The differential section of the transaxle has the same components as the differential gears in a RWD axle and basically operates in the same way. The drive pinion gear is connected to the transmission's output shaft and the ring gear is attached to the differential case. The pinion and ring gear set are helical gears that provide for a multiplication of torque.

Some transaxles route power from the transmission's through two helical-cut gears to a transfer shaft. A helical-cut pinion gear attached to the opposite end of the transfer shaft drives the differential ring gear and carrier **(Figure 2)**. The differential assembly then drives the axles and wheels.

Rather than use helical-cut or spur gears in the final drive assembly, some transaxles use a simple planetary gear set for its final drive **(Figure 3)**. The transmission's output is connected to the planetary gear set's sun gear. The final drive sun gear is splined to the front carrier and rear ring gear of the transmission's gear set **(Figure 4)**. Typically, a shaft is used to connect the carrier and ring gear to the final drive's sun gear. Doing this allows for input to the final drive unit when a forward and reverse gear is operating.

**Figure 1.** A helical-gear differential for a transaxle.

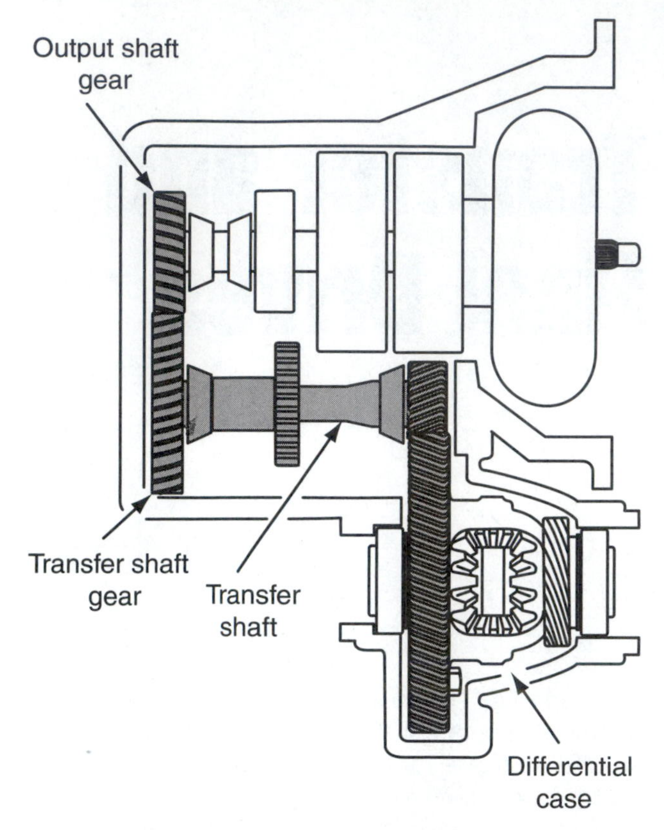


**Figure 2.** The entire final drive system for a transaxle with a transfer gear and shaft.

The final drive sun gear meshes with the final drive planetary pinion gears, which rotate on their shafts in the planetary carrier. The pinion gears mesh with the ring gear, which is splined to the transaxle case. The planetary carrier is part of the differential case, which contains typical differential gearing: two pinion gears and two side gears. The rotating planetary pinion gears drive the planetary carrier and differential case.

The ring gear of a planetary final drive assembly has lugs around its outside diameter, which fit into grooves machined inside the transaxle housing. As you recall, when the ring gear is held and input is sent to the sun gear, forward gear reduction takes place. This gear reduction is the final drive gear ratio.

Chain-drive final drive assemblies use a multiple-link chain **(Figure 5)** to connect a drive sprocket, connected to the transmission's output shaft, to a driven sprocket **(Figure 6)**, which is connected to the differential case. This design allows for remote positioning of the differential within the transaxle housing. Final drive gear ratios are determined by the size of the driven sprocket compared to the drive sprocket. The driven sprocket is attached to the differential case, which provides differential action for the drive wheels.

**Figure 3.** A planetary final drive and output shaft assembly.

## Differentials

Although the differential can be part of the rear drive axle and typically is covered in detail in a manual transmission course, the differential may be in the transaxle and be the cause of a problem in a transaxle. The basics of how a differential works are important for both automatic and manual transmission specialists. The differential **(Figure 7)** is a geared mechanism located between two driving axles. It rotates the driving axles at different speeds when the vehicle is turning and at the same speed when the vehicle is traveling in a straight line. The drive axle assembly directs driveline torque to its driving wheels, the gear ratio between the pinion and ring gear is used to increase torque and improve driveability, the gears within the differential serve to establish a state of balance between the forces or torques between the drive wheels, and allows the drive wheels to turn at different speeds as the vehicle turns a corner.

In gear-type final drives, transmission output is delivered to the differential assembly on the pinion gear. The pinion teeth engage the ring gear; therefore, as the pinion turns, so does the ring gear.

The ring gear is fastened to the differential carrier assembly. Holes machined through the center of the differential carrier support the differential pinion shaft. The pinion shaft is retained in the housing case by clips or a specially designed bolt. Two beveled pinion gears and thrust washers are mounted on the differential pinion shaft. In mesh with the differential pinion gears are two axle side gears splined internally to mesh with the external splines on the left and right axle shafts. Thrust washers are placed between the differential pinions, axle side gears, and differential case to prevent wear on the inner surfaces of the differential case.

In operation, the rotating differential carrier causes the pinion shaft and pinion gears to rotate end over end

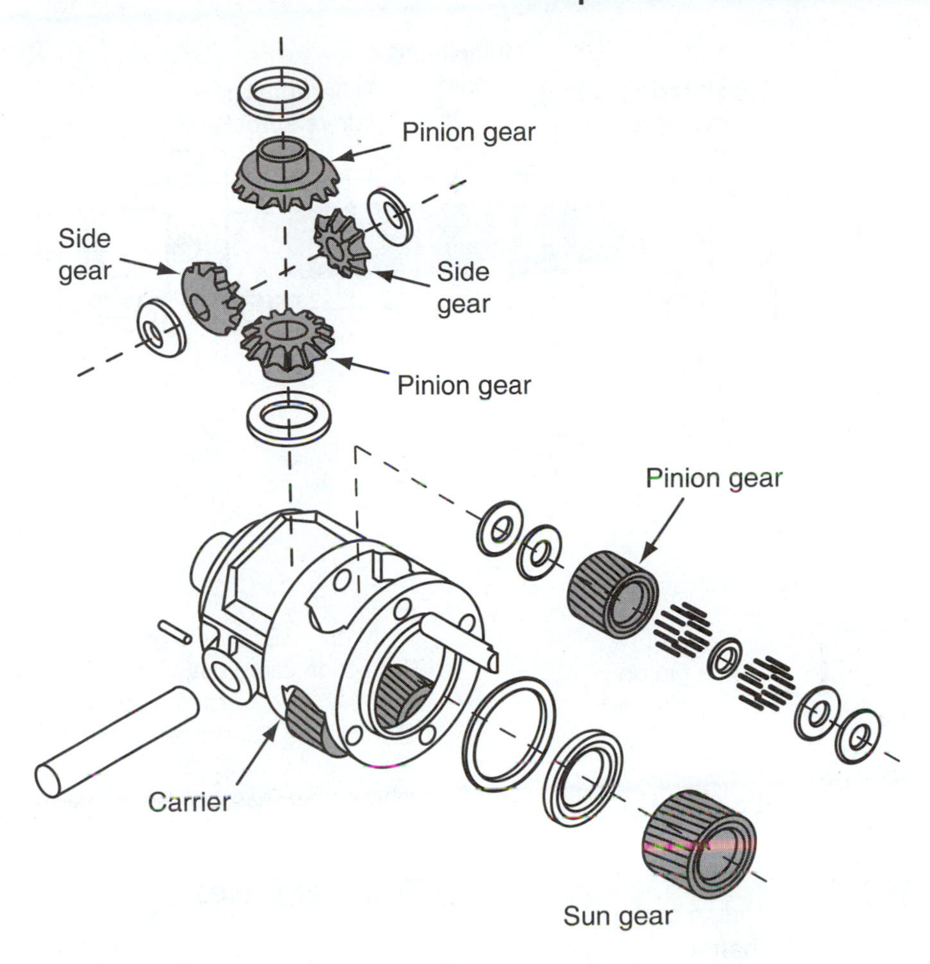


**Figure 4.** The components of a final drive unit that utilizes a planetary gearset.

with the case. Because the pinion gears are in mesh with the side gears, the side gears and axle shafts are also forced to rotate.

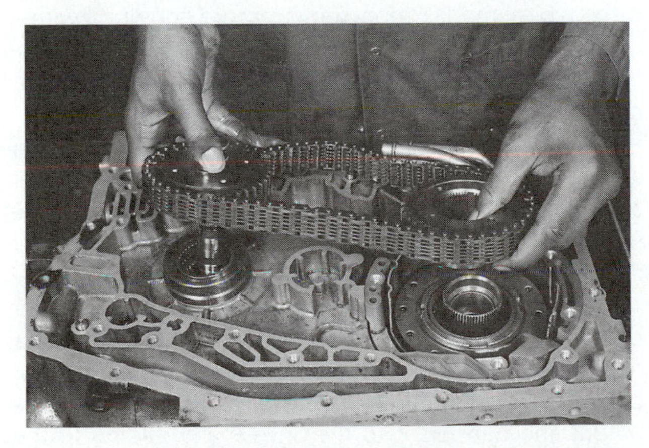

**Figure 5.** A multiple-link chain that connects to connect a drive sprocket, connected to the transmission's output shaft, to a driven sprocket, which is connected to the differential case.

When a car is moving straight ahead, both drive wheels are able to rotate at the same speed. Engine power comes in on the pinion gear and rotates the ring gear. The differential case is rotated with the ring gear. The ring gear carries around the pinion shaft and pinion gears and all of the gears rotate as a single unit. Each side gear rotates at the same speed and in the same plane as does the case and they transfer their motion to the axles. Each axle rotates at the same speed and the vehicle moves straight.

As the vehicle goes around a corner, the inside wheel travels a shorter distance than the outside wheel. The inside wheel must therefore rotate slower than the outside wheel. In this situation, the differential pinion gears will "walk" forward on the slower turning or inside side gear. As the pinion gears walk around the slower side gear, they drive the other side gear at a greater speed. An equal percentage of speed is removed from one axle and given to the other; however, an equal amount of torque is applied to each wheel.

When one of the driving wheels has little or no traction, the torque required to turn the wheel without traction is very low. The wheel with good traction is, in effect,

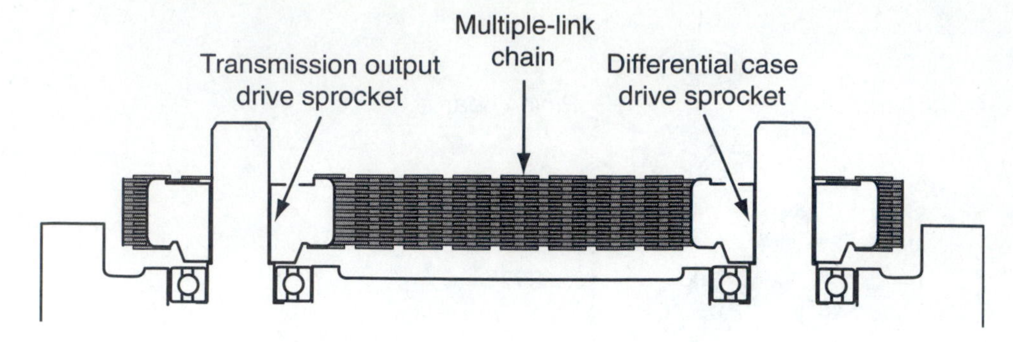


**Figure 6.** The chain setup for a final drive unit.

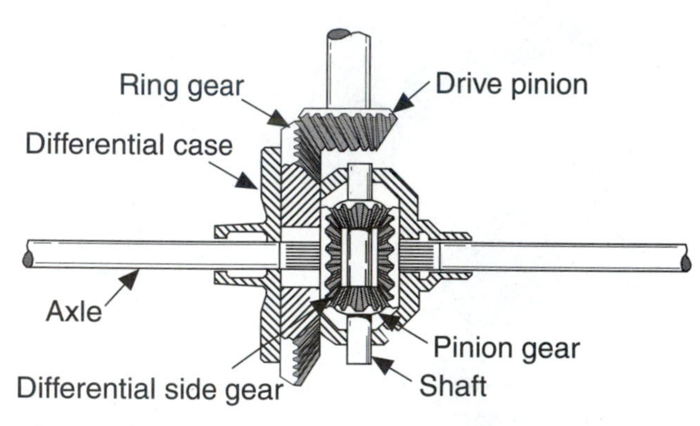


**Figure 7.** Major components in a basic differential.

holding the axle gear on that side stationary. This causes the pinions to walk around the stationary side gear and drive the other wheel at twice the normal speed but without any vehicle movement. With one wheel stationary, the other wheel turns at twice the speed shown on the speedometer.

## 4WD DESIGN VARIATIONS

4WD vehicles normally have some sort of transfer case to distribute the transmission's output to two or four drive wheels. Some transfer cases are mounted directly onto the transmission housing; others are separate and mounted to the frame between the transmission and the axles.

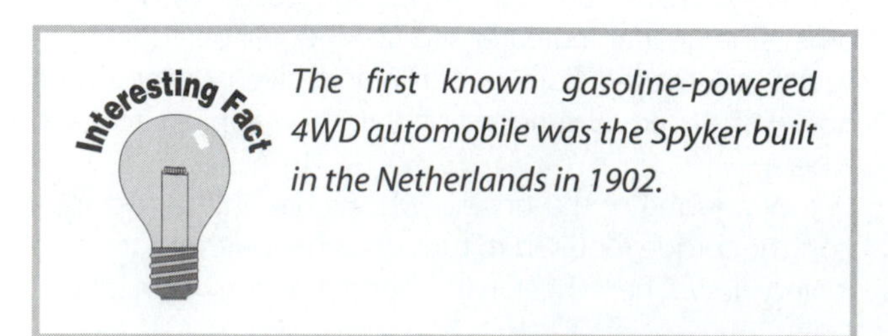


*The first known gasoline-powered 4WD automobile was the Spyker built in the Netherlands in 1902.*

With 4WD, engine power can flow to all four wheels. This action can greatly increase a vehicle's traction when traveling in adverse conditions and also can improve handling, as side forces generated by the turning of a vehicle or by wind gusts will have less of an effect on a vehicle that has power applied to the road on four wheels.

### Transfer Cases

4WD is most useful when a vehicle is traveling off the road or in deep mud or snow. However, some high-performance cars are equipped with 4WD to improve the handling characteristics of the car. Most of these cars are FWD models converted to 4WD. Normally, 4WD cars are modified by adding a transfer case **(Figure 8)**, a rear drive shaft, and a rear axle with a differential. Although this is the typical modification, some cars are equipped with a center differential or clutch assembly in place of the transfer case. These allow the rear and front wheels to turn at different speeds.

The typical 4WD system on trucks and SUVs consists of a front-mounted, longitudinally positioned engine, either an automatic or manual transmission, front and rear drive shafts, front and rear drive axle assemblies, and a transfer case.

The transfer case is usually mounted to the side or rear of the transmission. When a drive shaft is not used to connect the transmission to the transfer case, a chain or gear drive within the transfer case, receives the transmission's output and transfers it to the drive shafts leading to the front and rear drive axles **(Figure 9)**.

The drive shafts from the transfer case shafts connect to differentials at the front and rear drive axles. As on 2WD vehicles, these differentials are used to compensate for road and operating conditions by altering the speed of the wheels connected to the axles.

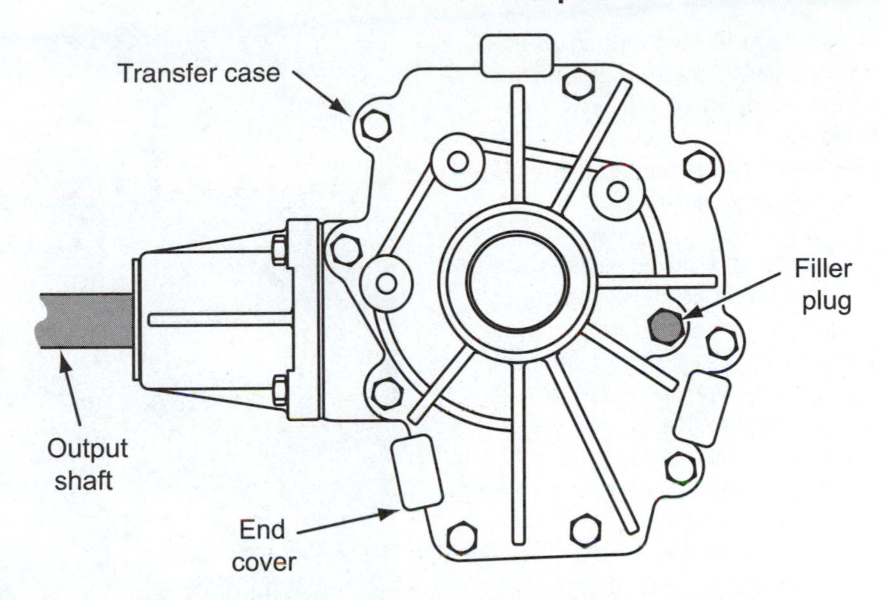


**Figure 8.** A typical transfer case for an FWD vehicle converted to 4WD or AWD.

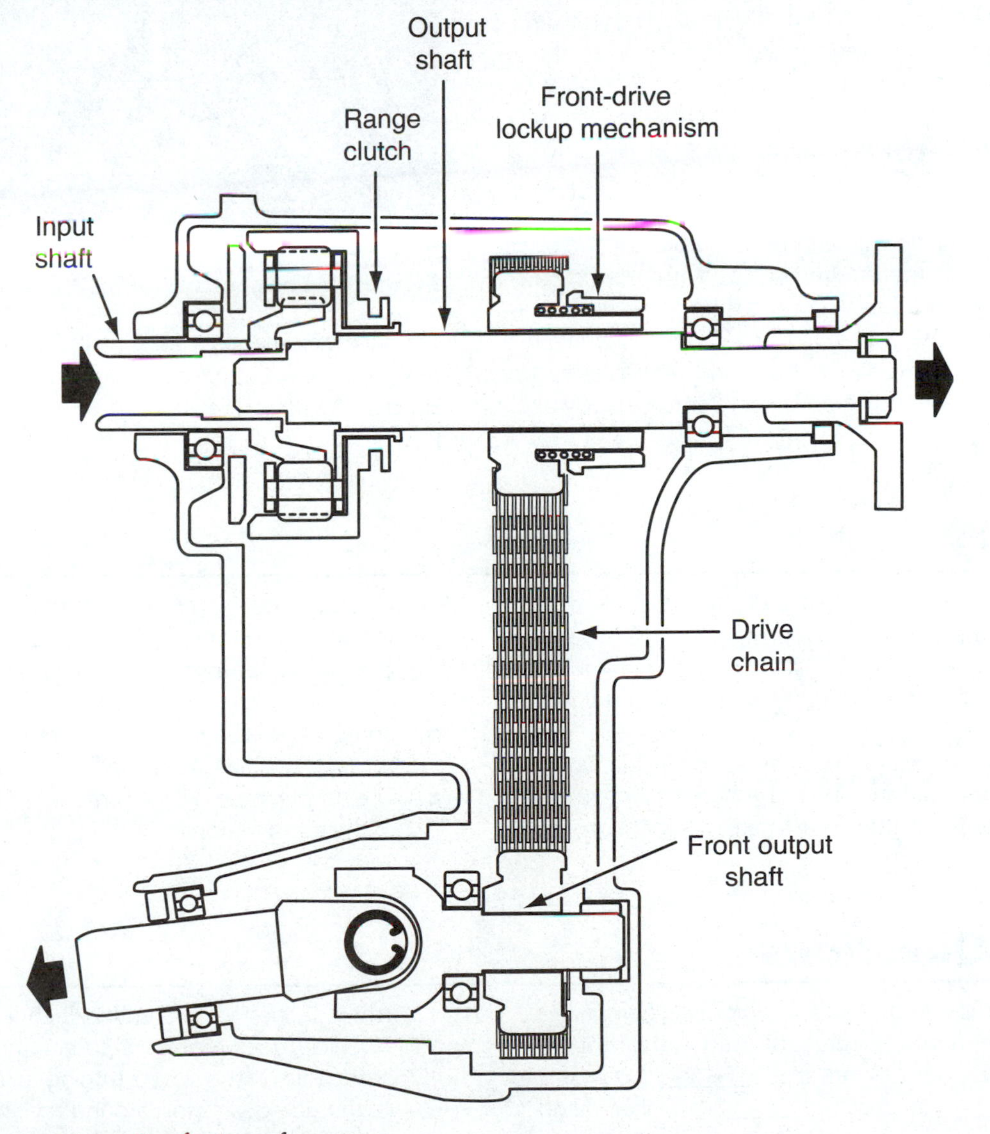


**Figure 9.** Basic components of a transfer case.

An electric switch or shift lever, located in the passenger compartment, controls the transfer case so that power is directed to the axles selected by the driver. Power can typically be directed to all four wheels, two wheels, or none of the wheels. On many vehicles, the driver can also select a low-speed range for extra torque while traveling in very adverse conditions.

Although most 4WD trucks and utility vehicles are design variations of basic RWD vehicles, most passenger cars equipped with 4WD are based on FWD designs. These modified FWD systems consist of a transaxle and differential to drive the front wheels, plus some type of mechanism for connecting the transaxle to a rear driveline. In many cases, this mechanism is a simple clutch or differential. Some vehicles are fitted with a compact transfer case bolted to the front-drive transaxle. A drive shaft assembly carries the power to the rear differential. The driver can switch from 2WD to 4WD by pressing a dashboard switch **(Figure 10)**. This switch activates a solenoid vacuum valve, which applies vacuum to a diaphragm unit in the transfer case. The linkage of the diaphragm unit linkage locks the output of the transaxle to the input shaft of the transfer case.

The driver cannot select between 2WD or 4WD in an All-Wheel-Drive (AWD) system. These systems always drive four wheels. AWD vehicles are not designed for off-road operation; rather, they are designed to increase vehicle performance in poor traction situations, such as icy or snowy roads. AWD allows for maximum control by transferring a large portion of the engine's power to the axle with the most traction. Most AWD designs use a center differential to split the power between the front and rear axles. On some designs, the center differential locks automatically or the driver can manually lock it with a switch. AWD systems may also use a viscous coupling to allow variations in axle speeds.

**Figure 10.** The controls for a 4WD system.

## Summary

- A transaxle uses helical gears, planetary gears, or a chain-type final drive unit when the engine is transversely mounted. Hypoid final drive gear assemblies are normally found in vehicles with a longitudinally placed engine.
- As the vehicle goes around a corner, the inside wheel travels a shorter distance than the outside wheel. The inside wheel must therefore rotate slower than the outside wheel. An equal percentage of speed is removed from one axle and given to the other; however, an equal amount of torque is applied to each wheel.
- Some transaxles route power from the transmission through two helical-cut gears to a transfer shaft.
- The transfer case is usually mounted to the side or rear of the transmission.

## Review Questions

1. What is the most important function of a differential?
2. The four common configurations used as the final drives on FWD vehicles are the __________ gear, __________ gear, __________ gear, and __________ __________.
3. When discussing differentials, Technician A says that as a vehicle goes around a corner, the differential side gears walk on the slower turning or inside pinion gear. As the side gears walk around the slower pinion gear, they drive the other pinion gear at a greater speed.

Technician B says that an equal percentage of speed is removed from one axle and given to the other; however, an equal amount of torque is applied to each wheel. Who is correct?

A. Technician A only
B. Technician B only
C. Both Technician A and Technician B
D. Neither Technician A nor Technician B

4. Technician A says that the output shaft connects the output of a driving member to the final drive unit. Technician B says that the input shaft connects the impeller of the torque converter to the front driven member in the transmission. Who is correct?

A. Technician A only
B. Technician B only
C. Both Technician A and Technician B
D. Neither Technician A nor Technician B

5. Why might a chain drive be used in a transaxle's final drive unit?

# Section 7

## Transmission Service



## SECTION OBJECTIVES

After you have read, studied, and practiced the contents of this section, you should be able to:

- Remove and reinstall transmission and torque converter.
- Inspect converter flexplate, attaching parts, pilot, pump drive, and seal areas.
- Measure torque converter end play and check for interference; check stator clutch.
- Inspect and repair the case.
- Inspect, measure, and replace oil pump assembly.
- Inspect, measure, clean, and replace valve body.
- Inspect, adjust, repair, and replace the governor assembly.
- Measure shaft end play or preload.
- Inspect and replace shafts.
- Inspect and measure planetary gear assembly.
- Inspect, measure, and replace bushings, thrust washers, and bearings.
- Inspect oil delivery seal rings, ring grooves, and sealing surface areas.
- Inspect and reinstall parking pawl assembly.
- Inspect, measure, repair, adjust, or replace transaxle final drive components.
- Inspect and replace bands and drums.
- Adjust bands, internally and externally.
- Inspect clutch assemblies.
- Measure clutch pack clearance.
- Air test operation of clutch and servo assemblies.
- Inspect roller and sprag clutch assemblies.
- Inspect servo and accumulator bore assemblies.
- Assemble transmission/transaxle.

*Things are always changing for automatic transmission specialists. The evolution of automatics brought more complex hydraulic systems, transaxles, torque converter clutches, and electronic controls. Automatic transmissions are units that shift automatically and do not require a clutch. A new trend in transmission design is a transmission that works without a clutch but has some sort of manual shifting. These are called sequential manual transmissions and their diagnosis and service will be added to the duties of an automatic transmission specialist.*

# Transmission/ Transaxle Removal

## Introduction

Removal of the transmission is required to do any service to the inside of the transmission. The exact procedures for transmission removal vary with the different vehicle models. However, there are general guidelines that should be followed when removing a transmission from a FWD vehicle and a RWD vehicle.

**You Should Know**

*Be sure to wear safety glasses or goggles when working under the vehicle and when handling ATF. ATF, rust, and dirt can cause serious damage to your eyes.*

Removing the transmission from a RWD car is more straightforward than removing one from a FWD model, as there is typically one cross member, one driveshaft, and easy access to cables, wiring, cooler lines, and bellhousing bolts. Transmissions in FWD cars can be more difficult to remove as you may need to disassemble or remove large assemblies such as engine cradles, suspension components, brake components, splash shields, or other pieces that would not usually affect RWD transmission removal.

### RWD VEHICLES

The following is a list of components that are typically removed or disconnected while removing an transmission from a RWD vehicle. This list is typical and is arranged in a suggested order of events. Some vehicles will require more than this, while others will require less.

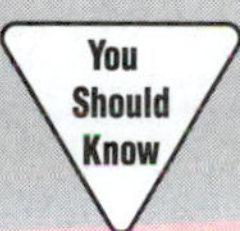


*Safe removal of transmissions and transaxles requires the purchase or fabrication of tools to help support the engine in the chassis and to lift and carry the transmission away from the vehicle.*

1. Battery ground cable
2. Transmission oil pan
3. Torque converter access plate
4. Torque converter drain plug
5. Transmission dipstick tube
6. Transmission cooler lines
7. Speedometer cable
8. Vacuum hose to modulator
9. Electrical connectors to solenoids
10. Electrical connectors to sensors
11. Gear selector linkage
12. Throttle pressure linkage
13. Kickdown linkage or electrical connector to switch
14. Neutral safety switch
15. Reverse lamp switch
16. Starter motor
17. Exhaust heat shields
18. Electrical connectors to oxygen sensors
19. Exhaust pipes and catalytic converters
20. Drive shaft
21. Torque converter to flexplate bolts or nuts
22. Transmission mounts
23. Cross member
24. Bellhousing to engine bolts

**You Should Know** *If you plug the cooler line fittings on the housing and the lines themselves, you will prevent the frustration of having ATF drip down your neck or into your eyes while removing the transmission.*

## Procedure

The exact procedure for removing a transmission varies with each year, make, and model of vehicle; always refer to the service manual for the procedure you should follow. Normally the procedure begins with placing the vehicle on a hoist.

Once the vehicle is in position, disconnect the negative battery cable and place it away from the battery. Disconnect and remove any transmission linkages connected to the engine. Also, remove the ATF dipstick.

Then raise the vehicle and disconnect the parts of the exhaust system that may interfere with transmission removal. Disconnect all electrical connections at the transmission **(Figure 1)**. Make sure you place these away from the transmission so they are not damaged during transmission removal or installation.

Remove the torque converter inspection plate or dust cover **(Figure 2)**. Place an index mark on the converter and the flexplate to ensure the two will be properly mated during installation. Using a flywheel turning tool, rotate the flywheel until some of the converter-to-flexplate bolts are exposed. Loosen and remove the bolts. Then rotate the flywheel until more bolts are accessible. Remove them and continue the process until all mounting bolts are removed. Once the bolts are removed, slide the converter back into the transmission.

Place a drain pan under the transmission **(Figure 3)** and drain the fluid from the transmission. Once the fluid is out, place the oil pan back onto the transmission and keep it in place with three or four bolts.

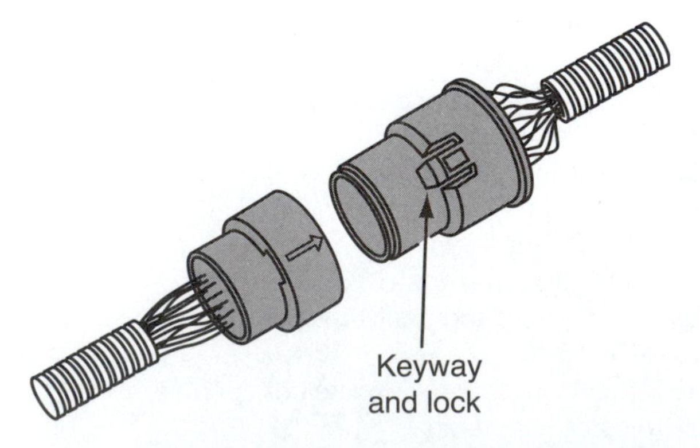


**Figure 1.** Be sure to examine the electrical connectors before attempting to separate them. Most have locks that must be unlocked.

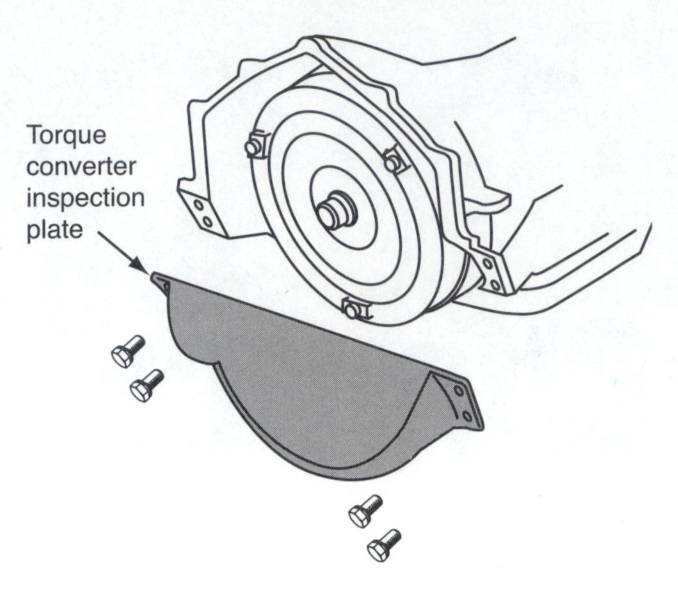


**Figure 2.** Location of a typical torque converter (and flex plate) inspection cover and bolts.

**Figure 3.** Place a drain pan under the transmission and drain the fluid from the transmission.

Disconnect the shift and other linkages from the transmission. Then remove the speedometer cable (if there is one). Disconnect the cooler lines from the transmission **(Figure 4)**. Cap the ends of the lines to prevent leakage and dirt from entering the cooling system. Disconnect the brackets for the transmission vent and dipstick tube. Then remove the vent and dipstick tubes **(Figure 5)**.

Move the drain pan under the rear of the transmission. Then, with a paint stick or chalk, make reference marks on rear universal joint and the yoke at the rear axle **(Figure 6)**. Then unbolt the joint from the yoke. Carefully pull the slip yoke of the drive shaft out of the transmission and remove the drive shaft. To prevent fluid leakage at the output shaft, install a transmission plug **(Figure 7)**.

Place a transmission jack under the transmission **(Figure 8)** and secure the transmission to it. The use of a

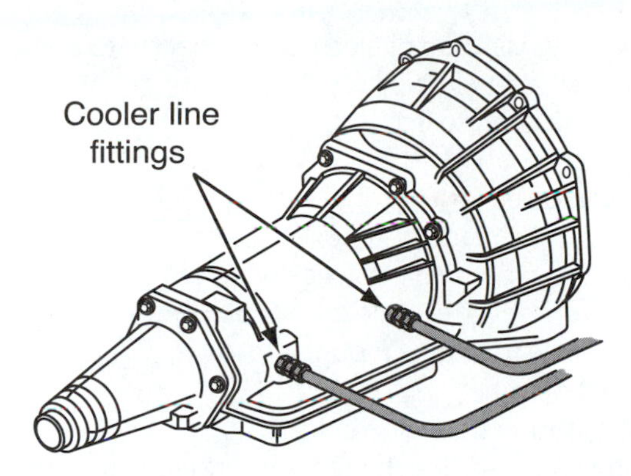


**Figure 4.** The cooler lines should be disconnected at the transmission and carefully pushed out of the way.

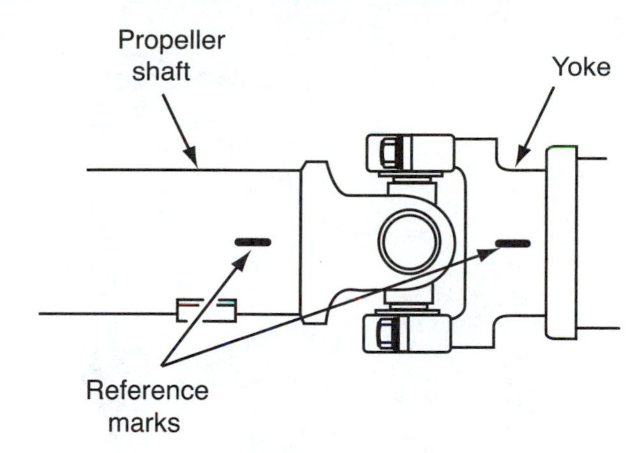


**Figure 6.** Place reference marks on the rear of the drive shaft and its mounting yoke before removing a drive shaft.

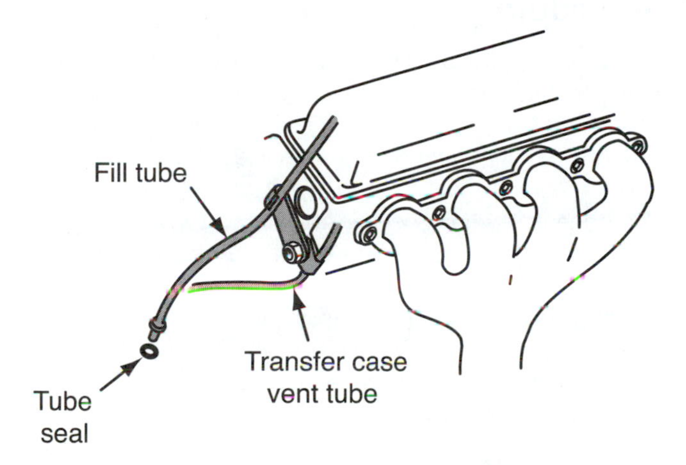


**Figure 5.** Disconnect the brackets for the transmission vent and dipstick tube and then remove the vent and dipstick tubes.

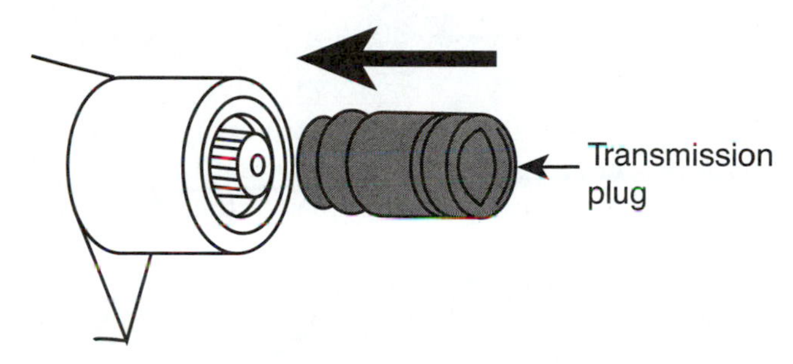


**Figure 7.** After the drive shaft has been pulled out the transmission, install a transmission plug into the rear of the transmission.

transmission jack allows for easier access to parts hidden by cross members or hidden in the space between the transmission and the vehicle's floor pan. Using a chain or other holding fixture, secure the transmission to the jack's pad. Using two chains in an "X" pattern around the transmission works well to secure the transmission to the jack. If the transmission begins to slip on the jack while you are removing it, never try to catch it. Let it fall.

Remove the transmission mounting bolts. Then, remove the cross member at the transmission. After the mounts are free from the transmission, lower the transmission slightly so you can easily access the top transmission-to-engine bolts. Loosen and remove these bolts. Then, remove the remaining transmission-to-engine bolts.

Move the transmission away from the engine until the unit is clear of the alignment dowels at the rear of the engine. Then slowly lower it **(Figure 9)**. Make sure the converter hub and any associated shafts have a clear path

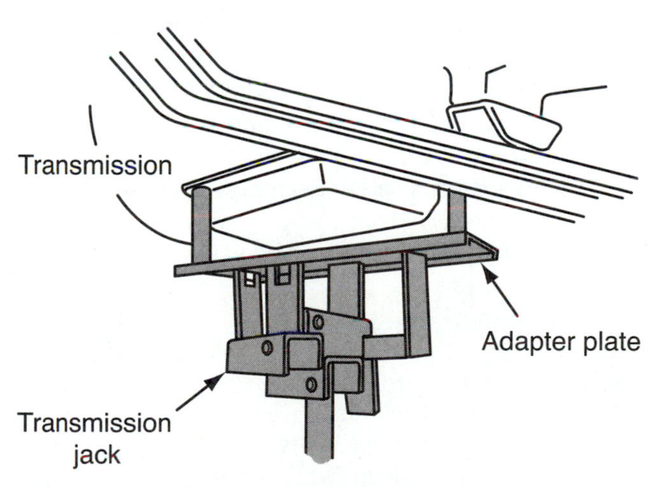


**Figure 8.** Place a transmission jack under the transmission and secure the transmission to it.

while lowering the transmission. Once the transmission is out of the vehicle, carefully move it to the work area and mount it to a stand.

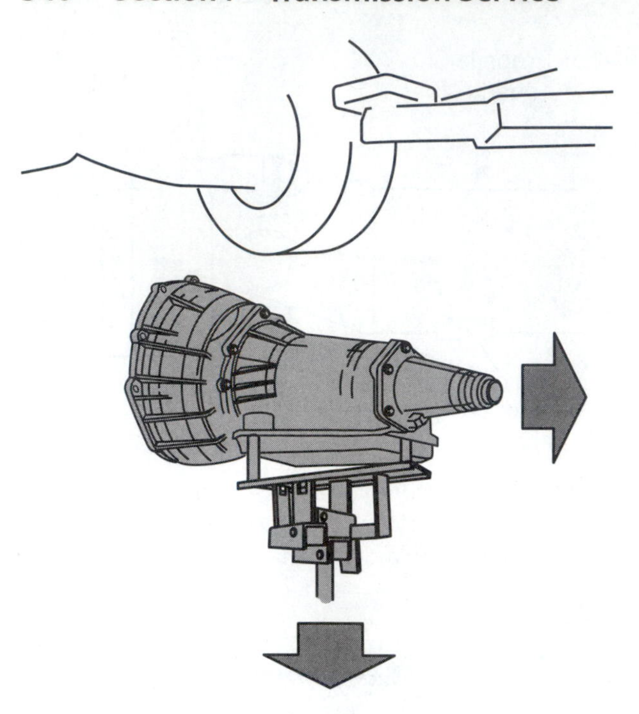

**Figure 9.** Move the transmission away from the engine until the unit is clear of the alignment dowels at the rear of the engine and the converter hub has a clear path while lowering the transmission.

## FWD VEHICLES

Obviously, the procedure for removing a transmission from a FWD vehicle is different. Because all of the components are confined to a very small area, the exact items that need to be removed or disconnected will vary from vehicle to vehicle. The following is a list of components that are typically removed or disconnected. This list is arranged in a suggested order of events.

1. Battery ground cable
2. Underhood electrical connectors to transaxle
3. Front wheels
4. Electrical connectors at the wheel brake units
5. Brake calipers
6. Steering knuckles
7. Drive axles
8. Transmission oil pan
9. Transaxle to engine brackets
10. Torque converter access plate
11. Torque converter drain plug
12. Transmission dipstick tube
13. Transmission cooler lines
14. Speedometer cable
15. Electrical connectors to solenoids
16. Electrical connectors to sensors
17. Gear selector linkage
18. Throttle pressure linkage
19. Kickdown linkage or electrical connector to switch
20. Neutral safety switch
21. Reverse lamp switch
22. Starter motor
23. Exhaust heat shields
24. Electrical connectors to oxygen sensors
25. Exhaust pipes and catalytic converters
26. Torque converter to flexplate bolts or nuts
27. Transaxle to engine mount
28. Cross member
29. Bellhousing to engine bolts

On some vehicles, the recommended procedure may include removing the engine with the transaxle. Always refer to the service manual before proceeding to remove the transaxle. You will waste much time and energy if you do not check the manual first.

## Procedure

Begin removal by placing the vehicle on a lift. However, before raising the vehicle, take a look around the engine bay to see if any interference will occur between the firewall and engine components—such as distributors, fans and fan shrouds, fuel lines, exhaust systems, or electrical components—when the transmission is removed. If any causes for interference are found, these problems should be corrected before continuing. Also, at this time, any bellhousing bolts, wiring, or TV cables accessible from above should be removed. Before raising a FWD vehicle to begin transaxle removal, a support fixture should be attached to the engine **(Figure 10)**.

**You Should Know** *Always begin the transmission removal procedure by disconnecting the battery ground cable. This is a safety related precaution to help avoid any electrical surprises when removing starters or wiring harnesses. It also is possible to send voltage spikes, which may kill the PCM if wiring is disconnected when the battery is still connected.*

Raise the vehicle to a comfortable height and remove all but the three or four corner bolts of the transmission pan, depending on the shape of the pan.

Place a large drain pan under the transaxle. Carefully remove bolts from one side of the pan. Back off the bolt or bolts on the other side just enough to allow the pan to drop slightly as you pry it loose. Be careful, as some pans will come loose without being pried and if you loosened the pan too much, you may have a large mess to clean up! When the fluid stops draining, replace the pan with a minimum of bolts taking care not to lose the remaining bolts.

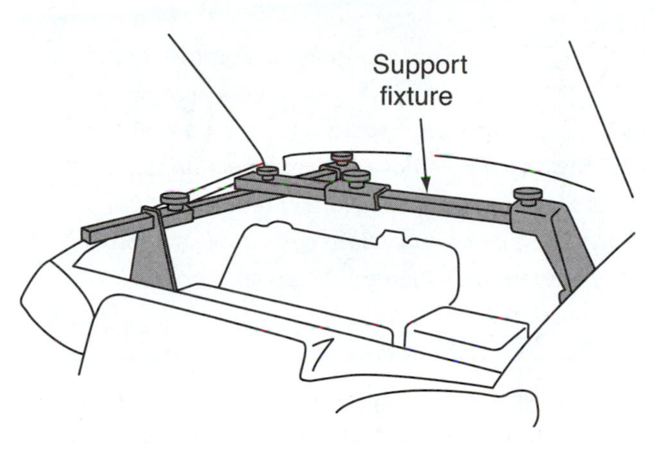


Figure 10. A typical engine support fixture for a FWD vehicle.

> **You Should Know** *To control the mess, some technicians disconnect a cooler line at the radiator and place the end of the line in a drain pan. The engine is then cranked and fluid pumped into the drain pan. This can be done to move most of the fluid out of the transmission.*

To remove FWD driveshafts, you must first loosen the large nut that retains the outer CV joint, which is splined shaft to the hub **(Figure 11)**. It is recommended that this nut be loosened with the vehicle on the floor and the brakes applied, as this reduces possible damage to the CV joints and wheel bearings.

Now raise the vehicle and remove the front wheels. Tap the splined CV joint shaft with a soft faced hammer to see if it is loose. Most will come loose with a few taps. Many cars use an interference fit spline at the hub and you will need a special puller for this type CV joint **(Figure 12)**;

Figure 11. It is recommended that the axle nut be loosened with the vehicle on the floor and the brakes applied.

Figure 12. Use the correct type of puller and other tools to remove the drive axle from the hub and bearing assembly.

the tool pushes the shaft out, and on installation pulls the shaft back into the hub.

The lower ball joint must now be separated from the steering knuckle. The ball joint **(Figure 13)** will either be bolted to the lower control arm or the ball joint will be held into the knuckle with a pinch bolt. Once the ball joint is loose, the control arm can be pulled down and the knuckle can be pushed outward to allow the splined CV joint shaft to slide out of the hub **(Figure 14)**. The inboard joint can then be pried out or it will simply slide out. Some transaxles have retaining clips that must be removed before the inner joint can be removed.

The speedometer drive gears may need to be removed before pulling out the driveshafts on some vehicles. The inner CV joint, on some cars, may have a flange type mounting. These must be unbolted for removal of the shafts. In some cases, the flange mounted driveshafts may be left attached to the wheel and hub assembly and only

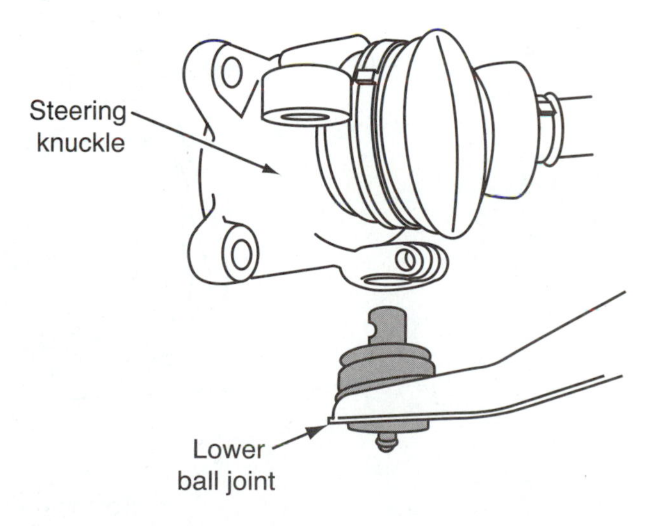


Figure 13. Typical ball joint to steering knuckle attachment.

**Figure 14.** The knuckle assembly is moved out of the way to allow the stub shaft to be pulled out of the hub.

unbolted at the transmission flange. The free end of the shafts should be supported and placed out of the way. Doing this will greatly decrease the amount of time needed to remove and install the transmission.

Now, the shift linkages, vacuum hoses, electrical connections, speedometer drives, and control cables should be disconnected **(Figure 15)**. The exhaust system also may need to be lowered or partially removed. The inspection cover between the transaxle and engine should be removed to allow access to the torque converter bolts. There will be three to six bolts or nuts securing the converter to the flexplate, depending on the application.

Mark the position of the converter to the flexplate to help maintain balance or runout. The crankshaft will need to be rotated to remove all of the converter bolts. This can be done with a long ratchet and socket on the front crankshaft bolt, or with a flywheel turning tool.

**You Should Know** *To prevent accidental damage, pay attention to the proper location of all mounting bolts and hardware when the transmission is removed. For example, on some torque converters, using longer than normal flywheel-to-converter bolts can dimple the torque converter clutch friction surface, causing the converter clutch to chatter.*

Carefully remove the cooler lines by holding the case fitting with one wrench and loosening the line nut with a line wrench. Doing this ensures that you will not twist the steel lines, which will damage them and restrict their flow.

With the transmission jack supporting the transmission, remove the transaxle mounts. If the car is equipped with an engine cradle that will separate, remove the half of the cradle that allows for transaxle removal.

Now remove the starter. The starter wiring may be left connected but you will need to hang the starter with heavy mechanics wire to avoid damage to the cables. You also

**You Should Know** *Never allow the starter to hang by the wires attached to it. The weight of the starter can damage the wires or worse break the wire and allow the motor to fall, possibly on you or someone else. Always securely support the starter and position it out of the way after you have unbolted it from the engine.*

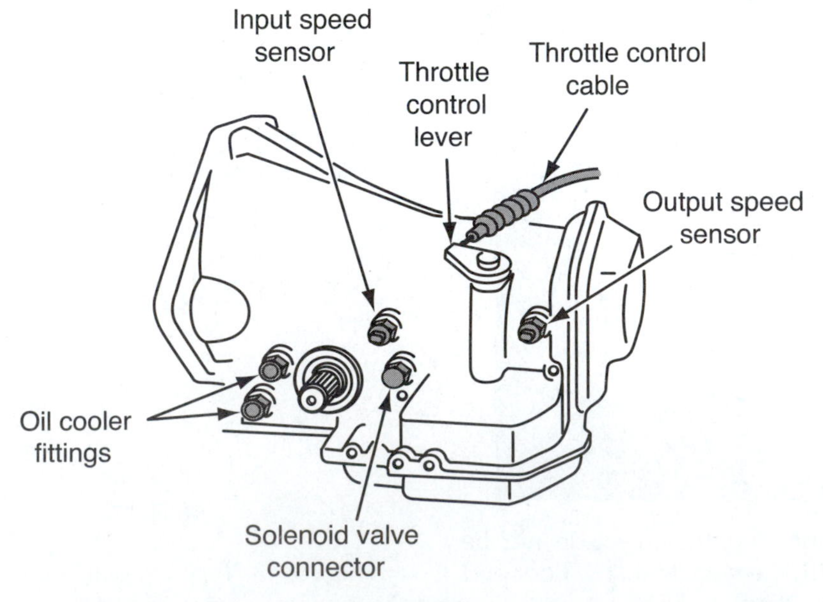


**Figure 15.** Location of the various switches, connectors, and levers on a typical transaxle.

can completely remove the starter from the vehicle to get it totally out of the way.

Now pull the transaxle away from the engine. It may be necessary to use a pry bar between the transaxle and engine block to separate the two units. Make sure the converter comes out with the transmission. This prevents bending the input shaft, damaging the oil pump, or distorting the drive hub. After separating the transaxle from the engine, retain the torque converter in the bellhousing. This can be simply done by bolting a small combination wrench to a bellhousing bolthole across the outer edge of the converter.

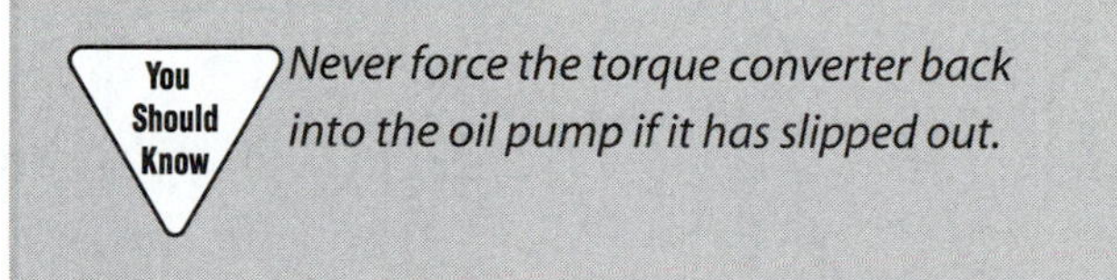


*Never force the torque converter back into the oil pump if it has slipped out.*

## Summary

- Normally, transmission removal begins with placing the vehicle on a hoist so that you can easily work under the vehicle and under the hood.
- Always disconnect the negative battery cable and place it away from the battery before beginning to remove the transmission or transaxle.
- Disconnect and remove anything that may get in the way when removing the transmission or transaxle.
- Unbolt the torque converter from the flexplate.
- Place a transmission jack under the transmission and secure the transmission to it before loosening and removing mounts and transmission-to-engine bolts.
- When lowering the transmission, make sure the converter hub and any associated shafts have a clear path.
- On some vehicles, the recommended removal procedure may include removing the engine with the transaxle.
- The drive or axle shafts must be removed before a transaxle can be removed.
- Engine support bars must be installed before attempting to remove the transaxle in most FWD vehicles.

## Review Questions

1. What is the best way to drain the fluid from a transmission?
2. What type of special tool may be needed to unbolt the torque converter from the flexplate?
3. What is the typical procedure for removing the axle shafts from a FWD vehicle?
4. *True or False*: On some vehicles, the engine must be removed before the transaxle can be removed.
5. Technician A says that a transmission jack is a good workstand for overhauling a transmission after it has been removed from the vehicle. Technician B says that the engine in most FWD vehicles needs to be supported when removing the transaxle. Who is correct?
   A. Technician A only
   B. Technician B only
   C. Both Technician A and Technician B
   D. Neither Technician A nor Technician B

# Chapter 29 Torque Converters

## *Introduction*

After the transmission has been removed, the torque converter should be inspected. The torque converter should be checked for internal interference, incorrect end play, and condition of the stator one-way clutch.

> **You Should Know** *It is common practice to replace all lockup-type converters whenever a transmission is internally serviced. The clutch discs wear and the seals harden from heat. There is no way to physically check these so to be safe, the converters are replaced.*

## INSPECTION

Before removing the torque converter from the transmission, measure the depth of the converter in the transmission case **(Figure 1)**. To do this, hold a straightedge across the bellhousing and measure in to the pilot hub or a drive pad on the converter. Record this dimension and save it for use when installing the converter.

### Flexplate

Remove the torque converter **(Figure 2)** and inspect the drive studs or lugs used to attach the converter to the flexplate. These are designed to hold the converter firmly to the flexplate and to keep the converter in line with the engine. Damaged studs or lugs will cause runout, misalignment, and vibration problems and can result in damage to the bushings of the oil pump. If the threads are damaged slightly, they can be cleaned up with a thread file, tap, or die set. However, if they are badly damaged, the converter should be replaced. The converter also should be replaced if the studs or lugs are loose or damaged. Also, check the shoulder area around the lugs and studs for cracked welds or other damage. If any damage is found, the converter should be replaced. An exception to this is when the internal threads of a drive lug are damaged. These often can be

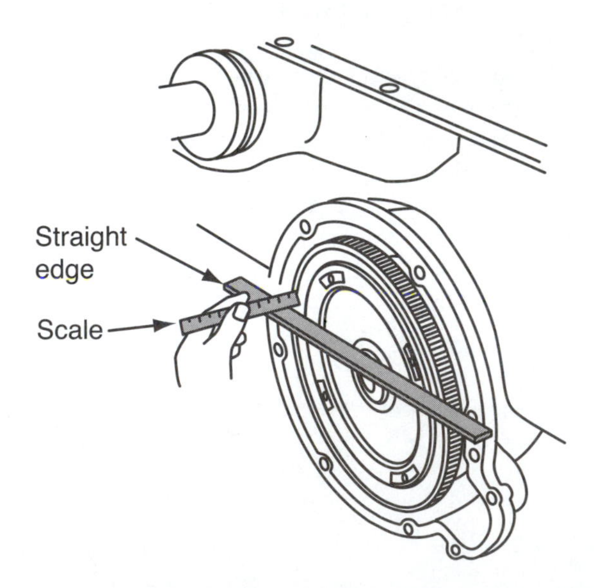


**Figure 1.** After the transmission and engine have been separated, measure and record the depth of the converter into the housing. This measurement will be used during reassembly to ensure the torque converter and oil pump are properly assembled.

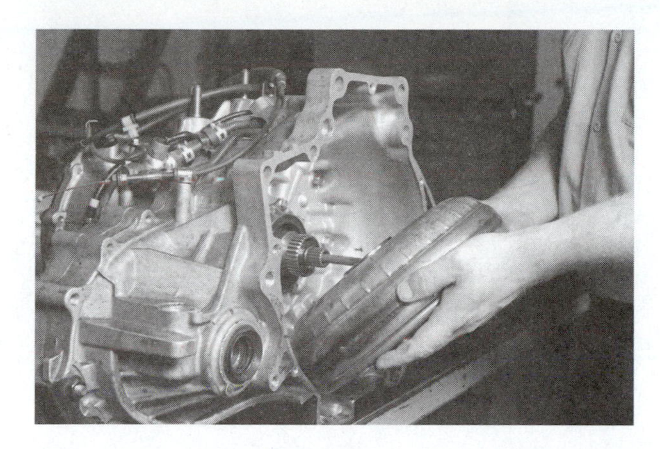

Figure 2. Remove the torque converter and inspect the drive lugs or studs.

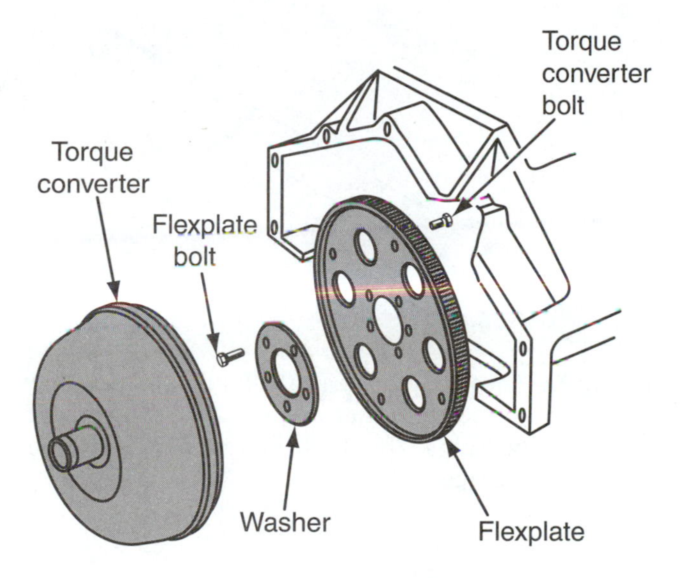


Figure 3. Carefully inspect the flexplate and torque converter mounting hardware.

repaired by tapping the threads or by installing a threaded insert. Also, inspect the converter attaching bolts or nuts and replace them if they are damaged.

Also, inspect the flexplate for warpage and cracks **(Figure 3)**. Pay attention to the condition of the teeth on the starter ring gear. Check the flexplate for excessive runout and elongated boltholes. Replace the flexplate if there is evidence of damage. Also, inspect the converter attaching bolts. Replace any worn bolts with suitable equivalents. When inspecting the flexplate, check the starter ring gear for looseness and damage. Pay attention to the welds that secure the ring gear to the plate. It is common for these and/or the areas around them to crack.

Misalignment can cause many problems. It is important to check for excessive runout of the flexplate and the converter's hub. Check flexplate runout with a dial indicator. Mount the indicator on the engine block and set its plunger on the flexplate **(Figure 4)**. Rotate the flexplate and observe the readings. Typically, the maximum allowable runout is 0.03 inch.

To check the runout of the converter's hub, mount the converter onto the flexplate. Tighten the retaining bolts to the specified torque. Position a dial indicator on the engine block. Set the plunger on the hub **(Figure 5)**. Rotate the torque converter and observe the readings. Again, the typical maximum allowable runout is about 0.03 inch. If the runout is excessive, remove the torque converter and set it at a new position on the flexplate. Check the runout again. If the runout is now within specifications, mark its position on the flexplate. If repositioning the converter on the flexplate does not correct hub runout, the converter should be replaced.

Because of the weight of torque converters, the above check is not always reliable. To get an accurate check of torque converter runout, mount the converter on two vee-blocks. Set the dial indicator so it can read lateral runout of the converter housing. To do this, place the indicator's

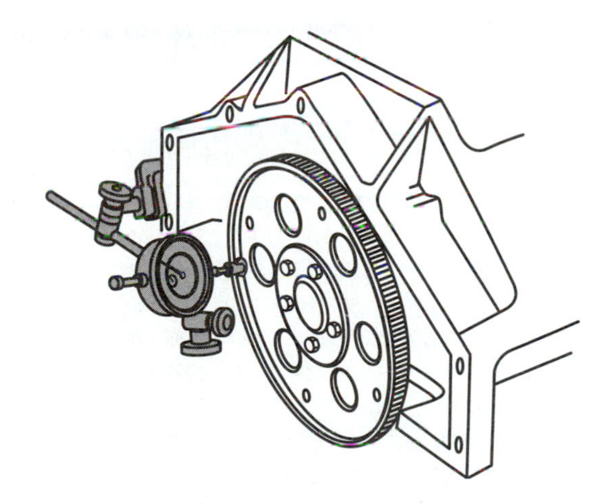

Figure 4. Setup for checking flexplate runout.

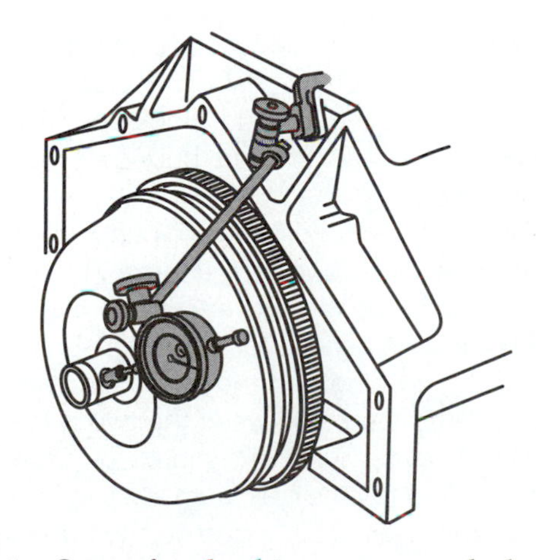

Figure 5. Setup for checking converter hub runout.

plunger against the top of the converter housing's flywheel mounting surface. Rotate the housing and observe the runout. To check radial runout, place the dial indicator on the outside end of the pump hub. Rotate the housing and observe the runout. Then move the indicator on the hub and toward the housing. Again, rotate and measure the runout.

Check the flexplate for cracks at the crankshaft mounting boltholes. The best way to do this is to remove the plate and hold it up toward a bright light. If there are cracks, the light will shine through the cracks. If there is any flexplate damage, it should be replaced. Before reinstalling the flexplate, check the service manual to make sure the attaching bolts are reusable or if they should be replaced. Also make sure the flexplate is installed in the correct direction and that all spacers are in place.

## Torque Converter

A complete inspection of the torque converter should be done any time the torque converter is removed, especially if the oil pump is damaged or if the customer's complaint is related to the torque converter. Check the pilot of the converter for wear and other damage. Also, check the area around the pilot for cracks. If the area around the pilot has dimples and looks like it has contacted the flywheel bolts, the converter has ballooned. If the converter has ballooned or has other damage, it should be replaced and the cause corrected.

Check the drive hub of the torque converter. It should be smooth and not show any signs of wear. If the hub is worn, carefully inspect the oil pump drive and replace the torque converter. Light scratches, burrs, nicks, or scoring marks on the hub surface can be polished with fine crocus cloth. Be careful not to allow dirt to enter into the converter when polishing the hub. Use a rag to cover the opening in the hub. Dirt and dust that enter the converter can cause the converter to wear rapidly. After polishing, clean the hub with solvent and a clean, lint-free rag. If the hub has deep scratches or other major imperfections, the converter should be replaced.

In general, a torque converter should be replaced if there is fluid leakage from its seams or welds, or if there are loose drive studs, worn drive stud shoulders, stripped drive stud threads, a heavily grooved hub, or excessive hub runout **(Figure 6)**.

Transaxles that do not drive the oil pump directly by the torque converter use a drive shaft **(Figure 7)** that fits into a support bushing inside the converter's hub **(Figure 8)**. This bushing should be checked for wear. Measure the inside diameter of the bushing and the outside diameter of the drive shaft. The difference between the two is the amount of clearance. This measurement should be compared to specifications. Normally, the maximum allowable clearance is 0.004 inch. If the clearance is excessive, the bushing should be replaced.

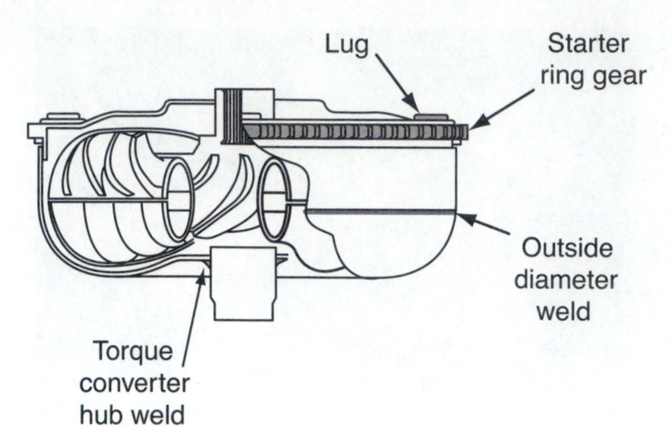


**Figure 6.** Some of the areas of a torque converter that need to be carefully checked.

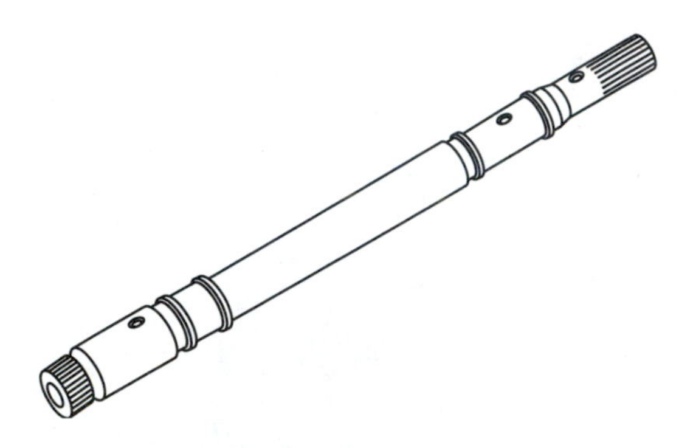

**Figure 7.** The pump drive shaft should be inspected in the areas that ride on the bushings and bearings. The shaft seals should always be replaced.

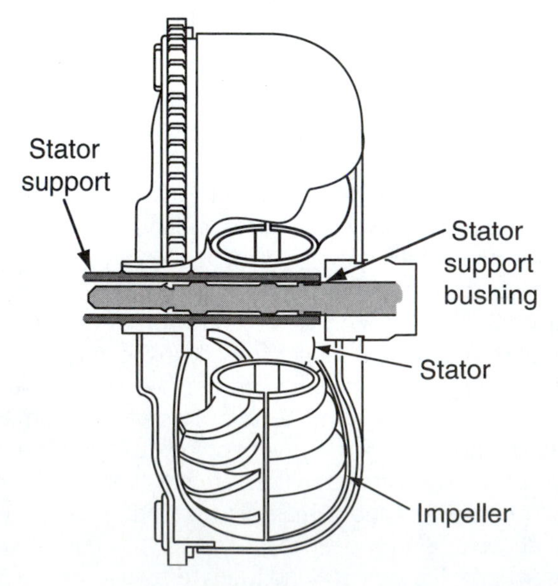


**Figure 8.** Typical stator support and oil flow in a torque converter.

## Stator One-Way Clutch Diagnosis

The operation of the stator one-way clutch inside the torque converter is critical to overall effectiveness of the torque converter. If a problem occurs in this clutch assembly, either the clutch will fail to lock when rotated in either direction or it will fail to unlock when rotated in either direction. Although these problems are similar, they affect efficiency at opposite ends of the engine's operating speeds. However, in either case, fuel economy will be affected.

When the stator clutch does not lock, there is a disruption in vortex flow and a loss of torque multiplication in the torque converter. This problem will cause sluggish low speed performance but will not affect high speed operation when the stator is supposed to free wheel.

A vehicle with a constantly locked stator will have good low speed and poor high speed performance. Torque multiplication will always occur as will speed reduction. A vehicle with a constantly locked stator will show signs of overheating. If you suspect a locked stator, check for a bluish tint on the hub of the converter. This discoloration typically results from overheating. It is normal for some blue to be evident at the spot where the hub was welded to the housing. However, if the most of the hub is blue, the converter has overheated.

To check the one-way clutch with the converter on a bench, insert a finger or long-jawed snapring pliers into the splined inner race of the clutch. Attempt to turn the inner race in both directions. You should be able to turn the race freely in one direction and feel lockup in the opposite direction. If the clutch rotates freely in both directions or if the clutch is locked in both directions, the converter should be replaced.

Because this check does not put a load on the clutch assembly, it does not totally check the unit. Some manufacturers recommend the use of a special tool set that holds the inner race and exerts a measurable amount of torque on the outer race, thereby allowing the technician to observe the action of the clutch when it is under a load **(Figure 9)**.

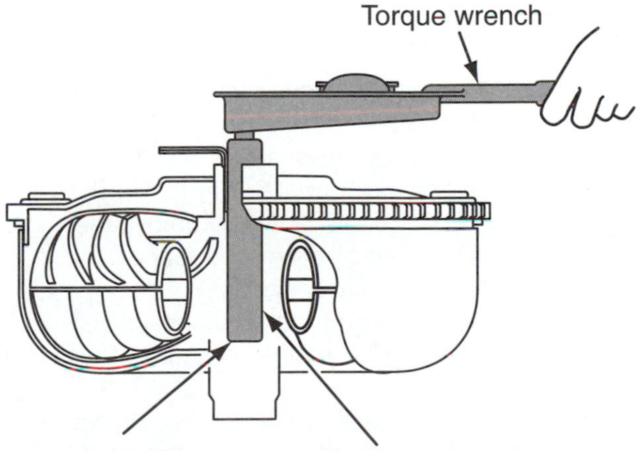

Torque wrench
Splined into stator clutch inner race
Converter clutch torqueing tool

**Figure 9.** Checking the one-way clutch with a special driver tool and torque wrench.

## Internal Interference Checks

Internal converter parts hitting each other or hitting the housing also may cause noises. To check for any interference between the stator and turbine, place the converter face down on a bench. Then install the oil pump assembly. Make sure the oil pump drive engages with the oil pump. Insert the input shaft into the hub of the turbine. Hold the oil pump and converter stationary, and then rotate the turbine shaft in both directions **(Figure 10)**. If the shaft does not move freely and/or makes noise, the converter must be replaced.

To check for any interference between the stator and the impeller, place the transmission's oil pump on a bench and fit the converter over the stator support splines **(Figure 11)**. Rotate the converter until the hub engages with the oil pump drive. Then hold the pump stationary and rotate the converter in a counterclockwise direction. If the converter does not freely rotate or makes a scraping noise during rotation, the converter must be replaced.

**You Should Know**

*During this check, some converters will go into the pump deep enough to rub on the pump's housing. This is typically not a problem, as there will be adequate clearance between the two after the converter is bolted to the flywheel.*

**Figure 10.** Insert the input shaft through the pump and into the hub of the turbine. Then, hold the oil pump and converter stationary, and rotate the shaft in both directions. If the shaft does not move freely and/or makes noise, the converter must be replaced.

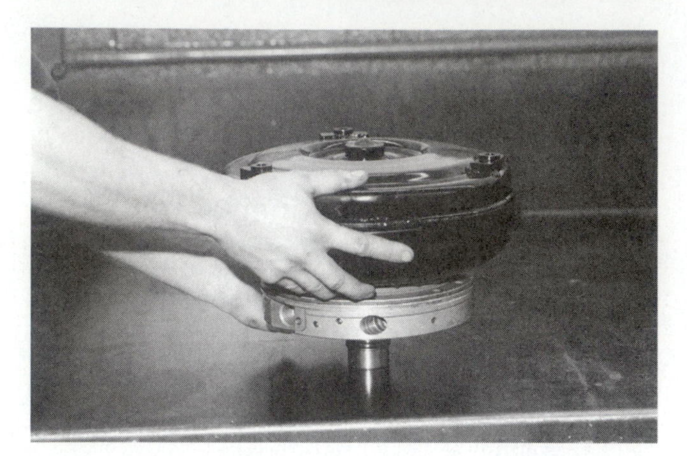

**Figure 11.** Hold the pump stationary and rotate the converter in a counterclockwise direction. If the converter does not freely rotate or makes a scraping noise during rotation, the converter must be replaced.

## End Play Check

The special tools required to check the internal end play of a torque converter are typically part of the essential tool kit recommended by each manufacturer. However, these specialty tools can be individually purchased through specialty tool companies. Basically, the special tools are a holding tool and a dial indicator with a holding fixture. The holding tool is inserted into the hub of the converter and, once bottomed, it is tightened in place. This locks the tool into the splines of the turbine. The dial indicator is fixed onto the hub **(Figure 12)**. The amount indicated on the dial indicator, as the tool is lifted up, is the amount of end play inside the converter **(Figure 13)**. If this amount exceeds specifications, replace the converter.

## Converter Leakage Tests

If the initial visual inspection suggested the converter has a leak, special test equipment can be used to determine if the converter is leaking. This equipment **(Figure 14)** uses compressed air to pressurize the converter. Leaks are found in much the same way as tire leaks are; the converter is submerged in water and the trail of air bubbles leads to the source of leakage.

Although some specialty shops will rebuild a converter, nearly all technicians replace the converter when it is faulty or damaged. This is especially true of converters with a clutch. Rebuilding a converter is not a normal task for an automatic transmission specialist and is done only at specialty shops. Because this procedure requires special equipment and knowledge, do not attempt to repair a faulty converter.

**Figure 12.** To check the end play of a torque converter, insert the holding tool into the hub of the converter and mount a dial indicator onto the converter. Set the plunger of the dial indicator so that it can read the movement of the tool.

## Oil Pump Seal

If the back side of the torque converter is wet, it is very likely that the pump seal is bad. There are many possible causes for seal leaks; these must also be looked at and corrected when the new seal is installed **(Figure 15)**. The following list describes many of the causes of pump seal leakage:

- The seal has a missing or damaged spring.
- The lip of the seal is cut or damaged.
- The pump bore is worn, scratched, or damaged.
- The pump seal bore is off the centerline of the crankshaft.
- The pump drain hole is too small or is restricted.
- The pump bushing is worn, loose, or missing.
- The hub seal area is worn, rusted, or pitted.
- The hub has excessive runout.
- The hub is cracked or machined rough.
- The hub is the incorrect length.

Figure 13. Lift up on the tool and observe the reading on the indicator. This is the amount of end play inside the converter.

- The hub or converter is out of balance.
- Excessive crankshaft end clearance (should be no more than 0.006 inch).
- Crankshaft pilot bore is worn.
- Crankshaft pilot sleeve is missing.
- The flex plate spacer is broken or missing.
- The flex plate is broken or has excessive runout.
- The converter housing is loose.
- The dowels in the converter housing are worn or missing.
- Replacement seal has incorrect elastomer or is missing an auxiliary lip.
- The inside or outside diameter of the seal is incorrect.
- The elastomer of the seal is too thick and is restricting the drain hole.
- The helix direction of the seal is incorrect.
- The seal was installed upside down.
- The seal was damaged during installation by the use of a defective or wrong driver.
- The seal was damaged by careless installation of the torque converter.

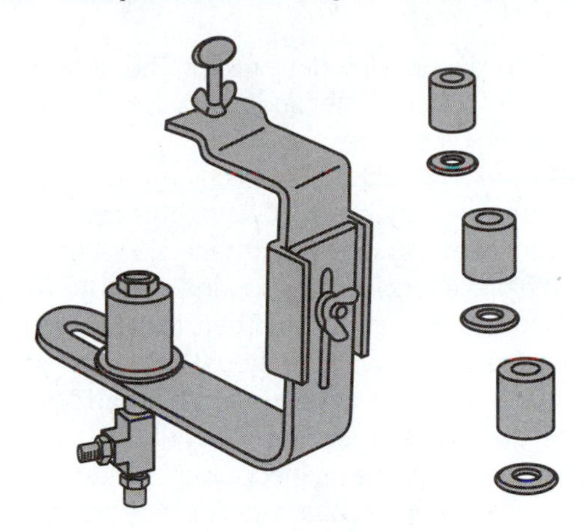

Figure 14. Converter leakage test kit.

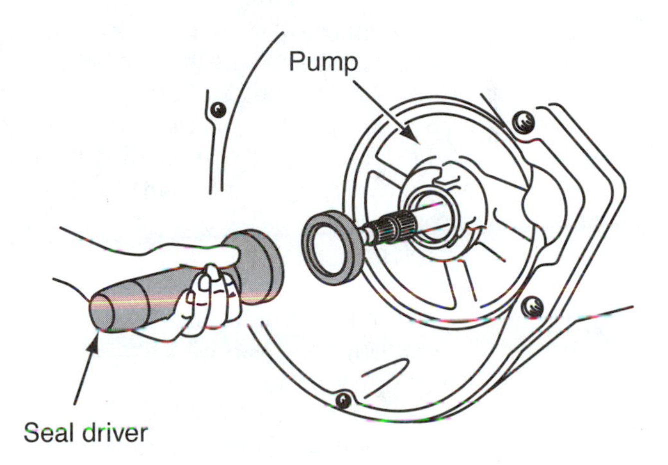


Figure 15. The converter/pump seal must be installed with the proper driver in order to seal.

## CLEANING

A torque converter without a clutch should be flushed any time it will be reused, if there was contaminated fluid in the transmission, or if the transmission overheated. There are many small corners inside a converter for dirt to become trapped in. Leftover debris in the converter can lead to converter and transmission damage. Remember, the fluid that circulates through the converter circulates through the transmission next. Converters are typically cleaned by flushing the inside of the housing with solvent.

Flushing with solvent is not recommended for most converters with a clutch. It is recommended that clutch-type converters be replaced rather than flushed or cleaned.

Flushing removes all debris and sludge from the inside of the converter. Flushing is done either by a machine or by hand. Some transmission shops use a torque converter flushing machine, which pumps a solvent through the converter as it is rotated by the machine. This method keeps the fluid moving inside the converter. The moving fluid is quite efficient at picking up any dirt present. The dirty solvent is pumped out of the converter and new solvent is added during the flushing procedure.

In shops without a flushing machine, you can clean the inside of converters by hand. However, this method is risky and not very effective, so it is not generally recommended. Basically, the procedure involves pouring about two quarts of clean solvent into the converter. Then, forcefully rotate and vibrate the converter. This action should dislodge any trapped debris. It also is helpful to use the input shaft and spin the turbine while the fluid is inside. The solvent is drained and the process repeated until the drained solvent is clean.

Because any amount of dirt can destroy a transmission, most rebuilders replace the torque converter with a transmission overhaul. There is no true way of knowing how much dirt remains in the converter after cleaning and flushing it. The cleaning process may loosen up the debris, which will break down and contaminate the fluid once the torque converter starts spinning and gets hot.

Some transmission shops may cut the torque converter shell in half, then clean the parts, examine them for wear, and replace any that are worn or broken. The shell is then welded back together. This is a job only for shops that are equipped to do this the right way.

After the converter has been flushed, disconnect it from the machine. If the converter has a drain plug, invert and drain the complete assembly. Converters without drain holes or plugs can be drilled, with a 1/8-inch drill bit, between the top end of the impeller fin dimples. This hole will act as an air bleed to maximize flushing. After flushing and draining of the converter, the bleeder hole is sealed with a closed-end pop rivet covered with sealant.

## TORQUE CONVERTER REPLACEMENT

Extra care should always be taken when replacing a torque converter. Size and fit are not the only important variables. Nor does size alone determine the stall speed of the converter. Even if the converter has exactly the same stall speed, it may have a different torque ratio and should not be used. Always check and double check the part or model number of the torque converter you are removing and compare it to the one you are going to install. Converters are typically identified by a sticker or a number code stamped into the converter housing.

When replacing a torque converter, never use an impact wrench on the torque-converter bolts. Impact wrenches can drive the bolts through the cover, which will warp the inside surface and prevent proper clutch engagement or it will damage the clutch's pressure plate.

Always perform an end-play check and check the depth of the torque converter in the bell housing **(Figure 16)** before reinstalling a torque converter or installing a new unit.

Correct installation of a torque converter requires that the converter's hub be fully engaged with the transmission.

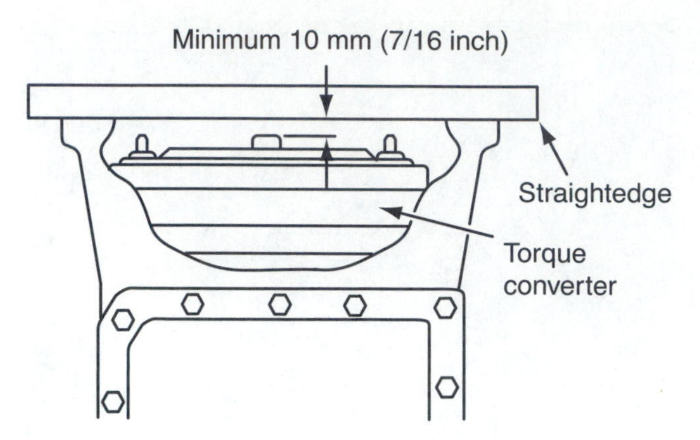


**Figure 16.** When installing a torque converter, check its installed depth in the bell housing.

To help in seating the torque converter into the transmission, push on the converter while rotating it **(Figure 17)**. Care should be taken not to damage the pump seal when installing the converter. To verify full engagement, place a straightedge across the bell housing or transaxle flange. Then measure from the straightedge to the pilot. Compare your readings to those specified by the manufacturer. If your reading is less than the specifications, the torque converter is not properly engaged in the transmission.

**You Should Know**

*Many OBD-II systems have an operational mode that allows for converter clutch breakin. The purpose of this mode is to allow the system to readjust to a new clutch in the torque converter. Whenever a torque converter is replaced in a vehicle equipped with OBD-II, set the system into the clutch breakin mode. Doing this will allow the system to realize things have changed and will allow the converter clutch to work more efficiently.*

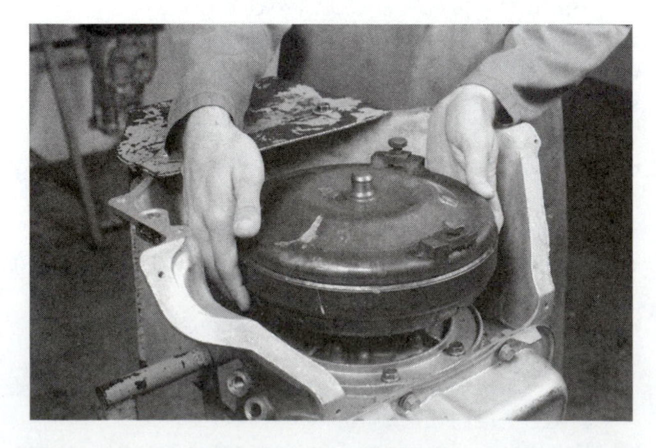

**Figure 17.** Make sure the converter's hub is fully engaged with the transmission.

## Converter Balance

When replacing a torque converter, make sure it is balanced. If the converter is not balanced, noise and vibrations can result, as well as damage to the transmission.

When reusing a torque converter, make sure it is positioned in its original position on the flexplate and the mounting bolts or nuts are tighten to specifications. Sometimes manufacturers install a slightly unbalanced torque converter in a precise position on the flexplate. This is done to offset or counter engine vibrations, and you have no way of knowing if this is what they did.

When installing a known balanced converter, it can be positioned anywhere on the flexplate as long as it is fully seated against it and the mounting bolts or nuts are properly torqued.

## Starter Ring Gear Replacement

The starter ring gear is most often part of the flexplate. Therefore, whenever the teeth of the ring gear are damaged, the entire flexplate is replaced. There are some transmissions equipped with a torque converter fitted with a ring gear around the outer circumference of the torque converter cover. The ring gear is welded to the front of the converter cover and replacement involves breaking or cutting the welds, then replacing the gear and welding it back on to the converter. This procedure is typically not recommended for torque converters fitted with a clutch, as the heat from welding can destroy the frictional surfaces inside the converter. In these cases, if the ring gear is damaged, the entire converter should be replaced.

# Summary

- A torque converter must be replaced:
  - If there is a stator clutch failure.
  - If there is internal interference.
  - If the transmission's front pump is badly damaged.
  - If the converter hub is severely damaged or scored.
  - If there are signs of external fluid leaks.
  - If the drive studs or lugs are damaged or loose.
  - If there are signs of overheating.
  - If heavy amounts of metal were found in the fluid.
  - If any damage indicates that the converter is no longer balanced.
- A torque converter without a clutch should be flushed any time it will be reused.
- Flushing with solvent is not recommended for most converters with a clutch.
- Correct installation of a torque converter requires that the converter's hub be fully engaged with the transmission.

# Review Questions

1. Why is it important to check the installed depth of a torque converter when installing it into a transmission?
2. What is indicated by the presence of oil on the back side of a torque converter?
3. When checking torque converter end play, Technician A says that the torque converter must be installed in the transmission. Technician B says that the torque converter end play is corrected by installing a thrust washer between the oil pump and the direct or front clutch. Who is correct?
   A. Technician A only
   B. Technician B only
   C. Both Technician A and Technician B
   D. Neither Technician A nor Technician B
4. Technician A says that a converter that had a tight fit at the pilot hub could hold the converter drive hub and inner gear too far into the pump, causing cover scoring. Technician B says that a front pump bushing that has too much clearance may allow the gears to run off center, causing them to wear into the crescent and/or the sides of the pump body. Who is correct?
   A. Technician A only
   B. Technician B only
   C. Both Technician A and Technician B
   D. Neither Technician A nor Technician B
5. Technician A says that a seized one-way stator clutch will cause the vehicle to have good low speed operation but poor high speed performance. Technician B says that a freewheeling or nonlocking one-way stator clutch will cause the vehicle to have poor acceleration. Who is correct?
   A. Technician A only
   B. Technician B only
   C. Both Technician A and Technician B
   D. Neither Technician A nor Technician B

# Chapter 30 Disassembly

## *Introduction*

The disassembly of a transmission and transaxle are similar; therefore, disassembly and the inspection of both transaxles and transmissions can safely be grouped together.

### DISASSEMBLY

Before disassembling a transmission, determine the causes of any leakage. If leakage is the reason for the tear-down, determine the path of the leakage before cleaning the area around the seals **(Figure 1)**. At times, the leak-age may be from sources other than a seal. Leakage could be from worn gaskets, loose bolts, cracked housings, or loose line connections.

Inspect the outside sealing area of the seal to see if it is wet or dry. If it is wet, determine if the oil is running out or is merely a lubricating film. Check both the inner and outer parts of the seals for wet oil.

When removing a seal, inspect the sealing surface or lips before cleaning it. Look for signs of unusual wear, warping,

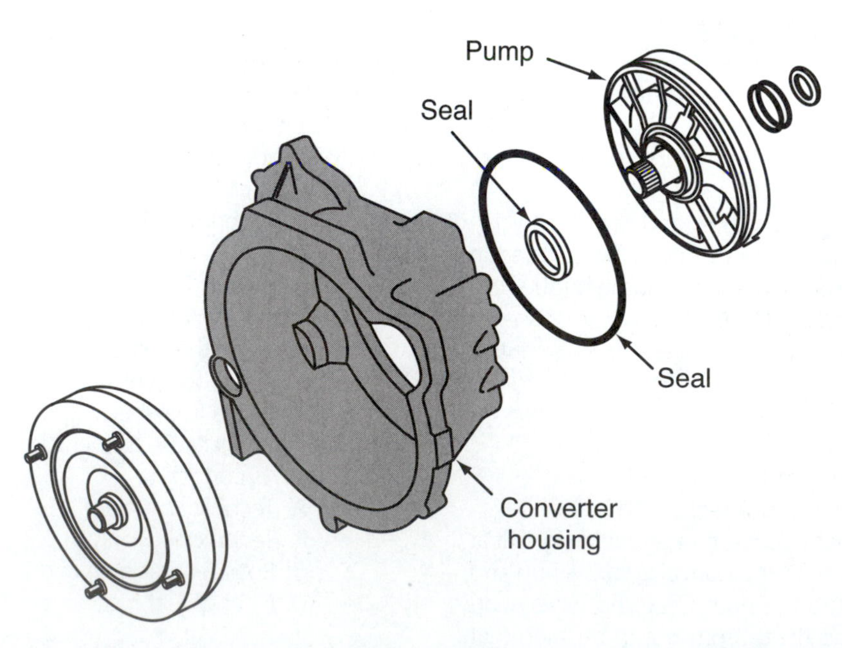


**Figure 1.** Signs of oil inside the converter housing can be caused by leaks at any of these seals shown in this drawing.

cuts and gouges, or particles embedded in the seal. On spring-loaded lip seals, make sure the spring is seated around the lip, and that the lip was not damaged when first installed. If the seal's lip is hardened, this was probably caused by heat from either the shaft or the fluid.

If the seal is damaged, check all shafts for roughness, especially at seal contact areas. Look for deep scratches or nicks that could have damaged the seal. Determine if a shaft spline, keyway, or burred end could have caused a nick or cut in the seal lip during installation. Inspect the bore into which the seal was fitted. Look for nicks and gouges that could create a path of oil leakage. A coarsely machined bore can allow oil to seep out through a spiral path. Sharp corners at the edges of a bore could have scored the metal case of the seal.

## Cleaning and Inspection

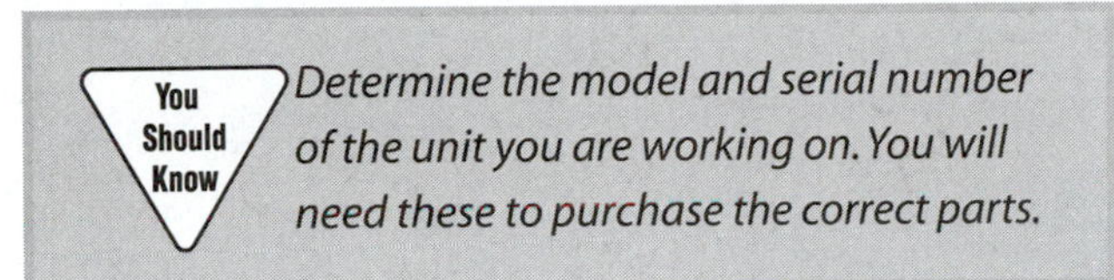

You Should Know *Determine the model and serial number of the unit you are working on. You will need these to purchase the correct parts.*

Before disassembling the transmission, make sure you clean away any dirt, undercoating, grease, or road grime on the outside of the case.

When cleaning transmission parts, avoid the use of solvents, degreasers, and detergents that can decompose the friction composites used in a transmission. It is best not to attempt to clean the friction members, as this will damage the parts. Use compressed air to dry components; do not wipe down parts with a rag. The lint from a rag can easily destroy a transmission after it has been rebuilt.

Some rebuilding shops use a parts washing machine to clean the case, converter housing, and extension housing. These parts washers use hot water and a special detergent that are sprayed onto the parts as they rotate inside the cleaner. The key to good cleaning with these machines is to use a very small amount of soap with very hot water. Many rebuilders simply clean the parts in a mineral spirits tank, where the parts are brushed and hand cleaned. No matter what type of cleaning procedure is followed, the transmission and parts should be rinsed with water then thoroughly dried with compressed air before reassembly.

You Should Know *Always wear safety goggles when using compressed air to dry something. The air pressure can easily move dirt, metal, or other debris from around the work area. If any of these get into your eye, it can cause permanent damage to your eye.*

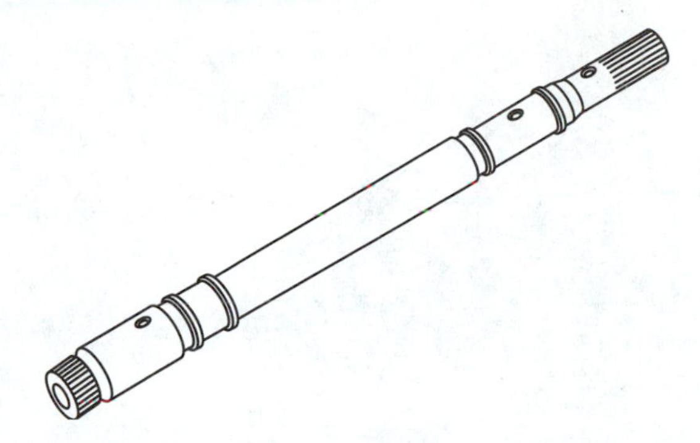

**Figure 2.** The pump drive shaft should be inspected in the areas that ride on the bushings and bearings.

After the torque convertor is removed, check the input shaft splines **(Figure 2)**, stator support splines, and the converter's pump drive hub for any wear or damage. Converters with direct drive shafts should be checked to be sure that no excessive play is present at the drive splines of the shaft or the converter. If any play is found in the converter, the converter or the shaft must be replaced.

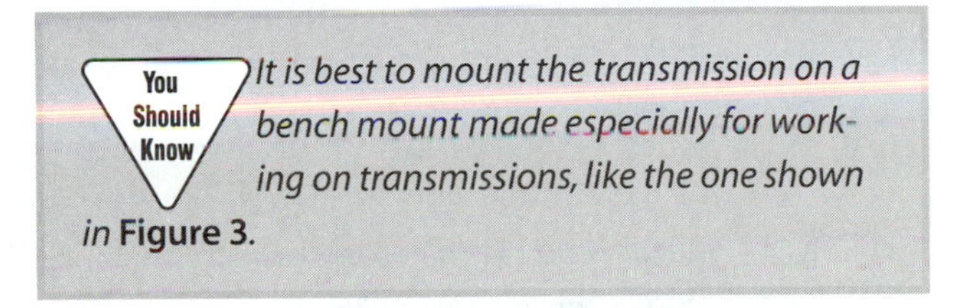

You Should Know *It is best to mount the transmission on a bench mount made especially for working on transmissions, like the one shown in* **Figure 3**.

Position the transmission so that the shaft centerline is vertical **(Figure 4)**. This allows the weight of the internal components to load the shafts toward the rear of the transmission. The end play of the shafts can now be checked. The transmission end play checks can provide the technician with information about the condition of internal bearings and seals as well as clues to possible causes of improper operation found during the road test.

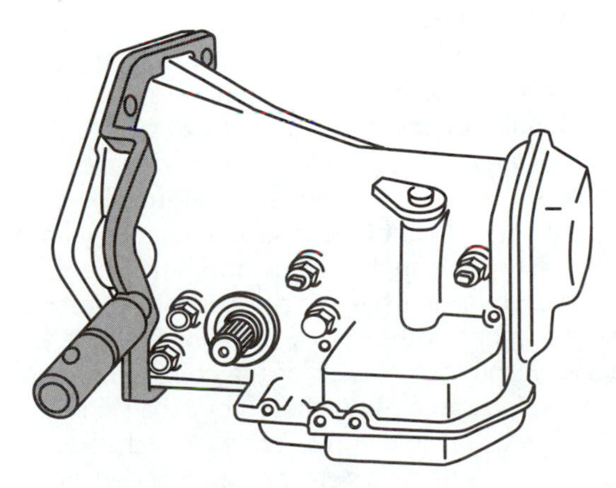

**Figure 3.** A typical transaxle holding fixture.

**Figure 4.** To measure the end play on many transmissions, the input shaft must be facing upward.

**Figure 5.** It is possible to measure end play on some transmissions with the input shaft in a horizontal position.

These measurements will also determine the thickness of the thrust washer(s) during reassembly. Thrust-washer thickness sets the end play of various components. Excessive end play allows clutch drums to move back and forth too much, causing the transmission case to wear. Assembled end play measurements should be between minimum and maximum specifications but preferably at the low end of the specifications.

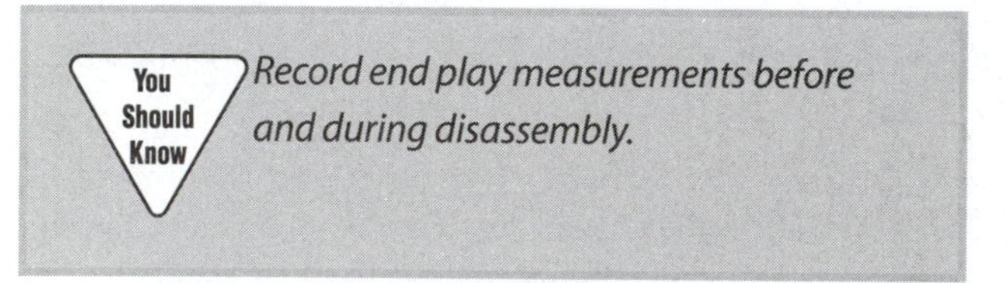


*Record end play measurements before and during disassembly.*

Most GM transmissions can be checked by mounting a dial indicator to read the input shaft movement. Zero the dial indicator and lift upward on the input shaft. Most Chrysler and some Ford transmissions require you remove the oil sump pan and valve body and pry the input shell upward. Chrysler and Ford transmissions may be measured horizontally **(Figure 5)**, but you may want to center the output shaft with a slip yoke for a more accurate reading.

Input shaft end play is measured with a dial indicator. The dial indicator should be solidly clamped to the converter housing and the plunger positioned so that it is centered on the end of the input shaft **(Figure 6)**. Move the input shaft in and out of the case and note the reading on the indicator. Compare this reading with the specifications for that transmission. If the end play is incorrect, it should be corrected during assembly by installing a thrust washer, selective snapring, or spacer of the correct thickness between the input shaft and the output shaft. Some transmissions require additional end play checks during disassembly. The manufacturer's recommended procedure for checking end play should always be followed.

**Figure 6.** Input shaft end play is measured with a dial indicator mounted to the converter housing and its plunger positioned so that it is centered on the end of the input shaft.

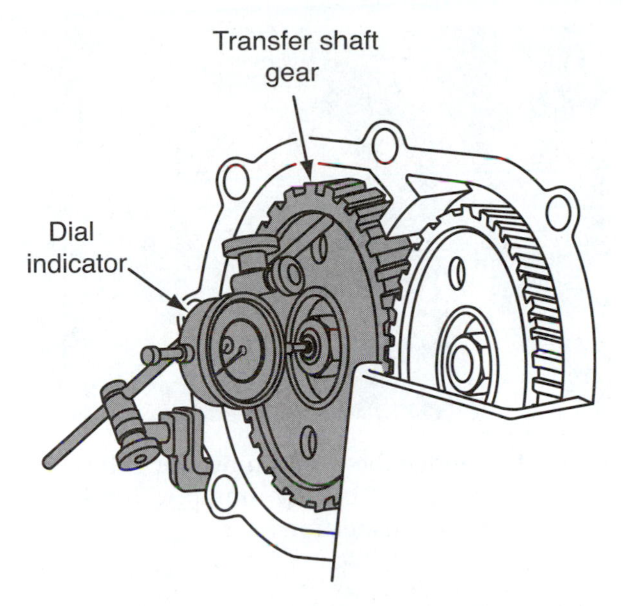


**Figure 7.** Some transaxles require additional end play checks during assembly and disassembly.

An output shaft or final drive end play check may be required **(Figure 7)**. These settings provide long gear life and eliminate engagement "clunk." All measurements taken during this phase of teardown should be used as a guide during the rebuild. This information can identify worn parts and will be used to obtain the shims necessary to correct undesired end play. If the end play is excessive, the thrust washers or bearings or the sealing rings and grooves may be worn. During rebuild, end play settings should be kept to the minimum allowable amount. Once the end play checks have been made and recorded, disassembly can begin.

**You Should Know** *When working on automatic transmissions, there is no such thing as being too clean. All dirt, grease, and other materials should be cleaned off any parts that are going to be reused.*

As you remove the various assemblies from the transmission, place them in the order that they were removed. The correct position of thrust washers, snaprings, and even bolts and nuts is of great importance and should be noted. Unless the transmission suffered a major breakdown or was overheated, the internal components of the transmission will be fairly clean due to the nature of the fluid. A light cleaning with mineral spirits or a simple blowing off with low-pressure air is all that is necessary for cleaning most parts for inspection.

Each subassembly should be completely disassembled and the parts thoroughly cleaned and checked for signs of excessive wear, scoring, overheating, chipping, or cracking. If there is any question as to the condition of a part, replace it.

## Basic Disassembly

The following is a general procedure for disassembling a transmission. Always refer to the manufacturer's recommended procedure for the specific transmission.

With the transmission mounted on a fixture, remove the oil pan. Carefully inspect the debris in the pan. Doing this will help you determine what else should be carefully examined.

Remove the converter housing from the transmission case **(Figure 8)**, if it is a separate piece. Remove all externally mounted solenoids **(Figure 9)** with their gaskets, O-rings, and seals. Pay attention to the wiring harness and connectors; check them for fraying, corrosion, or other damage. If the transmission has a vacuum modulator, unscrew and remove it. Remove the speedometer drive assembly with its gear and O-ring.

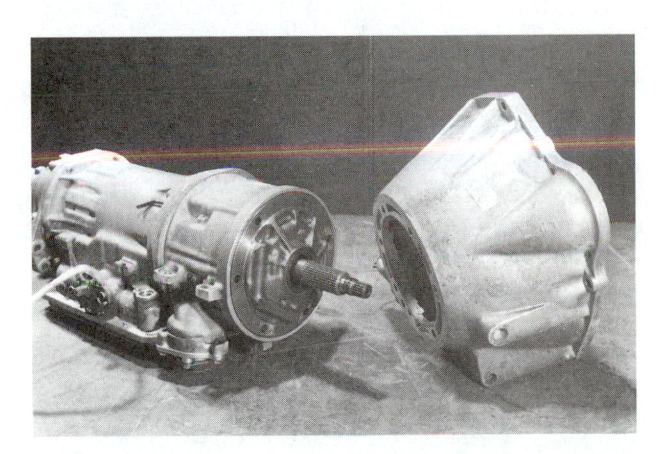

**Figure 8.** Remove converter housing from the transmission case, if it is a separate unit.

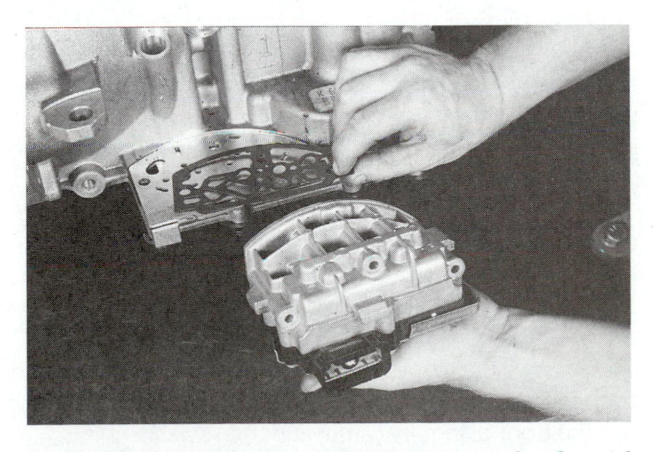

**Figure 9.** Remove all externally mounted solenoids with their seals and gaskets.

**Figure 10.** Unbolt and remove valve body from the case.

> **You Should Know** *The magnets inside electronic shift control solenoids will attract any iron that is floating around the inside of the transmission. Thoroughly clean or replace these solenoids during a transmission rebuild.*

Unbolt and remove the valve body from the case **(Figure 10)**. Then remove the manual valve from the valve body to prevent it from dropping out. Back off the servo piston stem locknut and tighten the piston stem to prevent the front clutch drum from dropping out when removing the front pump.

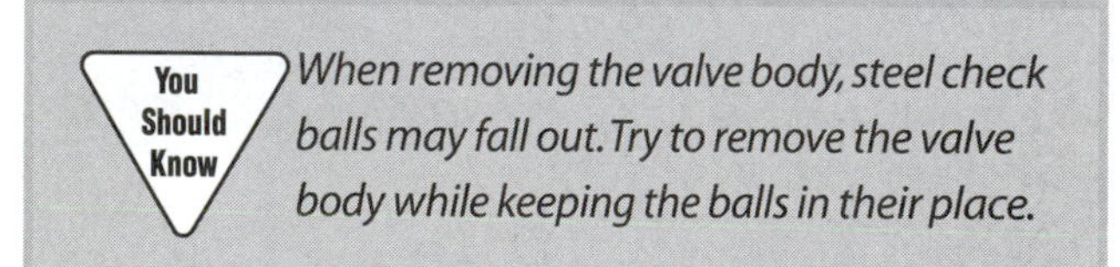

> **You Should Know** *When removing the valve body, steel check balls may fall out. Try to remove the valve body while keeping the balls in their place.*

Using the correct puller, remove the front pump from the case **(Figure 11)**. Lay this assembly to the side for further inspection. Remove the front clutch thrust washer and bearing race. Now, back off the front brake band servo piston stem to release the band. Remove the brake band strut and front brake band. The drum and band may be removed together.

Remove the front and rear clutch assemblies **(Figure 12)**. Make sure the input shaft is pointed up. This prevents the front clutch from falling out. Note the positions of the front pump thrust washers and rear clutch thrust washer, if the transmission has them.

Remove the rear clutch hub, front planetary carrier, and connecting shell. Note the positions of the thrust bearings and the front planetary carrier and thrust washer.

Remove the output shaft snapring **(Figure 13)**. Often it will be easier to remove the snapring if the carrier is removed first. Remove the carrier snapring and remove the carrier. Now, remove the output shaft snapring.

**Figure 11.** Remove the front pump using the correct puller and put the pump to the side for further inspection later.

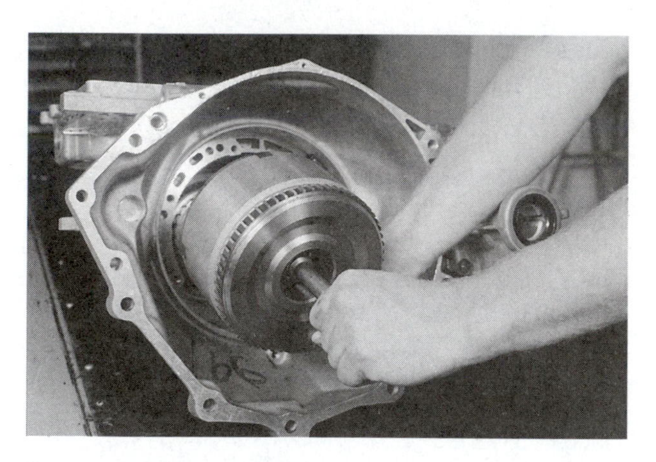

**Figure 12.** Remove the front and rear clutch assemblies. Pay attention to positions of the thrust washers.

Remove the rear connecting drum from the housing. Then, using a screwdriver, remove the large retaining snapring of the rear brake assembly. Tilt the extension housing upward and remove the rear brake assembly.

Then, unbolt and remove the extension housing. Be careful not to lose the parking pawl, spring, and retaining washer. Remove the governor **(Figure 14)** with its attachments (such as the oil distributor, thrust washer, and needle bearing assembly). Now, pull out the output shaft.

Remove the inner race of the one-way clutch, thrust washer, piston return spring, and thrust ring. Using compressed air, remove the rear brake piston and front servo.

Servo and accumulator pistons and seals are held in their bores by springs—sometimes very strong springs—and retaining rings **(Figure 15)**. Remove the retaining ring and pull the assembly from its bore **(Figure 16)**. Check the

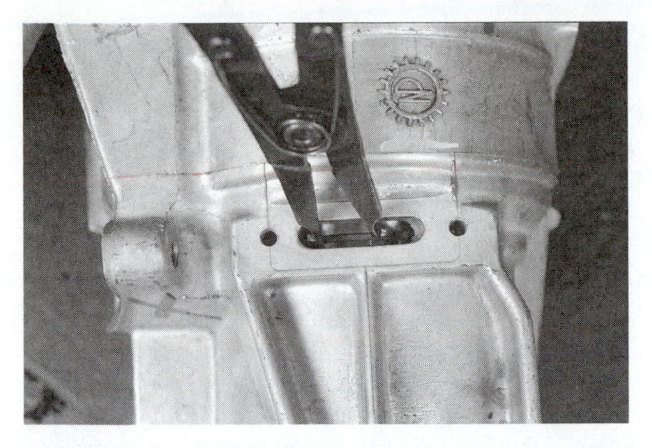

Figure 13. On some transmissions, it is necessary to loosen the snapring for the output shaft before the extension housing can be removed.

Figure 15. To remove servo and accumulator assemblies, the cover's retaining ring must be removed.

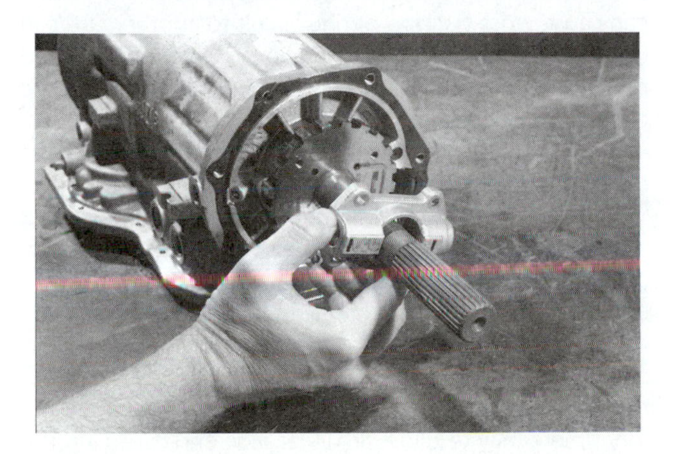

Figure 14. Remove the governor assembly.

condition of the piston and springs. Cast iron seal rings may not need replacement but elastomer seals should always be replaced.

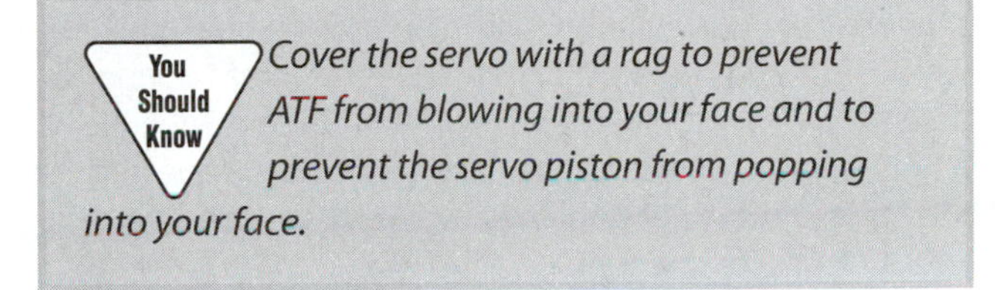


*Cover the servo with a rag to prevent ATF from blowing into your face and to prevent the servo piston from popping into your face.*

## TRANSMISSION CASE SERVICE

The transmission case should be thoroughly cleaned and all passages blown out. Make sure all electrical components have been removed from the case before cleaning it. After the case has been cleaned, all bushings, fluid passages, bolt threads, clutch plate splines, and the governor bore should be checked. The passages can be checked for restrictions and leaks by applying compressed air to each one. If the air comes out the other end, there is no restriction. To check for leaks, plug off one end of the passage and

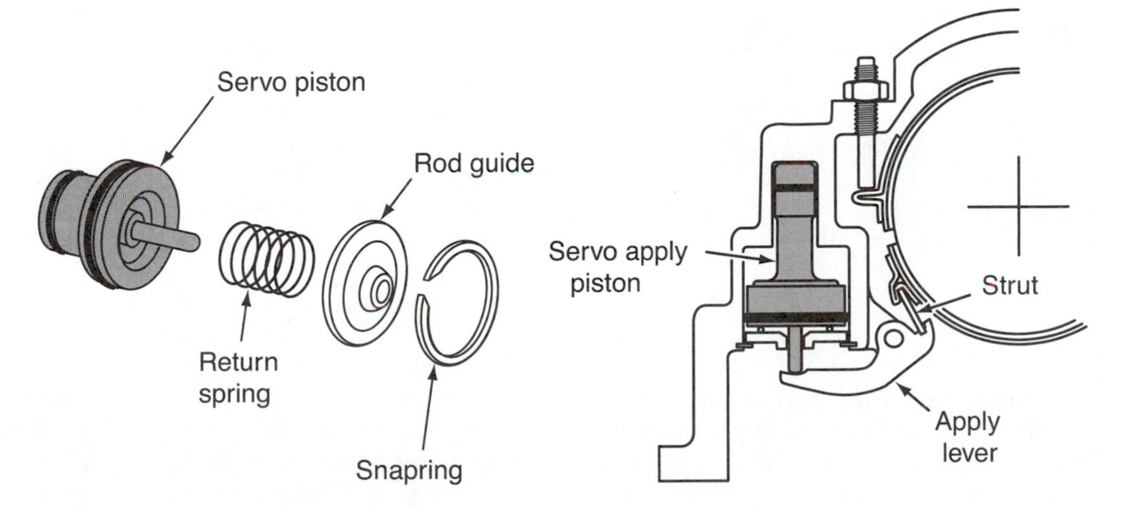


Figure 16. All servos and accumulators should be removed from their bores in the case and the bores carefully inspected.

apply air to the other. If pressure builds in that passage, there are probably no leaks in it.

The case of modern transmissions is made of aluminum. Aluminum is a soft material, which can easily be deformed, scratched, cracked, or scored. Special attention should be given to the following areas: clutch, oil pump, and servo and accumulator bores. All bores should be smooth. The servo piston also could hang up in a bore that is deeply scored. Check the fit of the servo piston in the bore without the seal, if possible, to be sure it has free travel. There should be no tight spots or binding over the whole range of travel. Any deep scratches or gouges that cause binding of the piston will require case replacement.

Case mounted accumulator bores are checked the same way as servo bores. The oil pump bore at the front of the case should be free of any scratches that would keep the O-ring from sealing the outer diameter of the pump to the front of the case. Case mounted hydraulic clutch bores are prone to the same problems as servo bores. Look for any scratches or gouges in the sealing area that would affect the rubber seals. It is possible to damage these areas during disassembly, so be careful with tools used during overhaul.

Sealing surfaces on the case should be inspected for roughness, nicks, or scratches where the seals ride **(Figure 17)**. Any problems found in servo bores, clutch drum bores, and governor support bores can cause pressure leakage. Imperfections in steel or cast iron parts usually can be polished out with crocus cloth. Care should be taken so as not to disturb the original shape of the bore. Use the crocus cloth inside clutch drums to remove the polish marks left by the cast iron sealing rings. This will help the new rings to rotate with the drum as designed. As a rule, all sealing rings, whether cast iron or Teflon, are replaced during overhaul, as this gives the desired sealing surface required for proper operation.

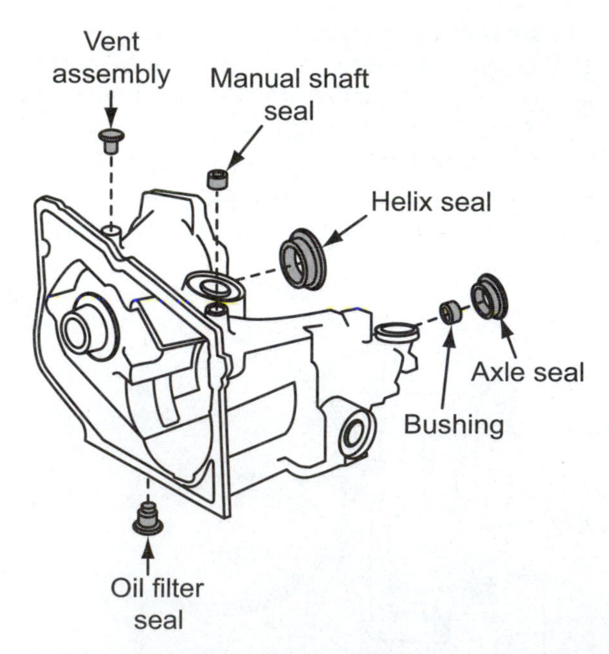


**Figure 17.** All sealing surfaces and bores of the transmission case should be carefully inspected for cracks, grooves, and scratches. (Courtesy of General Motors Corporation, Service Operations)

**You Should Know** *Crocus cloth is a very fine polishing paper. It is designed to remove very little metal; therefore, it is safe to use on critical surfaces. Never use sandpaper on sealing surfaces.*

Passages in the case guide the flow of fluid through the case. Although not that common, porosity in this area can cause cross-tracking of one circuit to another. This can cause bind up (two gears at once) or a slow pressure bleed off in the affected circuit. If this is suspected, try filling the circuit with solvent and watching to see if the solvent disappears or leaks away. If a leak is suspected, check each part of the circuit to find where the leak is.

Check the valve body mounting area for warpage with a straightedge and feeler gauge. This should be done in several locations so that any crossover from one worm track to another is evident. If there is a slight burr or high spot, it can be removed by flat filing the surface.

A long straightedge should be laid across the lower flange of the case to check for distortion. Any warpage found here may result in circuit leakage and cause a number of problems. Case warpage should be less than 0.002 inch. Cases with center supports should be checked with the support bolted in place to prevent case distortion.

Be sure to check all converter housing boltholes and dowel pins. Cracks around the boltholes indicate the case bolts were tightened with the case out of alignment with the engine block. The case should be replaced if the following problems are present: broken worm tracking, cracked case at the oil pump to case flange, case cracked at clutch housing pressure cavity, ears broken off the converter housing, or oil pan flange broken off the case. Although it is possible to weld the aluminum case, it is not possible to determine if the repair will hold. A transmission case is very thin and welding may distort the case.

If any of the bolts that were removed during disassembly have aluminum on the threads, the thread bore is damaged and should be repaired. Thread repair entails the installation of a thread insert, which serves as new threads for the bolt, or by retapping the bore. After the threads have been repaired, make sure you thoroughly clean the case.

The small screens found during teardown should be inspected for foreign material. These screens are used to prevent valve hangup at the pressure regulator and governor and must be in place. Most screens can be removed easily. Care should be taken when cleaning as some cleaning solvents will destroy the plastic screens. Low air pressure (approximately 30 psi) can be used to blow the screens out in a reverse direction.

Make sure the oil passage to a pressure fed bushing or bearing is open and free of dirt and foreign material. It does no good to replace a bushing without making sure there is good oil flow.

Vents are located in the pump body or transmission case and provide for equalization of pressures in the transmission. These vents can be checked by blowing low pressure air through them, squirting solvent or brake cleaning spray through them, or by pushing a small diameter wire through the vent passage. A clean, open passage is all you need to verify proper operation.

## Extension Housing

Check the extension housing for cracks, especially around the case mounting surface and the pad that attaches to the transmission mount. Using a straightedge and feeler gauge set, check the flatness of the mating surface. If there is 0.002 inch or more distortion, the surface may need to be resurfaced or the housing replaced. Minor problems may be corrected by filing down the surface. However, filing should be done only when it is necessary. To file the surface, select the largest single-cut file available. Place the file across one end of the surface and pull or draw the file across the surface to the opposite end. Lift the file off the surface and place it back at its original position, then draw the file to the other end. Repeat this process until the surface is corrected. Never move the file from side to side.

Carefully inspect all threaded and nonthreaded bores in the housing. All damaged threaded bores should be repaired by running a tap through the bore or by installing threaded inserts. If any condition exists that cannot be adequately repaired, the case should be replaced.

At the rear of the extension housing is the slip-yoke bushing. This bushing will normally wear to one side due to loads imparted on it during operation. Oil feed holes at this bushing must be checked to make sure oil can get to this bushing. Often the speedometer drive gear is responsible for throwing oil back to the rear bushing. A sheared or otherwise inoperative speedometer gear could cause the extension housing bushing to fail. Always make sure this bushing is aligned correctly during replacement or premature failure can result.

# Summary

- Before disassembling the transmission, identify the causes of any leakage.
- When removing a seal, inspect the sealing surface before cleaning it.
- Check the end play of the shafts before disassembling the transmission or transaxle.
- If the end play is incorrect, it should be corrected during assembly.
- The magnets inside electronic shift control solenoids will attract any iron that is floating around the inside of the transmission. Thoroughly clean or replace these solenoids during a transmission rebuild.
- Once the transmission has been disassembled into its various subassemblies, each subassembly should be disassembled, cleaned, inspected, and reassembled.
- Check all sealing surfaces, bushings, fluid passages, bolt threads, clutch plate splines, and bores in the transmission housing.

# Review Questions

1. If during disassembly a seal is found damaged, what should you do?
2. How should a transmission be cleaned?
3. Why should an end play check be done before disassembling a transmission?
4. An output shaft or final drive end play check should be made on either a transmission or transaxle. Technician A says that this is done to provide long gear life and to eliminate engagement "clunk." Technician B says that end play should be measured during teardown, as this helps you to obtain the correct shims necessary to correct undesired end play. Who is correct?
   A. Technician A only
   B. Technician B only
   C. Both Technician A and Technician B
   D. Neither Technician A nor Technician B

5. A transmission case is being inspected. Technician A says that the case should be replaced because there is a crack in the flange for the mounting of the oil pump. Technician B says that the case should be replaced because there was aluminum on the threads of the bolts that hold the extension housing to the case. Who is correct?

   A. Technician A only
   B. Technician B only
   C. Both Technician A and Technician B
   D. Neither Technician A nor Technician B

# Chapter 31 Servicing Pumps

## Introduction

The oil pump **(Figure 1)** of an automatic transmission should be carefully inspected during any transmission overhaul, especially when low line pressure was measured during a pressure test. Carefully remove the oil pump assembly. Some transmissions require the use of a special puller to remove the oil pump from the transmission case **(Figure 2)**. Never pry the pump out; it is easy to damage the case when doing this.

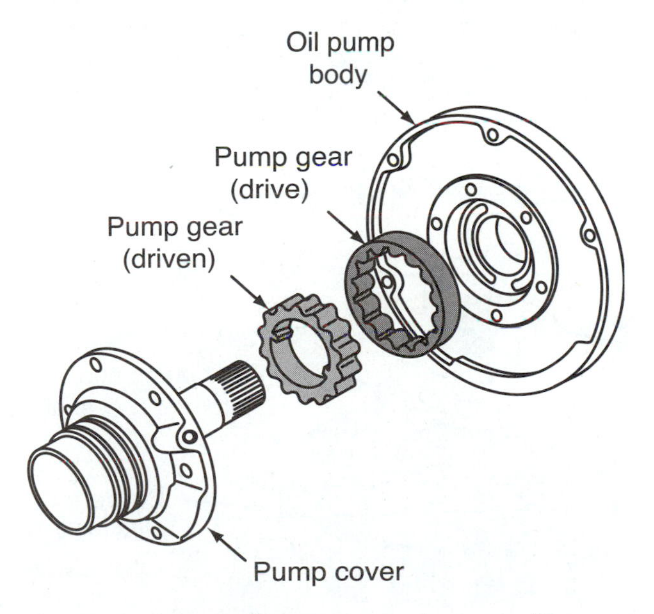


**Figure 1.** A typical gear-type oil pump. (Courtesy of General Motors Corporation, Service Operations)

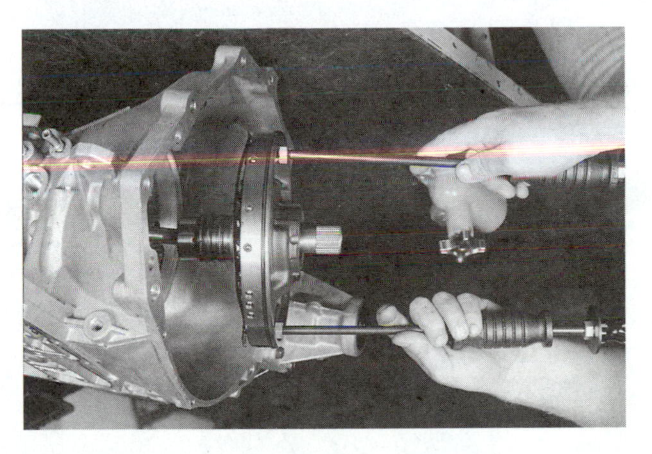

**Figure 2.** An oil pump being removed from a transmission housing with a special puller.

## DISASSEMBLY

Begin disassembly by removing the front pump bearing race, front clutch thrust washer, gasket, and O-ring. Inspect the pump bodies, pump shaft, and ring groove areas. Then unbolt and separate the pump bodies **(Figure 3)**.

Before doing anything with the gears or rotors, mark the alignment of the gears. If acceptably worn gears are reinstalled in a position other than their wear pattern, excessive noise will result. When the pump halves have been separated, look at the relationship between the inner and outer gears. Most gears will have a mark on them indicating the top side of the gear **(Figure 4)**.

If no mark is present, you should use a nondestructive-type marker to be sure you install the gears in their original position **(Figure 5)**. This ensures that the converter drive hub will mate correctly with the inner gear.

Figure 3. Unbolt and separate the pump bodies.

Figure 5. Marking the location of the outer and inner gears so that they can be properly meshed during reassembly.

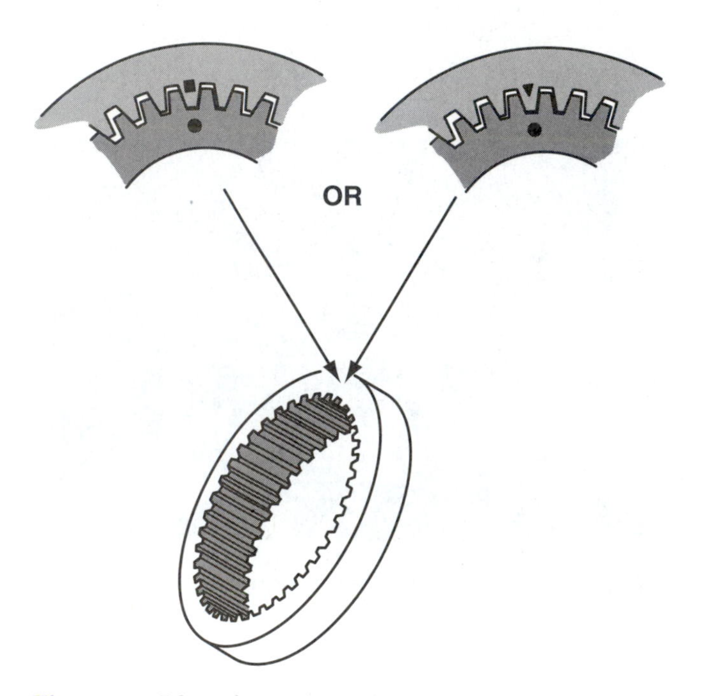


Figure 4. Identification marks on oil pump gear sets.

Another way of ensuring proper position of the gears is to observe the existing wear pattern on the gears as they are being removed and make a note of this for use during reassembly. If the inner gear is installed upside down, there will not be enough free movement, and that can result in one or more of the following problems: broken inner gear, broken flexplate, pump cover scoring, or broken transmission case ears.

## INSPECTION

Inspect the pump bore for scoring on both its bottom and sides. A converter that had a tight fit at the pilot hub could hold the converter drive hub and inner gear too far into the pump, causing cover scoring. A front pump bushing that has too much clearance may allow the gears to run off center, causing them to wear into the crescent and/or the sides of the pump body.

> **You Should Know** *Some service manuals will call for a maximum front bushing wear of 0.004 inch; however, a dimension of less than that will avoid wear as well as keep the pressure losses at the bushing to a minimum.*

The stator shaft should be inspected for looseness in the pump cover **(Figure 6)**. The shaft's splines and bushing also should be carefully looked at. If the splines are distorted, the shaft and the pump cover should be replaced. Because the bushings control oil flow through the converter and cooler, their fit must be checked **(Figure 7)**. Bushings must

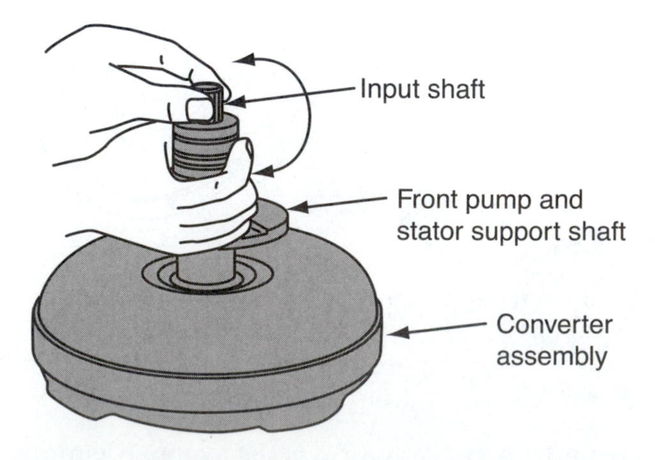


Figure 6. Checking the play between the splines of the oil pump and the input shaft.

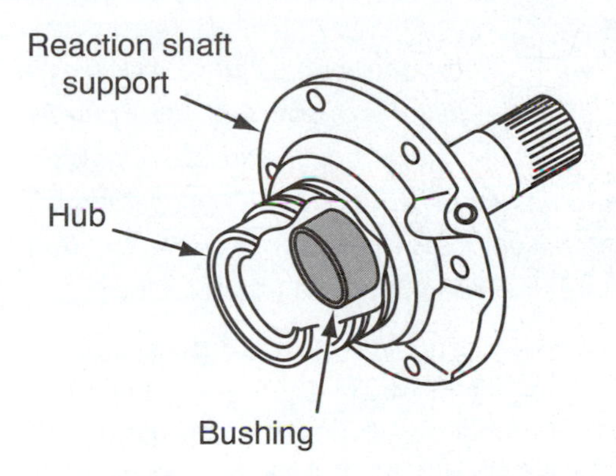


Figure 7. Location of a stator shaft bushing.

Figure 9. Measure between the outer gear and the pump housing. Replace the pump if the measurements exceed specifications.

be tight inside the shaft and provide the input shaft with a 0.0005- to 0.003- inch clearance.

Inspect the gears and pump parts for deep nicks, burrs, or scratches **(Figure 8)**. Examine the pump housing for abnormal wear patterns. The fit of each gear into the pump body, as well as the centering effect of the front bushing, controls oil pressure loss from the high pressure side of the pump to the low pressure input side. Scoring or body wear will greatly reduce this sealing capability.

Figure 10. Measure between the outer gear teeth and the crescent. Replace the pump if the clearance exceeds specifications.

> **You Should Know** *All pumps, valve bodies, and cases should be checked for warpage and should be flat filed to take off any high spots and burrs prior to reassembly.*

On positive displacement pumps, use a feeler gauge to measure the clearance between the outer gear and the pump pocket in the pump housing **(Figure 9)**. If possible, also check the clearance between the outer pump gear teeth and the crescent **(Figure 10)**, and between the inner gear teeth and the crescent. Compare these measurements to the specifications. Use a straightedge and feeler gauge to check gear side clearance **(Figure 11)**, and compare the clearance to the specifications. If the clearance is excessive, replace the pump.

Figure 8. Inspect the gears and all internal surfaces for defects and visible wear.

Variable displacement vane-type pumps require different measuring procedures. However, the inner pump rotor to converter drive hub fit is checked in the same way as described for the other pumps. The pump rotor, vanes, and slide are originally selected for size during assembly at the factory **(Figure 12)**. Changing any of these parts during overhaul can destroy this sizing and possibly the body of the pump. You must maintain the original sizing if any parts are found to need replacement. These parts are available in select sizes for just this reason.

**Figure 11.** Place the pump flat on a bench. Place the straightedge across the pump. Using a feeler gauge set, measure the clearance between the straightedge and the gears. Replace the pump if the measurements exceed specifications.

> **You Should Know** *Some early pump vanes cracked as a result of excessive load. These pumps did not have enough vanes to handle the hydraulic loads encountered during operation. Updated pumps with more vanes cured the problem.*

The vanes are subject to edge wear, as well as cracking and subsequent breakage. The outer edge of the vanes should be rounded with no flattening **(Figure 13)**. These pumps have an aluminum body and cover halves; therefore, any scoring indicates that they should be replaced.

Inspect the reaction shaft's seal rings. If the rings are made of cast iron, check them for nicks, burrs, scuffing, or uneven wear patterns. Replace them if they are damaged **(Figure 14)**. Make sure the rings are able to rotate in their grooves. Check the clearance between the reaction shaft support ring groove and the seal ring. If the seal rings are the Teflon full-circle type, cut them out and use the required special tool to replace them.

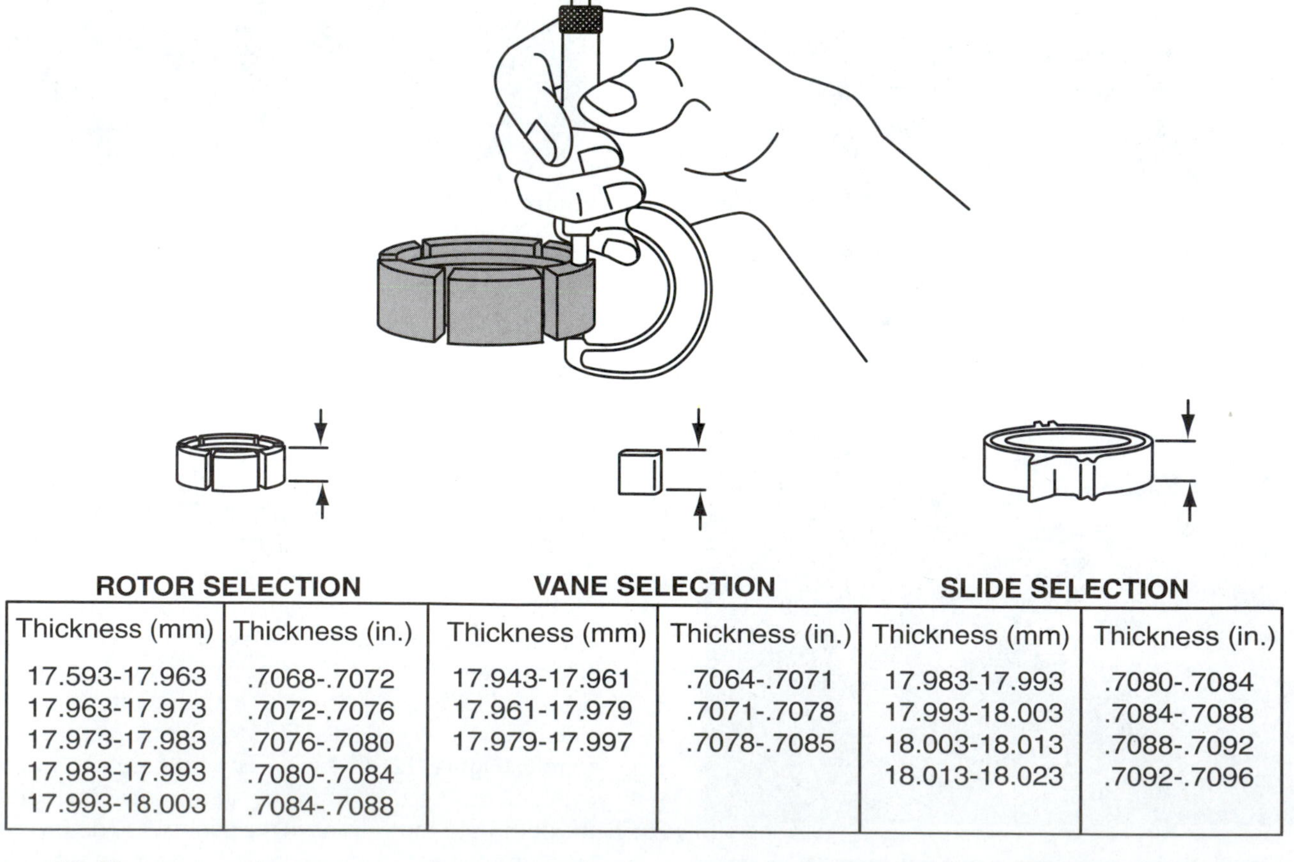

| ROTOR SELECTION | | VANE SELECTION | | SLIDE SELECTION | |
|---|---|---|---|---|---|
| Thickness (mm) | Thickness (in.) | Thickness (mm) | Thickness (in.) | Thickness (mm) | Thickness (in.) |
| 17.593-17.963 | .7068-.7072 | 17.943-17.961 | .7064-.7071 | 17.983-17.993 | .7080-.7084 |
| 17.963-17.973 | .7072-.7076 | 17.961-17.979 | .7071-.7078 | 17.993-18.003 | .7084-.7088 |
| 17.973-17.983 | .7076-.7080 | 17.979-17.997 | .7078-.7085 | 18.003-18.013 | .7088-.7092 |
| 17.983-17.993 | .7080-.7084 | | | 18.013-18.023 | .7092-.7096 |
| 17.993-18.003 | .7084-.7088 | | | | |

**Figure 12.** Vane-type oil pump measurements and selection chart. (Courtesy of General Motors Corporation, Service Operations)

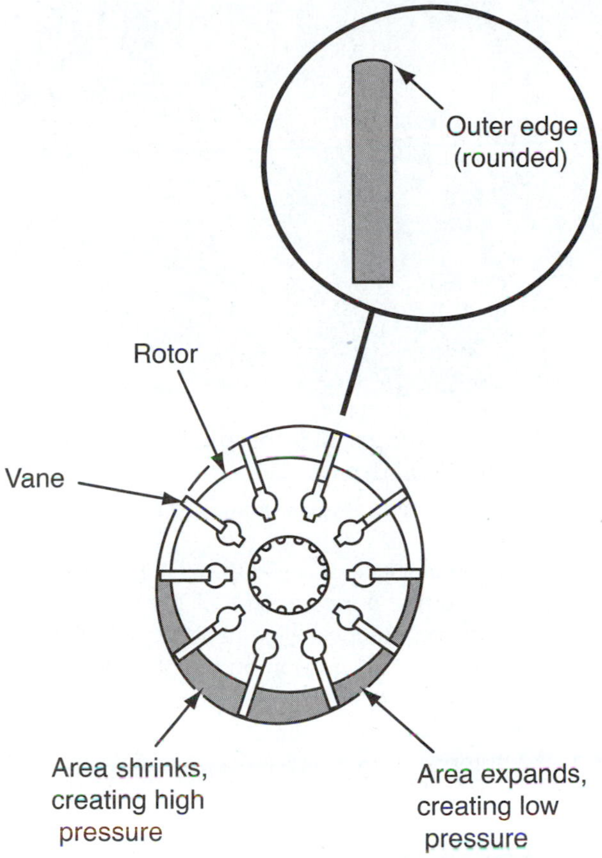

**Figure 13.** Placement of vanes in the pump's rotor.

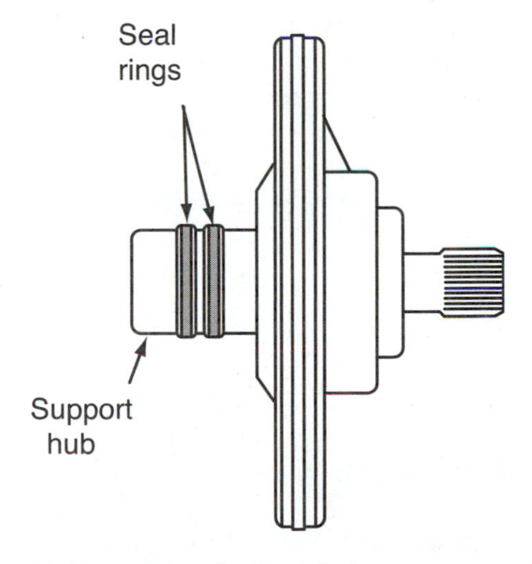


**Figure 14.** Location of oil seal rings on a typical oil pump assembly.

## REASSEMBLY

Assemble the pump by bolting the two halves together and torquing the bolts to specifications **(Figure 15)**. The outer area of most pumps utilizes a rubber seal **(Figure 16)**. Check the fit of the new seal by making sure the seal sticks out a bit from the groove in the pump. If it does not, it will leak. The seal at the front of the pump is always replaced during overhaul. Most of these seals are the metal clad lip seal type. Care must be taken to avoid damage to the seating area when removing the old seal.

Check the area behind the seal to make sure the drain back hole is open to the sump. If this hole is clogged, the new seal will possibly blow out. The drain back hole relieves pressure behind the seal. A loose fitting converter drive hub bushing can also cause the front pump seal to blow out.

Place a layer of clean ATF in two or three spots around the oil pump gasket and position it onto the transmission case. Tighten the pump attaching bolts to specifications in the specified order **(Figure 17)**. After the bolts are tight, check the rotation of the input shaft. If the shaft does not

**Figure 15.** Torque the two bodies of the pump together.

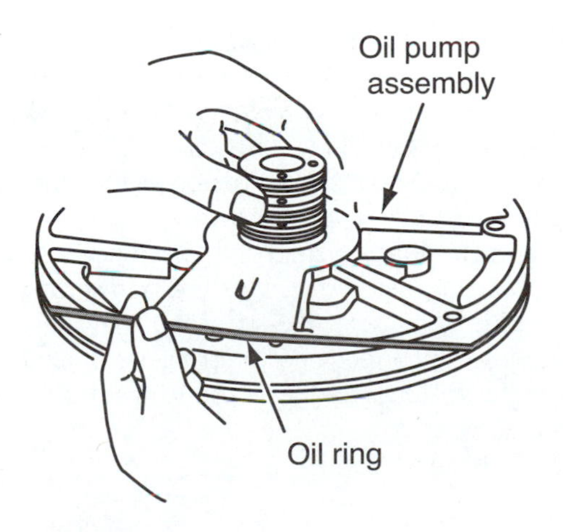


**Figure 16.** An outer oil pump seal.

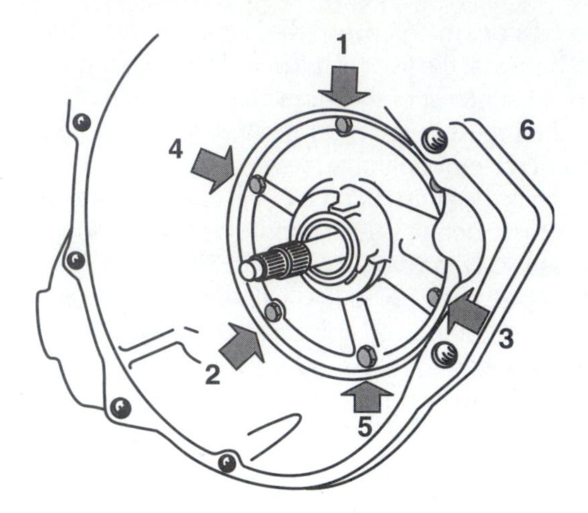


Figure 17. Oil pump bolts must be torqued in the sequence recommended by the manufacturer.

Figure 18. The pump seal is installed with a hammer and seal driver.

rotate, disassemble the transmission to locate the misplaced thrust washer or misaligned friction plate.

The pump seal can be installed with a hammer and seal driver **(Figure 18)**. Apply a very thin coating of RTV sealant around the outside surface of the seal case when installing the seal. Place some transmission fluid in the pocket of the pump housing and install the gears into the housing according to their alignment marks. Align and install the reaction shaft support and tighten the bolts to the specified torque. Make sure the pump is not binding after you have tightened it by using the torque converter to rotate the pump.

## Summary

- The pump should be carefully inspected during any transmission overhaul and especially when low line pressure is measured during a pressure test.
- Before doing anything with the gears or rotors, mark the alignment of the gears.
- A scored cover may be caused by insufficient clearance between the converter's drive hub and the pilot hub. A front pump bushing that has too much clearance may allow the gears to run off center, causing them to wear into the crescent and/or the sides of the pump body.
- On positive displacement pumps, check the clearances between the outer gear and the pump pocket, outer pump gear teeth and the crescent, and inner gear teeth and the crescent.
- A pump should be assembled with all new seals and gaskets.

## Review Questions

1. Why should you make alignment marks in a pump's gears before removing them?
2. Name three things that can happen if the inner gear of a pump is installed upside down.
3. When servicing a variable displacement vane-type pump, Technician A says that the pump rotor, vanes, and slide have selective sizes and may destroy the pump if the correct ones are not used. Technician B says that the outer edge of the vanes should be flat. Who is correct?
   A. Technician A only
   B. Technician B only
   C. Both Technician A and Technician B
   D. Neither Technician A nor Technician B
4. Technician A says that the gears in a gear-type oil pump should be replaced if the outer edges of the teeth are worn flat. Technician B says that all parts of gear-type pumps are selectively sized. Who is correct?
   A. Technician A only
   B. Technician B only
   C. Both Technician A and Technician B
   D. Neither Technician A nor Technician B
5. Technician A says that a converter that had a tight fit at the pilot hub could hold the converter drive hub and

inner gear too far into the pump and allow the gears to run off center, causing them to wear into the crescent and/or the sides of the pump body. Technician B says that if the splines of the stator shaft are distorted, the shaft and the pump cover should be replaced. Who is correct?

A. Technician A only
B. Technician B only
C. Both Technician A and Technician B
D. Neither Technician A nor Technician B

# Chapter 32 General Hydraulic System Service

## *Introduction*

Regardless of how a transmission is electronically controlled, its valve body is the control center. The action of the valve body is merely influenced by the various pressure and shift solenoids. The valve body is the main focus of this chapter on hydraulic system service. Another hydraulic device that may or may not be used, the governor, also is discussed.

## GOVERNOR SERVICE

If tests suggest there is a governor problem, it should be removed, disassembled, cleaned, and inspected. Some governors are mounted internally and the transmission must be removed to service the governor. Others can be serviced by removing the extension housing or oil pan, or by detaching an external retaining clamp and then removing the unit **(Figure 1)**.

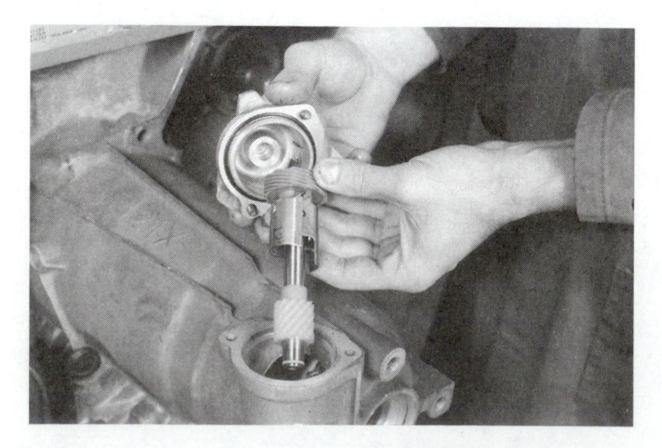

**Figure 1.** A typical governor assembly.

Improper shift points can be caused by a faulty governor or governor drive gear system. However, many transmissions do not rely on the hydraulic signals from a governor; rather, they rely on the electrical signals from sensors, such as speed and load sensors. Faulty electrical components and/or loose connections can cause improper shift points.

If the transmission has a shaft-mounted governor, it is driven by the output shaft and can be accessed by removing the extension housing **(Figure 2)**. Some transaxles require complete disassembly of the transaxle to access the governor. Other transmissions may have a protrusion off the side of the extension housing that contains the governor. These governors are typically driven by a gear and are accessible by removing the governor cover from the protrusion.

**Figure 2.** If the governor is driven directly by the output shaft, it may be accessed by removing the extension housing from the transmission.

## Disassembly

To disassemble a typical gear-driven governor, remove the governor cover by carefully prying it out of its bore. Once the cover has been removed, remove the primary governor valve from its bore in the governor housing. Then remove the secondary valve retaining pin, secondary valve spring, and valve.

Prior to disassembling the governor, it is wise to check the action of the governor valve by moving the weights. With the weights held to the shaft, the exhaust port of the valve should be open. The amount the port is opened can be measured with a feeler gauge. Typically, the port should be open at least 0.020 inch. With the weights held in their fully extended position, the exhaust port should be closed and the inlet port opened. Again, the amount the port is opened can be measured with a feeler gauge. Typically, the port should be open at least 0.020 inch. When moving the weights, pay attention to their movement. They should move freely and return to their rest position without much effort.

Thoroughly clean and dry the governor parts. Test the valve in its bore in the governor housing, it should move freely in the bore without sticking or binding. Also, check the valve for any signs of burning or scoring and replace it, if necessary. Inspect the springs for a loss of tension and burningmarks and replace if necessary. Make sure you check the ports of the governor for any buildups that may restrict fluid flow.

## Reassembly

To reassemble a shaft-mounted governor, place the spring around the secondary valve and insert them into the secondary valve bore. Then insert the retaining pin into the governor housing pinholes. Now, install the primary valve into the governor housing. The governor cover should then be driven in place with a new seal. Make sure you lubricate the seal with ATF before driving the cover into the bore.

> **You Should Know** *Never interchange components of the primary and secondary governors. Also, note that the flat faces of the primary valve must face outward when it is installed.*

To reassemble a shaft-driven governor, use a press and install the drive gear. Then install the weights, springs, valve, and thrust cap. Insert new retaining pins through the thrust cap and weights and crimp both ends of the pins to prevent them from working out.

If the governor assembly was removed from the governor support and parking gear, be sure to tighten the bolts to specifications with a torque wrench. After assembly, install the governor and torque the bolts to specifications. Overtightening can cause the valve to stick. Some transmissions use a drive ball on the output shaft, which locks the governor to the shaft **(Figure 3)**. Make sure it is in place when installing the governor.

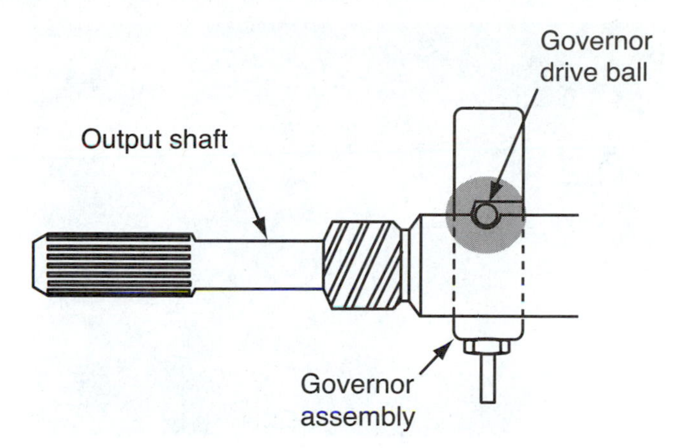


**Figure 3.** Governor drive ball in output shaft.

## VALVE BODY REMOVAL

If the pressure test indicated a problem associated with the valves in the valve body, a thorough cleaning in fresh solvent, careful inspection, and the freeing-up and polishing of the valves may correct the problem. Sticking valves and sluggish valve movements are caused by poor maintenance, use of the wrong type of fluid, and/or overheating the transmission. The valve body of most transmissions can be serviced when the transmission is in the vehicle; however, typically it is serviced when the transmission has been removed for other repairs.

### In-Vehicle Service

Typically, to remove a valve body from a transmission when it is in the vehicle, begin by draining the fluid and removing the oil pan. Then, disconnect the manual and throttle lever assemblies **(Figure 4)**. Carefully remove the detent spring and screw assemblies. Loosen and remove the valve body screws. Before lowering the valve body and separating the assembly, hold the assembly with the valve body on the bottom and the transfer and separator plates on top. Doing this will reduce the chances of dropping the check balls in the valve body **(Figure 5)**. Lower the valve body and note where these steel balls are located in the valve body, then remove them and set them aside, along with the various screws.

> **You Should Know** *To avoid spending hours crawling on the floor looking for lost parts, place your hand or fingers over spring-loaded valves or plugs when removing them.*

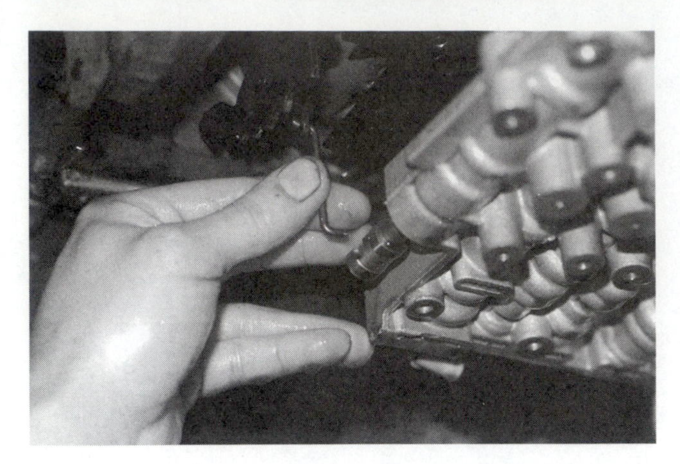

**Figure 4.** Before attempting to remove the valve body, disconnect the linkage from the manual valve.

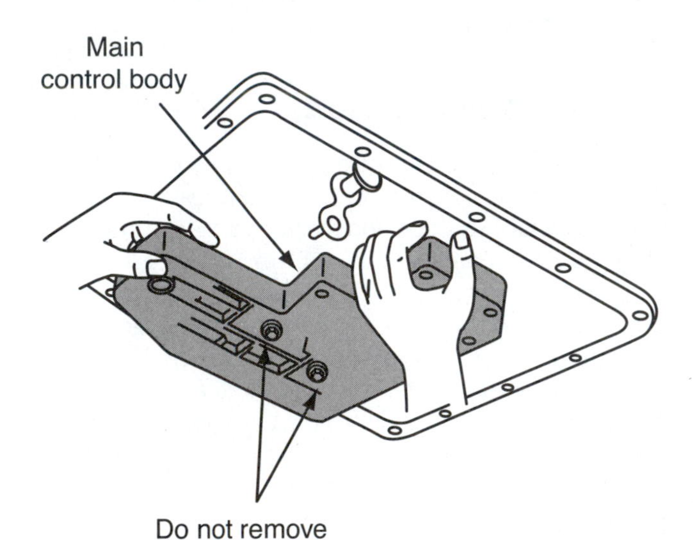


**Figure 5.** When removing a valve body from the transmission, remove only the bolts that are necessary to lower the valve body. Also, be careful not to lose any springs and check balls.

## Bench Service

When removing a valve body from a transmission on a bench, make sure the transmission is positioned so the valve body can be lowered while you keep the check balls in place.

There are many different designs and configurations used in today's transmissions. Each of these requires unique steps for removal and installation. It is extremely important that you refer to the service manual for these specifics. Valve body service procedures for some common transmissions follow.

**Figure 6.** The main parts of a typical Chrysler valve body.

## Chrysler Transaxles

The valve body assembly in Chrysler's 41TE transaxle is comprised of the valve body, a transfer plate, and a separator plate **(Figure 6)**. These control fluid flow to the TCC, the solenoid/pressure switch assembly, and the various hydraulic apply devices.

To remove the valve body, move the manual valve lever into low gear. Loosen and remove the valve body retaining bolts. Then, using a screwdriver, push the park rod rollers out of the guide bracket. Carefully pull the manual shaft from its bore and then remove the valve body assembly from the case.

Once removed, the valve body can be disassembled. This begins with the removal of the TR sensor assembly. The accumulator retaining plate and accumulator are then removed.

The transfer plate can now be unbolted and separated from the valve body. Remove the oil screen and overdrive clutch check valve from the separator plate. Then remove the separator plate. Then remove the thermal valve and check balls from the valve body. Keep track of their location and count the balls as they are removed. Disassemble, inspect, and clean the valve body components as needed.

To install the valve body, guide the park rod rollers into the guide bracket while positioning the valve body. Install the valve body mounting bolts and tighten them to specifications **(Figure 7)**.

## Ford Transmissions

Removing the valve body from a Ford 4R70W transmission is a rather simple process. The connectors to the various solenoids mounted to or around the valve body must be disconnected. Once they are, the wiring harness can be removed by disconnecting it from its main connector. After the harness is removed, the shift and TCC solenoids can be removed. The retaining bolt for the manual valve detent lever spring and the spring are removed **(Figure 8)**. After

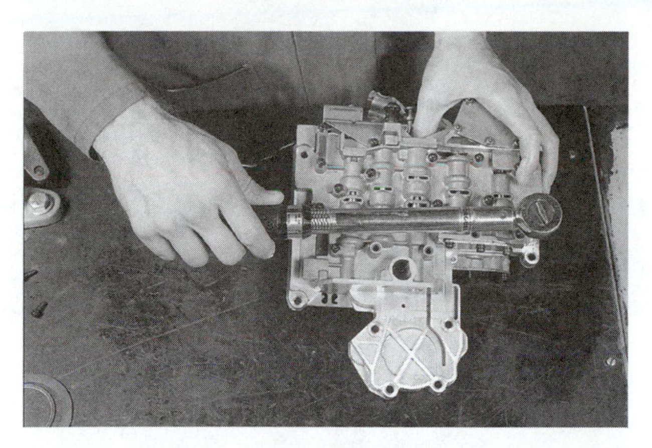

Figure 7. Bolt the valve body together and tighten the bolts to specifications.

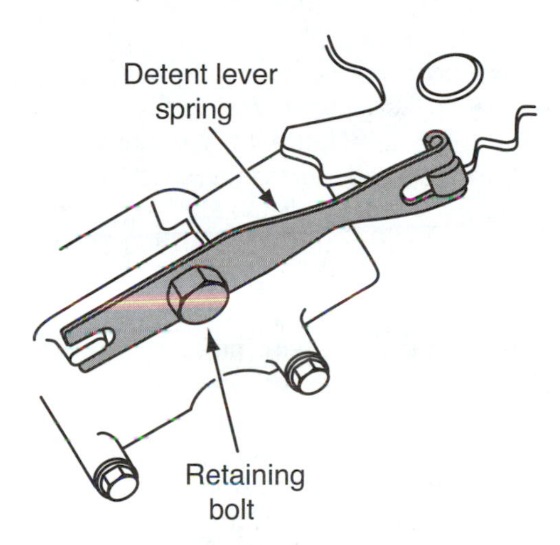


Figure 8. The detent lever spring for the manual valve.

the valve body is unbolted from the case, remove and discard the pump outlet screen.

Remove the separator plate and discard the gaskets. Note the location of the check balls **(Figure 9)** and remove them before cleaning the valve body. Once the valve body is serviced, install the check balls and the separator plate. Install a new pump outlet screen and gaskets during reassembly.

To ensure proper alignment of the valve body, this transmission has two alignment bolts. Position the valve body gasket and valve body onto the case using these alignment bolts as a guide. Loosely install the valve body retaining bolts. Then install the manual valve detent lever spring and tighten it in place.

Now tighten the valve body retaining bolts according to the specified sequence and to the correct torque. Install the shift and TCC solenoids and reconnect the wiring harness to the main connector and to the solenoids.

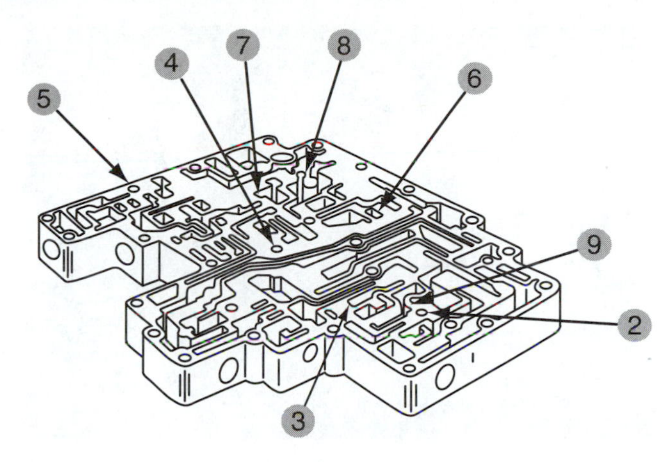


Figure 9. Location of the check balls in a 4R70W transmission.

## General Motors' Transmissions

The removal process of the valve body in a GM 4L60-E begins with disconnecting the electrical connections to the various switches and solenoids mounted to or around the valve body. The order in which the solenoids are removed is important. Some retaining bolts and brackets are covered by the solenoids. Remove the PWM solenoid to gain access to the retaining bolts for the TCC solenoid **(Figure 10)**. Once the TCC solenoid is removed, the electrical harness retaining bolts can be removed.

Now unbolt and remove the **transmission fluid pressure switch** assembly **(Figure 11)**. This switch assembly contains five different pressure switches and is connected to five different hydraulic circuits. It needs to be carefully inspected for any damage or debris.

The manual detent spring can now be unbolted and removed. Now the remaining valve body mounting bolts can be removed. Once loosened, the valve body should be

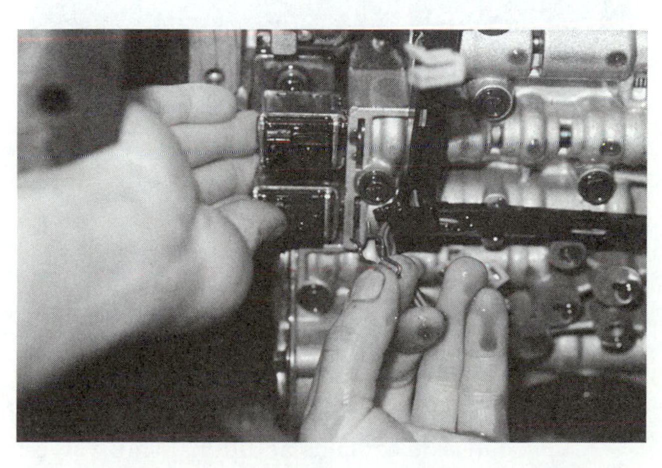

Figure 10. To remove the solenoids, remove the retaining clip and pull the solenoids out.

**Figure 11.** Remove the fluid pressure switch and carefully check it for damage and dirt.

lowered slightly to disconnect the manual valve's linkage. This allows for the removal of the valve body.

Installation is in the reverse order of removal. Again, it is important that the solenoids be installed according to the sequence. All of the valve body mounting bolts should be loosely installed as components are installed, until all of the bolts are in their proper location. Then the bolts should be tightened in the specified order and to the specified torque.

## Honda Transaxles

The valve body assembly in a Honda transaxle is comprised of a main valve body, the regulator valve body, the servo valve body, and the accumulator body. The main valve body contains the pump gears and the manual, modulator, shift, servo control, TCC control, and cooler check valves. The shift valves are controlled by fluid flow directed by shift solenoids.

The regulator valve body contains the regulator, TCC timing, and relief valves. The servo valve body **(Figure 12)** contains additional shift valves, two forward speed accumulators, and the servo valve. The accumulator or servo secondary valve body **(Figure 13)** contains two forward speed accumulators and a lubrication check valve.

**Figure 12.** The servo valve body in a Honda transaxle contains additional shift valves: two forward speed accumulators, and the servo valve.

**Figure 13.** The servo secondary or accumulator valve body contains two forward speed accumulators and a lubrication check valve.

To remove this valve body, the ATF feed pipes must be removed first. Then the fluid strainer and servo detent plate should be removed. Now press down on the accumulator cover while loosening its mounting bolts. Loosen the bolts in a staggered pattern. The cover is spring loaded and the threads in the valve body will strip if the cover is not evenly removed.

Now the servo valve body, servo separator plate, accumulator valve body, and regulator valve body can be removed. After these are removed, the stator shaft and shaft stop should be removed. Now unhook the detent spring from the detent arm and remove the detent arm shaft, detent arm, and control shaft.

Note the location of the cooler check ball and spring and remove them. Once these are removed, the main valve body can be unbolted and removed. When removing the valve body, be prepared to catch the TCC control valve and spring. These are held in position by the valve body.

Remove the pump gears, noting which side of the gears faces up. Now the main separator plate and its dowels can be removed. Clean and inspect all parts. Installation of the valve body assembly follows the reverse order of the removal procedure.

## VALVE BODY SERVICE

If previous tests suggest a problem with only one or two valves, start your inspection at those valves. Doing this will not only save you time but also will reduce the chance of something being misplaced or ruined during a total disassembly. If the transmission had heavily contaminated fluid, the entire valve body should be inspected and cleaned, or replaced.

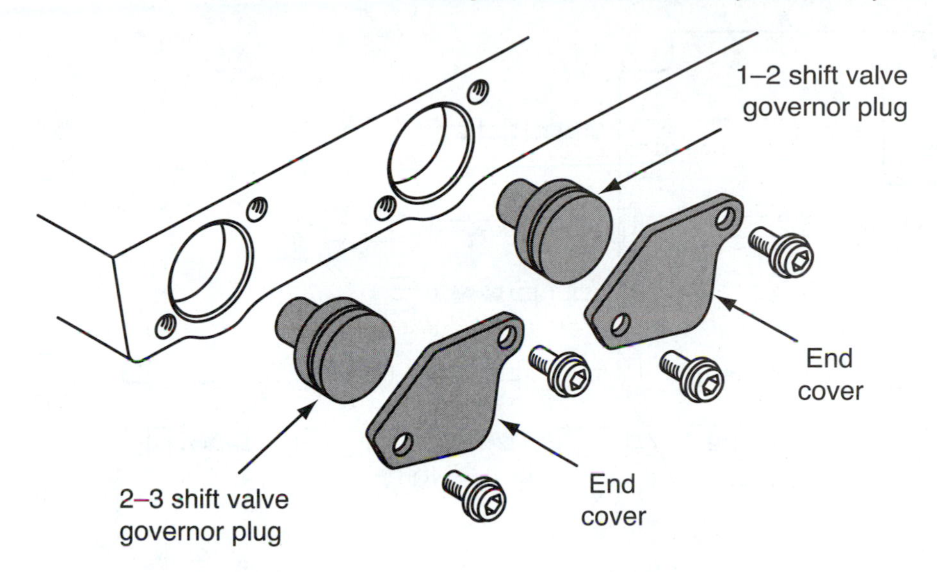


**Figure 14.** Examples of the covers used to retain valves and their springs in the bores of a valve body.

## Disassembly

A valve body contains many valves. These valves are typically held in their bores by a plug or cover plate. The cover plates are bolted or screwed to the valve body **(Figure 14)**. Plugs can be held in place in a number of different ways, all of which can be released by pressing the plug slightly into the bore **(Figure 15)**.

Begin disassembly by removing the manual shift valve from the valve body. Then remove all of the valves and springs from the valve body. Make sure that you keep all springs and other parts with their associated valves.

It is important that you keep track of the position of the springs in relationship to the valves **(Figure 16)**. You can draw diagrams on a piece of paper or use an instant camera to photograph the valve body. Both of these make an excellent reference for reassembly.

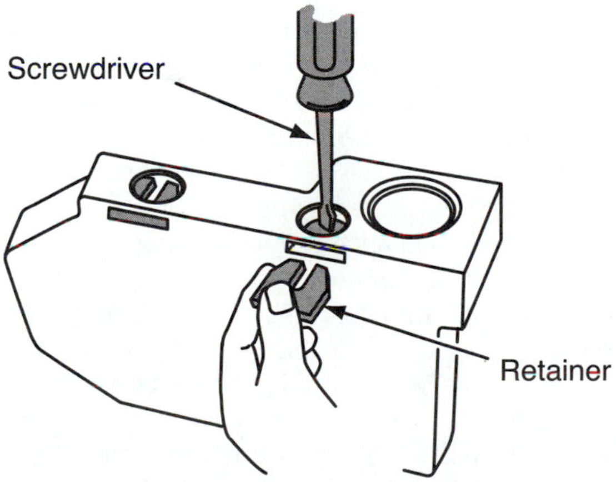

**Figure 15.** To remove or install the retainer for a valve retaining plug, the plug is depressed and the retainer pulled out or inserted into its groove.

**You Should Know**

*Another trick is to use the cardboard sheet included in every gasket set. Fold this sheet in 1-inch pleats like an accordion and lay it on your bench with the slick side of the cardboard facing up. Then follow these steps:*

1. *Take the valves and springs out of the valve body and lay them in the different grooves in the sequence they were removed.*
2. *Clean the valve body castings.*
3. *Clean the valves and springs. Do not put them back on the cardboard; rather, put them directly into their bores in the valve body.*

## Cleaning and Inspection

Remove the check balls and springs. Note their exact location and count them as they are removed. Compare your count to the number given in the service manual. This will ensure that all have been removed and some will not be lost during cleaning. Never use a magnet to remove the check balls. The balls may become magnetized and tend to collect particles in the fluid. Continue disassembly by removing the rest of the valves and springs.

After the valves and springs have been removed, soak the valve body and separator and transfer plates in mineral spirits or lacquer thinner, then wash the parts off with water. Thoroughly clean all parts and make sure all passages within the valve body are clear and free of debris.

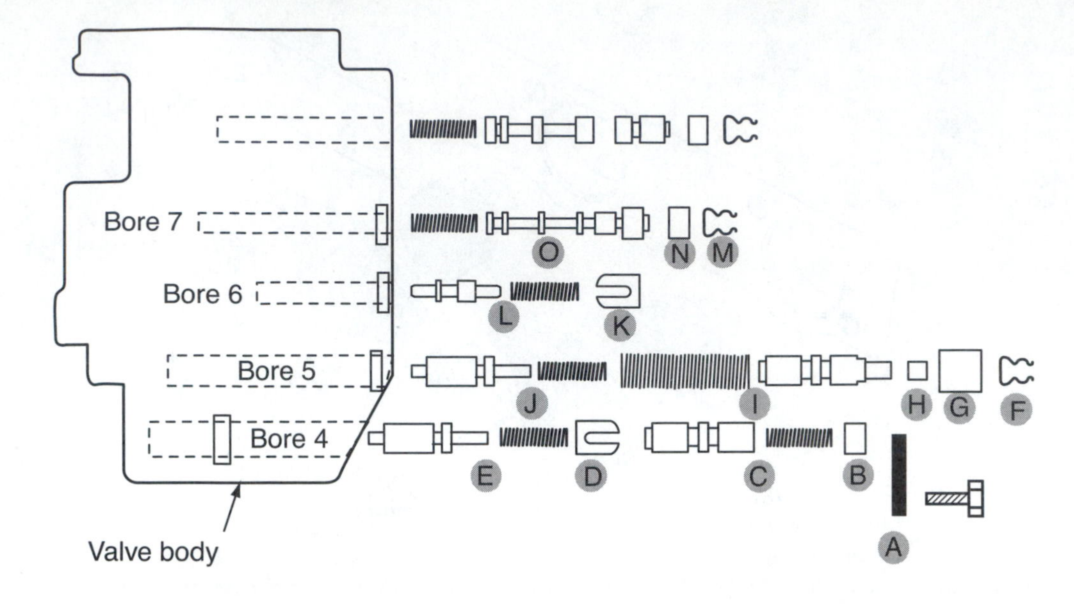


| Bore 4 | Bore 5 | Bore 6 | Bore 7 |
|---|---|---|---|
| A Spring retainer plate | F Clip | K Retainer plate | M Clip |
| B Bore plug | G Sleeve | L TV limit valve and spring | N Bore plug |
| C Orifice control valve and spring | H Plug | | O 1–2 shift valve and spring |
| D Spring retainer plate 1 | I 3–4 shift valve and spring | | |
| E 2–3 capacity modulator valve and spring | J 3–4 TV modulator valve and spring | | |

**Figure 16.** Typical location of some of the valves and springs in a valve body.

Carefully blow-dry each part with compressed air. Never wipe the parts with a rag or paper towel. Lint from either will collect in the valve body passages and cause shifting problems.

> **You Should Know** *Wear safety glasses when drying parts with compressed air. The overspray of solvent and other cleaners can damage your eyes and can lead to blindness.*

Check the separator plate for scratches or other damage. Scratches or score marks can cause oil to bypass correct oil passages and can lead to system malfunction. If the plate is defective in any way, it must be replaced. Check the oil passages in the upper and lower valve bodies for varnish deposits, scratches, or other damage that could restrict the movement of the valves. Make sure all fluid drain back openings are clear and have no varnish or dirt buildup. Check all of the threaded holes and related bolts and screws for damaged threads; replace as needed.

If the valve body has an oil strainer, make sure it is clean. This can be checked by pouring clean ATF through the strainer and observing the fluid flow out of the strainer. If there is poor flow, the strainer should be replaced.

Examine each valve for nicks, burrs, and scratches. If the valve lands have worn areas, this means the valve has been rubbing in its bore. Make sure each valve properly fits into its respective bore. To do this, hold the valve body vertically and install an unlubricated valve into its bore. Let the valve fall of its own weight into the valve body until the valve stops. Then place your finger over the valve bore and turn the valve body over. The valve should again drop by its own weight. If the valve moves freely under these conditions, it will operate freely with fluid pressure. Repeat this test on all valves.

If the valves do not move freely, the problem may be corrected by polishing the valve lands if the valves are steel. If the valves are aluminum, the valve body should be replaced. To polish a valve, use a polishing (Arkansas) stone or crocus cloth. Let the cloth soak in ATF before using it. Evenly rotate the valve on the polishing material; make sure this does not round the edges of the valve. Polish the valves only enough to ensure free movement in the bore.

After the valve is polished, it must be thoroughly cleaned to remove all of the cleaning and abrasive materials. After the valve has been recleaned, it should be tested in its bore again.

If the valve still cannot move freely in its bore, the valve body should be replaced. Individual valve body parts are usually not available. Individual valves are lapped to a particular valve body and therefore if any parts need to be replaced, the entire valve body must be replaced.

The valve bores need attention as well. Roll up a half sheet of ATF-soaked 600+ grit sandpaper. Insert the paper into the valve's bore. Turn the roll of paper so it unrolls and expands to the size of the bore. Then twist the paper while moving it in and out of the bore. Then clean the entire valve body in solvent and dry it with compressed air.

Although desirable to have the valves move freely in their bores, excessive wear is a problem. There should never be more than 0.001 inch clearance between the valve and its bore. If either is worn, the entire valve body needs to be replaced.

Check each spring for signs of distortion. If any spring is damaged, the valve body should be replaced. To check the springs lay them on their side and roll them. If they roll true, they are not distorted. If they wobble, they are.

With a straightedge laid across the sealing surface of the valve body, use a feeler gauge to check its flatness. If it is slightly warped, it can be flat filed. Be very careful when doing this. Keep the file flat and always file in one direction. If the surface is warped beyond repair, the valve body must be replaced.

## Reassembly

After the valve body has been cleaned, it should air dry. Then it should be dipped into a pan of clean ATF. While it is soaking in ATF, locate the installation specifications in the service manual.

Lubricate all parts with clean ATF. Then install the valves and associated springs into their bores. It is important that you place the valve retaining plugs or caps in the correct bore and in the correct direction **(Figure 17)**. Once

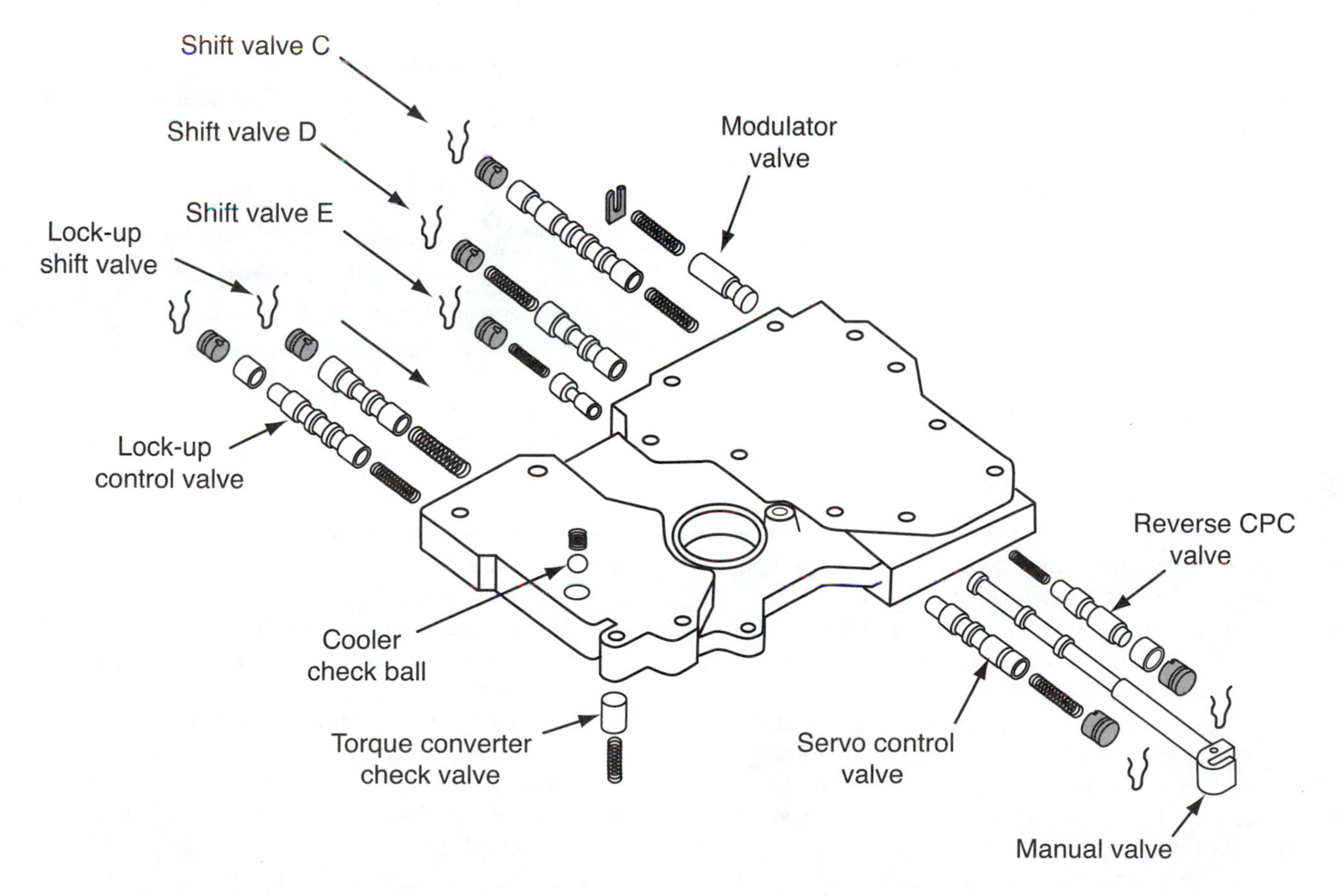


**Figure 17.** The plugs used to retain the valves in a valve body have designated bores and must face the correct direction.

> **You Should Know** *When assembling a valve body, always have your notes, photograph, and the correct service manual close for quick reference. Many valves and springs look similar; however, each has its own bore and purpose.*

the cap is in position, carefully depress it with a small screwdriver and install the retaining clip.

Install all check balls and springs in their correct location. If you have any doubts as to where they should be placed, refer to the service manual. Count the check balls as you install them to make sure you have inserted all of them.

Before beginning to install the valve body, check the new valve body gasket to make sure it is the correct one by comparing it to the old gasket. If the gasket appears to be the correct one, lay it over the separator plate and hold it up to a light, making sure no oil holes are blocked. Also check to make sure the gasket seals off the worm tracks and will not allow the fluid to go where it should not go. Then install the bolts to hold valve body sections together and the valve body to the case. Tighten the bolts to the torque specifications to prevent valve body warpage and possible leakover. Overtorquing also can cause the bores to distort, which would not allow the valves to move freely once the valve body is tightened to the transmission case.

Many transmissions use bolts of various lengths to secure the valve body to the case. It is important the correct length bolt is used in each bore. It is so important that service manuals list the size and location of the mounting bolts **(Figure 18)**.

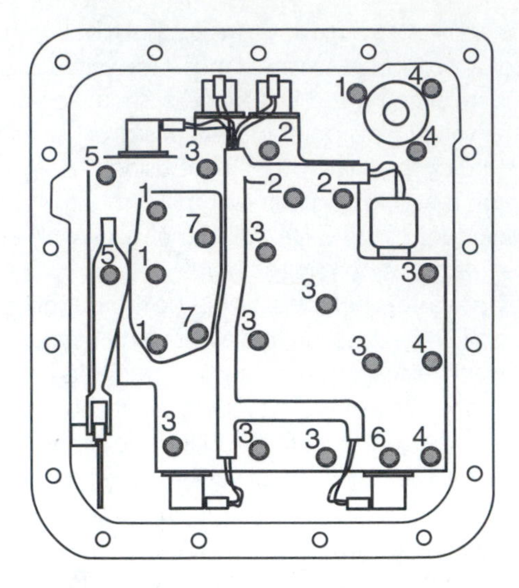

Each numbered bolt location corresponds to a specific bolt size and length, as indicated by the following:

1. M6 X 1.0 X 65.0
2. M6 X 1.0 X 54.4
3. M6 X 1.0 X 47.5
4. M6 X 1.0 X 35.0
5. M8 X 1.25 X 20.0
6. M6 X 1.0 X 12.0
7. M6 X 1.0 X 18.0

**Figure 18.** When installing the valve body mounting bolts, make sure you use the correct bolt size and length in the correct location. Refer to the service manual for guidance.

> **You Should Know** *The primary cause of valve sticking is the over-tightening of the valve body bolts. Always be careful when handling a valve body; they are very precise components.*

## Summary

- If tests suggest there was a governor problem, it should be removed, disassembled, cleaned, and inspected.
- Improper shift points are typically caused by a faulty governor, governor drive gear system, faulty electrical components, and/or loose connections.
- If there is a problem associated with the valve body, a thorough cleaning, careful inspection, and/or freeing up and polishing of the valves may correct the problem.
- Some valve bodies are actually many separate assemblies. Each one should be inspected and cleaned.
- When removing a valve body and/or separating its units, pay attention to the location and count of the check balls.
- Valves are held in their bores by a plug, cover plate, and/or retaining clips.
- After the valves and springs have been removed from the valve body, soak the valve body and separator and transfer plates in mineral spirits.
- Examine each valve for nicks, burrs, and scratches.
- If any spring is damaged, the valve body should be replaced.
- Check the flatness of all sealing and mounting surfaces of the valve body.
- Lubricate all parts with clean ATF.

## Review Questions

1. When removing scratches in a valve, Technician A uses a flat file to remove the scratch. Technician B uses a sand blaster or glass bead machine to polish the surface of the valve. Who is correct?
   A. Technician A only
   B. Technician B only
   C. Both Technician A and Technician B
   D. Neither Technician A nor Technician B
2. Technician A says that overtorqueing the hold-down bolts of the valve body can cause the valves to stick in their bore. Technician B says that flat filing the surface of the valve body will allow the valve body to seal properly and will therefore allow the valves to move freer in their bores. Who is correct?
   A. Technician A only
   B. Technician B only
   C. Both Technician A and Technician B
   D. Neither Technician A nor Technician B
3. When assembling a valve body after cleaning it, Technician A lubricates all parts with clean ATF. Technician B says that it is important that the valve retaining plugs or caps be placed in the correct bore and in the correct direction. Who is correct?
   A. Technician A only
   B. Technician B only
   C. Both Technician A and Technician B
   D. Neither Technician A nor Technician B
4. Technician A says that if diagnosis suggests a problem with only one or two valves, visual inspection of the valve body should start at those valves. Technician B says that if the transmission had heavily contaminated fluid, the entire valve body should be inspected and cleaned or replaced. Who is correct?
   A. Technician A only
   B. Technician B only
   C. Both Technician A and Technician B
   D. Neither Technician A nor Technician B
5. Technician A says that if a valve cannot be cleaned well enough to move freely in its bore, the valve body should be replaced. Technician B says that if there is even the slightest bit of damage or varnish buildup in a valve's bore, the entire valve body must be replaced. Who is correct?
   A. Technician A only
   B. Technician B only
   C. Both Technician A and Technician B
   D. Neither Technician A nor Technician B
6. Technician A says that if the governor pressure is slow to build, early upshifts will result. Technician B says that if line pressure is higher than normal, early shifts will take place. Who is correct?
   A. Technician A only
   B. Technician B only
   C. Both Technician A and Technician B
   D. Neither Technician A nor Technician B
7. Which of the following is NOT a common cause for sticking valves and sluggish valve movements?
   A. Overtorqued value body bolts
   B. A faulty pressure control solenoid
   C. The use of the wrong type of fluid
   D. Overheating the transmission
8. Technician A says that if aluminum valves are scored or otherwise damaged, the individual valve or entire valve body should be replaced. Technician B says that problems rarely result from excessive clearance between the valve and its bore. Who is correct?
   A. Technician A only
   B. Technician B only
   C. Both Technician A and Technician B
   D. Neither Technician A nor Technician B

# Gears and Shafts

## *Introduction*

The overall reliability of a transmission depends on the structural integrity of the shafts and gears. Therefore these parts, as well as all thrust washers, bearings, and bushings **(Figure 1)**, should be carefully inspected and replaced if damaged or worn.

### THRUST WASHERS, BUSHINGS, AND BEARINGS

A thrust washer is designed to support a **thrust load** and keep parts from rubbing together, such as planetary gearsets **(Figure 2)** and transfer and final drive assemblies **(Figure 3)**. **Selective thrust washers** come in various thicknesses to take up clearances and adjust end play at many different locations. Thrust washers and thrust bearings are used wherever rotating parts must have their end play maintained. To control the end play of nonrotating parts, selective shims or spacers are used.

Flat thrust washers and bearings have a fixed thickness and are used throughout a transmission. Typically, these are numbered by the manufacturer for easy identification **(Figure 4)**. Thrust washers should be inspected for scoring, flaking, and wear. Flat thrust washers also should be checked for broken or weak tabs **(Figure 5)**. These tabs are critical for holding the washer in place. On metal thrust washers, the tabs may appear cracked at the bend; however, this is a normal appearance. Plastic thrust washers will not show wear unless they are damaged. The only way to check their wear is to measure the thickness with a micrometer and compare them to a new part. All damaged and worn thrust washers and bearings should be replaced.

Proper thrust washer thicknesses are important to transmission operation. After following the recommended procedures for checking the end play of various components, refer to the manufacturer's chart for the proper thrust plate thickness for each application.

Use a petroleum jelly-type lubricant to hold thrust washers in place during assembly. If they move, end play may be affected. Besides petroleum jelly, there are special greases designed just for automatic transmission assembly that may work fine.

> **You Should Know** *Never use white lube or chassis lube. These greases will not mix in with ATF and can plug up orifices, passages, and even hold check balls off their seats.*

#### Bearings

All bearings should be checked for roughness before and after cleaning. Carefully examine the inner and outer races, and the rollers, needles **(Figure 6)**, or balls for cracks, pitting, **etching**, or signs of overheating.

Sprag and roller clutches should be inspected in the same way as bearings. Check their operation by attempting to rotate them in both directions **(Figure 7)**. If working properly, they will allow rotation in one direction only. Also, visually inspect each spring of the clutch unit.

#### Bushings

Bushings should be inspected for pitting and scoring. Always check the depth of installed bushings and the

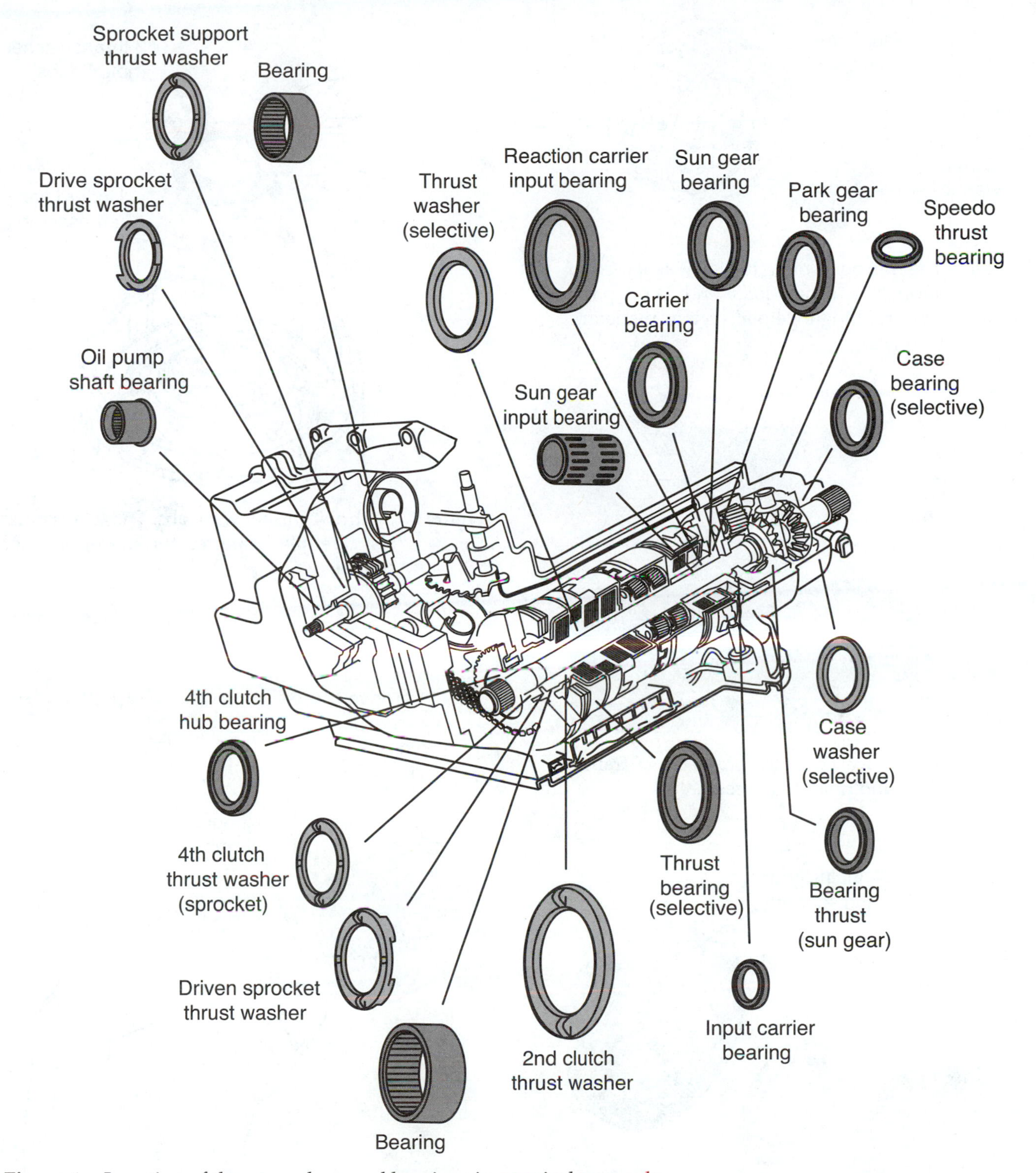


**Figure 1.** Location of thrust washers and bearings in a typical transaxle.

direction of their oil groove, if so equipped, before you remove them. Many bushings used in the planetary gearing and output shaft areas have oiling holes in them. Be sure to line these up correctly during installation, or you may block off oil delivery and destroy the gear train. If any damage is evident on the bushing, it should be replaced.

Bushing wear can be visually checked as well as checked by observing the lateral movement of the shaft that fits into the bushing. Any noticeable lateral movement indicates wear and the bushing should be replaced. Normally, bushings must

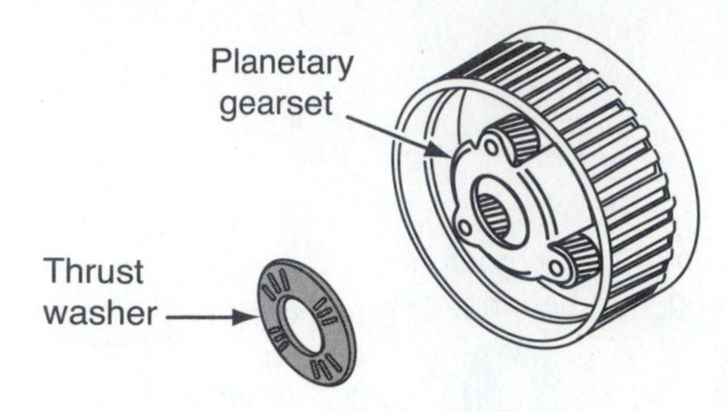


**Figure 2.** The purpose of a thrust washer is to support a thrust load and keep parts from rubbing together, such as planetary gearsets.

**Figure 3.** Thrust washers also are used in transfer shaft and final drive assemblies.

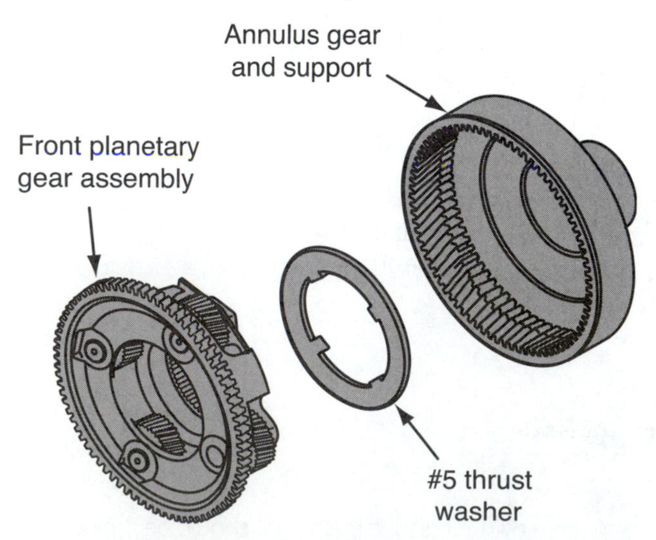


**Figure 4.** This is the #5 thrust washer for a transaxle. Flat thrust washers and bearings are numbered by the manufacturer for easy identification.

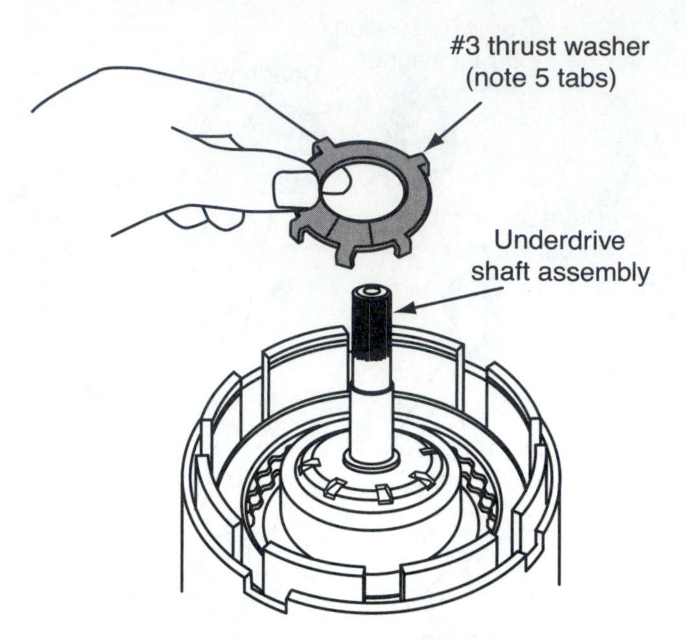


**Figure 5.** A thrust washer with tabs. These tabs must be carefully inspected for cracks or other damage.

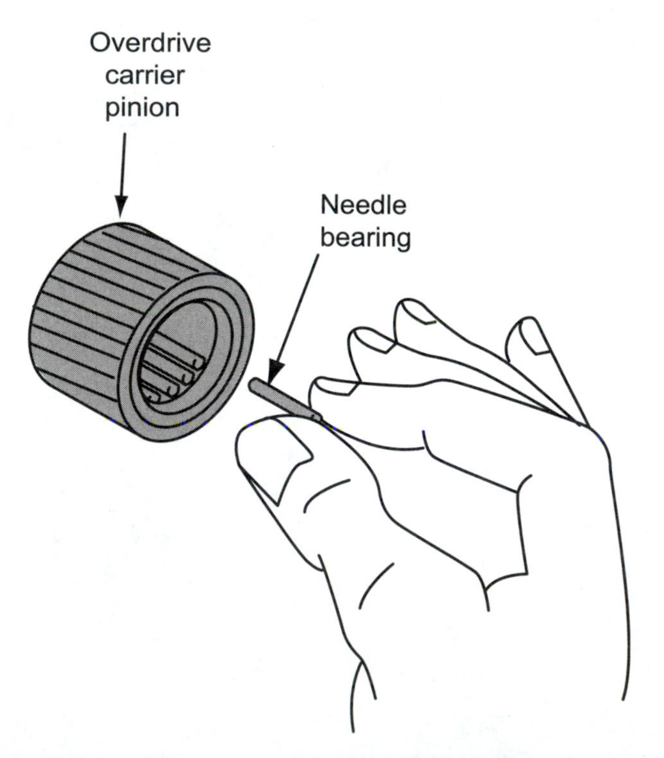


**Figure 6.** Needle bearings are often located inside the small pinion gears. Coating the inside of the gear with petroleum jelly before the needle bearings are positioned will help to keep them in place.

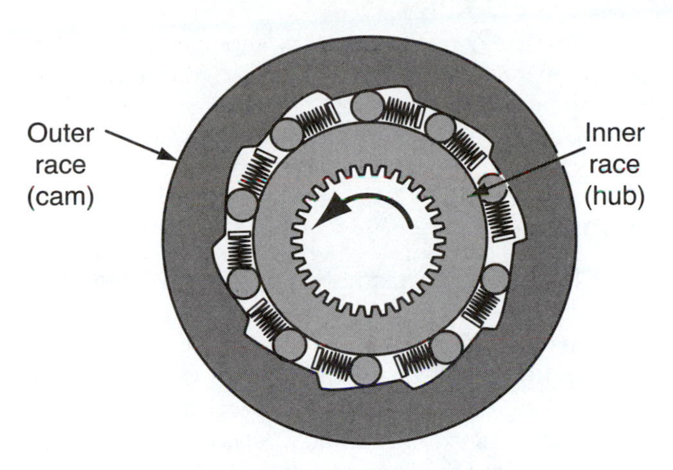


Figure 7. To check the action of a one-way clutch, hold the inner race and rotate the clutch in both directions. The clutch should rotate smoothly in one direction and lock in the other direction.

fit the shafts they ride on with about a 0.0015–0.003-inch clearance. You can check this fit by measuring the inside diameter of the bushing and the outside diameter of the shaft with a dial caliper or micrometer **(Figure 8)**. This is a critical fit throughout the transmission, especially at the converter drive hub, where 0.005 inch is the desired fit.

Most bushings are press-fit into a bore. They are normally driven out of the bore with a properly sized bushing tool. Some bushings can be removed with a slide hammer fitted with an expanding or threaded fixture that grips to the inside of the bushing. Another way to remove bushings is to carefully cut one side of the bushing and collapse it. Once collapsed, the bushing can be easily removed with a pair of pliers. Small-bore bushings located in areas where it is difficult to use a bushing tool can be removed by tapping the inside bore of the bushing to match those of a bolt that fits into the bushing. After the bushing has been tapped, insert the bolt and use a slide hammer to pull the bolt and bushing out of its bore.

All new bushings should be installed with the proper bushing driver and a press. These tools prevent damage to the bushing and allow for proper seating of the bushing in its bore.

## PLANETARY GEAR ASSEMBLIES

Inspection of the planetary gearset is important, as the type of wear or damage often will identify the cause or the cause of other problems.

### General Gear Wear

All new gear teeth have slight imperfections, but these normally disappear during break-in as the teeth are oiled and become polished. When a lack of lubrication or other factors cause a gear to fail, a thorough visual inspection can normally determine the cause of failure.

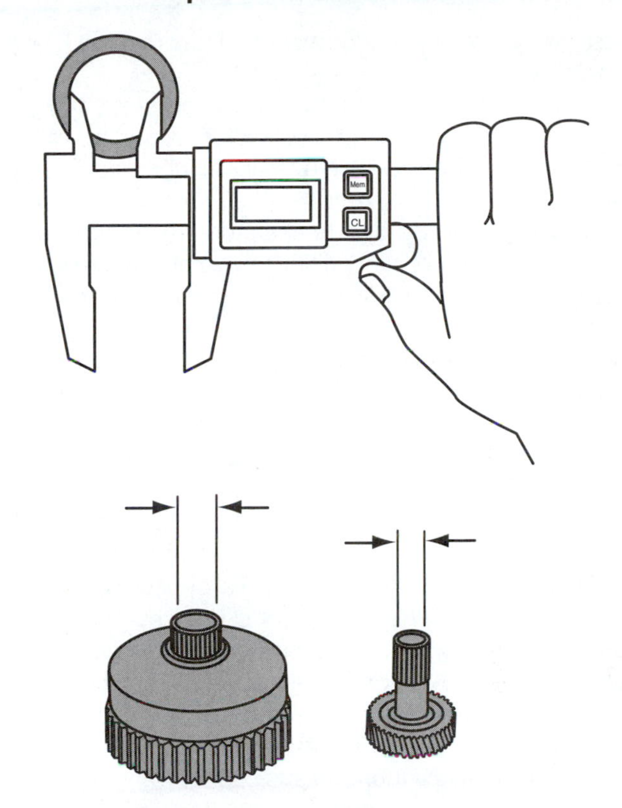

Figure 8. The inside diameter of bushings should be measured for wear with a caliper or a telescoping gauge and a micrometer.

The teeth of a normally worn gear have a polished surface that extends the full length of the tooth from the bottom to the tip of the tooth. Excessive wear and/or grooves on the teeth are normally caused by fine particles carried in the lubricant or embedded in the tooth surfaces. The usual sources of these fine particles are metal particles from gear teeth, abrasives left in the gear case, or sand and scale from the housing casting.

Scratches on the surface of the teeth also can be caused by suspended or embedded particles. However, this normally occurs when the particles are larger than the fine abrasives that cause excessive wear. These particles are normally metal pieces from the gears themselves.

If the contact surface of the teeth is worn smoothly, the gears have been overloaded and metal has been removed by the sliding pressure of the meshing gears.

Rolling is the result of overloading and sliding, which leaves a burr on the edge of the teeth. Insufficient bearing support results in rolling of the metal because of the sliding pressure.

**Peening** is the result of excessive backlash, which causes the teeth to contact others with so much impact they look as if they were beaten with a hammer. The impact forces the lubricant out of the teeth and the gears mesh without a protecting layer of oil, causing the heavy contact.

A wavy surface or "fish scales" on the teeth at right angles to the direction of the sliding action is called "rippling." This may be caused by a lack of lubrication, heavy loads, or vibration.

**Scoring** is caused by a temperature rise and thinning or rupture of the lubricant's film because of extreme loads. Pressure and sliding action heats the gear and permits the metal surface to transfer from one tooth to the face of another. As the process continues, chunks of metal loosen and gouge the teeth in the direction of the sliding motion **(Figure 9)**.

Pitting is a condition normally associated with a thin oil film, possibly because of high oil temperatures. A very small amount of pitting causes a gray appearance on the teeth.

Spalling is a condition that starts with fine surface cracks and eventually results in large flakes or chips coming off the tooth face. This typically is caused by improperly hardened teeth. Spalling may occur on only one or two teeth, but the chips may then damage the other teeth.

Corrosive wear results in an erosion of the tooth surfaces by acid. The acid is the result of moisture combining with impurities in the lubricant and contaminants in the air. Normally, the surfaces initially become pitted and then become chipped or spalled.

Evidence of high heat or burning is usually caused by failure of the lubricant or by a lack of lubrication. During high stress, friction causes excessive heat and the temperature limits of the metal are exceeded. Burned gear teeth become extremely brittle and are easily broken.

Broken teeth can be caused by high impact forces, overloading, fatigue, or defective manufacturing processes. If the break shows fresh metal over the break, an impact overload was probably the cause. If the break shows an area in the center of fresh metal with the edges dark and old looking, the breakage was because of fatigue, which started with a fine surface crack.

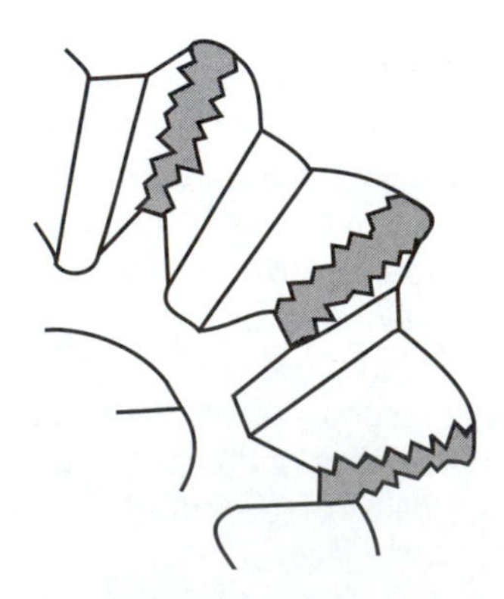

**Figure 9.** Scoring is caused by a temperature rise and thinning or rupture of the lubricant's film because of extreme loads.

Cracks are normally caused by improper heat-treating during manufacturing or improperly machined tooth dimensions. Most cracks resulting from improper heat-treating are extremely fine and will not show up until a gear has been used for some time.

## Inspection

All planetary gear teeth should be inspected for chips or stripped teeth. Any gear that is mounted to a splined shaft needs to be checked for mutilated or shifted splines **(Figure 10)**. Helical gears have many advantages over straight-cut gears, such as providing low operating noise, but the end play of the individual gears must be checked. The helical cut of the gears makes them thrust to one side during operation, and this can put a lot of load on the thrust washers.

Look first for obvious problems like blackened gears or pinion shafts. These conditions indicate severe overloading and require that the carrier be replaced. Occasionally, the pinion gear and shaft assembly can be replaced individually. When looking at the gears themselves, a bluish condition can be a normal condition, as this is part of a heat-treating process used during manufacture. Check the planetary pinion gears for loose bearings. Check each gear individually by rolling it on its shaft to feel for roughness or binding of the needle bearings. Wiggle the gear to be sure it is not loose on the shaft. Looseness will cause the gear to whine when it is loaded. Inspect the gears' teeth for chips or imperfections, as these also will cause whine.

Check the gear teeth around the inside of the front planetary ring gear. Check the fit between the front planetary carrier to the output shaft splines. Remove the snapring and thrust washer from the front planetary ring gear. Examine the thrust washer and the outer splines of the front drum for burrs and distortion. The rear clutch friction discs must be able to slide on these

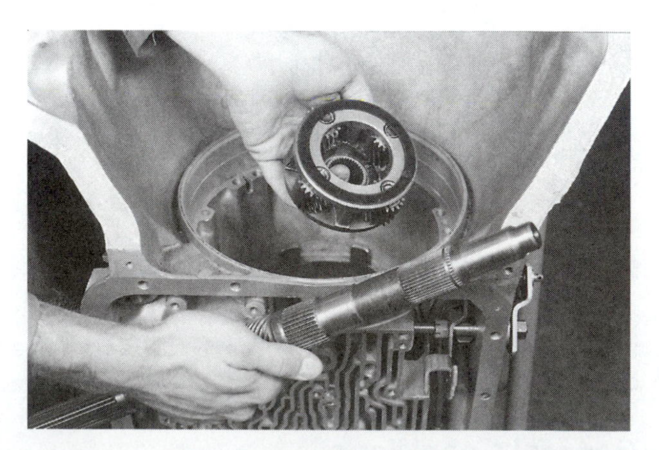

**Figure 10.** When a planetary member is splined to a shaft, the splines in the gear and on the shaft should be checked for mutilated or shifted splines.

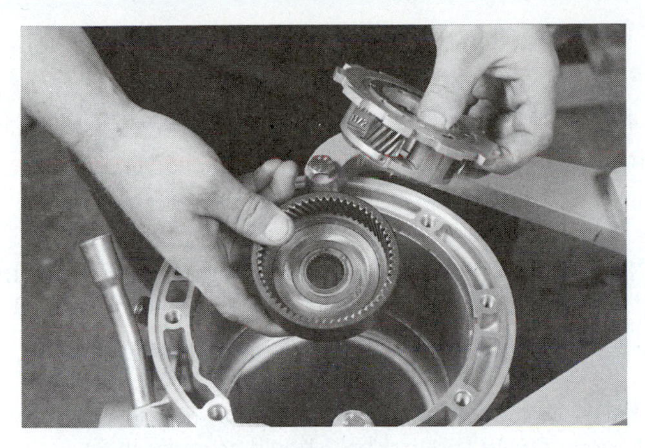

Figure 11. With the carrier and ring gear separated, carefully inspect both of them.

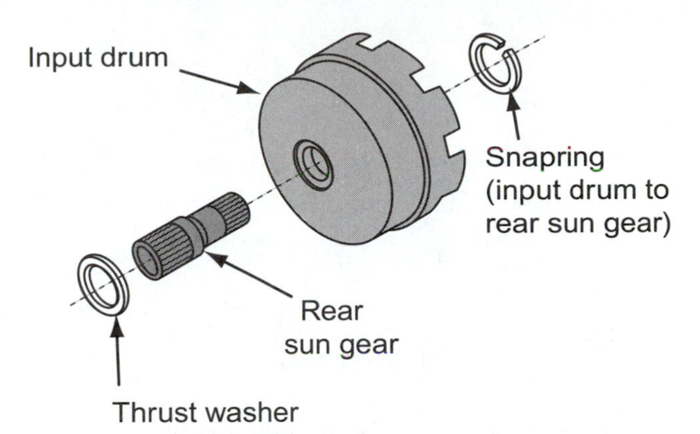


Figure 13. The fit of all drums onto the splines of their mating shafts should be checked.

splines during engagement and disengagement. With the snapring removed, the front planetary carrier can be removed from the ring gear **(Figure 11)**. Check the planetary carrier gears for end play by placing a feeler gauge between the planetary carrier and the planetary pinion gear **(Figure 12)**. Compare the end play to specifications. On some Ravigneaux units, the clearance at both ends of the long pinion gears also must be checked and compared to specifications.

Check the splines of the sun gear. Sun gears should have their inner bushings inspected for looseness on their respective shafts. Also, check the fit of the sun shell to the sun gear **(Figure 13)**. The shell can crack where the gears mates with the shell. The sun shell also should be checked for a bell-mouthed condition where it is tabbed to the clutch drum. Any variation from a true round should be considered junk and should not be used. Look at the tabs and check for the best fit into the clutch drum slots. This involves trial fitting the shell and drum at all the possible combinations and marking the point where they fit the tightest. A snug fit here will eliminate bell mouthing because of excess play at the tabs. It also can reduce engagement noise in reverse, second, and fourth gears. This excess play allows the sun shell tabs to strike the clutch drum tabs as the transmission shifts from first to second or when the transmission is shifted into reverse.

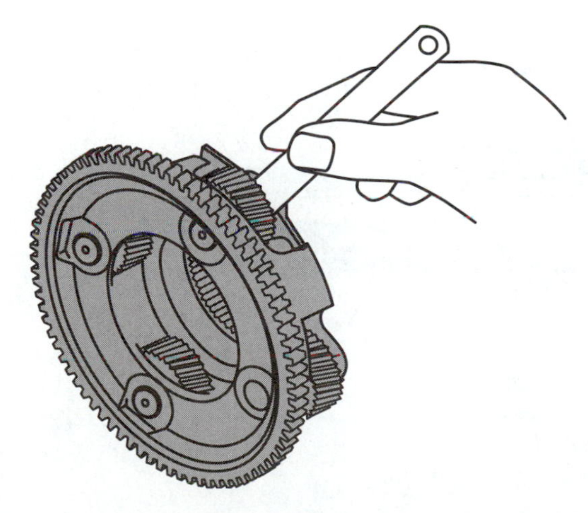

Figure 12. The clearance between the pinion gears and the planetary carrier should be checked and compared to specifications.

The gear carrier should have no cracks or other defects. Replace any abnormal or worn parts. Check the thrust bearings for excessive wear and, if required, correct the input shaft thrust clearance by using a washer with the correct thickness. Determine the correct thickness by measuring the thickness of the existing thrust washer and comparing it to the measured end play. Now move the gear back and forth to check its end play. Some shop manuals will give a range for this check but, if none can be found, you can figure about 0.007 to 0.025 inch as an average amount. All the pinions should have about the same end play.

## SHAFTS

Carefully examine the area on all shafts that rides in a bushing, bearing, or seal. Check the entire length of the shaft for signs of overheating and other damage. Inspect the splines for wear, cracks, or other damage **(Figure 14)**. A quick way to determine spline wear is to fit the mating splines and check for lateral movement.

Shafts are checked for scoring in the areas where they ride in bushings. As the shaft is a much harder material than the bushings, any scoring on the shaft would indicate a lack of lubrication at that point. The affected bushing should appear worn into the backing metal. Because shaft-to-bearing fit is critical to correct oil travel throughout the transmission, a scored shaft should be replaced. Lubricating oil is carried through most shafts; therefore, an internal inspection for debris is necessary. A blocked oil delivery

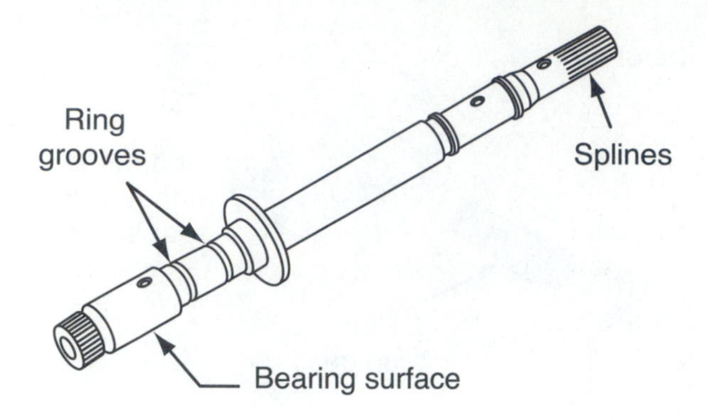


Figure 14. All shafts, including their splines and ring grooves, should be carefully inspected for wear or other damage.

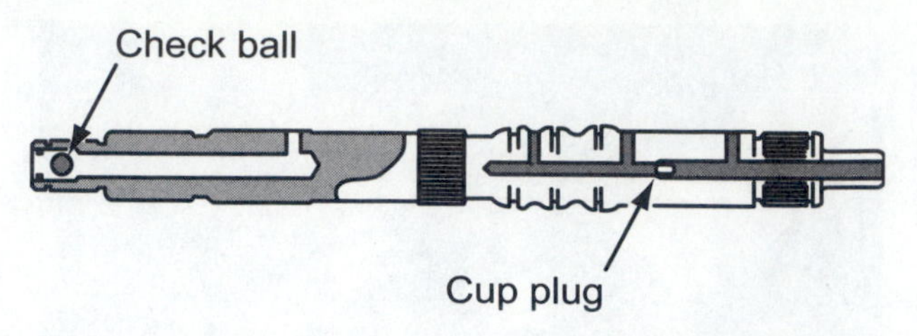


Figure 15. If the turbine shaft is fitted with a check ball, make sure it is able to seat and unseat.

hole can starve a bushing, resulting in a scored shaft. The internal oil passage of a shaft may not be able to be visually inspected; only observation during cleaning will give an indication of the openness of the passage. Washing the shaft passage out with a solvent and possibly running a piece of small diameter wire through the passage will dislodge most particles. Be sure to check that the ball that closes off the end of the shaft, if the shaft is so equipped, is securely in place. A missing ball could be the cause of burned planetary gears and scored shafts because of a loss of oil pressure. Any shaft that has an internal bushing should be inspected as described earlier. Replace all defective parts as necessary.

Shafts should be checked for wear in the ring groove area. Make sure there is no step wear in the groove and that the sides and bottom are square. Also, make sure the groove is not too wide for the ring. If a 0.005 inch feeler gauge will fit in the groove with the ring in place, the groove is worn and the shaft should be replaced.

Often shafts are supported by bushings and journal areas on the shafts should be free of noticeable wear. Small scratches can be removed with 320-grit emery cloth. Grooved or scored shafts require replacement. The splines should not show any sign of waviness along their length. Check drilled shafts to be sure the bore is open and free of any foreign material. Wash out the shaft with solvent and run a small diameter wire through the shaft to dislodge any particles. Then, wash out the shaft once more and blow it out with compressed air.

If the shaft has a check ball, such as the 4L60 turbine shaft **(Figure 15)**, be certain the ball seats in the correct direction. This particular check ball controls oil flow direction to the converter. Some shafts have a ball pressed into one end to block off one end of the shaft. This is used to hold oil in the shaft so the oil is diverted through holes in the side of the shaft. These holes supply oil to bushings, one-way clutches, and planetary gears. If the ball does not fully block the end of the shaft, oil pressure can be lost, causing failure of these components. Some shafts may be used to support another shaft **(Figure 16)**, as in the GM

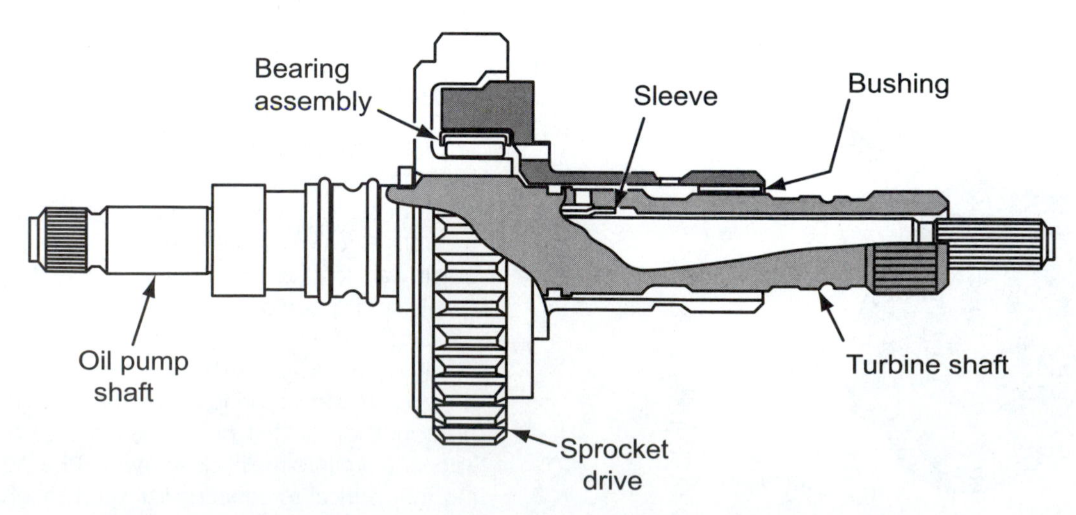


Figure 16. Some transmission shafts support another shaft through the bushings fitted to the inside diameter of one shaft. These bushings should be carefully inspected.

4L30. The output shaft uses the rear of the input shaft to center and support itself. The small bushing found in the front end of the output shaft always should be replaced on this transmission during rebuild. If the input shaft pilot is worn or scored, a replacement shaft will be necessary.

All hubs, drums, and shells should be carefully examined for wear and damage. Especially look for nicked or scored band application surfaces on drums, worn or damaged lug grooves in clutch drums, worn splines, and burned or scored thrust surfaces. Minor scoring or burrs on band application surfaces can be removed by lightly polishing the surface with a 600-grit crocus cloth. Any part that is heavily scored or scratched should be replaced.

## PARKING PAWL

The parking pawl assembly **(Figure 17)** can be inspected after the transmission is disassembled, or on some transmissions when the transmission is still in the vehicle. Examine the engagement lug on the pawl; make sure it is not rounded off. If the lug is worn, it may allow the pawl to slip out or not fully engage in the parking gear. Most parking pawls pivot on a pin. This also needs to be checked to make sure there is no excessive looseness at this point. The spring that pulls the pawl away from the parking gear also must be checked to make sure it can hold the pawl firmly in place. In addition, check the position and seating of the spring to make sure it will remain in that position during operation.

The push rod or operating shaft **(Figure 18)** must provide the correct amount of travel to engage the pawl to the gear. Make sure the shaft is not bent or that the pivot holes in the internal shift linkage are not worn oblong. Also, make sure the bushing or sleeve that supports the manual shaft is in good condition **(Figure 19)**.

The parking lock gear, like any other gear, needs to be carefully checked **(Figure 20)**. Make sure you look carefully at the edges of each tooth. The teeth must be able to hold the parking pawl securely. If the gear is damaged or worn, replace it.

Any components found unsuitable should be replaced. It should be noted that the components that make up the parking lock system are the only parts holding the vehicle in place when parked. If they do not function correctly, the car may roll or even drop into reverse when the engine is running, causing an accident or injury. Replace any questionable or damaged parts. Then install the housing and tighten the bolts to specifications.

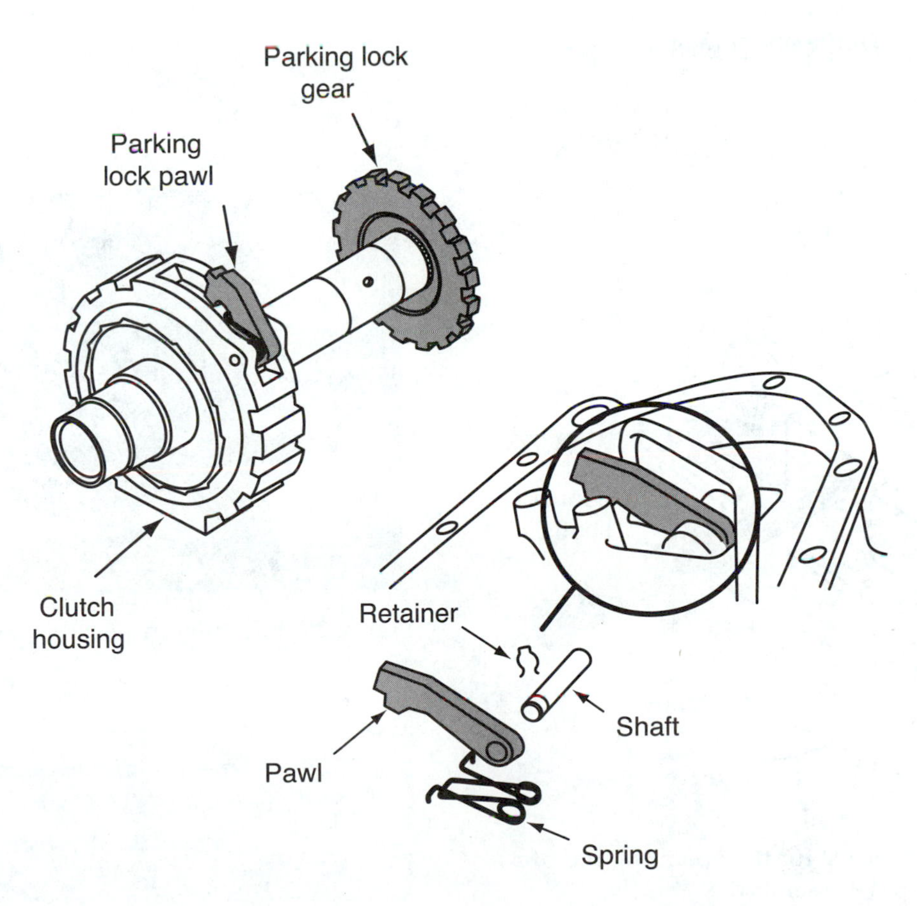


**Figure 17.** Each part of the parking pawl assembly should be carefully inspected.

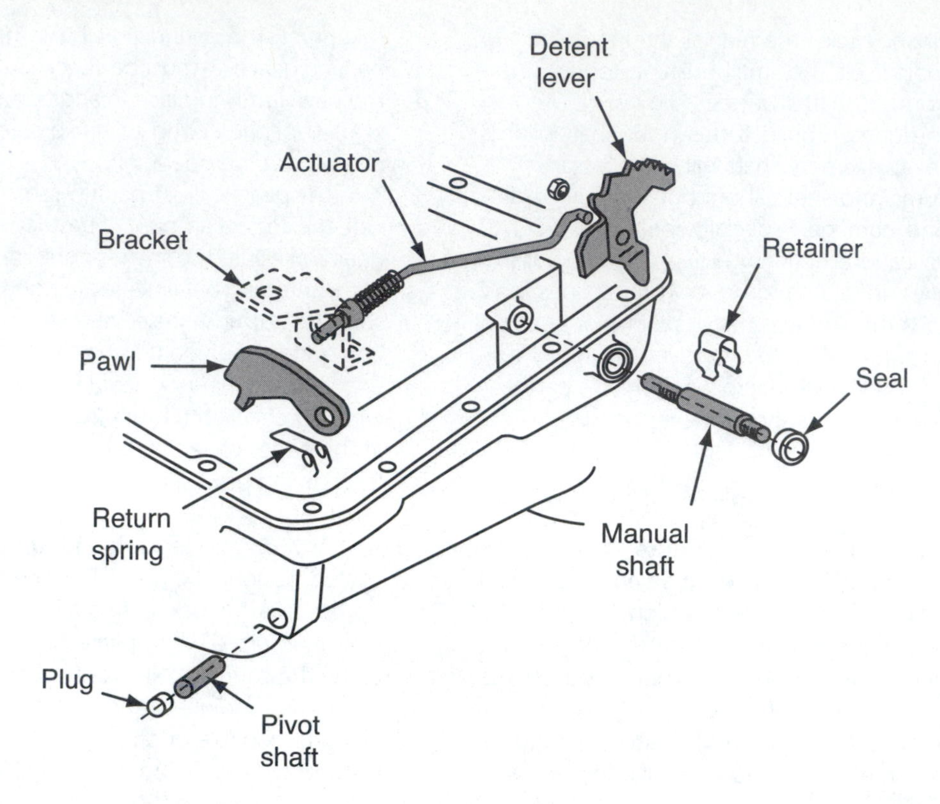


**Figure 18.** The push rod or operating shaft for the parking pawl must provide the correct amount of travel to engage the pawl to the gear. Make sure the shaft is not bent or that the pivot holes in the internal shift linkage are not worn oblong.

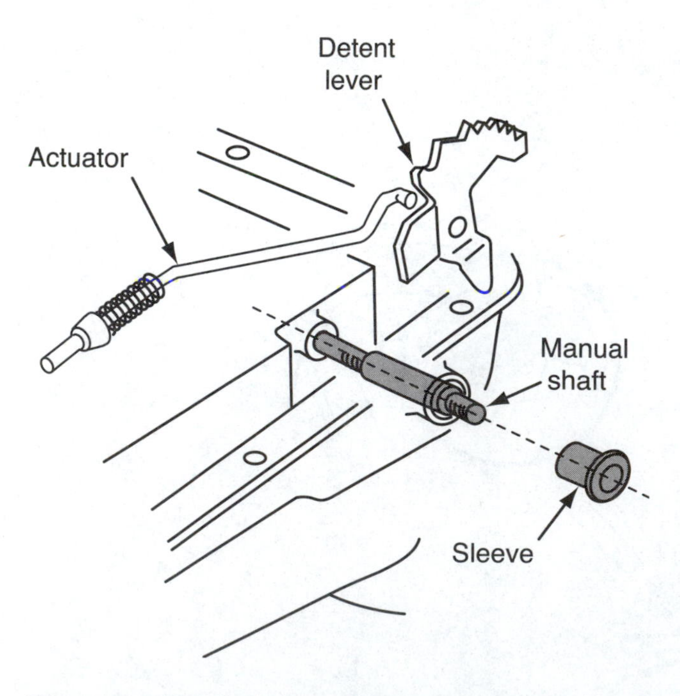


**Figure 19.** A typical sleeve for the operating rod of a parking pawl assembly.

**Figure 20.** Carefully inspect the parking gear.

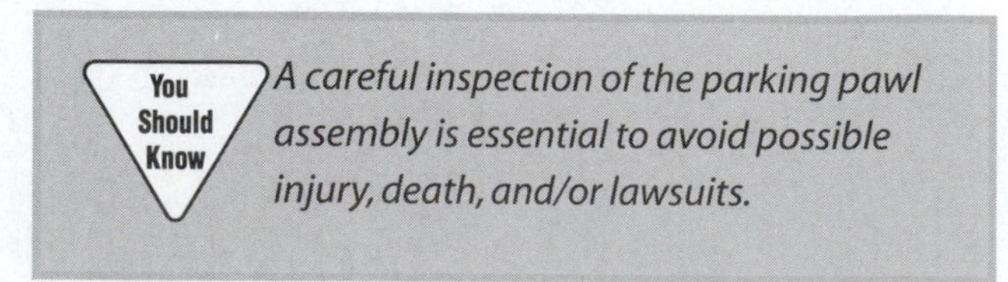


*A careful inspection of the parking pawl assembly is essential to avoid possible injury, death, and/or lawsuits.*

## Summary

- Thrust washers support a thrust load, set end play, and keep parts from rubbing together, preventing premature wear on parts.
- All bearings should be checked for wear, damage, and roughness.
- Sprag and roller clutches should be checked by attempting to rotate them in both directions.
- Bushings should be checked for wear, damage, and pitting and scoring.
- The installed depth and location of the oil holes of bushings should be checked.
- The teeth of a normally worn gear have a polished surface that extends the full length of the tooth from the bottom to the tip of the tooth.
- The pattern of wear or damage on a gear's tooth indicates probable problems.
- All planetary gear teeth should be inspected for chips or stripped teeth.
- Any gear mounted to a splined shaft should be checked for mutilated or shifted splines.
- Check the end play of the planetary carrier gears.
- Carefully examine the area on all shafts that rides in a bushing, bearing, or seal.
- Check the entire length of the shaft for signs of overheating and other damage.
- Inspect the splines for wear, cracks, or other damage.
- Shafts should be checked for wear in the ring groove area.
- All hubs, drums, and shells should be carefully examined for wear and damage.
- The entire parking pawl assembly should be inspected.

## Review Questions

1. When inspecting a planetary gearset, Technician A says that blackened pinion shafts indicate severe overloading. Technician B says that bluish gears indicate overheating. Who is correct?
   A. Technician A only
   B. Technician B only
   C. Both Technician A and Technician B
   D. Neither Technician A nor Technician B
2. Technician A says that loose planetary pinion gear bearings will cause the gear to whine when it is loaded. Technician B says that damaged teeth on a planetary pinion gear will cause the gear to whine. Who is correct?
   A. Technician A only
   B. Technician B only
   C. Both Technician A and Technician B
   D. Neither Technician A nor Technician B
3. Technician A removes a bushing by carefully cutting one side of the bushing and collapsing it. Once collapsed, the bushing can be easily removed with a pair of pliers. Technician B removes small-bore bushings by tapping the inside bore of the bushing with threads that match a selected bolt, which fits into the bushing. After tapping the bushing, the bolt is inserted and a slide hammer used to pull the bolt and bushing out of its bore. Who is correct?
   A. Technician A only
   B. Technician B only
   C. Both Technician A and Technician B
   D. Neither Technician A nor Technician B
4. Technician A says that bushings can be heated with a torch to remove them. Technician B says that bushings can be removed with a slide hammer and the correct attachment. Who is correct?
   A. Technician A only
   B. Technician B only
   C. Both Technician A and Technician B
   D. Neither Technician A nor Technician B
5. When checking a planetary gearset, Technician A says that the end clearance of the pinion gears should be checked with a feeler gauge. Technician B says that the end clearance of the long pinions in a Ravigneaux gear set should be checked at both ends. Who is correct?
   A. Technician A only
   B. Technician B only
   C. Both Technician A and Technician B
   D. Neither Technician A nor Technician B
6. Technician A says that a blocked oil delivery passage will cause a shaft to score. Technician B says that if a shaft is fitted with a check ball and the check ball does not seat properly, low oil pressure will result. Who is correct?
   A. Technician A only
   B. Technician B only
   C. Both Technician A and Technician B
   D. Neither Technician A nor Technician B
7. Technician A says that thrust washers should be inspected for scoring, flaking, and wear through to the base material. Technician B says that plastic thrust washers will not wear unless they are damaged. Who is correct?

A. Technician A only
B. Technician B only
C. Both Technician A and Technician B
D. Neither Technician A nor Technician B

8. When inspecting a parking pawl assembly, Technician A says that it must fit loosely in its bracket and tightly in the pawl. Technician B says that the engagement lug on the pawl must be square in order to fully engage into the parking gear. Who is correct?
   A. Technician A only
   B. Technician B only
   C. Both Technician A and Technician B
   D. Neither Technician A nor Technician B
9. Technician A checks bushing wear by observing the lateral movement of the shaft that fits into the bushing. Any noticeable lateral movement indicates wear and the bushing should be replaced. Technician B checks for bushing wear by measuring the inside diameter of the bushing and the outside diameter of the shaft with a dial caliper or micrometer. Who is correct?
   A. Technician A only
   B. Technician B only
   C. Both Technician A and Technician B
   D. Neither Technician A nor Technician B
10. Technician A says that all hubs, drums, and shells should be carefully examined for wear and damage. Technician B says minor scoring or burrs on band application surfaces of a drum can be removed by lightly polishing the surface with a 200-grit crocus cloth. Who is correct?
    A. Technician A only
    B. Technician B only
    C. Both Technician A and Technician B
    D. Neither Technician A nor Technician B

# Chapter 34 Transfer and Final Drives

## Introduction

The final drive unit **(Figure 1)** in a transaxle typically is serviced along with the transmission unit. Those procedures depend on the design of the final drive unit. Preventative maintenance procedures vary with the manufacturer not the design.

### Checking the Fluid Level

Some manufacturers recommend periodic final drive fluid changes, and all require fluid level checks. Be sure to use the proper lubricant whenever you change or top-up the fluid. Hypoid final drive gears and limited slip differentials both require special lubricants. Make sure the lubricant meets the manufacturer's specifications.

On most FWD models, the transmission and final drive assembly are lubricated with the same fluid and share the same fluid reservoir. Check the fluid level at the fill plug or dipstick. The fluid should be even with the bottom of the fill plug opening unless otherwise specified.

If the transmission and final drive have separate fluid reservoirs, check each individually and refill them with the proper lubricant.

**Figure 1.** A final drive unit.

### DRIVE CHAINS

The drive chains used in some transaxles should be inspected for side play and stretch. These checks are made during disassembly and should be repeated during reassembly. Typically, very little deflection is allowed.

Chain deflection is measured between the centers of the two sprockets. Deflect the chain inward on one side until it is tight **(Figure 2)**. Mark the housing at the point of maximum

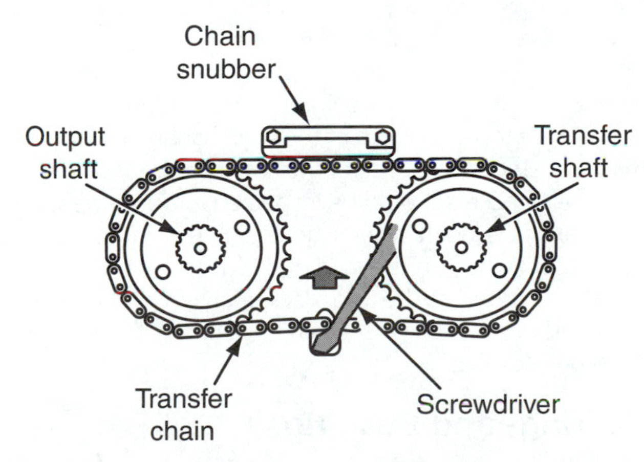


**Figure 2.** When measuring the slack of the drive chain, outwardly deflect the chain, and make a mark to that point of deflection.

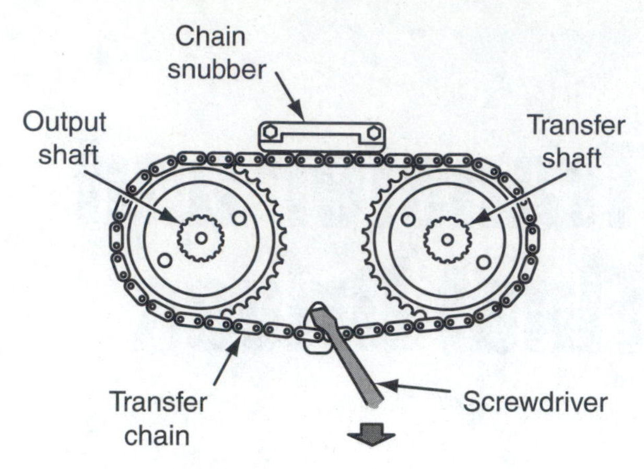


**Figure 3.** Continue measuring the slack of the drive chain by deflecting the chain inwardly. Mark the point of deflection. The distance between the outward and inward marks is the amount of chain slack.

**You Should Know** *According to the major supplier of drive chains, it is impossible to accurately measure chain stretch without expensive gauges. Therefore, they suggest that drive chains be replaced every 50,000 miles or during an overhaul of a transaxle or transfer case.*

deflection. Then deflect the chain outward on the same side until it is tight **(Figure 3)**. Again, mark the housing in line with the outside edge of the chain at the point of maximum deflection. Measure the distance between the two marks. If this distance exceeds specifications, replace the drive chain.

Check each link of the chain by pushing and pulling the links away from the pin that holds them together. All of the links should move very little and each move the same amount. Check each link carefully.

Check for an identification mark on the chain during disassembly. These can be painted or dark colored links, and indicate the top or bottom of the chain. During reassembly, make sure the correct side of the chain is up.

The sprockets should be inspected for tooth wear and for wear at the point where the chain rides. A slightly polished appearance on the face of the gears is normal. If the chain was found to be too slack, it may have worn the sprockets. Also, check the chain snubber or tensioner for damage and wear.

## Bearings and Bushings

The bearings and bushings for the sprockets need to be checked for damage. The radial needle thrust bearings must be checked for any deterioration and damage. The

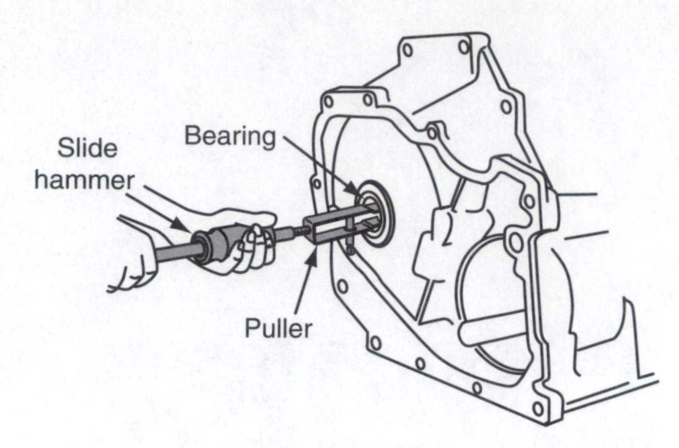


**Figure 4.** The bearings and bushings in a chain drive assembly should be removed with a puller.

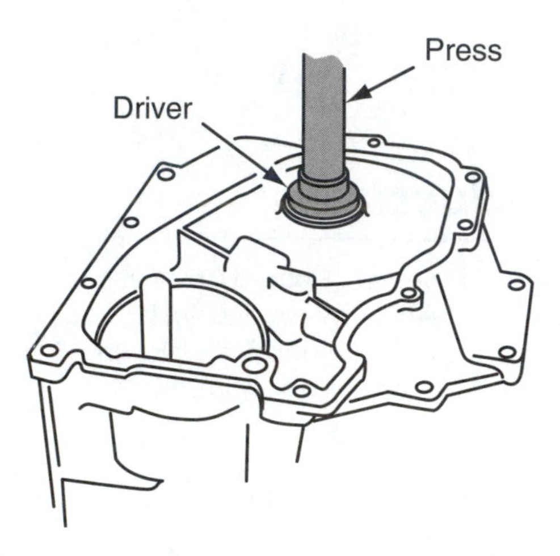


**Figure 5.** The bearings and bushings in a chain drive assembly should be installed with a driver and a hydraulic press.

bushings should be checked for any signs of scoring, flaking, or wear. Replace any defective parts.

Typically, the bearings and bushings are removed with a puller **(Figure 4)** and installed with a driver and a press **(Figure 5)**.

The removal and installation of some chain drive assemblies requires that the sprockets be spread slightly apart. The sprockets must be spread the right amount. If they are spread too far, they will not be easy to install or remove.

The shafts and gears of the assembly have numerous seals and thrust washers **(Figure 6)**. The seals must be replaced whenever the unit is disassembled.

## TRANSFER GEARS

Some transaxles use gears, instead of a drive chain, to move or transfer the output of the transmission to the final drive unit **(Figure 7)**. These shafts and gears must be carefully inspected and replaced if they are damaged.

**Figure 6.** Thrust washers for the output and transfer shafts.

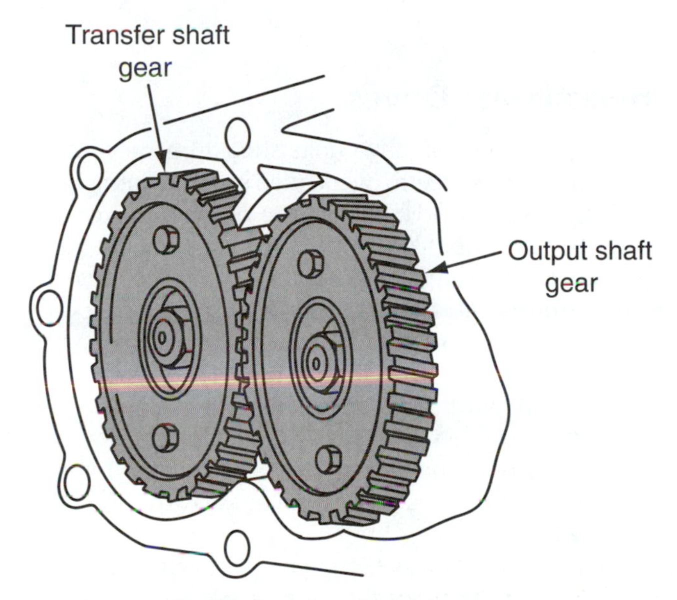


**Figure 7.** The transfer shafts and gears must be carefully inspected and replaced if they are damaged.

**Figure 8.** To remove and install the transfer shaft gear, a holding tool must be used to stop the transfer gear from turning when loosening or tightening the retaining nut.

To remove and install the transfer gear, a holding tool is used to stop the transfer gear from turning while loosening or tightening the retaining nut **(Figure 8)**. The nut is typically tightened to 200 ft.-lb. To remove the transfer shaft from the transaxle case, the retaining snapring must be removed, and then the shaft can be pulled out with its bearing **(Figure 9)**. The shaft's bearing must be pressed on and off the shaft. A puller must be used to remove the transfer gear from the output shaft after its retaining bolt has been removed.

Behind the transfer shaft gear is a selective shim used to provide correct meshing of the teeth of the transfer shaft gear and the output shaft transfer gear and to control transfer shaft end play. End play is measured with a dial indicator. To ensure good contact with the indicator's plunger and the end of the transfer shaft, Chrysler recommends that a steel check ball be placed between the

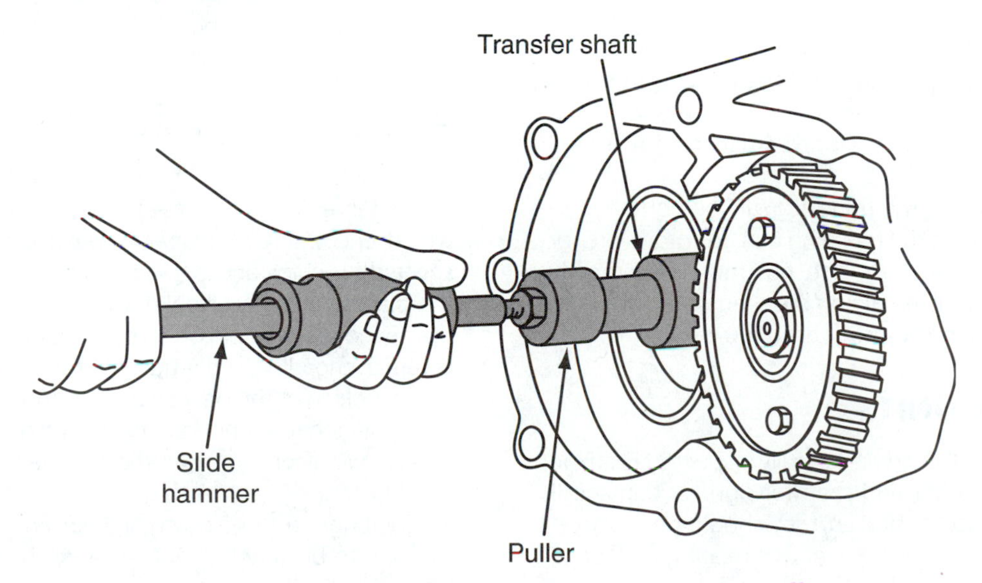


**Figure 9.** The transfer shaft can be pulled out with its bearing with a suitable puller tool.

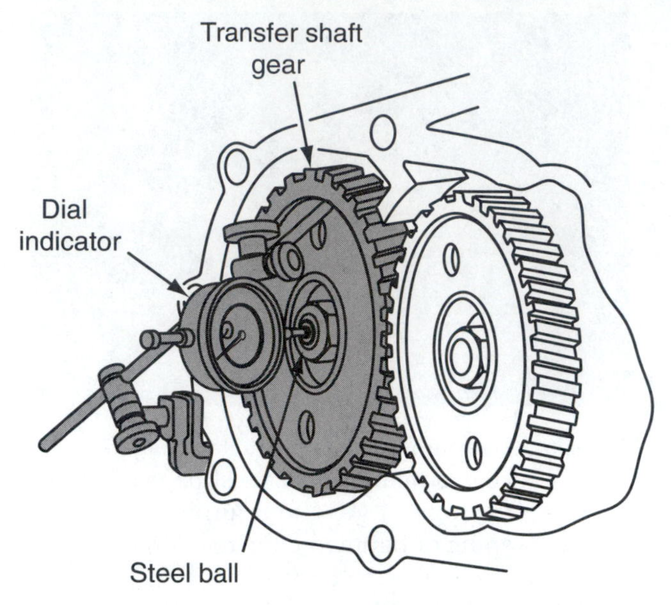


**Figure 10.** Chrysler recommends that a steel check ball coated in heavy grease be placed between the plunger tip and the shaft to ensure good contact with the indicator's plunger and the end of the transfer shaft.

plunger tip and the shaft **(Figure 10)**. To hold the ball in place during the check, coat the ball in heavy grease. If the end play is not within specifications, select a thrust washer with the correct thickness and install it behind the transfer gear.

The output shaft also should be checked for end play. If the end play is incorrect, a different size shim should be installed behind the output shaft transfer gear. Because the gear is splined to the shaft, a thrust bearing is not needed here. After end play has been corrected and the output shaft gear reinstalled, the turning torque of the shaft should be measured. If the turning torque is too high, a slightly thicker shim should be installed. If the turning torque is too low, a slightly smaller shim should be installed.

It is important that the transfer gears be tightened to specifications. Normally the retaining bolt has provisions for staking the nut to the shaft. The nut must be properly staked so the gears will remain on the shafts and will maintain the correct bearing adjustments **(Figure 11)**.

## FINAL DRIVE UNITS

Final drive units should be carefully inspected. Examine each gear, thrust washer, and shaft for signs of damage. If the gears are chipped or broken, they should be replaced. Also, inspect the gears for signs of overheating or scoring on the bearing surface of the gears.

**Figure 11.** Stake the output gear's retaining nut to the output shaft.

### Helical Gear Drives

Helical-type final drive units should be checked for worn or chipped teeth, overloaded tapered roller bearings, and excessive differential side gear wear. Excessive play in the differential is a cause of engagement clunk **(Figure 12)**. Be sure to measure the clearance between the side gears and the differential case and to check the fit of the gears on the gear shaft. Proper clearances can be found in the appropriate shop manual. The side bearings of some final drive units are preloaded with shims. Select the correct size shim to bring the unit into specifications. With a torque wrench, measure the amount of rotating torque. Compare your readings against specifications **(Figure 13)**.

If the bearing preload and end play are fine as is the condition of the bearings, the parts can be reused. However, always install new seals during assembly. New bearings require that preload be set to the specifications for a new bearing. Used bearings should be set to the amount found during teardown or about one half the preload of a new bearing.

### Planetary Gear Assemblies

Planetary-type final drives **(Figure 14)** also are checked for the same differential case problems that the helical type would encounter. All planetary gear teeth should be inspected for chips and damage. Pay particular attention to the planetary carriers. Look for obvious problems such as blackened gears or pinion shafts. These conditions indicate severe overloading and require that the carrier be replaced. When looking at the gears, a bluish condition can be a normal condition resulting from the heat-treating process used during manufacture. Check the planetary pinion gears for loose bearings. Check each gear by rolling it on its shaft to feel for roughness or binding of the needle bearings. Wiggle the gear to be sure it is not loose on the shaft, which can cause the gear to whine when it is loaded. Also, inspect the

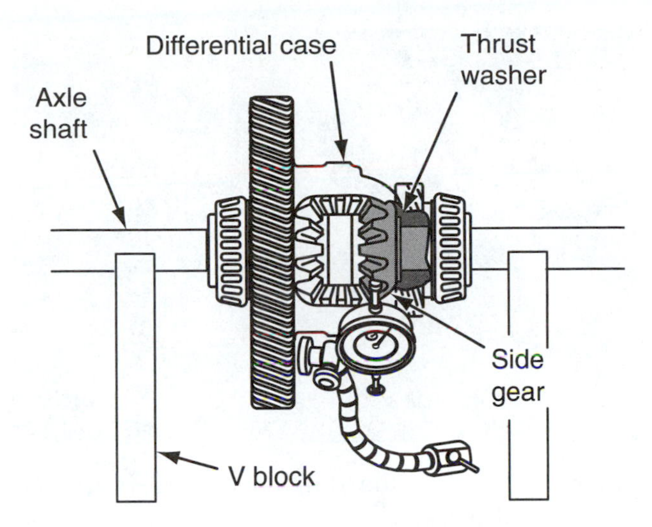


**Figure 12.** The backlash of the side gears in a final drive unit should be checked prior to assembling the transaxle.

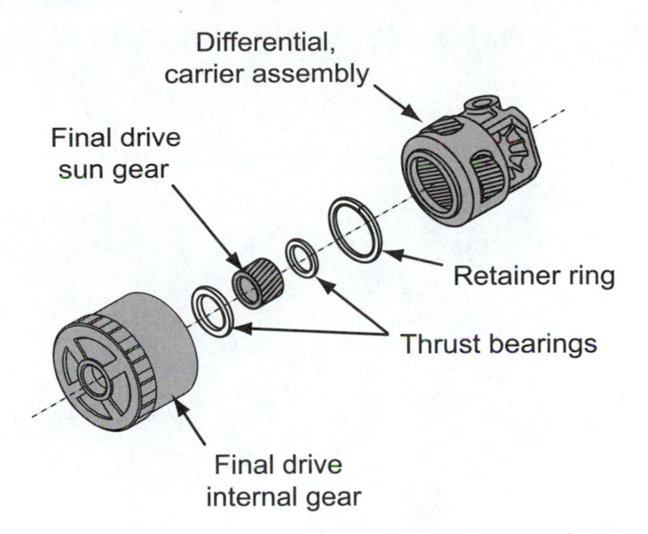


**Figure 14.** Planetary-type final drives should be checked in the same basic way as helical-type final drives.

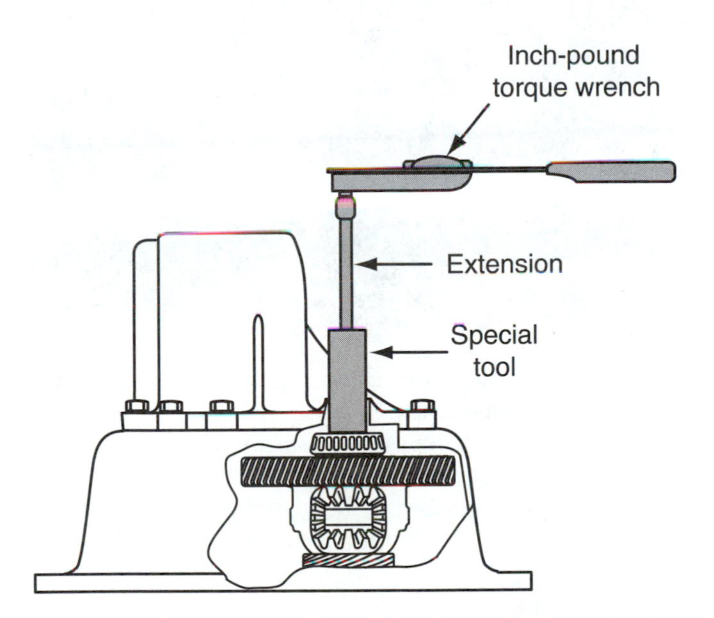


**Figure 13.** Using an inch-pound torque wrench, the turning torque of the transaxle assembly should be checked after assembly.

**Figure 15.** The end play of the planetary pinion gears should be checked with a feeler gauge.

gears' teeth for chips or imperfections, as these also will cause whine.

Check the gear teeth around the inside of the planetary ring gear. Check the carrier assembly for cracks or other defects. Check the end play of the pinion gears by placing a feeler gauge between the planetary carrier and each pinion gear **(Figure 15)**. Compare the end play to specifications. If the end play is too low, the pinion assembly must be removed and the correct thrust washer installed. If there is excessive end play, the differential assembly must be replaced.

Any problems normally result in the replacement of the carrier as a unit as most pinion bearings and shafts are not sold as separate parts. To disassemble a typical final drive unit, place the final drive unit into the transaxle's oil pan. Doing this will lessen the chances of losing bearings while disassembling the unit. Make sure the pan is clean.

With a pin punch and hammer, remove the differential pinion shaft retaining pin **(Figure 16)**. Then remove the retaining ring from the end of the output shaft. This ring must not be reused! Now pull the output shaft from the differential carrier and remove the final drive sun gear.

Remove the differential pinion shaft retaining pin **(Figure 17)**, then the differential pinion shaft, pinion gears, and thrust washers, and place them to one side. Keep these

**Figure 16.** Using a pin punch and hammer, remove the differential pinion shaft retaining pin.

**Figure 17.** Remove the differential side gears and thrust washers.

**Figure 18.** Install the pinion gear needle bearing spacer, thrust washer, and pinion gear onto the planet pinion gear pin. Then install the needle bearings, one at a time, into the top and bottom of the planet pinion gear.

parts together and in the order they were assembled. Then remove the side gears and thrust washers.

Using a screwdriver, carefully remove the final drive carrier retaining ring. Now remove the planet pinion pins, then the planet pinion gears, thrust washers, needle bearings, and the bearing spacers. Inspect the needle bearings, thrust washers, pinion gears, and planet pinion pins. To complete the disassembly, remove the final drive sun gear to carrier thrust bearing. Inspect all parts and replace all damaged or worn parts.

To begin reassembly, install the pinion gear needle bearing spacer, thrust washer, and pinion gear onto the planet pinion gear pin. Then install the needle bearings, one at a time, into the top and bottom of the planet pinion gear **(Figure 18)**. Use transjel or petroleum jelly to help hold the bearings in position.

Coat the sun gear-to-carrier thrust washer with transjel or petroleum jelly. Then install the thrust bearing onto the final drive carrier. Remove the pinion pin from the pinion gears. Install the planet pinion gear thrust washers and gears into the final drive carrier. Make sure the gears are in the same direction as they were when they were removed. Also, be careful not to let the bearings fall out. Install the planet pinion gear pins into the carrier. Then, install the final drive carrier retaining ring. With a feeler gauge, check the end play of the pinion gears. If the end play is too low, the planet pinion assembly must be removed and the correct thickness thrust washers installed. If the end play is too high, the differential assembly must be replaced.

Install the differential side gears and thrust washers into the carrier. Apply transjel or petroleum jelly onto the thrust washers for the pinion gears, and then install them into the carrier. Align the bores of the pinion gears with the bore in the carrier for the pinion pin, and then install the pin. Make sure the retaining pin hole is aligned with the retaining pin

Figure 19. Install the pinion pin's retaining pin.

bore. Install the pinion pin's retaining pin **(Figure 19)**. Insert the sun gear into the carrier, making sure the gear is placed in the correct direction. Install the output shaft into the final drive assembly. To complete assembly, install a new retaining snapring onto the end of the output shaft.

## Adjusting Side Gear End Play

Ring gear and side bearing adjustments are typically made with the differential case assembled and out of the transaxle case.

The procedure that follows is typical for measuring and adjusting the side gear end play in a transaxle's final drive. Always refer to your service manual before proceeding to make these adjustments on a transaxle.

1. Install the correct adapter into the differential bearings **(Figure 20)**.
2. Mount a dial indicator to the ring gear with the plunger resting against the adapter.

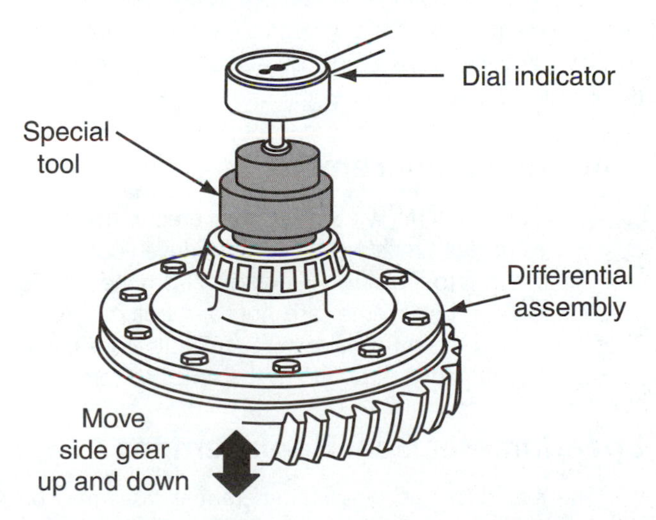


Figure 20. Tool setup for measuring side gear end play.

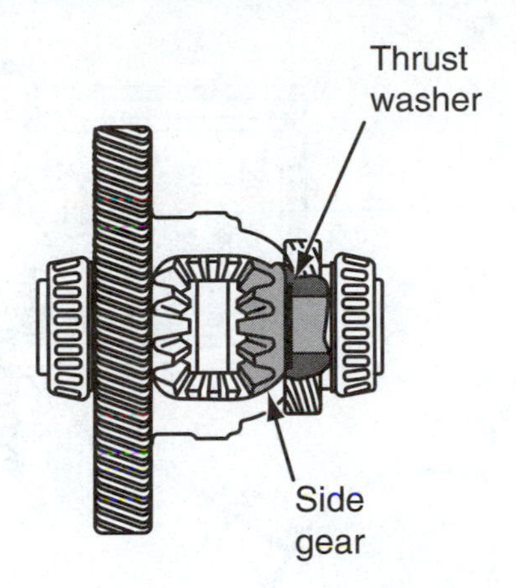


Figure 21. Location of thrust washers.

3. With your fingers or a screwdriver, move the ring gear up and down.
4. Record the measured end play.
5. Measure the old thrust washer with the micrometer.
6. Install the correct size thrust washer **(Figure 21)**.
7. Repeat the procedure for the other side.

## Adjusting Bearing Preload

The following procedure is typical for the measurement and adjustment of the differential bearing preload in a transaxle. Always refer to your service manual before proceeding to make these adjustments on a transaxle.

1. Remove the bearing cup and existing shim from the differential bearing retainer.
2. Select a gauging shim that will allow for 0.001- to 0.010-inch end play.
3. Install the gauging shim into the differential bearing retainer **(Figure 22)**.
4. Press in the bearing cup.
5. Lubricate the bearings and install them into the case.
6. Install the bearing retainer.
7. Tighten the retaining bolts.
8. Mount the dial indicator with its plunger touching the differential case **(Figure 23)**.
9. Apply medium pressure in a downward direction while rolling the differential assembly back and forth several times.
10. Zero the dial indicator.
11. Apply medium pressure in an upward direction while rotating the differential assembly back and forth several times.
12. The required shim to set preload is the thickness of the gauging shim plus the recorded end play.
13. Remove the bearing retainer, cup, and gauging shim.
14. Install the required shim.
15. Press the bearing cup into the bearing retainer.
16. Install the bearing retainer and tighten the bolts.

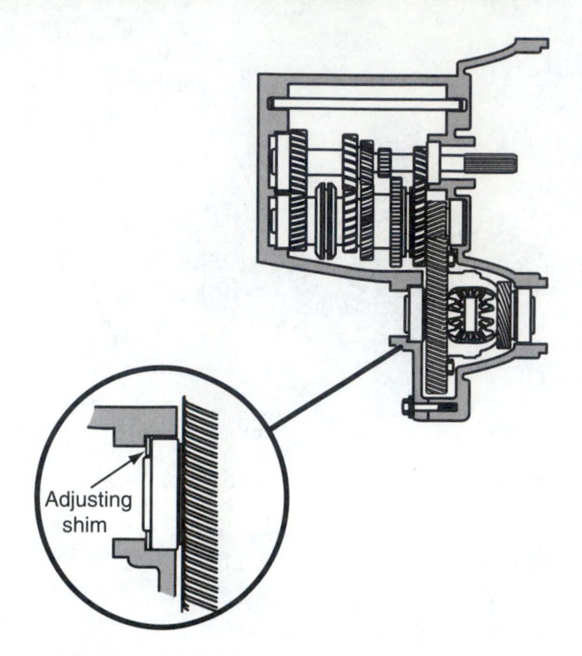


**Figure 22.** Typical location of preload shim.

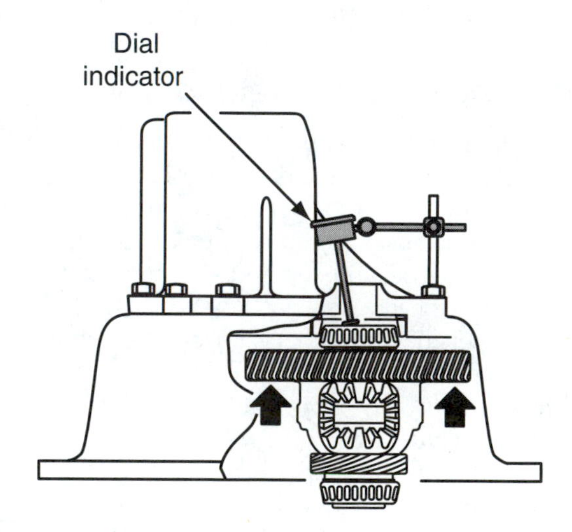


**Figure 23.** Setup for measuring differential bearing preload.

17. Check the rotating torque of the transaxle. If this is less than specification, install a thicker shim. If the torque is too great, install a slightly thinner shim.
18. Repeat the procedure until desired torque is reached.

## PART REPLACEMENT

Planetary-type final drives, like helical final drives, are available in more than one possible ratio for a given type of transaxle, so care should be taken to ensure that the same gear ratios are used during assembly. This is not normally a problem when overhauling a single unit; however, in a shop where many transmissions are being repaired, it is possible to mix up parts, causing problems during the rebuild.

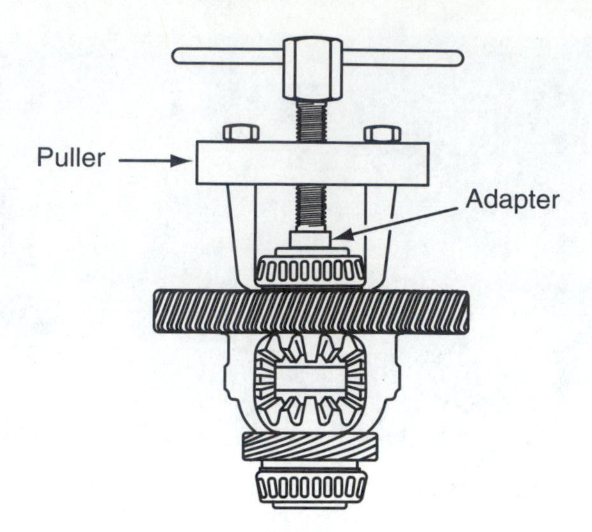


**Figure 24.** Typical setup for removing side bearings from a differential case.

### Side Bearings

The side bearings of most final drive units must be pulled off and pressed onto the differential case. Always be sure to use the correct tools for removing and installing the bearings.

### Final Drive Bearing Replacement

The final drive unit is positioned either in the transaxle case or in a separate housing mounted to the transaxle case. Regardless of its location, it is supported by bearings. Normally tapered roller bearings are used. These bearings are removed with pullers **(Figure 24)** and installed with a press and driver. When replacing the bearings, make sure the bearing seats are free from nicks and burrs. These defects will not allow the bearing to seat properly and will give false end play measurements. If the bearings were replaced, end play and gear clearances must be checked and corrected before installing the final drive unit into the housing.

### Ring Gear Replacement

The ring gear of many transaxles is riveted to the differential case. The rivets must be drilled, then driven out with a hammer and drift to separate the ring gear from the case. To install a new ring gear to the case, nuts and bolts are used. These nuts and bolts must be of the specified hardness and should be tightened in steps and to the specified torque.

### Speedometer Gear Replacement

The final drives of transaxle differential cases may be fitted with a speedometer gear pressed onto the differential case and under one side bearing. These gears are pulled off and pressed onto the case.

## Summary


- The last set of gears in the drive train is the final drive.
- On most FWD cars, the final drive is located within the transaxle. Some FWD cars with longitudinally mounted engines locate the differential and final drive in a separate case that bolts to the transmission.
- There are four common configurations used as the final drives on FWD vehicles: helical gear, planetary gear, hypoid gear, and chain drive. Hypoid final drive gear assemblies are only found in vehicles with a longitudinally placed engine.
- The differential section of the transaxle has the same components as the differential gears in a RWD axle and basically operates in the same way.
- The drive pinion gear is connected to the transmission's output shaft and the ring gear is attached to the differential case. The pinion and ring gear provide for torque multiplication.


## Review Questions

1. Technician A says that most FWD final drive units require that ring and pinion backlash and pinion depth be set to specifications. Technician B says that ring gear and side gear bearing adjustments are required on all FWD final drive units. Who is correct?
   A. Technician A only
   B. Technician B only
   C. Both Technician A and Technician B
   D. Neither Technician A nor Technician B
2. When checking the end play of a transfer gear assembly, Technician A says that if the end play of the output transfer shaft is incorrect, a different size thrust washer should be installed behind the output shaft transfer gear. Technician B says that after end play of the output shaft has been corrected, the turning torque of the shaft should be measured. If the turning torque is too high, a slightly thinner shim should be installed. If the turning torque is too low, a slightly thicker shim should be installed. Who is correct?
   A. Technician A only
   B. Technician B only
   C. Both Technician A and Technician B
   D. Neither Technician A nor Technician B
3. When servicing a final drive unit, Technician A checks gear and bearing end play whenever new bearings are installed in the unit. Technician B reuses the bearings, seals, and thrust washers if the bearing preload and end play are fine and so is the condition of the bearings. Who is correct?
   A. Technician A only
   B. Technician B only
   C. Both Technician A and Technician B
   D. Neither Technician A nor Technician B
4. When inspecting a FWD final drive assembly, Technician A says that a drive chain that is too loose should be shortened by removing a pair of links in the chain. Technician B says that the drive sprockets should be replaced if the gear teeth are polished or show any other signs of wear. Who is correct?
   A. Technician A only
   B. Technician B only
   C. Both Technician A and Technician B
   D. Neither Technician A nor Technician B
5. When replacing the ring gear in a FWD final drive unit, Technician A drills out the ring gear retaining rivets on some final drive units to remove the ring gear from the carrier. Technician B uses nuts and bolts of the specified hardness to fasten a ring gear to the differential carrier. Who is correct?
   A. Technician A only
   B. Technician B only
   C. Both Technician A and Technician B
   D. Neither Technician A nor Technician B

# Reaction and Friction Units

## *Introduction*

To provide the various forward speed gears and a reverse, transmissions use compound gears controlled by friction and reaction devices. Although diagnosis can lead you to the device that is not holding or driving properly, recognition of those suspected parts becomes difficult when the transmission is apart. Clutch and brake application charts are a must during diagnosis and transmission overhaul.

During inspection and disassembly of friction units and their hub or drum, pay close attention to which planetary gear set member each is attached. Keep in mind that some planetary members are connected to each other and that as one member rotates it rotates the other at the same speed. Sometimes the drum or housing for a clutch also is the drum for a brake band **(Figure 1)**. Often this means the same planetary member can be driven or held to achieve different gears.

Servos and accumulators control the application of bands and clutches to achieve a desirable shift feel. There are many different types and designs of each. Each has an important function in the operation of the various designs of transmissions and transaxles. Fortunately, service for all servos and accumulators is very similar.

## BAND SERVICE

Prior to overdrive automatic transmissions, most bands operated in a free condition during most driving conditions. This means the band was not applied in the cruising gear range. However, many overdrive automatic transmissions use a band in the overdrive cruise range, which puts an additional load and subsequently causes additional wear on the band. For this reason, a thorough inspection of the bands is very important **(Figure 2)**.

The friction material used on clutches and bands is absorbent, and this characteristic can be used to tell if there is much life left in the lining. Simply squeeze the lining with your fingers to see if any fluid appears. If fluid appears, the lining can still hold fluid and has some life left in it. Strap- or flex-type bands should never be twisted or flattened out; this may crack the lining and lead to flaking of the lining.

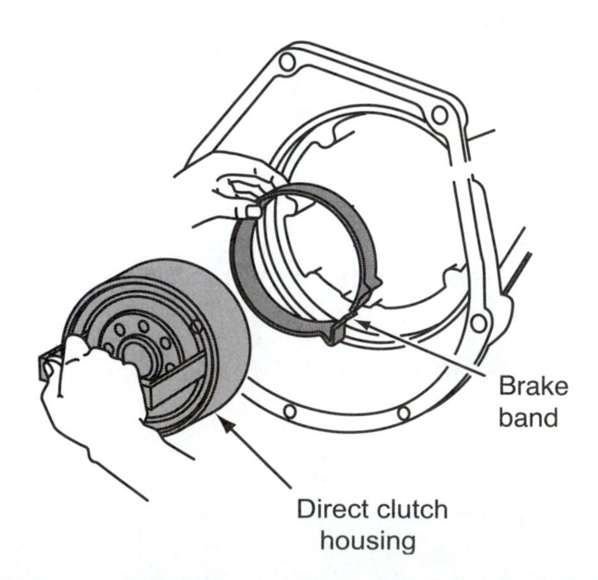


**Figure 1.** The drum for this clutch also serves as the clamping surface for the band. The tool shown in the clutch assembly is used to pull the assembly out of the transmission housing.

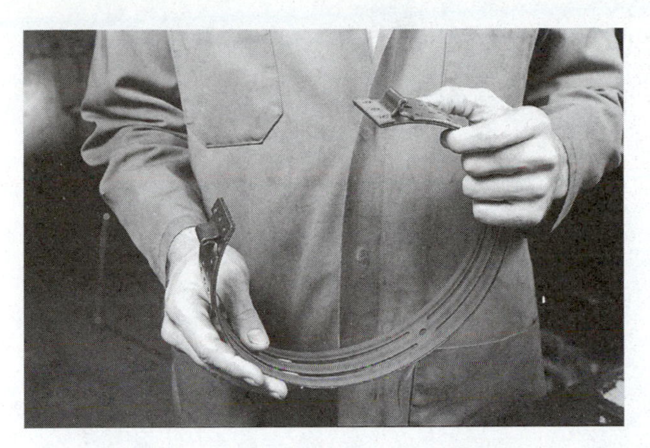
**Figure 2.** The friction material of the bands must be carefully inspected.

**Figure 4.** The surface of the drum should be checked for discoloration, scoring, glazing, and distortion.

## Inspection

Check the bands for chipping, cracks, burn marks, glazing, and nonuniform wear patterns and flaking. A twisted band will show tapered wear on the lining. If the friction material is blackened, this is caused by an excessive buildup of heat. High heat may weaken the bonding of the lining and allow the lining to come loose from the metal portion of the band. If any of these defects are apparent, the band should be replaced. On double-wrap bands, check both segments of the lining.

Make sure to check the band lugs for damage. Also carefully check the band struts **(Figure 3)**, levers, and anchors for wear. Replace any worn or damaged parts.

The drum surface **(Figure 4)** should be checked for discoloration, scoring, glazing, and distortion. Cast iron drums that are not scored can generally be cleaned up by sanding the running surface with 180-grit emery paper in the drum's normal direction of rotation. A polished surface is not desirable on cast iron drums.

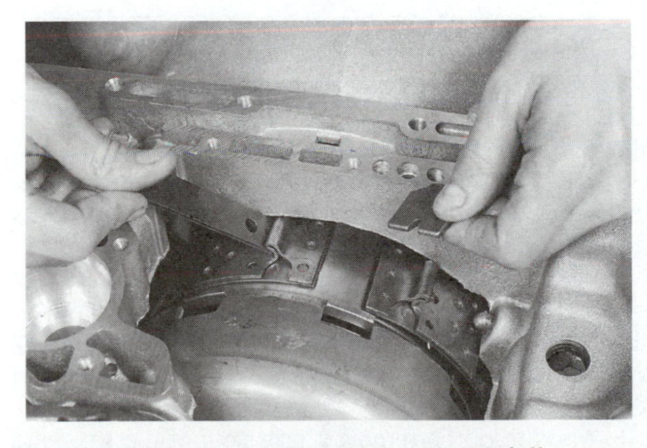
**Figure 3.** When removing the band, carefully inspect the band strut.

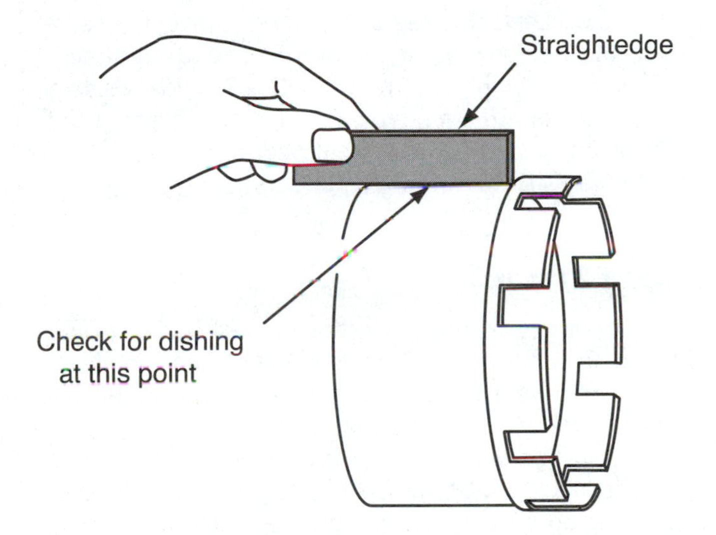

**Figure 5.** Check the flatness of the drum with a straightedge.

Check the drum for flatness across the surface where the band runs **(Figure 5)**. Any dishing will cause the band to distort as it attempts to get a full grip on the drum. Band distortion weakens the bond of the friction material to the band and will cause the friction material to flake. A dished drum should be replaced. Check the service manual for maximum allowable tolerances.

## Band Adjustments

After the band assembly has been installed around its drum, the band needs to be adjusted. Band adjustment is also part of a "transmission tuneup" on some models. Many transmissions have provisions for externally adjusting the band running clearance. Some transmissions have no provisions for band adjustment other than selectively sized servo apply pins and struts. Serious damage to the transmission can result if the specified band adjustment procedure is not followed.

To adjust the bands on transmissions with an adjustment screw, loosen the locknut on the adjuster screw and back it off about five turns. Backing off the locknut allows for tightening the screw to a specified torque, which simulates a fully applied band. After torquing, the adjuster screw is backed off a specified number of turns. To hold the adjustment, the locknut is generally tightened to 30–35 ft.-lb. with the adjuster screw held stationary.

The timing of band application has a lot to do with shift feel. This is why there are many different tightening torques for the various transmissions. The torque setting and number of turns the adjuster is backed off provide the proper clearance and grip for the different bands used. Additionally, the pitch of the threads on the adjuster screws varies. This is why you cannot use one particular adjustment sequence for all transmissions.

Not all band adjusters are on the exterior of the transmission case. Some transmissions have a band adjustment screw inside the oil pan. This typically is for a low wearing band and adjustment is typically made during a fluid and filter change when the pan is removed.

## MULTIPLE-FRICTION DISC ASSEMBLIES

Two types of multiple-friction disc assemblies will be found in a transmission: rotating drum and case held (grounded). Both are serviced in the same way but require slightly different inspection procedures.

All friction disc and steel plate packs are held in place by snaprings. It is common to use the same diameter snapring on more than one clutch in a single transmission. However, the rings can be of differing thickness and must be kept with the clutch pack during disassembly. This thickness variation is used to set the clearance in the clutch pack. The snaprings may also have a distinct shape, which will only be effective when used in the correct groove.

### Disassembly

Using a screwdriver, remove the clutch retaining plate snapring and remove the steel clutch pressure plate **(Figure 6)**. Now remove the disc assembly. The thickest plate or backing plate may be a selectively sized part. Keep the friction discs and steel plates in the order they were in the drum **(Figure 7)**.

Using a clutch spring compressor tool, compress the clutch return springs. Then remove the clutch hub retainer snapring, retaining plate, and springs. Most retainers have small tabs that prevent you from removing the snapring without using a compressor.

There are many types of clutch spring compressor tools available. The many different locations and depths the

**Figure 6.** Removing large snapring to disassemble clutch pack.

clutch springs may have in their drums dictates having more than one simple compressor **(Figures 8 and 9)**.

With the compressor installed on the spring retainer, compress the spring and retainer just enough to allow the snapring to be removed. Pushing the retainer down too much may bend or distort it. Make sure the snapring is not partially caught in its groove when releasing the spring compressor. Also be very careful when releasing the spring compressor tool; some springs have very high tensions and can injure you if they get a chance to fly out.

With the retainer removed, a single large coil spring or multiple small coil springs will be exposed. Note the number and placement of the multiple springs for use during assembly.

If coil springs are not used, a Belleville- or disc-type return spring is used **(Figure 10)**. This type of clutch will have a heavy pressure plate with one rounded side in the bottom of the clutch pack. The released position of this type of spring has very little or no outward force on the retaining snapring; therefore, the snapring can be removed without the use of a compressor. The snapring for this application could be selective, so remember where the snapring goes.

Some clutches have two or more snaprings. Pay attention to their placement in the clutch. Additionally, a wavy snapring is sometimes used to retain the Belleville spring. These are used to give some cushion to the application of the clutch.

For easy removal of the clutch piston, mount the clutch on the oil pump. Then lift the piston out of the bore. If it will not come out, apply pressurized air into the piston.

Use an air nozzle with a rubber tip and apply air pressure at the feedhole in the drum, or, by placing the drum on the clutch support, to its normal feed passage **(Figure 11)**. Wrap a rag around the clutch drum to catch the piston and the fluid overspray.

With the piston removed, take note of the seals and their position. This is important with lip seals; the lip should face the direction the fluid comes from. Lip seals installed in

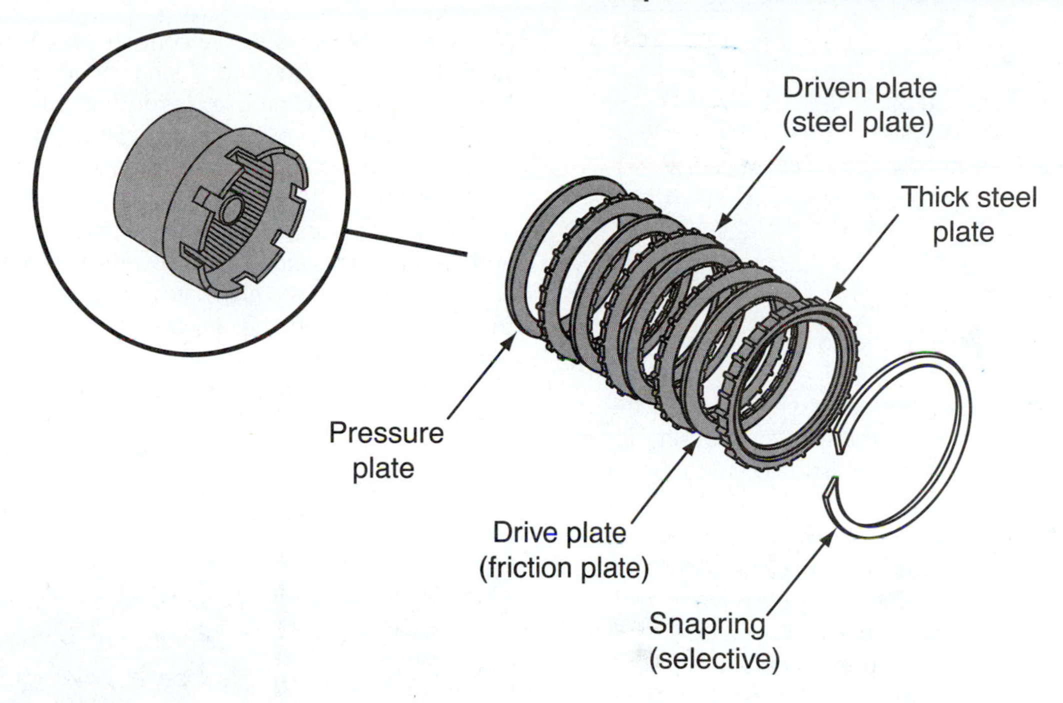


**Figure 7.** When removing the clutch discs from the drum, keep them in the same order.

**Figure 8.** A spring compressor tool for a multiple friction disc assembly in a rotating drum.

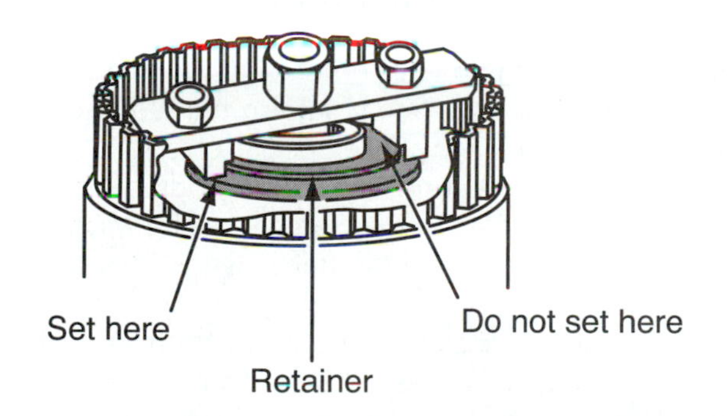


**Figure 10.** A clutch assembly fitted with a Belleville spring.

**Figure 9.** A spring compressor tool for a multiple friction disc assembly that is set in the transmission housing.

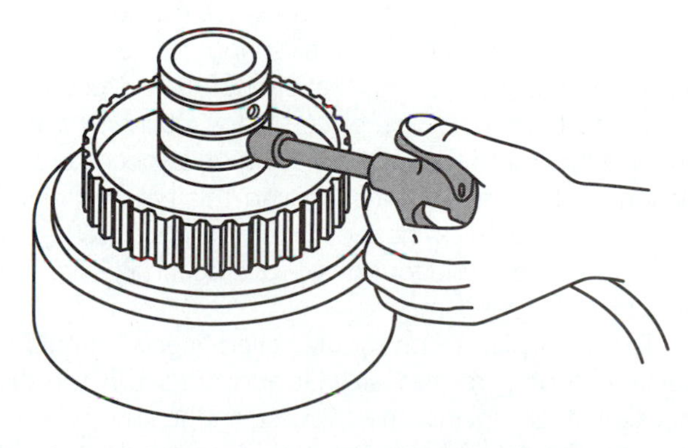

**Figure 11.** Use an air nozzle with a rubber tip and apply air pressure at the feed hole in the drum to pop out the piston from the drum.

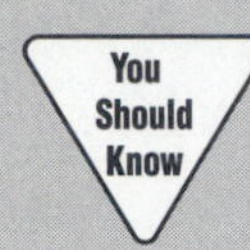

*Air pressure should be reduced to 25–30 lbs. to avoid expelling the piston at a high rate of speed, which could cause injury.*

the wrong direction will not hold pressure. The reuse of seals is not recommended. They should be replaced during overhaul.

After the clutch pack has been disassembled, check the clutch drum, clutch housing, or transmission housing for defects. The splined area in the transmission housing should be carefully inspected for breaks, chips, and other damage. If the splines are damaged, the case should be replaced. The bushings of the clutch drums and housings should be checked. These should be replaced if they are worn or damaged. The splines inside of the drum should be inspected. If they are damaged, the drum or housing should be replaced. The outside surface of the housing should be inspected for signs of overheating and other defects. The flatness of the surface also should be checked.

## Inspection and Cleaning

Clean the components of the clutch assembly. Make sure all clutch parts are free of varnish, burned disc facing material, or steel filings. Take special care to wash out any foreign material from the inside of drums and the hub disc splines. If left in, the material can be washed out by ATF and sent through the transmission. This can ruin the rebuild.

The clutch splines must be in good shape with no excessively rounded corners or shifted splines. Test their fit by trial fitting each new clutch disc on the splines. Move the discs up and down the splines to check for binding. If they bind, this can cause dragging of the discs during a time when they should be free floating. Replace the hubs if the discs drag during this check.

Check the spring retainer. It should be flat, not distorted, at its inner circumference. Check all springs for height, cracks, and straightness. Any springs that are not the correct height or are distorted should be replaced. Many retainers have the springs attached to them by crimping. This speeds up production at the assembly line. Turning this type of retainer upside down is a quick check of spring length. Closely examine the Belleville spring for signs of overheating or cracking, and replace it if it is damaged.

The steel plates, commonly called "steels," should be checked to be sure they are flat and not worn too thin. Check all steels against the thickest one in the pack or a new one. The steel plates also should be checked for flatness by placing one plate on top of the other and checking the space on the inside and outside diameters. Clutch plates must not be warped or cone shaped. Also, check the steel plates for burning and scoring and for damaged driving lugs. Check the grooves inside the clutch drum and the fit of the steel plates. If the plates pass inspection, remove the polished surface finish and the steels will be ready for reuse.

Many transmission rebuilders replace all friction discs and steel plates as a part of the overhaul process. This can actually save time as it eliminates inspection time. There is then no doubt of whether any discs or plates were of questionable condition.

Close inspection of the friction discs is simple **(Figure 12)**. The discs will show the same types of wear as bands will. Disc facings should be free of chunking, flaking, and burned or blackened surfaces. Discs that are stripped of their facing have been overheated and subjected to abuse. In some cases, the friction discs and steel plates can be welded together. This occurs when the facing comes off the disc because of extreme heat. As the facing comes off, metal-to-metal contact is made between the discs and steel plates. The friction that results from the clutch as it tries to engage can cause the discs and plates to fuse together. This may lock the clutch in an engaged condition and can cause a number of problems.

You Should Know

*New friction discs should not be used with used steel plates unless the steel plates are deglazed by sanding their faces with 320-grit emery paper.*

If the discs do not show any signs of deterioration, squeeze each disc to see if fluid is still trapped in the facing material. If fluid comes to the surface, the disc is not glazed.

**Figure 12.** Carefully examine the pressure plate and the friction discs.

Glazing seals off the surface of the disc and prevents it from holding fluid. Holding fluid is basic to proper disc operation. Fluid stored in the friction material cools and lubricates the facings as it transfers heat to the steel plates and also carries heat away, as some oil is spun out of the clutch pack by centrifugal force. This helps avoid the scorching and burning of the disc.

Clutch discs must not be charred, glazed, or heavily pitted. If a disc shows signs of flaking friction material or if the friction material can be scraped off easily, replace the disc. A black line around the center of the friction surface also indicates that the disc should be replaced. Examine the teeth on the inside diameter of each friction disc for wear and other damage.

Wave plates are used in some clutch assemblies to cushion the application of the clutch **(Figure 13)**. These should be inspected for cracks and other damage. Never mix wave plates from one clutch assembly to another. As an aid in assembly, most wave plates will have different identifying marks.

The clutch pistons **(Figure 14)** are checked for cracks, warpage, and fit in their bores. Carefully examine the seal ring grooves and inside diameter of the piston for cracks, nicks, and burrs. Excessive pump pressures can accelerate groove wear. The excess pressure forces the seal rings against the sides of the grooves so hard that fluid cannot get between the ring and groove to lubricate the ring. Excessive pressure can be caused by a stuck pressure regulator valve problem.

The reverse side of the pump cover is the clutch support that incorporates seal rings for the fluid circuits leading to the clutch drums. The seal rings fit loosely into the grooves in the clutch support and rely on pump oil pres-

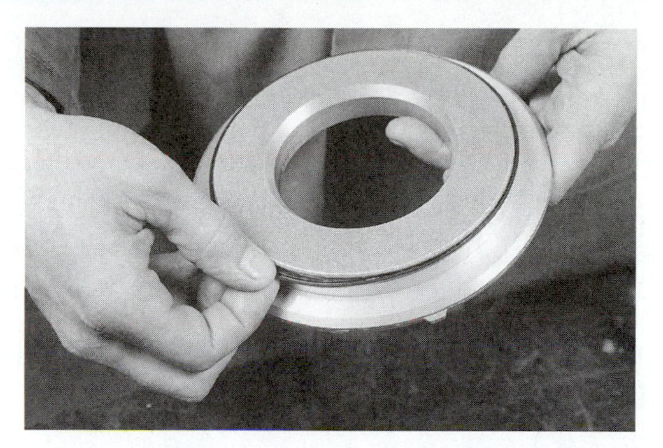

**Figure 14.** Carefully inspect the piston in each multiple-disc assembly.

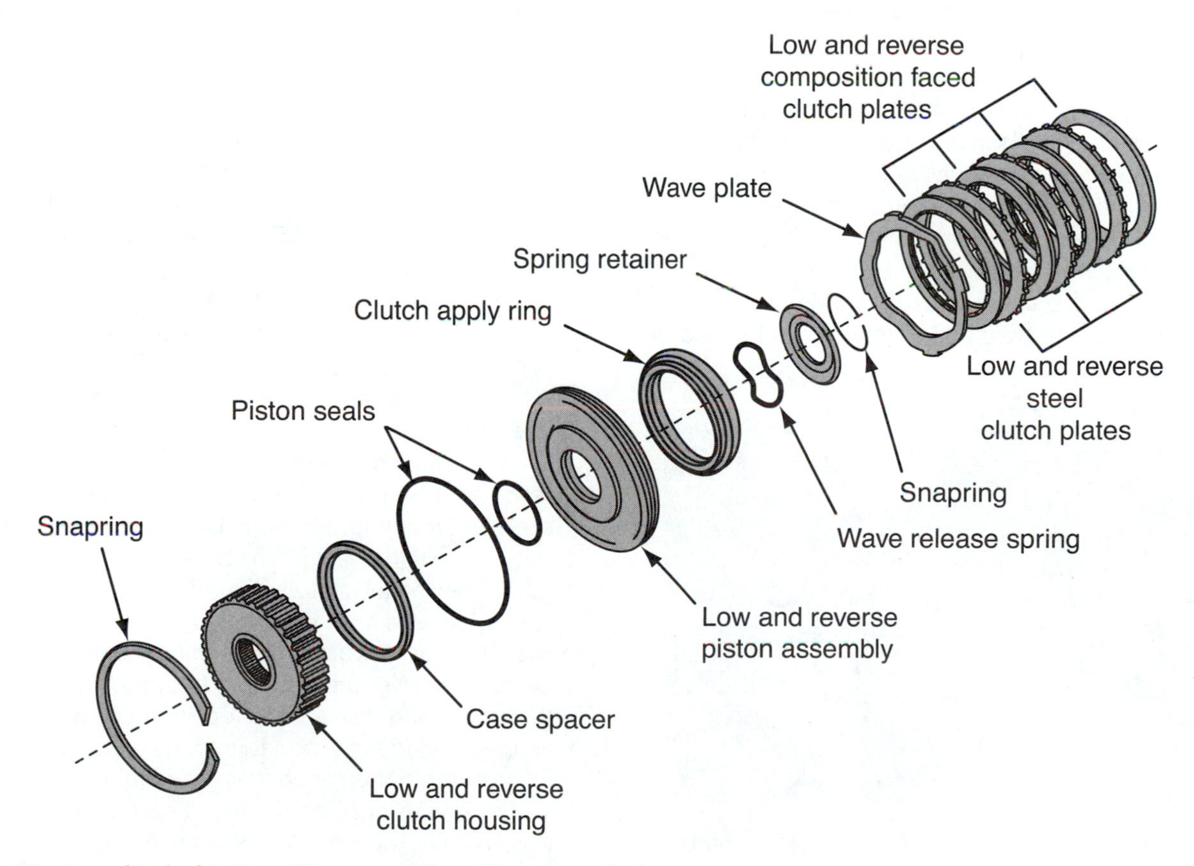


**Figure 13.** A multiple friction disc assembly with a waved plate.

sure to push them against the side of the groove to make the seal. The seal rings should be checked for side play in the grooves and for proper fit into the drum. Check the grooves for burrs, step wear, or pinched groove conditions. It should be noted that these seals rotate with the drum. Any condition that hinders rotation will cause the ring seals to bind, resulting in drum wear. This can destroy the drum if it is not corrected.

Carefully inspect aluminum pistons, which may have hairline cracks that will cause pressure leakage during use. This could cause clutch slippage, as hydraulic force is lost as the piston is pushed against the discs. If burned discs were found in the clutch pack, be sure to check for defects and cracks in the piston.

Stamped steel pistons have replaced most aluminum pistons because they are cheaper to produce. Aluminum pistons require a casting process followed by machining, whereas stamped pistons only require a simple spot weld to finish them. Cracks in stamped pistons are more evident than in aluminum ones. Also, look for any separation at the points of the spot weld.

It is common to find annular check balls in clutch pistons. These balls allow for air release when the bore is being filled with ATF and also allow for a quick release of pressure, which prevents residual pressure when the clutch is released. Inspect each check ball to be sure it is free in its bore. A fine wire can be pushed into the bore, followed by a spray of cleaner, to remove any stubborn deposits. The bore should then be blown out with compressed air.

A check ball may also be located in the drum **(Figure 15)** and should be checked. However, it is often very difficult to use the same cleaning techniques as used on pistons. A quick way to determine if the check ball is free is to shake the clutch drum to hear the relief check ball rattle. If the check ball does not rattle, replace the drum.

The ability of the check ball to seal is important. Pour clean solvent into the bore and observe the other end of the bore. If fluid leaks out, the ball is not seating and the piston or drum should be replaced.

**Figure 15.** Check the movement of the check balls in the clutch pistons and drums.

Examine the outside surface of the drum for glazing. Glazing can be removed with emery cloth. Also, check the drum's cylinder walls for deep scratches and nicks.

Inspect the front clutch bushing for wear and scores. If the bushing is worn, replace it. Also, inspect any bushings found in the clutch drums for excessive wear, scoring, or looseness in the drum bore. Replace as needed.

## Clutch Pack Reassembly

Begin assembly of a clutch unit by gathering the new seals and other new parts that may be necessary. Prior to installation, all clutch discs and bands are to be soaked in the type of transmission fluid that will be used in the transmission. The minimum soak time is one half hour. Be sure all discs are submerged in the fluid and that both sides are coated.

Before fitting the new rubber seals to the piston, check them against the old seals. This will ensure correct sizing and shape of the new seal. Most overhaul kits include more seals than are required to complete the job. This is because changes in transmission design may dictate the use of a seal or gasket of different design or size. Therefore, both the old design seal and the new design seal will be included in the kit. The rebuilder must check to be sure the correct seal is being used. This also holds true for cast iron and Teflon ring seals.

Once the correct seals are chosen, they can be installed on the piston. Remember to position lip seals so they face the direction that the fluid pressure comes into the drum. Seals should be lubricated with automatic transmission fluid, transjel, or petroleum jelly. Never use chassis lube, "white lube," or motor oil. These will not melt into the transmission fluid as it heats up, but they will clog filters and orifices or cause valves and check balls to stick.

Manufacturers switched to Teflon seals for these positions because they help reduce the drum bore wear. During an overhaul, they may be replaced with hook-end type steel rings.

Care must be taken to be sure the ends of scarf cut Teflon seals are installed correctly. These rings also must be checked for fit into the drum. They should have a snug—but not too tight—fit into the bore.

Some manufacturers use an endless type of Teflon seal. These seals must be installed with special sleeves and pushing tools to avoid overstretching the seal. The seals are first pushed over the installing sleeve to their location in the groove, and then a sizing tool is slipped over the sleeve and seal ring to push the seal into the groove.

Assemble the piston, being careful not to allow the seal to kink or become damaged during installation. The piston can now be installed. Several methods may be used to aid piston installation. Lathe-cut seals can be helped into their bores by using a thin feeler gauge mounted on a handle. These are available from most automotive tool suppliers. Pistons with lathe-cut seals are installed by first positioning the piston in the bore of the clutch drum. Then slowly work the piston down in the bore until resistance is felt. Using the feeler-type seal installer, work your way around the outer circumference of the seal using a downward action followed by a clockwise pulling motion as you push the seal back into the groove in the piston.

Occasionally a piston will not allow access to the outer seal area. A large chamfered edge is at the top of the seal bore to allow the seal to be worked into the bore without the help of any special tools. The piston can be installed by rotating the piston as you push down. Use even pressure. Uneven pressure also can cause the ring seal to be pushed out and/or tear.

**You Should Know** *A wax stick can be used to coat lathe cut seals for installation. This is available under the trade name "Door Ease®." If Door Ease is used, do not coat the drum bore with ATF. The Door Ease works fine by itself.*

Pistons with lip seals require a more delicate installation. The lips can be bent back or torn unless proper caution is taken during installation. The basic shape of a lip seal makes it necessary to use an installation tool. The lip must be pushed back toward the piston body in order to allow the seal to enter the bore. Lip seals will often stick in snapring grooves as you try to slip the piston and seals into the drum. Piano wire installers can be used to roll lip seals back away from the snapring grooves or the bore of the drum **(Figure 16)**. The round cross-section of the wire prevents cutting or tearing of the seal lip during installation.

While holding the piston as squarely in the drum as possible, work the tool around the lip to allow the seal to enter into the bore or around the center of the drum. Do not apply too much downward force while working the seal into the bore. The piston will fall into place once the lip is fully inserted.

Pistons with multiple seals will require special care to avoid damaging the other seals as you work on one seal. Multiple seal piston installation is made simple by using plastic ring seal installers. These rings compress the seals back into the piston grooves so they will not hang up. Often two installers are used at the same time.

Regardless of the type of tool used to install the piston seals, always take your time to avoid tearing or rolling the new seals during installation.

When working with stamped steel pistons, install the ring spacer on the top side of the piston, making sure it has the correct thickness for the application.

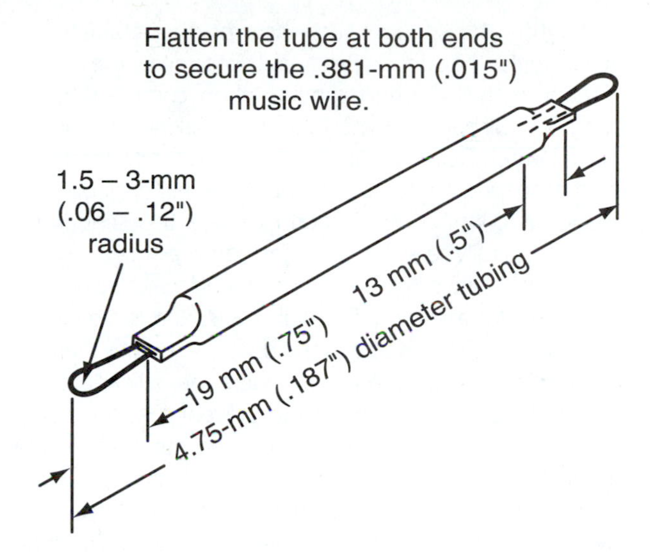


**Figure 16.** Piano wire-type piston installation tool.

Once the piston seals enter the bore, push the piston all the way down until it stops. Then, lift the piston up slightly. Push it back down to the bottom of the bore to be sure it is all the way down. After the piston is installed, rotate the piston by hand to ensure that there is no binding.

Reassemble the springs and retainer after the piston has been installed. Place the single or multiple spring set onto the piston. Make sure the loose spring sets are spaced as they were during teardown. Set the retainer plate over the springs. Make sure the retainer is facing the correct direction **(Figure 17)**. On some transmissions, if the retaining ring and springs are faced in the wrong direction, the clutch assembly will work well during an air check, but it will not apply the clutches during use.

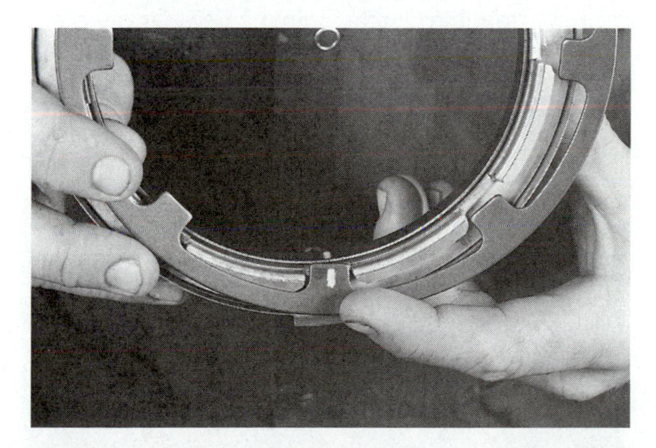

**Figure 17.** On many transmissions, the friction disc pack's snapring and retainer should be indexed before they are removed and then returned to the same relationship during reassembly.

Position the spring compressor on the retainer plate and compress the springs. Be careful not to allow the retainer to catch in the snapring groove when compressing the spring. This will bend or distort the plate, making it unsuitable for use. Remember to compress the springs only enough to get the snapring into its groove. Use snapring pliers to expand or contract the snapring. Once the snapring is installed and fully seated, release the compressor.

Clutches with a Belleville return spring may not require a compressor for assembly. The Belleville spring is merely laid on the piston and centered in the bore. Sometimes there is a wire ring inserted on the piston where the Belleville spring touches the piston. This is used to prevent the steel spring from chafing the aluminum piston. If left out, there will be too much end play, and piston damage will occur.

The large snapring that retains the Belleville spring is now inserted in its groove. This snapring may be either flat or of wavy construction. There also may be a plastic spacer ring used under the snapring to center the spring. Be certain the snapring seats firmly against the drum. Then install the pressure plate on top of the Belleville spring.

Continue clutch assembly by stacking the clutch pack. Begin by installing the pressure plate with the dish facing outward. If a wavy plate or cushion spring is used, it will normally be installed next to the piston. Alternately stack the steels and friction discs until the correct number of plates has been installed **(Figure 18)**. Place the backing plate (the thickest steel plate) on top and install the retainer ring in its groove. Assemble the remaining clutch packs.

Not all models of a transmission use the same number of discs and steels in their clutches. Most overhaul kits supply enough discs and steels to rebuild all models. They are likely to have more discs and steels than are required. This is why a technician should always note how many discs and steels were in each pack while disassembling the transmission. Trying to fit all of the supplied discs and steels may result in clutches with no free play or no room for the snapring. If these problems come up when stacking the pack, refer to a service manual. It may list the number of discs and steels for the model you are building. After the plates are installed, position the retainer plate and install the retaining snapring.

**You Should Know** *Steel plates and friction discs from the same transmission may look the same but may have different thicknesses. Compare the new plates and discs to the ones removed from each clutch during teardown. Incorrect disc and steel thicknesses can cause headaches, if you are not aware of this fact. The new disc facings should match the type that was removed in both thickness and grooving if correct clutch engagement is to be maintained.*

## Clearance Checks

The clearance of a clutch pack is critical for proper transmission operation. Excessive clearance will cause delayed gear engagements, whereas too little clearance causes the clutch to drag. Adjusting the clearance of multiple-friction disc assemblies can be done with the large outer snapring in place.

With the clutch pack and pressure plate installed, use a feeler gauge to check the distance between the pressure plate and the outer snapring **(Figure 19)**. Clearance also can be measured between the pressure plate and the uppermost friction disc. If the clutch pack has a waved snapring, place the feeler gauge between the flat pressure plate and the wave of the snapring farthest away from the pressure plate. Compare the distance to specifications.

Clearances also can be checked with a dial indicator and hook tool. The hook tool is used to raise one disc from its

**Figure 18.** Alternately stack the steels and friction discs until the correct number of plates has been installed.

**Figure 19.** Measuring clutch clearance with a feeler gauge.

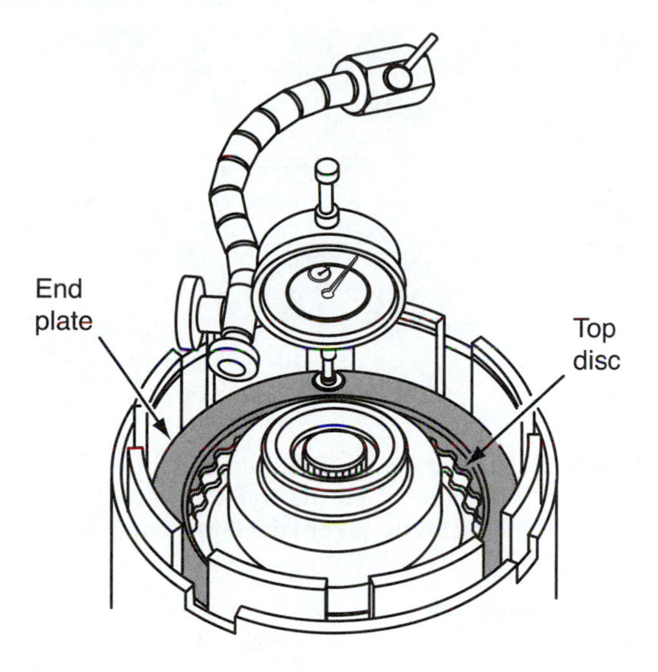


| | Service limit | |
|---|---|---|
| 1st | 0.65–0.85 mm | (0.026–0.033 in.) |
| 2nd | 0.65–0.85 mm | (0.026–0.033 in.) |
| 3rd | 0.40–0.60 mm | (0.016–0.024 in.) |
| 4th | 0.40–0.60 mm | (0.016–0.024 in.) |
| Low-hold | 0.80–1.00 mm | (0.031–0.039 in.) |

**Figure 20.** Dial indicator setup for measuring clutch clearances.

downward position and the amount that it is able to move is recorded on the dial indicator. This represents the clearance.

Another way to measure clearance is to mount the clutch drum on the clutch support and use 25–35 psi of compressed air through the oil pump body channels to charge the clutch. Clearance can be measured by mounting a dial indicator so that it reads backing plate movement as the air forces the piston to apply the clutch **(Figure 20)**.

If the clearance is greater than specified, install a thicker snapring to take up the clearance. If the clutch clearance is insufficient, install a thinner snapring. Another way to adjust clutch clearance is to vary the thickness of the clutch pressure plate. By using a pressure plate of the desired thickness, you can obtain adequate clutch clearance.

## Air Testing

After the clearance of the clutch pack is set, perform an air test on each clutch assembly **(Figure 21)**. This test will verify that all of the seals and check balls in the hydraulic component are able to hold and release pressure.

**Figure 21.** To test a clutch, apply air to the hole in the test plate designated for that clutch.

Air checks also can be made with the transmissions assembled. This is the absolute best way to check the condition of the circuit, because there are very few components missing from the circuit. Testing with the transmission assembled also allows for testing of the servos.

Install the oil pump assembly with its reaction shaft support over the input shaft and slide it into place on the front clutch drum. When the clutch drum is mounted on the oil pump, all components in the circuit can be checked. If the clutch cannot be checked this way, blocking off the apply ports with your finger and applying air pressure to one clutch apply port will work.

**You Should Know** *FWD transaxles that use a chain to connect the torque converter to the input shaft do not have a common shaft for the clutches and the oil pump. Therefore, it is impossible to air test the drums when they are mounted on the oil pump. Instead, the drums should be tested when they are mounted on the front sprocket support.*

To conduct an air test with the transmission assembled, invert the entire assembly and place it in an open vise or transmission support. Pour clean ATF into the circuits that will be tested. Try to get as much fluid in the circuit as you can. Then install an air test plate. On many transmissions, it is necessary to remove the valve body to mount the test plate. Then air test the circuit using the test hole designated for that clutch. Be sure to use low pressure air (25–35 psi) to avoid damage to the seals.

When applying air pressure, you may notice some escaping air at the metal or Teflon seal areas. This is normal but not necessarily desirable, as these seals have a controlled amount of leakage designed into them. Basically, the less a seal leaks, the better a seal it is. There should be no air escaping from piston seals. The clutch should apply with a dull but positive thud. It should also release quickly without binding. Examine the check ball seat for evidence of air leakage.

## CLUTCH VOLUME INDEX

Many electronically controlled transmissions regulate shift feel by controlling the volume of fluid used to apply or release a clutch or brake. The computer monitors gear ratio changes by monitoring the input and output speed of the transmission. By comparing the signal from the turbine or input speed sensor to the signal from the output speed sensor, the computer can determine the operating gear. The computer also monitors the **Clutch Volume Index (CVI)**. CVIs can be monitored with a scan tool and the readings compared to charts given in the service manual. Keep in mind the CVI lets you know how much fluid is needed to apply or release a friction component. When it is out of spec, a problem is indicated.

Based on its constant calculation of gear ratio, the computer can monitor how long it takes to change a gear **(Figure 22)**. If the time is too long, more fluid is sent to the apply device. The volume of the fluid required to apply the friction elements is continuously updated as part of the transmission's adaptive learning. As friction material wears, the volume increases.

## SERVOS AND ACCUMULATORS

On some transmissions, the servo and accumulator assemblies are serviceable with the transmission in the vehicle **(Figure 23)**. Others require the disassembly of the transmission. Internal leaks at the servo or clutch seal will cause excessive pressure drops during gear changes.

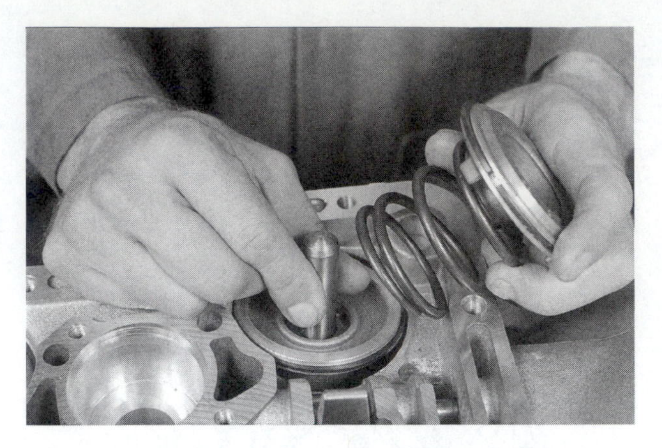

**Figure 23.** The servos in some transmissions are serviceable when the transmission is in the vehicle and is contained in its own bore.

Before disassembling a servo, carefully inspect the area to determine the exact cause of the leakage. Do this before cleaning the area around the seal. Look at the path of the fluid and identify possible sources. These sources could be worn gaskets, loose bolts, cracked housings, or loose line connections.

Inspect the outside area of the seal. If it is wet, determine if the oil is leaking out or if it is merely a lubricating film of oil. When removing the servo, continue to look for causes of the leak. Check both the inner and outer parts of the seal for wet oil, which means leakage. When removing the seal, inspect the sealing surface, or lips, before washing.

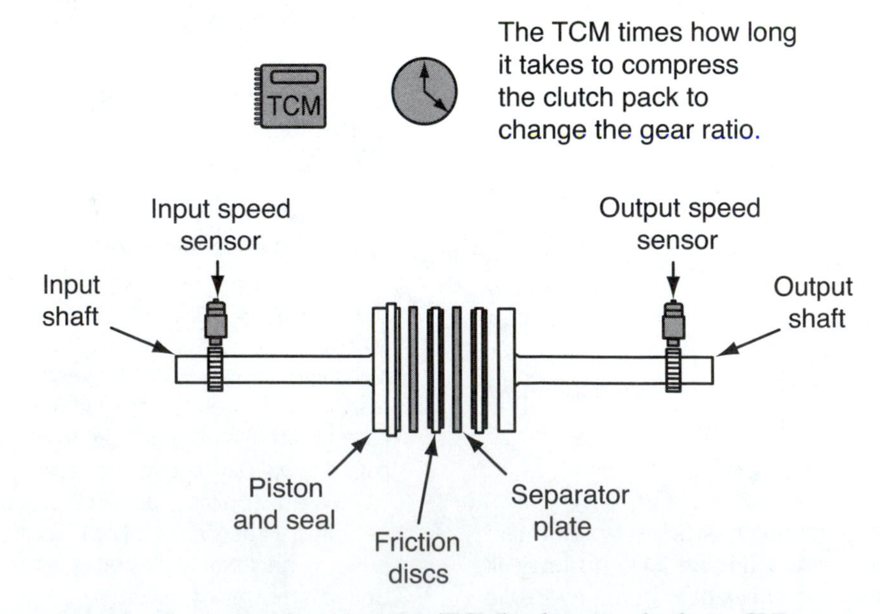


**Figure 22.** Things considered by the system's computer (TCM) when it calculates CVI.

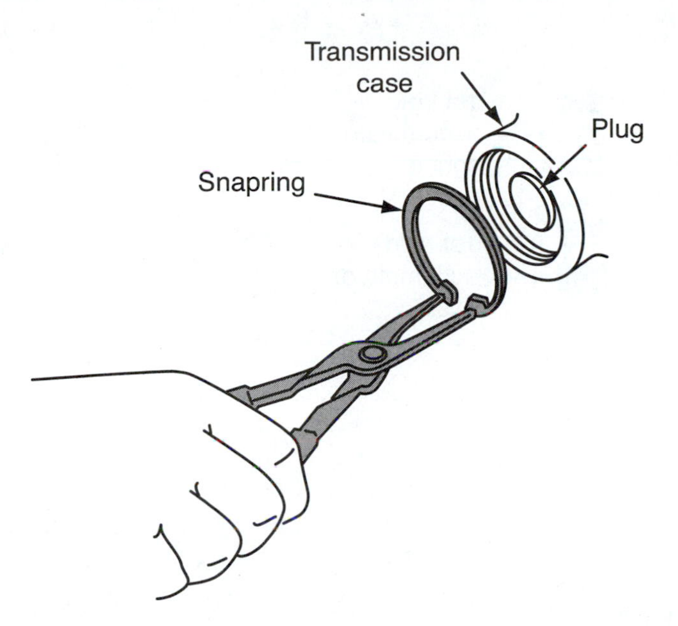


**Figure 24.** Servos and accumulators are normally retained by a snapring.

Look for unusual wear, warping, cuts and gouges, or particles embedded in the seal.

Band servos and accumulators are basically pistons with seals in a bore held in position by springs and retaining snaprings **(Figure 24)**. Remove the retaining rings and pull the assembly from the bore for cleaning. Cast iron seal rings may not need replacement, but rubber and elastomer seals should always be replaced. Most technicians will replace all seals.

Also, check the condition of the pins, piston, and springs. Look for signs of wear and damage. Also, check the fit of the pins and pistons in the case. The bores in the case should not allow the pins and pistons to wobble. If they do, they are worn or the bores in the case are worn.

## Accumulators

To disassemble an accumulator, remove the accumulator plate snapring or unbolt the cover **(Figure 25)**. After removing the accumulator plate, remove the spring and pistons. If rubber seal rings are installed on the piston, replace them. Lubricate the accumulator piston ring and install it on the piston. Lubricate the accumulator cylinder walls and install the accumulator piston and spring. Then, install the accumulator plate and retaining snapring.

Many accumulator pistons can be installed upside down. This results in free travel of the piston, or excessive compression of the accumulator spring. Note the direction of the piston during the teardown, as you will not always find a good reference to follow during reassembly. It is quite common for manufacturers to mate servo piston assemblies with accumulators. This takes up less space in the transmission case and, because they have the same basic shape, can reduce some of the machining during manufacture.

## Servos

A servo **(Figure 26)** is disassembled in a similar fashion. The servo's piston, spring, piston rod, and guide should be cleaned and dried. Check the servo piston for cracks, burrs, scores, and wear. Aluminum servo pistons should be carefully checked for cracks and their fit on the guide pins. Cracked pistons will allow for pressure loss, and being loose on the guide pin may allow the piston to bind in its bore. The seal groove should be free of nicks or any imperfection that might pinch or bind the seal. Clean up any problems with a small file.

## Seals

Most original equipment servo seals **(Figure 27)** are of the Teflon type. These seals will exhibit no feeling of drag in the bore because of the slipperiness of the Teflon. A majority of replacement transmission gasket sets will supply cast iron hook end seal rings in place of the Teflon seals. When installed, the cast iron seals will have a noticeable drag as the servo piston is moved through the bore. This is not a problem and in some cases may even improve operation.

Check the cast iron seal rings to make sure they are able to turn freely in the piston groove. These seal rings are not typically replaced unless they are damaged.

Some servo pistons have a molded rubber seal. This seal is usually replaced during overhaul as the rubber is subject to deterioration.

## Inspection

Inspect the servo spring for cracks. Also, check the area where the spring rests against the case or piston. The spring may wear a groove; make sure the piston or case material has not worn too thin. Some servos have a steel chafing plate in order to eliminate this problem.

Inspect the servo cylinder for scores or other damage. Move the piston rod through the piston rod guide and check for freedom of movement. Check band servo components for wear and scoring. Replace all other components as necessary, and then reassemble the servo assembly.

## Assembly

Servos and accumulators must be assembled in the correct sequence. Incorrect assembly can result in dragging bands and harsh shifting. If the servo pin has Teflon seals, these will need to be cut off with a knife and new Teflon seals installed with the correct sleeve and sizing tools.

When reassembling the servo, lubricate the seal with ATF and install it on the piston rod. On spring-loaded lip seals, make sure the spring is seated around the lip and that the lip is not damaged during installation. Lubricate and

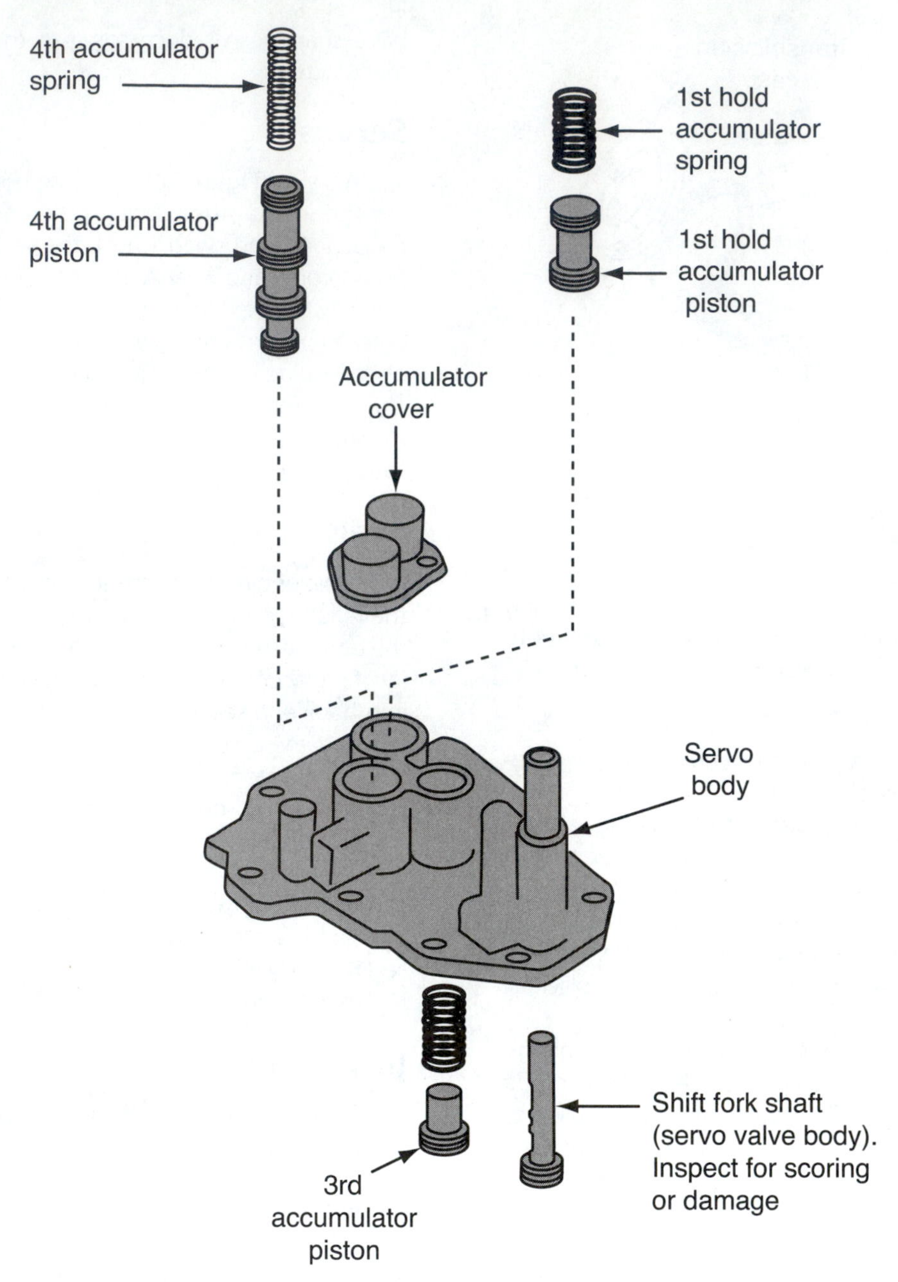


**Figure 25.** An accumulator is disassembled by initially removing the accumulator plate snapring or by unbolting the cover.

install the piston rod guide with its snapring into the servo piston. Then, install the servo piston assembly and return spring into the cylinder.

## Selective Servo Apply Pins

Transmissions without a band adjusting screw use selective servo apply pins, which maintain the correct clearance between the band and the drum. These servo apply pins must be checked for correct length to ensure proper stroking action of the servo piston, as well as shift timing. The travel needed to apply a band relates to the timing of band application. Although valving and orificing determines true shift timing, the adjustment of the servo pin completes the job by providing the correct clearance of the band to the drum. To select the correct apply pin, most transmissions require the use of special tools that simulate a fully applied band.

To select the servo apply pin in a 4L80-E transmission, two special tools and a torque wrench are needed. A gauge pin **(Figure 28)** is placed into the bore for the servo. Then the checking tool **(Figure 29)** is positioned over the bore

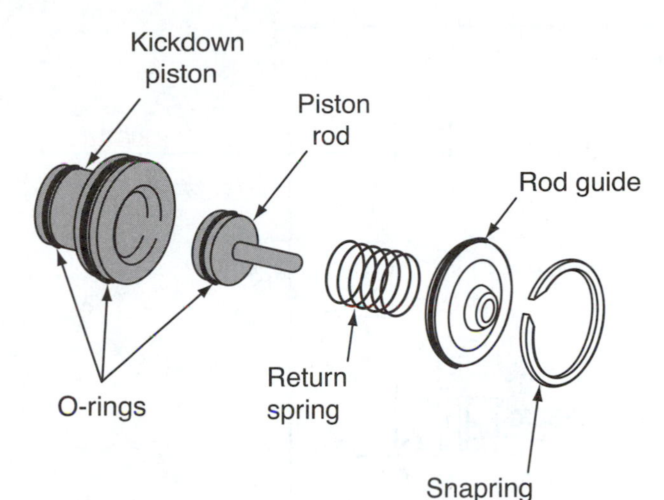


**Figure 26.** A typical servo assembly.

**Figure 27.** Inspect all servo seals.

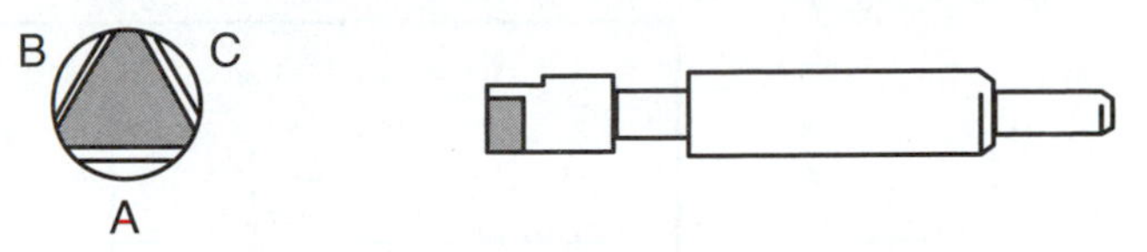


**Figure 28.** A special gauge pin that is placed into the bore for the servo to determine which selective servo pin should be used.

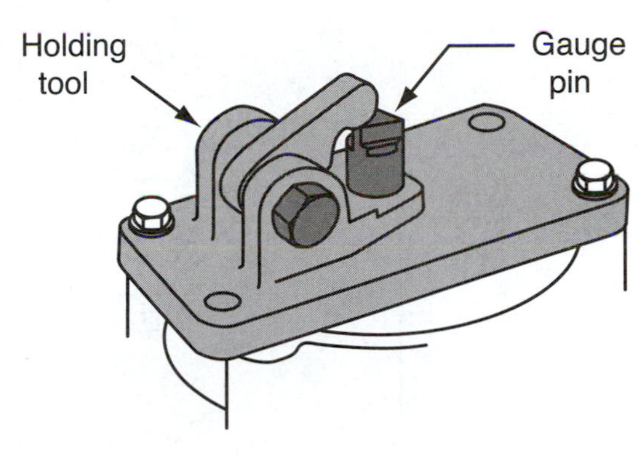


**Figure 29.** The checking tool is positioned over the bore and gauge pin with its hex nut facing the linkage for the parking pawl.

with its hex nut facing the linkage for the parking pawl. The checking tool is fastened with two servo cover bolts that are torqued to a specified amount. The hex nut on the checking tool is then torqued to 25 lbs. per ft. and the exposed part of the pin gauge is studied to determine the proper apply pin length. A chart in the service manual **(Figure 30)** relates the appearance of the exposed part of the pin gauge to the correct pin length.

If a replacement band, drum, or even case has been used, an apply pin check must be made. Because these pins are selective, you will have to start by checking the length of the pin already in the transmission. If it is too long or too short, a new length pin must be installed. These pins are available in the parts departments of most dealerships and parts stores.

## ONE-WAY CLUTCH ASSEMBLIES

One-way clutches are holding devices. They allow a gear set member to rotate in one direction only and are used to ground or effectively hold a member. One-way clutches also are used to transmit torque from one member to another. In these cases, the inner race is splined to one member of the gear set. The outer race is splined to another member.

Because they are purely mechanical in nature, one-way clutches are relatively simple to inspect and test. The durability of these clutches relies on constant fluid flow during operation. If a one-way clutch has failed, a thorough inspection of the hydraulic fluid feed circuit to the clutches must be made to determine if the failure was because of fluid starvation. The rollers and sprags ride on a wave of fluid when they overrun. Because most of these clutches spend most of their time in the overrunning state, any loss of fluid can cause rapid failure of the components.

Sprags, by design, produce the fluid wave effect as they slide across the inner and outer races, making them somewhat less prone to damage. Rollers, because of their spinning action, tend to throw off fluid, which allows more chance for damage during fluid starvation. During the check of the hydraulic circuit, take a look at the feedholes in the races of the clutch. Use a small diameter wire and spray carburetor cleaner or brake cleaner to be certain the feedholes are clear. Push the wire through the feedholes and spray the cleaner into them. Blowing through them with compressed air after cleaning is recommended.

One-way clutches should be inspected for wear or damage, spline damage, surface finish damage, and damaged

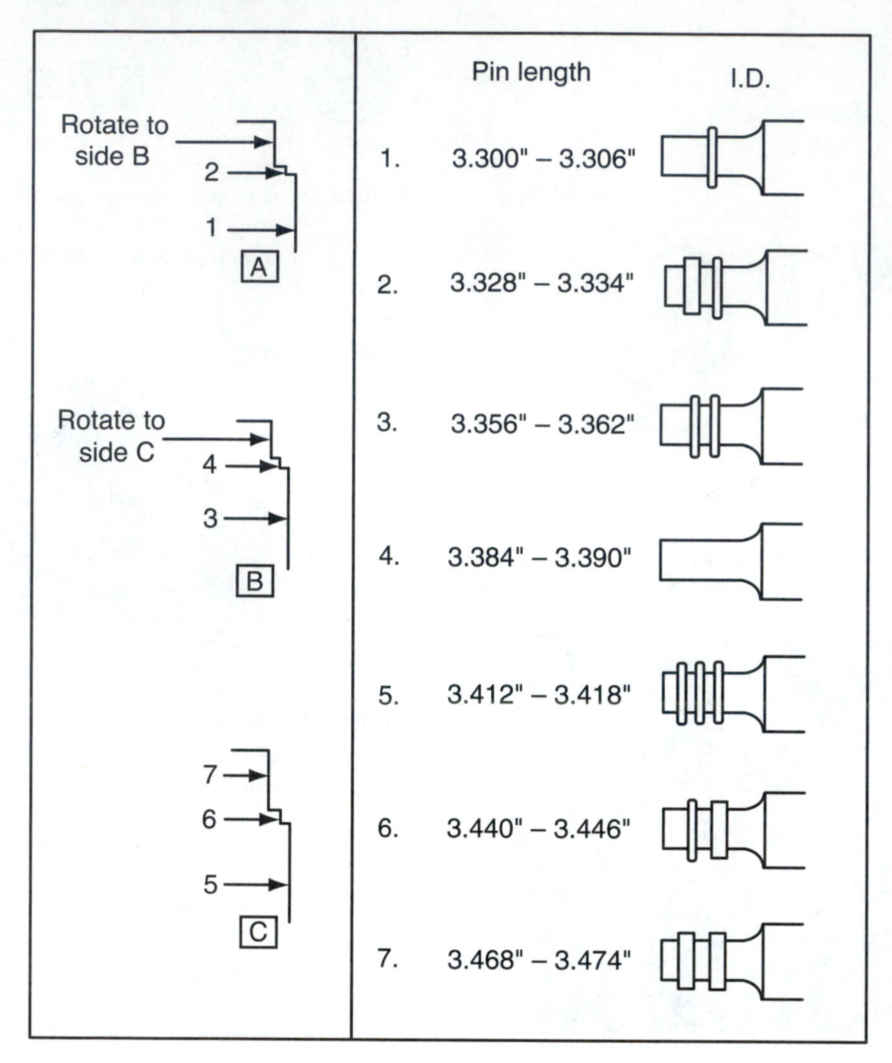


**Figure 30.** The exposed part of the gauge pin should be compared to the chart given in the service manual to determine the correct pin length.

retainers and grooves. Some clutch units have a plastic race that needs to be carefully inspected for chafing or other damage.

## Roller Clutches

Roller clutches should be disassembled for inspection. The surface of the rollers should have a smooth finish with no evidence of any flatness. Likewise, the race should be smooth and show no sign of brinnelling, as this indicates severe impact loading. This condition also may cause the roller clutch to "buzz" as it overruns.

All rollers and races that show any type of damage or surface irregularities should be replaced. Check the folded springs for cracks, broken ends, or flattening out. All of the springs should have approximately the same shape. Replace all distorted or otherwise damaged springs. The cam race of a roller clutch may show the same brinnelled wear, because of impact overloading. The cam surface, like the smooth race, must be free of all irregularities.

## Sprag Clutches

Sprag clutches cannot easily be disassembled; therefore, a complete and thorough inspection of the assembly is necessary. Pay particular attention to faces of the sprags. If the faces are damaged, the clutch unit should be replaced. Sprags and races with scored or torn faces are an indication of dry running and require the replacement of the complete unit.

## Installation

Once the one-way clutches are ready for installation, verify that they overrun in the proper direction. In some cases, it is possible to install these clutches backward; this would allow them to lock in the wrong direction. This would result in some definite driveability problems. One way to make sure you have installed the clutch in the correct direction is to determine the direction of lockup before installing the clutch **(Figure 31)**, then study the

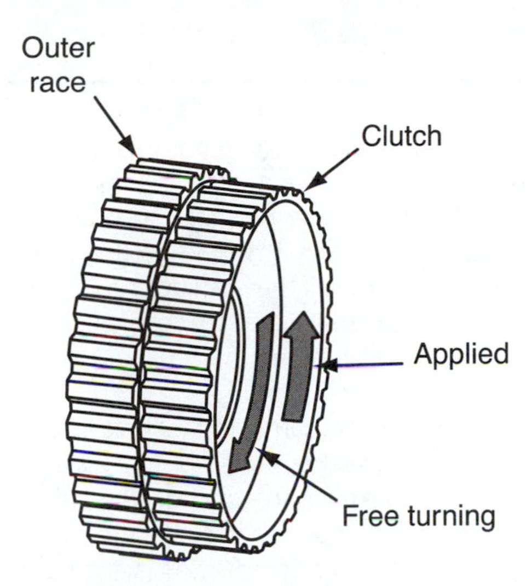


**Figure 31.** One way to make sure you have installed the clutch in the correct direction is to determine the direction of lockup before installing the clutch. The operation of a one-way clutch is checked in the same way.

transmission's powerflow and match the direction of the clutch with it. Most one-way clutches have some marking that indicates which direction the clutch should be set **(Figure 32)**.

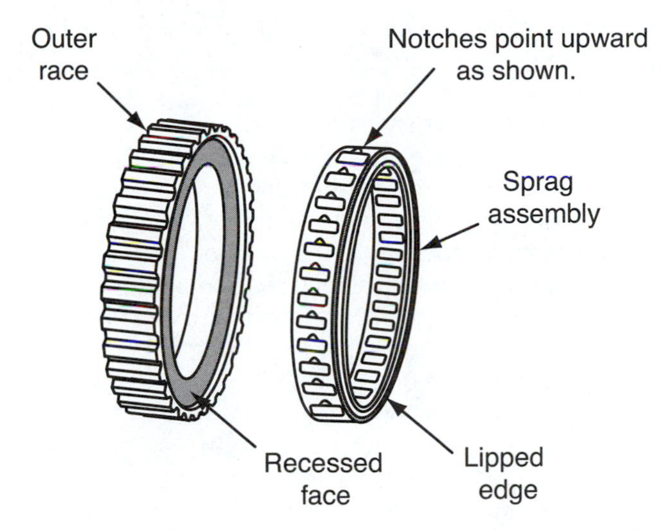


**Figure 32.** Most one-way clutches have some marking that indicates in which direction the clutch should be set.

## Summary

- Bands can be checked for usefulness by squeezing the lining to see if any fluid appears.
- Bands should be checked for wear, chipping, cracks, burn marks, glazing, and nonuniform wear patterns and flaking.
- Band struts, levers, and anchors should be checked for wear and damage.
- The drum surface should be checked for discoloration, scoring, glazing, and distortion.
- A band should be adjusted after it has been installed.
- All friction disc and steel plate packs are held in place by snaprings, which may differ in thickness.
- The splines in the clutch drum, clutch housing, or transmission housing should be checked for defects.
- The return spring assembly must be checked for distortion, overheating, and installed height.
- The clutch pistons should be checked for cracks, warpage, and their fit in their bore.
- When assembling a multiple disc pack, install the discs in an alternating pattern.
- The clearance check of a clutch pack is critical; excessive clearance will cause delayed gear engagements, whereas too little clearance causes the clutch to drag.
- After the clearance of the clutch pack is set, perform an air test on each clutch assembly to verify that all of the seals and check balls in the hydraulic component are able to hold and release pressure.
- Many electronically controlled transmissions regulate shift feel by regulating the volume of fluid used to apply or release a clutch or brake. The computer also monitors the Clutch Volume Index (CVI), which indicates how much fluid is needed to apply or release a friction component.
- Servos and accumulators control the application of bands and clutches to achieve a desirable shift feel.
- Internal leaks at the servo or clutch seal will cause excessive pressure drops during gear changes.
- Check the condition of the servo and accumulator pins, pistons, and springs.
- Inspect the servo or accumulator spring for possible cracks.
- Inspect the servo cylinder for scores or other damage.
- Transmissions without a band adjusting screw use selective servo apply pins, which maintain the correct clearance between the band and the drum.
- The rollers in roller clutches should have a smooth finish with no evidence of any flatness and the folded springs should not have cracks, broken ends, or be flattened out.
- If the faces of the sprags in a sprag-type clutch are damaged, the clutch unit should be replaced.

## Review Questions

1. True or False: Transmissions without a band adjusting screw use selective servo apply pins to maintain the correct clearance between the band and the drum.
2. Which of the following is *not* a good reason to replace a band?
   A. The band is chipped.
   B. The friction material is glazed.
   C. The friction material is soaked with fluid.
   D. The friction material has random wear patterns.
3. When conducting an air test on a transaxle, Technician A notices escaping air at the metal or Teflon seals. He proceeds to replace them. Technician B says that the clutch should apply with a slight delay, then a dull thud. Who is correct?
   A. Technician A only
   B. Technician B only
   C. Both Technician A and Technician B
   D. Neither Technician A nor Technician B
4. Technician A says that all adjustable bands have their locknut and adjusting screw on the outside of the transmission case. Technician B says that some bands do not have adjusting screws. Who is correct?
   A. Technician A only
   B. Technician B only
   C. Both Technician A and Technician B
   D. Neither Technician A nor Technician B
5. Technician A says that transmissions originally equipped with Teflon seals must be refitted with Teflon seals during an overhaul. Technician B says that a press is needed for the installation of Teflon seals. Who is correct?
   A. Technician A only
   B. Technician B only
   C. Both Technician A and Technician B
   D. Neither Technician A nor Technician B
6. Technician A says that incorrect assembly of servos and accumulators can result in dragging bands and harsh shifting. Technician B says that internal leaks at the servo or clutch seal will cause excessive pressure drops during gear changes. Who is correct?
   A. Technician A only
   B. Technician B only
   C. Both Technician A and Technician B
   D. Neither Technician A nor Technician B
7. When checking clutch discs, Technician A says that the steel plates should be replaced if they are worn flat. Technician B says that the friction discs should be squeezed to see if they can hold fluid. If they hold fluid and look okay, they are serviceable. Who is correct?
   A. Technician A only
   B. Technician B only
   C. Both Technician A and Technician B
   D. Neither Technician A nor Technician B
8. Technician A says that friction discs that have blackened surfaces have been subject to overheating. Technician B says that if the friction facing has been burn or worn off, the discs can become welded together and can cause the vehicle to creep when the transmission is in neutral. Who is correct?
   A. Technician A only
   B. Technician B only
   C. Both Technician A and Technician B
   D. Neither Technician A nor Technician B
9. Technician A says that excessive groove wear in a clutch piston can be caused by excessive pump pressures. Technician B says that excessive wear, scoring, and grooving in the bore of a clutch can be caused by a stuck pressure regulator valve. Who is correct?
   A. Technician A only
   B. Technician B only
   C. Both Technician A and Technician B
   D. Neither Technician A nor Technician B
10. Technician A says that if a one-way clutch has failed, a thorough inspection of the hydraulic fluid feed circuit to the clutch must be made to determine if the failure was because of fluid starvation. Technician B says that if the race of a roller-type one-way clutch has random hard spots, it is likely that the roller clutch will "buzz" as it overruns. Who is correct?
    A. Technician A only
    B. Technician B only
    C. Both Technician A and Technician B
    D. Neither Technician A nor Technician B

# Reassembly and Testing

## *Introduction*

After each subunit of the transmission is assembled, reassembly of the transmission can begin. There are many variances to this procedure based on the transmission model. Always refer to the appropriate service manual when assembling a transmission.

### GASKETS

Gaskets are used to create a seal between two flat surfaces **(Figure 1)**. This seal must be able to prevent fluid leaks when undergoing changes in pressure and temperature. Whenever a transmission is overhauled, new gaskets should be used throughout the transmission.

When disassembling a transmission, keep the old gaskets. These will be used to select the correct gaskets for reassembly. Rebuilding kits often come with a few different gaskets for the same part. To help select the correct gasket, compare the old gasket with the new one.

When installing a gasket, make sure both surfaces are clean and flat. Any imperfections in a sealing surface should be corrected before installing a new gasket. These imperfections will prevent the gasket from providing a good seal. To remove minor scratches, use a fine flat file. Make sure you do not remove more metal than is necessary. Also, make sure the surface is flat after you have filed away the scratches.

To ensure flatness, spread a piece of medium-grade (approximately 300-grit) sandpaper on the surface. Then set the sealing surface of the object on the paper. With an even amount of pressure, move the object in a figure-eight pattern around the surface of the sandpaper. When the sandpaper leaves a mark on the entire surface, the surface is flat. Stop sanding and clean it off.

Oil pans typically are made of stamped steel **(Figure 2)**. The thin steel tends to become distorted around the attaching bolt holes. These distortions can prevent the pan from



**Figure 1.** A typical application of a gasket in an automatic transmission.

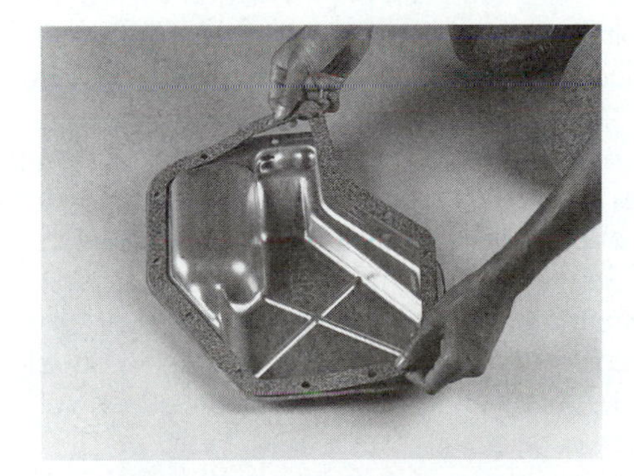

**Figure 2.** A transmission oil pan shown with its gasket.

fitting tightly against the transmission case. To flatten the pan, place the mounting flange of the pan on a block of wood and flatten one area at a time with a ball-peen hammer.

Transmission gaskets should not be installed with any type of liquid adhesive or sealant, unless specifically noted by the manufacturer. If any sealer gets into the valve body, severe damage can result. Also, sealant can clog the oil filter. If a gasket is difficult to install, a thin coating of transmission assembly lube can be used to hold the gasket in place.

One type of gasket that presents unique installation problems is a cork gasket. These gaskets tend to change shape and size with changes in humidity. If a cork gasket is slightly larger than it should be, soak it in water and lay it flat on a warm surface. Allow it to dry before installing it. If the gasket is slightly smaller, soak it in warm water prior to installation.

You Should Know *Another way to make a cork gasket grow is to lay it on a flat, clean, and hard surface, then strike it with a hammer all the way around until it is the correct size.*

Whenever you are installing a cork gasket, make sure you tighten the attaching bolts or nuts in a staggered pattern to the specified torque so that the gasket material is evenly squeezed between the two surfaces. If too much torque is applied, the gasket may split.

## SEALS

Four types of seals are used in automatic transmissions: O-ring, square-cut (lathe-cut), and lip seals, and sealing rings **(Figure 3)**. These seals are designed to stop fluid from leaking out of the transmission and to stop fluid from moving into another circuit of the hydraulic circuit. The latter ensures proper shifting of the transmission.

O-ring and square-cut seals are used to seal nonrotating parts. These seals can provide a good seal at high pressure points within the transmission. Common points that are sealed with these seals are oil pumps, servos, clutch pistons, speedometer drives, and vacuum modulators.

When installing a new O-ring or square-cut seal, coat the entire seal with assembly lube. Make sure you do not stretch or distort the seal when you work it into its holding groove. Some stretching may be necessary to work the seal over a shaft or fitting; however, do not stretch it more than is needed. After a square-cut seal is installed, double-check it to make sure it is not twisted. The flat surface of the seal should be parallel with the bore.

Lip seals are used around rotating shafts **(Figure 4)** and apply pistons. Those used to seal a shaft typically have a metal flange around their outside diameter. The shaft rides on the lip seal at the inside diameter of the assembly. The rigid outer diameter provides a mounting point for the seal and is pressed into a bore. Once pressed into the bore, the outer diameter prevents fluid from leaking into the bore and the inner lip prevents leakage past the shaft.

Piston lip seals are set into a machined groove on the piston. This type of lip seal is not housed in a rigid metal flange. They are designed to be flexible and provide a seal when the piston moves up and down. When the piston moves, the lip flexes up and down. Any distortion or damage to the seal will allow fluid to escape. If fluid escapes from around a piston, the piston will not move with the force and speed it should.

The most important thing to keep in mind when installing a lip seal is to make sure the lip is facing the correct direction. The lip always should be aimed toward the source of pressurized fluid. If installed backward, fluid under pressure will easily leak past the seal. Also, remember to make sure the surfaces to be sealed are clean and not damaged.

Teflon or metal sealing rings are commonly used to seal servo pistons, oil pump covers, and shafts **(Figure 5)**. These rings may be designed to provide for a seal, but they also may allow a controlled amount of fluid leakage. This leakage may be used to lubricate some shaft bushings. Sealing rings are solid rings or are cut. Cut sealing rings are one of three designs: open end, butt end, or locking end.

Solid sealing rings are made of a Teflon-based material and are commonly used and never reused. To remove them, cut the seal after it has been pried out of its groove. Installing a new ring requires special tools. These tools allow you to stretch the seal when pushing it into position. Once in position, the tool holds the seal in place until the components are assembled together. Never attempt to install a solid seal without the proper tools. These seals are soft and are easily distorted and damaged.

Open-end sealing rings fit loosely into a machined groove. The ends of the rings do not touch when they are installed. This type of ring is typically removed and installed with a pair of snapring pliers. The ring should be expanded just enough to move it off or onto the shaft.

Butt-end sealing rings are designed so that their ends butt up or touch each other once the seal is in place. This type of seal can be removed with a small screwdriver. The blade of the screwdriver is used to work the ring out of its groove. To install this type of ring, use a pair of snapring pliers and expand the ring to move it into position.

Locking-end rings may have hooked ends that connect or ends that are cut at an angle. These seals are removed and installed in the same way as butt-end rings. After these rings are installed, make sure the ends are properly positioned and touching.

Before final assembly, it is important to verify that the case, housing, and parts are clean and free from dust, dirt,

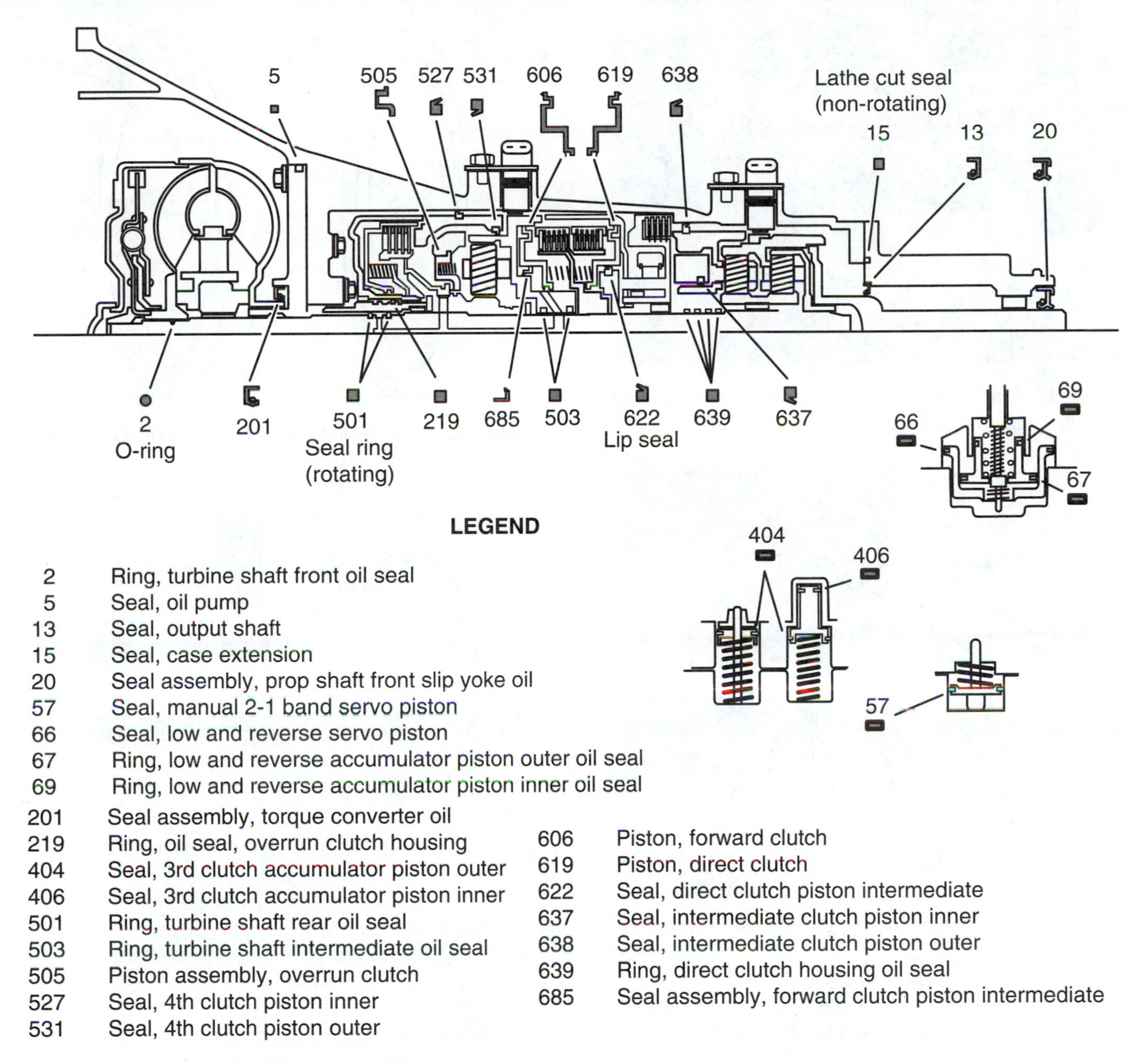


**Figure 3.** Location of the various seals used in today's transmissions.

and foreign matter. Have a tray of clean ATF available for lubricating parts. Also have some assembly lube for securing washers during installation.

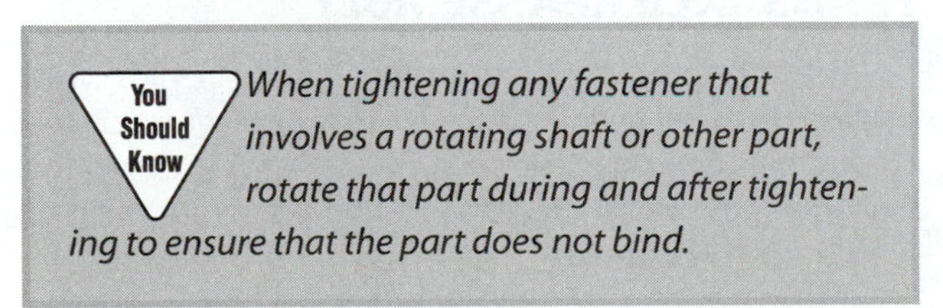


*When tightening any fastener that involves a rotating shaft or other part, rotate that part during and after tightening to ensure that the part does not bind.*

Soak bands and clutches in the fluid for at least 15 minutes before installing them. All new seals and rings should be installed before beginning final assembly.

## Seal Installation

All seals should be checked in their own bore. They should be slightly smaller or larger (+ or − 3%) than their groove or bore. If a seal is not the proper size, find one that is. Do not assume that because a particular seal came with the overhaul kit it is the correct one.

Never install a seal when it is dry. The seal should slide into position and allow the part it seals to slide into it. A dry seal is easily damaged during installation. The following are some guidelines to follow when installing a seal in an automatic transmission.

1. Install only genuine seals recommended by the manufacturer of the transmission.

Figure 4. Lip seals are commonly used around rotating shafts, such as axle shafts.

Figure 5. Teflon seals are commonly used to seal servo pistons, oil pump covers, and shafts.

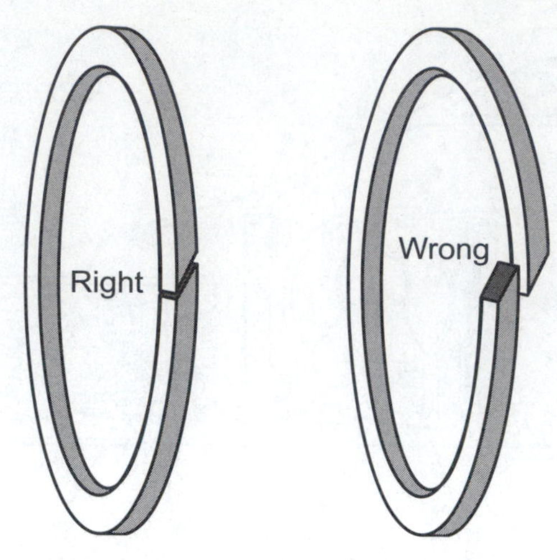


Figure 6. Installation of Teflon (scarf-cut) oil seal rings.

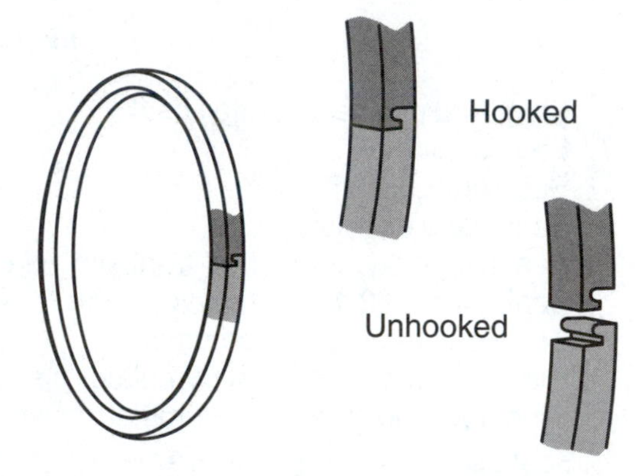


Figure 7. Installation of cast iron oil seal rings.

2. Use only the proper fluids as stated in the appropriate service manual.
3. Keep the seals and fluids clean and free of dirt.
4. Before installing seals, clean the shaft and/or bore area. Carefully inspect these areas for damage. File or stone away any burrs or bad nicks and polish the surfaces with a fine crocus cloth, then clean the area to remove the metal particles.
5. Lubricate the seal, especially any lips, to ease installation.
6. All metal sealing rings should be checked for proper fit. Because these rings seal on their outer diameter, they should be inserted in their bore and feel tight. If the seal has some form of locking ends, these should be interlocked prior to trying the seal in its bore **(Figures 6 and 7)**. The fit of the rings in their shaft groove should also be checked. If the ring can move laterally in the groove, the groove is worn and this will cause internal fluid leaks. To check the side clearance of the ring, place the ring into its groove and measure the clearance between the ring and the groove with a feeler gauge. Typically, the clearance should not exceed 0.003 inch.
7. When checking the clearance, look for nicks in the grooves and for evidence of groove taper or stepping. If the grooves are tapered or stepped, the shaft will need to be replaced. If there are burrs or nicks in the grooves, they can be filed away.
8. Always use the correct driver when installing a seal and be careful not to damage the seal during installation. When possible, press the seal into position. This typically prevents the garter spring from moving out of position during installation.

## TRANSAXLE REASSEMBLY

The assembly of a transaxle typically begins with installing the final drive unit and transfer gear set. Install the output shaft, gear, selective shim, washer, and bolt. Hold the shaft with the special holding tool and tighten the bolt to specifications **(Figure 8)**.

Install the transfer shaft, transfer shaft seals, bearing snapring, selective shim, bearing retainer, and a new shaft nut. Then install the transfer gear, washer, and nut. Hold the shaft with a special tool and tighten the nut to specifications.

**Figure 8.** Install the output gear. While holding the shaft, tighten the bolt to specifications.

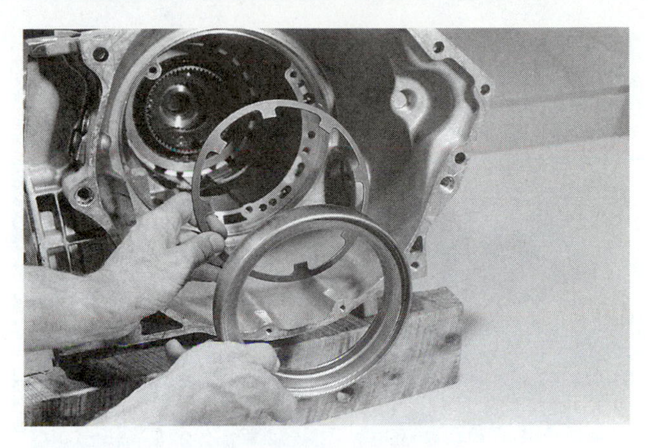

**Figure 9.** The return spring and retainer for a multiple-friction disc pack.

Assemble the drive and driven sprockets with their chain link. Then install the assembly with the appropriate thrust washers. Make sure they are facing the correct direction and the tabs on the washers are aligned. Slightly rotate the shafts when lowering the chain assembly to ensure that the sprockets are fully seated on them.

Check the drive sprocket end clearance. To do this, determine if the machined bolt hole surfaces on the driven sprocket support are above or below the case machined surface. If they are above the surface, place a depth micrometer on the machined bolt hole surface and measure the distance to the case's machined surface. If they are below the case's surface, place the depth micrometer on the case's surface and measure the distance to the machined bolt hole surface. Install the correct thrust washer. Repeat this procedure for the drive sprocket.

Install a new chain cover gasket or apply a 1/8-inch bead of RTV on the chain cover. Then install the cover and tighten the bolts to specifications.

Prior to installing the final drive unit, lubricate all bearings, bushings, and seals with clean ATF. Then position the correct size thrust washers and bearings onto the final drive assembly and install the unit. Petroleum jelly can be used to hold the washers and bearings in place while positioning the unit.

Tap the seal for the manual shaft into the case. Start the manual shaft through the seal and slide the manual detent lever onto the shaft. Then slide the shaft through the park rod actuating lever and tap it into the bore in the case. Install a new lock pin through the case hole. Make sure it is aligned with the groove in the shaft. Install new roll pins. Install the parking pawl assembly. Then install the parking assist lever and snapring.

The rear planetary gearset and accompanying friction and reaction devices should be installed next. When installing a friction disc pack, make sure the individual parts are put in their proper order. Typically, the rear carrier assembly is installed first, followed by the clutch pack, the reaction plate flat snapring, and the reaction plate. This assembly is followed by another clutch pack, a reaction plate, return springs, and a clutch retainer **(Figure 9)**.

On some transaxles, a band is used with the rear planetary gearset. If this is the case, the shell is installed onto the output gear with a thrust washer. The band **(Figure 10)** is then installed in the case. Make sure the band is properly aligned with its anchor pin.

The front planetary gear set is now installed with its appropriate friction and reaction devices. Make sure all thrust washers and seals are lubricated and correctly placed. Make sure the band is aligned to its anchor and strut assembly. Make sure all overrunning clutches are placed in the correct location. Once the planetary unit is in place, check and correct endplay. Install a new oil pump seal, using the correct seal installing tool. Then install the oil pump with a new gasket **(Figure 11)**. Tighten the oil pump's attaching bolts to specifications.

Identify the correct accumulator and piston spring for each accumulator. Then install the accumulator assemblies with their retaining rings.

**Figure 10.** When installing a band, make sure it is properly aligned with its anchor pin.

**Figure 11.** Install the oil pump with a new seal and gasket and tighten the oil pump's attaching bolts to specifications.

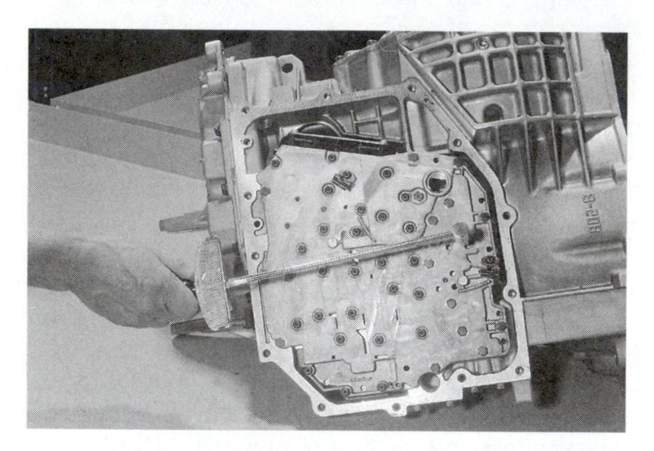

**Figure 12.** Install and tighten the valve body retaining bolts to specifications and according to the order prescribed by the manufacturer.

Then install the valve body assembly. Push the park rod rollers away from the guide bracket when positioning the valve body. Install and tighten the valve body retaining bolts to specifications and according to the correct order **(Figure 12)**.

Connect the manual valve rod to the manual valve. Place the detent lever in the park position and install the retaining clip. Connect the lever link for the TV bracket and install the bracket onto the valve body. Install the solenoid assembly onto the valve body. Connect them and the inhibitor switch to their wiring harness.

Install a new oil filter. Then, install the O-rings onto the oil filter and press the filter into the case. Install the oil pan with a new gasket. Tighten the bolts to specification.

Install the governor assembly, with a new O-ring on its cover. Install the speedometer gear or speed sensor rotor into the governor cover. Then tighten the governor cover bolts to specifications. Install the solenoid assembly with a new gasket and all electrical switches on the outside of the transaxle case. Install dipstick tube and oil filler tube assemblies. Install new Teflon seals onto the pump drive shaft and install the shaft **(Figure 13)** and torque converter.

**Figure 13.** Lubricate the pump drive shaft and carefully install it.

## TRANSMISSION REASSEMBLY

Before reassembly, make sure the case, housing, and parts are clean and free from dust, dirt, and foreign matter (use an air gun). Have a tray available with clean ATF for lubricating parts. Also have a jar or tube of Vaseline for securing washers during installation. All new seals and rings should have been installed before beginning final assembly.

Install a new shaft seal in the case and the output shaft **(Figure 14)**. Then install the rear planetary gearset and drum. Make sure all thrust washers are installed in their correct locations. Install the rear band and anchor assembly and tighten the adjustment screw.

On some models, the output shaft, bearing, and appropriate gauging shims are placed into the transmission housing. The output shaft washer and bolt are then installed. While holding the output shaft and gear assembly, torque

**Figure 14.** Install the output shaft with the governor into the case.

the output shaft nut to specifications. Then install a dial indicator and check the travel of the output shaft as it is pushed and pulled. Remove the gauging shims and install the correct sizes of service shims, output shaft gear, washer, and nut. Torque the output shaft nut to specifications. Using an inch-pound torque wrench, check the turning torque of the output shaft and compare this reading to specifications.

Install the thrust ring, piston return spring, thrust washer, and one-way clutch inner race into the case. Torque the bolts into the inner race from the rear of the case to specifications.

> **You Should Know** *Make sure the return spring is centered onto the race before tightening.*

Install new seals onto the rear piston. Then lubricate them and install the rear piston into the case. After determining the correct number of friction and steel plates, install the steel dished plate first. Make sure the steel disc is facing the correct direction, then install the steel and friction plates, the retaining plate, and the snapring **(Figure 15)**.

> **You Should Know** *When installing the snaprings, make sure the taper faces up or toward the outside of the shaft. This provides a good gripping surface for the snapring pliers during installation and for removal the next time the transmission is taken apart.*

Using a suitable blow gun with a rubber tapered tip, air check the rear brake operation. After the rear brake has been completely assembled, measure the clearance between the snapring and the retainer plate. Select the proper thickness of retaining plate that will give the correct ring to plate clearance if the measurement does not meet the specified limits.

Before installing the rear extension housing, assemble the parking pawl pin, washer, spring, and pawl, and any other assembly that is enclosed by the extension housing. Be sure they are assembled properly. Install the housing and tighten the bolts to specifications.

Place the rear clutch assembly, front clutch drum, turbine shaft, and thrust washer into the housing. Locate and align the rear clutch over its hub. Gently move the rear clutch and turbine shaft around, rotating the assembly to engage the teeth of the friction discs with the rear clutch hub. Align the direct clutch assembly over the front clutch hub. Move the input shaft back and forth, rotating it so the front clutch friction discs engage with the front clutch hub.

**Figure 15.** Insert the disc pack, starting with the steel reaction plate followed by alternating friction and steel discs.

Position the thrust washer to the back of the rear planetary carrier. Install the planetary carrier and thrust washer into the housing. Install the front thrust washer and the drive shell assembly, engaging the common sun gear with the planetary pinions in the rear planetary carrier. Assemble the front planetary gear assembly into the front planetary ring gear. Make sure the planetary pinion gear shafts are securely locked to the planetary carrier.

Install the one-way sprag into the one-way clutch outer race with the arrow on the sprag facing the front of the transmission. Some overrunning clutches do not have an arrow. For these, the manufacturers recommend that an index mark be made on the clutch and the transmission case **(Figure 16)**. This index mark should be made at the nonthreaded bore in the clutch.

Install the connecting drum with the sprag by rotating the drum clockwise using a slight pressure and wobbling to align the plates with the hub and sprag assembly. The connecting drum should now be free to rotate clockwise only. This check will verify that the sprag is correctly installed and

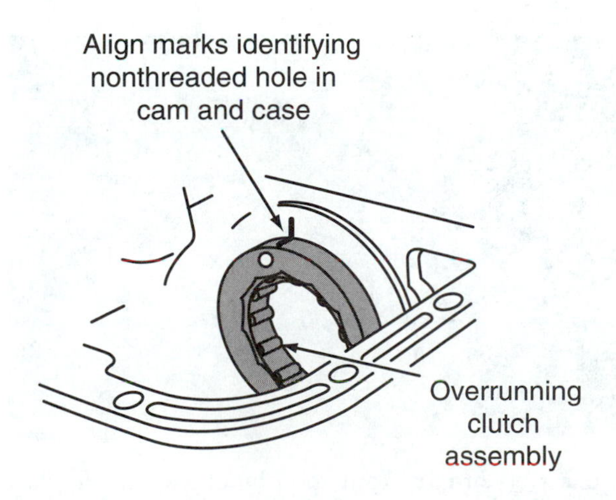


**Figure 16.** Aligning an overdrive clutch to the case.

operative. Now install the rear internal gear and the shaft's snapring. Secure the thrust bearing with petroleum jelly and install the rear planet carrier and the snapring.

> **You Should Know** *This snapring may be thinner than the clutch drum snapring, so be sure you are using the correct size. Should you encounter trouble in having sufficient space to install the snapring into the drum groove, pull the connecting drum forward as far as possible. This will give you sufficient groove clearance to install the drum snapring.*

Total end play should now be checked **(Figure 17)**. Set the transmission case on end, front end up. Be sure the thrust bearings are secured with petroleum jelly. Pick up the complete front clutch assembly and install it into the case. Be sure all parts are seated before proceeding with the measurement. Using a dial indicator or caliper, measure the distance from the rear hub thrust bearing to the case. Next, measure the pump with the front bearing race and gasket installed. Tolerance should fall within specifications. If the difference between the measurements is not within tolerance, select the proper size front bearing race. If it is necessary to change the front bearing race, be sure and change the front clutch thrust washers the same amount.

Install the brake band servo. Use extreme care not to damage the O-rings. Lubricate around the seals. Then install and torque the retainer bolts to specifications. Loosen the piston stem. Install the brake band and strut and finger tighten the band servo piston stem just enough to keep the band and strut snug or from falling out. Do not adjust the band at this time. Air check for proper performance.

Place some petroleum jelly in two or three spots around the oil pump gasket and position it on the housing. Next, align the pump and install it with care. Tighten the pump attaching bolts to specifications in the specified order. Then check the rotation of the input shaft. If the shaft does not rotate, disassemble the transmission to locate the misplaced thrust washer.

Before installing the converter housing, check the bolt holes alignment. Install the converter housing and torque the retaining bolts to specifications **(Figure 18)**.

Now, adjust the band, after you check to make sure that the brake band strut is correctly installed. Before proceeding with the installation of the valve body assembly, perform a final air check of all assembled components. This will ensure that you have not overlooked the tightening of any bolts or damaged any seals during assembly.

Assemble the parking pawl assembly. Place the assembly into its position and install the extension housing with a new gasket, then tighten the attaching bolts to the proper specifications.

Install the valve body. Be sure the manual valve is in alignment with the selector pin. Tighten the valve body attaching bolts to the specified torque.

Before installing the vacuum modulator valve, measure the depth of the hole in which it is inserted. This measurement determines the correct rod length to ensure proper performance. Refer to the service manual to determine the correct rod length based on your measurements. You should note that the actual rod size is slightly longer than the measurement taken. If you do not have the correct chart, it is fairly safe to simply add 0.070 inch to the measurement taken.

Before installing the solenoids, check to verify that they are operating properly. Connect a solenoid checker to each solenoid. When the solenoids are activated, you should hear the solenoid click on. You can also check the integrity of a solenoid with an ohmmeter. If the solenoid is good, install it. If the solenoid does not check out good, replace it.

Install the kickdown, inhibitor, manual lever position, and/or transmission range switch. Again, check the operation of the switches. This is best done by connecting an

**Figure 17.** With the front and planetary units in place, check the total end play of the assembly.

**Figure 18.** Install the converter housing and torque the retaining bolts to specifications.

ohmmeter across the switch and moving the switch through its different settings.

Before installing the oil pan, check the alignment and operation of the control lever and parking pawl engagement. Make a final check to be sure that all bolts are installed in the valve body. Install the oil pan with a new gasket. Torque the bolts to specifications.

Lubricate the oil pump's seal and the converter neck before installing the converter. Install it and make sure the converter is properly meshed with the oil pump drive gear.

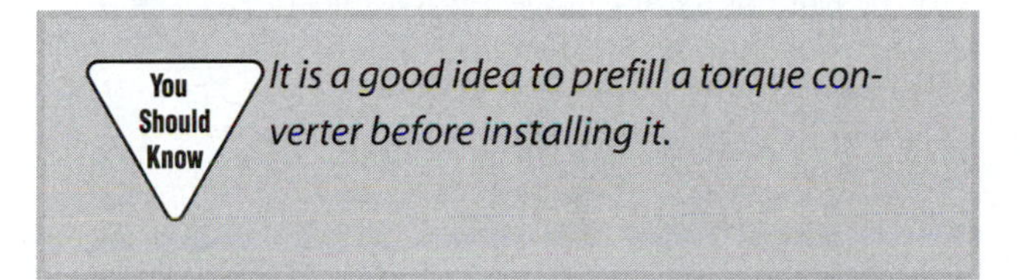


*It is a good idea to prefill a torque converter before installing it.*

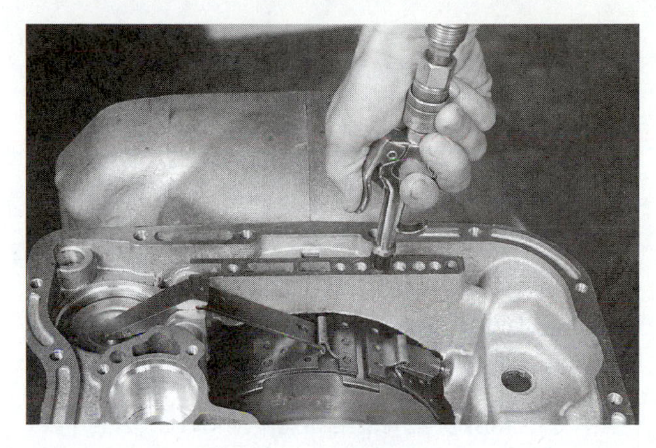

**Figure 19.** Air test the entire transmission before installing the oil filter and oil pan.

## Air Testing

Air checks are the absolute best way to check the condition of a hydraulic circuit, because there are very few components missing from the circuit. Testing with the transmission assembled also allows for testing the servos as well.

With the transmission assembled, pour clean ATF into the circuits that will be tested. Try to get as much fluid in the circuit as you can. Then locate the test ports in the housing **(Figure 19)** or install an air test plate. It may be necessary to remove the valve body in order to mount the test plate. Then air test the circuit using the test hole designated for that clutch. Be sure to use low pressure compressed air (25–35 psi) to avoid damage to the seals. Higher pressures may blow the rubber seals out of the bore or roll them on the piston.

When applying air pressure, you may notice some escaping air at the metal or Teflon seal areas. This is normal but not necessarily desirable, as these seals have a controlled amount of leakage designed into them. There should be no air escaping from the piston seals. The clutch should apply with a dull but positive thud. It also should release quickly without any delay or binding. Examine the check ball seat for evidence of air leakage.

# Summary

- Gaskets prevent fluid leaks when undergoing changes in pressure and temperature.
- When installing a gasket, make sure both surfaces are clean and flat.
- Transmission gaskets should not be installed with liquid adhesives or sealants unless specifically noted by the manufacturer.
- Four types of seals are used: O-ring, square-cut (lathe-cut), lip, and sealing rings.
- Teflon or metal sealing rings are used to seal servo pistons, oil pump covers, and shafts.
- Solid sealing rings are made of a Teflon-based material and are never reused.
- Coat all parts with ATF. Soak bands and clutches in the fluid for at least 15 minutes before installing them.
- The assembly of a transaxle typically begins with installing the final drive unit and the transfer gear set.
- Prior to reassembly, check the service manual for the order and procedure for installing the different components and units.
- Air checks are the best way to check the condition of a hydraulic circuit.

# Review Questions

1. What do the input shaft and the output shaft normally connect to?
2. Technician A says that seals should be kept clean and free of dirt, before and during installation. Technician B says that most seals should be installed dry. Who is correct?

   A. Technician A only
   B. Technician B only
   C. Both Technician A and Technician B
   D. Neither Technician A nor Technician B

3. When discussing transmission housings, Technician A says that good surfaces are necessary to provide a good seal between the mating sections or parts. Technician B says that proper mating surfaces are necessary to keep shafts properly aligned. Who is correct?
   A. Technician A only
   B. Technician B only
   C. Both Technician A and Technician B
   D. Neither Technician A nor Technician B
4. Technician A says that end play checks should be taken before the transmission is disassembled. Technician B says that end play checks should be taken after the transmission is reassembled. Who is correct?
   A. Technician A only
   B. Technician B only
   C. Both Technician A and Technician B
   D. Neither Technician A nor Technician B
5. Technician A says that if sealants get into the valve body, severe damage can result. Technician B says that sealants can clog the oil filter. Who is correct?
   A. Technician A only
   B. Technician B only
   C. Both Technician A and Technician B
   D. Neither Technician A nor Technician B

# Chapter 37 Install Transmission/ Transaxle

## Introduction

The transmission is now ready for installation into the vehicle. Use the reverse of the removal procedures. Never bolt the converter onto the flexplate and then try to install the transmission onto the converter. Remember to follow the correct fluid filling procedure and check the service manual for transmission relearn instructions.

### INSTALLATION

Transmission installation is generally a reverse of the removal procedure. Care must be taken to avoid destroying the new or rebuilt transmission during installation. A quick check of the following will greatly simplify your installation and reduce the chances of destroying something during installation.

- Make sure the block alignment pins (dowels) are in their appropriate bores and are in good shape, and that the alignment holes in the converter housing are not damaged **(Figure 1)**. Also, make sure expansion plugs at the rear of the engine are not leaking.
- Make sure the pilot hole in the crankshaft is smooth and not out-of-round. This will allow the converter to move in and out on the flexplate.
- Make sure the pilot hub of the converter is smooth and cover it with a light coating of chassis lubricant to prevent chafing or rust.
- Make sure the converter's drive hub is smooth and coat it with transjel.
- Secure all wiring harnesses out of the way to prevent their being pinched between the converter housing and engine block. If the wires get pinched, not only will there be a large electrical short but also you may destroy the car's computer.

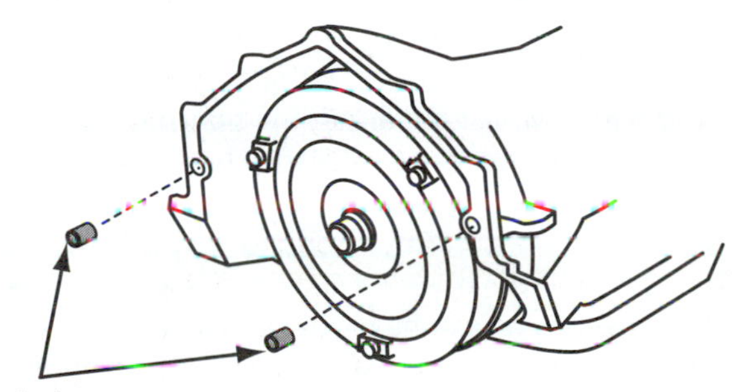


**Figure 1.** Dowel pins help position the transmission to the engine.

- Flush out the oil cooler lines and the cooler itself to remove any material from before that could damage the transmission. Likewise, the converter should be flushed. It is recommended that clutch-type converters be replaced, as it is not possible to tell how much the unit has been damaged.
- Always perform an end play check and check the overall height before reinstalling a torque converter or installing a fresh unit out of the box.
- Pour one quart of the recommended fluid into the converter before mounting the converter to the transmission. This will ensure that all parts in the converter have some lubrication before startup.

Slide the converter into the transmission, making sure that all drives are engaged. Double-check this by using the height dimension **(Figure 2)** you measured during transmission removal. Older transmissions only have the input

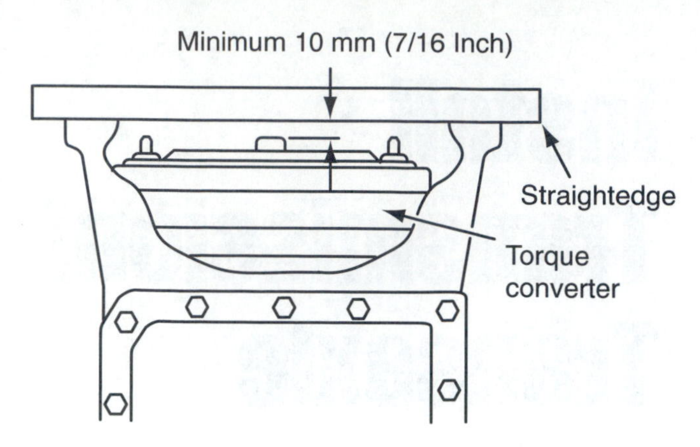


**Figure 2.** Check the installation of the torque converter before attempting to install the transmission.

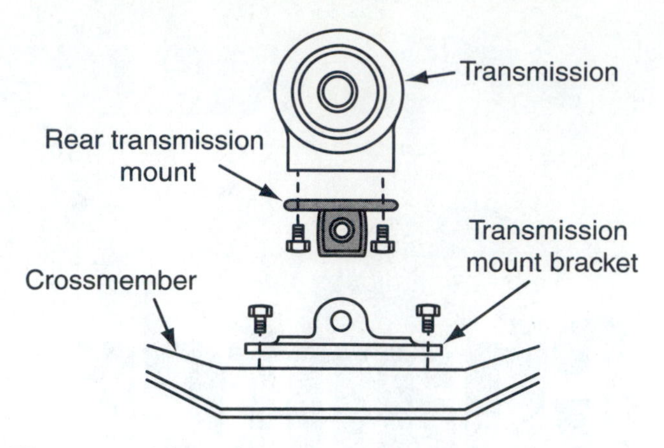


**Figure 3.** After the rear transmission mount is attached to the cross member, the transmission jack can be removed.

shaft stator splines and the oil pump into which the torque converter must engage, whereas later models will have those as well as (possibly) a direct drive shaft or an oil pump drive shaft.

Secure the converter in the transmission as you did during removal. Then transfer the transmission to the jack and move it under the car. Raise the transmission to get close alignment to the engine block. If the transmission has a full circle converter housing, you will need to align the converter drive studs or bolt pads before you push the transmission against the block. Make sure to set the torque converter onto the flexplate in alignment with the marks made during removal. Once the converter is aligned, be sure the block dowel pins line up with the converter housing then push the transmission against the engine block. Check to be sure that nothing is caught between the block and converter housing. Start two converter housing bolts across from each other and slowly tighten them. Check for free rotation of the torque converter when doing this. Then, install the rest of the bolts and torque them to specifications.

On FWD cars, if the engine is held in place with a support bar, the transmission jack may be removed at this time. If the car has a split cradle-type subframe, it should be installed now.

On RWD vehicles, you may need to leave the transmission jack in place while the exhaust cross-overs or frame cross-members are installed. Do not connect the gearshift linkage until the transmission is mounted to the cross-member and the transmission jack is removed. Once the rear mount is attached to the cross-member **(Figure 3)**, the jack may be removed.

Install the converter drive bolts (or nuts). You should notice when installing these fasteners that the converter has some noticeable fore and aft movement. This amount varies with different transmissions but is generally between 1/8 to 1/4 inch. This is normal and necessary as it allows the converter to move on the flexplate and also allows for a noninterference fit at the oil pump drive gear. If there was no movement, premature pump failure will result.

**You Should Know** *Never use an impact wrench on linkage nuts, cooler line fittings, or torque converter bolts. Impact wrenches can drive the bolts through the cover, which will warp the inside surface and prevent proper clutch apply or may damage the clutch-pressure plate.*

The cooler lines should be installed into their bores **(Figure 4)**. Then tighten the fittings, making sure you do not twist or distort the lines. The remaining components: starter, throttle cables, electrical connections, and dipstick tube can now be installed.

When connecting the manual shift linkage **(Figure 5)**, pay attention to the condition of the plastic bushings. If

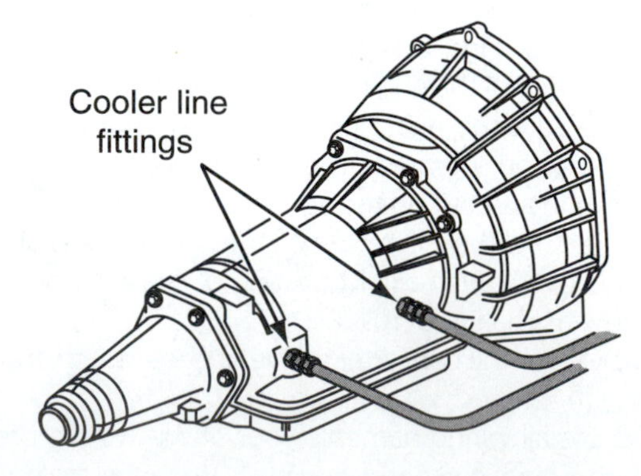


**Figure 4.** When connecting the cooler lines, be careful not to twist or distort the lines.

> **You Should Know** *After the transmission has been installed, make sure everything is properly realigned. Repositioning the transmission an inch or less can have a big effect on the manual shift linkage adjustment.*

these are worn or missing, hard or inaccurate linkage movement will result. Replace the bushings if needed.

Some other areas that require special attention during installation are the grounding straps and any rubber tubes. Often, these are overlooked during transmission removal and installation and can cause problems if not checked. Any rubber tube is suspect, as they are exposed to heat, which causes them to crack. New parts should be used if necessary. The ground straps must be in good condition and free of corrosion as these provide an electrical ground path to the body of the car during operation. Failure to clean or attach these straps or cables also can result in poor signals to the PCM, voltage spikes that can damage the PCM, electrolysis through the fluid that welds internal transmission components, or even fused manual linkage cables as the current looks for a path to flow.

On RWD vehicles, the driveshaft must be installed using the marks you made during removal **(Figure 6)**. Be sure to coat the slip yoke with ATF before sliding it into the extension housing. The drive shafts of FWD cars are installed in the reverse manner in which they were removed and the related components (speedometer gears, strut lower ball joint bolts, etc.) should be installed just as they were before they were removed. Use a new nut on the outer CV joint stud, if the bolt is the self-locking type. The torque of this nut is critical. Torquing the nut should be done with the car's tires on the ground and/or with the brakes held. Air-type impact tools should not be used, as they can damage the wheel bearing or hub.

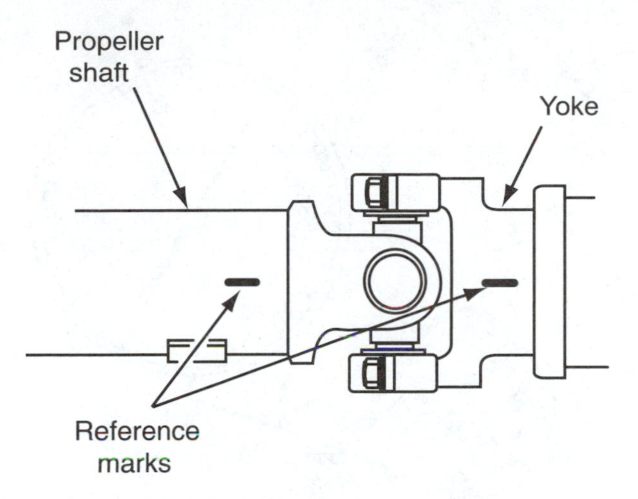


**Figure 6.** When installing the driveshaft, make sure the shaft and yoke are aligned.

Connect the battery cables. Then add about one half of the total quantity of the proper ATF to the transmission. This amount varies, but an average amount is four quarts. Some technicians also will connect a pressure gauge to the transmission **(Figure 7)** and take pressure readings during the initial operation of the fresh transmission. Some imported transmission manufacturers require this be done, because line pressures must be set after startup. Connecting the pressure gauges also will give you an indication of whether or not there is sufficient oil pressure to road test the car.

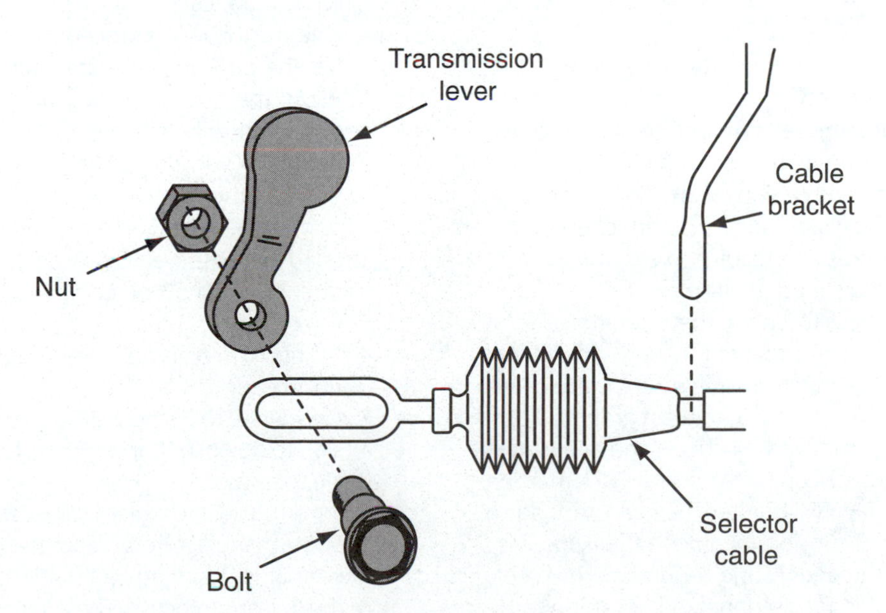


**Figure 5.** The connection for the gear selector cable to the shift lever on a transaxle.

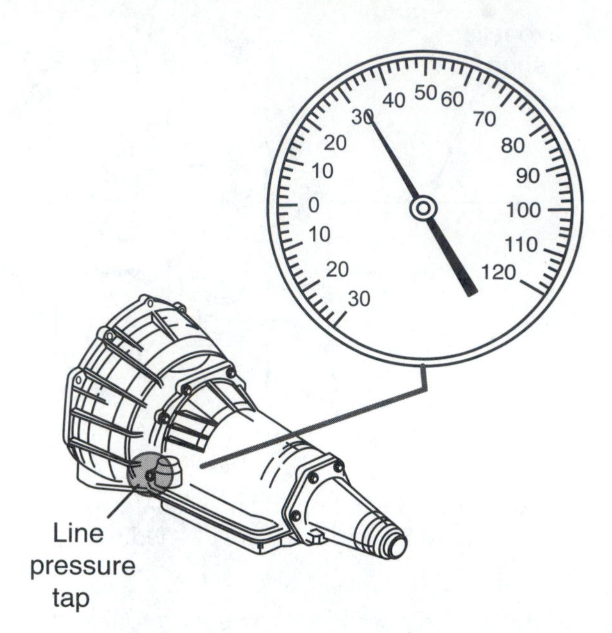


**Figure 7.** Connecting a pressure gauge to the transmission before road testing it is recommended by many technicians and some transmission manufacturers.

> **You Should Know** *On governor-equipped transmissions, put the transmission into reverse as soon as the engine starts. Many transmissions do not send fluid to the governor in reverse, and putting the transmission in reverse prevents the possibility of moving debris from the rebuild into the governor, which would cause it to stick or seize.*

Put the vehicle on a lift, then start the engine and apply the brakes. Move the shift selector through all of the gear ranges. Then place the selector into PARK. Check the fluid level and add ATF until you reach the ADD or COLD mark. Then inspect the transmission for any signs of leakage and loose bolts. If a pressure gauge was attached to the transmission and the pressure reading were fine, disconnect the gauge after you shut down the engine.

Road test the vehicle to check the operation of the transmission and anything else that may have been affected by your work. If any adjustments are necessary, make them. The road test should cover at least 20 miles in order to warm up the transmission completely. Recheck and adjust the fluid level. It should be between the ADD and FULL mark on the dipstick. DO NOT OVERFILL! After the road test, again visually inspect the transmission for signs of leakage. Also carefully look around the engine and transmission to see if any wires, hoses, or cables are disconnected or positioned in an undesirable spot.

> **You Should Know** *If the transmission is computer controlled, check the service manual before taking a road test. Some automatic transmissions require that a "learning procedure" be followed, which includes various driving conditions. Because you need to road test and teach the transmission, you may as well do both at the same time.*

## COMPUTER RELEARN PROCEDURES

Vehicles equipped with engine or transmission computers may require a relearn procedure after the battery has been disconnected. Many vehicle computers memorize and store vehicle operation patterns for optimum driveability and performance. When the vehicle battery is disconnected, this memory is lost. The computer will use default data until new data from each key start is stored. As the computer memorizes vehicle operation for each new key start, driveability is restored. Vehicle computers may memorize vehicle's operation patterns for forty or more key starts. Always refer to the service manual, as some transmissions have the ability to learn quickly when certain conditions are met.

Customers often complain of problems during the relearn stage because the vehicle acts differently than before it was serviced. Depending on the vehicle and how it is equipped, the following complaints may exist: harsh or poor shift quality, rough or unstable idle, hesitation or stumble, rich or lean running, and poor fuel economy. These complaints should disappear after a number of drive cycles. To reduce the possibility of complaints, after any service that requires that the battery be disconnected, the vehicle should be road tested. If a specific relearn procedure is not available, the following procedure may be used:

1. Set the parking brake and start the engine in P or N. Allow the engine to warm up to normal operating temperature or until the electric cooling fan cycles on.
2. Allow the vehicle to idle for about a minute in the N position, then move the gear selector into the D position and allow it to idle in gear for one minute.
3. Road test the vehicle. Accelerate at normal throttle openings (20–50 percent) until the vehicle shifts into top gear.
4. Maintain a cruising speed with a light to medium throttle.
5. Decelerate to a stop. Make sure you allow the transmission to downshift, and use the brakes to bring the vehicle to a stop.
6. If a driveability problem still exists, repeat the sequence.

Some manufacturers recommend a specific relearn procedure that is designed to establish good driveability during the relearn process. These procedures are especially important for all vehicles equipped with an electronically controlled converter and/or transmission. Always complete the procedure before returning the vehicle to the customer.

Chryslers with 41TE and 42LE transaxles are relearned by first warming up the transaxle to normal operating temperatures by allowing the vehicle to idle, then operating the vehicle and maintaining a constant throttle opening during upshifts. This sets the transaxle into the upshift relearn process.

Accelerate the vehicle with a throttle opening of 10–50 degrees. Accelerating the vehicle from a stop to 45 mph with a moderate throttle is sufficient for this part of the procedure. Then operate the vehicle until the transaxle completes 1–2, 2–3, and 3–4 shifts at least twenty times.

Now operate the vehicle at a speed of less than 25 mph and force downshifts with a wide open throttle. Repeat this at least eight times.

Operate the vehicle above 25 mph in fourth gear and force downshifts with a wide open throttle. Repeat this at least eight times. The forced downshifts allow the computer to relearn kickdown operation.

Ford Motor Company also specifies a relearn procedure for some of their transmissions and transaxles. All of the specific procedures begin with a fluid check and warming the ATF to normal temperatures. The procedure has two segments: an idle relearn and a drive relearn. The idle relearn procedure begins with starting the vehicle in P with all accessories off and the parking brake set. Move the gear selector to N and allow the engine to idle for one minute. Move the gear selector to D and again allow the engine to idle for one minute. After the idle relearn procedure is complete, the drive relearn procedure can begin. The drive relearn procedures are specific for the different transmissions.

If the vehicle is equipped with 4R70W or AX4S transmission, road test the vehicle. With the gear selector in overdrive, moderately accelerate the vehicle to 50 mph for a minimum of 15 seconds. The transmission should be in fourth gear at the end of that time. Then hold the speed with a steady throttle and lightly apply and release the brake for about 5 seconds. Then stop and park the vehicle with the gear selector in the D position for a minimum of 20 seconds. Repeat this procedure five times.

If the vehicle is equipped with an E4OD transmission, put the gear selector in the D position and depress the O/D cancel button (the LED should light). Then moderately accelerate to 40 mph for a minimum of 15 seconds. Hold the throttle steady and depress the O/D cancel button (LED should go out). Accelerate to 50 mph. The transmission should be in fourth gear at this time. Hold that speed for 15 seconds, then lightly apply and release the brake. Maintain 50 mph when applying the brake. Then stop and park the vehicle with the gear selector in the D position for a minimum of 20 seconds. Repeat this procedure five times.

## Summary

- Before installing a transmission or transaxle, the torque converter, flexplate, cooler, and wiring should be checked again to avoid problems during and after installation.
- The torque converter must be properly aligned to the flexplate before attempting to seat the transmission against the engine.
- Care must be taken not to strip or damage the cooler lines when connecting them.
- Make sure all ground straps are reconnected and are in good shape.
- Vehicles equipped with engine or transmission computers may require a relearn procedure after the battery has been disconnected.
- Always refer to the service manual for the proper relearn procedure.

## Review Questions

1. Why do some manufacturers require that pressure readings be taken on a freshly overhauled transmission before it is road tested?
2. What can result from dirty or disconnected ground straps?
3. When discussing computer relearn procedures, Technician A says that this type of computer strategy is designed to optimize driveability. Technician B says that the computer on some cars needs to learn about the vehicle before it can control the systems properly. Who is correct?
   A. Technician A only
   B. Technician B only
   C. Both Technician A and Technician B
   D. Neither Technician A nor Technician B
4. After completing a transmission overhaul, Technician A checks the service manual for the relearn procedures before taking the vehicle out on a road test. Technician B connects a scan tool to the electronic control system and checks for DTCs before operating the transmission. Who is correct?

A. Technician A only
B. Technician B only
C. Both Technician A and Technician B
D. Neither Technician A nor Technician B

5. Which of the following probably will NOT occur while the vehicle's PCM is going through its relearn process?

A. The engine will not start.
B. The engine will hesitate or stumble during acceleration.
C. The transmission will shift hard and abruptly.
D. The engine will idle poorly.

# Appendix A

## ASE PRACTICE EXAM FOR AUTOMATIC TRANSMISSIONS AND TRANSAXLES

1. When diagnosing a no engine braking condition in manual low operation, Technician A suspects a defective oil pump. Technician B suspects a bad drive link. Who is correct?
   A. Technician A only
   B. Technician B only
   C. Both Technician A and Technician B
   D. Neither Technician A nor Technician B
2. Which of the following is the MOST likely cause of slipping in all forward gear ranges?
   A. Faulty governor
   B. Clogged oil filter
   C. Faulty band or clutch
   D. Sticking valve in the valve body
3. When diagnosing the cause of no forced downshifts during full throttle operation, Technician A suspects a misadjusted manual linkage. Technician B suspects a dirty valve body and valves. Who is correct?
   A. Technician A only
   B. Technician B only
   C. Both Technician A and Technician B
   D. Neither Technician A nor Technician B
4. When diagnosing the cause of gear slipping in second gear only, Technician A suspects a worn or damaged clutch assembly. Technician B suspects a damaged drive link. Who is correct?
   A. Technician A only
   B. Technician B only
   C. Both Technician A and Technician B
   D. Neither Technician A nor Technician B
5. When conducting a pressure test, Technician A says that the cause of low pressure in all operating ranges could be a clogged filter. Technician B says that the cause of high pressure in all operating ranges could be a defective throttle valve. Who is correct?
   A. Technician A only
   B. Technician B only
   C. Both Technician A and Technician B
   D. Neither Technician A nor Technician B
6. When diagnosing the cause of no torque converter clutch engagement on a transmission that seems to shift fine, Technician A suspects a damaged clutch pressure plate. Technician B suspects a severely worn input clutch. Who is correct?
   A. Technician A only
   B. Technician B only
   C. Both Technician A and Technician B
   D. Neither Technician A nor Technician B
7. When diagnosing the cause of transmission overheating, Technician A checks for contaminated fluid. Technician B suspects a damaged flexplate. Who is correct?
   A. Technician A only
   B. Technician B only
   C. Both Technician A and Technician B
   D. Neither Technician A nor Technician B
8. When diagnosing the cause of sluggish acceleration, Technician A suspects a faulty governor. Technician B checks the condition of the fluid. Who is correct?
   A. Technician A only
   B. Technician B only
   C. Both Technician A and Technician B
   D. Neither Technician A nor Technician B
9. Technician A says that slippage in first gear only can be caused by a faulty VSS. Technician B says that a defective one-way clutch can cause slippage in all gears. Who is correct?
   A. Technician A only
   B. Technician B only
   C. Both Technician A and Technician B
   D. Neither Technician A nor Technician B

10. Which of the following statements about transmission solenoid valves is NOT true?
    A. They can be checked with an ohmmeter.
    B. They can be checked with a lab scope.
    C. They can be checked with air pressure.
    D. They can be checked by applying current to them and listening for their movement.
11. The vehicle experiences an intermittent second gear start. Technician A says that a bad one-way clutch may be the cause of the problem. Technician B says that low governor pressure may be the cause of the problem. Who is correct?
    A. Technician A only
    B. Technician B only
    C. Both Technician A and Technician B
    D. Neither Technician A nor Technician B
12. Technician A says that vibration problems can be caused by an unbalanced torque converter. Technician B says that a faulty output shaft can cause vibration problems. Who is correct?
    A. Technician A only
    B. Technician B only
    C. Both Technician A and Technician B
    D. Neither Technician A nor Technician B
13. Technician A says that a ballooned torque converter can cause damage to the oil pump. Technician B says that a ballooned torque converter is caused by excessive pressure in the torque converter. Who is correct?
    A. Technician A only
    B. Technician B only
    C. Both Technician A and Technician B
    D. Neither Technician A nor Technician B
14. When discussing end play checks, Technician A says that these checks should be taken before the transmission is disassembled. Technician B says that these checks should be taken after the transmission is reassembled. Who is correct?
    A. Technician A only
    B. Technician B only
    C. Both Technician A and Technician B
    D. Neither Technician A nor Technician B
15. When checking a planetary gear set, Technician A says that the end clearance of the pinion gears should be checked with a feeler gauge. Technician B says that the end clearance of the long pinions in a Ravigneaux gear set should be checked at both ends. Who is correct?
    A. Technician A only
    B. Technician B only
    C. Both Technician A and Technician B
    D. Neither Technician A nor Technician B
16. Technician A says that a blocked oil delivery passage will cause a shaft to score. Technician B says that if a shaft is fitted with a check ball and the check ball does not seat properly, low oil pressure will result. Who is correct?
    A. Technician A only
    B. Technician B only
    C. Both Technician A and Technician B
    D. Neither Technician A nor Technician B
17. When discussing computer relearn procedures, Technician A says that this type of computer strategy is designed to optimize driveability. Technician B says that the computer on some cars needs to learn about the vehicle before it can control the systems properly. Who is correct?
    A. Technician A only
    B. Technician B only
    C. Both Technician A and Technician B
    D. Neither Technician A nor Technician B
18. When removing scratches on a valve, Technician A uses fine emery cloth to remove the scratches. Technician B uses a sand blaster or glass bead machine to polish the valve. Who is correct?
    A. Technician A only
    B. Technician B only
    C. Both Technician A and Technician B
    D. Neither Technician A nor Technician B
19. Which of the following is the LEAST likely cause for the transmission not upshifting and operating only in first gear?
    A. Improper fluid level or condition
    B. Faulty valve body
    C. Defective oil pump
    D. Faulty governor
20. Technician A says that overtorquing the valve body retaining bolts can cause the valves to stick in their bore. Technician B says that flat filing a valve body's surface will allow the valve body to seal properly and will allow the valves to move freer in their bores. Who is correct?
    A. Technician A only
    B. Technician B only
    C. Both Technician A and Technician B
    D. Neither Technician A nor Technician B
21. When discussing the results of an oil pressure test, Technician A says that when the fluid pressures are high, internal leaks, a clogged filter, low oil pump output, or a faulty pressure regulator valve is indicated. Technician B says that if the fluid pressure increased at the wrong time, an internal leak at the servo or clutch seal is indicated. Who is correct?

A. Technician A only
B. Technician B only
C. Both Technician A and Technician B
D. Neither Technician A nor Technician B

22. Technician A says that an air test can be used to check servo action. Technician B says that an air test can be used to check for internal fluid leaks. Who is correct?
    A. Technician A only
    B. Technician B only
    C. Both Technician A and Technician B
    D. Neither Technician A nor Technician B
23. When checking clutch discs, Technician A says that the steel plates should be replaced if they are worn flat. Technician B says that the friction discs should be squeezed to see if they could hold fluid. If they hold fluid and look okay, they are serviceable. Who is correct?
    A. Technician A only
    B. Technician B only
    C. Both Technician A and Technician B
    D. Neither Technician A nor Technician B
24. When diagnosing noises apparently from a transaxle assembly, Technician A says that a knocking sound at low speeds is probably caused by worn CV joints. Technician B says that a clicking noise heard when the vehicle is turning is probably caused by a worn or damaged outboard CV joint. Who is correct?
    A. Technician A only
    B. Technician B only
    C. Both Technician A and Technician B
    D. Neither Technician A nor Technician B
25. Technician A says that low engine vacuum can cause a vacuum modulator to sense a load condition that is not present. Technician B says that poor engine performance can cause delayed shifts through the action of the TV assembly. Who is correct?
    A. Technician A only
    B. Technician B only
    C. Both Technician A and Technician B
    D. Neither Technician A nor Technician B
26. Technician A says that delayed shifting can be caused by worn planetary gear set members. Technician B says that delayed shifts or slippage may be caused by leaking hydraulic circuits or sticking spool valves in the valve body. Who is correct?
    A. Technician A only
    B. Technician B only
    C. Both Technician A and Technician B
    D. Neither Technician A nor Technician B
27. When discussing proper band adjustment procedures, Technician A says that on some vehicles the bands can be adjusted externally with a torque wrench. Technician B says that a calibrated inch-pound torque wrench is normally used to tighten the band adjusting bolt to a specified torque. Who is correct?
    A. Technician A only
    B. Technician B only
    C. Both Technician A and Technician B
    D. Neither Technician A nor Technician B
28. When checking the condition of a car's ATF, Technician A says that if the fluid has a dark brownish or blackish color and/or a burned odor, the fluid has been overheated. Technician B says that if the fluid has a milky color, this indicates that engine coolant has been leaking into the transmission's cooler. Who is correct?
    A. Technician A only
    B. Technician B only
    C. Both Technician A and Technician B
    D. Neither Technician A nor Technician B
29. When assembling a transmission, Technician A coats the steel clutch discs with petroleum jelly. Technician B soaks the friction discs in clean ATF before installing them. Who is correct?
    A. Technician A only
    B. Technician B only
    C. Both Technician A and Technician B
    D. Neither Technician A nor Technician B
30. Technician A says that end play often is corrected by selective snaprings. Technician B says that end play often is corrected by selective thrust washers. Who is correct?
    A. Technician A only
    B. Technician B only
    C. Both Technician A and Technician B
    D. Neither Technician A nor Technician B
31. Technician A says that delayed shifts can be caused by a faulty TP sensor. Technician B says that delayed shifts can be caused by an open shift solenoid. Who is correct?
    A. Technician A only
    B. Technician B only
    C. Both Technician A and Technician B
    D. Neither Technician A nor Technician B
32. When checking end play, Technician A measures the movement of the shaft with a dial indicator. Technician B uses a spring compressor to get a maximum reading. Who is correct?
    A. Technician A only
    B. Technician B only
    C. Both Technician A and Technician B
    D. Neither Technician A nor Technician B
33. The back side of a torque converter was wet. Technician A says that this may be caused by excessive torque converter hub runout. Technician B says that this can be caused by insufficient input shaft end play. Who is correct?
    A. Technician A only
    B. Technician B only
    C. Both Technician A and Technician B
    D. Neither Technician A nor Technician B

34. Which of the following is the LEAST likely cause for ATF coming out of the transmission vent or the dipstick tube?
    A. Defective pressure regulator valve
    B. Plugged drain holes in the valve body
    C. Fluid at too high a level
    D. Defective oil cooler
35. Technician A says that abnormal noises from a transmission will never be caused by faulty clutches or bands. Technician B says that abnormal noises from a transmission are typically caused by a faulty torque converter. Who is correct?
    A. Technician A only
    B. Technician B only
    C. Both Technician A and Technician B
    D. Neither Technician A nor Technician B
36. Technician A says that abnormal noises and vibrations can be caused by damaged or worn gears, damaged clutches, and/or damaged bands. Technician B says that abnormal noises can be caused by a bad oil pump or contaminated fluid. Who is correct?
    A. Technician A only
    B. Technician B only
    C. Both Technician A and Technician B
    D. Neither Technician A nor Technician B
37. Technician A says that vibration problems can be caused by an unbalanced torque converter. Technician B says that vibration problems can be caused by a faulty output shaft. Who is correct?
    A. Technician A only
    B. Technician B only
    C. Both Technician A and Technician B
    D. Neither Technician A nor Technician B
38. Which of the following is the LEAST likely cause for a buzzing noise?
    A. Improper fluid level or condition
    B. A defective oil pump
    C. A defective flexplate
    D. A damaged planetary gearset
39. Technician A says that the best way to diagnose noise problems is to take a road test and pay attention to operating gear, speed, and the conditions at which the noise occurs. Technician B says that noise diagnosis should begin with putting the vehicle in gear and applying the brake. If the noise or vibration is not evident, the problem is undoubtedly in the driveline or output of the transmission. Who is correct?
    A. Technician A only
    B. Technician B only
    C. Both Technician A and Technician B
    D. Neither Technician A nor Technician B
40. After soaking valve body parts in mineral spirits, Technician A wipes the parts off with a paper towel. Technician B blow dries each part with compressed air. Who is correct?
    A. Technician A only
    B. Technician B only
    C. Both Technician A and Technician B
    D. Neither Technician A nor Technician B

# Appendix B

## NATEF CORRELATIONS

### A. General Transmission and Transaxle Diagnosis

A.1 Identify and interpret transmission concern; ensure proper engine operation; determine necessary action. See Chapter 14

A.2 Research applicable vehicle and service information, such as transmission/transaxle system operation, vehicle service history, service precautions, and technical service bulletins. See Chapter 4

A.3 Locate and interpret vehicle and major component identification numbers (VIN, vehicle certification labels, calibration decals). See Chapter 4

A.4 Diagnose unusual fluid usage, level, and condition concerns; determine necessary action. See Chapter 14

A.5 Perform pressure tests; determine necessary action. See Chapters 20 and 23

A.6 Perform stall test; determine necessary action. See Chapter 17

A.7 Perform lockup converter system tests; determine necessary action. See Chapter 17

A.8 Diagnose electronic, mechanical, hydraulic, vacuum control system concerns; determine necessary action. See Chapters 8, 12, 15, and 16

A.9 Diagnose noise and vibration concerns; determine necessary action. See Chapter 14

A.10 Diagnose transmission/transaxle gear reduction/multiplication concerns using driving, driven, and held member (power flow) principles. See Chapters 9, 14, 20, and 22

### B. Transmission and Transaxle Maintenance and Adjustment

B.1 Inspect, adjust, or replace throttle (TV) linkages or cables, manual shift linkages or cables; transmission range sensor; check gear select indicator (as applicable). See Chapters 11, 14, and 19

B.2 Service transmission; perform visual inspection; replace fluids and filters. See Chapters 14 and 18

### C. In Vehicle Transmission and Transaxle Repair

C.1 Inspect, adjust or replace (as applicable) vacuum modulator; inspect and repair or replace lines and hoses. See Chapters 11, 14, and 23

C.2 Inspect, repair, and replace governor assembly. See Chapters 11, 20, and 32

C.3 Inspect and replace external seals and gaskets. See Chapters 26 and 36

C.4 Inspect extension housing, bushings, and seals; perform necessary action. See Chapters 18, 26, and 36

C.5 Inspect, leak test, flush, and replace cooler, lines, and fittings. See Chapters 14 and 18

C.6 Inspect and replace speedometer drive gear, driven gear, vehicle speed sensor (VSS), and retainers. See Chapters 16 and 33

C.7 Diagnose electronic transmission control systems using a scan tool; determine necessary action. See Chapters 15 and 16

C.8 Inspect, replace, and align powertrain mounts. See Chapters 9, 14, and 18

### D. Off-Vehicle Transmission and Transaxle Repair

*1. Removal, Disassembly, and Reinstallation*

D.1.1 Remove and reinstall transmission and torque converter (rear-wheel drive). See Chapters 28 and 37

D.1.2 Remove and reinstall transaxle and torque converter assembly. See Chapters 28 and 37

D.1.3 Disassemble, clean, and inspect transmission/transaxle. See Chapter 30

D.1.4 Inspect, measure, clean, and replace valve body (includes surfaces and bores, springs, valves, sleeves, retainers, brackets, check balls, screens, spacers, and gaskets). See Chapters 20 and 32

D.1.5 Inspect servo bore, piston, seals, pin, spring, and retainers; determine necessary action. See Chapters 25 and 35

D.1.6 Inspect accumulator bore, piston, seals, spring, and retainer; determine necessary action. See Chapters 25 and 35

D.1.7 Assemble transmission/transaxle. See Chapter 36

*2. Oil Pump and Converter*

D.2.1 Inspect converter flexplate, attaching parts, pilot, pump drive, and seal areas. See Chapters 10 and 29

D.2.2 Measure torque converter end play and check for interference; check stator clutch. See Chapters 10 and 29

D.2.3 Inspect, measure, and replace oil pump assembly and components. See Chapters 11 and 31

*3. Gear Train, Shafts, Bushings, and Case*

D.3.1 Measure end play or preload; determine necessary action. See Chapters 9 and 33

D.3.2 Inspect, measure, and replace thrust washers and bearings. See Chapters 26 and 33

D.3.3 Inspect oil delivery seal rings, ring grooves, and sealing surface areas. See Chapters 26, 30, and 33

D.3.4 Inspect bushings; replace as needed. See Chapters 26, 30, and 36

D.3.5 Inspect and measure planetary gear assembly (includes sun, ring gear, thrust washers, planetary gears, and carrier assembly); determine necessary action. See Chapters 24 and 33

D.3.6 Inspect case bores, passages, bushings, vents, and mating surfaces; determine necessary action. See Chapters 9 and 30

D.3.7 Inspect transaxle drive, link chains, sprockets, gears, bearings, and bushings; perform necessary action. See Chapters 27 and 34

D.3.8 Inspect, measure, repair, adjust, or replace transaxle final drive components. See Chapters 27 and 34

D.3.9. Inspect and reinstall parking pawl, shaft, spring, and retainer; determine necessary action. See Chapters 26 and 33

*4. Friction and Reaction Units*

D.4.1 Inspect clutch drum, piston, check balls, springs, retainers, seals, and friction and pressure plates; determine necessary action. See Chapters 25 and 35

D.4.2 Measure clutch pack clearance; determine necessary action. See Chapters 25 and 35

D.4.3 Air test operation of clutch and servo assemblies. See Chapters 25 and 35

D.4.4 Inspect roller and sprag clutch, races, rollers, sprags, springs, cages, and retainers; replace as needed. See Chapters 25 and 35

D.4.5 Inspect bands and drums; determine necessary action. See Chapters 25 and 35

# Appendix C

## USCS AND METRIC CONVERSIONS

Linear Measurements
- 1 meter (m) = 39.37 inches (in.)
- 1 centimeter (cm) = 0.3937 inch
- 1 millimeter (mm) = 0.03937 inch
- 1 inch = 2.54 centimeters
- 1 inch = 25.4 millimeters
- 1 mile = 1.6093 kilometers

Area (Square) Measurements
- 1 square inch = 6.452 square centimeters
- 1 square centimeter = 0.155 square inch

Volume Measurements
- 1 cubic inch = 16.387 cubic centimeters
- 1000 cubic centimeters = 1 liter (l)
- 1 liter (l) = 61.02 cubic inches
- 1 gallon = 3.7854 liters

Weight Measurements
- 1 ounce = 28.3495 grams
- 1 pound = 453.59 grams
- 1000 grams = 1 kilogram
- 1 kilogram = 2.2046 pounds

Temperature Measurements
- 1 degree Fahrenheit (F) = $\frac{9}{5}$ C + 32 degrees
- 1 degree Celsius (C) = $\frac{5}{9}$ (F − 32 degrees)

Pressure Measurements
- 1 pound per square inch (psi) = 0.07031 kilogram (kg) per square centimeter
- 1 kilogram per square centimeter = 14.22334 pounds per square inch
- 1 bar = 14.504 pounds per square inch
- 1 pound per square inch = 0.06895 bar

Torque Measurements
- 10 pounds (lbs.) per foot = 13.558 Newton meters (N•m)
- 1 N•m = 0.7375 lb.-ft.
- 1 lb.-ft. = 0.138 m kg
- 1 cm kg = 7.233 ft.-lb
- 10 cm kg = 0.98 N•m

# Appendix D

## FRACTIONS OF INCHES TO DECIMAL AND METRIC EQUIVALENTS

| Fraction | Decimal | Metric |
|---|---|---|
| $^{1}/_{64}$ | 0.015625 | 00.39688 |
| $^{1}/_{32}$ | 0.03125 | 00.79375 |
| $^{3}/_{64}$ | 0.046875 | 01.19062 |
| $^{1}/_{16}$ | 0.0625 | 01.58750 |
| $^{5}/_{64}$ | 0.078125 | 01.98437 |
| $^{3}/_{32}$ | 0.09375 | 02.38125 |
| $^{7}/_{64}$ | 0.109375 | 02.77812 |
| **$^{1}/_{8}$** | **0.125** | **03.1750** |
| $^{9}/_{64}$ | 0.140625 | 03.57187 |
| $^{5}/_{32}$ | 0.15625 | 03.96875 |
| $^{11}/_{64}$ | 0.171875 | 04.36562 |
| $^{3}/_{16}$ | 0.1875 | 04.76250 |
| $^{13}/_{64}$ | 0.203125 | 05.15937 |
| $^{7}/_{32}$ | 0.21875 | 05.55625 |
| $^{15}/_{64}$ | 0.234375 | 05.95312 |
| **$^{1}/_{4}$** | **0.250** | **06.35000** |
| $^{17}/_{64}$ | 0.265625 | 06.74687 |
| $^{9}/_{32}$ | 0.28125 | 07.14375 |
| $^{19}/_{64}$ | 0.296875 | 07.54062 |
| $^{5}/_{16}$ | 0.3125 | 07.93750 |
| $^{21}/_{64}$ | 0.328125 | 08.33437 |
| $^{11}/_{32}$ | 0.34375 | 08.73125 |
| $^{23}/_{64}$ | 0.359375 | 09.12812 |
| **$^{3}/_{8}$** | **0.375** | **09.52500** |
| $^{25}/_{64}$ | 0.390625 | 09.92187 |
| $^{13}/_{32}$ | 0.40625 | 10.31875 |
| $^{27}/_{64}$ | 0.421875 | 10.71562 |
| $^{7}/_{16}$ | 0.4375 | 11.11250 |
| $^{29}/_{64}$ | 0.453125 | 11.50937 |
| $^{15}/_{32}$ | 0.46875 | 11.90625 |
| $^{31}/_{64}$ | 0.484375 | 12.30312 |
| **$^{1}/_{2}$** | **0.500** | **12.7000** |
| $^{33}/_{64}$ | 0.515625 | 13.09687 |
| $^{17}/_{32}$ | 0.53125 | 13.49375 |
| $^{35}/_{64}$ | 0.546875 | 13.89062 |
| $^{17}/_{16}$ | 0.5625 | 14.28750 |
| $^{37}/_{64}$ | 0.578125 | 14.68437 |
| $^{19}/_{32}$ | 0.59375 | 15.08125 |
| $^{39}/_{64}$ | 0.609375 | 15.47812 |
| **$^{5}/_{8}$** | **0.625** | **15.8750** |
| $^{41}/_{64}$ | 0.640625 | 16.27187 |
| $^{21}/_{32}$ | 0.65625 | 16.66875 |
| $^{43}/_{64}$ | 0.671875 | 17.06562 |
| $^{11}/_{16}$ | 0.6875 | 17.46250 |
| $^{45}/_{64}$ | 0.703125 | 17.85937 |
| $^{23}/_{32}$ | 0.71875 | 18.25625 |
| $^{47}/_{64}$ | 0.734375 | 18.65312 |
| **$^{3}/_{4}$** | **0.750** | **19.05000** |
| $^{49}/_{64}$ | 0.765625 | 19.44687 |
| $^{25}/_{32}$ | 0.78125 | 19.84375 |
| $^{51}/_{64}$ | 0.796875 | 20.24062 |
| $^{13}/_{16}$ | 0.8125 | 20.63750 |
| $^{53}/_{64}$ | 0.828125 | 21.03437 |
| $^{27}/_{32}$ | 0.84375 | 21.43125 |
| $^{55}/_{64}$ | 0.859375 | 21.82812 |
| **$^{7}/_{8}$** | **0.875** | **22.22500** |
| $^{57}/_{64}$ | 0.890625 | 22.62187 |
| $^{29}/_{32}$ | 0.90625 | 23.01875 |
| $^{59}/_{64}$ | 0.921875 | 23.41562 |
| $^{15}/_{16}$ | 0.9375 | 23.81250 |
| $^{61}/_{64}$ | 0.953125 | 24.20937 |
| $^{31}/_{32}$ | 0.96875 | 24.60625 |
| $^{63}/_{64}$ | 0.984375 | 25.00312 |
| **$^{2}/_{2}=1$** | **1.000** | **25.40000** |

# Appendix E

## GENERAL TORQUE SPECIFICATIONS

**NOTE: The values in this chart should only be used when manufacturer's specifications are NOT available. Also, the values are only valid when SAE 10 oil is used to lubricate the threads of the bolt.

| Bolt Diameter in Inches | Torque: lb-ft. SAE 2 | SAE 5 | SAE 8 |
|---|---|---|---|
| 1/4 | 7 | 10 | 14 |
| 5/16 | 14 | 21 | 30 |
| 3/8 | 24 | 37 | 52 |
| 7/16 | 39 | 60 | 84 |
| 1/2 | 59 | 90 | 128 |
| 9/16 | 85 | 130 | 184 |
| 5/8 | 117 | 180 | 255 |
| 3/4 | 205 | 320 | 450 |
| 7/8 | 200 | 515 | 730 |
| 1 | 300 | 775 | 1090 |

Torque: kg cm* kg m

| Bolt Diameter in Millimeters | Property Class: 4.6 | 4.8 | 5.6 | 5.8 | 6.6 | 6.8 | 6.9 | 8.8 | 10.9 | 12.9 |
|---|---|---|---|---|---|---|---|---|---|---|
| 6 | 49* | 63* | 61* | 79* | 74* | 95* | 103* | 126* | 172* | 206* |
| 8 | 119* | 153* | 148* | 178* | 178* | 230* | 250* | 306* | 417* | 500* |
| 10 | 235* | 303* | 294* | 379* | 353* | 455* | 495* | 606* | 8.2 | 10 |
| 12 | 411* | 529* | 427* | 662* | 616* | 7.9 | 8.6 | 10.5 | 14 | 17 |
| 14 | 654* | 8.4 | 8.2 | 10.5 | 10 | 12 | 13 | 17 | 23 | 27 |
| 16 | 10 | 13 | 12 | 16 | 15 | 20 | 21 | 26 | 36 | 43 |
| 18 | 14 | 18 | 17 | 23 | 21 | 27 | 30 | 36 | 49 | 59 |
| 22 | 27 | 35 | 34 | 44 | 41 | 52 | 57 | 70 | 95 | 114 |

# Bilingual Glossary

**Abrasion** Wearing or rubbing away of a part.
***Erosión*** *cuando una pieza se desgasta.*

**Accumulator** A device used in automatic transmissions to cushion the shock of shifting between gears, providing a smoother feel inside the vehicle.
***Acumulador*** *Dispositivo que se usa en las transmisiones automáticas para amortiguar los cambios y provoca que se sienta el movimiento más suave dentro del vehículo.*

**Actuator** A control device that delivers mechanical action in response to an electrical signal.
***Biela de accionamiento*** *Dispositivo de control que envía una acción mecánica en respuesta a una señal eléctrica.*

**Adaptive Learning** The ability of a computer to monitor the drivers' habits and the operating conditions of its system and make adjustments to its program to correct for them.
***Aprendizaje adaptable*** *Habilidad de una computadora para inspeccionar los hábitos del conductor y las condiciones operacionales de su sistema y hacer ajustes a su programa para corregirlos.*

**Adhesives** Chemicals used to hold gaskets in place during the assembly of an engine. They also aid the gasket in maintaining a tight seal by filling in the small irregularities on the surfaces and by preventing the gasket from shifting because of engine vibration.
***Adhesivos*** *Químicos que se usan para detener el empaque en su lugar durante el montaje del motor. También ayudan al empaque a mantenerse firmemente sellado al llenar las pequeñas irregularidades en la superficie y al prevenir que el empaque se mueva con la vibración del motor.*

**Aeration** The process of mixing air into a liquid.
***Ventilación*** *Proceso de mezclar aire con un líquido.*

**Alignment** An adjustment to a line or made to bring into a line.
***Alineación*** *Ajuste a una línea o hecho para regular la posición de una línea.*

**Alternating Current** Electrical current that changes direction between positive and negative.
***Corriente alterna*** *Corriente eléctrica que cambia de dirección entre positivo y negativo.*

**Ammeter** The instrument used to measure electrical current flow in a circuit.
***Amperímetro*** *Instrumento que se usa para medir el flujo de la corriente eléctrica en un circuito.*

**Ampere** The unit for measuring electrical current; usually called an amp.
***Amperio*** *Unidad para medir la corriente eléctrica.*

**Amplitude** The height of a waveform.
***Amplitud*** *Altura de una onda.*

**Analog Signal** A voltage signal that varies within a given range (from high to low, including all points inbetween).
***Señal análoga*** *Señal de voltaje que varía en un cierto campo (de alto a bajo, incluyendo todos los puntos intermedios.)*

**Annulus gear** Another name for the ring gear of a planetary gear set.
***Palanca circular*** *Otro nombre para el eje de anillo de un conjunto de ejes planetarios.*

**Apply devices** Devices that hold or drive members of a planetary gear set. They may be hydraulically or mechanically applied.
***Dispositivo de ajuste*** *Dispositivos que detienen o conducen a los miembros de un conjunto de ejes planetarios. Se pueden ajustar hidráulica o mecánicamente.*

**ATF** Automatic Transmission Fluid.
***LTA*** *Líquido para la transmisión automática.*

**AWG** American Wire Gauge System. The system used to designate wire size.
**SAMA** *Sistema americano del manómetro para alambre. El sistema que se usa para designar el tamaño del alambre.*

**Axial** Parallel to a shaft or bearing bore.
***Axial*** *Paralelo a un eje o a la barrena del cojinete.*

**Axis** The centerline of a rotating part, a symmetrical part, or a circular bore.
***Árbol*** *Línea central de una parte giratoria, una parte simétrica, o una barrena circular.*

**Axle** The shaft or shafts of a machine upon which the wheels are mounted.
***Eje*** *La flecha o flechas de un motor en donde se montan las ruedas.*

**Axle housing** Designed in the removable carrier or integral carrier types to house the drive pinion, ring gear, differential, and axle shaft assemblies.
***Campana del eje*** *Designada en el portador removible o en los tipos de cargador integral para albergar el piñón de accionamiento, el eje de anillo, el diferencial y los montajes de las flechas. (The dictionary doesn't have drive pinion but pinion drive, which means transmisión por engranaje. I'm supposing that the main word here is pinion in the translation.)*

**Axle ratio** The ratio between the rotational speed (rpm) of the driveshaft and that of the driven wheel; gear reduction through

the differential, determined by dividing the number of teeth on the ring gear by the number of teeth on the drive pinion.
***Proporción del eje*** *Proporción entre la velocidad de giro (rmp) de un eje de transmisión o de una llanta impulsada.*

**Axle shaft** A shaft on which the road wheels are mounted.
***Flecha*** *Eje en donde se montan las ruedas para la carretera.*

**Backlash** The amount of clearance or play between two meshed gears.
***Contragolpe*** *Cantidad de espacio libre o de juego entre dos engranajes enlazados.*

**Balance** Having equal weight distribution. The term is usually used to describe the weight distribution around the circumference and between the front and back sides of a wheel.
***Balanceo*** *Tener la distribución equitativa de peso. El término se usa generalmente para describir la distribución del peso alrededor de la circunferencia y entre el frente y los lados traseros de una rueda.*

**Balance valve** A regulating valve, which controls a pressure of just the right value to balance other forces acting on the valve.
***Válvula de balance*** *Válvula reguladora que controla la presión de un valor exacto balancea otras fuerzas ejercidas sobre la válvula.*

**Ball bearing** An antifriction bearing consisting of a hardened inner and outer race with hardened steel balls, which roll between the two races, and which supports the load of the shaft.
***Cojinete de bola*** *Cojinete antifricción que consiste de carriles interior y exterior duros con bolas de acero endurecidas, los cuales ruedan entre las dos carriles, y lo cual sostiene el peso del eje.*

**Ballooning** A condition in which the torque converter has been blown up like a balloon, caused by excessive pressure in the converter.
***Aterrizaje brusco o aerostación*** *Condición en la que el convertidor de torsión se ha inflado como globo a causa de presión excesiva en el convertidor.*

**Ball-type valve** A valve that uses the movement of a ball to control fluid flow.
***Válvula tipo esfera*** *Válvula que usa el movimiento de una esfera para controlar el flujo del líquido.*

**Band** A steel band with an inner lining of friction material. Device used to hold a clutch drum at certain times during transmission operation.
***Banda*** *Banda de acero forrada con material de fricción. Aparato que se usa para detener el tambor del cloche a ciertos momentos durante la operación de la transmisión.*

**Bearing** The supporting part, which reduces friction between a stationary and rotating part or between two moving parts.
***Cojinete*** *Parte que sostiene, la cual reduce la fricción entre una pieza estacionaria y una giratoria o entre dos piezas movibles.*

**Bearing cage** A spacer that keeps the balls or rollers in a bearing in proper position between the inner and outer races.
***Canasta del cojinete*** *Espaciador que mantiene las esferas o rodillos en un cojinete en la posición apropiada entre los carriles interiores o exteriores.*

**Bearing caps** In the differential, caps held in place by bolts or nuts that, in turn, hold bearings in place.
***Casco del cojinete*** *En el diferencial, cascos que se mantienen en su lugar con pernos o tuercas que, a su vez, mantienen en su lugar a los cojinetes.*

**Bearing cone** The inner race, rollers, and cage assembly of a tapered roller bearing. Cones and cups always must be replaced in matched sets.
***Conos del cojinete*** *Carril interno, rodillos y canasta de ensamblado de un cojinete de rodillo graduado. Los conos y las tazas deben siempre remplazarse en conjuntos iguales.*

**Bearing cup** The outer race of a tapered roller bearing or ball bearing.
***Copa del cojinete*** *Carril exterior de un cojinete de rodillo graduado.*

**Bearing race** The surface upon which the rollers or balls of a bearing rotate. The outer race is the same thing as the cup, and the inner race is the one closest to the axle shaft.
***Carril del cojinete*** *Superficie en donde giran los rodillos o las bolas de un cojinete. El carril exterior es lo mismo que la copa, y el carril interior es el más cercano al eje o flecha.*

**Belleville spring** A tempered spring steel cone-shaped plate used to aid the mechanical force in a pressure plate assembly.
***Resorte Belleville*** *Placa en forma cónica de acero con resorte graduado que se usa para ayudar a la fuerza mecánica en un ensamblado de placa a presión.*

**Bevel gear** A form of spur gear that has its teeth cut at an angle.
***Engranaje en el bisel*** *Forma de un engranaje espolonado con dientes cortados en un ángulo.*

**Bolthead** The part of a bolt that the socket or wrench fits over in order to torque or tighten the bolt.
***Cabeza de tornillo o perno*** *Parte del perno que encaja en el que el manquito o la llave para torcer o apretar el perno.*

**Bolt shank** The smooth area on a bolt from the bottom surface of the head to the start of the threads.
***Mango del perno*** *Área lisa de un perno desde la superficie inferior de la cabeza hasta donde empieza el enroscado.*

**Bolt torque** The turning effort required to offset resistance as the bolt is being tightened.
***Torsión del perno*** *Esfuerzo de torsión requerido para contrabalancear la resistencia mientras se aprieta el perno.*

**Burnish** To smooth or polish by the use of a sliding tool under pressure.
***Bruñido*** *Alisar o lustrar usando una herramienta corrediza bajo presión.*

**Burr** A featheredge of metal left on a part being cut with a file or other cutting tool.
***Rebaba*** *Canto biselado de metal en una parte cortada con lima o con otra herramienta.*

**Bus** A common connector used as an information source for the vehicle's various control units.
***Barra colectora*** *Conector común que sirve como fuente de información para los varios controles del vehículo.*

**Bushing** A cylindrical lining used as a bearing assembly made of steel, brass, bronze, nylon, or plastic.
***Camisa*** *Forro cilíndrico que se usa como ensamblaje del cojinete y hecho de acero, latón, bronce, nailon o plástico.*

**C-clip** A C-shaped clip used to retain the drive axles in some rear axle assemblies.
***Pinza C*** *Pinza o grapa en forma de C que se usa para detener los ejes de manejo en algunos ensamblados de eje trasero.*

**Cage** A spacer used to keep the balls or rollers in proper relation to one another. In a constant-velocity joint, the cage is an open metal framework that surrounds the balls to hold them in position.
***Caja o jaula*** *Espaciador que mantiene las esferas o rodillos en relación apropiada el uno con el otro. En articulación de velocidad constante la caja es una armazón abierta de metal que rodea las esferas para mantenerlas en su posición.*

**Cantilever** A projecting lever or beam that is supported on only one end.

***Ménsula o cantilever*** *Palanca saliente o viga que sólo está sostenida en un extremo.*

**Cardan Universal Joint** A nonconstant velocity universal joint consisting of two yokes with their forked ends joined by a cross. The driven yoke changes speed twice in 360 degrees of rotation.
***Junta cardánica universal*** *Junta universal de velocidad no constante que consiste de dos grapas con terminales bifurcadas unidas por una cruz. La varilla de maniobra cambia de velocidad dos veces en un ángulo de rotación de 360 grados.*

**Case-harden** To harden the surface of steel. The carburizing method used on low-carbon steel or other alloys to make the case or outer layer of the metal harder than its core.
***Templar*** *Endurecer una superficie de acero. El método de carburización que se usa en acero bajo en carbon u otras aleaciones para hacer la caja o la superficie exterior del metal más duro que su centro.*

**Castellate** Formed to resemble a castle battlement, as in a castellated nut.
***Almenar*** *Formado para que parezca el almenaje de un castillo, como en un perno almenado.*

**Castellated nut** A nut with six raised portions or notches through which a cotter pin can be inserted to secure the nut.
***Perno almenado*** *Perno con seis porciones elevadas o muescas mediante las cuales puede insertarse una chaveta de dos patas para asegurar el perno.*

**Caustic** Something that has the ability to eat away at something through chemical action.
***Cáustico*** *Substancia que tiene la habilidad de destruir otra cosa mediante una acción química.*

**Centrifugal clutch** A clutch that uses centrifugal force to apply a higher force against the friction disc as the clutch spins faster.
***Cloche centrífugo*** *Cloche que usa fuerza centrífuga para aplicar mayor fuerza contra el disco de fricción mientras el cloche gira más rápido.*

**Centrifugal force** The force acting on a rotating body, which tends to move it outward and away from the center of rotation. The force increases as rotational speed increases.
***Fuerza centrífuga*** *Fuerza sobre un cuerpo giratorio, la cual tiende a moverlo hacia fuera y alejado de su centro de rotación. La fuerza aumenta al aumentar la velocidad giratoria.*

**Chamfer face** A beveled surface on a shaft or part that allows for easier assembly. The ends of FWD drive shafts are often chamfered to make installation of the CV joints easier.
***Cara del bisel*** *Superficie biselada en un eje o parte que permite un ensamblado más fácil. Las terminales de los ejes de TD generalmente están chaflanadas o biseladas para hacer más fácil la instalación de las juntas de velocidad constante.*

**Chase** To straighten up or repair damaged threads.
***Cincelar*** *Enderezar o reparar las roscas dañadas.*

**Check ball valve** A valve that uses a check ball to open and close a fluid circuit.
***Válvula de esferas de retén*** *Válvula que usa una esfera de retén para abrir y cerrar un circuito de flujo.*

**Circuit breaker** A resettable circuit protection device that automatically opens in response to high current.
***Cortacircuitos*** *Mecanismo reprogramable de protección del circuito que se abre automáticamente respondiendo a alta corriente.*

**Circlip** A split steel snapring that fits into a groove to hold various parts in place. Circlips are often used on the ends of FWD drive shafts to retain the constant velocity joints.
***Abrazadera en forma circular*** *Anillo de acero que se abre de golpe y cabe en una abierta que queda en un acanalado o ranura para detener varias partes en su lugar. Las abrazaderas en forma circular generalmente se usan en las terminales de los ejes TD para retener las juntas de velocidad constante.*

**Clearance** The space allowed between two parts, such as between a journal and a bearing.
***Holgura o juego*** *Espacio permitido entre dos partes, tal como entre el muñón y el cojinete.*

**Clutch** A device for connecting and disconnecting the engine from the transmission or for a similar purpose in other units.
***Cloche, embrague*** *Pieza para conectar y desconectar el motor de la transmisión o para un propósito similar en otras unidades.*

**Clutch packs** A series of clutch discs and plates installed alternately in a housing to act as a driving or driven unit.
***Fardo del cloche*** *Serie de discos y placas del cloche instaladas alternadamente en la campana que actúa como unidad impulsada o manejada.*

**Clutch slippage** Engine speed increases but increased torque is not transferred through to the driving wheels because of clutch slippage.
***Gasto no medido del cloche*** *La velocidad del motor aumenta pero no se transfiere el aumento de torsión a través de las ruedas impulsadas dado al gasto no medido del cloche.*

**Coefficient of friction** The ratio of the force resisting motion between two surfaces in contact to the force holding the two surfaces in contact.
***Coeficiente de fricción*** *Relación del movimiento de la fuerza de resistencia entre dos superficies en contraste de la fuerza de mantiene dos superficies en contacto.*

**Coil spring** A heavy wirelike steel coil used to support the vehicle weight when allowing for suspension motions. On FWD cars, the front coil springs are mounted around the MacPherson struts. On the rear suspension, they may be mounted to the rear axle, to trailing arms, or around rear struts.
***Resorte espiral*** *Espiral de acero parecido al alambre pesado que se usa para sostener el peso del vehículo al propio tiempo que se permiten los movimientos de la suspensión. En automóviles TD los resortes de espiral frontales se montan alrededor de los postes MacPherson. En la suspensión trasera se montan en el eje trasero a los brazos remolcados, o alrededor de los postes traseros.*

**Coil preload springs** Coil springs are made of tempered steel rods formed into a spiral that resists compression, located in the pressure plate assembly.
***Resortes espirales antecarga*** *Resortes espirales hechos de acero templado formados en espiral que resiste la compresión. Localizados en el ensamblaje de placa a presión.*

**Compound** A mixture of two or more ingredients.
***Compuesto*** *Mezcla de dos o más ingredientes.*

**Complete Circuit** An electrical circuit that includes a path that connects the positive and negative terminals of the electrical power source.
***Circuito completo*** *Circuito eléctrico que incluye un curso que conecta las terminales positivas y negativas de una fuente de energía eléctrica.*

**Composition gasket** A gasket made of more than one material, such as rubber and cork.
***Junta de composición*** *Junta hecha de más de un material, tal como hule y corcho.*

**Computer** An electronic device that receives information, stores information, processes information, and communicates information.
***Computadora u ordenador*** *Aparato electrónico que recibe, almacena, procesa y transfiere información.*

**Concentric** Two or more circles having a common center.
***Concéntrico*** *Dos o más círculos que tienen un centro común.*

**Conductor** A material that allows electrical current to easily flow through it.
***Conductor*** *Material que permite el fácil flujo de la corriente eléctrica.*

**Constant velocity joint (also called CV joint)** A flexible coupling between two shafts that permits each shaft to maintain the same driving or driven speed regardless of operating angle, allowing for a smooth transfer of power. The constant velocity joint consists of an inner and outer housing with balls in between, or a tripod and yoke assembly.
***Junta de velocidad constante (también llamada junta vc)*** *Acoplamiento flexible entre dos ejes que permite que cada eje mantenga la misma velocidad de movimiento o manejo sin importar el ángulo de operación, permitiendo una suave transferencia de energía. La junta de velocidad constante consiste en un campana interior y exterior con esferas en medio, o un ensamblado de trípode o de varilla de maniobra.*

**Control arm** A suspension component that links the vehicle frame to the steering knuckle or axle housing and acts as a hinge to allow up-and-down wheel motions. The front control arms are attached to the frame with bushings and bolts and are connected to the steering knuckles with ball joints. The rear control arms attach to the frame with bushings and bolts and are welded or bolted to the rear axle or wheel hubs.
***Brazo de control*** *Componente de suspensión que une el armazón del vehículo a las charnelas de cambio o campana del eje y actúa como bisagra para permitir los movimientos hacia arriba y hacia abajo. Los brazos de control frontales están unidos a la armazón con manguitos y pernos y están soldados o atornillados al eje trasero o a los cubos.*

**Controlled load servo** A type of servo that has two pistons and allows for the quick release of a band during shifting.
***Servomando de carga controlada*** *Tipo de servo que tiene dos pistones y permite el disparo rápido de una banda durante los cambios.*

**Converter capacity** An expression of a torque converter's ability to absorb and transmit engine torque with a limited amount of slippage.
***Capacidad de conversión*** *Expresión de la habilidad de un convertidor de torsión para absorber y transmitir torsión del motor a una cantidad limitada de resbalamiento.*

**Corrosion** Chemical action, usually caused by an acid, that eats away (decomposes) a metal.
***Erosión*** *Acción química generalmente causada por un ácido, que se come (descompone) un metal.*

**Coupling** A connecting means for transferring movement from one part to another; may be mechanical, hydraulic, or electrical.
***Acoplamiento*** *Manera de conectarse para transferir movimiento de una parte a otra, ya sea mecánica, hidráulica o eléctrica.*

**Coupling phase** Point in torque converter operation where the turbine speed is 90 percent of impeller speed and there is no longer any torque multiplication.
***Fase de acoplamiento*** *Punto en una operación de convertidor de torsión en donde la velocidad de la turbina es el 90% de la velocidad de impulso y ya no hay multiplicación de torsión.*

**Cover plate** A stamped steel cover bolted over the service access to the manual transmission.
***Placa de cubierta*** *Cubierta troquelada de acero atornillada sobre el servicio de acceso de la transmisión de mano o estándar.*

**Crocus cloth** A very fine polishing paper. It is designed to remove very little metal; therefore it is safe to use on critical surfaces.
***Arpillera*** *Papel lustre muy fino. Está diseñado para quitar muy poco metal, por lo que se puede usar en superficies críticas.*

**Current** The flow of electrons through a conductor.
***Corriente*** *El flujo de electrones a través de un conductor.*

**Cycle** One set of changes in a signal that repeats itself several times.
***Ciclo*** *Juego de cambios en una señal que se repite varias veces.*

**Damper** A device used to reduce or eliminate vibrations.
***Amortiguador*** *Mecanismo que se usa para reducir o eliminar las vibraciones.*

**Default mode** A mode of operation that allows for limited use of the transmission in the case of electronic or serious mechanical failure.
***Modo de omisión*** *Modo de operación que permite el uso limitado de la transmisión en caso de una falla electrónica o una falla mecánica seria.*

**Deflection** Bending or movement away from normal due to loading.
***Desvío*** *Doblez o movimiento fuera del vencido normal por carga.*

**Degree** A unit of measurement equal to 1/360th of a circle.
***Grado*** *Unidad de medición igual a 1/360avo. de un círculo.*

**Detent** A small depression in a shaft, rail, or rod into which a pawl or ball drops when the shaft, rail, or rod is moved. This provides a locking effect.
***Retén*** *Pequeña depresión en un eje, carril, o varilla en la que cae un trinquete o bola cuando se mueve el eje, carril o varilla. Esto proporciona un efecto de cierre.*

**Detent mechanism** A shifting control designed to hold the manual transmission in the gear range selected.
***Mecanismo de retén*** *Un control de cambio de velocidad diseñado para detener la transmisión de mano en la serie de velocidad seleccionada.*

**Diagnosis** A systematic study of a machine or machine parts to determine the cause of improper performance or failure.
***Diagnóstico*** *Estudio sistemático de un motor o de partes de un motor para determinar la causa de su mal funcionamiento o falla.*

**Dial indicator** A measuring instrument with the readings indicated on a dial rather than on a thimble as on a micrometer.
***Indicador de cuadrante*** *Instrumento de medición con las lecturas indicadas en la esfera en lugar de en el manguito como en un micrómetro.*

**Differential** A mechanism between drive axles that permits one wheel to run at a different speed than the other when turning.
***Diferencial*** *Mecanismo entre los ejes motrices que permite que una rueda corra a una velocidad diferente que la otra mientras se da vuelta.*

**Differential action** An operational situation in which one driving wheel rotates at a slower speed than the opposite driving wheel.
***Acción diferencial*** *Situación operativa en donde una rueda en movimiento gira a una velocidad menor que la rueda en movimiento opuesta.*

**Differential case** The metal unit that encases the differential side gears and pinion gears, and to which the ring gear is attached.

***Caja del diferencial*** *Unidad de metal que encajona los engranajes laterales del diferencial y los engranajes de piñón, y en el que está adjunto el engranaje en anillo.*

**Differential drive gear** A large circular helical gear driven by the transaxle pinion gear and shaft and that drives the differential assembly.
***Engranaje de accionamiento diferencial*** *Engranaje espiral circular grande movido por el engranaje de piñón del transeje y la flecha y que mueve el ensamblado diferencial.*

**Differential housing** Cast iron assembly that houses the differential unit and the drive axles. This also is called the rear axle housing.
***Caja o campana del diferencial*** *Ensamblado de hierro fundido que aloja la unidad del diferencial y los ejes de manejo. También se le llama caja o campana del eje trasero.*

**Differential pinion gears** Small beveled gears located on the differential pinion shaft.
***Engranaje de piñón del diferencial*** *Pequeños engranajes biselados que se encuentran en el eje del piñón del diferencial.*

**Differential pinion shaft** A short shaft locked to the differential case. This shaft supports the differential pinion gears.
***Eje del piñón del diferencial*** *Eje corto trabado en la caja del diferencial. Este eje sostiene los engranajes de piñón del diferencial.*

**Differential ring gear** A large circular hypoid-type gear enmeshed with the hypoid drive pinion gear.
***Engranaje de anillo del diferencial*** *Engranaje grande de tipo hipoide circular enredado con el engranaje de piñón de manejo hipoide.*

**Differential side gears** The gears inside the differential case that are internally splined to the axle shafts and that are driven by the differential pinion gears.
***Engranajes laterales del diferencial*** *Engranajes dentro de la caja del diferencial que está internamente entablillada a los ejes, y a los cuales los activan los engranajes del piñón del diferencial.*

**Digital Signal** A voltage signal that has only two values—on or off.
***Señal digital*** *Señal de voltaje que tiene sólo dos valores—prendido o apagado.*

**Diode** A semiconductor that allows current to flow through it in one direction only.
***Diodo*** *Semiconductor que permite que fluya la corriente a través de él en una sola dirección.*

**Direct Current** A type of electrical power used in mobile applications. A unidirectional current of substantially constant value.
***Corriente directa*** *Tipo de energía eléctrica que se usa en las aplicaciones movibles. Una corriente unidireccional de valor constante y considerable.*

**Direct drive** One turn of the input driving member compared to one complete turn of the driven member, such as when there is direct engagement between the engine and drive shaft, where the engine crankshaft and the drive shaft turn at the same rpm.
***Acoplamiento directo*** *Vuelta del órgano motor de admisión comparada con una vuelta completa del miembro accionado. Tal como cuando hay un engrane directo entre el motor y el eje conductor en donde el cigüeñal del motor y el eje conductor giran con las mismas revoluciones por minuto.*

**Disengage** When the operator moves the clutch pedal toward the floor to disconnect the driven clutch disc from the driving flywheel and pressure plate assembly.
***Desengranar*** *Cuando el operador mete el pedal del cloche para desconectar el disco del cloche accionado del volante de manejo y del ensamblado de la placa de presión.*

**Distortion** A warpage or change in form from the original shape.
***Distorción*** *Alabeamiento o cambio en forma de su figura original.*

**DLC** The data link connector. This is the connector used to connect into a vehicle's computer system for the purpose of diagnostics. Prior to J1930 this commonly was referred to as the ALDL.
***CTD*** *Conector de transmisión de datos. Conector que se usa para conectarse en el sistema de computadora de un vehículo para hacer diagnóstico. Antes del J1930 se le conocía comúnmente como ALDL.*

**DMM** The acronym for a digital multimeter.
***MMD*** *Sigla para un multímetro digital.*

**Double-wrap band** A transmission band that is split with overlapping ends.
***Banda de arrollamiento doble*** *Banda de la transmisión que está dividida con terminales sobrepuestas.*

**Dowel pin** A pin inserted in matching holes in two parts to maintain those parts in fixed relation one to another.
***Clavija*** *Chaveta que se mete en agujeros adaptados en dos partes para mantener dichas partes en relación fija la una con la otra.*

**Downshift** To shift a transmission into a lower gear.
***Bajar la velocidad*** *Meter el cambio a la transmisión a bajo embrague.*

**Drive line** The universal joints, drive shaft, and other parts connecting the transmission with the driving axles.
***Línea de accionamiento*** *Las juntas universales, el eje conductor y otras partes que conectan a la transmisión con los ejes motrices.*

**Driveline torque** Relates to the rear-wheel driveline and is the transfer of torque between the transmission and the driving axle assembly.
***Torsión de la línea de accionamiento*** *Se relaciona con la línea trasera de accionamiento y es la transferencia de torsión entre la transmisión y el ensamblado del eje motriz.*

**Drive pinion gear** One of the two main driving gears located within the transaxle or rear driving axle housing. Together the two gears multiply engine torque.
***Engranaje del piñón de accionamiento*** *Uno de los dos piñones conductores principales que se encuentran dentro del eje transversal o transeje o de la caja del eje motriz trasero. Entre ambos engranajes multiplican el par motor.*

**Drive shaft** An assembly of one or two universal joints connected to a shaft or tube; used to transmit power from the transmission to the differential. Also called the propeller shaft.
***Eje conductor*** *Ensamblado de una o dos juntas universales conectadas a un eje o tubo; se usa para transmitir potencia de la transmisión al diferencial. También se le llama eje de transmisión.*

**Driven gear** The gear meshed directly with the driving gear to provide torque multiplication or reduction or a change of direction.
***Engranaje de accionamiento*** *Engranaje endentado directamente con el piñón conductor para proporcionar aumento de torsión, reducción o cambio de dirección.*

**Drop forging** A piece of steel shaped between dies while hot.
***Troquelar*** *Una pieza de acero que se le da forma con dados mientras está caliente.*

**Drum** A cylinder shaped device that is attached to or houses components. A friction unit is normally used to stop and hold a rotating drum.
***Cilindro desgranador*** *Aparato en forma de cilindro que une o guarda los componentes. Una unidad de fricción generalmente se usa para parar y detener un cilindro desgranador giratorio.*

**DSO** Digital storage scope.
***AAD*** *Alcance de almacenamiento digital.*

**DTC** The acronym for Diagnostic Trouble Code.
***CAD*** *Sigla para el código de averías de diagnóstico*

**Dual band** Another name for a double-wrap band.
***Doble banda*** *Otro nombre para la banda de arrollamiento doble.*

**Duty cycle** The variation in length of time the solenoid is energized per cycle.
***Ciclo de rendimiento*** *La variación en el tiempo que el selenoide se carga por ciclo.*

**DVOM** The acronymn for digital volt/ohm meter A tool that combines the voltmeter, ohmmeter, and ammeter together in one diagnostic instrument and provides a digital reading.
***Medidor digital de voltios y ohmios*** *Herramienta que combina el voltímetro, ohmímetro y amperímetro en un solo instrumento de diagnóstico y proporciona una lectura digital.*

**Dry friction** The friction between two dry solids.
***Fricción seca*** *Fricción entre dos sólidos secos.*

**Dynamic** In motion.
***Dinámico*** *En movimiento.*

**Dynamic balance** The balance of an object when it is in motion; for example, the dynamic balance of a rotating drive shaft.
***Balance dinámico*** *Balance de un objeto cuando está en movimiento; por ejemplo, el balance dinámico de un eje conductor giratorio.*

**Dynamic pressure** Pressure that changes and/or causes something to move as pressurized fluid flow is present.
***Presión dinámica*** *Presión que cambia y/o causa que algo se mueva con la presencia del flujo del líquido a presión.*

**Dynamic seal** A seal used between two parts that move in relation to each other.
***Obturación dinámica*** *Obturación o burlete que se usa entre dos partes que se mueven una en relación con la otra.*

**Eccentric** One circle within another circle, wherein both circles do not have the same center or a circle mounted off center. On FWD cars, front-end camber adjustments are accomplished by turning an eccentric cam bolt that mounts the strut to the steering knuckle.
***Excéntrica*** *Un círculo dentro de otro en donde ambos no tienen el mismo centro o un círculo montado fuera del centro. En los vehículos TD los ajustes de la inclinación frontal se logran al dar vuelta al perno de leva excéntrico que monta el poste a la charnela de dirección.*

**EEPROM** An electrically erasable programmable read only memory chip.
***CMPSLBE*** *Chip de memoria programable de sólo lectura que se borra electrónicamente.*

**Elastomer** Any rubberlike plastic or synthetic material used to make bellows, bushings, and seals.
***Elastómero*** *Cualquier material sintético o de plástico parecido al hule que se usa para hacer fuelles, manguitos y burletes.*

**Electricity** The type of energy caused by the flow of electrons from one atom to another. It is the release of energy as one electron leaves the orbit of one atom and jumps into the orbit of another.
***Electricidad*** *Tipo de energía que causa el flujo de electrones de un átomo a otro. Es el desprendimiento de energía cuando un electrón sale de la órbita de un átomo y brinca a la órbita de otro.*

**Electromagnetism** A principle of using electricity to align the electrons in some material to give it magnetic properties.
***Electromagnetismo*** *Principio del uso de la electricidad para alinear los electrones en un material para darle propiedades magnéticas.*

**End clearance** Distance between a set of gears and their cover, commonly measured on oil pumps.
***Juego o holgura terminal*** *Distancia entre un conjunto de engranajes y su cubierta, comúnmente medido en las bombas de aceite.*

**Endplay** The amount of axial or end-to-end movement in a shaft because of clearance in the bearings.
***Juego longitudinal*** *Cantidad de movimiento del eje o de extremo a otro en un eje dado al juego entre los cojinetes.*

**Engage** When the vehicle operator moves the clutch pedal up from the floor, this engages the driving flywheel and pressure plate to rotate and drive the driven disc.
***Embrague*** *Cuando el operador de un vehículo sube el pedal del cloche, esto engrana el volante y la placa de presión a girar y a accionar el disco de mando.*

**Engagement chatter** A shaking, shuddering action that takes place as the driven disc makes contact with the driving members. Chatter is caused by a rapid grip and slip action.
***Traqueteo del engranaje*** *Acción de sacudida y estremecimiento que se lleva a cabo cuando el disco de accionamiento hace contacto con el órgano motor. El traqueteo lo causa la acción rápida de sujeción y deslizamiento.*

**Engine load** The physical resistance placed on an engine's crankshaft.
***Carga del motor*** *Resistencia física ejercida sobre el cigüeñal de un motor.*

**Engine torque** A turning or twisting action measured in foot-pounds or kilogram-meters developed by the engine.
***Par motor*** *Acción de dar vueltas y torcerse desarrollada por el motor y que se mide en pies y libras o en kilogramos y metros.*

**Engine vacuum** The low pressure formed by an engine's pistons.
***Vacío del motor*** *Baja presión que forman los pistones de un motor.*

**EPC Solenoid** Electronic Pressure Control solenoid, used to control and maintain mainline pressure in some transmissions.
***Solenoide CPE*** *Solenoide de control de presión electrónica, se usa para controlar y mantener la presión de las líneas principales en algunas transmisiones.*

**Etching** A discoloration or removal of some material caused by corrosion or some other chemical reaction.
***Mordiente*** *Mancha o eliminación de algún material causa da por la corrosión o alguna otra reacción química.*

**Extension housing** An aluminum or iron casting of various lengths that encloses the transmission output shaft and supporting bearings.
***Caja de extensión*** *Vaciado o colada de aluminio o hierro de varias longitudes que encierra el árbol motor de la transmisión y los cojinetes portadores.*

**External gear** A gear with teeth across the outside surface.
***Engranaje externo*** *Engranaje con dientes a través de la superficie exterior.*

**Externally tabbed clutch plates** Clutch plates are designed with tabs around the outside periphery to fit into grooves in a housing or drum.
***Placas de embrague de aleta externa*** *Las placas de embrague se diseñan con aletas alrededor de la periferia exterior para hacerlas caber en hendiduras en una caja o cilindro.*

**Extreme-pressure lubricant** A special lubricant for use in hypoid-gear differentials; needed because of the heavy wiping loads imposed on the gear teeth.
***Lubricante de presión extrema*** *Lubricante especial para usarse en los diferenciales de engranaje hipoide; se necesita por las cargas de barrido pesadas que se gravan en los dientes del engranaje.*

**Fail-safe valve** The valve, found in some hydraulic circuits, that allows limited transmission operation when a component or components have failed.
***Válvula de seguridad contra fallo*** *`Válvula, encontrada en algunos circuitos hidráulicos, permite la operación limitada de la transmisión cuando fallan uno o más componentes.*

**Fatigue** The buildup of natural stress forces in a metal part that eventually causes it to break. Stress results from bending and loading the material.
***Fatiga*** *Acumulación de fuerzas naturales de tensión en la parte de un metal que eventualmente causa que se rompa. La tensión resulta cuando dobla y se carga el material.*

**Feeler gauge** A metal strip or blade finished accurately with regard to thickness used for measuring the clearance between two parts; such gauges ordinarily come in a set of different blades graduated in thickness by increments of 0.001 inch.
***Galga para huelgos*** *Tira metálica terminada con precisión para medir y anchura se usa para medir el juego o huelgo entre dos partes; tales galgas ordinariamente vienen en un juego con diferentes hojas graduadas por grosor en incrementos de 0.001 pulgada.*

**Final drive ratio** The ratio between the drive pinion and ring gear.
***Porcentaje de accionamiento final*** *Porcentaje entre el piñón de accionamiento y el engranaje circular.*

**Fixed displacement pump** Fluid pumps that maintain a particular amount of fluid flow.
***Bomba de desajuste fija*** *Bomba de líquido que mantiene una cierta cantidad de flujo de líquido.*

**Fixed-type constant-velocity joint** A joint that cannot telescope or plunge to compensate for suspension travel. Fixed joints are always found on the outer ends of the drive shafts of FWD cars. A fixed-joint may be of either Rzeppa or tripod type.
***Junta de velocidad constante tipo fija*** *Junta que no puede condensarse o hundirse para compensar por el desplazamiento de la suspensión. Las juntas fijas siempre se encuentran en las terminales exteriores del eje conductor de los vehículos TD. Una junta fija puede ser de tipo Rzeppa o tripoide.*

**Flange** A projecting rim or collar on an object for keeping it in place.
***Reborde*** *Borde proyectado u collarín en un objeto para mantenerlo en su lugar.*

**Flat rate** Flat rate is a pay system in which technicians are paid for the amount of work they do. Each job has a flat rate time.
***Tarifa o precio fijo*** *Tarifa fija es un sistema de pago en la cual al técnico se le paga por la cantidad de trabajo que desempeñe. Cada trabajo tiene un tiempo de precio fijo.*

**Flexplate** A lightweight flywheel used only on engines equipped with an automatic transmission. The flexplate is equipped with a starter ring gear around its outside diameter and also serves as the attachment point for the torque converter.
***Placa encorvada*** *Volante ligero que se usa sólo en los motores equipados con una transmisión automática. La placa encorvada está equipada con un engranaje circular de encendido alrededor del diámetro exterior y también sirve como punto de apoyo para el convertidor del par motor.*

**Flow-directing valves** Valves that direct pressurized fluid to the appropriate apply device to cause a change in gear ratios.
***Válvulas reguladoras del flujo*** *Válvulas que regulan el flujo a presión hacia el mecanismo de fijación apropiado para provocar un cambio en la relación de transmisión.*

**Fluid coupling** A device in the power train consisting of two rotating members; transmits power from the engine, through a fluid, to the transmission.
***Acoplamiento hidráulico*** *Aparato en motor que consiste de dos miembros giratorios; transmite potencia del motor a la transmisión a través de un líquido,.*

**Fluid drive** A drive in which there is no mechanical connection between the input and output shafts, and power is transmitted by moving oil.
***Accionamiento hidráulico*** *Accionamiento en el que no hay una conexión mecánica entre los ejes de entrada y salida, y la potencia la transmite el aceite en movimiento.*

**Flywheel** A heavy metal wheel that is attached to the crankshaft and rotates with it; helps smooth out the power surges from the engine power strokes; also serves as part of the clutch and engine-cranking system.
***Volante*** *Rueda de metal pesada que se sujeta al cigüeñal y gira con él; ayuda a suavizar las sobrecargas causadas por los golpes de potencia del motor; también sirve como parte del cloche y del sistema de arranque del motor.*

**Flywheel ring gear** A gear, fitted around the flywheel, that is engaged by teeth on the starting-motor drive to crank the engine.
***Engranaje redondo del volante*** *Engranaje que va alrededor del volante y que se embraga con dientes en el encendido para arrancar el motor.*

**Force** Any push or pull exerted on an object; measured in pounds and ounces, or in Newton meters (N·m) in the metric system.
***Fuerza*** *Cualquier empujón o jalón que se le da a un objeto; se mide en libras y onzas, o en newtons y metros en el sistema métrico.*

**Four-wheel-drive** On a vehicle, driving axles at both front and rear, so that all four wheels can be driven. Also known as 4WD.
***Tracción cuatro ruedas*** *En un vehículo son los ejes motrices traseros como delanteros para que las cuatro ruedas puedan accionarse. También se le conoce como T4R o 4WD.*

**Free wheel** To turn freely and not transmit power.
***Ir en un punto muerto*** *Rotar libremente sin transmitir potencia.*

**Freewheeling clutch** A mechanical device that will engage the driving member to impart motion to a driven member in one direction but not the other. Also known as an "overrunning clutch."
***Cloche de rotación libre*** *Aparato mecánico que accionará el órgano motor para impartir movimiento a un miembro accionado en una dirección pero no en otra. También se le conoce como "cloche de sobremarcha".*

**Frequency** The number of complete cycles that occur within a given period of time.
***Frequencia*** *Número de ciclos completos que suceden en un cierto período de tiempo.*

**Friction** The resistance to motion between two bodies in contact with each other.
***Fricción*** *Resistencia al movimiento entre dos cuerpos en contacto el uno con el otro.*

**Friction bearing** A bearing in which there is sliding contact between the moving surfaces. Sleeve bearings, such as those used in connecting rods, are friction bearings.
***Cojinete de fricción*** *Cojinete en donde hay un contacto de deslizamiento entre las superficies en movimiento. Los cojinetes de camisa, tales como los que se usan en bielas de conexión, son cojinetes de fricción.*

**Friction disc** In the clutch a flat disc, faced on both sides with friction material and splined to the clutch shaft. It is positioned between the clutch pressure plate and the engine flywheel. Also called the clutch disc or driven disc.
***Disco de fricción*** *En el cloche es un disco plano, revestido por ambos lados con material de fricción y acuñado en el eje del cloche. Está colocado entre la placa de presión del cloche y el volante. También se conoce como disco del cloche o disco accionador.*

**Friction facings** A hard-molded or woven asbestos or paper material that is riveted or bonded to the clutch or driven disc.
***Caras de fricción*** *Material de asbestos o papel moldeado duro o entretejido que se remacha o pega al disco impulsador del cloche.*

**Front pump** Pump located at the front of the transmission. It is driven by the engine through two dogs on the torque converter housing. It supplies fluid whenever the engine is running.
***Bomba anterior*** *Bomba localizada en el frente de la transmisión. La impulsa el motor a través de dos perros de apriete en la campana del convertidor del par motor. Proporciona líquido cuando está corriendo el motor.*

**Front-wheel-drive (FWD)** The vehicle has all drive train components located at the front.
***Tracción delantera (TD)*** *Vehículo que tiene los componentes del tren de accionamiento localizados en el frente.*

**Fulcrum** The point at which a lever pivots.
***Punto de apoyo*** *Punto en el que gira una palanca.*

**Fuse** An electrical device used to protect a circuit against accidental overload or unit malfunction.
***Fusible*** *Aparato eléctrico que se usa para proteger un circuito contra la sobrecarga por accidente o por el fallo de la unidad.*

**Fusible Link** A type of fuse made of a special wire that melts to open a circuit when current draw is excessive.
***Enlace del fusible*** *Tipo de fusible hecho con un alambre especial que se derrite para abrir un circuito cuando es excesiva la carga de corriente.*

**FWD** Abbreviation for front-wheel-drive.
***TD*** *Abreviación para la tracción delantera.*

**Galling** Wear caused by metal-to-metal contact in the absence of adequate lubrication. Metal is transferred from one surface to the other, leaving behind a pitted or scaled appearance.
***Excoriación superficial por abrasión*** *Gasto causado por el contacto de metal con metal cuando no hay lubricación adecuada. El metal se transfiere de una superficie a otra dejando una apariencia cicatrizada o escamosa.*

**Gasket** A layer of material, usually made of cork, paper, plastic, composition, or metal, or a combination of these, placed between two parts to make a tight seal.
***Junta elástica*** *Capa de material, generalmente hecha de corcho, papel, plástico, composición o metal, o una combinación de éstas, colocada entre dos partes para hacer una obturación hermética.*

**Gasket cement** A liquid adhesive material, or sealer, used to install gaskets.
***Cemento de junta elástica*** *Material adhesivo líquido, u obturador, que se usa para instalar juntas elásticas.*

**Gear** A wheel with external or internal teeth that serves to transmit or change motion.
***Engranaje*** *Rueda con dentadura externa o interna que sirve para transmitir o cambiar movimiento.*

**Gear lubricant** A type of grease or oil blended especially to lubricate gears.
***Lubricante de engranaje*** *Tipo de grasa o aceite mezclado especialmente para lubricar engranajes.*

**Gear ratio** The number of revolutions of a driving gear required to turn a driven gear through one complete revolution. For a pair of gears, the ratio is found by dividing the number of teeth on the driven gear by the number of teeth on the driving gear.
***Relación de transmisión*** *Número de revoluciones que requiere el aparato de mando para dar vueltas a un engranaje activado a través de una vuelta completa. Para un par de engranajes, el porcentaje o relación se encuentra al dividir el número de dientes en el engranaje activado entre el número de dientes del aparato de mando.*

**Gear reduction** When a small gear drives a large gear, there is an output speed reduction and a torque increase, which results in a gear reduction.
***Reducción de engranajes*** *Cuando un engranaje pequeño acciona uno grande, hay una reducción de velocidad de salida y aumenta el par motor, lo cual resulta en una reducción de engranajes.*

**Gear-type pump** A fixed displacement pump that consists of two gears in mesh.
***Bomba de tipo engranaje*** *Bomba fija de desajuste que consiste en dos engranajes endentados.*

**Gear whine** A high-pitched sound developed by some types of meshing gears.
***Plañido del engranaje*** *Sonido de tono alto que desarrollan algunos tipos de engranajes endentados.*

**Gearshift** A linkage-type mechanism by which the gears in an automobile transmission are engaged and disengaged.
***Caja de cambios*** *Mecanismo de tipo enlace en donde los engranajes en una transmisión automática se embragan y desembragan.*

**Glitches** Abnormal, slight movements of a waveform on a lab scope. These can be caused by circuit problems or noise in the circuit.
***Interferencias*** *Movimientos anormales y pequeños de una onda en un campo de aplicación de laboratorio. Pueden ser causadas por problemas o ruidos en el circuito.*

**Governor pressure** The transmission's hydraulic pressure that is directly related to output shaft speed. It is used to control shift points.
***Presión reguladora*** *Presión hidráulica de la transmisión que se relaciona directamente con la velocidad del eje motor. Se usa para controlar puntos de cambio.*

**Governor valve** A device used to sense vehicle speed. The governor valve is attached to the output shaft.
***Válvula reguladora*** *Aparato que se usa para sentir la velocidad del vehículo. La válvula reguladora está sujeta al eje motor.*

**Grade marks** Marks on fasteners that indicate strength.
***Marcas de referencia*** *Marcas en los sujetadores para indicar la fuerza.*

**Graphite** Very fine carbon dust with a slippery texture used as a lubricant.
***Grafito*** *Polvo muy fino de carbón con textura deslizante que se usa como lubricante.*

**Guide rings** Rings built into a torque converter to direct the vortex flow and provide for smooth and turbulence-free fluid flow.
***Corona directriz*** *Anillos hechos en un convertidor de par motor para dirigir la circulación en remolino y proporcionar un flujo de líquido suave y libre de turbulencia.*

**Hard gasket** A type of gasket that is used between two smooth surfaces.
***Junta plástica dura*** *Tipo de junta plástica que se usa entre superficies para alisarlas.*

**Heat exchanger** A heat exchanger also may be called an intercooler and is used to transfer heat from one object to another. Heat is exchanged because of a law of nature in that the heat of an object will always attempt to heat a cooler object.
***Termopermutador*** *Al termopermutador también se le puede llamar interenfriador, y se usa para transferir calor de un objeto a otro. El calor se intercambia debido a una ley natural que establece que el calor de un objeto siempre tenderá a calentar un objeto más frío.*

**Heat treatment** Heating, followed by fast cooling, to harden metal.
***Tratamiento térmico*** *Calentamiento seguido por enfriamiento rápido para endurecer un metal.*

**Helical gear** A gear with teeth that are cut at an angle or are spiral to the gear's axis of rotation.
***Engranaje de dentadura helicoidal*** *Engranaje con dientes que están cortados en ángulo o que son espirales al eje de rotación del engranaje.*

**Hook-end seals** A metal sealing ring that has small hooks at each end.
***Obturadores de ganchos terminales*** *Obturador redondo metálico que tiene pequeños ganchos en cada extremo.*

**Hub** The center part of a wheel, to which the wheel is attached.
***Cubo*** *Parte central de una rueda en la cual se fija la rueda.*

**Hydraulic press** A piece of shop equipment that develops a heavy force by use of a hydraulic piston-and-jack assembly.
***Prensa hidráulica*** *Pieza de equipo de taller que desarrolla una fuerza pesada con el uso de un ensamblaje hidráulico de pistón y gato.*

**Hydraulic pressure** Pressure exerted through the medium of a liquid.
***Presión hidráulica*** *Presión ejercida a través de un medio líquido.*

**Hybrid valve** A spool valve that relies on spring tension and hydraulic force for movement.
***Válvula híbrida*** *Válvula de carrete que depende de la tensión del resorte y la fuerza hidráulica para el movimiento.*

**ID** Inside Diameter.
***DI*** *Diámetro interior.*

**Ignitability** The characteristic of a material that enables it to spontaneously ignite.
***Inflamabilidad*** *Característica de un material que le permite una inflamabilidad espontánea.*

**Impact sockets** Heavier walled sockets designed for use with an impact wrench.
***Clavija bipolar de choque*** *Clavijas bipolares de pared más pesadas que se diseñan para usarse con una llave de tuercas de choque.*

**Impedance** The operating resistance of an electrical device.
***Impedencia*** *Resistencia de operación de un aparato eléctrico.*

**Impeller** The pump or driving member in a torque converter.
***Soplante*** *Bomba u órgano motor en un convertidor del par motor.*

**Increments** Series of regular additions from small to large.
***Incrementos*** *Series de adiciones regulares de pequeño a grande.*

**Index** To orient two parts by marking them. During reassembly, the parts are arranged so the index marks are next to each other. Used to preserve the orientation between balanced parts.
***Indexar*** *Orientar dos partes al marcarlas. Durante el reensamblado las partes se colocan de tal manera que la indiciación de una está seguida por las otras. Se usa para conservar la orientación entre las partes balanceadas.*

**Induction** The process of producing electricity through magnetism rather than direct flow through a conductor.
***Inducción*** *Proceso de producir electricidad por medio de magnetismo en lugar de un flujo directo mediante un conductor.*

**Input shaft** The shaft carrying the driving gear by which the power is applied, as to the transmission.
***Eje de entrada*** *Eje que lleva el aparato de mando por medio del cual se aplica la energía, como a la transmisión.*

**Inspection cover** A removable cover that permits entrance for inspection and service work.
***Cubierta de inspección*** *Cubierta removible que permite la entrada para inspección y servicio.*

**Insulator** A material that does not allow for current to flow through it easily.
***Aislante*** *Material que no permite que fluya la corriente a través de él.*

**Integral** Built into, as part of the whole.
***Integral*** *Incorporado, como parte de un entero.*

**Interaxle differential** Another name for the center differential unit used in some 4WD vehicles.
***Diferencial de entreeje*** *Otro nombre para la unidad del diferencial de centro que se usa en algunos vehículos de T4R.*

**Internal gear** A gear with teeth pointing inward, toward the hollow center of the gear.
***Engranaje de dentadura interior*** *Engranaje con dentadura apuntando hacia adentro, hacia el centro hueco del engranaje.*

**Isolator springs** Springs used in converter clutches to absorb the normal torsional vibrations of the engine.
***Resortes aisladores*** *Resortes que se usan en los cloches conmutadores para absorber las vibraciones torsionales normales del motor.*

**IX-pump** A name given sometimes to a rotor-type fluid pump because of its internal/external design.
***Bomba IX*** *Nombre dado algunas veces a una bomba de líquido tipo rotor por su diseño interno-externo.*

**Jam nut** A second nut tightened against a primary nut to prevent it from working loose. Used on inner and outer tie-rod adjustment nuts and on many pinion-bearing adjustment nuts.
***Contratuerca*** *Segunda tuerca apretada contra la tuerca primaria para prevenir que se suelte. Se usa en tuercas de ajuste de varillas de amarre interior o exteriormente y en muchas tuercas de ajuste del cojinete del piñón.*

**Journal** A bearing with a hole in it for a shaft.
***Muñequilla*** *Cojinete con un agujero para el eje.*

**Key** A small block inserted between the shaft and hub to prevent circumferential movement.
***Chaveta*** *Bloque pequeño metido entre el eje y el cubo para prevenir movimiento circunferencial.*

**Keyway** A groove or slot cut to permit the insertion of a key.
***Cerradura o chavetero*** *Hendidura o ranura cortada para permitir insertar una llave o chaveta.*

**Kickdown** Forced downshift.
***Cambio a velocidad menor*** *Meter forzadamente una velocidad más baja.*

**Knock** A heavy metallic sound usually caused by a loose or worn bearing.
***Detonación*** *Sonido metálico pesado que generalmente lo causa un cojinete suelto o gastado.*

**Lands** The raised area on a spool valve.
***Terrenos o fundos*** *Área levantada en una válvula de carrete.*

**Pressure sensors** Sensors used to monitor fluid pressure.
***Detectores de presión*** *Detectores que se usan para mantener la presión de los fluidos.*

**PROM** Programmable Read Only Memory; contains all the information about the vehicle and its systems so that the control system can make decisions that are based on that vehicle.
***MPSL*** *Memoria programable de sólo lectura que contiene toda la información sobre el vehículo y sus sistemas para que el sistema de control pueda tomar decisiones que se basen en ese vehículo.*

**Prussian blue** A blue pigment; in solution, useful in determining the area of contact between two surfaces.
***Azul de Prusia*** *Pigmento azul; en una solución es útil para determinar el área de contacto entre dos superficies.*

**Psi** Abbreviation for pounds per square inch, a measurement of pressure.
***psi*** *Abreviatura de las libras por pulgada cuadrada, una medida de la presión.*

**Pulsation** To move or beat with rhythmic impulses.
***Pulsación*** *Mover o golpear con impulsos rítmicos.*

**Pulse width** The length of time something is energized.
***Duración del impulso*** *El período de tiempo durante el cual algo está cargado.*

**PWM** Pulse width modulated, a term used to define the operation of a solenoid that is turned on and off by a computer to control its output.
***MDI Modulación de la duración del impulso*** *Término para definir la operación de un solenoide que prende y apaga una computadora para controlar su rendimiento.*

**Race** A channel in the inner or outer ring of an antifriction bearing in which the balls or rollers roll.
***Carril*** *Canal en el anillo o corona interior o exterior de un cojinete antifricción en donde ruedan las esferas o los rodillos.*

**Radial** The direction moving straight out from the center of a circle, perpendicular to the shaft or bearing bore.
***Radial*** *La dirección que se sale directa del centro del círculo. Perpendicular al eje o a la perforación del cojinete.*

**Radial clearance (Radial displacement)** Clearance within the bearing and between balls and races perpendicular to the shaft.
***Huelgo radial (Desplazamiento radial)*** *Huelgo entre los cojinetes y entre las esferas y carriles perpendiculares al eje.*

**Radial load** A force perpendicular to the axis of rotation.
***Carga radial*** *Fuerza perpendicular al eje de rotación.*

**RAM (Random Access Memory)** A type of memory in a computer that allows for the storage of collected data.
***RAM Memoria de acceso aleatorio*** *Tipo de memoria en una computadora que permite el almacenamiento de colecciónes de datos.*

**Ratio** The relation or proportion that one number bears to another.
***orcentaje o relación*** *Relación o proporción que un número eva sobre otro.*

**avigneaux gear train** A compound gear set that combines o planetary units with a common ring gear.
***en de engranaje Ravigneaux*** *Juego de engranaje mpuesto que combina dos unidades satélites con un engranaje ılar común.*

**ction area** The area around a spool valve that allows surized fluid to move the valve.
***a de reacción*** *Área alrededor de una válvula de carrete que ite que el flujo a presión mueva la válvula.*

**Reaction members** The members of a gear set that are held stationary so the other members can react to them.
***Piezas de reacción*** *Piezas de un juego de engranaje que se mantienen fijas para que las otras piezas les puedan responder.*

**Reactivity** The characteristic of a material that enables it to react violently with water and other materials.
***Reactividad*** *Característica de un material que le permite reaccionar violentamente con el agua y otros materiales.*

**Reactor** Another name for the stator in a torque converter.
***Motor de reacción*** *Otro nombre para el estator en el conversor del par motor.*

**Reamer** A round metal cutting tool with a series of sharp cutting edges; enlarges a hole when turned inside it.
***Máquina de escarear*** *Herramienta circular para cortar metal con una serie de orillas afiladas; agranda un orificio cuando se voltea dentro de él.*

**Reference voltage sensors** An electrical device that responds to changes by altering the voltage (reference voltage) it receives.
***Detectores del voltaje de referencia*** *Aparato eléctrico que responde a los cambios al alterar el voltaje (voltaje de referencia) que recibe.*

**Relay** An electromagnetic device used to control a high current circuit with a low current control circuit.
***Relé*** *Aparato electromagnético que controla un circuito de alta corriente con un circuito de control de baja corriente.*

**Relay valve** A spool valve used to control the direction of fluid flow without affecting the pressure of the fluid.
***Válvula de relé*** *Válvula de carrete que controla la dirección del flujo del líquido sin afectar la presión del líquido.*

**Relief valve** A valve used to protect against excessive pressure in the case of a malfunctioning pressure regulator.
***Válvula de seguridad*** *Válvula que protege contra una presión excesiva en caso de un regulador de presión con fallas.*

**Resistance** An electrical term for something that opposes current flow.
***Resistencia*** *Término eléctrico para algo que se opone al flujo de corriente.*

**Retaining ring** A removable fastener used as a shoulder to retain and position a round bearing in a hole.
***Corona de sujeción*** *Sujetador desmontable que se usa como reborde para detener y colocar un cojinete redondo en un orificio.*

**RFI** Radio frequency interference, which is an unwanted voltage signal that rides on a signal.
***IRF Interferencia de radiofrecuencia*** *Es la señal de voltaje no deseada que monta una señal.*

**Rheostats** A type of variable resistor. Rheostats provide for varying amounts of voltage and current from the tap and have two connections—one to the fixed end of a resistor and one to a sliding contact with the resistor. Turning the control moves the sliding contact away from or toward the fixed end tap, increasing or decreasing the resistance.
***Reostatos*** *Tipo de resistencia variable. Los reostatos proporcionan cantidades variadas de voltaje y corriente de su enchufe y tienen dos conexiones—una para la terminal fija de un resistor y otra para el contacto deslizante con la resistencia. Al darle vuelta al control retira el contacto deslizante o lo acerca hacia el enchufe de la terminal fija, aumentando o disminuyendo la resistencia.*

**Right-to-Know Laws** Laws requiring employers to provide employees with a safe workplace as it relates to hazardous materials and information about any and all hazards the employees may face when performing their job.



**Lapping** The process of fitting one surface to another by rubbing them together with an abrasive material between the two surfaces.
***Recubrimiento o solape*** *Proceso de que una superficie quede en otra al rozarlas o frotarlas juntas con un material abrasivo entre ambas superficies.*

**Lash** The amount of free motion in a gear train, between gears, or in a mechanical assembly, such as the lash in a valve train.
***Coletazo*** *Cantidad de movimiento libre en un tren de engranajes, entre los engranajes, o en un ensamblado mecánico, tal como el coletazo en un tren de válvula.*

**Lathe-cut seal** Another name for a square cut seal designed to withstand axial movement.
***Obturador cortado en el torno*** *Otro nombre para el obturador cortado en cuadrado que está diseñado para resistir el movimiento del eje.*

**Linkage** Any series of rods, yokes, and levers, and so on— used to transmit motion from one unit to another.
***Acoplamiento o enlace*** *Cualquier serie de bielas, varillas de maniobra y balancines o palancas. Se usa para transmitir movimiento de una unidad a otra.*

**Lip seals** Type of seal used to seal parts that have axial movement.
***Obturadores de reborde*** *Tipo de obturadores que se usa para sellar las partes que tienen movimiento del eje.*

**Loads** Devices that use electricity to perform work, such as lights and motors.
***Receptores*** *Aparatos que usan electricidad para realizar trabajo, tales como las luces y los motores.*

**Lock pin** Used in some ball sockets (inner tie-rod end) to keep the connecting nuts from working loose. Also used on some lower ball joints to hold the tapered stud in the steering knuckle.
***Perno de cierre*** *Usado en algunos cojinetes esféricos (terminal de la varilla de amarre interno) para asegurar que los pernos de conexión no se suelten. También se usan en rótulas inferiores para detener la borna o contrete escalonado en la charnela de cambio.*

**Locking ring** A type of sealing ring that has ends that meet or lock together during installation. There is no gap between the ends when the ring is installed.
***Anillo de cierre*** *Tipo de anillo obturador que tiene terminales que se cierran durante la instalación. No hay espacio intermedio entre las terminales cuando se instala el anillo.*

**Locknut** A second nut turned down on a holding nut to prevent loosening.
***Contratuerca*** *Segunda tuerca apretada en una tuerca sujetadora para prevenir que se suelte.*

**Lockplates** Metal tabs bent around nuts or bolt heads.
***Placas de retención*** *Salientes metálicas que se doblan alrededor de tuercas o en cabezas de pernos.*

**Lockwasher** A type of washer which, when placed under the head of a bolt or nut, prevents the bolt or nut from working loose.
***Arandela de seguridad*** *Tipo de arandela que, cuando se coloca debajo de la cabeza de un perno o de una tuerca, previene que se suelten los pernos o tuercas.*

**Lubricant** Any material, usually a petroleum product such as grease or oil, which is placed between two moving parts to reduce friction.
***Lubricante*** *Cualquier material, generalmente un producto de petróleo tal como grasa o aceite, que se coloca entre dos partes en movibles para reducir la fricción.*

**Magnetism** A force between two poles of opposite potential, caused by the alignment of electrons.
***Magnetismo*** *Fuerza entre dos polos de voltaje opuesto causada por la alineación de los electrones.*

**Mainline pressure** The hydraulic pressure that operates apply devices and is the source of all other pressures in an automatic transmission. It is developed by pump pressure and regulated by the pressure regulator.
***Presión de la tubería principal*** *Presión hidráulica que hace funcionar aparatos de aplicación y que es la fuente de todas las demás presiones en una transmisión automática. Se desarrolla por la presión de la bomba y es regulada por el regulador de presión.*

**Main oil pressure regulator valve** Regulates the line pressure in a transmission.
***Válvula principal reguladora de la presión del aceite*** *Regula la presión de la línea en una transmisión.*

**Manual shift valve** A valve used to manually select the operating mode of the transmission; it is moved by the gearshift lever.
***Válvula de cambio manual*** *Válvula que se usa para seleccionar manualmente el modo de funcionamento de la transmisión; es movida por la palanca de la caja de cambios.*

**MAP Sensor** A Manifold Absolute Pressure sensor used to measure the pressure in an area and compare it to the pressure around it.
***Detector PAC*** *Detector de presión absoluta del colector se usa para medir la presión en un área y para compararla con la presión a su alrededor.*

**Material Safety Data Sheets (MSDS)** Information sheets containing chemical composition and precautionary information for all products that can present a health or safety hazard.
***Hojas de datos de seguridad del material (HDSM)*** *Hojas informativas que contienen la composición química e información de precaución para todos los productos que pudieran presentar un peligro de salud o de seguridad.*

**Mechanical Advantage** Based on the principles of levers, force can be increased by increasing the length of the lever or by moving the pivot or fulcrum.
***Ventaja mecánica*** *Basado en los principios de las palancas, la fuerza puede incrementarse al aumentar el largo de la palanca o al mover el pivote o el punto de apoyo.*

**Meshing** The mating, or engaging, of the teeth of two gears.
***Endentado*** *Acople o engrane de los dientes en dos engranajes.*

**Micrometer** A precision measuring device used to measure small bores, diameters, and thickness. Also called a mike.
***Micrómetro*** *Aparato de medición precisa que se usa para medir perforaciones pequeñas y grosores.*

**MIL** The malfunction indicator lamp for a computer control system. Prior to J1930, the MIL commonly was called a Check Engine or Service Engine Soon lamp.
***LIF*** *Lámpara indicadora de fallo para un sistema de control de computadora. Antes del J1930, la LIF se conocía comúnmente como lámpara para revisar motores o lámpara indicadora de servicio inmediato.*

**Mineral spirits** An alcohol-based liquid that leaves no surface residue after it dries.
***Espíritu de petróleo*** *Líquido con base de alcohol que no deja residuos en la superficie cuando se seca.*

**Misalignment** When bearings are not on the same centerline.
***Desalineado*** *Cuando los cojinetes no están en la misma línea de centro.*

**Modulator** A vacuum-diaphragm device connected to a source of engine vacuum. It provides an engine load signal to the transmission.
***Modulador*** *Aparato de diafragma al vacío conectado a una fuente del vacío del motor. Proporciona una señal de carga del motor a la transmisión.*

**Mounts** Made of rubber to insulate vibrations and noise when they support a power train part, such as engine or transmission mounts.
***Montajes o soportes*** *Hechos de caucho para aislar las vibraciones y ruidos mientras cargan una parte del tren de energía, tales como los soportes del motor o de la transmisión.*

**Multiple disc** A clutch with a number of driving and driven discs as compared to a single plate clutch.
***Pluridisco*** *Cloche con un número de discos directrices y activos o accionados comparado con un cloche de una placa.*

**Multiplexing** An electrical system in which voltage signals or information can be shared with two or more computers.
***Transmisión simultánea o multiplexación*** *Sistema eléctrico en el que las señales del voltaje o de información pueden compartirse entre dos o más computadoras.*

**Needle bearing** An antifriction bearing using a great number of long, small-diameter rollers. Also known as a quill bearing.
***Cojinete de agujas*** *Cojinete antifricción que usa un gran número de rodillos largos y de diámetro pequeño. También llamado cojinete de puas.*

**Needle check valve** A valve that utilizes a straight rod to open and close a fluid passage.
***Válvula de revisión de agujas*** *Válvula que utiliza una biela recta para abrir y cerrar un pasaje de líquido.*

**Needle deflection** Distance of travel from zero of the needle on a dial gage.
***Aguja de desviación*** *Distancia de movimiento de la aguja a partir del cero en una galga de cuadrante.*

**Neutral-start switch** A switch wired into the ignition switch to prevent engine cranking unless the transmission shift lever is in neutral or the clutch pedal is depressed.
***Interruptor de encendido en neutral*** *Interruptor conectado al interruptor de arranque para prevenir el arranque del motor a menos que la palanca de cambios de la transmisión esté en neutral o el pedal del cloche no esté metido.*

**Newton meter (N·m)** Metric measurement of torque or twisting force.
***Newton-metro (N-m)*** *Medida métrica de la fuerza del par motor o de torsión.*

**Nominal shim** A shim with a designated thickness.
***Calzo para ajuste nominal*** *Calzo con un grosor designado.*

**Nonhardening** A gasket sealer that never hardens.
***Sin endurecedor*** *Obturador de junta elástica que nunca se endurece.*

**O-ring** A type of sealing ring, usually made of rubber or a rubber-like material. In use, the O-ring is compressed into a groove to provide the sealing action.
***Anular*** *Tipo de anillo obturador que está hecho generalmente de caucho o de un material parecido al caucho. Cuando se usa, el anillo se prensa dentro de una hendidura para proporcionar una acción de obturación.*

**Occupational Safety and Health Administration (OSHA)** The government agency charged with ensuring safe work environments for all workers.
***Administración de Salud y Seguridad del Trabajo (ASSO)*** *Agencia gubernamental encargada de asegurar ambientes de trabajo seguros para todos los trabajadores.*

**OD** Outside diameter.
***DE*** *Diámetro exterior.*

**Ohm** A unit of measured electrical resistance.
***Ohmio*** *Unidad para medir la resistencia eléctrica.*

**Ohmmeter** The meter used to measure electrical resistance.
***Ohmiómetro*** *Medidor que se usa para medir la resistencia eléctrica.*

**Ohm's law** A basic law of electricity expressing the relationship between current, resistance, and voltage in any electrical circuit.
***Ley del ohmio*** *Ley o principio básico de electricidad que expresa la relación entre la corriente, la resistencia y el voltaje en cualquier circuito eléctrico.*

**Oil circuit** A schematic drawing showing the paths of fluid flow and the valves necessary to perform a specific function, also called a flow diagram.
***Circuito de aceite*** *Dibujo esquemático que muestra las trayectorias del flujo de líquido y las válvulas necesarias para realizar una función específica, también llamado un diagrama de fluido.*

**Oil seal** A seal placed around a rotating shaft or other moving part to prevent leakage of oil.
***Retén*** *Obturador colocado alrededor del eje giratorio u otra parte en movimiento para prevenir el goteo de aceite.*

**One-way clutch** See Sprag clutch.
***Cloche unidireccional*** *Vea el cloche patín o de calzo.*

**Open** An electrical circuit that has a break in the wire or is not complete.
***Abierto*** *Circuito eléctrico que tiene un interruptor en el alambre o que no está completo.*

**Open-end seals** A seal whose ends do not meet when it is installed.
***Obturadores de terminales abiertas*** *Obturador con terminales que no se unen cuando se instalan.*

**Orifice** A small opening designed to restrict flow.
***Orificio*** *Abertura pequeña diseñada para restringir el flujo.*

**Outer bearing race** The outer part of a bearing assembly on which the balls or rollers rotate.
***Carril del cojinete exterior*** *Parte exterior del ensamblado de cojinete en el cual giran las esferas o rodillos.*

**Output Driver** Located in the processor and operated by the digital commands of the computer, an output driver is an electronic on/off switch used to control the ground circuit of a specific actuator.
***Órgano motor de salida*** *Situado en el procesador y lo operan los mandos digitales de la computadora, el órgano motor de salida es un interruptor electrónico de encendido y apagado que se usa para controlar el circuito de tierra de una biela de accionamiento específica.*

**Output shaft** The shaft or gear that delivers the power from a device, such as a transmission.
***Eje de salida*** *Eje o engranaje que lleva la potencia que sale de un mecanismo, tal como una transmisión.*

**Overall ratio** The product of the transmission gear ratio multiplied by the final drive or rear axle ratio.
***Relación total*** *Producto de la relación de engranaje de la transmisión multiplicado por la relación del órgano motor final o del eje trasero.*

**Overcenter spring** Another name for a Belleville-type return spring.
***Resorte sobre centro*** *Otro nombre para el resorte de retroceso tipo Belleville.*

**Overdrive ratio** Identified by the decimal point indicating less than one driving input revolution compared to one output revolution of a shaft.
***Relación del órgano de salida total*** *Se identifica por el punto decimal que indica menos que una revolución del impulsor de entrada comparada con una revolución de salida de un eje.*



**Overrun coupling** A freewheeling device to permit rotation in one direction but not in the other.
***Acoplamiento de rebase*** *Aparato de volante libre para permitir la rotación en una dirección pero no en la otra.*

**Overrunning clutch** A device consisting of a shaft or housing linked together by rollers or sprags operating between movable and fixed races. As the shaft rotates, the rollers or sprags jam between the movable and fixed races. This jamming action locks together the shaft and housing. If the fixed race should be driven at a speed greater than the movable race, the rollers or sprags will disconnect the shaft.
***Cloche de sobremarcha o de rotación libre*** *Aparato que consiste de un eje o caja unidos por rodillos o calzos que funcionan entre carriles móviles y fijos. Mientras gira el eje, los rodillos o calzos se traban entre los carriles móviles y fijos. Este acto de trabamiento junta el eje con la caja. Si el carril fijo debe operarse a una velocidad mayor que el carril móvil, los rodillos o calzos desconectarán el eje.*

**Oxidation** Burning or combustion; the combining of a material with oxygen. Rusting is slow oxidation, and combustion is rapid oxidation.
***Oxidación*** *Cocción o combustión; la combinación de un material con el oxígeno. La herrumbre es oxidación lenta, y la combustión es la oxidación rápida.*

**Parallel Circuit** An electrical circuit that allows current to flow in various branches without affecting the operation of the other circuits.
***Circuito paralelo*** *Circuito eléctrico que permite que fluya la corriente en varias ramificaciones sin afectar la operación de los otros circuitos.*

**Pascal's Law** The law of fluid motion.
***Ley de Pascal*** *Ley del movimiento del líquido.*

**Pawl** A lever that pivots on a shaft. When lifted, it swings freely and when lowered locates in a detent or notch to hold a mechanism stationary.
***Linguete*** *Palanca que se apoya en un eje. Cuando se levanta se mece libremente, y cuando se baja se coloca en una abolladura o hendidura para mantener estacionario un mecanismo.*

**PCM** The powertrain control module of a computer control system. Prior to J1930, the MIL was commonly called an ECA, ECM, or one of many acronyms used by the various manufacturers.
***Módulo de control del motor (CM)*** *de un sistema de control de computadora. Antes del J1930, la LIF se llamaba comúnmente ECA, ECM, o una de las muchas siglas que usaban varios fabricantes.*

**PCS** A Pressure Control Solenoid.
***SCP*** *Solenoide de control de la presión.*

**Pinion gear** The smaller of two meshing gears.
***Piñón diferencial*** *El más pequeño de los engranajes endentados.*

**Pitch** The number of threads per inch on any threaded part.
***Paso*** *Número de roscas por pulgada en cualquier parte enroscada.*

**Pivot** A pin or shaft upon which another part rests or turns.
***Pivote o espiga*** *Chaveta o eje sobre el cual descansa o gira otra parte.*

**Planet carrier** In a planetary gear system, the carrier or bracket that contains the shafts upon which the pinions or planet gears turn.
***Cargador satélite*** *En un sistema de engranaje planetario, el cargador o la abrazadera que contiene los ejes en los que giran los piñones o engranajes satélites.*

**Planet pinions** In a planetary gear system, the gears that mesh with, and revolve about, the sun gear; they also mesh with the ring gear.
***Piñones planetarios*** *En un sistema de engranaje satélite, los engranajes que se endentan y giran alrededor del engranaje solar; también se endentan con el engranaje anular.*

**Planetary gear set** A system of gearing that is modeled after the solar system. A pinion is surrounded by an internal ring gear, and planet gears are in mesh between the ring gear and pinion around which all revolves.
***Juego de engranaje planetario*** *Sistema de engranaje que se parece al sistema solar. Un piñón está rodeado por un engranaje anular interno, y los engranajes planetarios están endentados entre el engranaje anular y el piñón en el cual todo gira a su alrededor.*

**Poppet valves** A valve shaped like a "T" and used to open and close a fluid circuit.
***Válvulas de retención*** *Válvula en forma de "T" que se usa para abrir y cerrar un circuito hidráulico.*

**Porosity** A statement of how porous or permeable to liquids a material is.
***Porosidad*** *Enunciado de cuan poroso o impermeable a los líquidos es un material.*

**Potentiometers** A type of variable resistor. Potentiometers provide a varying voltage signal from the tap and have three connections–one at each end of the resistance and one connected to a sliding contact with the resistor. Turning the control moves the sliding contact away from one end of the resistance and toward the other end.
***Potenciómetros*** *Tipo de resistor variable. Los potenciómetros proporcionan una señal de voltaje variable de su tomacorriente, y tienen tres conexiones una en cada terminal de resistencia y otra conectada a un contacto de deslizamiento con el resistor. Al mover el control retira el contacto de deslizamiento de la terminal de la resistencia, pero la mueve hacia la otra terminal.*

**Power** A measure of work being done.
***Potencia*** *Medida del trabajo que se realiza.*

**Preload** A load applied to a part during assembly so as to maintain critical tolerances when the operating load is applied later.
***Carga previa*** *Carga aplicada a una pieza durante ensamblado para mantener tolerancias críticas cuando se apli más tarde la carga operacional.*

**Press fit** Forcing a part into an opening that is slightly sm than the part itself to make a solid fit.
***Encastre a presión*** *Forzar una pieza en una abertura e pequeña que la misma pieza para conseguir un ajuste sólido.*

**Pressure** Force per unit area, or force divided by area measured in pounds per square inch (psi) or in kilopascal the metric system.
***Presión*** *Fuerza por área de unidad, o fuerza dividida p Generalmente se mide en libras por pulgada cuadrad kilopascales (kPa) en el sistema métrico.*

**Pressure plate** That part of the clutch that against the friction disc; it is mounted on and rot flywheel.
***Placa de presión*** *Parte del cloche que ejerce p disco de fricción; va montada en el volante y gira co*

**Pressure-regulating valves** Valves used pressure of the fluid by not allowing it to build b amount.
***Válvulas reguladoras de presión*** *Válvul presión de un líquido al no permitir que aumen especificada.*

***Leyes de derecho de información*** *Leyes que requieren que los patrones proporcionen a sus empleados un lugar de trabajo seguro y se refiere a los materiales peligrosos e información sobre cualquier peligro que pueda enfrentar el empleado mientras realiza su trabajo.*

**Ring gear** The outside internal gear of a planetary gear set.
***Engranaje anular*** *Engranaje interno exterior de un juego de engranaje satélite.*

**Roller bearing** An inner and outer race upon which hardened steel rollers operate.
***Cojinetes de rodillos*** *Carriles interiores y exteriores en los cuales operan rodillos de acero endurecido.*

**Roller clutch** A type of overrunning clutch that uses rollers to engage or disengage.
***Cloche de rodillos*** *Tipo de cloche de sobremarcha que utiliza rodillos para embragar y desembragar.*

**Rollers** Round steel bearings that can be used as the locking element in an overrunning clutch or as the rolling element in an antifriction bearing.
***Rodillos*** *Cojinetes redondos de acero que pueden usarse como elemento de bloqueo en un cloche de sobremarcha o como el elemento de rodillo en un cojinete antifricción.*

**Rotary flow** A fluid force generated in the torque converter that is related to vortex flow. The vortex flow leaving the impeller is not only flowing out of the impeller at high speed but also is rotating faster than the turbine. The rotating fluid striking the slower turning turbine exerts a force against the turbine that is defined as rotary flow.
***Flujo rotativo*** *Fuerza del flujo que se genera en el convertidor del par motor que se relaciona al flujo de remolino. El flujo de remolino que sale del propulsor no solamente está saliendo del propulsor a velocidad alta pero también está girando más rápido que la turbina. El flujo rotativo que golpea a la turbina en movimiento más lento ejerce una fuerza contra la turbina que se define como flujo rotativo.*

**Rotor-type pump** A fluid pump that uses an inner and outer rotor inside a housing to move the fluid.
***Bomba de tipo rotativo*** *Bomba hidráulica que usa una rueda móvil interna y externa dentro de una caja para mover el flujo.*

**RTV sealer** Room temperature vulcanizing gasket material, which cures at room temperature; a plastic paste squeezed from a tube to form a gasket of any shape.
***Obturador VATA*** *Material de junta elástica vulcanizable a temperatura ambiente, la cual se cura a la temperatura de la habitación; una pasta de plástico que sale de un tubo para formar una junta plástica con cualquier forma.*

**Runout** Deviation of the specified normal travel of an object. The amount of deviation or wobble a shaft or wheel has as it rotates. Runout is measured with a dial indicator.
***Descentramiento*** *Desviación de una trayectoria normal específica de un objeto. La cantidad de desviación o giro excéntrico que tiene un eje mientras gira. El descentramiento se mide con un indicador de cuadrante.*

**Scan tool** A microprocessor designed to communicate with a vehicle's on-board computer in order to perform diagnosis and troubleshooting.
***Herramienta exploratoria*** *Microprocesador diseñado para comunicarse con la computadora instalada en el vehículo para realizar diagnósticos y localización de averías.*

**Scarf-cut rings** A name for Teflon locking-end seals.
***Anillos de corte escarpado*** *Nombre para los obturadores de terminales cerradas hechos de teflón.*

**Schematics** Wiring diagrams used to show how circuits are constructed.
***Diseños*** *Diagramas de alambrados que muestran cómo se construyen los circuitos.*

**Score** A scratch, ridge, or groove marring a finished surface.
***Estría*** *Araño, reborde, o hendidura que estropean el acabado de una superficie.*

**Scuffing** A type of wear in which there is a transfer of material between parts moving against each other; shows up as pits or grooves in the mating surfaces.
***Desgaste abrasivo*** *Tipo de desgaste en el que hay una transferencia de material entre las partes que se mueven una contra la otra; se muestra como depresiones o hendiduras en las superficies acopladas.*

**Screw pitch gauge** A tool used to check the threads per inch (pitch) of a fastener.
***Calibrador de paso del tornillo*** *Herramienta que se usa para revisar las roscas por pulgada (paso) de un sujetador.*

**Seal** A material, shaped around a shaft, used to close off the operating compartment of the shaft, preventing oil leakage.
***Obturador*** *Material que se le da forma alrededor del eje, y que se usa para cerrar el compartimiento operacional del eje, previniendo así el goteo de aceite.*

**Sealer** A thick, tacky compound usually spread with a brush, which may be used as a gasket or sealant to seal small openings or surface irregularities.
***Sellador*** *Compuesto grueso y pegajoso que generalmente se extiende con un cepillo, y que puede usarse como una junta elástica o tapador para obturar aberturas pequeñas o irregularidades de la superficie.*

**Seat** A surface, usually machined, upon which another part rests or seats; for example, the surface upon which a valve face rests.
***Montaje*** *Superficie, generalmente labrada o maquinada, en la que reposa o descansa otra pieza; por ejemplo, la superficie en la que descansa la cara de una válvula.*

**Semiconductor** A material that is not a good conductor or a good insulator but can function as either when certain conditions exist.
***Semiconductor*** *Material que no es un buen conductor o un buen aislante pero que puede funcionar como cualquiera de los anteriores cuando existen ciertas condiciones.*

**Separator plate** A plate found in valve body assemblies and is used to seal off some fluid passages as well as to control and direct fluid flow through other passages.
***Placa desconectadora*** *Placa que se encuentra en los ensamblados del cuerpo de la válvula y que se usa para obturar algunos pasajes de líquido como también para controlar y dirigir el flujo del líquido por otros pasillos.*

**Serial data** The communications to and from the computer.
***Datos en serie*** *Comunicación que entra y sale de la computadora.*

**Series circuit** An electrical circuit that has only one path for current flow.
***Circuito en serie*** *Circuito eléctrico que sólo tiene una trayectoria para el flujo de la corriente.*

**Servo** A device that converts hydraulic pressure into mechanical movement, often multiplying it. Used to apply the bands of a transmission.
***Servo o servomando*** *Aparato que convierte la presión hidráulica en movimiento mecánico, multiplicándolo generalmente. Se usa para fijar las bandas de la transmisión.*

**Sheaves** The sides of a pulley.
***Rolfanas*** *Los lados de una polea.*

**Shift feel** The quality of a shift into another forward gear.
***Tacto de deslizamiento*** *La calidad de un enbregue hacia otro eje delantero.*

**Shift lever** The lever used to change gears in a transmission. Also, the lever on the starting motor that moves the drive pinion into or out of mesh with the flywheel teeth.
***Balancín de la transmisión o palanca de volteo*** *Palanca o balancín que se usa para meter los cambios en la transmisión. También la palanca en el motor de arranque que mete o saca el piñón de accionamiento del endentado con los dientes del volante.*

**Shift schedule** Best described as a three-dimensional graph that plots engine speed and load, as well as other operating conditions. Certain parts of the graph have designated gear ranges. When the conditions fall into a range, the computer causes the transmission to shift into that gear.
***Tabulación de cambios*** *Mejor descrito como una gráfica tridimensional que traza la velocidad y la carga del motor, como también las condiciones de operación. Ciertas partes de la gráfica tienen designadas relaciones de variación de velocidades. Cuando las condiciones caen en una relación, la computadora obliga a la transmisión a meter ese cambio.*

**Shift valve** A valve that controls the shifting of the gears in an automatic transmission.
***Válvula de cambios*** *Válvula que controla los cambios de los engranajes en una transmisión automática.*

**Shim** Thin sheets used as spacers between two parts, such as the two halves of a journal bearing.
***Calzo para ajuste*** *Hojas delgadas que se usan como separadores entre dos piezas, tales como dos mitades del cojinete radial.*

**Short** An electrical problem that results from an unwanted path to ground or to power.
***Corto*** *Problema eléctrico que resulta de una trayectoria no deseada hacia tierra o hacia la energía.*

**Side clearance** The clearance between the sides of moving parts when the sides do not serve as load-carrying surfaces.
***Huelgo lateral*** *Huelgo o separación entre los lados de piezas en movimiento cuando los lados no funcionan como superficies de arrastre de cargas.*

**Simpson gear train** A compound gear set in which two planetary units work together through a common sun gear.
***Juego de engranajes de Simpson*** *Juego de engranajes compuesto en donde dos unidades satélite trabajan juntas mediante un engranaje planetario común.*

**Single-wrap band** A solid, one-piece brake band.
***Banda de arrollamiento sencillo*** *Banda de frenos sólida de una pieza.*

**Sine wave** A waveform of a single frequency alternating current.
***Onda senoidal*** *Onda de corriente alterna de frecuencia sencilla.*

**Sinusoidal** The term sinusoidal means the wave is shaped like a sine wave.
***Sinusoidal*** *El término sinusoidal significa que la onda tiene forma de onda senoidal.*

**Sliding-fit** Where sufficient clearance has been allowed between the shaft and journal to allow free running without overheating.
***Ajuste suave*** *Cuando se ha dado el huelgo suficiente entre el eje y la muñequilla para permitir el funcionamiento sin sobrecalentarse.*

**Slip yoke** The transmission end of a drive shaft that fits into the extension housing and onto the splines of the output shaft.
***Horquilla de articulación colectora*** *La terminal de la transmisión de un eje motor que queda dentro de la caja o campana de ampliación y dentro de ranuras del eje de salida.*

**Snapring** Split spring-type ring located in an internal or external groove to retain a part.
***Anillo de resorte*** *Anillo abierto tipo resorte que se encuentra en la hendidura interna o externa para retener una pieza.*

**Soft gasket** A type of gasket that is used to seal two irregular or distorted surfaces.
***Junta plástica suave*** *Tipo de junta plástica que se usa para sellar dos superficies irregulares o distorsionadas.*

**Solenoids** An electromagnetic device that has a moveable core. The movement of the core causes some mechanical change or action to another device.
***Solenoides*** *Aparato electromagnético que tiene un núcleo movible. El movimiento del núcleo produce un cambio mecánico o acción a otro aparato.*

**Spalling** A condition in which the material of a bearing surface breaks away from the base metal.
***Exfoliación*** *Condición en donde el material de la superficie del cojinete se separa del metal base.*

**Speed reduction** The result of meshing two dissimilar sized gears. When a small gear drives a larger gear, speed reduction takes place.
***Reducción de velocidad*** *Resultado de endentar dos engranajes de distinto tamaño. Cuando un engranaje pequeño acciona un engranaje grande, se lleva a cabo la reducción de velocidad.*

**Spindle** The shaft on which the wheels and wheel bearings mount.
***Pivote*** *Eje en el que se montan las ruedas y los cojinetes de las ruedas.*

**Spline** Slot or groove cut in a shaft or bore; a splined shaft onto which a hub, wheel, or gear with matching splines in its bore is assembled and the two must turn together.
***Ranura*** *Muesca o hendidura cortada en el eje o en el orificio; un eje con ranura en el que se ensambla un cubo, rueda, engranaje etc., con ranuras emparejadas en el orificio para que las dos puedan dar vuelta simultáneamente.*

**Split lip seal** Typically, a rope seal sometimes used to denote any two-part oil seal.
***Obturador de reborde abierto*** *Generalmente un obturador de cable que a veces se usa para denotar cualquiera de las dos partes de un retén.*

**Split-band** Another name for a double-wrap band.
***Banda dividida*** *Otro nombre para una banda de arrollamiento doble.*

**Split pin** A round split spring steel tubular pin used for locking purposes; for example, locking a gear to a shaft.
***Chaveta hendida*** *Chaveta redonda tubular de acero con resorte dividiso que se usa para detener; por ejemplo, sujetar un engrajane al eje.*

**Spool valve** A cylindrically shaped valve with two or more valleys between the lands. Spool valves are used to direct fluid flow.
***Válvula de carrete*** *Válvula en forma cilíndrica con dos o más valles entre sus fundos. Las válvulas de carrete se usan para dirigir el flujo del líquido.*

**Sprag clutch** A member of the overrunning clutch family using a sprag to jam between the inner and outer races used for holding or driving action.
***Cloche de patín o de calzo*** *Miembro de la familia del cloche de sobremarcha que usa una calza para hacer una obstrucción entre los carriles internos y externos que se usan para las acciones mantenedora y de control.*

**Spur gear** A gear with teeth that are cut straight across the gear.

***Engranaje recto*** *Engranaje con dientes que están cortados rectamente a través del engranaje.*

**Stall** A condition in which the engine is operating and the transmission is in gear but the drive wheels are not turning because the turbine of the torque converter is not moving.
***Calado del motor*** *Condición en donde el motor está prendido y la transmisión está en engranaje pero las ruedas de accionamiento no están girando ya que no se está moviendo la turbina del convertidor del par motor.*

**Stall speed** Represents the highest engine speed attained with the turbine stopped (at stall).
***Velocidad del calado del motor*** *Representa la velocidad más alta que el motor logra con una turbina en detención (en calado del motor).*

**Static** A form of electricity caused by friction.
***Estática*** *Forma de electricidad causada por la fricción.*

**Static pressure** Pressure that is present and does not change.
***Presión estática*** *Presión que está presente y que no cambia.*

**Stator** The centrally located finned wheel in a torque converter. It receives transmission fluid from the turbine and directs it back to the impeller. The stator also makes torque multiplication in a torque converter possible.
***Estator*** *Rueda con aletas situada en el centro de un convertidor del par motor. Recibe flujo de la transmisión de la turbina y la conduce de regreso al propulsor. El estator también hace posible la multiplicación de torsión en el convertidor del par motor.*

**Stress** The force to which a material, mechanism, or component is subjected.
***Tensión*** *Fuerza a la que se somete un material, mecanismo o componente.*

**Sun gear** The central gear in a planetary gear system around which the rest of the gears rotate. The innermost gear of the planetary gear set.
***Engranaje planetario*** *Engranaje central en un sistema de engranaje satélite alrededor del cual giran el resto de los engranajes. El primer engranaje del juego de engranaje satélite.*

**Teardown** A term often used to describe the process of disassembling a transmission.
***Desequipo*** *Término generalmente usado para describir el proceso de desensamblar una transmisión.*

**Temper** To change the physical characteristics of a metal by applying heat.
***Templar*** *Cambiar las caracteríticas físicas de un metal al aplicarle calor.*

**Tension** Effort that elongates or "stretches" a material.
***Tensión*** *Fuerza efectiva que alarga o extiende un material.*

**Thermistor** A type of variable resistor. Thermistors are designed to change in value as its temperature changes. Thermistors are used to provide compensating voltage in components or to determine temperature.
***Termistor o resistencia térmica*** *Los termistores se usan para proporcionar voltaje de compensación en los componentes o para determinar la termperatura.*

**Thread chaser** A device, similar to a die, that is used to clean threads.
***Peine*** *Aparato similar a un dado que se usa para limpiar roscas.*

**Thread pitch** The number of threads in one inch of threaded bolt length. In the metric system, thread pitch is the distance in millimeters between two adjacent threads.
***Rosca*** *Número de roscas en una pulgada del largo de un tornillo roscado. En el sistema métrico la rosca es la distancia en milímetros entre dos enroscados adyacentes.*

**Threaded insert** A threaded coil that is used to restore the original thread size to a hole with damaged threads.
***Casquillo enroscado*** *Bobina enroscada que se usa para restaurar el tamaño original de la rosca en un orificio con roscas dañadas.*

**Throttle position (TP) sensor** A potentiometer used to monitor changes in throttle plate opening. The position of the throttle plate determines the voltage drop at the sensor's resistor and the resultant voltage signal is sent to a computer system.
***Detector de posición del acelerador (PA)*** *Potenciómetro que se usa para verificar los cambios en la apertura de la placa de aceleración. La posición de la palanca de aceleración determina la caída de voltaje al resistor del detector y la señal de voltaje resultante se envía a la computadora del sistema.*

**Throttle pressure** A signal pressure that varies with engine load.
***Presión del acelerador*** *Presión de señal que varía con la carga del motor.*

**Throttle valve** A valve that responds to throttle position and/or engine load.
***Válvula de mariposa*** *Válvula que responde a la posición del acelerador y/o a la carga del motor.*

**Thrust bearing** A bearing designed to resist or contain side or end motion as well as reduce friction.
***Cojinete de empuje*** *Cojinete diseñado para resistir o contener el movimiento lateral o trasero como también para reducir la fricción.*

**Thrust load** A load that pushes or reacts through the bearing in a direction parallel to the shaft.
***Carga de empuje*** *Carga que empuja o reacciona mediante el cojinete en una dirección paralela al eje.*

**Thrust washer** A washer designed to take up end thrust and prevent excessive endplay.
***Arandela o wacha de empuje*** *Arandela diseñada para levantar el empuje trasero y prevenir el juego longitudinal excesivo.*

**Tolerance** A permissible variation between the two extremes of a specification or dimension.
***Tolerancia*** *Variación permitida entre los dos extremos de una especificación o dimension.*

**Torque** A twisting motion, usually measured in ft.-lbs.
***Par motor*** *Movimiento de torsión que generalmente se mide en pies-libras o N-m.*

**Torque capacity** The ability of a converter clutch to hold torque.
***Capacidad del par motor*** *Habilidad de un cloche convertidor para mantener el par motor.*

**Torque converter** A turbine device utilizing a rotary pump, one or more reactors (stators) and a driven circular turbine or vane whereby power is transmitted from a driving to a driven member by hydraulic action. It provides varying drive ratios; with a speed reduction, it increases torque.
***Convertidor del par motor*** *Aparato de turbina que utiliza una bomba rotativa, uno o más reactores (estatores) y una turbina cilíndrica impulsada o paleta en donde se transmite la energía de un miembro conducido o accionador por medio de una acción hidráulica. Proporciona relaciones de transmisión variables; aumenta el par motor al disminuir la velocidad.*

**Torque curve** A line plotted on a chart to illustrate the torque personality of an engine. When the engine operates on its torque curve it is producing the most torque for the quantity of fuel being burned.
***Curva del par motor*** *Una linea que ilustra gráficamente la personalidad del par motor.*

**Torque multiplication** The result of meshing a small driving gear and a large driven gear to reduce speed and increase output torque.
***Multiplicación del par motor*** *Resultado del endentado de una rueda motriz pequeña y rueda conducida grande para reducir la velocidad e incrementar el par motor de salida.*

**Torque steer** An action felt in the steering wheel as the result of increased torque.
***Dirección del par motor*** *Acción que se siente en el volante como resultado del aumento del par motor.*

**Toxicity** The term used to describe how poisonous a substance is.
***Toxicidad*** *Término que describe cuan venenosa que es una sustancia.*

**Traction** The gripping action between the tire tread and the road's surface.
***Tracción*** *Acción de sujeción entre la superficie de rodadura y la superficie del camino.*

**Transfer case** An auxiliary transmission mounted behind the main transmission. Used to divide engine power and transfer it to both front and rear differentials, either full-time or part-time.
***Caja de reenvío*** *Transmisión auxiliar montada detrás de la transmisión principal. Se usa para dividir la potencia del motor y transferirla a los diferenciales frontales y traseros, ya sea permanentemente o por tiempo parcial.*

**Transfer plate** Another name for the separator plate found in some valve body assemblies.
***Planta de transbordo*** *Otro nombre para la placa separadora que se encuentra en algunos ensamblajes del cuerpo de la bobina.*

**Transmission Range (TR) sensor** Informs the TCM of the gear selected by the driver.
***Detector de la variación de velocidades (VV)*** *Informa al MCT (múltiplex por compresión en el tiempo) del engranaje que selecciona el conductor.*

**Transverse** Power train layout in a front-wheel drive automobile extending from side to side.
***Transversal*** *Motor bosquejado en un automóvil de tracción delantera que se extiende de lado a lado.*

**Turbine** A finned wheel-like device that receives fluid from the impeller and forces it back to the stator. The turbine transmits engine torque to the transmission's input shaft.
***Turbina*** *Aparato en forma de rueda con aletas que recibe el flujo del impulsor y lo hace retroceder al estator.*

**Two-way check valve** A ball-type check valve without a return spring. This type of valve is used where pressure from two different sources is sent to the same outlet port.
***Válvula de comprobación en las dos direcciones*** *Válvula de comprobación tipo esfera sin un muelle de desembrague. Este tipo de válvula se usa donde la presión de dos diferentes fuentes se envía a una misma válvula de escape.*

**U-joint** A four-point cross connected to two U-shaped yokes that serves as a flexible coupling between shafts.
***Junta en U*** *Cruz conectada a dos horquillas de articulación en forma de U que sirve como acoplamiento flexible entre los ejes.*

**United States Customary System (USCS)** A system of weights and measures used in places that do not use the metric system, such as the United States. The USCS system is also known as the English or inch system.
***Sistema Consuetudinario de los Estados Unidos (SCEU)*** *Sistema de pesos y medidas usado en lugares que no usan el sistema métrico, tal como en los Estados Unidos. El sistema de SCEU también se conoce como sistema inglés o sistema de pulgadas.*

**Universal joint** A mechanical device that transmits rotary motion from one shaft to another shaft at varying angles.
***Junta universal*** *Aparato mecánico que transmite un movimiento giratorio de un eje a otro eje en ángulos variados.*

**Upshift** To shift a transmission into a higher gear.
***Subir la velocidad*** *Meter el cambio de la transmisión a una velocidad más alta.*

**Vacuum** Any pressure lower than atmospheric pressure.
***Vacío*** *Cualquier presión menor que la presión atmosférica.*

**Vacuum modulator** A load-sensing device that regulates fluid pressure in response to a vacuum signal, which varies with throttle opening and vehicle load.
***Modulador de vacío*** *Aparato detector de carga que regula la presión del flujo en respuesta a una señal de vacío que varía con la apertura del acelerador y la carga del vehículo.*

**Vacuum transducer** An electronic device that changes vacuum signals into electrical signals.
***Transductor de medida al vacío*** *Aparato electrónico que cambia las señales de vacío en señales eléctricas.*

**Valley** The area on a spool valve that is between the valve's lands.
***Zona*** *Área de una válvula de carrete que está ente los fundos de la válvula.*

**Valve body** Main hydraulic control assembly of a transmission containing the components necessary to control the distribution of pressurized transmission fluid throughout the transmission.
***Cuerpo de válvula*** *Ensamblaje principal del control hidráulico de una transmisión que contiene los componentes necesarios para controlar la distribución del flujo a presión a través de la transmisión.*

**Vane-type pump** A type of fluid pump that uses several sliding vanes that seal against a slide in the housing to push fluid out.
***Bomba tipo álabe*** *Tipo de bomba hidráulica que usa varios álabes de deslizamiento que obturan contra un deslizamiento en la campana para sacar el líquido.*

**Variable capacity pumps** A classification of a fluid pump whose output is controllable.
***Bombas de capacidad variable*** *Clasificación de una bomba hidráulica de la cual se puede controlar la salida.*

**Variable force motor (VFM)** An electrohydraulic actuator made of a variable force solenoid and a regulating valve. It controls mainline pressure by moving a pressure regulator valve against spring pressure.
***Motor de fuerza variable (MFV)*** *Actuador electrohidráulico hecho de un solenoide de fuerza variable y de una válvula reguladora. Controla la presión de la línea principal al mover una válvula reguladora de presión contra la presión del muelle.*

**Vehicle identification number (VIN)** The number assigned to each vehicle by its manufacturer, primarily for registration and identification purposes.
***Número de identificación del vehículo (NIV)*** *Número asignado por el fabricante a cada vehículo, principalmente con el propósito de registro e identificación.*

**Vehicle speed sensor (VSS)** A sensor used to track the current speed and the total miles traveled by a vehicle. This input is used by many computer systems in the vehicle.
***Detector de la velocidad del vehículo (DVV)*** *Detector que se usa para deducir la velocidad actual y el millaje total recorrido por un vehículo. Esta admisión la usan varios sistemas de la computadora en el vehículo.*

**VFS (Variable Force Solenoid)** Used to maintain line pressure.

***SFV*** *Solenoide de fuerza variable se usa para mantener la presión de la línea.*

**Vibration** A quivering, trembling motion felt in the vehicle at different speed ranges.
***Vibración*** *Movimiento de vibración y temblor que se percibe en un vehículo a diferentes regímenes de velocidad.*

**Viscosity** The resistance to flow exhibited by a liquid. Thick oil has greater viscosity than thin oil.
***Viscosidad*** *Resistencia a un flujo que exhibe un líquido. El aceite grueso tiene mayor viscosidad que el aceite delgado.*

**Viscous clutch** A clutch assembly that engages and disengages in response to the viscosity of the fluid in the clutch assembly.
***Cloche viscose*** *Ensamblaje de cloche que se embraga y desembraga en respuesta a la viscosidad del líquido en el ensamblaje del cloche.*

**Volatile** The characteristic of a material that allows it to vaporize quickly.
***Volátil*** *Característica de un material que le permite evaporizarse rápidamente.*

**Volt** A unit of measurement of electromotive force. One volt of electromotive force applied steadily to a conductor of one ohm resistance produces a current of one ampere.
***Voltio*** *Unidad de medida de la fuerza electromotriz. Un voltio de fuerza electromotriz aplicada constantemente a un conductor de resistencia de un ohmio produce una corriente de un amperio.*

**Voltage** Electrical pressure formed by the attraction of electrons to protons.
***Voltaje*** *Presión eléctrica que se forma por la atracción de los electrones a los protones*

**Voltage drop** The voltage lost by the passage of electrical current through resistance.
***Baja de voltaje*** *Voltaje perdido por el pasaje de la corriente eléctrica a través de la resistencia.*

**Voltage-generating sensors** Sensors that are capable of producing their own input voltage signal. This varying voltage signal enables the computer to monitor and adjust for changes in the computerized control system.
***Detectores de generación de voltaje*** *Detectores que son capaces de producir su propia señal de voltaje de entrada. Esta señal de voltaje variable permite a la computadora controlar y adaptar en caso de cambios en un sistema de control computarizado.*

**Voltmeter** A tool used to measure the voltage available at any point in an electrical system.
***Voltímetro*** *Herramienta que se usa para medir el voltaje existente en cualquier punto en un sistema eléctrico.*

**Vortex flow** Recirculating flow between the converter impeller and turbine that causes torque multiplication.
***Flujo de remolino*** *Flujo de recirculación entre la rueda de aletas del convertidor y la turbina que produce la multiplicación del par motor.*

**Wet-disc clutch** A clutch in which the friction disc (or discs) is operated in a bath of oil.
***Cloche de discos sumergidos*** *Cloche en el que el disco o discos de fricción se opera(n) en un baño de aceite.*

**Work** The product of a force and the distance through which it acts.
***Trabajo*** *Producto de una fuerza y la distancia por la que actúa.*

**Yoke** In a universal joint, the drivable torque-and-motion input and output member, attached to a shaft or tube.
***Horquilla de articulación*** *En una junta universal es la pieza de accionamiento de entrada y de salida de torsión y huelgo, que está sujeta a un eje o tubo.*

# Index

**T**